Springer Series in
SOLID-STATE SCIENCES 115

Springer
Berlin
Heidelberg
New York
Barcelona
Hong Kong
London
Milan
Paris
Singapore
Tokyo

Springer Series in
SOLID-STATE SCIENCES

126 **Physical Properties of Quasicrystals**
Editor: Z.M. Stadnik

127 **Positron Annihilation in Semiconductors**
Defect Studies
By R. Krause-Rehberg and H.S. Leipner

128 **Magneto-optics**
Editors: S. Sugano and N. Koyima

129 **Introduction to Computational Materials Science**
From Ab Initio to Monte Carlo Methods
By K. Ohno, K. Esfarjani, and Y. Kawazoe

Volumes 1–125 are listed at the end of the book.

Jagdeep Shah

Ultrafast Spectroscopy of Semiconductors and Semiconductor Nanostructures

Second Enlarged Edition
With 256 Figures

Springer

Lucent Technologies
Bell Labs Innovations

Dr. Jagdeep Shah
Bell Laboratories
Lucent Technologies
101 Crawfords Corner Road
Holmdel, NJ 07733, USA
e-mail: jags@bell-labs.com

Series Editors:

Professor Dr., Dres. h. c. Manuel Cardona
Professor Dr., Dres. h. c. Peter Fulde*
Professor Dr., Dres. h. c. Klaus von Klitzing
Professor Dr., Dres. h. c. Hans-Joachim Queisser
Max-Planck-Institut für Festkörperforschung, Heisenbergstrasse 1, D-70569 Stuttgart, Germany
* Max-Planck-Institut für Physik komplexer Systeme, Nöthnitzer Strasse 38
D-01187 Dresden, Germany

Professor Dr. Roberto Merlin
Department of Physics, 5000 East University, University of Michigan
Ann Arbor, MI 48109-1120, USA

Professor Dr. Horst Störmer
Dept. Phys. and Dept. Appl. Physics, Columbia University, New York, NY 10023 and
Bell Labs., Lucent Technologies, Murray Hill, NJ 07974, USA

ISSN 0171-1873

ISBN 3-540-64226-9 2nd Edition Springer-Verlag Berlin Heidelberg New York

ISBN 3-540-60912-1 1st Edition Springer-Verlag Berlin Heidelberg New York

Library of Congress Cataloging-in-Publication Data.
Shah, J. (Jagdeep) Ultrafast spectroscopy of semiconductors and semiconductor nanostructures/Jagdeep Shah. – 2nd enl. ed. p. cm. – (Springer series in solid-state sciences, 0171-1873; v. 115) "Lucent Technologies." Includes bibliographical references and index. ISBN 3-540-64226-9 (alk. paper) 1. Semiconductors–Spectra. 2. Nanostructure materials–Spectra. 3. Laser pulses, Ultrashort. 4. Laser spectroscopy. I. Lucent Technologies. II. Title. III. Series. QC611.6.O6 S52 1998 537.6'22'0287–ddc21 98-46049

Typesetting: K + V Fotosatz, Beerfelden
Cover concept: eStudio Calamar Steinen
Cover production: *design & production* GmbH, Heidelberg

SPIN: 10673261 57/3144/ba - 5 4 3 2 1 0 – Printed on acid-free paper

To
my wife, Shobha
and
our children, Roopak and Raina

Preface

The field of ultrafast spectroscopy of semiconductors and their nanostructures continues to be an active field of research. Exciting new developments have taken place since the first edition of this book was completed in 1995. This revised edition includes a discussion of many of these recent developments in the field. This is accomplished by adding a chapter on *Recent Developments* at the end of the book. This approach was selected to provide a discussion of results while they are still relatively recent. Results published before the end of May 1998 were considered for inclusion in this book. The objective of this revised edition remains the same as before: to provide a cohesive discussion of the many diverse contributions of ultrafast spectroscopy to the field of semiconductors. Extensive cross-references are made to earlier chapters in order to accomplish this goal.

The chapter on *Recent Developments* begins with a brief discussion of new lasers, new techniques of ultrafast spectroscopy and novel nanostructures. This is followed by a section on *Coherent Spectroscopy* where some of the most interesting recent developments have taken place. These include observation of quantum kinetic effects, effects that require going beyond the mean-field approach of the semiconductor Bloch equations, coherent control of populations and current in semiconductors, exciton-continuum interactions, and many diverse aspects of coherent spectroscopy including studies of microcavities, Bragg structures, quantum dots and quantum wires. The next section deals with *Ultrafast Emission Dynamics*, with emphasis on resonant dynamics from quantum wells and microcavities. Recent developments clearly identify and isolate the coherent part of resonant emission using the techniques of phase-sensitive detection that were only beginning to be employed when the first edition of the book was completed. The next section discusses ultrafast dynamics in the incoherent regime. The discussion here includes recent studies of dynamics in quantum wires and dots, investigation of high-density effects and femtosecond transport dynamics using the recently developed technique of THz detection with ultrabroad bandwidth. A concerted effort is made to provide cross-references to earlier chapters in order to put the new developments in proper perspective, and Chap. 9 provides a list of more than 500 references.

I am grateful to Bell Laboratories, Lucent Technologies for permission to publish this book and for providing an intellectually stimulating environment conducive to successful and productive research in a rapidly advancing field. I would like to thank many colleagues, both within and outside

Bell Laboratories, with whom I have collaborated and interacted over the years. I would also like to thank Dr. H. Lotsch and the editorial staff of Springer-Verlag for help on many different aspects of this book. Last but not the least, I wish to express my appreciation to my wife and children for their encouragement, understanding and support.

Holmdel, New Jersey
November 1998

Jagdeep Shah

Preface to the First Edition

It is well known that linear optical spectroscopy of semiconductors has provided invaluable information on many diverse aspects of semiconductors such as electronic band structure, phonons, plasmons, single-particle spectra, and defects. These are impressive contributions, but optical spectroscopy can do much more by exploiting its additional unique strengths that make it a preferred technique for obtaining fundamental new information about non-equilibrium, nonlinear and transport properties of semiconductors. Optical excitation has the ability to generate non-equilibrium carrier and exciton distributions, and optical spectroscopy provides the best means of determining such distribution functions. When these unique strengths are combined with ultrashort laser pulses, and with spatial imaging techniques and special structures, optical spectroscopy becomes a powerful tool for investigating a wide diversity of phenomena related to relaxation and transport dynamics in semiconductors. Ultrafast spectroscopy of semiconductors has been an active field of research and has led to many new insights into phenomena of fundamental importance in semiconductor physics and in many optoelectronic and electronic devices.

The objective of this book is to provide a cohesive discussion of the many diverse contributions of ultrafast spectroscopy to the field of semiconductors. The field is so active and extensive that an exhaustive treatment of all the research activities would be nearly impossible. The emphasis of the book is, therefore, threefold: basic concepts and foundations, the historical perspective, and a discussion of selected experimental results, including recent developments, which have led to fundamental new insights into many diverse aspects of semiconductor physics. It is hoped that such an approach will be useful not only to researchers and graduate students entering into this active and vibrant field of research, but also to many researchers who are already active in the field.

The book begins with an introductory chapter on basic concepts in semiconductors, ultrafast lasers and ultrafast spectroscopy techniques, and the interpretation of results. This chapter discusses how carrier relaxation dynamics as well as transport dynamics can be investigated by using ultrafast spectroscopy, and how the relaxation dynamics can be divided into four temporally overlapping regimes. The first measurements of dynamics investigated the last (and the slowest) of these regimes because of the available time resolution and techniques. However, the chapters in the book are arranged not in an historical order, but in the order of occurrence of events in a semiconductor following photoexcitation by an ultrashort pulse.

Chapter 2, on *Coherent Spectroscopy* discusses the coherent regime in which the excitation is in phase with the electromagnetic radiation creating it. Some of the many exciting phenomena discussed in this chapter include exciton dephasing, the AC Stark effect, quantum beats of excitons, interaction-induced effects, which make the coherent response of semiconductors so different from that of systems of non-interacting atoms, coherent oscillations, Bloch oscillations, coherent spectroscopy of free carriers, coherent phonons, and terahertz spectroscopy of semiconductors and their nanostructures. The destruction of the coherence created by an ultrafast laser by various scattering processes provides important new information about some of the fastest dynamical processes in semiconductors, and is also discussed in Chap. 2.

Once the coherence is destroyed, the distribution function of carriers or excitons is still likely to be non-thermal. The initial relaxation of non-thermal carriers and the scattering processes that lead to a thermalized hot carrier distribution function, are discussed in Chap. 3, which is on *Initial Relaxation of Photoexcited Carriers*. The hot electron-hole plasma relaxes towards the lattice temperature primarily by interacting with phonons, a subject which is discussed in Chap. 4, *Cooling of Hot Carriers*. Chapter 4 also discusses how the large number of optical phonons generated in a typical experiment lead to significant non-equilibrium phonon populations that strongly reduce the rate of energy loss to the lattice by the now well-known hot phonon effect. Time-resolved light scattering techniques provide direct information about the dynamics of these photoexcited phonons. Various aspects of the phonon dynamics are described in Chap. 5, *Phonon Dynamics*.

Excitonic effects have a profound influence on the optical properties of semiconductors. Various aspects of excitonic dynamics have been investigated in recent years via ultrafast optical spectroscopy. These include exciton formation and relaxation dynamics, exciton recombination dynamics, and exciton spin dynamics. The physical processes involved in this dynamics are quite different from those involved in carrier relaxation. These topics are considered in Chap. 6, *Exciton Dynamics*.

Ultrafast spectroscopy provides information not only about carrier, exciton, and phonon relaxation dynamics but also about transport in semiconductors. This is obtained either by using spatial imaging techniques or by using the technique of optical markers. Combining these with ultrafast lasers allows one to investigate the dynamics of transport on femtosecond and picosecond time scales. A variety of interesting results have also been obtained on the dynamics of tunneling, a fundamental quantum-mechanical phenomenon. These results are discussed in Chap. 7, *Carrier Tunneling in Semiconductor Nanostructures*. The dynamics of a variety of other transport phenomena in semiconductor nanostructures is discussed in Chap. 8, *Carrier Transport in Semiconductor Nanostructures*.

I am grateful to Bell Laboratories for permission to publish this book and for providing an intellectually stimulating environment conducive to successful and productive research in a rapidly advancing field. I would like to thank many colleagues, both within and outside Bell Laboratories, with whom I have collaborated and interacted over the years. I have learned much from such interactions, and the perspective I have gained over the years through my research and my interactions with colleagues has been extremely valuable in writing this book. A list of such colleagues would be too long to be included here, but I would like to thank Prof. B. Deveaud, Prof. T. Elsaesser, Prof. E. Göbel, Dr. J. A. Kash, Prof. W. Pötz, Prof. L. J. Sham, Prof. D. G. Steel, and Dr. A. Vinattieri for reading the preliminary drafts of individual chapters and providing me with their comments and perspectives on them. I would also like to thank Prof. M. Cardona for his encouragement in this project, and Dr. H. Lotsch of Springer-Verlag for encouragement and help in many different aspects of this book. Last but not least, I wish to express my appreciation to my wife and children for their encouragement, understanding, and support.

Holmdel, New Jersey
January 1996

Jagdeep Shah

Contents

1. Introduction

Optical spectroscopy is a powerful technique for investigating electronic and vibrational properties of a variety of systems, and has provided extensive information and insights into the properties of atoms, molecules and solids. In semiconductors, the techniques of absorption, reflection, luminescence and light-scattering spectroscopies have provided invaluable information about such diverse aspects as the electronic band structure, phonons, coupled phonon-plasma modes, single-particle excitation spectra of electrons and holes, and properties of defects, surfaces and interfaces.

These are essential contributions to our understanding of semiconductors; but optical spectroscopy can do much more. Optical spectroscopy has additional unique strengths which makes it capable of providing fundamental information about *nonequilibrium, nonlinear and transport properties of semiconductors*. These strengths, which have been exploited since the 1960s and have been combined with the picosecond and femtosecond laser pulses to provide new insights into completely different aspects of semiconductors, can be classified into four groups: (1) photoexcitation generates excitations (electrons, holes, excitons, phonons, etc.) with non-equilibrium distribution functions, (2) optical spectroscopy provides the best means of determining the distribution functions of these excitations, and hence determining the dynamics of the relaxation of these excitations, (3) combined with spatial imaging and specially-designed structures, optical spectroscopy provides the ability to investigate the transport of excitations, and the dynamics of the transport, in semiconductors and their nanostructures, and (4) optical techniques provide the ability to investigate the nonlinear properties, including coherent effects, in semiconductors, and thus provide insights into yet different aspects of semiconductors, such as many-body effects, coherent effects and dephasing phenomena.

This book concentrates on the study of these nonequilibrium phenomena in semiconductors using ultrafast optical spectroscopy. This chapter provides an introduction to the book by discussing some basic concepts in semiconductors (Sect. 1.1), and surveying some techniques of generating ultrashort laser pulses (Sect. 1.2) and commonly used techniques of ultrafast spectroscopy (Sect. 1.3). Basic concepts in the interpretation of data obtained by ultrafast spectroscopy are discussed in Sect. 1.4.

1.1 Semiconductors: Basic Concepts

1.1.1 Band Structure

The periodicity of the semiconductor lattice introduces the concepts of the reciprocal lattice and the Brillouin zone, and allows one to specify the energy levels in the semiconductor in terms of the energy vs. momentum picture. The energy band structure of the semiconductor forms the basis of understanding most of the optical phenomena discussed in this book. Many textbooks [1.1–8] and specialized books on band structure calculations [1.9, 10] give a detailed discussion of the band structure of semiconductors. We will give a brief qualitative description of the basic ideas in the following sections.

(a) Bulk Semiconductors. Intrinsic semiconductors are characterized by a band gap separating the occupied valence-band and the empty conduction-band states at low temperatures. The energy band gaps range from very small values (close to zero) for semiconductors like HgCdTe and PbSnTe to several eV for semiconductors like ZnS. All semiconductors can be classified either as direct gap semiconductors with the valence band maximum and the conduction-band minimum occurring at the same point in the Brillouin zone, or indirect-gap semiconductors in which these extrema occur at different points in the Brillouin zone. The optical properties of direct-gap semiconductors differ considerably from those of indirect-gap semiconductors [1.11]. Although many of the concepts discussed in this book can be applied to all direct gap semiconductors and can perhaps be modified to apply also to indirect-gap semiconductors, we will be concerned primarily with compound III–V semiconductors and their nanostructures. GaAs is a prototype of this class of semiconductors and we will discuss it extensively.

Figure 1.1 shows a part of the band structure of bulk GaAs [1.9, 10, 12]. GaAs [1.13, 14] has a direct band gap located at the center of the Brillouin zone. The conduction band has subsidiary minima at higher energies at or near other symmetry points in the Brillouin zone. The subsidiary valleys play an important role in the high-field transport in GaAs [1.5, 7]. The Γ valley is spherically symmetric with a parabolic energy dispersion with wavevector (E vs k) relationship at low energies (an isotropic effective mass of 0.067 m_0). The electron states in the Γ valley are two-fold degenerate (including spin) and are primarily s-like near the minimum, with an increasing admixture of p-character at higher energies. The electron band structure near the X and the L minima is highly anisotropic.

The valence band in GaAs is four-fold degenerate at the zone center and is split into the Heavy-Hole (HH) and the Light-Hole (LH) bands at larger k values. The degeneracy between the HH and the LH bands at $k = 0$ can be lifted by perturbations such as stress or quantum confinement (Sect. 1.1.1 b). The valence band states are p-like and the valence band shows anisotropy and warping. There is also a spin-split-off valence band at lower energies.

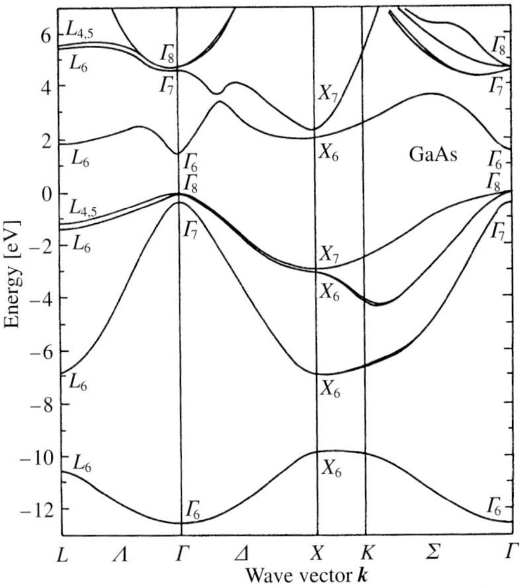

Fig. 1.1. Band structure of GaAs showing various conduction and valence bands in the vicinity of the direct band gap [1.10]

This complicated band structure must be used to understand many of the experiments performed on GaAs. However, for simple illustrative purposes, and some of the experiments, a simplified band structure considering the lowest conduction band and parabolic valence bands may be adequate.

(b) Quantum Wells. Epitaxial growth techniques [1.15 – 18] such as Molecular Beam Epitaxy (MBE) now allow growth of two dissimilar materials with nearly the same lattice constant on each other with high-quality interfaces. Such growth leads to planar quasi-two-Dimensional (2D) structures. Examples include GaAs/AlGaAs and InGaAsP/InP systems. If a semiconductor with a smaller band gap (GaAs) is sandwiched between layers of a larger-band-gap semiconductor (AlGaAs), and if the thickness of the smaller-band-gap material is comparable to or smaller than the carrier de Broglie wavelength in the semiconductor, then quantum confinement effects become important and lead to a profound change in the properties of the semiconductor [1.19 – 23]. The changes result from the fact that carriers are confined along the growth (z) direction but are free to move in the x and the y directions. Figure 1.2 shows the spatial electron potential profile along z of a type-I quantum well structure in which both electrons and the holes are confined to the same semiconductor. Strong interband transitions are possible in type I-quantum wells because there is a strong overlap between electron and hole wave functions. Also shown are the schematic of the two lowest electron wave functions. A Multiple Quantum Well (MQW) structure consisting of N such periods can be treated as N independent wells if the barrier height and thickness are sufficiently large so that there is negligible overlap between the wavefunctions of adjacent quantum wells.

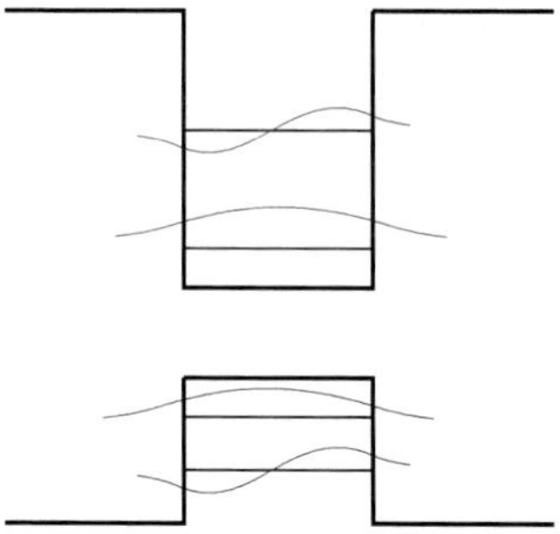

Fig. 1.2. Schematic of the spatial potential profile and wavefunctions for the two lowest electron and hole levels in a type-I quantum-well structure

The conduction and valence bands in quantum wells break up into several subbands, and the degeneracy between the HH and the LH valence band at $k = 0$ is lifted. Detailed discussion of the quantum well band structure is available in the literature [1.19–22]. The electrons have a simple subband structure, but the hole subbands are quite complicated. As an example, the calculated valence band structure of a 100 Å GaAs/AlGaAs quantum well as a function of $k_{||}$, the wavevector parallel to the interface planes, is shown in Fig. 1.3 [1.24]. Note that the $k_{||} = 0$ HH effective mass along $k_{||}$ is smaller than the corresponding LH mass.

If the quantum wells are undoped but the barriers are doped with donors (modulation-doping [1.25]), then the electrons transfer into the well and change the potential profile. The first modulation-doped heterostructures were

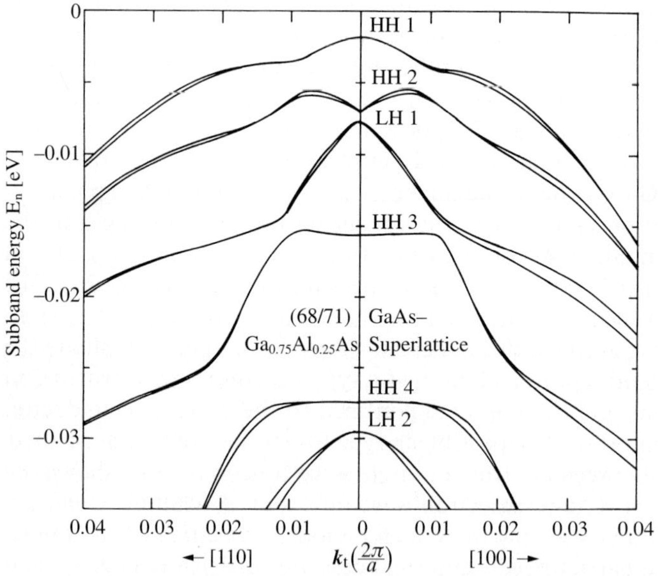

Fig. 1.3. Calculated valence band structure of a 192 Å thick GaAs (68 atomic layers) quantum well with 201 Å (71 atomic layers) thick $x = 0.25$ AlGaAs barriers [1.24]

MBE-grown GaAs/AlGaAs structures with modulated silicon doping reported by *Dingle* et al. [1.25]. Modulation-doped structures can have very high carrier mobility because the donor impurities are spatially removed from the carriers and the ionized impurity scattering is significantly reduced. The high-mobility 2D electron gas has many interesting transport and optical properties [1.26, 27].

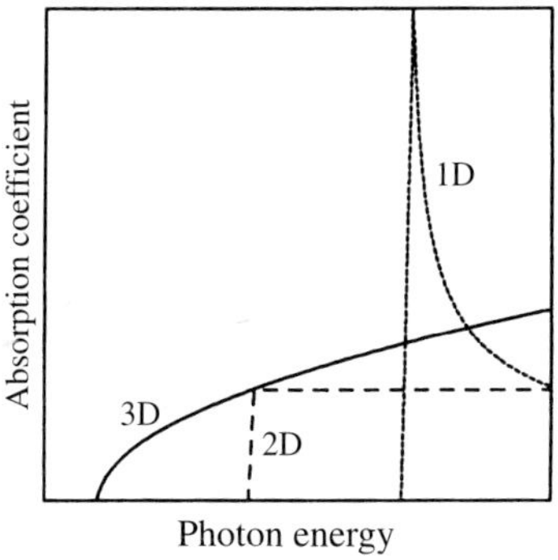

Fig. 1.4. Schematic comparison of the calculated absorption coefficients for a bulk semiconductor with parabolic bands, and the corresponding 2D (quantum well) and 1D (quantum wire) semiconductors with infinite barriers, without considering the excitonic effects

(c) Lower-Dimensional Systems. The dimensionality of the semiconductor structure can be further reduced either during the growth or by subsequent processing. The 1 and 0 dimensional systems (quantum wires and quantum boxes) are attracting considerable attention. The additional quantum confinement brings about remarkable changes in many properties. Figure 1.4 compares the calculated absorption coefficients for a bulk semiconductor with parabolic bands and the corresponding 2D (quantum well) and 1D (quantum wire) semiconductors with infinite barriers, without considering the excitonic effects.

1.1.2 Excitons

Coulomb interaction between electrons and holes has a profound influence on the interband optical properties of semiconductors. It introduces a hydrogenic series of exciton bound states converging to the electron-hole continuum. The matrix elements for the interband absorption of the continuum are also strongly affected by the Coulomb interaction [1.28, 29]. Excitons in semiconductors have been discussed in many excellent books and review chapters [1.30–37]. In Chap. 6 on dynamics of excitons, we will discuss the bound states of excitons, exciton dispersion, the coupled photon-exciton modes (polaritons), and various relaxation processes influencing exciton relaxation and dynamics. The important role of excitons in the coherent regime (Sect. 1.5) will be treated in Chap. 2.

1.1.3 Phonons in Semiconductors

The periodicity of the semiconductor lattice also allows the description of the quantized vibrational modes of the lattice in terms of phonon energy vs. wavevector diagrams or phonon dispersion relations [1.1, 4, 5, 7]. GaAs has more than one atom per unit cell so that both acoustic and optical phonon modes are present in GaAs. The acoustic phonons can be classified in terms of Longitudinal (LA) and Transverse (TA) Acoustic phonons with different velocities. There are also Longitudinal (LO) and Transverse (TO) Optical modes with a splitting at $q = 0$ because of the polar interaction in GaAs. Inelastic neutron scattering provides an experimental technique for measuring the dispersion relations. Neutron scattering results for the dispersion relations of the phonons in GaAs [1.38] are shown in Fig. 1.5.

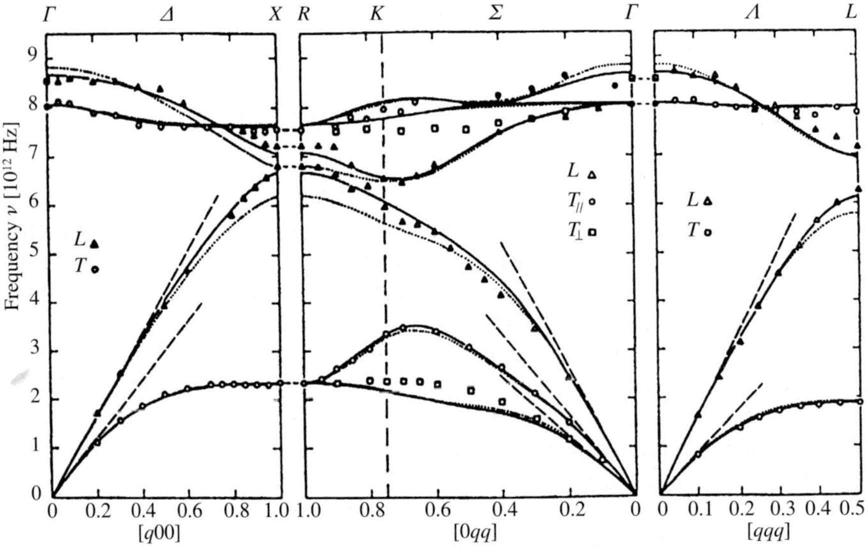

Fig. 1.5. Room-temperature dispersion curves of acoustic and optical phonons in bulk GaAs obtained from inelastic neutron scattering data [1.38]

A reduction in the dimensionality of the semiconductor also leads to a substantial change in the phonon modes. Different kinds of phonon modes, such as confined, propagating and interface modes, are present in quantum wells. The acoustic and optic phonon modes in quantum wells and quantum wires have been discussed quite extensively in the literature [1.39, 40] and will not be discussed here. We will, however, refer to the complicated phonon modes in reduced dimensionality structures during our discussion of experimental results, especially in Chap. 5 on phonon dynamics.

1.1.4 Scattering Processes in Semiconductors

The dynamics of electrons, holes, excitons and phonons is influenced by their interaction with each other, as well as with defects and interfaces of the system. There are once again extensive discussions of these interactions in the literature. In the following subsections, we present essentially a catalog of important interactions, with some representative references for further information.

(a) Carrier-Phonon Interactions. Interaction of carriers with phonons plays a major role in the exchange of energy and momentum between carriers and the lattice, and hence determines the relaxation of photoexcited semiconductors as well as the transport properties of semiconductors. Textbooks as well as books on high-field transport in semiconductors have extensive discussions on these processes [1.3, 5–7, 41–44]. The scattering rates are derived using the Fermi golden rule and are also discussed in detail in these references.

GaAs-like semiconductors are intermediate between the covalent semiconductors like Si and Ge, and ionic semiconductors like CdS. Polar optical phonons, therefore, play an important role in carrier-phonon scattering processes in GaAs-like semiconductors. Low energy electrons, because of their s-like wavefunctions, interact with optical phonons only through polar coupling. The matrix elements of this *Fröhlich interaction* varies inversely with the wavevector q of the phonon being emitted or absorbed. Holes also interact with LO phonons via Fröhlich interaction. Since the Fröhlich interaction rate is expected to vary as $m^{1/2}$, the heavier holes might be expected to interact more strongly with the LO phonons than the lighter electrons. However, the complicated band structure of holes leads to an effective reduction of the rate by a factor of 2 [1.45–49].

The electron polar optical mode scattering rate (Fröhlich interaction) is given by [1.3, 5]:

$$\begin{aligned} W(E) = {} & W_0(\hbar\omega_{\mathrm{LO}}/E)^{1/2}\{n(\hbar\omega_{\mathrm{LO}})\sinh^{-1}(E/\hbar\omega_{\mathrm{LO}}) \\ & +[n(\hbar\omega_{\mathrm{LO}})+1]\sinh^{-1}[(E/\hbar\omega_{\mathrm{LO}})-1]\} , \end{aligned} \tag{1.1}$$

where $n(\hbar\omega_{\mathrm{LO}})$ is the phonon occupation number and

$$W_0 = \frac{e^2(2m\hbar\omega_{\mathrm{LO}})^{1/2}}{4\pi\hbar^2}\left(\frac{1}{\varepsilon_\infty}-\frac{1}{\varepsilon_{\mathrm{s}}}\right) . \tag{1.2}$$

Here $\hbar\omega_{\mathrm{LO}}$ is the LO phonon energy, m is the electron mass, and ε_∞ and ε_{s} are the optical and static permittivities. The calculated electron-LO phonon normalized scattering rates for the Fröhlich interaction for electrons in a simple parabolic band with effective mass m are depicted in Fig. 1.6 for the lattice at 300 K. Values for electrons in the Γ valley of GaAs can be obtained by using $W_0 = (125\ \mathrm{fs})^{-1}$.

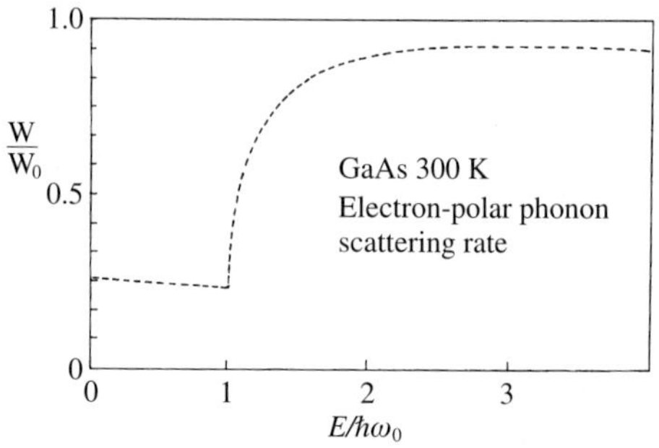

Fig. 1.6. Rate for polar optical mode scattering of Γ valley electrons in GaAs as a function of the energy of the electron

High-energy electrons, electrons in subsidiary valleys and holes can also interact with optical phonons through *non-polar optical deformation potential scattering* whose matrix elements are independent of the phonon wavevector. Coupling to both the LO and the TO phonons can occur through this mechanism. This interaction is also responsible for *intervalley scattering* of electrons.

Interaction with both LA and TA phonons can also be important. *Deformation potential* and *piezoelectric scattering mechanisms* play an important role, with the former dominating in most cases. If the carrier has sufficient energy to emit an optical phonon, then the optical phonon scattering rate is generally considerably higher than for the acoustic phonons. Phonon emission processes are independent of the phonon occupation number and hence temperature. Phonon absorption, on the other hand, is proportional to the phonon occupation number and hence is strongly temperature dependent.

(b) Carrier-Carrier Scattering. Carrier-carrier scattering determines the exchange of energy between carriers, and is primarily responsible for the thermalization of photoexcited non-thermal carriers (Sect. 1.1.5). This process is mediated by Coulomb interaction and includes electron-electron, hole-hole and electron-hole scattering. The difference in the masses between electrons and holes reduces the energy exchange between these two species. The long-range Coulomb interaction diverges in the absence of screening. Screening is generally included by calculating the frequency and wavevector dependent dielectric function of the semiconductor with a proper treatment of the plasmon-phonon coupling [1.50]. This may present a difficult challenge in many cases. Some aspects of this problem are discussed in Chap. 4.

(c) Spin Relaxation Processes. In addition to the energy and momentum relaxation processes discussed above, spin relaxation processes can also play an important role in the dynamics of photoexcited systems. This aspect can be investigated by studying the variation of the optical spectra for different polarizations. Exciton spin relaxation processes will be discussed in Chap. 6.

(d) Scattering Processes Specific to Quantum Wells. The scattering processes in quasi-two-dimensional semiconductors are similar to those discussed above for bulk semiconductors, but the scattering rates are modified by several factors such as different band structure, phonon modes and density-of-states in these semiconductors. One major difference in the carrier-phonon scattering rates results from the fact that only $k_{||}$, the momentum parallel to the interface planes, is conserved. There is a large literature on electron-phonon scattering rates in quantum wells [1.51, 52].

Some scattering processes specific to quantum wells are also important. Inter-subband scattering, capture of carriers from the barriers into quantum wells, and real space transfer of carriers from quantum wells into the barriers play an important role in specific situations. Tunneling of carriers between quantum wells and thermionic emission over the barriers contribute to the transport of carriers in MQW structures, and may well determine the speed of quantum well devices based on MQW. We will discuss these transport-related processes in Chap. 7 on tunneling and Chap. 8 on perpendicular transport.

1.1.5 Carrier Relaxation: Four Regimes

After a semiconductor in thermodynamic equilibrium is excited by an ultrashort pulse, it undergoes several stages of relaxation before it returns once again to the thermodynamic equilibrium. Investigation of these relaxation processes is a major theme of this book, but before we get into details, it is useful to provide a bird's eye view of these relaxation processes. The carrier relaxation can be classified into four *temporally overlapping* regimes:

(a) Coherent Regime. Excitation of a semiconductor with an ultrashort laser pulse creates excitation in the semiconductor with well-defined phase relationship within the excitation and with the electromagnetic field that created the excitations. The laser pulse can create either real or virtual excitation. This coherent regime exhibits many interesting phenomena which are elegant manifestations of basic quantum mechanics in semiconductors. Coherences in atomic and molecular systems and defects in solids have been investigated extensively, but coherence of intrinsic states in semiconductors has been explored only in recent years. This is because the scattering processes that destroy the coherence are extremely fast in semiconductors and pico- and femtosecond techniques are required to study the coherent regime in semiconductors. Chapter 2 will discuss many of the interesting recent investigations of coherence in semiconductors, as well as investigation of various scattering processes that destroy the coherence.

(b) Non-Thermal Regime. In the case of real excitation, the distribution of the excitation (free electron-hole pairs or excitons) after the destruction of coherence through dephasing is very likely to be non-thermal, i.e. the distribution function can not be characterized by a temperature. Investigation of this regime provides information about various processes (such as carrier-carrier or exciton-exciton scattering) that bring the non-thermal distribution to a hot, thermalized distribution. Non-thermal carriers and excitons can be probed by optical spectroscopy, and are discussed in Chaps. 3 and 6, respectively.

(c) Hot-Carrier Regime. Carrier-carrier (or exciton-exciton) scattering is primarily responsible for redistributing the energy within the carrier (exciton) system, and leads to a thermalized distribution function of carriers (excitons), i. e. a distribution that can be characterized by a temperature. The temperature can be, and usually is, higher than the lattice temperature, and may be different for different sub-systems (electrons, holes, excitons). The thermalization times depend strongly on many factors such as carrier density. Typically, the electrons and holes thermalize among themselves in hundreds of femtoseconds, while the electrons and holes achieve a common temperature in a couple of picoseconds. The thermalized electron-hole pairs reach lattice temperatures in hundreds of picoseconds through interaction with various phonons in the semiconductor. This interaction may lead to a large, wavevector-dependent population of non-equilibrium phonons. Therefore, this regime is described by various systems such as electrons, holes and non-equilibrium phonons which are not in equilibrium with each other.

Investigation of this hot carrier regime focuses on the rate of cooling of carriers to the lattice temperature and leads to information concerning various carrier-phonon, exciton-phonon, and phonon-phonon scattering processes. These will be discussed in Chaps. 4, 5 and 6.

(d) Isothermal Regime. At the end of the hot-carrier regime, all the carriers, phonons and excitons are in equilibrium with each other, i.e. can be described by the same temperature, that of the lattice. However, there is still an excess of electrons and holes compared to the thermodynamic equilibrium. These excess electron-hole pairs (or excitons) recombine either radiatively or non-radiatively and return the semiconductor to the thermodynamic equilibrium. We will not be concerned with this last regime in this book, although radiative recombination of excitons will be discussed in Chap. 6 because it contributes to the initial dynamics of excitons in 2D systems.

These temporally-overlapping regimes are schematically illustrated in Fig. 1.7 with some typical processes that occur in each relaxation regime [1.53]. The time scale for each event depends very strongly on parameters such as band structure, excess energy, the nature of the excitation (free carrier vs exciton), the density of excitation, the lattice temperature and so on. Many of these processes will be discussed in subsequent chapters.

It should be emphasized that many of the physical processes leading to relaxation in the different regimes are occurring simultaneously. For example, the processes that destroy coherence may also contribute to thermalization of carrier distribution functions, and emission of phonons may occur while the electron and holes are thermalizing to a hot distribution. The non-thermal carrier distribution function created by a femtosecond pulse is influenced by the dephasing of coherent polarization during the pulse. Nonetheless, this description in terms of four relaxation regimes does provide a convenient

Four relaxation regimes
in photoexcited semiconductors

0 ──────── Ultrashort laser pulse

Coherent Regime (≤200 fs)
- Momentum scattering
- Carrier-carrier scattering
- Intervalley scattering ($\Gamma \rightarrow L, X$)
- Hole-optical-phonon scattering

Non-thermal regime (≤2 ps)
- Electron-hole scattering
- Electron-optical-phonon scattering
- Intervalley scattering $(L, X \rightarrow \Gamma)$
- Carrier capture in quantum wells
- Intersubband scattering ($\Delta E > \hbar\omega_{LO}$)

Hot-excitation regime (~1–100 ps)
- Hot-carrier-phonon interactions
- Decay of optical phonons
- Carrier-acoustic-phonon scattering
- Intersubband scattering ($\Delta E < \hbar\omega_{LO}$)

Isothermal regime (≥100 ps)
- Carrier recombination

Time

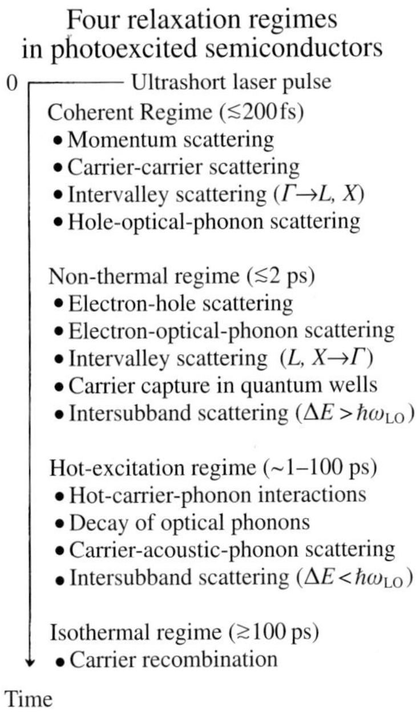

Fig. 1.7. Four temporally-overlapping relaxation regime in photoexcited semiconductors, with some typical scattering and relaxation processes for each regime [1.108]

framework for describing and discussing the dynamics of relaxation in semiconductors.

1.1.6 Carrier Transport

As we discussed at the beginning of this chapter, one of the strengths of optical spectroscopy is its ability of investigating the dynamics of carrier transport in semiconductors. For lateral transport (parallel to the surface), this is accomplished by monitoring linear or nonlinear optical properties at different points on the surface. Spatial resolution down to a small fraction of a micrometer can be achieved by spatial imaging techniques and recent advances in near-field scanning optical microscopy [1.54, 55]. For perpendicular transport (normal to the surface), the most commonly used technique is to incorporate *optical markers* in the semiconductors [1.56]. Optical markers are thin layers of semiconductors which have specific spectral properties that can be monitored to study the arrival or departure of carriers in that spatial region. There is a unique relationship between the spatial position and the spectral properties of the optical marker region. Chapters 7 and 8 will discuss how the techniques of optical markers and ultrafast optical spectroscopy can be combined to investigate the dynamics of carrier transport and tunneling in semiconductor nanostructures.

1.2 Ultrafast Lasers

Several techniques have been developed in the past thirty years to generate ultrashort pulses. Ideally one would like to have ultrashort laser pulses of desired pulse width and pulse shape, wavelength, pulse energy, and repetition rate. A single laser source obviously cannot cover the entire range of desired parameters, and different approaches have been developed. Development of ultrafast lasers is an area of continued intense activity and with a large literature [1.57–59]. We present a very brief summary, with some selected references.

Research on the generation of ultrashort laser pulses started almost immediately after the first demonstration of lasers. Techniques for switching the quality factor (Q) of the laser cavity typically generated nanosecond pulses [1.60]. A laser oscillates in many longitudinal modes supported by the cavity and the gain spectrum. The technique of mode-locking, in which these modes are locked in phase, was soon developed [1.60] by using a saturable absorber in the laser cavity. This and many other mode-locking techniques are now being used to generate ultrashort pulses [1.57–59].

Enormous progress has been made in the generation of ultrashort pulses since the first solid-state modelocked lasers which did not provide tunability. Organic dye lasers, with wide gain spectra and the potential for femtosecond pulses, were invented in the late 1960s, and rapidly developed into versatile tools for spectroscopy because of their ultrashort pulse widths and tunability in the visible and the near-IR range. Techniques have also been developed to amplify and compress these short pulses, to generate different frequencies using nonlinear second harmonic generation and nonlinear mixing, and to generate ultrashort pulses of white-light continuum by passing an intense ultrashort pulse through a liquid [1.61]. New solid state materials with wide gain spectra have been developed and have been used to generate pulses shorter than 10 fs directly from the laser oscillators without having to resort to pulse amplification and compression [1.62–72]. Optical Parametric Oscillators (OPO) pumped by such solid-state lasers [1.73–78] have increased the tunability range. These solid-state lasers have practically replaced the dye lasers as the basic ultrafast spectroscopy tool, and pumping these lasers with semiconductor diode lasers has made it possible to think of compact ultrafast sources. The availability of ultrafast lasers with a wide range of pulse widths, wavelengths, pulse energy, and pulse repetition rate makes it possible not only to investigate a broad range of physical phenomena using ultrafast optical spectroscopy, but also investigate practical applications of ultrafast technology.

1.3 Ultrafast Spectroscopy Techniques

Many techniques have been developed to investigate linear and nonlinear optical properties of semiconductors using ultrafast lasers. We provide a brief description of important aspects of some of the more commonly used techniques.

1.3.1 Pump-Probe Spectroscopy

This is the most common form of ultrafast spectroscopy. In its simplest form (*degenerate* pump-probe spectroscopy), the output pulse train from an ultrafast laser is divided into two. The semiconductor sample under investigation is excited by one pulse train (pump) and the changes it induces in the sample are probed by the second pulse train (probe), which is suitably delayed with respect to the pump by introducing an optical delay in its path. Some property related to the probe (e.g. reflectivity, absorption, Raman scattering, luminescence) is then monitored to investigate the changes in the sample produced by the pump. For suitably thin samples, the time resolution in pump-probe spectroscopy is limited only by the pulse width of the laser or the jitter between the lasers, if two lasers are used. In *non-degenerate* pump-probe spectroscopy, one uses two synchronized lasers at different wavelengths, or a laser and a synchronized white light continuum. This increases the versatility of the technique enormously by allowing a determination of the changes induced at photon energies different from that of the pump.

Many different pump-probe spectroscopy techniques have been developed based on different techniques of linear optical spectroscopy. We discuss them briefly in the following sections.

(a) Transmission/Reflection Spectroscopy. Perhaps the simplest of the pump-probe spectroscopy techniques is the measurement of the transmitted or reflected probe. The probe is typically much weaker than the pump, and the spot diameter of the probe at the sample is ideally considerably smaller than that of the pump so that it probes a region of uniform photoexcited density. A schematic diagram of a generic set-up for a two-beam nonlinear experiment is depicted in Fig. 1.8. For pump-probe transmission spectroscopy, one generally measures the *change* in the transmitted probe pulse energy induced by the pump as a function of the time delay between the pump and the probe pulses by chopping the pump beam and using a lock-in amplifier. Some experimental situations require that the pump and the probe beams are chopped at different frequencies, and the change in the transmitted probe is detected at, for example, the sum of the two chopping frequencies. The time resolution is provided at the sample by the delay between the pump and the probe, so the transmitted probe is detected with a slow detector, and can be spectrally resolved if desired. The data are typically presented in the form of normalized Differential Transmission (DT) given by $\Delta T/T_0 = (T - T_0)/T_0$, the change in transmission $\Delta T = T - T_0$ induced by the pump pulse divided by the transmission of the probe in the absence of the pump, T_0. The Differential Transmission *Spectrum* (DTS) can be obtained either by varying the wavelength of the probe laser or by spectrally analyzing the transmitted probe energy with a spectrometer if

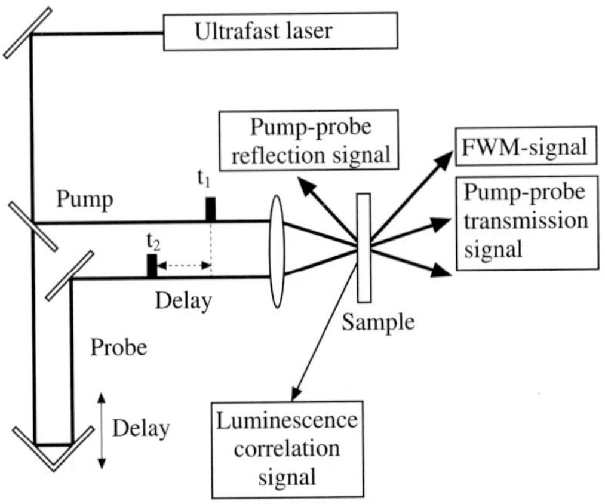

Fig. 1.8. Schematic for a generic setup for a two-beam nonlinear experiment. This setup can be used for pump-probe transmission or reflection spectroscopy, FWM spectroscopy, or luminescence correlation spectroscopy as shown (see the text). If the laser source is sufficiently intense, then white light continuum can be generated in a liquid cell and used as a probe pulse for pump-probe transmission or reflection measurements

a broadband ultrafast source (e.g. a white light continuum) is used as the probe. The use of a multichannel detector such as a diode array or a CCD camera is generally preferable in the latter case. Although one generally measures the total transmitted probe energy, the transmitted pulse shape can be measured by a suitable fast detector or a nonlinear gating technique (Sect. 1.3.3c). Different possibilities of processing the nonlinear pump-probe or FWM (Four-Wave-Mixing) signals (Sect. 1.3.2) are illustrated in Figs. 1.9 and 1.10.

The strength of this pump-probe spectroscopy technique is its ability to obtain time-resolution limited by the laser pulse width. It is also easy to detect the transmitted signal, but if one wants to measure very small differential transmission, careful attention has to be paid to minimizing all the noise com-

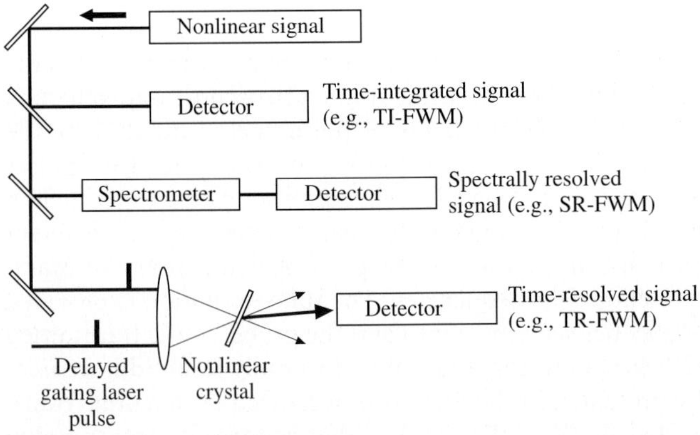

Fig. 1.9. Schematic experimental arrangements for analyzing the nonlinear pump-probe or FWM signals in different ways: Time-Integrated (TI) to measure the spectrally- and temporally-integrated nonlinear signals, Time-Resolved (TR) to measure the temporally resolved (spectrally-integrated) nonlinear signal, and Spectrally-Resolved (SR) to measure spectrally-resolved (but time-integrated) signals

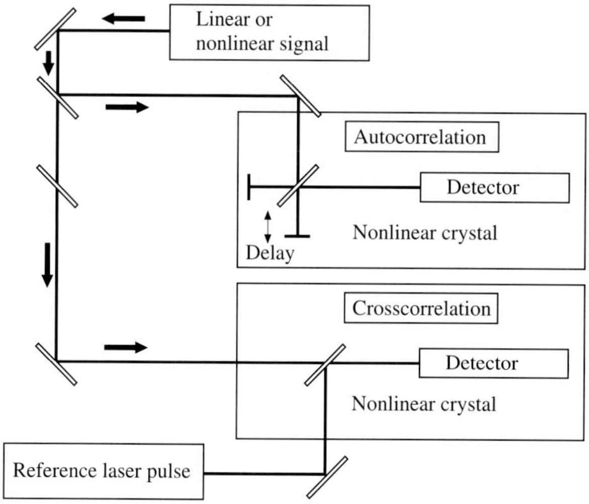

Fig. 1.10. Schematic experimental arrangement for interferometric auto- and cross-correlation measurements of linear or nonlinear signals

ponents. In the absence of any other noise, photon shot noise becomes the limiting factor.

A variation of the pump-probe transmission/reflection technique is the Electro-Optics Sampling Technique (EOS) which can be used with Transmitted (TEOS) or Reflected (REOS) signals [1.79, 80]. In this case, the two polarization components of the transmitted or reflected beam are analyzed and detected separately to provide additional information about the nature of the changes induced in the sample by the pump.

(b) Raman Spectroscopy. In this form of pump-probe spectroscopy, the changes induced by the pump are measured by the change in the Raman scattering signal generated by the probe laser. In degenerate pump-probe Raman spectroscopy, the two beams are derived from the same laser and are generally cross-polarized to distinguish the Raman signals generated by the pump and the probe. If the two lasers have different wavelengths, the signals can be distinguished spectrally. The time-integrated Raman signal of the probe, or the total energy of the Raman-scattered probe pulse, is measured as a function of the time delay between the pump and the probe. Both vibrational and electronic Raman signals have been used in this form of time-resolved Raman spectroscopy to monitor the dynamics of photoexcited phonons [1.81] and electronic excitations [1.82]. Application of this technique for investigating intersubband scattering and phonon dynamics are discussed in Chaps. 3 and 5, respectively.

1.3.2 FWM Spectroscopy

In the simplest form of FWM spectroscopy, the two-beam Degenerate Four-Wave-Mixing (DFWM), the pump photons with wavevector $\boldsymbol{q}_1$ generate a coherent polarization in the sample, and the probe photons with wavevector $\boldsymbol{q}_2$ arrive at the sample after a delay time τ_d. If τ_d is smaller than the dephasing time of the polarization, an interference grating is produced from which $\boldsymbol{q}_2$ can be self-diffracted along the phase matched direction $\boldsymbol{q}_d = 2\boldsymbol{q}_2 - \boldsymbol{q}_1$. The diffracted energy (i.e. time-integrated FWM signal) is measured as a function of τ_d using lock-in detection with a single or double chopper as described above.

In a three-beam degenerate FWM (Fig. 1.11), the sample is excited by three beams traveling along $\boldsymbol{q}_1$, $\boldsymbol{q}_2$, and $\boldsymbol{q}_3$ arriving at the sample at times t_1, t_2 and t_3, respectively. τ_{21}, τ_{31}, and τ_{32} ($\tau_{ij} = t_i - t_j$) define the delays between the three beams. The diffracted total energy along the phasematched direction $\boldsymbol{q}_d = \boldsymbol{q}_3 + \boldsymbol{q}_2 - \boldsymbol{q}_1$ is measured as a function of various time delays to obtain different information about the system being investigated. The ability to individually control the parameters of three beams lends greater versatility to the three-beam FWM compared to two-beam FWM discussed in the previous paragraph. For example, in the transient grating experiments, $\tau_{12} = 0$ [1.83], and the diffracted signal as a function of τ_{13} provides information about the population decay time (T_1) of the system. In the phase-conjugation configuration, two of the beams travel in opposite directions, a geometry that was used to study disorder in semiconductor quantum wells [1.84] as we will discuss in Chap. 2. Information about the nature of broadening (homogeneous or inhomogeneous) [1.85], spectral relaxation [1.85], and non-Markovian relaxation [1.86–89] can also be obtained from three-beam FWM.

Although most of the FWM experiments measure the total diffracted energy, the temporal profile of the diffracted pulse can be measured to obtain addi-

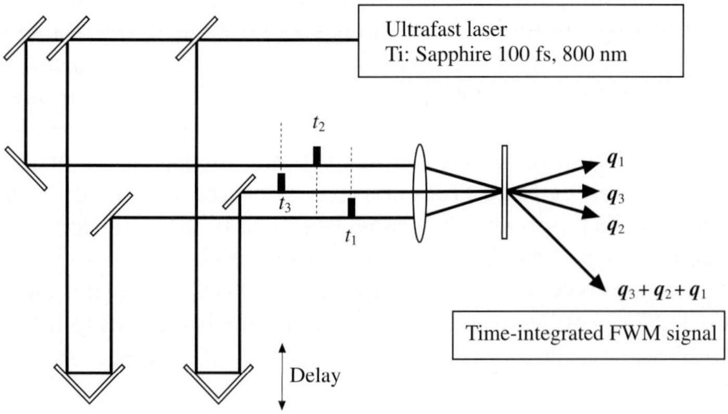

Fig. 1.11. Schematic experimental arrangement for degenerate three-beam FWM measurements

tional information. Early measurements on photon echoes [1.90] from defects in solids with long dephasing times resolved the echo signal in time using an appropriately fast detector and usual electronics. With short dephasing times in semiconductors, one needs to use a streak camera or nonlinear upconversion techniques (Sect. 1.3.3c) [1.91] to obtain high temporal resolution. Such Time-Resolved FWM (TR-FWM) experiments [1.92, 93] have provided additional information about semiconductors. Similarly, only a specific spectral component of the diffracted signal is measured in the Spectrally-Resolved FWM (SR-FWM) which also provides considerable additional information about the system. Furthermore, the time-dependent amplitude and the phase of the diffracted pulse can be measured using interferometric techniques [1.94, 95] to provide important additional information (Sect. 1.3.4). We will discuss examples of such measurements in Chap. 2. Experimental setups for some of these FWM techniques are schematically illustrated in Figs. 1.9 and 1.10.

Since optical transitions follow various selection rules, different combinations of linear polarizations or circular polarizations can be employed in the FWM experiments to obtain different information about the semiconductor under investigation. For example, if the two beams are co-linearly polarized, they can create a density grating, but cross-linearly polarized beams can produce orientational grating. Similarly, different combinations of circular polarizations can be used to investigate two-photon coherences.

Finally, we note that pump-probe spectroscopy and FWM spectroscopy are only two particular forms of nonlinear spectroscopy. Many other nonlinear spectroscopy techniques have been developed [1.96, 97] and applied to many different physical systems, including semiconductors. For example, using two independently tunable ultrafast lasers and adjusting the difference in the photon energies to correspond to the energy of a phonon in the system, it is possible to generate a coherent Raman excitation of the phonon, and to investigate anti-Stokes scattering from this excitation using an appropriate third beam. This technique of Coherent Anti-Stokes Raman Scattering (CARS) has been applied to various physical systems [1.96, 97], including semiconductors [1.98, 99], and some of the results on semiconductors using this technique will be discussed in Chap. 5 on phonon dynamics.

Most of the FWM experiments are performed in the transmission geometry, with the diffracted beam in the forward direction, as indicated in the experimental schematic (Fig. 1.8). *Honold* et al. [1.100] have proposed and demonstrated the use of FWM experiments in which the incident beams all come from one direction and the diffracted signal is measured in the backward direction (reflection geometry). This technique is particularly valuable for thin layers of sample (such as a single quantum well of GaAs grown on a GaAs substrate) for which the substrate is strongly absorptive at the wavelengths of interest and detection of FWM signal in the forward direction requires the removal of the substrate, possibly introducing strain in the thin layer of interest. Some of the experiments discussed above were performed in this geometry.

This brief discussion shows that the various forms of FWM spectroscopy provide a powerful means of investigating the coherent response of matter.

Investigation of the coherent regime of semiconductors using these techniques will be discussed in Chap. 2.

1.3.3 Luminescence Spectroscopy

Different techniques have been employed to measure time-resolved luminescence spectra of semiconductors with picosecond and femtosecond time resolutions. These luminescence techniques are used not only to investigate carrier relaxation processes, but also carrier transport processes in semiconductor nanostructures. Chapters 3, 4, 6, 7 and 8 provide examples of application of these techniques.

(a) Direct Detection. The simplest technique is to measure luminescence dynamics directly with a fast detector and fast electronics. This includes the use of fast photodiodes which can provide time resolution in the picosecond range for intense pulses, but are not well-suited for low-level signal detection. Fast channel-plate photomultipliers with time-to-amplitude conversion electronics provide time resolution of tens of picoseconds, and a very high dynamic range. Perhaps the best technique for direct measurement is the use of streak cameras. Considerable progress has been made recently in the technology of streak cameras which can now provide sub-picosecond time resolution in single-shot operation, and about 10 ps in synchroscan mode for low-level repetitive signals. Two-dimensional streak cameras which provide spectral and temporal information simultaneously have become very attractive and versatile tools for time-resolved luminescence spectroscopy in 10 ps time range. Better time resolution are not routinely available with this technique and the dynamic range is limited to about two orders-of-magnitude, but a factor of 2 – 3 improvement in time resolution may be possible with special efforts.

(b) Correlation Spectroscopy. This is really a form of pump-probe spectroscopy in which two pulse trains delayed by τ_d excite the semiconductor and the *time-integrated* luminescence spectra are measured as a function of τ_d [1.101 – 104]. If the luminescence signal is a nonlinear function of the excitation density, then the signal with two excitation pulses is not simply the sum of the signal with each beam exciting the sample separately. In practice, one chops the two beams at two different frequencies, and the nonlinear correlation signal is measured at the sum frequency. The advantages of the technique are its simplicity, and high time resolution limited only by the laser pulse width. The disadvantage is the complexity in the interpretation of the results.

(c) Upconversion Spectroscopy. Luminescence upconversion spectroscopy provides a means for measuring low-level luminescence signal with time resolution limited by the laser pulse width [1.91, 105, 106] and has been applied to a wide variety of investigations in semiconductors [1.107, 108]. The output of an ultrafast laser is divided into two beams and the sample is excited by one

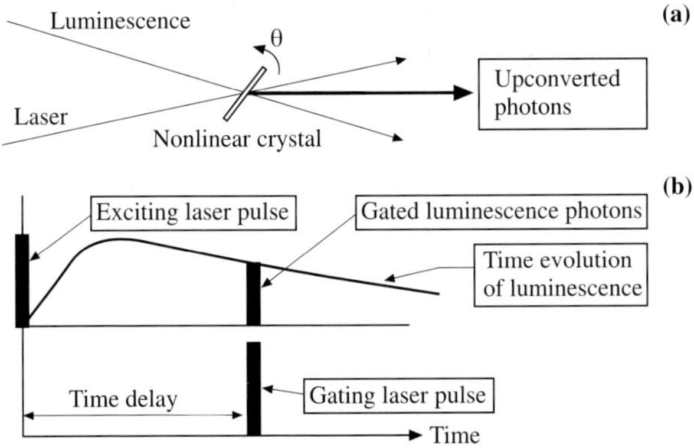

Fig. 1.12. **(a)** Schematic of the luminescence upconversion setup using either an ultrafast single laser for both exciting the sample and gating the luminescence in the nonlinear crystal, or one laser for exciting the semiconductor (resonantly or non-resonantly) and a synchronized ultrafast laser at a different wavelength for gating the luminescence in the nonlinear crystal; **(b)** Schematic showing how sum frequency generation acts as a gate

of these two beams. The luminescence from the sample is collected in the usual manner and focused on a nonlinear crystal. The second laser beam is suitably delayed and also focused on the nonlinear crystal to overlap the luminescence spot. A schematic of the upconversion setup is shown in Fig. 1.12. The angle between the two beams and the angle of the nonlinear crystal are adjusted for phasematched sum-frequency generation at a given luminescence photon energy. The sum-frequency signal is generated only during the gating laser pulse, as illustrated in Fig. 1.12. The time resolution is limited by the laser pulse width, and group velocity dispersion in the nonlinear crystal [1.91]. The latter can be made small compared to the laser pulse width by using a sufficiently thin nonlinear crystal. Since the time resolution is obtained at the nonlinear crystal, the sum-frequency signal is dispersed by a spectrometer and detected in a time-integrated manner by photon counting electronics. Time evolution of luminescence at a given photon energy can be obtained by scanning the relative delay between the two laser beams. Time-resolved luminescence spectra can be obtained by keeping the delay fixed and scanning the angle of the nonlinear crystal and the spectrometer in a synchronized manner.

If the luminescence wavelength is close to the laser wavelength, the optimum angle for sum frequency generation may be quite close to the optimum angle for the second-harmonic generation of the gating pulse. Under these conditions, the nonlinear crystal generates a strong second harmonic of the gating pulse even when the crystal is adjusted for optimum sum-frequency generation. One of the drawbacks of the upconversion technique is that this strong second-harmonic signal often prevents the measurement of the weak sum-frequency signal from resonant or near-resonant luminescence. One technique to avoid this is to use two synchronized lasers at different wavelengths, one to resonantly excite the sample and the other as a gating pulse on the nonlinear crystal [1.109–111]. The time resolution in such cases is determined not only by the pulse width and crystal parameters, but also by the time jitter between the pulses of the two lasers. For dye lasers, it is difficult to reduce the jitter to less than a couple of picoseconds. However, the jitter in the recently-developed optical parametric oscillators pumped by solid-state lasers can be reduced to less than 100 fs, allowing such two-color upconversion schemes to provide 100 fs time resolution [1.111]. Application of this technique to exciton luminescence will be discussed in Chap. 6.

Other forms of nonlinear frequency-mixing techniques (e.g. difference frequency mixing) can also be employed, and different crystals covering a wide range of wavelengths can be used. This technique provides a versatile, high-sensitivity technique with laser-pulsewidth-limited time resolution. Although the ultimate time-resolution of the technique is better than that of a streak camera by more than two-orders of magnitude, the sensitivity in most cases is lower than that of the best streak cameras.

1.3.4 Interferometric Techniques

The period of an optical cycle near 8000 Å is approximately 3 fs which corresponds to a distance of 0.9 μm in air. Micropositioners with accuracy better than 0.1 μm are now available so that relative delays between two pulse trains can be controlled to a small fraction of the optical period. This opens up the possibility of measuring not only the intensity but also the amplitude and the phase of the transmitted or diffracted signals using interferometric techniques. This is schematically illustrated in Fig. 1.10. Such techniques can be used not only to characterize a short laser pulse, but also to obtain information about the system under investigation not available through the intensity measurements. These interferometric techniques are just beginning to be used in spectroscopic measurements on semiconductors. Some of these results will be discussed in Chap. 2.

1.3.5 Terahertz Spectroscopy

Ultrafast lasers can generate coherent mm- and submm-wave radiation by a variety of different techniques. These include photoexcitation of photoconductive gaps under bias [1.112, 113], sometimes with appropriate antenna structures [1.114–116], photoexcitation of semiconductor surfaces which have a built-in surface potential [1.117], photoexcitation of semiconductor surfaces under external bias [1.118], and photoexcitation of semiconductor nanostructures [1.119, 120]. This radiation can be collimated and transmitted over reasonable distances, and can be detected using optically-gated photoconductive antennas. By adjusting the delay between the THz signal and the gating pulse, information about the amplitude and the phase of the THz signals can be obtained. A variety of experiments in different physical systems have been performed [1.57, 121, 122]; typical experimental arrangements and some of recent results will be discussed in Chap. 2.

1.4 Interpretation of Results

The experimental techniques discussed above investigate the ultrafast response of a semiconductor by measuring either the nonlinear properties (e.g., pump-probe and FWM spectroscopy) or the linear properties (e.g., luminescence spectroscopy) following the photoexcitation. The most general approach to analyzing these dynamical results is to use quantum kinetic equations with non-equilibrium Green's functions [1.123 – 132]. Although considerable progress has been made in recent years in obtaining results using this approach, such calculations require immense effort. Most of the experiments have therefore been analyzed using simplified approaches which make several approximations. We discuss in this section some general considerations on how the results of ultrafast spectroscopy of semiconductors are analyzed.

1.4.1 FWM Spectroscopy

The nonlinear response of a semiconductor can, in most cases, be described by the third-order polarization induced by the optical fields. The diffracted signal in a FWM experiment measures only the coherent response of the semiconductor whereas the differential transmission or reflection signal is determined by the coherent as well as the incoherent dynamics. The response of a system is determined not only by the photoexcited populations (the diagonal components of the density matrix), but also by the photoexcited polarizations (the off-diagonal components of the density matrix). Atomic systems are usually analyzed using the independent two-level approximation of the Liouville equations, or the optical Bloch equations [1.96, 97, 133, 134]. In semiconductors the Coulomb interactions between various polarization components cannot be ignored and the optical Bloch equations must be suitably modified to take this into account. The coherent dynamics of semiconductors is generally analyzed by calculating the third order nonlinear polarization using these modified equations, known as the semiconductor Bloch equations [1.132]. This approach to understanding coherent dynamics probed by FWM spectroscopy will be discussed in Chap. 2.

1.4.2 Pump-Probe Spectroscopy

The dynamics in a pump-probe transmission experiment is determined by coherent and incoherent response of the system. If the time scale is comparable to or shorter than the dephasing times of the system, one must once again use the semiconductor Bloch equations to calculate the third-order nonlinear polarization in the appropriate direction. Just as the FWM signal corresponds to the third-order polarization radiating in the diffracted direction, the pump-probe signal corresponds to the third-order polarization radiating in the probe direction. The pump-probe signal can be viewed as the diffraction of the pump beam in the probe direction ($q_d = q_1 + q_2 - q_1 = q_2$). However, if the time scale of the experiment is long compared to the dephasing times, it may be possible to achieve great simplification by ignoring the coherent effects, i.e., by ignoring the off-diagonal components of the density matrix and considering only the diagonal components corresponding to the population of various states.

The effects in this incoherent regime can be qualitatively divided into two general classes: the first class encompasses *many-body effects* which result, for example, in a change of the energy band structure (e.g., band-gap renormalization or a change in the exciton binding energy), broadening of energy levels, or a change of the various matrix elements. The second class encompasses *occupation effects*, or changes in the optical properties or transition rates brought about by the nonequilibrium occupation of certain states (e.g., the interband absorption at a given photon energy is reduced if the final state is partially occupied by electrons). In general, these two effects are coupled because the many-body effects depend not only on the total density of the photoexcited carriers or excitons but also on their distribution functions.

The many-body effects depend strongly on the photoexcited density and the nature of the system being investigated (e.g., excitons vs free carriers). For excitons, many-body effects such as energy shifts, broadening and bleaching induced by screening and phase-space filling dominate even at moderate densities. For free carriers, on the other hand, important effects such as band-gap renormalization and changes in the optical matrix elements become important only at relatively high excitation density. If the density is high enough, both the many-body effects and the occupation effects must be considered.

Under certain experimental conditions, it may be a good approximation to assume that the many-body effects depend only on the photoexcitation density and are independent of time for the time scale of interest. Furthermore, for certain spectral regions, the change in the matrix elements for interband transitions may be small enough to be neglected. Under these conditions, the analysis of experiments is considerably simplified. If f_e and f_h are the photoexcited electron and hole distribution functions at energies E_e and E_h coupled by a photon of energy $h\nu$, then the change in the absorption coefficient at photon energy $h\nu$ is given by

$$\Delta\alpha(h\nu) = (1 - f_e - f_h)\,\alpha_0(h\nu) \ , \tag{1.3}$$

where $\alpha_0(h\nu)$ is the absorption coefficient of the unexcited semiconductor at $h\nu$. This is valid in a direct-gap semiconductor such as GaAs for free carriers with kinetic energies larger than the plasma frequency where $\alpha_0(h\nu)$ does not change significantly with photoexcitation. For sufficiently small $\Delta\alpha(h\nu)d$, where d is the thickness of sample over which the change in absorption is induced,

$$\mathrm{DTS}(h\nu) = [T(h\nu) - T_0(h\nu)]/T_0(h\nu) = -\Delta\alpha(h\nu)d \ , \tag{1.4}$$

where T and T_0 are the transmitted intensities at $h\nu$ with and without photoexcitation, and DTS is the differential transmission spectrum. With these simplifications, the results can simply be related to the distribution functions of electrons and holes as a function of time. We emphasize that the full complexity of the semiconductor Bloch equations and distribution function-dependent many-body effects must be considered in the general case, and the conditions for the validity of the simplified approach must be carefully scrutinized before applying it to a specific experiment.

1.4.3 Raman Spectroscopy

The first-order Raman measurements (involving a single phonon or a single electronic excitation) in general probes the excitation only at a certain well-defined wavevector determined by energy and momentum conservation (although the selection rules can be relaxed in the case of disorder or very short absorption lengths). Therefore the time-resolved pump-probe Raman signals are generally analyzed in terms of the temporal evolution of the occupation of this particular mode. The anti-Stokes intensity is directly proportional to this occupation and provides a good measure of the dynamics.

1.4.4 Luminescence Spectroscopy

Resonant luminescence is sensitive to both coherent and incoherent dynamics. In the coherent regime it may be difficult to distinguish between resonant luminescence and resonant Rayleigh scattering. This question is considered in Chap. 6 in connection with exciton luminescence. Once the coherence is lost, luminescence is sensitive to only the many-body and occupation effects, just like the pump-probe spectroscopy. Non-resonantly excited luminescence is almost always in this regime. The interpretation of luminescence results may in some cases be simpler than the interpretation of pump-probe results because the latter is a nonlinear technique while luminescence is a linear technique. For example, the interband luminescence of a given spin-state of an exciton is determined only by the population of that state, whereas the DTS at this exciton energy may be influenced by the population of excitons in other exciton states, and the population and the distribution function of the free carriers. Luminescence and pump-probe spectroscopy provide useful complementary information in general.

In the simplest case, corresponding to the case for DTS discussed above, the luminescence intensity at the photon energy $h\nu$ is given by

$$L(h\nu) = \alpha(h\nu) f_e f_h \cong \alpha_0(h\nu) f_e f_h \ . \tag{1.5}$$

The last approximation is valid if the change in the absorption coefficient induced by the photoexcitation is sufficiently small. In this case, the results are determined by the temporal evolution of the product of the appropriate electron and hole distribution functions. The pump-probe spectroscopy results, on the other hand, are determined by the *sum* of the appropriate electron and hole distribution functions (1.4).

The assumption that the absorption coefficient can be approximated by its unperturbed value is reasonable for photon energies that exceed the band-gap energy by the energy of the photoexcited plasma. Under these conditions, a simultaneous measurement of DTS and luminescence spectrum can provide a means of determining both f_e and f_h. In general, this is not possible close to the band gap because of excitonic effects as well as excitation-dependent absorption coefficient. The luminescence spectrum at energies which are several kT higher than the Fermi energy can be described in terms of a simple exponential with an effective temperature. The effective temperature is determined by the electron and hole temperatures, and equals the carrier temperature if the electrons and holes are at the same temperature. The accuracy of the temperature determined by such an analysis depends on the dynamics range of the signal, and is generally better for the case of luminescence spectra compared to the DTS results. For the case of a high photoexcitation density, the low-energy tail of the luminescence spectrum is influenced by indirect, plasma-assisted processes, as well as gain and band-gap renormalization.

1.4.5 Calculation of the Dynamics

It has become clear in recent years that dephasing of coherence during the generation process plays an important role in determining the initial photoexcited distribution function, and efforts are beginning [1.114, 116, 135–140] to include the influence of coherent polarization induced by photoexcitation in the calculations. However, most of the pump-probe transmission, reflection or Raman experiments and most of the luminescence experiments to-date have been analyzed in the simplified limit in which coherent effects are neglected and many-body effects are considered to be independent of the time-dependent distribution functions. In this limit, the dynamical response of a photoexcited semiconductor can be calculated within the framework of the semiclassical Boltzmann equation. Many different techniques, involving both analytical and numerical approaches, have been developed to solve the Boltzmann equation in the field of high-field transport in semiconductors [1.7, 43, 44].

Many of the physical processes that determine the dynamics of a semiconductor excited by an ultrashort laser pulse are the same as those that determine high-field transport in semiconductors. High-field transport and transient dynamics can be thought of as complementary ways of investigating some of the same physics. The scattering processes that determine the transport properties under applied electric field also determine the relaxation of non-equilibrium excitations produced by photoexcitation. Therefore, these approaches developed for high-field transport can easily be modified to investigate the dynamics following photoexcitation. We discuss two of these approaches based on the Boltzmann equation in the following subsections.

(a) Balance-Equations Approach. Forming the moments of the Boltzmann equation, one can obtain several equations relating the changes in the energy and momentum of a carrier with the external perturbations. This method was applied to the problem of non-ohmic transport in semiconductors using the electron-temperature model [1.3, 5, 7, 41, 42, 141] and was extended to include the effects of nonequilibrium phonon populations [1.142–144]. This technique was later applied to the study of the dynamics of photoexcited semiconductors [1.145–147]. This balance-equations approach provides a computationally simpler approach, but the simplicity is achieved through the use of simple band structure, and simple analytic distribution functions for carriers. This limits the validity of the approach to the hot-carrier relaxation regime (Sect. 1.1.5c). The direct numerical-integration methods (e.g., the Monte Carlo simulation method discussed in Sect. 1.4.5b) overcome these limitations and are applicable on the shorter time scales to the nonthermal relaxation regime (Sect. 1.1.5b). But these direct methods may be impractical for times longer than a few picoseconds because of the small time-steps used in such calculations. Thus the two approaches complement each other. A detailed discussion of the hot-carrier cooling using the balance equations approach has been recently given by *Pötz and Kocevar* [1.148], and will be discussed in Chap. 4.

(b) Monte Carlo Simulations. Monte Carlo calculations have provided a powerful means of simulating the dynamics of photoexcited semiconductors. This technique has so far been applied only to dynamics of free carriers, although the dynamics of exciton distribution function can also be investigated by the same technique. In the early days, the technique followed a single particle through a large number of two-step iterations. The first step is the free flight of the particle, terminated at time t selected on a random basis using a predetermined set of scattering probabilities. In the second step, the particle is scattered from the state at the end of this free flight into a new state determined by energy and momentum conservation of the particular scattering process which is randomly selected. Details of the technique can be found in recent review articles and books [1.7, 44, 149, 150].

The ensemble Monte Carlo approach, in which an ensemble of several thousand particles is followed at once, has been used in recent years. In contrast to the earlier technique of following a single particle through several thousand iterations, this technique allows one to include carrier-carrier scattering in the simulations. All dynamic averages are now computed as averages over this ensemble. The initial distribution of electrons and holes is calculated using the laser pulse shape and photon energy, the energy-band structure and optical-matrix elements for the semiconductor under investigation. The dynamics of each particle is then determined using the single-particle approach, and dynamical distribution functions of the particles are calculated by using averages over the ensemble. The measured transient response, such as temporal evolution of the transmitted probe intensity or the luminescence intensity, is then determined by using these distribution functions and the known or pre-determined optical matrix elements.

The ensemble Monte Carlo technique has become a very powerful method for quantitatively interpreting the results of ultrafast experiments in semiconductors. The technique has become very sophisticated with the inclusion of virtually all important scattering processes, and realistic band structure and phonon modes of semiconductors and their nanostructures. One of the advantages of this technique is that the relative importance of various scattering processes can be evaluated by selectively turning on and off these scattering processes. Some of the many useful results have been obtained using this technique will be discussed in Chaps. 3–6.

1.4.6 Current Trends

These computational techniques have made it possible to make detailed quantitative comparison with ultrafast experiments in semiconductors, and thereby have contributed much to our understanding of the dynamics of semiconductors following ultrafast photoexcitation. In spite of this success, efforts are continuing to remove the limitations of the current techniques. For example, the neglect of coherence during and immediately following the ultrafast excitation is certainly not justified from a conceptual point of view, and efforts are under way to evaluate the importance of these effects and incorporate them in a Monte Carlo calculation [1.114, 116, 137, 140]. The question of dynamic screening at very short times is also important and a real-space molecular-dynamics approach is being investigated to address this problem [1.7, 151, 152]. Theorists have long attempted to go beyond the semiclassical Boltzmann transport equations and consider quantum effects in high-field transport. The tremendous progress in developing mesoscopic devices and systems, and ultrafast spectroscopic tools to investigate femtosecond time scales, have provided further impetus to these investigations. There is much that we do not know, but progress is being made in understanding these complex phenomena.

1.5 Summary

This brief introduction to ultrafast dynamics in semiconductors clearly shows the complexity of the dynamical processes in semiconductors. Understanding these processes through ultrafast optical spectroscopy has been a challenging and exciting field for the past several years. The remaining chapters discuss some of the understanding and insights developed as a result of such ultrafast optical investigations.

2. Coherent Spectroscopy of Semiconductors

The temporal regime during and immediately following photoexcitation by an ultrashort pulse is the coherent regime in which the photoexcited excitations have retained some definite phase relationship among themselves and with the electromagnetic radiation creating them. Photoexcitation creates a macroscopic polarization (an ensemble average of the individual photoexcited dipole moments) in the system. This macroscopic polarization acts as a source term in Maxwell's equations, and determines the linear and nonlinear response of the system. Therefore, appropriate investigations of the linear and nonlinear response of the system can provide information about the induced macroscopic polarization and hence the coherent regime.

Such investigations have been performed in a variety of systems and have provided invaluable information on many dynamical processes. Coherent phenomena in atoms, molecules and defects in solids in the optical frequency domain have been investigated in detail since the invention of lasers in early 1960s [2.1 – 4] using both cw and transient optical spectroscopy. We focus here on the use of *transient spectroscopy* to investigate coherent phenomena in semiconductors. Since the scattering rates in semiconductors are extremely high (Chap. 1), the dephasing times in semiconductors are in the pico- and femtosecond regime. Therefore, an investigation of the coherent transient response of semiconductors requires the use of ultrafast lasers. With the recent availability of such lasers (Chap. 1), investigation of coherent transient spectroscopy of semiconductors has become a very active and exciting current field of research.

The motivation for investigating this regime is two-fold: the coherent regime in semiconductors is essentially unexplored and its investigation offers the opportunity of learning about some of the most fundamental quantum mechanical processes in solids. Secondly, since scattering processes lead to the eventual loss of coherence, investigation of this regime also offers the opportunity of learning about various scattering processes and scattering rates in semiconductors.

This chapter begins with a brief review of the basic concepts (Sect. 2.1) in the coherent regime. The optical Bloch equations for an ensemble of independent two-level systems with homogeneous and inhomogeneous broadening are introduced first using the Markovian approximation for the relaxation Hamiltonian. This is followed by a discussion of the importance of interaction in semiconductors and a discussion of the semiconductor Bloch equations which provide a comprehensive framework for discussing coherent phenomena in semiconductors.

The primary goal of this chapter is to present a discussion of some of the important experiments investigating the coherent response of semiconductors using ultrafast spectroscopy. Therefore, after the introductory discussion in Sect. 2.1, the remainder of the chapter (Sects. 2.2 through 2.12) is devoted to a discussion of experimental results. This is a very rapidly advancing field with some excellent recent reviews [2.5]. Our discussion is intended to be representative of the work done in the field rather than comprehensive. Most of the work on coherent spectroscopy of semiconductors has been concentrated on excitons [2.6] for two reasons. First, excitons are neutral and therefore have longer dephasing times than charged carriers (electrons and holes). Second, excitons have discrete energy levels and large oscillator strengths associated with them, and they act as "atoms" of semiconductors. Therefore, our discussion begins with excitons and then discusses free carriers and phonons. For excitons, we discuss many different topics including dephasing, AC Stark effect, quantum-beat spectroscopy, coherent oscillations of an electronic wavepacket, and Bloch oscillations in semiconductor superlattices. For free carriers, we discuss photon-echo experiments on intrinsic and modulation-doped quantum wells. Most of these experiments measure time-integrated FWM signals but we will also discuss recent results showing that femtosecond time-resolved, spectrally-resolved and interferometric FWM experiments, as well as coherent THz emission experiments, provide much additional information. As we discussed in the preface, the emphasis in this chapter, and throughout the book, is on III-V semiconductors and their nanostructures.

2.1 Basic Concepts

The coherent phenomena in atoms and molecules are generally analyzed for an ensemble of independent two-level systems [2.1, 2]. The *independent two-level model* assumes that the photon is nearly resonant with the transition between $|a\rangle$ and $|b\rangle$ and far off resonance with respect to all other transitions. It further assumes that there is no interaction between various two-level atoms or molecules making up the ensemble. The transition under consideration can be homogeneously broadened, inhomogeneously broadened or a combination of both.

The electronic states in a semiconductor are considerably more complicated than those in atoms, as discussed in Chap. 1. However, each exciton in a semiconductor (e.g., 1s, 2p etc. states of a HH or a LH exciton) can be considered as a two-level system in the simplest approximation. The continuum states of a semiconductor can be considered as inhomogeneously broadened in the momentum space in the absence of any interaction between the states. Therefore, in the simplest approximation one may be able to apply the independent two-level model to semiconductors. For this reason, we begin our discussion with the predictions of the independent two-level model in various cases.

2.1.1 An Ensemble of Independent Two-Level Systems

(a) The Density Matrix. The quantum mechanics of transition probabilities in a two-level system by a near-resonant excitation is well known [2.1, 2, 7, 8]. If the wavefunctions of the two states are known, then one can describe the transition probabilities as bilinear combinations of transition amplitudes. Since the expectation value of any observable involves such bilinear combinations, the density matrix method dealing directly with the bilinear combinations has been developed. The density matrix formalism facilitates the treatment of interacting quantum systems. We are interested not in a single two-level system but an ensemble of two-level systems. In such cases, the wavefunction of the ensemble of two-level systems is generally not known but certain statistical properties of the ensemble may be known. Such statistical properties are conveniently described in terms of the general density matrix operator:

$$\rho = \sum_j P_j \, |\Psi_j\rangle\langle\Psi_j| \ , \tag{2.1}$$

where P_j is that fraction of the systems which has the state vector $|\Psi_j\rangle$. The density matrix obeys the Liouville variant of the Schrödinger equation:

$$\mathrm{i}\hbar\dot{\rho} = [\mathscr{H}, \rho] \ , \tag{2.2}$$

where the square bracket is the commutator $\mathscr{H}\rho - \rho\mathscr{H}$ and $\dot{\rho} = \mathrm{d}\rho/\mathrm{d}t$. $\mathscr{H}$ is the Hamiltonian operator of the system given by

$$\mathscr{H} = \mathscr{H}_0 + \mathscr{H}_{\mathrm{int}} + \mathscr{H}_{\mathrm{R}} \ . \tag{2.3}$$

$\mathscr{H}_0$ is the Hamiltonian of the isolated two-level system, $\mathscr{H}_{\mathrm{int}}$ is the Hamiltonian describing the interaction between the radiation field and the two-level system and the relaxation Hamiltonian $\mathscr{H}_{\mathrm{R}}$ describes all the processes that return the ensemble to thermal equilibrium. The expectation value of an operator O is given by

$$\langle O \rangle = \mathrm{Tr}\{O\rho\} \ . \tag{2.4}$$

For a two-level system with the ground state with state vector $|a\rangle$ and energy E_a and an excited state with the state vector $|b\rangle$ and energy E_b (Fig. 2.1), the density operator can be written as

$$\rho = \begin{bmatrix} \rho_{bb} & \rho_{ba} \\ \rho_{ab} & \rho_{aa} \end{bmatrix} . \tag{2.5}$$

The diagonal elements of the density matrix represent the *probability* of finding the system in the two energy eigenstates; i.e. the populations in the two eigenstates. The off-diagonal elements represent the *coherence* intrinsic to a superposition state.

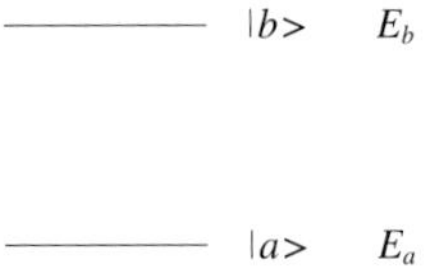

Fig. 2.1. Schematic of a two-level system consisting of the ground state with state vector $|a\rangle$ and energy E_a and an excited state with the state vector $|b\rangle$ and energy E_b

In a *closed* two-level system, the diagonal components are related by

$$\rho_{aa}+\rho_{bb}=1 \ , \tag{2.6}$$

since the sum of the populations in the lower and upper states is constant. Because of the complexity of electronic states in a semiconductor, the assumption of a closed system may be difficult to satisfy in a semiconductor.

(b) The Unperturbed Hamiltonian. The state vector $|\Psi_j\rangle$ obeys the time-dependent Schrödinger equation

$$\mathrm{i}\hbar\,|\dot{\Psi}_j\rangle=\mathscr{H}|\Psi_j\rangle \ . \tag{2.7}$$

For an isolated two-level system in the absence of any interactions $\mathscr{H}=\mathscr{H}_0$, and $\mathscr{H}_0$ has no explicit time-dependence so that for the system at position $\boldsymbol{R}$

$$\Psi_k(\boldsymbol{R},t)=\exp(-\mathrm{i}E_k t/\hbar)\,u_k(\boldsymbol{R}) \tag{2.8}$$

and

$$\mathscr{H}_0 u_k(\boldsymbol{R})=E_k u_k(\boldsymbol{R}) \ , \tag{2.9}$$

where $k=a$ or b. Thus, the unperturbed Hamiltonian $\mathscr{H}_0$ is given by

$$\mathscr{H}_0=\begin{bmatrix} E_b & 0 \\ 0 & E_a \end{bmatrix} . \tag{2.10}$$

For an ensemble of two-level systems, the state vector for the j-th system is

$$|\Psi_j(t)\rangle=C_{aj}(t)\,|a\rangle+C_{bj}(t)\,|b\rangle \tag{2.11}$$

and the density matrix can be put in the more familiar form

$$\rho=\sum_j P_j\begin{bmatrix} |C_{bj}|^2 & C_{bj}C_{aj}^* \\ C_{aj}C_{bj}^* & |C_{aj}|^2 \end{bmatrix} , \tag{2.12}$$

where P_j is the probability of being in the state j. If the state vectors for all j's are identical (i.e., amplitudes C_{aj} and C_{bj} are the same for all j's), but the phases of the coherent superposition are randomly distributed between 0 and

2π, then the off-diagonal elements of the density matrix vanish and there is no coherence in the ensemble. On the other hand, if there is a well-defined phase relationship for different j, the ensemble has coherence (see, for example, Ref. 2.2a, Chap. 7). The concepts of phase and coherence are central to coherent spectroscopy.

(c) The Interaction Hamiltonian. For an electric-dipole allowed transition, one generally neglects the electric quadrupole and magnetic dipole interactions (these are smaller by the fine structure constant $\approx 1/137$) in the interaction Hamiltonian. This is equivalent to assuming a zero wavevector for the radiation, a good assumption since the photon wavevector q and the characteristic length a_0 generally satisfy $qa_0 \ll 1$. In this dipole approximation,

$$\mathscr{H}_{\text{int}} = -\boldsymbol{d}\cdot\boldsymbol{E}(\boldsymbol{R},t) \ , \tag{2.13}$$

$\boldsymbol{d}$ is the dipole moment operator and $\boldsymbol{E}(\boldsymbol{R}, t)$ is the electric field of the light. The components of the dipole moment operator are given by

$$\boldsymbol{d}_{nk} = -e\int u_n^* \boldsymbol{r} u_k \mathrm{d}^3\bar{r} \equiv -e\boldsymbol{r}_{nk} \ , \tag{2.14}$$

where $\boldsymbol{r}$ is the electron coordinate with respect to the location of the nucleus at $\boldsymbol{R}$. The components of the operator $\mathscr{H}_{\text{int}}$ are given by

$$\varDelta_{nk} = -\boldsymbol{d}_{nk}\cdot\boldsymbol{E}(\boldsymbol{R},t) \ . \tag{2.15}$$

The diagonal components of $\boldsymbol{d}_{nk}$ and $\varDelta_{nk}$ are zero because $\boldsymbol{d}$ is an odd-parity operator.

For a monochromatic plane wave at angular frequency ω and linearly polarized in the direction $\boldsymbol{\varepsilon}$, the electric field can be written as the sum of two fields

$$\boldsymbol{E}(\boldsymbol{R},t) = \boldsymbol{E}^+(\boldsymbol{R},t) + \boldsymbol{E}^-(\boldsymbol{R},t) \tag{2.16}$$

with

$$\boldsymbol{E}^+(\boldsymbol{R},t) = (1/2)\,\boldsymbol{\varepsilon}\,\mathscr{E}_0 \exp\{\mathrm{i}\,[\boldsymbol{q}\cdot\boldsymbol{R} - (\omega t + \eta)]\} \tag{2.17}$$

and

$$\boldsymbol{E}^-(\boldsymbol{R},t) = \boldsymbol{E}^{+*}(\boldsymbol{R},t) \ , \tag{2.18}$$

where $\mathscr{E}_0$ is the (real) electric field amplitude, $\boldsymbol{\varepsilon}$ is the polarization vector, and η is a phase factor. In calculating the transition probabilities one generally makes the *rotating-wave approximation* in which the term with the rapidly varying phase factor $\exp[\mathrm{i}(\omega + \Omega)t]$ (where $\hbar\Omega = E_b - E_a$) and the large denominator $(\omega + \Omega)$, corresponding to $\boldsymbol{d}\cdot\boldsymbol{E}^-$, is neglected. In this approximation,

$$\Delta_{ba} = (\hbar/2)\chi_R \exp\{\mathrm{i}[\boldsymbol{q}\cdot\boldsymbol{R}-(\omega t-\eta)]\} = \Delta^*_{ab} \ , \tag{2.19}$$

where

$$\hbar\chi_{\mathrm{R}} = e(\boldsymbol{r}_{ba}\cdot\varepsilon)\,\mathscr{E}_0 \tag{2.20}$$

and the quantity χ_{R} is known as the *Rabi frequency* at resonance $\omega = \Omega$.
The matrix form of $\mathscr{H}_{\mathrm{int}}$ can thus be written as

$$\mathscr{H}_{\mathrm{int}} = \begin{bmatrix} 0 & \Delta_{ba} \\ \Delta^*_{\mathrm{ba}} & 0 \end{bmatrix} . \tag{2.21}$$

The linear and nonlinear response of the system to electromagnetic field is determined by the macroscopic polarization, as discussed at the beginning of this chapter. The macroscopic polarization is related to the macroscopic dielectric polarization density, and is given by, see (2.4),

$$P = N\mathrm{Tr}\{d\rho\} \ , \tag{2.22}$$

where N is the number density in the ensemble.

(d) The Relaxation Hamiltonian. The processes which bring the ensemble back to thermal equilibrium include recombination, collisions with phonons and interaction with other electronic states. The relative time scales of these processes, and their relation to other characteristic times in the system such as the laser pulse duration, determine the correct treatment of these relaxation processes.

Non-Markovian Behavior. The most general approach to the description of nonequilibrium properties of semiconductors excited by laser pulses is the quantum-kinetic equations approach based on non-equilibrium Green's functions [2.9, 10]. These methods have been applied to the description of excitation and relaxation processes in laser-excited semiconductors [2.11–21]. Numerical solutions of these equations have shown [2.22–24] that non-Markovian behavior becomes significant on time scales small compared to the dephasing times, a condition that is easily achieved for femtosecond photoexcitation not too far from a resonance. Under these conditions, the interband polarizations do not follow the pulse adiabatically, and the dephasing time is not an instantaneous function of its environment but depends on the "history" of the environment. These "memory effects" are under active theoretical investigation currently [2.25–30]. Solvent-dependent non-Markovian effects in large dye molecules have been experimentally investigated by *Bigot* et al. [2.31] and *Nibbering* et al. [2.32–34], and experimental investigation of non-Markovian effects in semiconductors is just beginning [2.26].

Markovian Approximation. A simpler approach to the problem is to make the Markovian approximation under which the relaxation times are determined by the instantaneous distribution and polarization functions, and hence are time dependent. This approximation forms the basis for the classical Boltzmann equation approach to transport in semiconductors and hence the Monte-Carlo simulations of carrier relaxation following excitation by a short optical pulse.

Relaxation-Time Approximation. The simplest approach to the problem is to assume that the dynamical self-energies in the nonequilibrium Green's function approach can be replaced by *constant* phenomenological transverse and longitudinal relaxation rates for the relevant relaxation processes in the problem and simplify the relaxation Hamiltonian accordingly. Analysis of most coherent experiments and the semiconductor Bloch equations (Sect. 2.1.2) are based on this assumption. In this approach, one approximates the relaxation Hamiltonian by

$$[\mathscr{H}_{\mathrm{R}}, \rho]_{bb} = -\rho_{bb}/T_1 \tag{2.23}$$

and

$$[\mathscr{H}_{\mathrm{R}}, \rho]_{ba} = -\rho_{ba}/T_2 \ , \tag{2.24}$$

where T_1 is the lifetime of the state b and $1/T_2$, the transverse relaxation rate, is the sum of the recombination rate $(1/T_1)$ and the pure dephasing rate. One can think of T_2 as the lifetime of the coherent superposition state.

The relaxation-time approximation is valid in the limit that the response of the medium in which the system under study is embedded is either very fast or very slow compared to the system-medium interaction. It can be shown [2.35, 36] that the linear absorption shape is Lorentzian (corresponding to a homogeneously broadened line) in the limit of very fast response of the medium, and a Gaussian (corresponding to an inhomogeneously broadened line) in the limit of very slow response of the medium. Stochastic models have been developed to treat the case of intermediate time scales [2.37–41] and have been applied to the nonlinear response of the system [2.35, 42]. These calculations have also been generalized to a Brownian oscillator model for the bath [2.35, 43]. In our simplified treatment, we will assume the limit of homogeneous or inhomogeneous line widths.

(e) Optical Bloch Equations. The coupled equations of motion for the polarization and the population of an ensemble of independent two-level systems are known as the *optical Bloch equations,* in analogy with the equations first derived by *Bloch* [2.44] for the spin systems. For simplification, we introduce a new notation and substitute $n = \rho_{bb}$, $1-n = \rho_{aa}$, and $p = \rho_{ba}$. Using the Liouville equation (2.2) and the definitions of T_1, T_2, Δ_{ba}, and Δ^*_{ba} given above, one gets [2.1, 3, 4]

$$\rho = \begin{bmatrix} n & p \\ p^* & 1-n \end{bmatrix} , \tag{2.25}$$

$$\dot{n}+g_r n+(\mathrm{i}/\hbar)(\Delta_{ba}p^*-p\Delta^*_{ba})=0\ , \tag{2.26}$$

$$\dot{p}+gp+(\mathrm{i}/\hbar)\Delta_{ba}(1-2n)=0\ , \tag{2.27}$$

where $g_r=1/T_1$ and $g=(\mathrm{i}\Omega+1/T_2)$ with $\Omega=(E_b-E_a)/\hbar$. Equations (2.26, 2.27) are known as the *optical Bloch equations*, and form the basis for analyzing coherent transient experiments in independent two-level systems. In the next two subsections (Sects. 2.1.1f, g), we analyze these equations and present the predictions of the optical Bloch equations developed above for homogeneously broadened and inhomogeneously broadened independent two-level systems.

(f) Formal Solutions of the Optical Bloch Equations. The coupled optical Bloch equations (2.26 and 27) cannot be solved analytically in the general case. One generally resorts to expanding the density matrix into a Taylor series in the incident field amplitudes and obtains a solution to the desired order. Numerical techniques have to be employed in the general case. We formally write the density n and polarization p as

$$n=n^{(0)}+n^{(1)}+n^{(2)}+n^{(3)}+\dots \tag{2.28}$$

and

$$p=p^{(0)}+p^{(1)}+p^{(2)}+p^{(3)}+\dots \tag{2.29}$$

with $n^{(0)}=0$ and $p^{(0)}=0$. For the usual initial conditions of $n=0$ and $p=0$, it can be shown that the odd powers of n and the even powers of p are zero and other low-order components of n and p are given by

$$\dot{p}^{(1)}+gp^{(1)}+(\mathrm{i}/\hbar)\Delta_{ba}=0\ , \tag{2.30}$$

$$\dot{n}^{(2)}+g_r n^{(2)}+(\mathrm{i}/\hbar)(\Delta_{ba}p^{*(1)}-p^{(1)}\Delta^*_{\mathrm{ba}})=0\ , \tag{2.31}$$

$$\dot{p}^{(3)}+gp^{(3)}-2(\mathrm{i}/\hbar)\Delta_{ba}n^{(2)}=0\ . \tag{2.32}$$

Thus, n (p) of order $j+1$ is influenced by $p(n)$ of order j.

(g) Analysis of Two-Beam, Degenerate FWM Experiment. In the simplest of the FWM experiments (the two-beam degenerate FWM configuration discussed in Sect. 1.3.2), the sample is excited by two pulse trains from the same modelocked laser. The second pulse train along $\boldsymbol{q}_z$ is delayed with respect to the first pulse train along $\boldsymbol{q}_1$ by a time delay τ_d. If the dephasing time T_2 is long compared to τ_d, then the polarizations of pulses 1 and 2 interfere to create a grating. Pulse 2 can be self-diffracted by this grating in the direction $\boldsymbol{q}_\mathrm{d}=2\boldsymbol{q}_2-\boldsymbol{q}_1$. In the TI-FWM experiment, one measures the total energy diffracted along $\boldsymbol{q}_\mathrm{d}$, either with or without spectrally resolving the signal, whereas in the TR-FWM experiment, one measures the temporal evolution of the diffracted signal along $\boldsymbol{q}_d$. TR-FWM in semiconductors has become possible only recently, so most experiments have measured TI-FWM signals. The experimental arrangement corresponding to both kinds of experiments have been discussed in Chap. 1 (Figs. 1.8, 9).

The equations of Sect. 2.1.1e have to be solved iteratively for two incident electric fields along $\boldsymbol{q}_1$ and $\boldsymbol{q}_2$ to obtain the third order polarization along the direction $\boldsymbol{q}_d$ which is related to the FWM signal along $\boldsymbol{q}_d$. Furthermore, the electric fields are not monochromatic plane waves but, typically, mode-locked laser pulses. If the pulse along $\boldsymbol{q}_2$ is delayed by time delay τ_d with respect to the pulse along $\boldsymbol{q}_1$, then the electric field can be expressed as

$$\begin{aligned}\boldsymbol{E}(\boldsymbol{R},t) = (1/2)\boldsymbol{\varepsilon}_1\, \mathscr{E}_1 A(\boldsymbol{R},t) \exp[\mathrm{i}(\boldsymbol{q}_1\cdot\boldsymbol{R}-\omega t)]\\ +(1/2)\boldsymbol{\varepsilon}_2\, \mathscr{E}_2 A(\boldsymbol{R},t-\tau_d) \exp[\mathrm{i}(\boldsymbol{q}_2\cdot\boldsymbol{R}-\omega t-\eta)]\end{aligned} \tag{2.33}$$

within the rotating-wave approximation. Here $\boldsymbol{\varepsilon}_1\,(\boldsymbol{\varepsilon}_2)$ is the direction of linear polarization for the first (second) field, $\mathscr{E}_1\,(\mathscr{E}_2)$ is the amplitude of the first (second) electric field, $A(\boldsymbol{R},\,t)$ is the electric field pulse shape at $\boldsymbol{R}$, and η is the relative phase factor between the two fields. ω corresponds to the peak of the mode-locked laser spectrum. Ignoring polarization effects on dipole matrix elements and the relative phase η, absorbing 1/2 in the field amplitude, taking $r_{ba} = \boldsymbol{r}_{ba}\cdot\boldsymbol{\varepsilon}$, and assuming $\boldsymbol{\varepsilon}_1 = \boldsymbol{\varepsilon}_2$

$$E(\boldsymbol{R},t) = \mathscr{E}_1 A(\boldsymbol{R},t) \exp[\mathrm{i}(\boldsymbol{q}_1\cdot\boldsymbol{R}-\omega t)] + \mathscr{E}_2 A(\boldsymbol{R},t-\tau_d) \exp[\mathrm{i}(\boldsymbol{q}_2\cdot\boldsymbol{R}-\omega t)] \tag{2.34}$$

and

$$\begin{aligned}\Delta_{ba}/\mathrm{i}\hbar &= (e/\mathrm{i}\hbar) r_{ba} \exp(-\mathrm{i}\omega t)\\ &\quad\times[\mathscr{E}_1 A(\boldsymbol{R},t) \exp(\mathrm{i}\boldsymbol{q}_1\cdot\boldsymbol{R}) + \mathscr{E}_2 A(\boldsymbol{R},t-\tau_d) \exp(\mathrm{i}\boldsymbol{q}_2\cdot\boldsymbol{R})]\\ &\equiv \exp(-\mathrm{i}\omega t) f(t)\ .\end{aligned} \tag{2.35}$$

If we define $p^{(1)} = p^{(1)}(t) \exp(-\mathrm{i}\omega t)$, then

$$\dot{p}^{(1)}(t) + G p^{(1)}(t) = f(t)\ , \tag{2.36}$$

which has the solution

$$p^{(1)} = \left\{\int_{-\infty}^{t} \mathrm{d}t' \exp[-G(t-t')]\right\} f(t') \exp(-\mathrm{i}\omega t)\ , \tag{2.37}$$

where $G = g - \mathrm{i}\omega$. The signal along $\boldsymbol{q}_d$ results from the diffraction of pulse 2 from a grating created by pulses 1 and 2, i.e., it is related to $p^{(3)}_{221^*}$. We therefore need to calculate only the $q_2 - q_1$ component of the density $n^{(2)}$, i.e., $n^{(2)}_{21^*}$ which, using (2.31), can be shown to be given by

$$n^{(2)}_{21^*} = \int_{-\infty}^{t} dt'' \exp[-g_r(t-t')] f_{21^*}(t'') \ , \tag{2.38}$$

where

$$f_{21^*}(t) = (e^2 |r_{ba}|^2/\hbar^2) \mathscr{E}_2 \mathscr{E}_1 \exp[i(q_2 - q_1)\cdot R]$$

$$\times \left\{ \begin{array}{l} A(R, t-\tau_d) \int_{-\infty}^{t} dt' A(R,t') \exp[-G^*(t-t') \\ \\ +A(R,t) \int_{-\infty}^{t} dt' A(R, t'-\tau_d) \exp[-G(t-t')] \end{array} \right\} \ . \tag{2.39}$$

The third-order polarization $p^{(3)}_{221^*}$ can be obtained from (2.32)

$$p^{(3)}_{221^*} = -2i(e|r_{ba}|/\hbar)^3 \exp(-gt) \mathscr{E}_1 \mathscr{E}_2^2 \int_{-\infty}^{t} dt''' \int_{-\infty}^{t'''} dt'' \int_{-\infty}^{t''} dt'$$

$$\times \{\exp(Gt''') \exp[-g_r(t'''-t'')] A(R, t'''-\tau_d)\}$$

$$\times \left\{ \begin{array}{l} A(R, t''-\tau_d) A(R,t') \exp[-G^*(t''-t')] \\ +A(R, t'-\tau_d) A(R,t'') \exp[-G(t''-t')] \end{array} \right\} \ . \tag{2.40}$$

This equation is equivalent to that of *Yajima* and *Taira* [Ref. 2.45, Eq. (9)]. In order to determine the signal at a given point in space and time, the Maxwell's propagation equations have to be solved in general. However, for an optically thin sample with thickness small compared to the wavelength of light, the signal diffracted signal along q_d can be approximated by the macroscopic polarization given by

$$P^{(3)}_{221^*} = N\mathrm{Tr}\{d\rho\} \tag{2.41}$$

with appropriate order and component of ρ.

The above discussion applies to a homogeneously-broadened system. An inhomogeneously-broadened system behaves quite differently because different homogeneous components of the inhomogeneous line evolve at different rates because of the difference in their energies. The macroscopic polarization of the ensemble can be lost by this interference between the different components, even if there is no dephasing of the individual components. If the transition under consideration is inhomogeneously broadened, rather than homogeneously broadened as we have assumed so far, then one needs to integrate over the inhomogeneous distribution $g(\Omega_0)$,

$$P^{(3)}_{221^*} = -eN\left(\int_0^\infty g(\Omega_0)\,\mathrm{d}\Omega_0\,|r_{ba}|\,p^{(3)}_{221^*} + \int_0^\infty g(\Omega_0)\,\mathrm{d}\Omega_0\,|r_{ab}|\,p^{(3)}_{221^*}\right) , \quad (2.42)$$

where Ω_0 is the center of the inhomogeneous distribution and

$$\int_0^\infty \mathrm{d}\Omega_0 g(\Omega_0) = 1 . \quad (2.43)$$

These equations can be numerically integrated to obtain the *temporal evolution* of the FWM signal (TR-FWM) in the $\boldsymbol{q}_\mathrm{d}$ direction

$$S^{(3)}_{221^*}(t) = |P^{(3)}_{221^*}|^2 \quad (2.44)$$

and the *time-integrated* (TI-FWM) signal as a function of time delay τ_d

$$I^{(3)}_{221^*}(\tau_\mathrm{d}) = \int_{-\infty}^{\infty} S^{(3)}_{221^*}(t)\,\mathrm{d}t . \quad (2.45)$$

(h) Analytic Solutions for Delta Pulses. The iterative equations can be solved numerically for a given pulse shape. However, we consider in this section a simpler case of pulses that can be described by the Dirac δ-functions in time. We further assume that the sample is thin and the propagation effects can be neglected, as discussed above, and the first pulse arrives at the sample at time $t = 0$ and the second pulse at time $t = \tau_\mathrm{d}$. Then for a homogeneously broadened independent two-level system, (2.34) can be rewritten as

$$E(\boldsymbol{R},t) = [\mathscr{E}_1\delta(t)\exp(\mathrm{i}\boldsymbol{q}_1\cdot\boldsymbol{R}) + \mathscr{E}_2\delta(t-\tau_\mathrm{d})\exp(\mathrm{i}\boldsymbol{q}_2\cdot\boldsymbol{R})]\exp(-\mathrm{i}\omega t) . \quad (2.46)$$

Equation (2.36) gives

$$p^{(1)} = (er_{ba}/\mathrm{i}\hbar)\{\mathscr{E}_1\exp(-Gt)\Theta(t) + \mathscr{E}_2\exp[-G(t-\tau_\mathrm{d})]\Theta(t-\tau_\mathrm{d})\}\exp(\mathrm{i}\omega t) , \quad (2.47)$$

where Θ is the Heaviside step function, and we recall that $G = (1/T_2) + \mathrm{i}(\Omega - \omega)$. Thus, at resonance ($\Omega = \omega$), the first order polarization is a sum of two damped oscillations at ω displaced by the time delay τ_d. This is shown schematically in Fig. 2.2a without the oscillating part. For the electric field (2.46), Eqs. (2.37 and 38) can be similarly integrated to yield

$$n^{(2)}_{21^*} = (e^2|r_{ba}|^2/\hbar^2)\,\mathscr{E}_2\mathscr{E}_1\left\{\begin{array}{l}\Theta(-\tau_\mathrm{d})\Theta(t)\exp(-g_\mathrm{r}t + G\tau_\mathrm{d})\\ +\Theta(\tau_\mathrm{d})\Theta(t-\tau_\mathrm{d})\exp[-g_\mathrm{r}(t-\tau_\mathrm{d}) - G^*\tau_\mathrm{d}]\end{array}\right\} . \quad (2.48)$$

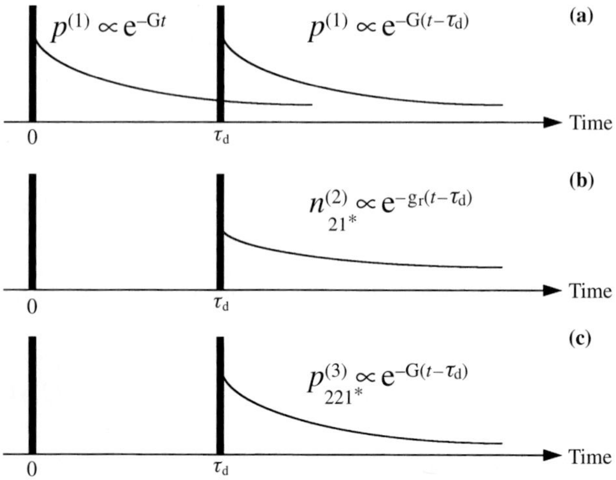

Fig. 2.2a–c. Schematic of the solution of the optical Bloch equation for two δ-function pulses delayed by time delay τ_d; **(a)** the first-order polarization, **(b)** the second-order population responsible for the third-order polarization in the diffraction direction, and **(c)** the third order polarization in the diffraction direction, as discussed in the text

There are two terms of which only one is non-zero for a given τ_d: the first term is non-zero only when $\tau_d < 0$, i.e., for *negative* time delays, whereas the second term is non-zero only for *positive* time delays ($\tau_d > 0$). This equation simply states that, after the arrival of the *second* pulse (pulse 2 if $\tau_d > 0$, and pulse 1 if $\tau_d < 0$), $n^{(2)}_{21^*}$ decreases exponentially with τ_d. For a given τ_d, $n^{(2)}_{21^*}$ decreases with the population decay constant T_1 as a function of time. This behavior is schematically illustrated in Fig. 2.2b for positive τ_d. Similar results are obtained for negative τ_d.

This process can be repeated to obtain an expression for the third-order polarization in the $2\boldsymbol{q}_2 - \boldsymbol{q}_1$ direction from (2.40):

$$p^{(3)}_{221^*} = -\mathrm{i}(e\,|r_{ba}|/\hbar)^3 \exp(-\mathrm{i}\omega t) \exp[\mathrm{i}(2\boldsymbol{q}_2 - \boldsymbol{q}_1)\cdot\boldsymbol{R}]\, \mathcal{E}_1 \mathcal{E}_2^2$$
$$\times \Theta(\tau_d)\Theta(t-\tau_d) \exp[-Gt + (G-G^*)\tau_d]\ . \qquad (2.49)$$

Therefore, at resonance $\omega = \Omega$, the third-order polarization in the direction $\boldsymbol{q}_d$ is non-zero only when $\tau_d \geq 0$ and $t \geq \tau_d$, and it decays with the time constant T_2, the dephasing time.

The expected behavior of $p^{(3)}_{221^*}$ as a function of time t is shown in Fig. 2.2c (without the oscillatory part) for a homogeneously-broadened independent two-level system at resonance ($\omega = \Omega$). The polarization is maximum at $t = \tau_d$ (the time of the arrival of the second pulse) and decays with a time constant T_2. The behavior of the third-order polarization as a function of τ_d is determined by integrating such curves over time.

The experiments are usually analyzed in two limits: (1) homogeneous broadening, when the levels are homogeneously broadened with line width $\Gamma_h = 2\hbar/T_2$ and (2) inhomogeneous broadening, when the inhomogeneous broadening Γ_{in} is much larger than Γ_h. We depart from the usual practice of defining Γ as a rate and define it as line width in energy units (e.g., meV). Based on the analysis given by *Yajima* and *Taira* [2.45], the expected behavior for the TI-FWM (2.45) and TR-FWM (2.44) can be summarized as follows.

Homogeneously Broadened System. In this case, one expects to see Free-Polarization Decay (FPD) following the arrival of the second pulse, provided τ_d is positive. For δ-function pulses, the TR-FWM will peak at $t = \tau_d$ and decay with a decay constant $\tau_{decay} = T_2/2$. For pulses of width τ_p, the FPD reaches a peak $t \sim \tau_d + \tau_p$, and then decays with $\tau_{decay} = T_2/2$. This behavior of TR-FWM is sketched in Fig. 2.3 a. The peak height depends on the polarization created by the first pulse that is present when the second pulse arrives, and is therefore proportional to $\exp(-2\tau_d/T_2)$. The TI-FWM as a function of time delay τ_d is simply given by the integration of the TR-FWM and is schematically shown in Fig. 2.3 b. TI-FWM signal decays as $\exp(-2\tau_d/T_2)$. These simple exponential decays can be used to determine the dephasing time T_2 provided the nature of line broadening is known.

Inhomogeneously Broadened System. For an inhomogeneously broadened system, the phases of the different frequency components excited within the spectral bandwidth of the laser evolve at different rates according to the time-dependent Schrödinger equation. Therefore, the macroscopic polarization created by the first pulse decays to zero within the laser pulse width τ_p if the inhomogeneous line width Γ_{in} is larger than the laser spectral width, and with a time constant inversely proportional to Γ_{in} otherwise. This is true even if there are no dephasing collisions. If the second pulse arrives at time τ_d short

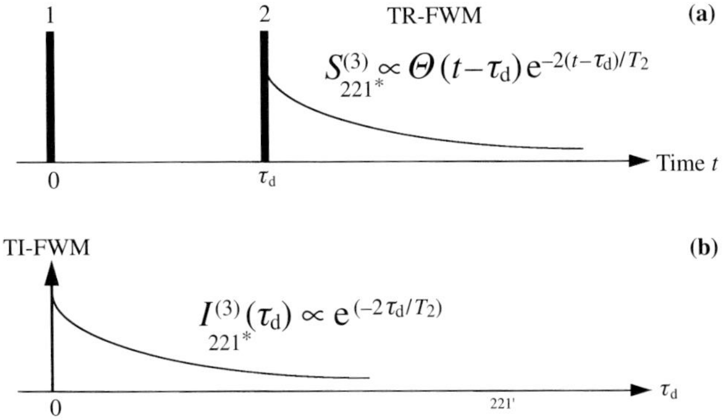

Fig. 2.3. Schematic representation of **(a)** the Time-Resolved FWM (TR-FWM) signal as a function of time t, and **(b)** the Time-Integrated FWM (TI-FWM) signal as a function of time delay τ_d for a homogeneously broadened two-level system for δ-function pulses

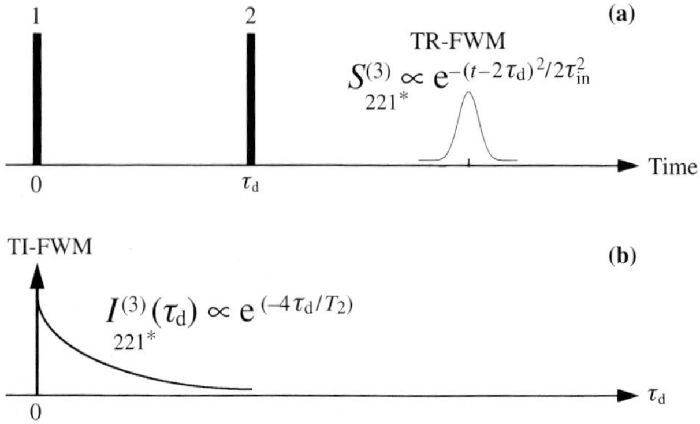

Fig. 2.4 a, b. Schematic representation of the FWM signals from an inhomogeneously broadened two-level system for δ-function pulses; (**a**) TR-FWM as a function of time for a given time delay τ_d, showing the photon echo at $2\tau_d$, and (**b**) TI-FWM signal as a function of the time delay τ_d. The width of the photon echo (τ_{in}) is related to the inhomogeneous linewidth Γ_{in}

compared to T_2, it reverses the phase evolution of the different frequency components, leading to a photon echo at time τ_d after the second pulse. This is the well-known photon echo first investigated by *Kurnit* et al. [2.46] for the case of ruby. The FWHM of the echo is determined by τ_p and Γ_{in}, and the height of the echo is proportional to $\exp(-4\tau_d/T_2)$. This TR-FWM signal is shown schematically in Fig. 2.4a. The TI-FWM is given by the time integration of the TR-FWM and decays as $\exp(-4\tau_d/T_2)$, as shown schematically in Fig. 2.4b. This exponential dependence can also be used to determine the T_2. Note that the decay time constant is $T_2/4$, a factor of two smaller than the decay constant of $T_2/2$ obtained for the case of homogeneous broadening.

It must be emphasized that since the nature of broadening (homogeneous or inhomogeneous) is not known a priori, there is an uncertainty in the value of T_2 by a factor of up to 2 in TI-FWM measurements unless some other information is used to determine the nature of broadening. The three-beam FWM technique proposed by *Weiner* et al. [2.47] provides one way of determining this information. Their technique utilizes the fact that the expected line shape of the TI-FWM signal in a three-beam experiment they proposed is symmetric for the homogeneously broadened system but asymmetric for the inhomogeneously broadened system. Time-resolved FWM provides another means of distinguishing between homogeneous and inhomogeneous cases. For the homogeneous case, TR-FWM signal peaks close to the end of the laser and decays with a time constant $T_2/2$. For the inhomogeneous case, TR-FWM comes at time τ_d after the arrival of the second pulse as discussed above. One must, however, be careful in applying these simple arguments, because interaction-induced effects change these simple temporal shapes as discussed in Sect. 2.7. Spectral shape of the line in absorption or emission spectrum also provides a means of determining the nature of broadening. A homogeneously-broadened line has a Lorentzian spectral shape, whereas an inhomogeneously-broadened line has a Gaussian spectral shape. If both types of broadening are comparable then the Voigt line shape must be considered [2.48], and the decay of the FWM signal has a time constant between $T_2/2$ and $T_2/4$.

2.1.2 Semiconductor Bloch Equations

The electron and hole states in a semiconductor span a wide range of energies and wavevectors. If we consider only the lowest conduction band and the highest valence band, each state has a well-defined energy and momentum. In the absence of Coulomb interaction, one can then consider these continuum electron-hole pair states as being *inhomogeneously broadened in the momentum space*. In this simplest picture, one would then expect to observe a photon-echo from this manifold of continuum states. TI-FWM experiments in the photon echo geometry have been performed using 6 fs pulses [2.49] and will be discussed later in this chapter (Sects. 2.10.2, 3).

It is well-known that Coulomb attraction between electrons and holes leads to important modifications to this simple picture. The Coulomb interaction results in a series of bound states of the electron-hole pair, the well-known hydrogenic exciton, and also alters the optical matrix elements for interband transitions involving the continuum states [2.50–53], as discussed in detail in Chap. 6. These changes induced by the Coulomb interaction have profound implications not only for the linear optical properties of semiconductors [2.52, 53], but also for the nonlinear properties of semiconductors [2.54–56]. In the simplest approximation, one can ignore all the continuum electron-hole pair states and consider only the crystal ground state and one excited state corresponding to the exciton bound state in which the oscillator strength is concentrated. If one ignores interaction between excitons, then a collection of excitons can be thought of as an ensemble of independent two-level systems. The two-level system may be homogeneously or inhomogeneously broadened depending on the nature of the sample under investigation.

Many FWM experiments on semiconductors have indeed been analyzed on the basis of this simple model, and much useful information about excitons and semiconductors has been obtained in this way. However, it should be clear that such an analysis can not adequately describe all the situations one might encounter in a real semiconductor. In particular, Coulomb interactions *between* excitons can not be ignored. These many-body interactions lead to important effects such as exciton bleaching and band-gap renormalization which have been investigated by incoherent nonlinear optical spectroscopy [2.56]. These interactions also have a profound influence on the coherent nonlinear response of semiconductors. It is clear that one has to go beyond the assumption of independent two-level systems for a true understanding of the nonlinear response of semiconductors.

A theoretical framework to include many-body Coulomb interaction in semiconductors has been developed in recent years [2.20]. The resulting equations, known as the *Semiconductor Bloch Equations* (SBE), reduce to the optical Bloch equations in independent two-level systems (2.26) when the Coulomb interaction between the excitons or the electron-hole pairs is set to zero.

The derivation of SBE is beyond the scope of this book, but a brief description is given here. In a spatially inhomogeneous system, one develops a real-space description of SBE [2.20] but we discuss a spatially homogeneous system in which a description of SBE in the momentum space is more appropriate. In this case the density matrix is given by [2.20]

$$\rho_k = \begin{bmatrix} n_{ek} & p_k \\ p_k^* & n_{hk} \end{bmatrix}, \tag{2.50}$$

where p_k describes the polarization induced by the coherent pump field in the momentum state indexed by k, and $n_{ek}(n_{hk})$ describe the non-equilibrium electron and hole distribution functions of state k. The coherent Hamiltonian, the sum of the Hamiltonians for the electronic system (including Coulomb interaction between the carriers) and its interaction with radiation, is

$$\mathscr{H}_{\text{coh}} = \begin{bmatrix} \varepsilon_{ek}^0 & \Delta_k^0 \\ \Delta_{\text{k}}^{0*} & \varepsilon_{hk}^0 \end{bmatrix} - \sum_{k' \neq k} V_{kk'} \rho_{k'}, \tag{2.51}$$

where $\varepsilon_{ek}^0 = (E_g/2) + \hbar^2 k^2/2m_e$ is the electron energy in the absence of Coulomb interaction, $\varepsilon_{hk}^0 = -(E_g/2) - \hbar^2 k_2/2m_h$ is the hole energy in the absence of Coulomb interaction, E_g is the energy band gap, $\Delta_k^0 = -\boldsymbol{d}_k \boldsymbol{E}(\boldsymbol{R}, t)$ with $\boldsymbol{d}_k = -e\boldsymbol{r}_k$ the dipole matrix element for the electron and hole states at k, and $m_e(m_h)$ is the electron (hole) effective mass (parabolic bands assumed, for simplicity). The diagonal components of the coherent Hamiltonian are given by

$$\varepsilon_{ik} = \varepsilon_{ik}^0 - \sum_{k \neq k'} V_{k,k'} n_{ik}, \quad i = \text{e or h} \tag{2.52}$$

and the off-diagonal components are given by

$$\Delta_k = \Delta_k^0 - \sum_{k' \neq k} V_{k,k'} p_{k'}. \tag{2.53}$$

These two equations show that there are two primary effects of introducing Coulomb interaction between pairs at different k values:

(1) The Coulomb interaction renormalizes the electron and hole energies and introduces excitons into the picture, as given by (2.52), and
(2) the Coulomb interaction renormalizes the interaction strength between the radiation field and the semiconductor as shown by the renormalized Rabi frequency in (2.53).

These two modifications have a profound influence on the nonlinear response of semiconductors.

Substituting these expressions into the Liouville equation (2.2), and using the same relaxation Hamiltonian as before yields the following coupled equations for the time evolution of the polarization and electronic distribution functions [2.20]:

$$(\dot{p}_k + g_k p_k) + (\mathrm{i}\Delta_k/\hbar)(1 - n_{ck} - n_{hk}) = 0 \ , \tag{2.54}$$

$$\dot{n}_{ek} + n_{ek}/T_{1e} + (\mathrm{i}/\hbar)(\Delta_k p_k^* - \Delta_k^* p_k) = 0 \ , \tag{2.55}$$

$$\dot{n}_{hk} + n_{hk}/T_{1h} + (\mathrm{i}/\hbar)(\Delta_k p_k^* - \Delta_k^* p_k) = 0 \ . \tag{2.56}$$

Here g_k is the generalization of g given following (2.26, 27) and is given by

$$g_k = \Omega_k + 1/T_2 \ . \tag{2.57}$$

These are the semiconductor Bloch equations which properly account for the Coulomb interaction between carriers in a semiconductor. Analytical solutions of these coupled equations are not possible for the general case. One has to resort to extensive numerical calculations to obtain results that can be compared with experiments. Such calculations can be very computation-intensive, taking many hours of CPU time on the fastest Cray available today. In spite of this, such calculations have been made and compared with experiments [2.22, 23, 57–64]. We will present a few such results when discussing some of the experiments in the following sections. In the remainder of this section, we will consider three special cases of SBE to obtain some physical insight into the kinds of solutions expected in limiting cases.

(a) No Coulomb Interaction: Inhomogeneously Broadened Independent Two-Level System in *k*-space. We note that SBE reduce to the usual optical Bloch equations for a two-level system in the absence of interaction. This is easy to see from (2.52, 53) by assuming that $V_{k,k'} = 0$. In this case we recover the optical Bloch equations (2.26, 27) for the independent two-level system for each k. This is the case of inhomogeneously-broadened independent two-level systems in the momentum space, corresponding to the continuum states in the valence and conduction bands of a semiconductor in the absence of any Coulomb interaction.

(b) Non-Interacting Excitons. Another simple case is that of a finite $V_{k,k'}$ with quasi-equilibrium assumption for carriers. Under this assumption, the time scales are assumed to be long enough that the carrier distribution functions can be represented by quasi-equilibrium Fermi-Dirac distributions. In this case the homogeneous part of the first of the SBE (2.54) corresponds to the usual Wannier equation for exciton, as shown in [2.20, 65].

(c) Interacting Excitons: Low Excitation Regime. For the case of weak ultrafast excitation, the dynamics is often described by ignoring the exchange interaction; i.e., one keeps the term Δ_k but ignores any product terms $\Delta_k n_k$ or $\Delta_k p_k^*$. Under these conditions, it can be shown [Ref. 2.20, pp. 224–227] that the optical response of a semiconductor is once again equivalent to that of an inhomogeneously broadened two-level system. However, in this case the Sommerfeld enhancement of the density of states of the continuum states due to Coulomb interaction must be properly taken into account. In other words, the absorption edge does not have a square root behavior but is given by the Elliott's formula (Chap. 6) [2.50].

(d) Interacting Excitons: Local Field Model. We now look for an iterative solution to the SBE under the assumption

$$\sum_{k' \neq k} V_{k,k'} p_{k'} = V p_{\mathrm{k}} \ . \tag{2.58}$$

Thus, the off-diagonal elements of the interaction and coherent Hamiltonians are given by

$$\Delta_k = \Delta_k^0 + V p_k \ . \tag{2.59}$$

Since $\Delta_k^0 = e\boldsymbol{r}_k \cdot \boldsymbol{E}(\boldsymbol{R}, t)$, the effect of the second term in (2.59) is equivalent to the statement that the total field is the sum of the electromagnetic field and the "local field" generated by the polarization of all other pairs. For this reason this is known as the *local-field model* [2.66].

For simplicity, we drop the index k and assume that the electron and hole scattering terms are the same so that $n_{\mathrm{e},k} = n_{\mathrm{h},k} = n$. Then the first- and third-order polarizations and the second-order density are given by

$$\dot{p}^{(1)} + g p^{(1)} + (\mathrm{i}/\hbar)\Delta^0 = 0 \ , \tag{2.60}$$

$$\dot{n}^{(2)} + g_{\mathrm{r}} n^{(2)} + (\mathrm{i}/\hbar)(\Delta^0 p^{*(1)} - p^{(1)}\Delta^{0*}) = 0 \ , \tag{2.61}$$

$$\dot{p}^{(3)} + g p^{(3)} - 2(\mathrm{i}/\hbar) n^{(2)}(\Delta^0 + V p^{(1)}) = 0 \ . \tag{2.62}$$

These equations are similar to (2.30–32), but with one important difference. In the absence of interaction, the third-order polarization (2.32) is driven by the bare Rabi frequency ($\chi_{\mathrm{R}}^0 = \Delta^0/\hbar$). This driving term is present only so long as the external field is present. In contrast, in the presence of interaction, the third-order polarization is driven by the renormalized Rabi frequency ($\chi_{\mathrm{R}} = \Delta/\hbar$) that includes the first-order polarization. Since $p^{(1)}$ builds up during the external field and decays with the dephasing time included in the constant g, the driving term can persist for times much beyond the duration of the external field. This has a profound influence on the nature of the nonlinear response of semiconductors.

To visualize these effects, we once again consider the case of δ-function pulses (Sect. 2.1.1h). The first-order polarization and the second-order density are still given by equations similar to those given earlier for the independent two-level model (2.47, 48). However, the third-order polarization in direction $\boldsymbol{q}_{\mathrm{d}} = 2\boldsymbol{q}_2 - \boldsymbol{q}_1$ is given by

$$p^{(3)}_{221^*} = -\mathrm{i}(e\,|r_{ba}|/\hbar)^3 \exp(-\mathrm{i}\omega t) \exp[\mathrm{i}(2\boldsymbol{q}_2 - \boldsymbol{q}_1)\cdot\boldsymbol{R}]\, \mathscr{E}_1 \mathscr{E}_2^2$$

$$\times \left\{ \begin{array}{l} \Theta(\tau_{\mathrm{d}})\Theta(t-\tau_{\mathrm{d}}) \exp[-Gt + (G-G^*)\tau_{\mathrm{d}}] \\ \times [1 + 2(\mathrm{i}\,V/\hbar g_{\mathrm{r}})]\{1 - \exp[-g_{\mathrm{r}}(t-\tau_{\mathrm{d}})]\} \\ +2(\mathrm{i}\,V/\hbar g_{\mathrm{r}})\Theta(-\tau_{\mathrm{d}})\Theta(t) \exp[-G(t-2\tau_{\mathrm{d}})]\,[1 - \exp(-g_{\mathrm{r}}t)] \end{array} \right\} . \tag{2.63}$$

For long recombination times T_1, this can be approximated by

$$p^{(3)}_{221^*} \cong -\mathrm{i}(e\,|r_{ba}|/\hbar)^3 \exp(-\mathrm{i}\omega t) \exp[\mathrm{i}(2\boldsymbol{q}_2 - \boldsymbol{q}_1)\cdot\boldsymbol{R}]\, \mathscr{E}_1 \mathscr{E}_2^2$$

$$\times \left[\begin{array}{l} \Theta(\tau_{\mathrm{d}})\Theta(t-\tau_{\mathrm{d}}) \exp[-Gt + (G-G^*)\tau_{\mathrm{d}}]\{1 + 2[\mathrm{i}\,V(t-\tau_{\mathrm{d}})/\hbar]\} \\ +2(\mathrm{i}\,Vt/\hbar)\Theta(-\tau_{\mathrm{d}})\Theta(t) \exp[-G(t-2\tau_{\mathrm{d}})] \ . \end{array} \right] . \tag{2.64}$$

These equations show that the third-order polarization can be written as the sum of three terms

$$p^{(3)}_{221^*} \equiv p^{(3)}_{221^*,\mathrm{N}^+} + p^{(3)}_{221^*,\mathrm{II}^+} + p^{(3)}_{221^*,\mathrm{II}^-} \ . \tag{2.65}$$

We see that the local-field approximation of SBE introduces two new terms, compare (2.49), proportional to the interaction potential V. The first Interaction-Induced (II) term ($p^{(3)}_{221^*,\mathrm{II}^+}$) is present for $\tau_{\mathrm{d}} > 0$ and adds to the normal term (N) for $\tau_{\mathrm{d}} > 0$ ($p^{(3)}_{221^*,\mathrm{N}^+}$) in the absence of interaction. The second II term ($p^{(3)}_{221^*,\mathrm{II}^-}$) leads to a finite third-order polarization in the $\boldsymbol{q}_{\mathrm{d}}$ direction for $\tau_{\mathrm{d}} < 0$, i.e., for *negative* time delays, in contrast to the N term for which the third-order polarization in the $\boldsymbol{q}_{\mathrm{d}}$ direction is zero for $\tau_{\mathrm{d}} < 0$. Both of these terms introduce profound changes in the nonlinear response of a semiconductor. We discuss these terms in some detail in the following two sections. Experimental observations are discussed in Sects. 2.7.1a, b. Local-field effects in atomic systems has also been investigated [2.67], but will not be discussed here.

Time-Resolved Signal. The macroscopic polarization $P^{(3)}$ along $\boldsymbol{q}_{\mathrm{d}}$ is given by (2.41) and can also be expressed as a sum of three terms corresponding to each of the three $p^{(3)}$ terms above. Since the N and the II terms are out-of-phase and since the terms for positive and negative τ_d are orthogonal because of the Heaviside step functions, the signal at time t along $\boldsymbol{q}_d$ can be expressed as

$$S^{(3)}_{221^*}(t) = |P^{(3)}_{221^*}|^2 = |P^{(3)}_{221^*,\mathrm{N}^+}|^2 + |P^{(3)}_{221^*,\mathrm{II}^+}|^2 + |P^{(3)}_{221^*,\mathrm{II}^-}|^2 \ . \tag{2.66}$$

Time-Integrated Signal. For a homogeneously broadened system in which T_1 is much longer than T_2, the time-integrated signal as a function of τ_d is given by

$$I^{(3),\mathrm{hom}}_{221^*}(\tau_d) \propto \Theta(\tau_d) \exp(-2\tau_d/T_2)[1+2(VT_2/\hbar)^2] + \Theta(-\tau_d) \exp(4\tau_d/T_2) \ . \tag{2.67}$$

There are thus two contributions to the signal at positive τ_d, and they both decay with the usual $\tau_{decay} = T_2/2$. In addition, there is also a contribution at negative τ_d arising solely from the interaction between pairs. This contribution has half the slope, i.e., the risetime is given by $T_2/4$. For an inhomogeneously broadened system, the contribution for $\tau_d < 0$ vanishes and the signal is given by [2.66]

$$I^{(3),\mathrm{inhom}}_{221^*}(\tau_d) \propto \Theta(\tau_d) \exp(-4\tau_d/T_2)\{1+(2VT_1/\hbar)^2[1-\exp(-\tau_d/T_1)]^2\} \ . \tag{2.68}$$

Figure 2.5 schematically presents the expected TR-FWM and TI-FWM for the two cases discussed above. Experimental observations are discussed in Sects. 2.7.1a, b.

(e) Interacting Excitons: Excitation-Induced Dephasing. In the discussion on interacting excitons, it is tacitly assumed that interaction between excitons does not change the relaxation rates. In fact, it is known for quite some time

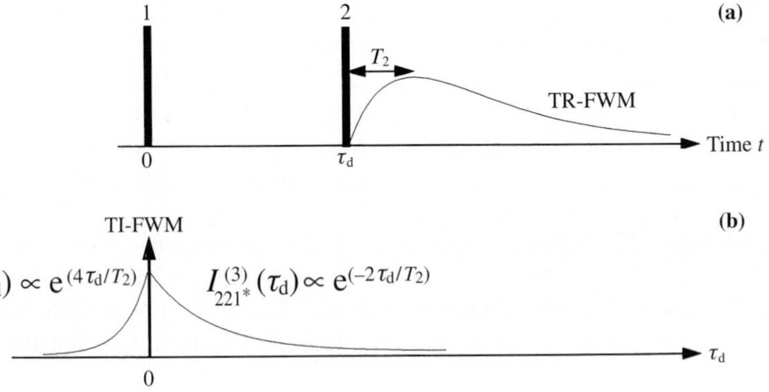

Fig. 2.5. Schematic representation of **(a)** the TR-FWM signal as a function of time t, and **(b)** the TI-FWM signal as a function of time delay τ_d for the local field model for homogeneously-broadened two-level system

[2.68 – 70] that interaction of excitons with free carriers and other excitons affects the exciton linewidth (homogeneous broadening). The influence of excitation-induced dephasing on nonlinear response of the system has recently been investigated by introducing density-dependent Hartree-Fock self energies in SBE [2.71] and by phenomenologically introducing an additional dephasing rate proportional to the excited state population [2.72, 73]. These analyses show that excitation-induced dephasing also introduces a delayed rise in the TR-FWM signal and a negative time-delay signal in TI-FWM; i.e., the effects on FWM signals are qualitatively similar to those introduced by local-field effects discussed in Sect. 2.1.2d. Note, however, that local-field effects are linear (i.e., do not depend on intensity) whereas excitation-induced dephasing is nonlinear (i.e., depends on intensity). Dependence of the measured FWM signals on intensity and on different polarizations of the fields may provide a technique for distinguishing between the two effects experimentally. Some experiments related to this are discussed in Sect. 2.7.1d.

2.1.3 Coherence Effects in Other Optical Experiments

The measured diffracted signal in a FWM experiment is purely coherent; i.e., the signal disappears as the coherence is lost through dephasing. In other experiments such as induced absorption or emission, strong signals persist even after the coherence is completely lost. Such incoherent signals have been the focus of considerable investigation and have provided invaluable information on many dynamical processes in semiconductors, as we will discuss in later chapters. One undesirable side effect of this has been to largely ignore the role of coherence in such experiments at ultrashort times. In the early days, such effects were termed "coherent artifacts" and were not paid much attention. It is clear, however, that coherent effects do play an important role [2.74] and provide important insights into the physics of the systems under investigation even in experiments in which incoherent effects dominate at long times. We briefly discuss such effects in pump/probe transmission experiments and luminescence experiments in this section.

(a) Coherence Effects in Pump/Probe Transmission Experiments. In a pump/probe transmission experiment, the sample is excited by two pulse trains from the same ultrafast laser. A pump beam traveling along $\boldsymbol{q}_1$ excites the sample and the change in the properties of the sample induced by this pump beam is investigated by measuring the change in the transmission of a much weaker probe beam traveling along $\boldsymbol{q}_2$ and delayed with respect to the pump pulse by a time delay τ_d. These experiments can be analyzed starting with the well known relation

$$\begin{aligned}\alpha &= \left\langle \frac{\mathrm{d}}{\mathrm{d}t}\,(\text{absorbed energy/volume}) \right\rangle / \text{energy flux} \\ &= \langle \boldsymbol{j}\cdot\boldsymbol{E}(\boldsymbol{R},t)\rangle / \text{energy flux}\ , \end{aligned} \tag{2.69}$$

where, in the electric dipole approximation,

$$j = \frac{\partial}{\partial t} \boldsymbol{P} = -\mathrm{i}\omega \boldsymbol{P} \ . \tag{2.70}$$

These equations can be solved in the thin sample limit, i.e., when $\alpha l \ll 1$ and $\alpha\lambda \ll 1$ where l being the effective length of the sample for which the absorption coefficient is α, and λ the wavelength of light. While the first inequality can be satisfied by appropriately selecting l, the second inequality is not well satisfied for typical semiconductors. In the thin sample limit, it can be shown that [2.75] the induced differential transmission signal ($\Delta I/I \cong \Delta\alpha/\alpha$) in direction $\boldsymbol{q}_2$ is given by

$$\mathrm{DTS}(t) \propto -\mathrm{Im}\{P^{(3)}_{121^*} E^*_2\} \ , \tag{2.71}$$

where the third-order polarization $P^{(3)}_{121^*}$ in direction $\boldsymbol{q}_2$ is obtained from the appropriate component of the density matrix using the relation given by (2.39), and E_2 is the electric field for the beam propagating in the direction $\boldsymbol{q}_2$. One generally measures the integrated differential transmission signal, given by the time integral of (2.71). This DTS signal can be considered as the diffraction of the pump in the direction of the probe by the grating created by the pump and the probe, and is the coherent component of the DTS signal. We will see that recent experiments (Sect. 2.8) give a clear indication of coherent signals in differential transmission experiments.

(b) Coherence in Luminescence and Light Scattering. Resonant emission is a linear property of the system. Resonant emission from an excited system may be resonant Rayleigh scattering or resonant luminescence. Conceptually, one can think of Rayleigh scattering as the coherent emission from the polarization of the excited ensemble, and luminescence as arising from the population of excited states. From a practical point of view, it is often difficult to distinguish between resonant Rayleigh scattering and resonant luminescence [2.76, 77], or between resonant Raman scattering and hot luminescence [2.78, 79]. In atomic physics, the role of coherence in resonant emission has been extensively explored and a detailed discussion has been given by *Loudon* [2.48] and *Mollow* [2.80]. Most of the concepts discussed in these discussions of two-level systems can be applied to semiconductors, with the caveat that interaction between excitons is strong in semiconductors and the independent two-level model should be applied with caution.

Some emission experiments in semiconductors in which coherent effects are important will be discussed in the section on quantum beats (Sect. 2.6) and in Chap. 6 on exciton dynamics.

2.1.4 Concluding Remarks

We have given a brief discussion of the basic concepts necessary to understand coherent spectroscopy experiments in semiconductors. In the remainder of this chapter, we will discuss some of the many interesting experiments performed in semiconductors over the last few years and discuss how these experiments can be understood using the concepts discussed above. We will first discuss coherent effects related to excitons, free carriers and phonons, and conclude with a discussion of coherent THz emission experiments.

2.2 Dephasing of Excitons

Dephasing of excitons in quantum wells was first investigated by *Hegarty* and *Sturge* [2.81] using the technique of cw and time-resolved (30 ps time resolution) resonant Rayleigh scattering. The intensity profile of the resonantly scattered Rayleigh signal was analyzed to determine the homogeneous linewidth (and hence the dephasing rate) of excitons as a function of exciton energy and temperature. The influence of various scattering processes on the dephasing of excitons was systematically investigated in a series of experiments by *Schultheis* et al. [2.69, 82–85] using the two-beam, degenerate FWM technique with picosecond laser pulses. With the laser tuned to the HH resonance, they measured the time-integrated FWM signal as a function of the delay between the two pulses (TI-FWM) and determined the exciton dephasing times by deconvolving their data with the pulse shape using optical Bloch equations for an Independent Two-Level (ITL) system. In most cases, there was an uncertainty of up to a factor of two in the determination of the dephasing rate because of uncertainty in the nature of broadening (homogeneous or inhomogeneous, Sect. 2.1.1h). They found that exciton dephasing times T_2 in bulk GaAs [2.84, 86] as well as in GaAs quantum wells [2.69, 70, 83] are in the picosecond range for a low excitation density at helium temperatures. With this as the starting point, they investigated the effects of temperature and collisions with other excitons and free carriers.

2.2.1 Exciton-Exciton and Exciton-Free-Carrier Collisions in Quantum Wells

These collision rates were determined by *Honold* et al. [2.70, 87] using the two-beam, time-integrated FWM technique, but with a third pulse of variable intensity from an independently tunable laser pre-exciting the sample 20 ps before the two weak pulses used to measure the FWM signals. The 20 ps time delay between the two pulses ensured that the exciton population excited by this third pulse was incoherent before the arrival of the other two pulses.

For measuring the influence of Exciton-Exciton (XX) collisions, the third pulse had the same photon energy as the other two pulses and hence resonantly excited the Heavy-Hole (HH) excitons. They determined the dephasing times T_2 by fitting their time-integrated FWM data by assuming that the excitons were inhomogeneously broadened ($T_2 = 4\tau_{\mathrm{decay}}$), and calculated the homogeneous linewidth $\Gamma_{\mathrm{h}} = 2\hbar/T_2$ ($= 1.316$ meV$/T_2$ [ps]). Their results on the variation of Γ_{h} with exciton density (Fig. 2.6) for a 120 Å wide quantum well, obtained using lasers with 2.6 ps pulses, show a linear dependence on the exciton density:

$$\Gamma_{\mathrm{h}}(n_{\mathrm{x}}) = \Gamma_{\mathrm{h}}(0) + \gamma_{\mathrm{xx}} a_{\mathrm{B}}^2 E_{\mathrm{b}} n_{\mathrm{x}} \ , \tag{2.72}$$

where a_{B} is the exciton Bohr radius, E_{b} is the exciton binding energy, and n_{x} is the exciton density per unit area. From the slope of best linear fit to the data, they obtain $\gamma_{\mathrm{xx}} = 1.5 \pm 0.3$.

Honold et al. [2.70] repeated the FWM measurements as a function of free-carrier density by tuning the pre-exciting laser photon energy to 8 meV above the band gap, thus pre-exciting free electrons and holes in the quantum wells. These results also show (Fig. 2.6) a linear increase of the homogeneous linewidth with the free-carrier density n. From the slope of the linear fit, they obtain $\gamma_{\mathrm{xeh}} = 11.5 \pm 2$, so that the exciton-free carrier interaction is about a factor of 8 stronger than the exciton-exciton interaction. They also presented argu-

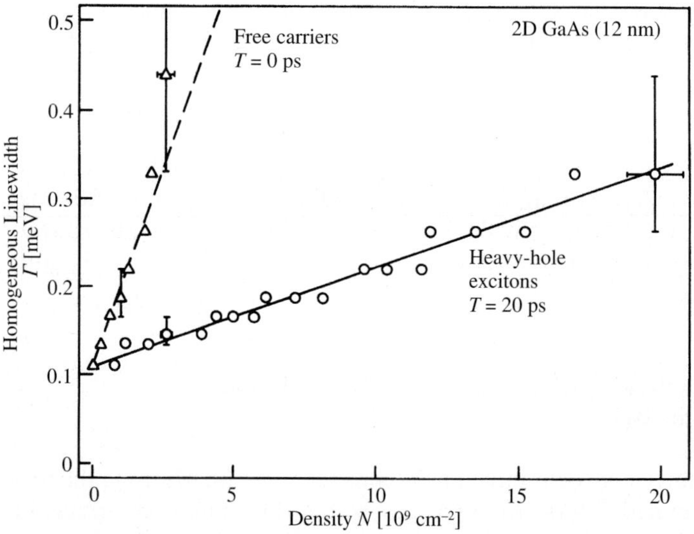

Fig. 2.6. Exciton linewidth deduced from an analysis of TI-FWM from a 120 Å GaAs quantum well at 2 K, as a function of the density of pre-excited electron-hole pairs and excitons. The linewidths were determined from best fit of the optical Bloch equations to the experimental curves, taking into account the inhomogeneous broadening of the exciton [2.70]

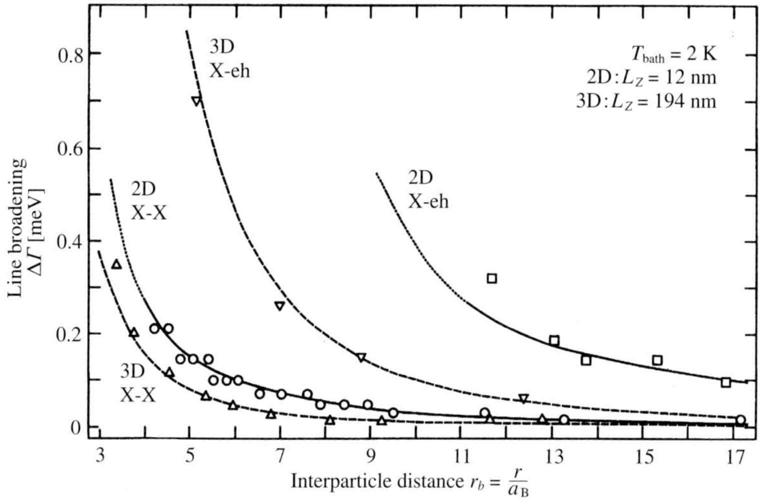

Fig. 2.7. Line broadening of two-dimensional (*circles* and *squares*; 120 Å quantum well) and three-dimensional (*triangles*) excitons at 2 K subjected to collisions with excitons (x-x) or free carriers (x-eh) as a function of the normalized interparticle distance [2.70]

ments that electron-exciton scattering is more efficient than hole-exciton scattering.

Exciton-exciton and exciton-free carrier scattering in bulk GaAs was investigated using luminescence techniques by *Leite* et al. [2.68]. *Schultheis* et al. [2.84, 85] used FWM techniques to determine the homogeneous width of excitons resulting from such collisions in 1900 Å-thick bulk GaAs film with $Al_{0.3}Ga_{0.7}As$ cladding layers. These results agreed quite well with the earlier luminescence studies [2.68]. Comparing these data with their results on GaAs quantum wells, *Honold* et al. [2.70] found (Fig. 2.7) that both the exciton-exciton and exciton-electron collisions were more effective in quantum wells than in the bulk when compared at equivalent interparticle distances. This result was attributed by *Honold* et al. [2.70] to relatively weaker screening in 2D systems as discussed by *Schmitt-Rink* et al. [2.56]. An additional consideration about exciton-exciton scattering is whether it depends on the spin state of the exciton. *Wang* et al. [2.71] have investigated this in connection with their study of interaction-induced dephasing (Sect. 2.7.1d), and found the HH XX scattering rate to be spin-independent for bulk GaAs. In contrast, the LH XX scattering rate shows a strong spin dependence as discussed recently by *Ferrio* et al. [2.88].

2.2.2 Exciton-Phonon Interactions

Honold et al. [2.87] also investigated the temperature dependence of the dephasing time and hence the homogeneous linewidth for $T<25$ K by using

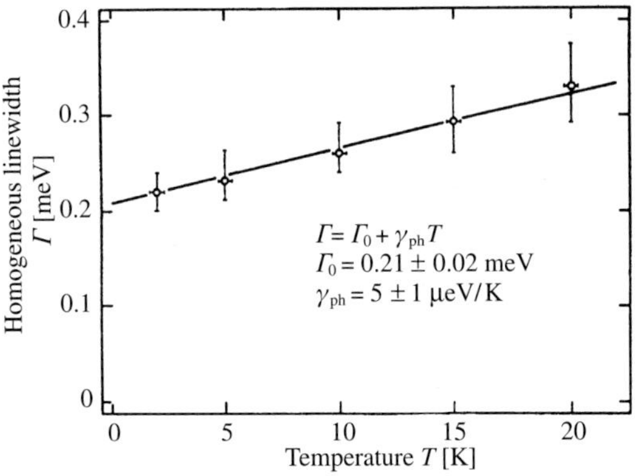

Fig. 2.8. Lattice-temperature dependence of the homogeneous linewidth of excitons in a 120 Å GaAs quantum well, deduced from TI-FWM decay rates assuming that excitons are homogeneously broadened [2.89]

two-beam TI-FWM experiments. Their results, obtained for the same 120 Å quantum well sample, reveal (Fig. 2.8) that the homogeneous linewidth increases linearly with increasing temperature. This can be expressed as

$$\Gamma_h(T) = \Gamma_h(0) + aT \ , \tag{2.73}$$

where $a = 5$ μeV/K. *Schultheis* et al. [2.69] investigated HH and LH line widths for several quantum well thicknesses and bulk GaAs to 80 K. The values for a ranged from 5 to 10 μeV/K for quantum well excitons, somewhat smaller than the 17 μeV/K deduced for bulk GaAs. The linear dependence on temperature was interpreted to imply the dominance of one-phonon scattering due to acoustic phonons. Note that there is an uncertainty of up to a factor of 2 in $\Gamma_h(0)$ and a determined from TI-FWM measurements, as discussed earlier. These values were obtained assuming homogeneously-broadened excitons ($\tau_{decay} = T_2/2$). If the line is inhomogeneously-broadened, as seems likely [2.70], then they [$\Gamma_h(0)$ and a] should be reduced by a factor of two.

These measurements were extended up to room temperature by *Kim* et al. [2.89] for a sample consisting of 100 Å GaAs quantum wells. Using 100 fs pulses, they measured the two-beam, TI-FWM signal as a function of the time delay τ_d between the two pulses and determined the decay constant τ_{decay} at several temperatures for many different intensities. At each temperature, they found a linear dependence of τ_{decay} on the excitation density and obtained a zero-density decay constant $\tau_{decay}(0)$ by extrapolating such curves to zero density. Figure 2.9 depicts the variation of $\tau_{decay}(0)$ with temperature T. The decay constant varies by more than a factor of 30 over a temperature range of 10 to 300 K.

Kim et al. [2.89] found that both acoustic and optical phonon scattering processes were necessary to fit this curve. Assuming that $T_2 = 2\tau_{decay}$ for the entire temperature range, an assumption that may be questionable at low temperatures where the homogeneous linewidth is probably smaller than the inho-

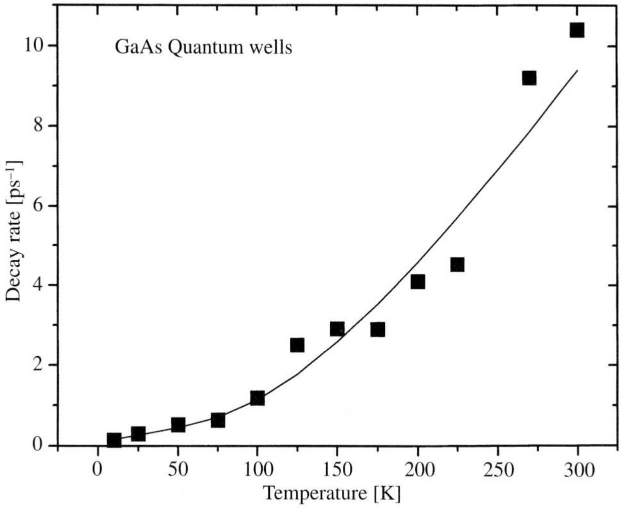

Fig. 2.9. Measured decay rates of TI-FWM from a 100 Å GaAs quantum well sample in the zero density limit as a function of the lattice temperature *T*. The *solid curve* is a least squares fit to (2.74) with parameters listed in the text [2.91]

mogeneous linewidth, they found that Γ_h can be fitted to an equation of the form

$$\Gamma_h(T) = \Gamma_h(0) + aT + bn(T) \ , \qquad (2.74)$$

where the first term is the temperature-independent contribution (scattering due to impurities, interface roughness, etc.), the second is the acoustic phonon contribution, and the last term is due to optical phonon scattering, primarily reflecting the optical phonon occupation factor $n(T) = [\exp(\hbar\omega_{LO}/kT) - 1]^{-1}$ where $\hbar\omega_{LO}$ is the optical phonon energy. Figure 2.9 shows the best fit to the data obtained by using the values $\Gamma_h(0) = 0.07$ meV, $a = 4.6$ μeV/K, and $b = 13.8$ meV. The value of a agrees well with 5 μeV/K obtained by *Honold* et al. [2.87]. Note that both analyses assumed homogeneous linewidth, an assumption that may not be valid at low temperatures where acoustic phonons make the most important contributions.

There also have been a number of calculations of the scattering rates and energy relaxation rates of $K_{||} = 0$ excitons as a function of temperatures. We will discuss some of these in Chap. 6 dealing with exciton dynamics. We note here that *Lee* et al. [2.90] have calculated the homogeneous widths of the excitons as a function of temperature and find $a = 3$ μeV/K and $b = 8$ meV, somewhat smaller than the values deduced above.

It should be remarked that earlier measurements of absorption linewidths [2.91] were fitted using only optical phonon interactions. Inhomogeneous broadening of the exciton absorption lines gave a large temperature-independent contribution at low temperature and made it difficult to obtain an accurate dependence at low and intermediate temperatures where acoustic phonons make the most important contributions. An advantage of the coherent FWM technique is that effects of inhomogeneous broadening are minimized. Thus, even with some uncertainty in the value of Γ_h at low temperatures, the TI-FWM measurements provide a much clearer picture of exciton-phonon interactions in semiconductors.

Finally, the zero-density value of $T_2 = 200$ fs obtained for excitonic dephasing at room temperature is quite important. It means that, even at low densities, TI-FWM from excitons will decay with $\tau_{decay} = 100$ fs for homogeneously broadened excitons and $\tau_{decay} = 50$ fs for inhomogeneously broadened excitons. Thus, at room temperature, the optical and acoustic phonons make a strong contribution to exciton dephasing that must be taken into account when considering density dependence of the dephasing rate.

2.2.3 Localized Excitons

The subject of disorder and localization has attracted considerable attention since the pioneering work of *Anderson* [2.92]. Semiconductors provide an ideal system for exploring the fundamental physics of these phenomena. Two kinds of systems have been used: the first is a mixed crystal of the type AlGaAs or CdSSe where the disorder is introduced by the random nature of the alloy. The second is a quantum-well system where roughness in the interface between the two constituents introduce well-width fluctuations within a single-quantum well. Since the confinement energies depend on the well width, quite strongly in thin wells, this leads to different exciton energies in different spatial regions. Both types of disorder have been investigated using ultrafast coherent spectroscopy.

(a) Quantum Wells. The work by *Hegarty* et al. [2.81, 93–97], using the techniques of resonant Rayleigh scattering and spectral hole burning, established that for an inhomogeneously broadened exciton in a GaAs quantum well, the states below the line center behave like localized states with finite activation energies whereas the states above the line center behave as delocalized states. This interpretation is consistent with the idea of a mobility edge separating the localized and delocalized states. This observation is not universally observed in all semiconductor quantum well systems. For example, recent results by *Hegarty* et al. [2.98] and *Albrecht* et al. [2.99] show that all excitons are localized in the InGaAs/InP quantum wells investigated by them.

These states in GaAs quantum wells have been further investigated by *Steel* and *coworkers* in recent years [2.100–106], in a series of experiments in the frequency and the time domain. We discuss here the time-domain results reported by *Webb* et al. [2.104, 105] and *Cundiff* et al. [2.107,108] on a sample with 96 Å GaAs quantum well and 98 Å $Al_xGa_{1-x}As$ ($x = 0.3$) barriers. The exciton absorption linewidth is $\cong 2.3$ meV (FWHM) and the luminescence peak is Stokes shifted by $\cong 1$ meV from the absorption peak. These authors used a three-beam FWM setup with counter-propagating beams over a wide range of densities extending to extremely low densities. At a density of 5×10^7 cm^{-2} excited with 1.6 ps laser pulses (spectral width $\approx$ 1 meV) centered 1 meV below the absorption line center, they observe a well-defined Stimulated Photon Echo (SPE) whose peak is located at a time corresponding to the delay time. The temporal width of the SPE corresponds to an inhomogeneous width of 2.5 meV, close to the observed FWHM in the absorption spectrum. From the decay of the time-integrated signal, *Webb* et al. [2.104,105] deduced a dephasing time of 68 ps, corresponding to $\Gamma_h = 0.019$ meV. Thus, the dephasing times of localized excitons are considerably longer than those for delocalized excitons discussed in Sects. 2.2.1, 2.

When the excitation intensity is increased to 5×10^9 cm^{-2}, the TR-FWM signal shows not only the SPE, but also a prompt signal characteristic of Free Polarization Decay (FPD). The SPE and FPD have different polarization dependence (Fig. 2.10). At low excitation levels and when all the beams are co-polarized, only the SPE signal is observed, whereas when beams 1 and 2 are cross-polarized, only the prompt FPD signal is observed. *Webb* et al. [2.104, 105] found that the SPE and FPD have the same spectral response, and that the dephasing time for the FPD signal was nearly constant across the exciton line whereas that for the SPE decreased at higher energies. Furthermore, the temperature dependence of the decay rates for the two emissions were found to be different, that for the FPD corresponding to interaction with acoustic phonons whereas that for the SPE corresponding to the migration between localized sites that was determined earlier in frequency-domain experiments [2.100, 103]. On the basis of these observations, *Webb* et al. argued that the simple picture of mobility edge separating localized and delocalized states is incomplete and that in reality these two kinds of states seem to co-exist and contribute to the FWM signal at the same photon energy. This possible interpretation is not universally accepted and more experiments and discussion on this interesting topic is likely. An alternate view of the disorder in quantum wells induced by well-width fluctuations has also been proposed by *Stolz* et al. [2.109].

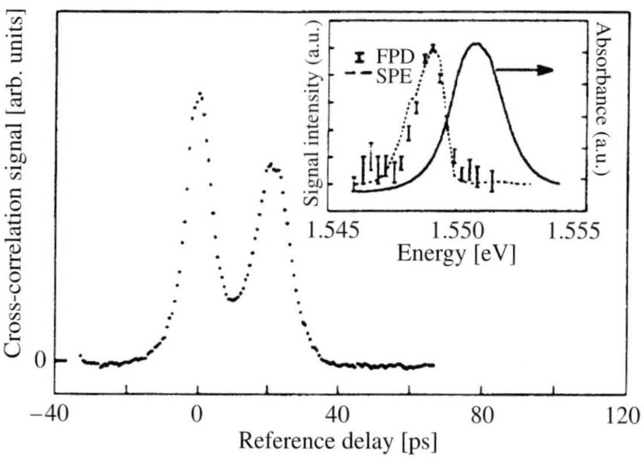

Fig. 2.10. TR-FWM signal showing both a stimulated photon echo (at 20 ps) and free polarization decay (prompt signal). The *inset* shows the spectral dependence of both signals and the sample absorbance. The sample was a 65 period 96 Å GaAs MQW at 1.8 K with an absorption linewidth of 2.35 meV [2.105]

(b) Mixed Crystals. We now discuss some results obtained [2.110–112] on disorder and localization induced by alloying in mixed crystal systems CdS_xSe_{1-x} and $Al_xGa_{1-x}As$. Although the linear optical spectra are very similar, the nature of disorder and its influence on exciton dephasing are very different in the two systems. In CdS_xSe_{1-x}, *Noll* et al. [2.110] investigated FWM signals with laser pulses of 7–10 ps and spectral width of less than 0.5 meV. They found exciton dephasing times of several hundred picoseconds, considerably longer than those obtained in CdSe. Comparison with the broad spectral line shape immediately shows that the spectrum is strongly inhomogeneously broadened. Time-resolved FWM measurements (Fig. 2.11) show a clear signature of photon echoes, as expected for an inhomogeneously-broadened two-

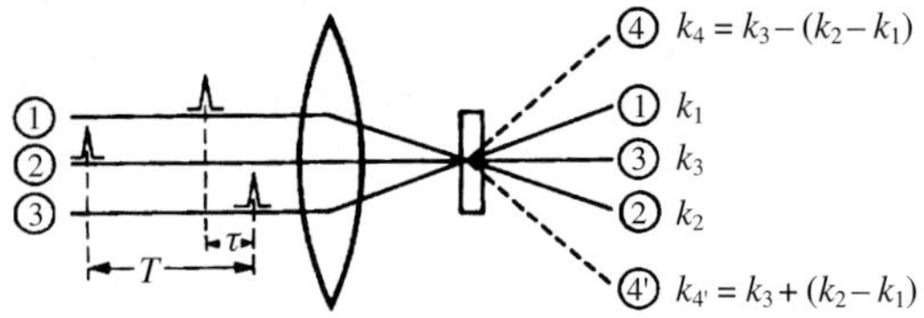

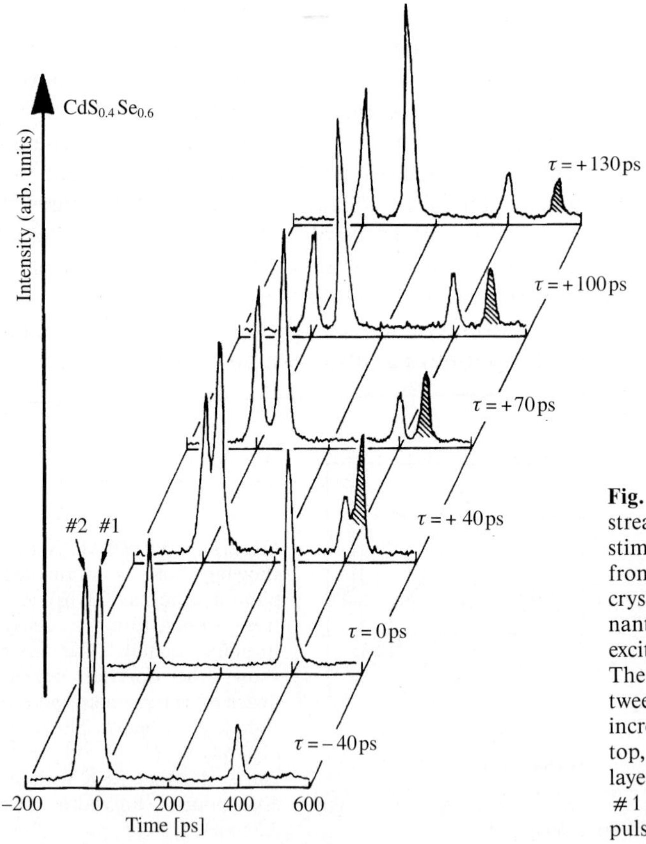

Fig. 2.11. Time-resolved streak camera traces of the stimulated photon echo from a CdS_xSe_{1-x} mixed crystal at 10 K for resonant excitation of localized excitons (E_{exc} = 2.0505 eV). The time separation between pulse #1 and #2 increases from bottom to top, and pulse #3 is delayed 400 ps from pulse #1 in all cases. The echo pulses are shaded [2.110]

level system from our discussion in Sect. 2.1.1. From the measurements of dephasing times for various excitation densities, temperatures and photon energies, *Noll* et al. [2.110] concluded that localization strongly reduces exciton-exciton and exciton-phonon scattering in CdS_xSe_{1-x}, and that energy relaxation induced by phonon-assisted hopping and recombination are the primary dephasing processes for the localized excitons.

In contrast to CdS_xSe_{1-x}, the dephasing times in $Al_xGa_{1-x}As$ [2.112] are approximately a few picoseconds, determined primarily by elastic disorder scattering. The contributions of exciton-phonon interaction and energy relaxation by hopping are small. They also conclude from their measurements that excitons are mobile in the sense that they are either delocalized or weakly localized as a whole with a localization radius larger than the exciton Bohr radius. The strong contribution of disorder to dephasing is a result of considerable short-range fluctuations in the disorder potential. These measurements show that considerable information about the nature of disorder can be obtained from coherent spectroscopy of semiconductors.

2.3 AC Stark Effect

The studies described above investigated properties of excitons in semiconductors and in particular their interactions with phonons and other excitations and with local potential fluctuations in the crystal (localization effects). In these cases, nonlinear optical spectroscopy allowed a determination of properties of excitons that could not be determined as well by linear spectroscopy. The intensities of the laser pulses used for these measurements were kept intentionally low so as not to perturb the excitonic states, and the laser was tuned close to the exciton resonance to maximize the FWM signals. A completely different regime can be explored if the laser is tuned to the transparent region of the semiconductor (well below the exciton resonance) and if strong optical pulses are used. One of the most interesting phenomena in this regime is that of AC Stark effect on excitonic states in semiconductors.

AC or dynamic Stark effect is well known in atomic physics and refers to the shift (splitting) of the atomic energy levels induced by a strong non-resonant (resonant) optical field [2.113]. The first such measurement in semiconductors was attempted by *Fröhlich* et al. [2.114, 115] in Cu_2O by using an intense CO_2 laser close to the 1s-2p exciton transition in the semiconductor. While there were clear indications of changes in the absorption spectrum from the ground state into the 2p state, a clear evidence of the shift of the transition was not obtained because of large inhomogeneous broadening of the transitions compared to the detuning of the CO_2 laser from the 1s-2p resonance. The first clear evidence for AC Stark effect in semiconductors was obtained in the case of GaAs quantum wells by *Mysyrowicz* et al. [2.116] and by *von Lehman* et al. [2.117].

Mysyrowicz et al. [2.116] excited a GaAs/AlGaAs MQW sample (with 100 Å quantum well and 100 Å barrier) at 15 K in its transparency region (i.e., sufficiently below the exciton absorption peak in the quantum well) with intense 100 fs pulses from an appropriately filtered amplified continuum, and measured the transmission spectra of the sample with a weak 100 fs continuum in the standard pump-and-probe configuration. Their results (Fig. 2.12) clearly show a shift in the exciton absorption peak to higher energy, and a slight bleaching of the exciton absorption during the pulse. Note that the shift exists only while the intense pump pulse is present. Since the lifetime of photoexcited excitons or free carriers is hundreds of picoseconds, much longer than pulse-width of 100 fs, these results demonstrate that the effect observed by *Mysyrowicz* et al. [2.116] is due to *virtual excitation*. A small residual change in the absorption spectrum persists for a long time and is related to a small density of real excitations that is probably created by the two-photon absorption of the intense excitation pulse. *Mysyrowicz* et al. [2.116] found no polarization dependence, and also reported interesting results for excitation very close to the exciton absorption peak.

These data were qualitatively explained [2.116] in terms of a dressed atom picture [2.113]. More sophisticated many-body theories more applicable to semiconductors followed soon, beginning with the work of *Schmitt-Rink* and *Chemla* [2.65] who explained the AC Stark shift in terms of the creation of virtual excitation that disappears when the exciting pulse ends. There was some initial controversy between this approach and that by *Combescot* and *Combescot* [2.118], but it was shown by *Zimmermann* [2.119–121] that the major results of the two approaches were identical. *Zimmermann* [2.120] also showed that the intense field should induce a shift of the continuum states as well and that the shifts of the continuum and the exciton are different for moderate detuning. A shift in the continuum absorption spectrum to higher energy was indeed observed by *Peyghambarian* et al. [2.122] in the case of CdS. These re-

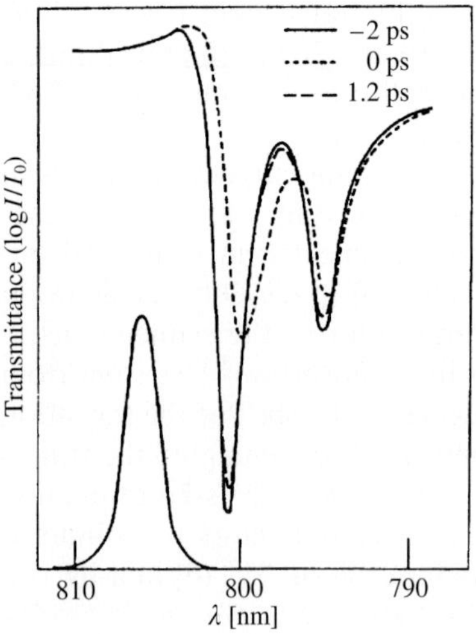

Fig. 2.12. Subpicosecond time-resolved transmission spectrum of a 100/100 Å GaAs-AlGaAs sample at 15 K recorded at three different delays after excitation by an intense pulse below the bandgap (pulse spectrum is also shown in the figure). The recovery of the absorption at $\tau_d = 1.2$ ps provides evidence for the optical Stark effect [2.116]

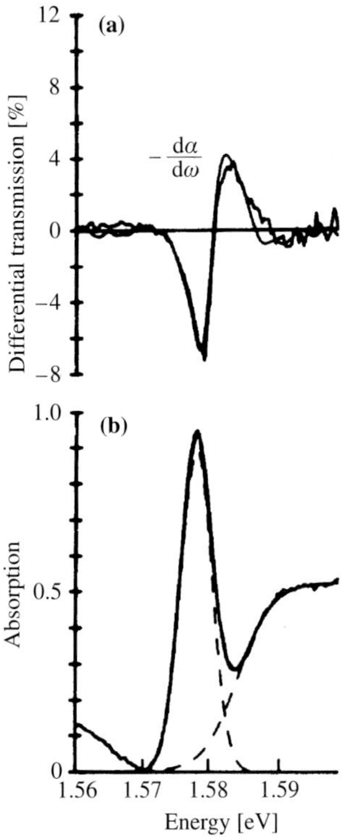

Fig. 2.13. **(a)** Differential transmission spectrum at $\tau_d = 0$ for a pump intensity of 30 MW/cm^2. The derivative of the experimental absorption spectrum in **(b)** matches the differential transmission spectrum **(a)**, demonstrating a pure Stark shift (shift without bleaching). 74 Å GaAs MQW sample at 35 K [2.123]

searchers also provided a detailed comparison with theory, including the relationship of the AC Stark effect to transient spectral oscillations (Sect. 2.3).

Knox et al. [2.123] reported the observation of a pure Stark shift, i.e., a shift in the absorption spectrum without any loss of oscillator strength. Their results (Fig. 2.13) reveal that for a low excitation intensity (30 MW/cm^2) and moderate detuning, the measured differential transmission spectrum can be fit well by the difference between the unperturbed absorption spectrum and an appropriately shifted absorption spectrum without any change in the spectral shape. This may appear surprising at first because virtual exciton pairs are also expected to bleach the exciton through phase-space-filling and screening effects. However, such a pure shift was predicted by *Zimmermann* [2.119–121] for certain excitation conditions for which the shift in the continuum states is larger than that in the exciton. For these conditions, the exciton binding energy increases and the increased oscillator strength resulting from this increase in the binding energy compensates the decrease in the exciton oscillator strengths caused by the reasons mentioned above. *Chemla* et al. [2.124] have reviewed experiments and theory of AC Stark effect with femtosecond pulses.

The observation of AC Stark effect in semiconductors, in addition to being of considerable fundamental interest, may also be of practical use in ultrafast modulators [2.125] because the change in the absorption lasts only as long as the exciting pulse is present. Therefore, the recovery time of such a modulator would be considerably faster than the recombination-rate-limited recovery time in modulators which rely on real excitations to produce modulation.

Finally we note two related phenomena. If the electromagnetic field is nearly resonant with the transition energy between two levels, then AC Stark *splitting* of the levels is expected [2.48, 113] instead of the AC Stark shift for non-resonant excitation discussed in this section. Dynamic Stark splitting is a well-known phenomenon in atomic systems [2.48, 113]. Dynamic or AC Stark splitting has not been reported in semiconductors. A time-domain analog of the dynamic Stark splitting is the Rabi oscillations. Evidence for Rabi oscillations in semiconductors has been reported recently by *Cundiff* et al. [2.126].

2.4 Transient Spectral Oscillations

The AC Stark effect investigates the influence of a strong laser pulse in the transparent region of the semiconductor on the exciton resonance for *positive time delays* ($\tau_d > 0$, probe pulse arriving at the sample *after* the pump pulse). Under the same experimental conditions, but for *negative time delays,* one observes the interesting phenomena of transient spectral oscillations. This was first reported by *Fluegel* et al. [2.127] for semiconductors and by *Brito Cruz* et al. [2.128] for dyes. Such transient oscillations have been observed for three different cases: (1) pumping in the continuum and observation in the spectral vicinity of the pump [2.127], (2) pumping below the exciton resonance and observation in the vicinity of the exciton resonance [2.129], and (3) pumping in the continuum and observation in the vicinity of the exciton resonance [2.129]. Although interband polarization of free carriers is involved in (1) and (3), we discuss these effects under this section of excitons because the observations are very similar.

The DTS in the vicinity of the above-bandgap pump, measured by *Sokoloff* et al. [2.129] for a 0.5 μm thick bulk GaAs at 15 K (Fig. 2.14), show clear spectral oscillations whose period increases as τ_d approaches zero. The increasing period, as τ_d approaches zero, is a general feature for all three cases. However, for cases (1) and (3) involving continuum excitation, the asymptotic shape of DTS is a symmetric peak indicative of spectral hole in case (1) and exciton bleaching in case (3). In contrast, for case (2) where the excitation is below the exciton energy, the asymptotic shape is dispersive, indicative of AC Stark effect. Thus, the transient spectral oscillations can be thought of as *precursors* to (1) spectral hole burning, (2) AC Stark effect, and (3) exciton bleaching resulting from phase-space filling and screening.

Lindberg and *Koch* [2.130] have obtained good fits to these oscillations by numerical simulations, but a qualitative discussion is adequate for our purposes. Since the dephasing time T_2 at the exciton is longer than the typical 100 fs pulses used for such experiments, the spectral oscillations in DTS near the exciton resonance can be primarily attributed to the pump-induced perturbation of the Free-Polarization Decay (FPD) of the polarization produced by the

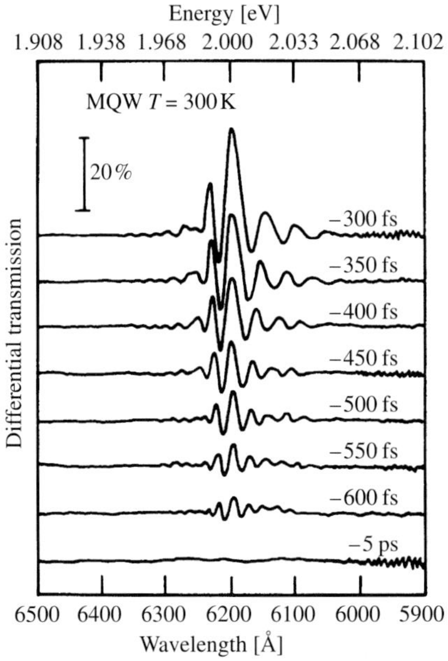

Fig. 2.14. Differential transmission spectrum measured at 300 K near the pump-photon energy in a 0.75 μm thick GaAs MQW sample with 100 Å quantum wells, at 300 K. The pump pulse is centered around 620 nm, has a 150 fs FWHM, and a peak intensity of 10^9 W/cm^2. Probe pulse arrives before the pump pulse (negative time delays) as indicated [2.129]

probe pulse. In contrast, the dephasing time of the continuum states is comparable or shorter than the typical 100 fs pulses used in these experiments. Therefore, the primary effect for DTS in the continuum states is the diffraction of the pump pulse in the probe direction by the grating produced by the pump and the probe. Thus, the effect at exciton resonance can be present even after the probe pulse is completely transmitted through the sample, so long as the pump pulse arrives at the sample within approximately the dephasing time after the probe pulse leaves the sample (negative time delay). In contrast, the effect at the continuum states is present only when the pump and the probe overlap in the sample. The description in terms of these two effects is, of course, an oversimplification and both terms must be treated in a unified manner for a complete description.

Joffre et al. [2.131] have discussed their results near the exciton in terms of the uncertainty principle. They have argued that when the pump-induced changes in the transmission are faster than T_2, monitoring the line or a part of it provides a temporal response limited by T_2. Therefore, in order to measure the true bleaching dynamics, one must integrate over a spectral width of the order of the inverse of the laser pulsewidth.

2.5 Exciton Resonance in FWM

The difference in the exciton and free-carrier dephasing times leads to another important effect, as discussed recently by *Kim* et al. [2.89]. The strong optical

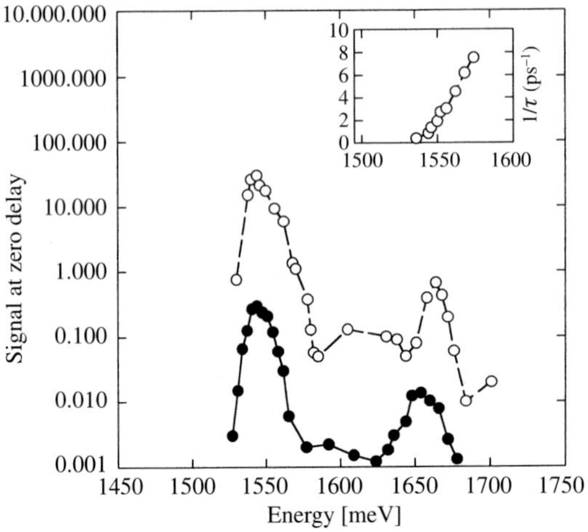

Fig. 2.15. Time-integrated FWM signal at the peak (close to the zero delay) from a GaAs quantum well sample at 10 K (*open circles*) and 300 K (*solid circles*) as a function of the central photon energy of the laser pulse. The 300 K data have been shifted up by the known band gap shift between 10 and 300 K. The excitation density is 2×10^{10} cm^{-2}. The data clearly show a strong resonance at the HH exciton. The *insert* shows the decay rate of the time-integrated FWM signal as a function of the excitation energy at 2×10^{9} cm^{-2} [2.91]

nonlinearity at the exciton resonance in quantum wells is well known and has led to the observation of FWM signals at room temperature with very weak light sources by *Miller* et al. [2.132, 133], and *Chemla* and *Miller* [2.134]. We discuss in this section the remarkably strong resonance in the FWM signals at the exciton resonance which cannot be explained simply by the strong optical nonlinearity at the exciton.

Kim et al. [2.89] measured the time-integrated FWM signals as a function of time delay τ_d for a 65 period 100 Å GaAs/100 Å $Al_{0.3}Ga_{0.7}As$ MQW sample using a Ti : Sapphire laser with 100 fs pulses. Data were obtained by varying temperature (10 to 300 K), excitation density (2×10^{8} to 4×10^{10} cm^{-2}) and photon energy (1.44 to 1.59 eV). The peak TI-FWM signal, occurring at $\tau_d \approx \tau_p$ (the laser pulsewidth), is plotted in Fig. 2.15 as a function of the laser-photon energy for the sample at 10 and 300 K. The data for 300 K have been appropriately shifted along the energy axis to account for the change in the band gap. The data show a strong peak at the $n = 1$ HH exciton, a weak peak at the $n = 2$ exciton and a very weak signal when the laser is tuned to the continuum. The ratio of the signal at the $n = 1$ exciton to that in the continuum is $\cong 300$, a remarkably large number.

In order to understand this behavior, *Schäfer* and *Schmitt-Rink* [2.135] performed a numerical simulation [2.89] of the semiconductor Bloch equations. They found that for long laser pulses with spectrum narrower than the exciton homogeneous linewidth, the calculated ratio for the FWM signal at the exciton and in the continuum is $\cong 4$, as expected from the oscillator strengths. For

70 fs pulses, on the other hand, the ratio is 1 if T_2 at the exciton and in the continuum is 800 fs. However, for 70 fs pulses, if T_2 (exciton) = 800 fs and T_2 (free carriers) = 80 fs, then the calculated ratio of the FWM signal is 300. This is because the signal with the long dephasing times dominates at long times. Thus, the correct way to consider the FWM signals is in the time domain and the primary cause of the large resonance in the FWM signal at the exciton is the long dephasing time of the exciton compared to that of the free carriers. This result shows that, even when the central wavelength of an ultrashort laser pulse is tuned to the continuum states, excitonic contributions to the FWM signal cannot be neglected because the tail of the laser spectrum may overlap the exciton, and although the number of excitons may be much smaller than that of the free pairs, the excitons may make a substantial contribution because of their longer dephasing times.

2.6 Quantum Beats of Excitons

We have so far considered systems that can be described reasonably well by a two-level model; i.e., when the photon energy is close to the energy difference between two levels and very different from the energies of other allowed transitions in the system. We now consider the case of a three-level system (Fig. 2.16a) which has a common ground state and two excited states that are close in energy. If the system is excited resonantly with a short laser pulse such that the spectral width of the laser is greater than the energy separation between the two excited states $|1\rangle$ and $|2\rangle$, then the laser pulse excites a coherent superposition state, i.e., a linear superposition of the two wavefunctions. This

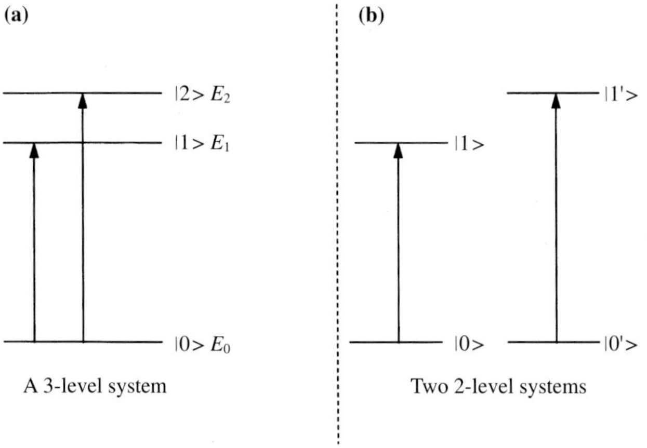

Fig. 2.16. **(a)** Schematic of a three-level system with two closely-spaced excited states and a common state. **(b)** Two two-level systems with close transition energies but no common state

is a non-stationary state whose time evolution is determined by the time-dependent Schrödinger equation. Since the energies of the eigenstates comprising the non-stationary state are different, the phases of the two eigenstates will evolve at a different rate. It can be shown that the spontaneous emission intensity from this system following the pulsed excitation oscillates with a period given by $h/\Delta E$ where ΔE is the energy separation between levels $|1\rangle$ and $|2\rangle$, and h is the Planck's constant. This is the phenomenon of quantum beats. Quantum beats have been observed in atoms [2.136], molecules [2.137] and nuclei [2.138]. It can also be shown that the third-order diffracted polarization signal from this system also oscillates with the same period so that TI-FWM signals would also show these quantum beats.

The term "quantum beats" has also been used to describe interference phenomena which arise when the two transitions *not* sharing a common ground or excited state are excited by a short laser pulse whose spectrum is wider than the difference in the transition energies (Fig. 2.16b). An example of such a system is a gas consisting of a mixture of two isotopes. Although such interference has also been considered a special case of "quantum beats" [2.139], there is no wavefunction superposition or quantum interference. The nonlinear or linear polarization emitted by such a system can show polarization interference effects which are difficult to distinguish from quantum beats. We discuss both types of beats in Sects. 2.6.1, 3, and experimental means of distinguishing them in Sect. 2.6.4.

Several general comments are appropriate before discussing some specific examples. Quantum beats from defect states in solids have been investigated for a number of years [2.46]. However, quantum beats of extended states in solids have been observed only recently, primarily because of short dephasing times of extended excitons in semiconductors. The first observations were made for excitons in AgBr by *Langer* et al. [2.140] and excitons in GaAs quantum wells by *Göbel* et al. [2.141]. It should be remarked that no beats are expected for the continuum states because they can be considered as inhomogeneously broadened two-level system in the momentum space as discussed above. Although the exciton wavefunction is made up of linear superposition of electron and hole wavefunctions at many different k-states, the concentration of the oscillator strength at the exciton energy allows a description of excitons in terms of a two-level system reasonable in the lowest approximation. Thus, closely-spaced exciton levels makes the observation of quantum beats possible. We now discuss some specific examples, with emphasis on III–V semiconductors.

2.6.1 Beats from Discrete Excitonic Islands

The confinement energies of the electron and holes in a quantum well depends on the thickness of the quantum well. Since the two interfaces of a quantum well are not perfect, the thickness of the quantum well is a function of the position in the plane of the quantum well. It is known for some time that a good-quality quantum well consists of islands whose average thickness differ by approximately a monolayer [2.142–145]. There may be atomic-scale fluctuations in the thickness within an island [2.145] but the average thickness within an island is constant, leading to a relatively narrow exciton line width in the luminescence spectrum. An excitation spot may encompass islands of two or three different average thickness, so that the luminescence spectrum may be composed of two or more well-separated lines.

Göbel et al. [2.141] investigated TI-FWM signals from two similar double quantum well samples of GaAs. Each sample consisted of 10 periods of a double quantum well structure separated by 150 Å $Al_{0.35}Ga_{0.65}As$ barriers. The double quantum well structure in each sample consisted of two quantum wells of nominal thickness 70 Å and 45 Å separated by a 48 Å $Al_{0.35}Ga_{0.65}As$ barrier. Sample A was grown with growth interruption at the interfaces and exhibited an exciton luminescence spectrum with three spectral features corresponding to islands with different thicknesses. Sample B was grown without growth-interruption and exhibited a single broad (3.4 meV) exciton luminescence feature. TI-FWM signals were measured using $\approx$ 500 fs pulses from a dye laser with the laser spectrum ($\approx$ 3 meV) centered at the PLE peak for sample B, and encompassing the two lower energy peaks in the PLE of sample A. Typical TI-FWM signals from these two samples at 10 K are depicted in Fig. 2.17. Sample B shows an exponentially decaying signal; the decay time is related to T_2 as discussed above. In contrast, the TI-FWM signal from sample A displays periodic oscillations superposed on the exponential decay. The period of the oscil-

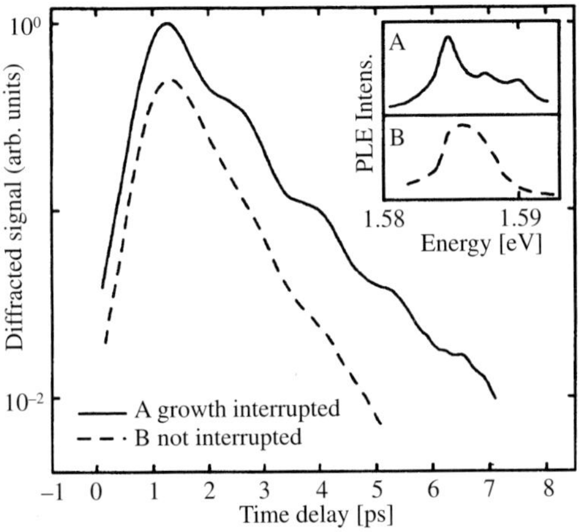

Fig. 2.17. Time-integrated FWM signal from two 70/48/45 Å $GaAs/Al_{0.35}Ga_{0.65}As/GaAs$ near the excitonic resonance of the 70 Å quantum well, at 1×10^9 cm^{-2} and 5 K. The *inset* shows the PhotoLuminescence Excitation (PLE) spectra for sample A (with growth interruption) which shows a broad peak, and sample B (no growth interruption) which shows three peaks due to monolayer-flat islands. Sample B was resonantly excited at the peak whereas sample A was excited with a spectrum overlapping the two low-energy PLE peaks [2.141]

lations is 1.3 ps, and corresponds to an energy splitting of 3.1 meV, somewhat larger than the 2.7 meV splitting observed in the PLE spectrum. These oscillations correspond to the beats between the two closely-spaced exciton levels.

These beats were analyzed in terms of the interference between the polarizations created at slightly different energies in different islands [2.141]. The exciton transitions were treated as inhomogeneously broadened and a reasonable fit to the TI-FWM signal was obtained by a straightforward calculation of the third-order polarization. This approach implicitly assumes that the beats are between two two-level systems without a common state; i.e., they are polarization beats. While this may seem reasonable because excitons in different islands do not have a common state, subsequent results by *Koch* et al. [2.146] reveal that the oscillations are actually quantum beats. However, the coupling mechanism between excitons in different islands is not well understood [2.146]. Techniques to distinguish between these two types of beats, based on TR-FWM [2.147, 148] and SR-FWM [2.149], will be discussed in Sects. 2.6.4a and b, respectively.

2.6.2 HH-LH Beats

Removal of the translational symmetry in the growth direction in quantum wells lifts the degeneracy between the HH and the LH valence bands at $k = 0$, and leads to well-separated HH and LH excitons in quantum wells. The energy separation between the excitons depend on quantum well parameters such as the well width and the barrier height. This is an intrinsic three-level system

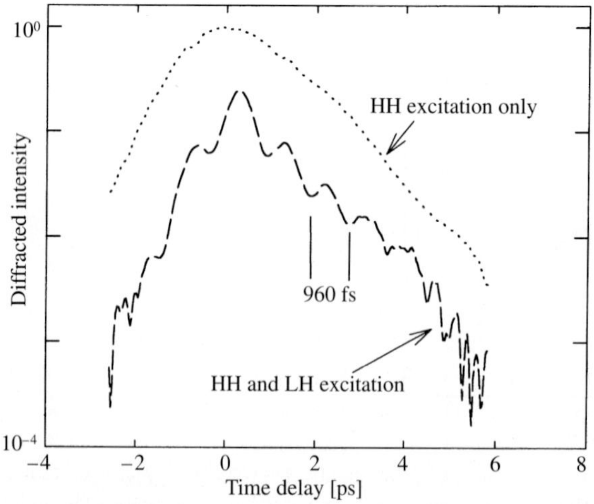

Fig. 2.18. Time-integrated FWM signal from 170/150 Å GaAs/$Al_{0.36}Ga_{0.64}As$ quantum well sample obtained at 5 K and 4×10^8 cm^{-2}; *dash-dot curve*: 1.5265 eV excitation (exciting primarily the HH exciton) showing smooth decay; *dashed curve*: 1.5310 eV, exciting both the HH and the LH excitons showing quantum beats [2.150]

with a common ground state. TI-FWM signals from such a system was first investigated by *Leo* et al. [2.150] and *Feuerbacher* et al. [2.151].

Figure 2.18 shows the TI-FWM signals from a 10 period MQW sample consisting of 170 Å GaAs wells separated by 150 Å $Al_{0.36}Ga_{0.64}As$ barriers [2.150]. FWM measurements were performed with a synchronously-pumped LDS-751 dye laser with $\approx$ 500 fs pulsewidth and 4.2 meV spectral width. When the laser is tuned to low energies so that its spectrum excites only the HH exciton, the TI-FWM signal for the sample at 10 K shows an exponential decay with $\tau_{decay} = 950$ fs. When the laser is tuned to an energy between the HH and the LH exciton, strong periodic modulation of the TI-FWM signal is clearly evident in the figure. The period of the oscillations is 960 ± 40 fs, in good agreement with the period of 980 fs calculated from the 4.2 meV HH-LH exciton splitting observed in the PLE spectrum.

The HH-LH beats can be analyzed on the basis of a three-level model that is a simple extension of the independent two-level model discussed in Sect. 2.1.1. According to this model, it can be shown that the intensity of the third-order diffracted polarization is given by [2.152]

$$I(\tau_d) = \Theta(\tau_d)\left(\frac{w_1^2}{2\gamma_1}+\frac{w_2^2}{2\gamma_2}+\frac{2w_1w_2(\gamma_1+\gamma_2)}{(\gamma_1+\gamma_2)^2+(\Delta E)^2}\right)$$
$$\times[w_1^2 e^{-2\gamma_1\tau_d}+w_2^2 e^{-2\gamma_2\tau_d}+2w_1w_2\cos(\Delta E\tau_d)e^{-(\gamma_1+\gamma_2)\tau_d}] \qquad (2.75)$$

for δ-function pulses. Here Θ is the Heaviside step function, $\gamma_{1,2}$ and $w_{1,2}$ are the dephasing rates and the spectral weights of the two transitions, ΔE is the energy separation between levels $|1\rangle$ and $|2\rangle$, and τ_d is the delay between the two pulses. This model provides a good description of the data and allows one to determine various dephasing rates by fitting the model calculations to the experimental observations. Figure 2.19 shows a comparison of the experimental data with a curve calculated from (2.75) using the following parameters: $w_1 = 1$, $w_2 = 1/3$, $\gamma_1(\gamma_2)$ corresponding to 0.35 (0.55) meV. This model gives a reasonably good agreement with the data despite its simplicity. It is interesting to note from (2.75) that for equal spectral weights and dephasing rates, this model predicts "perfect" quantum beats, i.e., *zero* FWM signal at certain delays with the peaks of the beats decaying with $2\gamma(=\gamma_1+\gamma_2)$.

There are two other interesting aspects of the observations. First, the treatment of the excitons as a non-interacting three-level system (in the same sense as a non-interacting two-level system discussed in Sect. 2.1.1) provides a good first-order description. However, note that the oscillatory signal is present for the negative time delays τ_d as well. As discussed in Sect. 2.7.1a, this negative τ_d signal can not be explained on the basis of the independent two-level model, and is a consequence of many-body interactions between excitons.

Another interesting observation is that the relative polarizations of the two beams in the two-beam degenerate FWM experiment have an important effect on the signal. The simple two- and three-level models discussed above ignore

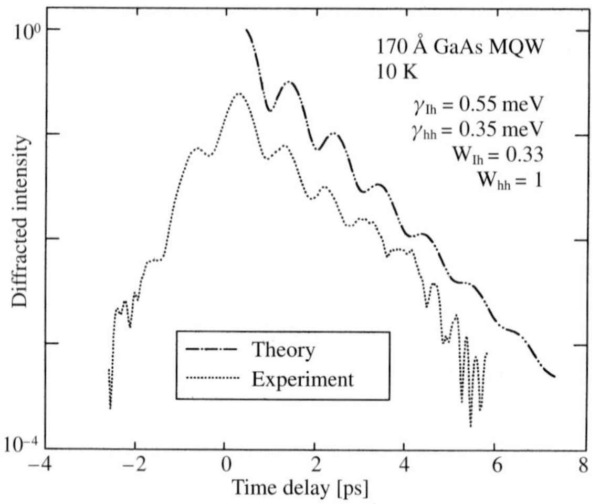

Fig. 2.19. Comparison of the experimental time-integrated FWM signal showing HH-LH quantum beats (*dashed line*) with the calculations based on a simple three-level model (*dash-dotted line*) [2.152]

the details of the crystal structure and symmetry, and hence can not predict the polarization dependence of the FWM signals. *Schmitt-Rink* et al. [2.153] considered a six-level model describing the $J = 1/2$ electrons and $J = 3/2$ holes and linearly polarized laser beams. They showed that (i) for co-polarized incident beams, the TI-FWM signal is linearly polarized parallel to the polarization of the two beams, whereas for cross-polarized incident beams the TI-FWM signal in the $2q_2 - q_1$ direction is polarized anti-parallel to beam 1. The HH and LH components are in-phase for co-linear and out-of-phase for cross-linear conditions, so that the quantum beats are out of phase for the two conditions. (ii) For all other incident polarizations, the TI-FWM signal is elliptically polarized, and that the beats disappear when the two beams are polarized at 45° with respect to each other. These conclusions are a consequence of the circular selection rules for the HH and LH exciton transition in III–V quantum wells.

These ideas were experimentally tested [2.153] by measuring two-beam TI-FWM signals from a 10 period, 170 Å GaAs quantum well sample with a HH-LH splitting of 4.2 meV. The sample at 12.5 K was excited with 100 fs pulses from a Ti:Sapphire laser tuned between the HH and LH excitons. The excitation density was $< 10^9$ cm^{-2}, and TI-FWM signals in both the $2q_2 - q_1$ and $2q_1 - q_2$ directions were measured without any analyzers. The results show pronounced quantum beats which are 180° out of phase for the co-linear and cross-linear geometry (Fig. 2.20), thus confirming the first prediction. The beats show a maximum for $\tau_d = 0$ for the co-linear case as expected. The experiments also show the unexpected result that the decay time depends strongly on the relative polarization directions. This aspect of the results has been in-

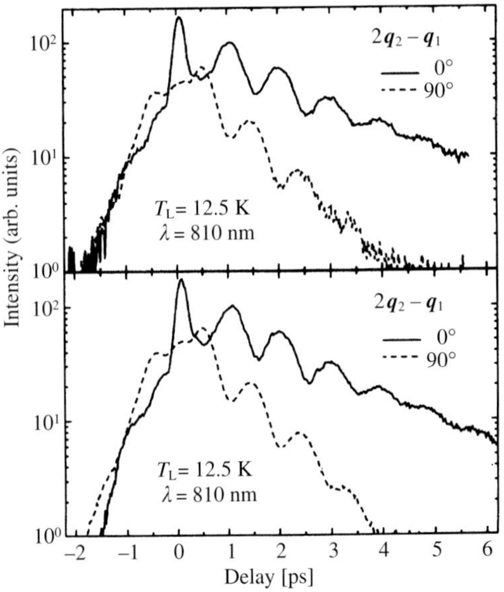

Fig. 2.20. Two-beam self-diffracted time-integrated FWM signal from 170/150 Å GaAs quantum well sample at 12.5 K and $<1\times10^9$ cm^{-2} excitation density for co-linear (*solid line*, 0°) and cross-linear (*dashed line*, 90°) polarizations, showing the π change in the phase of the oscillations due to quantum beats [2.153]

vestigated extensively by *Kuhl* and collaborators [2.154], and is discussed in more detail in Sect. 2.7.2d.

2.6.3 Quantum Beats of Magneto-Excitons

Bar-Ad and *Bar-Joseph* [2.155] investigated differential transmission in pump-probe experiments on a 100-period, stepped MQW GaAs sample consisting of adjacent layers of 30 Å GaAs and 100 Å $Al_{0.1}Ga_{0.9}As$ separated by 100 Å $Al_{0.3}Ga_{0.7}As$ barriers. The excitons were inhomogeneously broadened, with the HH exciton absorption line width of ≈ 6 meV at helium temperatures. They found that the differential transmission showed periodic oscillations as a function of time delay τ_d when a magnetic field was applied perpendicular to the interface planes. Figure 2.21 presents their results for $B = 4$ T for cross-linear (solid curve) and co-linear (dotted curve) polarizations of the pump and the probe beams. The two curves are out-of-phase, similar to the behavior of HH-LH quantum beats discussed in the previous section. These absorption quantum beats have an origin similar to the quantum beats in FWM experiments in that they also arise from the third-order optical nonlinearity in the direction $q_1 + q_2 - q_1$ where the pump and probe are along q_1 and q_2, respectively. In other words, the pump-probe results can be analyzed in terms of the diffraction of the pump (q_1) along the probe direction (q_2) by the grating ($q_2 - q_1$) created by the pump and the probe. Note that this corresponds to the coherent part of the differential transmission signal, and that the incoherent part of the signal persists after the beats (and hence coherence) disappear, as discussed in Sect. 2.1.3a. The period of the quantum beats varied with the magnetic field, and the beats disappeared for circular polarizations or when

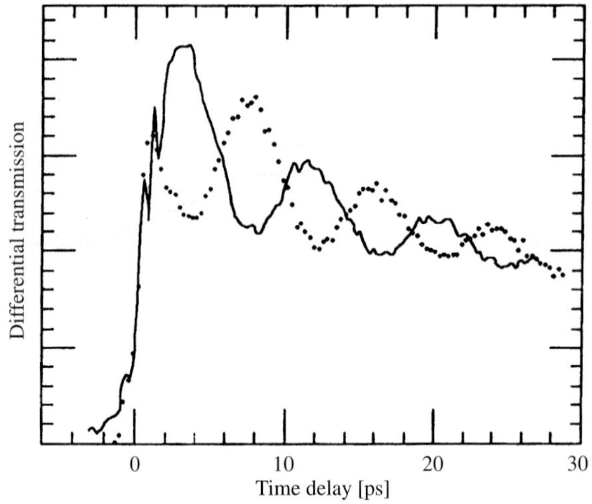

Fig. 2.21. Normalized differential transmission as a function of time delay τ_d for a stepped multiple quantum well sample (30 Å GaAs adjacent to 100 Å $Al_{0.1}Ga_{0.9}As$, surrounded on both sides with 100 Å $Al_{0.3}Ga_{0.7}As$ barriers) at $B = 4$ T normal to the interface planes. *Solid curve:* co-linear polarizations, *dashed curve:* cross-linear polarizations showing the π phase change in the magneto-exciton quantum beats [2.155]

the magnetic field was parallel to the interface planes. The beats were interpreted in terms of the magnetic field-induced spin splitting of the excitonic levels, and were used to measure the Landé g factor [2.155]. Spin splitting of excitons and spin-relaxation dynamics was also investigated by spectral hole burning and FWM techniques by *Wang* et al. [2.156].

2.6.4 Distinction Between Quantum and Polarization Beats

If the coherent laser field excites more than one electronic transition, interference phenomena, such as quantum beats discussed above, can be observed until the system is dephased. The microscopic nature of these beats can be quite different depending on whether the different transitions are independent or have some states in common. This was first discussed by *Lambert* et al. [2.157]. For example, in a three-level system with two closely-spaced excited states, a quantum mechanical interference due to the coupling of the transitions via a common state causes quantum beats. In contrast, for two closely spaced two-level system with no common state can also show oscillations due to the interference of the emitted fields from each two-level system at the detector, known as polarization interference.

It is not always easy to ascertain whether the observed beats are quantum beats or polarization beats, even though the origin of the two interference phenomena are quite different. We discuss in this section two techniques [2.148,

149] that have been developed to provide an experimental distinction between these.

(a) Distinction Using TR-FWM. *Koch* et al. [2.148] showed that the temporal evolution of the third-order polarizations along $2\boldsymbol{q}_2 - \boldsymbol{q}_1$ in a two-beam FWM (TR-FWM) experiment are different for the two cases:

$$P^{(3)}_{\mathrm{QB}}(t, \tau_\mathrm{d}) \propto P^{(1)*}(\tau_\mathrm{d}) P^{(1)}(t - \tau_\mathrm{d}) \tag{2.76}$$

for the quantum beats, and

$$P^{(3)}_{\mathrm{PI}}(t, \tau_\mathrm{d}) \propto P^{(1)*}(0) P^{(1)}(t - 2\tau_\mathrm{d}) \tag{2.77}$$

for the polarization interference, where

$$P^{(1)}(t) \propto \cos(\Delta E t/2h) \exp(\mathrm{i}\omega_1 t/2 + \mathrm{i}\omega_2 t/2) , \tag{2.78}$$

and ΔE is the difference in energy between the two transitions. On the basis of this analysis, it can be shown that the quantum beat signal for a given τ_d is maximum when $t = \tau_\mathrm{d} + nh/\Delta E$ for integer n. In contrast, the polarization beats will give a maximum signal for $t = 2\tau_\mathrm{d} \pm nh/\Delta E$, with the restriction that t exceeds τ_d. Furthermore, in the absence of dephasing the maximum in the polarization interference signal remains constant as a function of τ_d. In contrast, the maximum in the quantum beat signal oscillates with τ_d as a result of the first term of (2.77).

Koch et al. [2.148] used two different GaAs nanostructures to verify these predictions. They measured TR-FWM from a GaAs MQW sample with 150 Å quantum wells and a HH-LH splitting of 5.4 meV using 100 fs pulses from a mode-locked Ti:Sapphire laser. Figure 2.22a exhibits the TR-FWM signal in a two-beam self-diffraction experiment as a function of time following the arrival of the *first* pulse on the sample for various time-delays τ_d. For a given τ_d, the TR-FWM signal oscillates as a function of time t. The intensity at each of these maxima shows an oscillatory behavior as a function of τ_d, thus confirming that the signal is a result of quantum beats. The second nanostructures they used consisted of 40 periods of 80 Å GaAs quantum wells followed by 40 periods of 90 Å GaAs quantum wells. The laser spectral width was smaller than the HH-LH splitting in each region but larger than the energy difference between the HH excitons in the two regions (13.4 meV). The TR-FWM from this structure (Fig. 2.22b) displays periodic oscillations as a function of t, but the intensity at the maxima do not show oscillatory structure as a function of τ_d. Therefore, this signal is due to polarization interference. The decay of signals as a function of t and τ_d results from the dephasing that is ignored in (2.76, 77), for simplicity. Analysis of the results shows that t_m, the time t at which the signal has a maximum for a given τ_d, changes twice as rapidly for the polarization interference case compared to the quantum beat case, as expected.

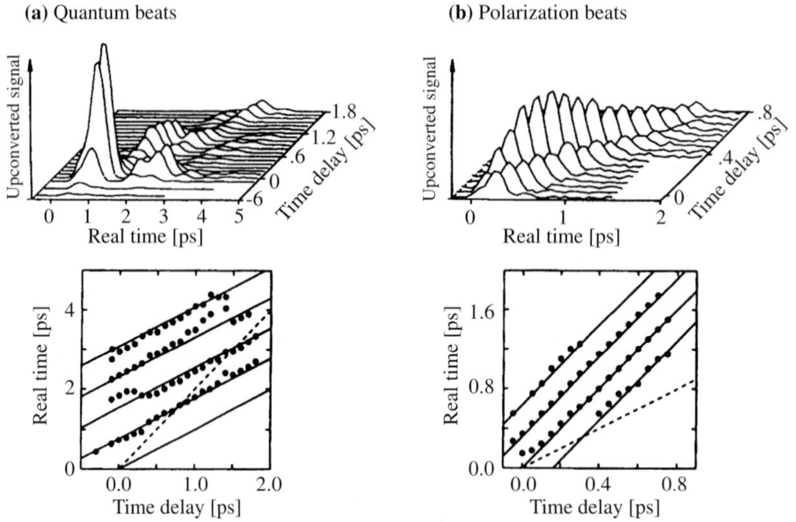

Fig. 2.22 a, b. The top traces show the Time-Resolved FWM (TR-FWM) signal as a function of time for different time delays between the beams: **(a)** from a GaAs MQW sample with 150 Å quantum wells at 5 K; **(b)** from a GaAs MQW sample with 40 periods of 80 Å quantum wells followed by 40 periods of 90 Å quantum wells at 5 K. The lower traces represent the corresponding plots of the peak position in real time as a function of time delays. Different behavior of the two systems demonstrates that TR-FWM can be used as a technique to distinguish between **(a)** quantum beats and **(b)** polarization interference (see the text) [2.148]

The oscillations in TI-FWM from monolayer-flat islands in GaAs quantum were analyzed as polarization beats [2.141]. However, application of these time-resolved techniques to this sample showed [2.146] that these oscillations are true quantum beats. Therefore, the excitons in different monolayer-flat islands can not be considered as two uncoupled two-level systems, and there is a sufficiently strong coupling between excitons in different islands to make the system equivalent to a three-level system. Various possible coupling mechanisms were discussed, but the origin of such coupling is not clear at this time [2.146].

(b) Distinction Using SR-FWM. *Lyssenko* et al. [2.149] have demonstrated that Spectrally-Resolved FWM (SR-FWM) also provides a technique for distinguishing between polarization interference and quantum beats. If two transitions with frequencies ω_1 and ω_2 are excited by transform-limited pulses with spectral width much smaller than ω_1 and ω_2, but larger than $|\omega_1 - \omega_2|$, then for τ_d larger than the pulse widths, they showed that for quantum beats

$$|P^{(3)}_{\mathrm{QB}}(\omega_d, \tau_d)|^2 = \left| \frac{R_1 \exp(\mathrm{i}\omega_1 \tau_d) + Q_{12} \exp(\mathrm{i}\omega_2 \tau_d)}{\omega_1 - \omega_d - \mathrm{i}\gamma_1} + \frac{R_2 \exp(\mathrm{i}\omega_2 \tau_d) + Q_{21} \exp(\mathrm{i}\omega_1 \tau_d)}{\omega_2 - \omega_d - \mathrm{i}\gamma_2} \right|^2 \quad (2.79)$$

and for polarization interference

$$|P_{\mathrm{PI}}^{(3)}(\omega_{\mathrm{d}}, \tau_{\mathrm{d}})|^2 = \left| \frac{R_1 \exp(\mathrm{i}\omega_1 \tau_{\mathrm{d}})}{\omega_1 - \omega_{\mathrm{d}} - \mathrm{i}\gamma_1} + \frac{R_2 \exp(\mathrm{i}\omega_2 \tau_{\mathrm{d}})}{\omega_2 - \omega_{\mathrm{d}} - \mathrm{i}\gamma_2} \right|^2 . \tag{2.80}$$

Here, ω_{d} is the detection frequency, $R_i = 2N_i\mu_i^4 \exp(-\gamma_i\tau_{\mathrm{d}})$, $Q_{ij} = N_i\mu_i^2\mu_j^2 \exp(-\gamma_j\tau_{\mathrm{d}})$, N_i is the density of excited levels, μ_i is the dipole matrix element, and γ_i is the dephasing rate. *Lyssenko* et al. [2.149] demonstrated that, for quantum beats, the SR-FWM signal as a function of time-delay shows only small changes of the phase and the amplitude when the detection frequency passes through one of the resonances. In contrast, for polarization interference, the phase of the oscillatory SR-FWM signal as a function of the time-delay changes by π and the amplitude goes through zero as the detection frequency is varied around one of the resonances. This provides a mean of distinguishing between the two types of oscillations.

Lyssenko et al. [2.149] investigated SR-FWM from a GaAs MQW sample with 116 Å quantum wells and 20 periods, using picosecond pulses with a coherence time of less than 200 fs. The SR-FWM signals in two-beam self-diffraction experiment are shown in Fig. 2.23a as a function of τ_{d} for various

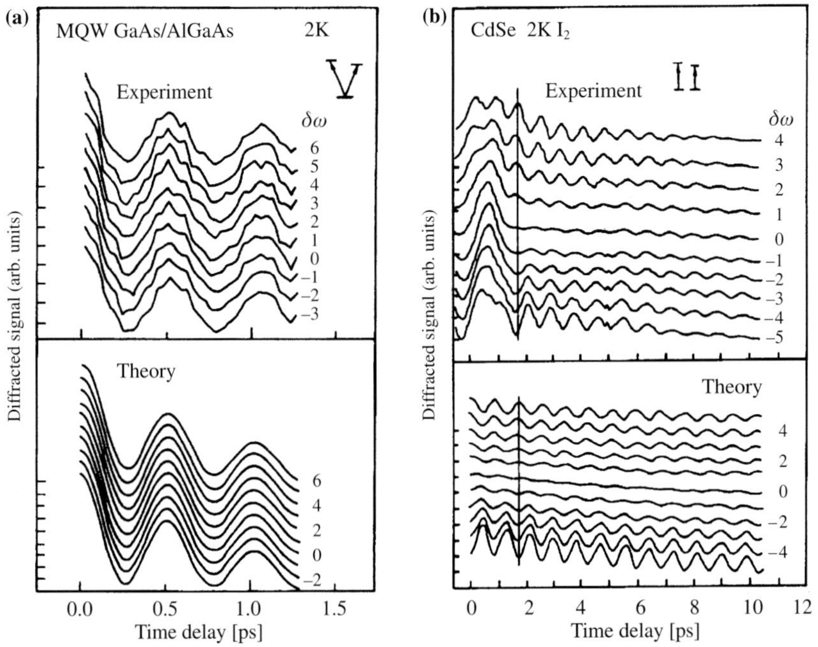

Fig. 2.23 a, b. Spectrally-resolved FWM signal as a function of time delay for various detunings. **(a)** MQW GaAs with 116 Å quantum wells near the HH-LH exciton resonances; $\delta\omega = \hbar(\omega_{\mathrm{d}} - \omega_1)/0.08$ meV. **(b)** CdSe near two bound exciton resonances; $\delta\omega = \hbar(\omega_{\mathrm{d}} - \omega_2)/0.08$ meV. All curves are normalized to their maximum value. The difference in the behavior provides another technique to distinguish between **(a)** quantum beats and **(b)** polarization interference [2.149]

detunings of the detection frequency from the frequency of the first transition in the units of 0.08 meV. The beats are present for all detunings and remain in phase (maxima and minima occurring at the same τ_d) as the detuning is varied. In contrast, the results on two closely-spaced bound excitons in CdSe (Fig. 2.23 b) show that the amplitude of the oscillations goes through a minimum for zero detuning, and that the phase of the oscillation changes as the detuning changes from positive to negative. These results reveal that the HH-LH beats in GaAs quantum wells are quantum beats, whereas the bound exciton beats in CdSe are due to polarization interference.

2.6.5 Propagation Quantum Beats

The discussion on quantum beats in this section implicitly assumes that (i) interaction between excitons can be ignored so that excitons can be treated within the framework of the independent two- or three-level model, and (ii) the sample under investigation is optically thin so that propagation effects can be ignored. In this section, we discuss some results within the independent two-level model when propagation effects are important. Interaction-induced effects are discussed in Sect. 2.7.

In the limit of optically thin sample, the electric field at any point in the sample can be approximated by the incident electric field of the laser pulse, and the optical Bloch equations adequately describe the response of the systems to the incident laser pulse. When the sample is not optically thin, the total electric field at any point in the crystal must be determined by using the Maxwell's equations. Therefore, the response of the system is determined by the solution of the coupled Maxwell-Bloch equations. The inclusion of the propagation effects introduced by Maxwell's equations has important ramifications for the transmitted pulse as well as for many nonlinear properties of the system. The general problem of pulse propagation in a dispersive medium has been investigated by a number of researchers [2.158, 159]. Under certain conditions [2.158] the electric field and the polarization of the medium can be expressed in terms of an integral of the Bessel functions of zero and first order. Numerical integration revealed [2.158, 160] that the intensity of a pulse transmitted through an optically thick medium exhibits oscillatory behavior. Unlike the quantum beats discussed above, the period of oscillations is not constant but increases at longer times. Polaritons [2.161 – 163] are the stationary coupled modes of excitons and photons in a crystal. The polariton description provides an alternate description of pulse propagation effects in an optically thick solid. We discuss here a striking example of "propagating quantum beats" reported by *Fröhlich* et al. [2.164] for the quadrupole polaritons in Cu_2O.

(a) Quantum Beats of Quadrupole Polaritons in Cu_2O. The 1s exciton in Cu_2O is dipole forbidden but is weakly allowed through quadrupole interaction. The polariton dispersion relation in the region of the resonance is given by the usual expression which includes the exciton dispersion, damping and oscillator strength. The weak quadrupole interaction leads to a very small longitudinal-transverse splitting of the polariton ($\Delta_{LT} \cong 10\ \mu eV$). The small oscillator strength leads to a long radiative recombination time constant. Furthermore, exciton-acoustic phonon scattering rate is strongly reduced at low temperatures because of the small number of final states. Therefore, the dephasing times of the quadrupole excitons is several hundred picoseconds, considerably longer than the excitons in GaAs discussed above.

Fröhlich et al. [2.164] investigated propagation of picosecond pulses through high-quality crystals of Cu_2O which were oriented using x-ray diffraction. The crystals at 1.8 K were excited with 30 ps dye laser pulses with a spectral width of 25 μeV. The temporal evolution of the transmitted pulse was measured using time-resolved photon counting system with microchannel-plate photomultiplier with a time resolution of 60 ps. When the laser was tuned to excite only one branch of the polariton, the transmitted pulse intensity showed a smooth decay with time. When the laser was tuned at the resonance so that both the branches of the polariton were coherently excited, the transmitted pulse revealed pronounced oscillatory behavior. The results for a 0.91 mm thick crystal of Cu_2O are exhibited in Fig. 2.24 [2.165]. The data, plotted on a log scale, display eight pronounced minima. The period of the oscillation increases with increasing delay. The solid lines are the result of a numerical fit to the data using Maxwell's equations with appropriate dispersion. From the fits to the data, they obtained an oscillator strength of 3×10^{-9} and a broadening of 0.65 μeV. The same parameters also give good fit to the data

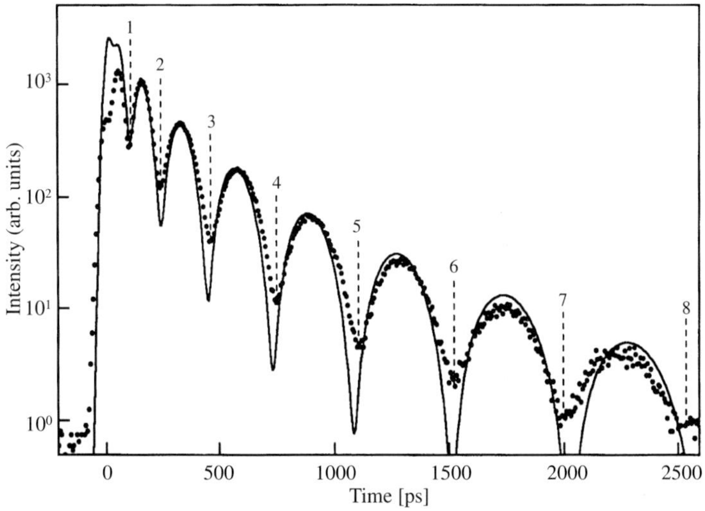

Fig. 2.24. Temporal evolution of the intensity of a 30 ps pulse (tuned to the 1s excitonic resonance) after propagation through 0.91 mm thick sample of Cu_2O. The *dots* are experimental and the *solid curve* is a theoretical fit [2.165]

on crystals of different thicknesses. Oscillator strengths in different directions were determined by measuring and analyzing pulse propagation along different crystal directions.

Fröhlich et al. [2.164] also presented an intuitive explanation for the observation of quantum beats of propagating polaritons in terms of the quantum interference between the two coherently excited polariton branches. A pair of pulses, one from each branch of the polaritons, propagating with the same group velocity experience a relative phase difference which depends on the difference in the frequency, difference in the wavevector, the group velocity, and the total distance traveled. For a given crystal thickness, minima in the transmitted pulse intensity occur corresponding to pairs of pulses with different group velocities. Since the frequency and wavevector differences decrease as the group velocity decreases, there is a corresponding increase in the beat period at longer times.

The results in Fig. 2.24 provide a striking demonstration of propagation effects in the transmitted pulse intensity. As mentioned above, propagation effects can also influence the nonlinear optical response on a medium. For example, *Pantke* et al. [2.166] have reported propagation effects on the third-order nonlinear polarization by measuring four-wave-mixing signals near the $n = 1$ and $n = 2$ A excitons in CdSe.

(b) Interferometric Measurements of Transmitted Pulse. The pulse distortion and oscillatory behavior of the transmitted pulse has also been observed in GaAs quantum wells by *Kim* et al. [2.167, 168] who investigated the temporal evolution of the transmitted pulse intensity as a function of total sample thickness and temperature. By cross-correlating the transmitted pulse with the laser in a nonlinear crystal, they were able to obtain time resolution of 100 fs, limited by the pulsewidth. They found that the pulse distortion was small for a MQW sample with a total quantum well thickness of 1500 Å. However, for samples with total thickness of ≈ 6000 Å, they found significant pulse distortion and aperiodic oscillatory behavior for the transmitted intensity. Figure 2.25 depicts the temporal evolution of the transmitted pulse after propagation through a 75-period multiple quantum well sample with 84 Å quantum wells (total thickness of 6300 Å) for sample temperatures of 10 and 100 K. Strong pulse distortion and aperiodic oscillations reminiscent of the results in Cu_2O are observed at 10 K. Increasing the temperature to 100 K decreases the dephasing time and reduces the pulse distortion.

Kim et al. [2.167, 168] proposed that the observations can be understood in terms of the interference between the incident field and the radiation from the induced dipoles. Since the induced polarization is 90° out-of-phase with the incident field, and the radiated field is an additional 90° out-of-phase, the radiated field is 180° out-of-phase with the incident field. This was confirmed by their interferometric measurements. For these measurements, they used a modified Michelson interferometer, as shown in the inset of Fig. 2.26. The sample was in one arm of the interferometer so that the pulse transmitted through the sample and an undistorted laser pulse were allowed to interfere at

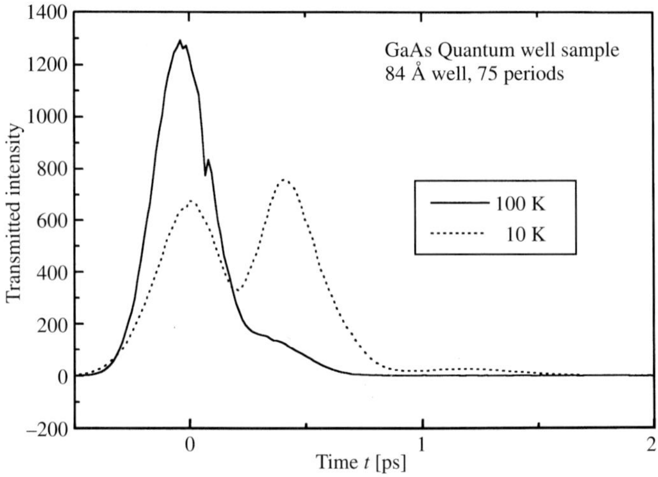

Fig. 2.25. Temporal evolution of the intensity of a 100 fs pulse transmitted through a 75 period, 84 Å GaAs multiple-quantum-well sample at 10 K and 100 K for an excitation density of 3×10^7 cm^{-2}. The transmitted pulse is strongly distorted at 10 K, but the distortion diminishes at 100 K because of a decrease in the dephasing time T_2 [2.167]

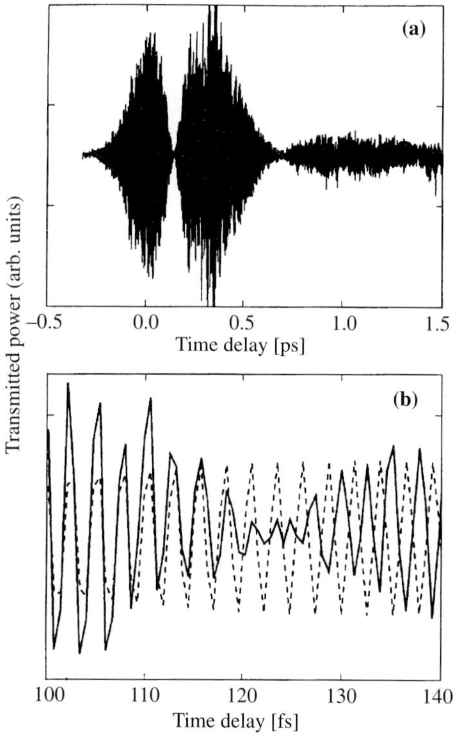

Fig. 2.26 a, b. Interferometric measurement of the temporal evolution of the transmitted pulse amplitude after propagation through a 75 period GaAs quantum well sample with 84 Å quantum wells (total thickness of 6300 Å) [2.167]

the detector. The electric fields of the two pulses add algebraically, and the measured intensity is the square of the total field at the detector. Figure 2.26 exhibits the temporal evolution of the measured intensity when the laser is tuned at the exciton resonance in the quantum well. The data show two clear nodes in the envelope, corresponding to the time at which the incident and the radiated fields cancel each other. An expanded view of the region near the first node [2.169] reveals that there is a 180° change in the phase at the node, a clear signature that the radiated field dominates beyond the node.

2.6.6 Concluding Remarks

We have discussed some experiments that clearly show the phenomenon of quantum beats in III–V semiconductors. Most of these experiments measured the beats in the two-beam FWM experiments. However, pump-and-probe transmission experiments also display beats as we have seen above. In addition to these nonlinear phenomena, linear phenomenon such as luminescence also reveal this quantum interference effect. We refer the reader to the work on AgBr by *Langer* et al. [2.140] and on CdS by *Stolz* et al. [2.170]. Recent reviews can be found in [2.76, 77]. Picosecond quantum beats resulting from Larmor precession of electrons [2.171] and femtosecond quantum beats in exciton radiation emitted from GaAs quantum wells [2.172, 173] have been reported recently. The latter are discussed in Chap. 6 on exciton dynamics.

2.7 Interaction-Induced Effects: Beyond the Independent-Level Approximation

Nearly all the nonlinear effects discussed so far in this chapter can be explained in terms of the independent two- or three-level models. Optical Bloch equations provide an excellent description of these phenomena. In the extremely short pulse limit, the independent two-level model makes several simple predictions for the two-beam self-diffracted FWM experiments in which the sample is excited by pulse 1 along $\boldsymbol{q}_1$ at time t_1, pulse 2 along $\boldsymbol{q}_2$ at time t_2 with time delay $\tau_\mathrm{d} = t_2 - t_1$, and the signal is measured along the direction $2\boldsymbol{q}_2 - \boldsymbol{q}_1$: (1) the TI-FWM is zero for negative time delays, peaks close to the zero delay ($\tau_\mathrm{d} = 0$), and decays with a single exponential time constant related to T_2 for positive time delays, (2) for a homogeneously-broadened system and $\tau_\mathrm{d} > 0$, the TR-FWM signal peaks close to $t = t_2$ (prompt free polarization decay signal) and decays exponentially with a time constant $T_2/2$, (3) for an inhomogeneously-broadened system and $\tau_\mathrm{d} > 0$, TR-FWM exhibits a delayed peak at $t = \tau_\mathrm{d}$ (the photon echo) whose temporal width reflects the inhomogeneous spectral width, and (4) no unusual polarization dependence is expected.

During the last few years, it has become quite clear that the nonlinear response of semiconductors deviates significantly from these simple predictions. For example, TI-FWM signals at negative time-delays have been observed [2.66, 174] in GaAs and InGaAs and their nanostructures. TR-FWM in high-quality homogeneously-broadened GaAs quantum wells revealed that the signal is not instantaneous, but shows a slow rise following t_2; the difference between the time the signal reached its peak and t_2 was found to be independent of the time-delay between the pulses but to depend on the dephasing time of the excitons [2.175]. The polarization properties were also anomalous. The relative polarizations of the two beams had a strong influence on the intensities as well as the decay time-constants of the TI-FWM signals [2.108, 176–178]. Furthermore, TR-FWM indicates free polarization-like prompt signals for certain polarizations of the beams even when the system was inhomogeneously broadened and was expected to lead to photon echoes [2.108].

These departures result from the complicated nature of electronic states in semiconductors and their nanostructures, and from the complex interactions between these states. Investigations of these departures from simple models provide an opportunity to explore these states and their interactions. It is therefore not surprising that considerable effort has been devoted in recent years to investigate these phenomena experimentally and theoretically. Experimentally, the techniques of two- and three-beam TI-FWM and TR-FWM have been applied; efforts are also beginning to investigate the amplitude as well as the phase of the nonlinear signals [2.167, 179–181]. Theoretically, the interactions between excitons are considered within the mean-field approximation using Semiconductor Bloch Equations (SBE) as already discussed in Sect. 2.1.2. There are, however, effects which can not be explained on the basis of the current SBE models, and various other interactions have been invoked or introduced phenomenologically into the optical Bloch equations. Excitonic molecules have played a large role in the linear and nonlinear optical properties of II–VI compounds because of their large binding energies. In very-high-quality GaAs quantum wells, biexcitons have also been shown to play an important role in III–V semiconductor nanostructures even though the binding energy of the biexciton is small (~ 1 meV). Also, density-dependent dephasing has been invoked [2.71] and incorporated into SBE [2.182] to explain results in bulk GaAs. Disorder-induced coupling between exciton states has also been invoked [2.177] to explain some of the polarization anomalies. The proceedings of a recent NATO Advanced Study Institute on coherent processes in semiconductors [2.6] provide an excellent overview of the current activities in this rapidly moving field. We discuss some of these aspects in the next few sections, once again emphasizing the experimental results and their interpretations.

2.7.1 Exciton-Exciton Interaction Effects

We have discussed in Sects. 2.1.2c–e how interaction between excitons lead to a renormalization of the energies and the Rabi frequency of excitonic transitions in semiconductors [2.18, 22, 65, 183]. The influence of these effects on the nonlinear response of the semiconductor can be qualitatively assessed by considering that the total field is the sum of the incident electromagnetic field and the Lorenz local field resulting from the induced polarization. It can be shown that these effects lead to the diffraction of polarization in addition to the usual diffraction of the electric field predicted by the independent-level model. Diffraction of polarization is expected to lead to TI-FWM signal for negative time-delays, and a slow rise in the Free Polarization Decay (FPD) signal in homogeneously-broadened systems (2.65 – 66). These effects have indeed been observed and are discussed in the following.

(a) Signal at Negative Time Delays. FWM signals at negative time-delays were first reported in two-beam self-diffracted TI-FWM measurements on GaAs and InGaAs quantum wells by *Leo* et al. [2.174]. The experiments on GaAs were performed on a 170 Å GaAs/AlGaAs multiple quantum-well sample as a function of temperature (5 – 70 K) and excitation intensity using a synchronously-pumped dye laser (500 fs pulsewidth) tuned 3 meV below the HH exciton resonance to avoid the strong influence of the HH-LH beats (Sect. 2.6.2). Figure 2.27 depicts the TI-FWM signals at three different temperatures in $2\boldsymbol{q}_2 - \boldsymbol{q}_1$ direction as a function of time-delay τ_d between the pulses. Positive (negative) time-delays correspond to pulse 2 arriving after (before) pulse 1. The figure shows that there is a substantial signal extending to negative time-delays much larger than the pulsewidth. The magnitude of the slope of the negative time-delay signal increases with increasing temperature, remaining approx-

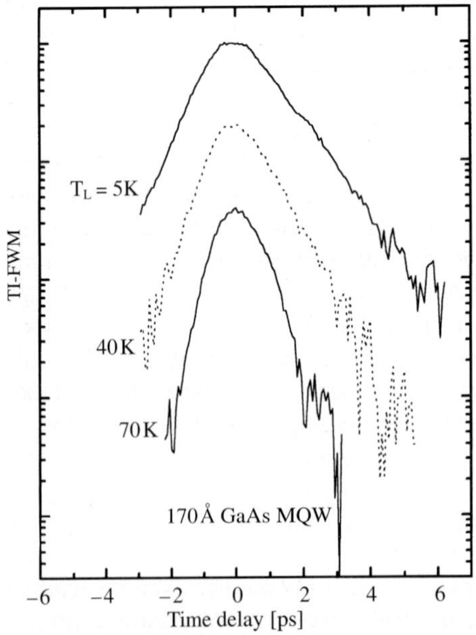

Fig. 2.27. Time-integrated, self-diffracted FWM signal from a 170 Å GaAs MQW sample as a function of time delay at three different temperatures, showing signal at negative time delays [2.174]

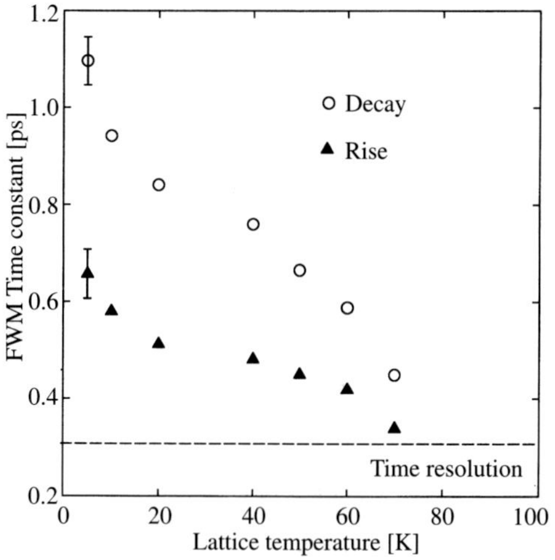

Fig. 2.28. Temperature dependence of the rise and decay time constants of the TI-FWM signals from the 170 Å GaAs MQW sample of Fig. 2.27 [2.174].

imately half that for the positive time-delay. The rise and decay times of the TI-FWM signals, obtained at various temperatures from plots such as Fig. 2.27, are plotted in Fig. 2.28. Similar results were obtained for 5 K at various intensities [2.152]. These results indicate that the rise time is half the decay time at low temperature and intensity where the results are not limited by the time resolution of the experiment. We recall from Sect. 2.1.2 that the prediction of the exciton-exciton interaction model is that for a homogeneously-broadened interacting two-level system, the TI-FWM signal rises as $(4\tau_d/T_2)$ for negative time-delays, and decays as $(-2\tau_d/T_2)$ for positive time-delays [2.66, 174, 183]. The experimental results agree with these expectations. The results on InGaAs at low intensities showed an asymmetric signal similar to those for GaAs discussed above. With increasing intensity, a pronounced dip developed at zero time delay, that was attributed to a subtle interference between the third-order and higher-order processes [2.66, 174].

We conclude this section by noting that other physical mechanisms, such as excitonic molecule and excitation-induced dephasing also predict signals at negative time delays. One can attempt to distinguish between different possibilities by a number of techniques, such as polarization dependence or spectrally resolved signals. We will discuss some of these in Sect. 2.7.2.

(b) Slow Rise of Time-Resolved FWM in Homogeneously-Broadened GaAs Quantum Wells. A second important consequence of the diffraction of polarization caused by exciton-exciton interaction is that a part of the TR-FWM signal from a homogeneously-broadened system is expected to rise slowly following the arrival of the second pulse on the sample. Experimental results confirming this prediction were obtained by *Kim* et al. [2.175, 184] who investigated a high-quality intrinsic GaAs/$Al_{0.3}Ga_{0.7}As$ MQW sample with 170 Å quantum wells and 10 periods. A Ti:Sapphire laser with transform-limited 100 fs

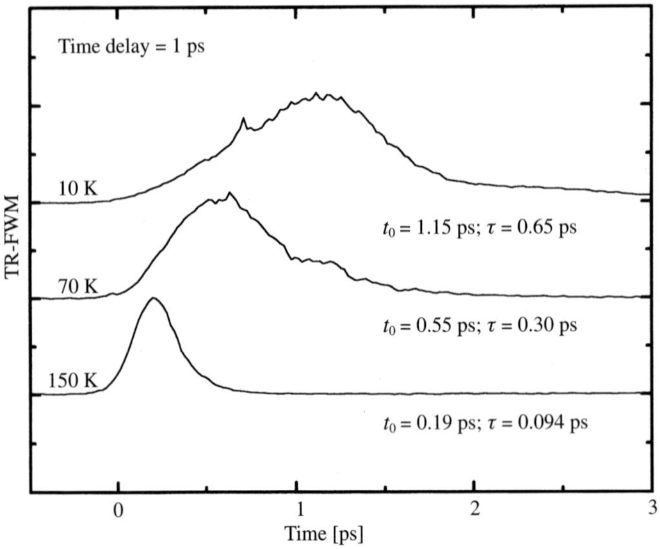

Fig. 2.29. Temporal evolution of the two-beam self-diffracted four-wave mixing (TR-FWM) signal from a nearly homogeneously-broadened 170 Å GaAs quantum well sample at a time delay $\tau_d = 1$ ps for three different temperatures. The peak position t_0 and the decay times τ are indicated for each curve. The signal peaks long after the end of the 100 fs pulse, and the peak is reached at shorter times with increasing temperature (decreasing T_2) [2.175]

pulses was used to perform TI- and TR-FWM measurements at various lattice temperatures, laser wavelengths and excitation densities. The sample was excited by two pulses along $\boldsymbol{q}_1$ and $\boldsymbol{q}_2$ with the second pulse delayed by a time delay τ_d with respect to the first. TR-FWM in the $2\boldsymbol{q}_2 - \boldsymbol{q}_1$ direction was measured using cross-correlation in a $LiIO_3$ nonlinear crystal. The zero of time was determined with an accuracy better than 20 fs by measuring the scattered laser light from the sample surface.

Figure 2.29 exhibits the TR-FWM signals for three different temperatures when the two beams are linearly co-polarized and $\tau_d = 1$ ps (pulse 2 arriving 1 ps after pulse 1 at the sample). The zero of the time axis corresponds to the arrival of the *second* pulse on the sample. The laser was tuned 4 meV below the HH transition energy, and the excitation density was calculated to be 1×10^{10} cm^{-2} from the measured spot diameter and the total absorbed power. The top curve at 10 K shows a gradually increasing signal which reaches a peak 1.15 ps after the arrival of the second pulse and then decays with a time constant $\tau_{decay} = 0.65$ ps. The decay constant $\tau_{decay} = 0.65$ ps is the same as the decay constant for the TI-FWM signal obtained under the same conditions.

These results clearly express that the diffracted signal increases slowly with time after the arrival of the second pulse on the sample. The peak of this signal occurs more than 10 pulsewidths after the second pulse has left the crystal. The fact that the peak position moves closer to $t = 0$ with increasing temperature shows clearly that this is not a photon echo. The temporal evolution has a long rise and a decay constant that agrees with the dephasing time. The shape of the temporal evolution is asymmetric and gives further support to the argument that this signal is not a photon echo. Finally, this assignment is also supported by the data obtained at various time delays τ_d for (a) co-linearly polarized beams, and (b) cross-linearly polarized beams (Fig. 2.30). We concentrate here on the positive τ_d. Except near $\tau_d = 0$, the TR-FWM response is characterized by a peak that remains *nearly fixed in time* regardless of the time-delay τ_d. (The additional peak at $t = \tau_d$ for large τ_d in the co-polarized signals has

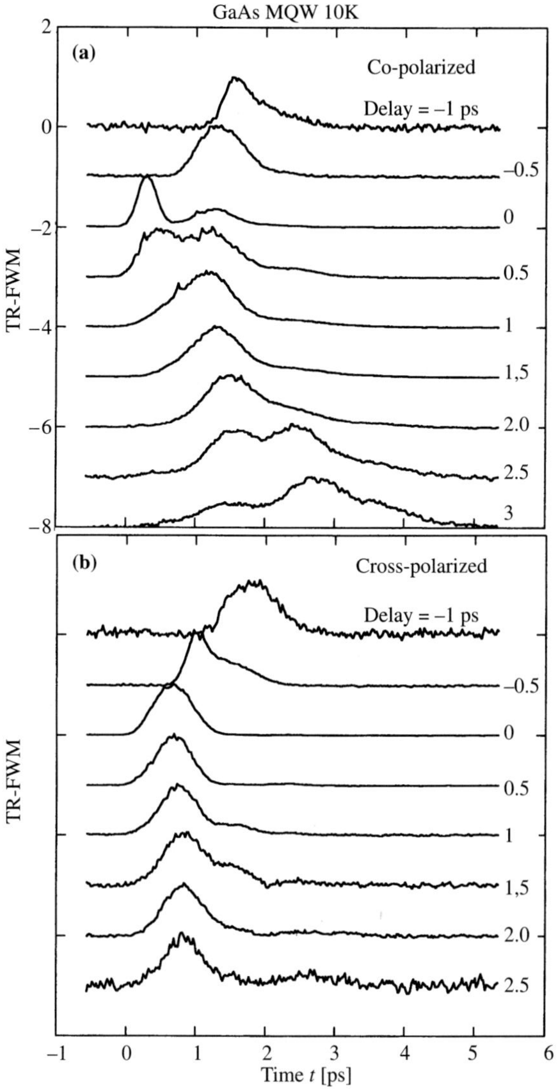

Fig. 2.30 a, b. Time-resolved FWM signal from a nearly homogeneously-broadened 170 Å GaAs quantum well sample at 10 K for different time-delays τ_d for **(a)** co-linearly polarized beams, and **(b)** cross-linearly polarized beams [2.175]

a different origin [2.175, 184]). The cross-polarized signal exhibits a peak at $t = 0.9$ ps, independent of the time-delay τ_d (and shown in Fig. 2.30b). The peak occurs at shorter time t because the decay time-constant τ_{decay} of the TI-FWM signal is shorter for cross-polarized case compared to co-polarized case [2.72, 154, 176, 177].

Detailed analysis of the data shows that the peak intensity of the TR-FWM signal is nearly *100 times* stronger than the signal at the end of the second pulse. Since the prompt signal arises from the diffraction of the field (as expected for the independent two-level model) whereas the delayed signal arises from the diffraction of the polarization, we conclude that the latter completely dominates the nonlinear response of high-quality GaAs quantum wells at low temperatures. In the simplified model of a k-independent interaction potential V between excitons (Sect. 2.1.2d), the ratio of the interaction-induced (II) signal to the normal (N) signal (2.66) expected for the independent two-level system is $(VT_2/\hbar)^2$ where T_2 is the dephasing time. The measured ratio of ≈ 100 and $T_2 = 1.3$ ps leads to a value of a few meV for V.

Weiss et al. [2.185] have also investigated TI-FWM and TR-FWM from GaAs quantum wells at 300 K at high excitation densities. The short dephasing times of excitons at 300 K (Sect. 2.2.2) makes it difficult to separate the effects of electric field and polarization diffraction in these measurements. However, from a detailed analysis of the results, they also conclude that interaction effects are necessary to explain their results.

(c) Interferometric Measurements of Frequency Dynamics. Photoexcitation of a system generates a macroscopic polarization in the system which radiates an electric field. The measurement of this field, superposed on any probe fields that may be present, provides information about the linear and nonlinear response of the system. Both the amplitude and the phase of the emitted coherent electromagnetic field are necessary to fully characterize the response of the system. Most of the experiments, however, have only measured the intensity of the emitted field and ignored the information contained in the phase. An example of the kind of information available by measuring the amplitude and the phase in a *linear* experiment was discussed in Sect. 2.6.5b. *Bigot* et al. [2.179] have performed the first interferometric measurements of the nonlinear response of a semiconductor. *Chemla* et al. [2.180] and *Bigot* et al. [2.181] have further analyzed and reviewed these results on the complex frequency dynamics in semiconductor nanostructures. We discuss some of their important results in this section. A third example of how phase dependent dynamics is important will be discussed in connection with the emitted THz radiation in Sect. 2.12.

Before discussing the results of *Bigot* et al. [2.179], it should be mentioned that the problem of characterizing the amplitude and the phase of a signal in the optical domain where an optical cycle is ≈ 2 fs long is very complex. Two general techniques have been discussed recently. The first uses the chronocyclic representation [2.186] as a way to display an ultrashort light pulse in a frequency-time space using the Wigner distribution function. In the second technique,

Frequency-Resolved-Optical-Gating (FROG) is used to measure the amplitude and the time evolution of an ultrashort pulse [2.187]. Recently [2.179, 180] these concepts have been applied to the coherent emission from excitons in GaAs quantum wells as a first attempt to obtain information about the complex phase and frequency dynamics in coherently excited semiconductors.

Bigot et al. [2.179] investigated the complex amplitude, phase and frequency dynamics of the nonlinear response of coherently excited excitons in GaAs quantum wells at 300 K. The experiments were performed with a nearly transform-limited (time-bandwidth product of 0.45) Gaussian pulse from a Ti : Sapphire laser operating near 800 nm with a pulse width of ≈ 80 fs corresponding to a spectral bandwidth of ≈ 22 meV. Extreme care was taken to fully characterize and stabilize the measurement system [2.179–181]. Temporal accuracy of 0.14 fs, corresponding to an average of 21 points within a 2.81 fs optical cycle, was achieved in the interferometric measurements. A 50-period 95 Å GaAs/45 Å $Al_{0.3}Ga_{0.7}As$ MQW sample was investigated at 300 K where the exciton is homogeneously broadened ($T_2 \approx 200$ fs [2.91]) by interactions with phonons (Sect. 2.2.2), and the HH and LH exciton energies were 1.467 and 1.482 eV, respectively.

The sample was excited along $\boldsymbol{q}_1$ by pulse 1 and along $\boldsymbol{q}_2$ by pulse 2 delayed by time with respect to pulse 1, and the diffracted signal was measured along $2\boldsymbol{q}_2 - \boldsymbol{q}_1$. In addition to the TI-FWM and the TR-FWM signals, *Bigot* et al. measured the spectrally-resolved FWM (SR-FWM or the power spectrum of the diffracted signal) as well as the first-order autocorrelation of the dif-

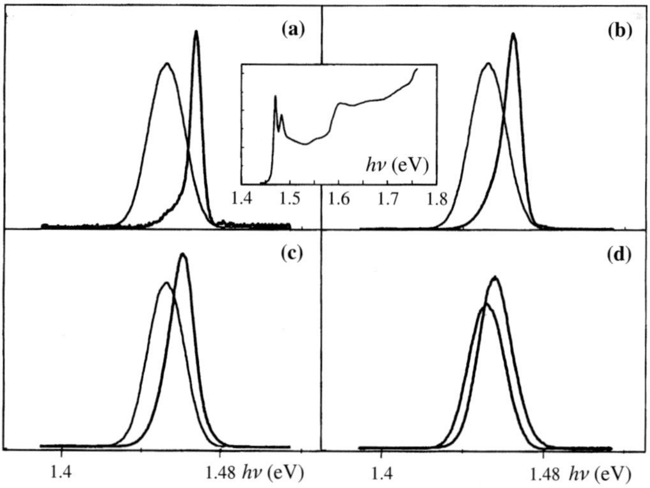

Fig. 2.31 a–d. Spectrally-resolved FWM (SR-FWM) or the power spectrum of the diffracted signal for a 95 Å GaAs MQW sample at 300 K for $\tau_d = 0$ and four different excitation densities: **(a)** 3×10^9 cm^{-2}; **(b)** 1.2×10^{10} cm^{-2} ; **(c)** 6×10^{10} cm^{-2}; **(d)** 3×10^{11} cm^{-2}. The laser spectrum is also shown as the *light curve*. The *inset* shows the absorption spectrum of the sample at 300 K [2.179, 180]

fracted signal $S^{\mathrm{I}}_{\mathrm{a}}(\tau,\tau_{\mathrm{d}}) = \int_{-\infty}^{\infty} \mathrm{d}u\,|P(u,\tau_{\mathrm{d}})+P(\tau-u,\tau_{\mathrm{d}})|^2$ and the first-order cross-correlation of the diffracted signal with the incident laser field $S^{\mathrm{I}}_{\mathrm{c}}(\tau,\tau_{\mathrm{d}}) = \int_{-\infty}^{\infty} \mathrm{d}u\,|P(u,\tau_{\mathrm{d}})+E_{\mathrm{l}}(\tau-u,\tau_{\mathrm{d}})|^2$. Here τ_{d} is the delay between the two pulses on the sample, τ is the delay between the two arms of the interferometer, and $E_{\mathrm{l}}(u)$ is the field of the reference laser which was derived from the same laser. $P(t,\,\tau_{\mathrm{d}})$ is the nonlinear polarization induced in the system so that the power spectrum (or spectrally-resolved FWM signal) is proportional to $|P(t,\,\tau_{\mathrm{d}})|^2$. The signal levels were found to be too weak to measure the second-order interferometric auto- and cross-correlation functions. In the absence of higher-order correlations, *Bigot* et al. determined the Differential Fringe Spacing, DFS(τ), defined as the difference between the laser fringe spacing and the FWM fringe spacing in first-order interferograms as a function of τ. For well-behaved amplitude functions and slowly-varying instantaneous frequency, DFS(τ) reproduces the time-dependent phase shifts $\Phi(\tau)$. These measurements were performed as a function of the excitation density between 3×10^9 and 3×10^{11} cm^{-2} and as a function of various detunings of the

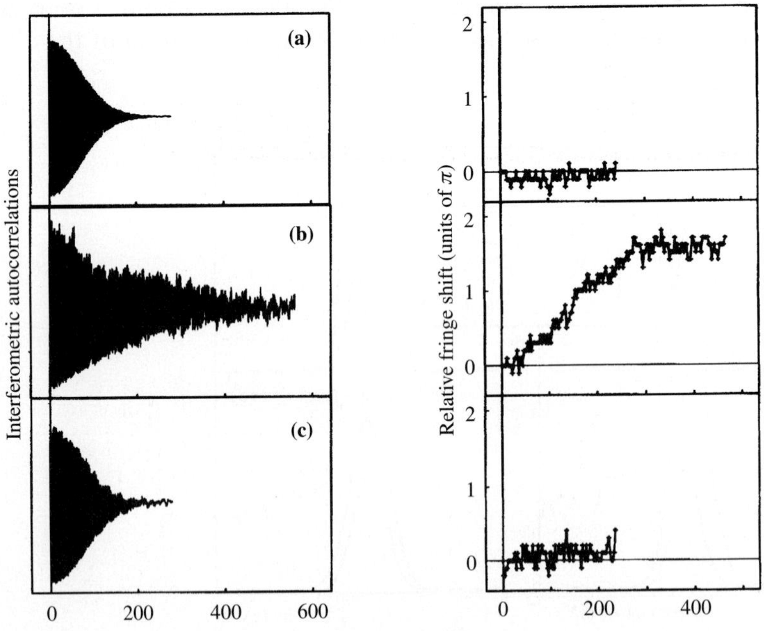

Fig. 2.32 a – c. *Left panel:* First-order interferometric autocorrelation traces at $\tau_{\mathrm{d}} = 0$ for **(a)** the laser, **(b)** FWM at 3×10^9 cm^{-2}, and **(c)** FWM at 3×10^{11} cm^{-2}. *Right panel:* the corresponding differential fringe spacing DFS(τ) for the 95 Å GaAs MQW sample under the same conditions as in Fig. 2.31 [2.179]

laser from the HH exciton energy, and allowed *Bigot* et al. to determine the complex dynamics of the FWM signals in the frequency-time domain.

Figure 2.31 exhibits the SR-FWM spectra of the sample at four different excitation densities. The laser was tuned below the HH exciton energy, as shown in the figure. With increasing excitation density, these time-integrated spectra evolve from an asymmetric line shape peaking at the HH exciton energy to a nearly symmetric line shape peaking close to the laser energy. The first-order interferometric autocorrelation and DFS(τ) for the laser for the FWM signals at the lowest and the highest densities are depicted in Fig. 2.32. DFS(τ) indicates that the instantaneous frequency of the FWM at high intensities is nearly the same as that of the laser. However, for low excitation densities, the interferogram is much longer and DFS(τ) shows a nonlinear behavior with τ. For small τ, the instantaneous frequency increases, with a slope approximately given by the difference between the laser and the exciton frequency. The slope then decreases and becomes practically zero. These experiments were analyzed using the semiconductor Bloch equations with a six-band model [2.180]. This complicated model requires extensive numerical simulations. However, under certain simplifying conditions such as low density, occupation of the lowest HH exciton, and no dephasing, it was shown that the time-dependent phase

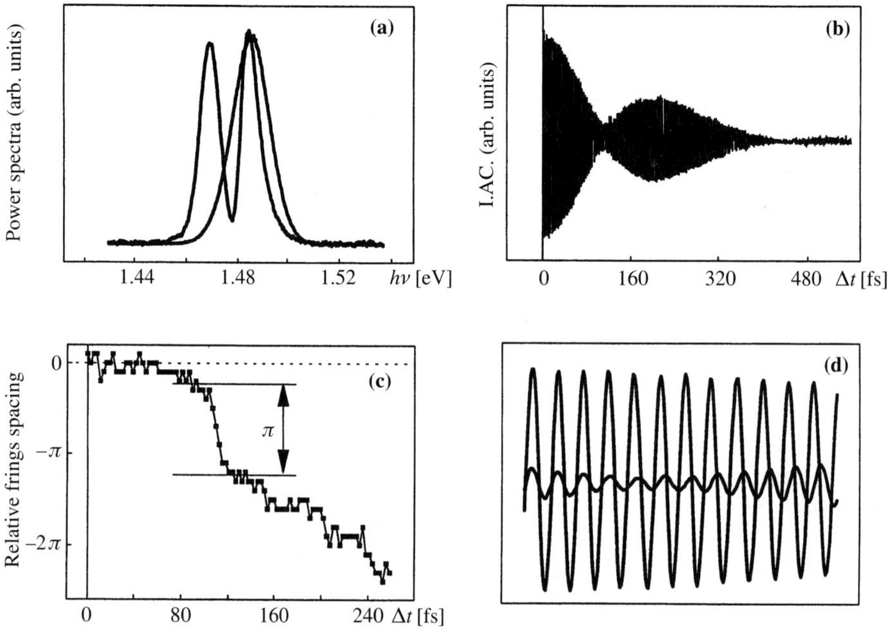

Fig. 2.33. **(a)** SR-FWM and the laser spectrum (light curve); **(b)** first-order interferometric autocorrelation, **(c)** differential fringe spacing DFS(τ); **(d)** details of the interferometric autocorrelation signal near $\tau = 0$, and close to the first node. Data are obtained on the 95 Å sample of GaAs (Fig. 2.31) under the same conditions, except that the laser is tuned to higher energy to excite the HH and the LH excitons equally [2.179, 180]

shift or the instantaneous frequency is determined by two factors: (1) the detuning of the laser from the exciton resonance, and (2) the relative strength of the exciton-exciton interaction compared to the Pauli blocking. Information about the latter was obtained by comparing the experiments with theory.

When these measurements were repeated with the laser tuned to a higher energy such that the HH and the LH excitons were equally excited (Fig. 2.33 a), the interferograms exhibited beating with a period corresponding to the HH-LH splitting (Fig. 2.33 b). This is expected from the results on HH-LH beating observed previously in TI-FWM [2.150] and TR-FWM [2.148]. Interestingly, DFS(τ) shows a phase change of π at the first node ($\tau \approx 120$ fs, Fig. 2.33 c, d). Figure 2.33 c, d further shows that the phase change at the node is not instantaneous but occurs over several (≈ 20) optical cycles or ≈ 50 fs.

These measurements show that investigations of the amplitude and the phase of the emitted radiation provides new information about the complex nonlinear response of the semiconductor to an ultrashort laser pulse. This technique provides another powerful tool in the arsenal of scientists investigating optical response of semiconductors.

(d) Excitation-Induced Dephasing. Semiconductor Bloch equations are the best formalism currently available for including the effects of interaction between excitons in the nonlinear response of semiconductors. However, various dephasing processes are still included in a phenomenological manner by introducing dephasing times (T_2) which are independent of the density of excitation. We have discussed in Sect. 2.2.1 that exciton-dephasing times as measured from TI-FWM experiments depend on exciton and free electron-hole pair densities, so that the approximation of density-independent dephasing times is valid only at very low densities.

Wang et al. [2.71] have recently shown that in high-quality bulk GaAs, the differential transmission spectrum in a pump-probe measurement at a relatively low exciton density (3×10^{15} cm^{-3}) is determined primarily by the broadening of the exciton resonance because of exciton-exciton interaction. By measuring the TI-FWM response at the HH exciton resonance for co-circularly polarized beams as a function of pre-pulse polarization and intensity, they showed that excitation-induced dephasing is *spin-independent* for HH excitons in strained bulk GaAs. Exciton-exciton scattering for LH excitons shows a strong dependence on the spin [2.90]. They also found that the ratio of the TI-FWM intensities for co-linear and cross-linear polarizations depends on excitation density. They argued that excitation-induced dephasing, i.e., a density-dependent T_2, gives a strong contribution to TI-FWM signal in the co-linearly polarized geometry and this contribution vanishes for the cross-linearly polarized geometry. The dephasing rate depends linearly on density at low densities, but saturates at higher density. Therefore, at higher densities, the contribution of the excitation-induced dephasing to TI-FWM signal diminishes and the ratio of co- to cross-linearly polarized signal diminishes.

These considerations were made more rigorous [2.71] by modifying the semiconductor Bloch equations to include excitation-induced dephasing. This was accomplished by including a density-dependent term in the imaginary part of the dynamically-screened Hartree-Fock self-energies. The calculations do indeed show that the ratio of co- to cross-linearly polarized signal decreases with density. However, the calculated change was much larger than the experimentally determined change in the ratio.

These calculations as well as the more physical semi-quantitative picture presented by *Wang* et al. [2.73] and *Steel* et al. [2.72] indicate that excitation-induced dephasing also leads to the diffraction of polarization, just as in the case of exciton-exciton interaction discussed in Sects. 2.1.2d and 2.7.1a, b. The question then arises as to which contribution to the negative time-delay signal and the slow rise of the signal for the positive time-delays is more important. Detailed polarization-dependent FWM experiments may be necessary to provide a definitive answer to this question. It was argued recently [2.188, 189] that excitation-induced dephasing is not a major factor in high-quality quantum wells of GaAs (see also Sect. 2.7.2b).

2.7.2 Biexcitonic Effects

The concept of an excitonic molecule or biexciton is similar to that of a positronium molecule [2.190–192]. When two excitons are close together, the two-exciton wavefunction can be either a symmetric or an anti-symmetric combination of the exciton wavefunctions. The symmetric combination has lower energy and forms the bound state of the excitonic molecule. The anti-symmetric state is not bound.

The binding energy of the biexciton depends on various parameters of the crystal. Since the biexciton binding energy is expected to be rather small (a fraction of a meV) in bulk III–V semiconductors (0.13 meV in GaAs [2.191, 193]), biexcitons have not played a major role in the optical response of these semiconductors. In contrast, the binding energies of the biexciton in II–VI and I–VII semiconductors are rather large so that biexcitons have played a major role in their linear and nonlinear properties of these semiconductors [2.194–203].

Since quantum confinement leads to an increase in the binding energy of the exciton in quantum wells compared to the bulk, the binding energy of the biexciton would also be expected to be larger in quantum wells. This expectation is indeed supported by calculations [2.192]. With increased binding energy and oscillator strength, biexcitons might be expected to play a bigger role in the linear and nonlinear optical properties of GaAs quantum wells, and this expectation is supported by recent experiments on III-V quantum wells. Properties of biexcitons in GaAs quantum wells using luminescence and other linear spectroscopy techniques have been investigated by a number of researchers [2.204–210]. The importance of biexcitons in the nonlinear response of GaAs quantum wells was first considered by *Feuerbacher* et al. [2.211] in relation to the observation of a negative time-delay signal. Other nonlinear optical studies include investigation of SR-FWM signals [2.188, 189, 211, 212], quantum beats in differential absorption [2.213] and TI-FWM signals [2.189, 212, 214, 215], pump-probe investigations of biexcitons [2.209, 216, 217] photon-echo studies using TR-FWM [2.218], theory of nonlinear response related to biexcitons [2.118, 189, 196, 197, 219–221], as well as recent detailed studies of polarization-dependent FWM signals [2.188, 189, 222, 223].

We have seen above (Sects. 2.1.2 and 2.7.1) that the interaction between excitons leads to a strong modification in the nonlinear response of a semiconductor. An extensive and successful formalism, in the form of semiconductor Bloch equations, has been developed to account for this interaction which makes semiconductors different from atomic systems. This formalism is based on a mean field approximation which is valid so long as the homogeneous line width of the excitons is larger than the binding energies of the biexciton. This condition is satisfied for many nonlinear optical experiments when densities are sufficiently high. However, the homogeneous line width of excitons in high-quality GaAs quantum wells is smaller than the biexciton binding energy at sufficiently low excitation density. The current formulation of semiconductor Bloch equations can not treat this case in which biexciton effects are important. The increasing amount of experimental information being accumulated on the influence of biexcitons in GaAs quantum wells is therefore being analyzed in terms of four- or five-level *optical* Bloch equations, with appropriate consideration for the selection rules for various transitions. While this phenomenological approach may be adequate in some cases, a microscopic model including a proper treatment of biexcitons must be developed for a true understanding of the nonlinear response of biexcitons. Efforts to develop a microscopic treatment of biexcitons are beginning [2.224, 225].

The current effort on nonlinear properties of biexcitons has three main objectives: (1) to understand the influence of biexcitons on negative time-delay TI-FWM signals, (2) to understand the influence of biexcitons on polarization-dependent properties of FWM signals, and (3) to understand the observation of quantum beats. We discuss in this section some results related to these topics, primarily concentrating on high-quality GaAs/AlGaAs quantum well samples in which the excitons are homogeneously broadened. For such samples, TI-FWM signals in both co-linear and cross-linear polarizations show nearly the same decay constant τ_{decay} [2.154, 177], and the photon-echo effects arising from inhomogeneous broadening, a subject of much recent discussion [2.72, 106–108, 154, 177, 218], are not important.

Figure 2.34 presents a schematic diagram of the exciton-biexciton level structure. The nonlinear response of such a system depends on various parameters such as the laser photon energy and its detuning from the exciton transition energy (E_{x}) , the spectral width of the laser and its relation to the biexciton binding energy Δ_{bx}, polarization vectors of the different beams in the FWM experiment, the homogeneous and inhomogeneous broadening of the exciton and the temperature. The selection rules in this system are such that an electromagnetic field with a given circular polarization can only induce transitions between the ground state and a particular exciton state or between the biexciton state and a particular exciton state. In addition, direct excitation

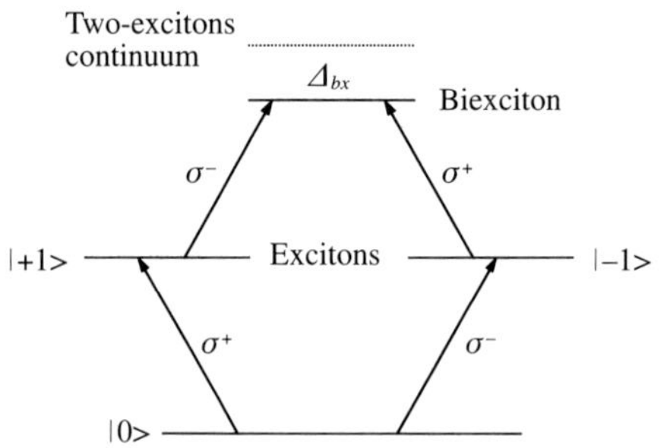

Fig. 2.34. Schematic diagram of the energy levels of biexcitons in GaAs quantum wells, showing the ground state $|0\rangle$, two (single-) exciton states $|+1\rangle$ and $|-1\rangle$ at energy E_x, the unbound two-exciton state at $2E_x$, and the bound state of the biexciton at $2E_x - \Delta_{bx}$, where Δ_{bx} is the biexciton binding energy. The selection rules for circularly-polarized σ^+ and σ^- fields are also shown

from the ground state to the biexciton state is possible by absorption of two photons with opposite circular polarizations. For two photons with the same photon energy, this two-photon absorption process peaks at $(E_x - \Delta_{bx}/2)$. These selection rules and transitions are shown in Fig. 2.34. The importance of biexcitonic contribution to the two photon absorption process in GaAs quantum wells was confirmed by recent pump-probe differential transmission measurements [2.209].

(a) TPC-Induced Biexcitonic FWM Signals. In the two-beam degenerate FWM experiment with pulse 1 along q_1 and pulse 2 along q_2 delayed by time delay τ_d, the polarization created by pulse 1 interferes with the field of pulse 2 and creates a grating which diffracts the field of pulse 2 along $2q_2 - q_1$. This is the usual *grating-induced* signal which is expected for only $\tau_d > 0$ for δ-function pulses in the absence of interaction between excitons as discussed in Sect. 2.1.1h. The biexcitons introduce an *additional process* by which FWM signals can be generated. The two-photon absorption process discussed above leads to Two-Photon Coherence (TPC) between the ground state and the biexciton state. A subsequent pulse can be diffracted from this TPC in a direction determined by momentum conservation. Biexcitons thus provide a mechanism for this *TPC-induced* FWM signal.

This TPC-induced TI-FWM signal for GaAs quantum wells was first considered by *Feuerbacher* et al. [2.211] to explain their results for negative time-delays in a two-pulse self-diffraction experiment with linearly polarized fields. They found a weak shoulder in SR-FWM along $2q_2 - q_1$ and different TI-FWM vs time-delay profiles for different laser-photon energies. *Pantke* et al. [2.212] also reported weak structures in the SR-FWM at $(E_x - \Delta_{bx})$ and $(E_x - \Delta_{bx}/2)$. Dependence on the polarization of the two beams was not investigated in either study. *Wang* et al. [2.188] and *Mayer* et al. [2.189] have recently reported extensive investigations of TI-FWM and SR-FWM for different combinations of linear or circular polarizations of the beams using two- and three-beam FWM experiments. These experiments provide the most conclusive evidence for the importance of biexcitons [2.188, 189] in the nonlinear response of GaAs quantum wells.

Wang et al. [2.188] investigated a GaAs MQW sample with 15 periods of 170 Å GaAs quantum well and 175 Å $Al_{0.3}Ga_{0.7}As$ barriers. The HH exciton

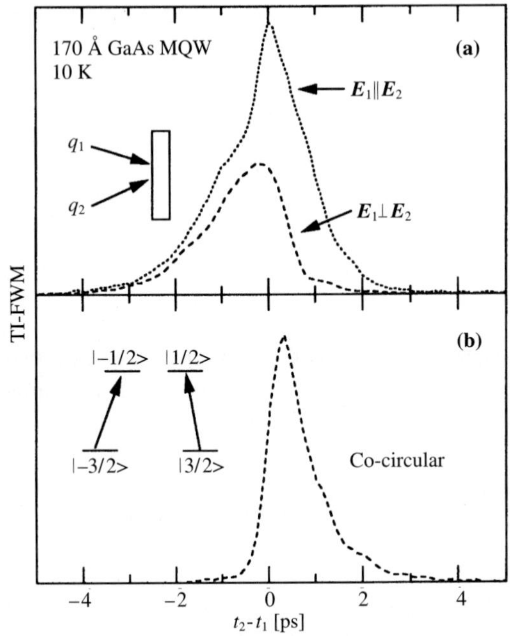

Fig. 2.35 a, b. Self-diffracted TI-FWM signal in the $2q_2 - q_1$ direction from a nearly homogeneously-broadened, 170 Å GaAs quantum-well sample at 10 K. **(a)** linearly polarized fields showing strong negative time-delay signals; *dotted:* co-linear, *dashed:* cross-linear; **(b)** co-circularly polarized fields showing no signals at large negative time delays [2.188]

was homogeneously broadened under the conditions of the experiment with absorption line width of 0.6 meV at 10 K. 400 fs pulses (4.5 meV spectral width) from a mode-locked Ti : Sapphire laser were used to excite the sample along $\boldsymbol{q}_1$ by pulse 1 and along $\boldsymbol{q}_2$ by pulse 2 arriving at the sample at time t_1 and t_2 respectively, so that the time delay $\tau_d = t_2 - t_1$. The TI-FWM and SR-FWM signals were measured along $2\boldsymbol{q}_2 - \boldsymbol{q}_1$. Strong negative time-delay ($\tau_d < 0$) TI-FWM signals were observed for both co- and cross-linearly polarized fields, as shown in Fig. 2.35 a. These signals are caused by the diffraction of pulse 1 by the TPC created by two photons of pulse 2 which arrives at the sample before pulse 1. The two photons with opposite circular polarizations required to create TPC are provided by the linearly polarized beam 2. Therefore, the TPC-induced signals are independent of the relative polarizations of the two beams. This interpretation can be tested by using circularly polarized beams. Since two photons of the same circular polarization can not create TPC, there should be no negative time-delay signal when the two pulses are circularly polarized. The results for co-circularly polarized beams in Fig. 2.35 b indeed reveal that there is no negative time-delay signal. No signal is expected, or observed, for positive or negative time-delay for cross-circular polarizations. These observations demonstrate that the TPC-induced signal from biexcitons makes a dominant contribution to TI-FWM at large negative time-delays.

The polarization-dependent study by *Wang* et al. [2.188] also distinguishes between the biexciton-related negative time-delay signal and exciton-exciton interaction-induced negative time-delay signal discussed in Sect. 2.7.1 a. The negative time-delay signal for linearly polarized case is clearly dominated by biexcitonic TPC as discussed above. For co-circularly fields (Fig. 2.35 b) this TPC-induced signal is absent. However, the signal at negative time-delays in Fig. 2.35 b is *not* pulse width limited and shows behavior similar to that discussed earlier for exciton-exciton interaction-induced negative time-delay signal (Sect. 2.7.1 a).

The nature of TPC-induced signal has been further investigated by *Wang* et al. [2.188, 209] and *Mayer* et al. [2.189] by using three-pulse TI-FWM and SR-FWM signals. The use of three-pulse configuration provides greater flexibility in exploring various selection rules. *Wang* et al. [2.188, 209] investigated the MQW sample of GaAs discussed above with three circularly polarized pulses with fields $\boldsymbol{E}_1$, $\boldsymbol{E}_2$ and $\boldsymbol{E}_3$ arriving along $\boldsymbol{q}_1$, $\boldsymbol{q}_2$, $\boldsymbol{q}_3$ at times t_1, t_2 and t_3, respectively. The FWM signals were measured along $\boldsymbol{q}_d = (\boldsymbol{q}_1 + \boldsymbol{q}_2 - \boldsymbol{q}_3)$ for various combinations of t_1, t_2, and t_3. The results for $t_1 = t_2$ and the sample at 10 K are depicted in Fig. 2.36. The solid curve is when all three fields are σ^+ polarized, and the dotted curve is when $\boldsymbol{E}_1$ and $\boldsymbol{E}_3$ are σ^+ polarized but $\boldsymbol{E}_2$ is σ^- polarized. The signal for $t_1 - t_3 < 0$ results from the diffraction of the third pulse ($\boldsymbol{E}_3^*$) from the TPC induced by pulses 1 and 2 ($\boldsymbol{E}_1 \cdot \boldsymbol{E}_2$), and corresponds to the negative time-delay signal in Fig. 2.35. The solid curve in Fig. 2.36 shows that the TPC-induced signal for $t_1 - t_3 < 0$ (negative time de-

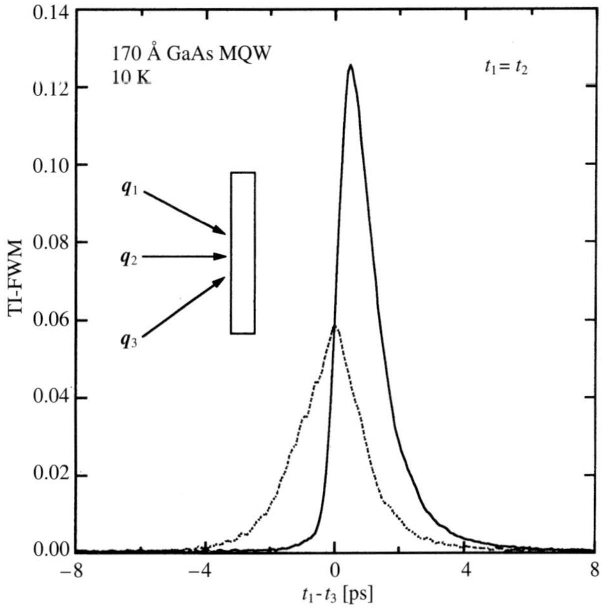

Fig. 2.36. Three-pulse TI-FWM signals along $\boldsymbol{q}_1 + \boldsymbol{q}_2 - \boldsymbol{q}_3$ for the 170 Å MQW GaAs (of Fig. 2.35) at 10 K and $t_1 = t_2$ as a function of the time-delay $t_1 - t_3$; *solid curve:* all three fields are co-circularly polarized; *dashed curve:* the second field has the circular polarization opposite to those of fields 1 and 3. Note the strong negative time-delay signal for the *dashed curve* [2.188]

lays) is absent when all three fields are σ^+. This is because selection rules do not allow creation of the TPC when E_1 and E_2 are both of the same circular polarization. However, when E_1 is σ^+ and E_2 is σ^-, biexcitonic TPC is created from which E_3 is diffracted. These selection rules conclusively prove the biexcitonic nature of TPC.

It is also interesting to note that the decay of the TPC-induced signals in Figs. 2.35a and 2.36 correspond to $T_2 = 1.8$ ps, comparable to $T_2 = 1.9$ ps for positive time-delays obtained from the solid curve in Fig. 2.36. This is in contrast to the hypothesis in some models [2.218, 226] that require different dephasing rates for biexcitons and excitons. All the results in Figs. 2.35 and 2.36 were obtained for a density of 10^9 cm^{-2}. Qualitatively the same results were obtained in the excitation density range 10^8 to 10^{10} cm^{-2}, except that increasing the density or the temperature reduces T_2 [2.70, 84] which would eventually destabilize the biexcitons when the homogeneous broadening of excitons exceeds the biexciton-binding energy, as discussed above. Finally, we note that *Ferrio* and *Steel* [2.227] have recently reported quantum coherence oscillations in GaAs quantum wells using interferometric technique, and interpreted the oscillation as the temporal evolution of the biexcitonic TPC.

(b) Grating-Induced Biexcitonic FWM Signals. Biexcitons also play an important role in grating-induced signal which normally occurs at positive time-delays in the two-pulse experiments and for some combination of delays in three-pulse experiments. In a two-beam self-diffracted experiment for $\tau_d > 0$, E_2 can be diffracted by the grating formed by $E_1^* \cdot E_2$ through both the exciton transition (at E_x) and the exciton-biexciton transition (at $E_x - \Delta_{bx}$). The relative strengths of the two signals depend on various parameters such as the relative oscillator strengths, the detuning of the laser from each transition, and the laser spectral width. A similar situation can be created in three-pulse experiments by appropriate selection of the time-delays.

Wang et al. [2.188] investigated the same GaAs MQW sample as above at 10 K for $t_2 - t_1 = 5$ ps. Figure 2.37 shows the TI-FWM signal as a function of $t_1 - t_3$ when E_1 and E_3 are σ^+ polarized but E_2 is σ^- polarized. The inset in Fig. 2.37 exhibits that the spectrum of the diffracted signal (SR-FWM) under the same conditions for $t_1 = t_3$ clearly has two peaks corresponding to E_x and $E_x - \Delta_{bx}$. For $|t_1 - t_3| < 5$ ps, excitation by pulses 1 and 3 creates a grating in the sample from which E_2 is diffracted along $q_1 + q_2 - q_3$ through the excitonic resonance (at E_x) as well as through the exciton-biexciton resonance (at $E_x - \Delta_{bx}$). The observation of two peaks, and the observation that the lower energy peak disappears when all fields are co-circularly polarized, confirms that exciton as well as biexcitons contribute to this grating-induced signal. From the peak separation, *Wang* et al. deduced $\Delta_{bx} = 1.2$ meV for the 170 Å GaAs quantum well, in agreement with their pump-probe differential absorption results [2.188].

There are several other points of interest. The symmetric line shape (Fig. 2.37) demonstrates [2.47] that the excitonic transition is homogeneously-broadened under the conditions of the experiment. Also, the signals in

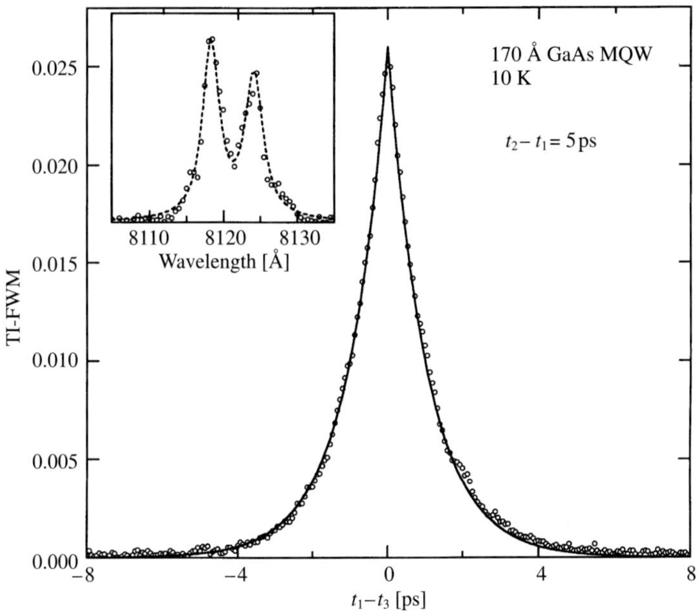

Fig. 2.37. Three-pulse TI-FWM signals for the 170 Å MQW GaAs (of Fig. 2.35) at 10 K for the $\sigma^+\sigma^-\sigma^+$ configurations of the fields for $t_2 - t_1 = 5$ ps as a function of $t_1 - t_3$. This corresponds to the diffraction of the field $\boldsymbol{E}_2$ at positive time delay by the grating formed by $\boldsymbol{E}_1$ and $\boldsymbol{E}_3$. The SR-FWM signal at $t_1 = t_3$ in the *inset* shows two peaks at E_x and $E_x - \Delta_{bx}$. The results show that biexciton contributes to the grating-induced signal and determine the biexciton-binding energy $\Delta_{bx} = 1.3$ meV [2.188] (*circles:* experimental; *solid curve:* fit to Lorentzian)

Fig. 2.37 can be generated by biexcitonic effects as well as by Excitation-Induced Dephasing (EID) discussed in Sect. 2.7.1d. However, EID is expected to give a signal only at the exciton energy. The fact that the signals at the exciton and the biexciton energies are nearly equal in strength (inset of Fig. 2.37) leads to the conclusion that excitation-induced dephasing is not important in high-quality GaAs quantum wells under the conditions investigated [2.188]. *Mayer* et al. [2.189] have also reached similar conclusions. Also, we note that exciton-exciton interaction generates grating-induced negative time-delay signals both at the exciton resonance (E_x) and at the exciton-biexciton resonance ($E_x - \Delta_{bx}$). Thus a double-peaked spectrum in the SR-FWM would be expected in this case as well. The relative strengths of the two peaks is determined by the relative strengths of the grating-induced and TPC-induced negative time-delay signals. For comparable dephasing times for excitons and biexcitons and appropriate polarizations of the beams, TPC-induced signal dominates the grating-induced negative time-delay signals for large negative time-delays, as already discussed above.

(c) Biexcitonic Quantum Beats. *Mayer* et al. [2.189] have re-visited the observation of quantum beats [2.214] with a period corresponding to Δ_{bx}. They have considered a 5 level model (Fig. 2.34) by phenomenologically adding an unbound two-exciton state to the usual 4 level model. For negative time-delay, they assign the quantum beats to the interference between the biexciton bound state and the two-exciton unbound state. Exciton-exciton interaction effects are introduced through the local field model. The 5 level model in presence of local field effects predicts the quantum beats for negative time-delays. Quantum beats are also observed at positive time-delays. They show that higher-order effects become important at relatively low exciton densities ($>10^9\ \mathrm{cm}^{-2}$). Since the strength of these beats increases with increasing density, they attribute the beats at positive time-delays to fifth-order effects.

(d) Polarization Anomalies. Polarization anomalies in the nonlinear response of semiconductor nanostructures have received considerable attention [2.72, 154, 218], as already discussed in Sects. 2.6.2 and 2.7.1. Different mechanisms have been proposed to account for the differences in the decay rates, signal strengths, and temporal behavior for different polarization configurations, and have been reviewed recently by *Kuhl* et al. [2.154] and *Steel* et al. [2.72]. The effect of various polarization configurations on the TI-FWM signals expected in three-pulse experiments have been considered in detail recently by *Mayer* et al. [2.189]. Their results indicate that the co-polarized signals are the strongest, and the others are reduced in intensity by factors of $2-3$ except the $\sigma^+\sigma^+\sigma^-$ configuration which is forbidden in the third order for all mechanisms they considered. *Wang* et al. [2.188] have suggested that the grating-induced signal at E_x vanishes in the cross-linear configuration in the limit that Δ_{bx} is much larger than the exciton line width. Further, they have also argued that the TPC-induced signal behaves like free polarization decay even in an inhomogeneously-broadened system under certain conditions. Detailed polarization-dependent measurements and further development of semiconductor Bloch equations to include biexcitonic effects would be necessary for a complete understanding.

2.8 Coherent Oscillations of an Electronic Wavepacket

We now turn to a discussion of a slightly more complicated semiconductor nanostructure, namely an asymmetric Double-Quantum-Well Structure (*a*-DQWS). Such a structure consists of two quantum wells of unequal thickness a and a' (with $a>a'$) separated by a barrier of thickness b. *a*-DQWS will be denoted by $a_L/b/a_R$ where a_L is the thickness of the left well, and a_R is the thickness of the right well, with all values in Å. A typical structure with the Wide Well (WW) on the left and the Narrow Well (NW) on the right ($a/b/a'$) is shown in Fig. 2.38. Primed and unprimed symbols are also used to denote the energy levels in the NW and the WW, respectively. Because of the ability to control the spacing between various electronic levels by applying an electric

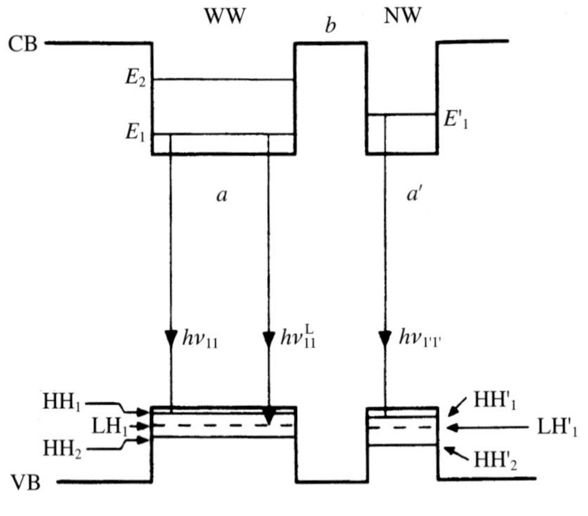

Fig. 2.38. Schematic of *a*-DQWS denoted as *a*/*b*/*a*′ where *a* is the width of the left quantum well, *b* is the barrier width and *a*′ is the width of the right quantum well. Unprimed and primed quantities refer to the WW and NW, respectively [2.236]

field in the growth direction, such a structure is enormously useful for investigating various tunneling processes. We will discuss this aspect of ultrafast spectroscopy of semiconductor nanostructure in Chap. 7, but discuss here a closely related phenomenon that occurs in the coherent regime.

We will adopt the convention that a reverse bias lowers the potential of the right well with respect to the left well and that all electron and hole energy levels are measured with respect to the center of the left well conduction and valence band potential profile. If a and a' are sufficiently different, the levels E_1 and E'_1 are well separated under flat-band condition, and their wavefunctions are primarily localized in the WW and the NW, respectively. With the application of reverse bias, the two levels come closer together, and anti-cross with a certain minimum spacing ΔE at the resonant electric field F_R. For $F = F_R$, the wavefunctions of E_1 and E'_1 are completely delocalized in the two wells. This behavior of the wavefunctions and the corresponding cw absorption spectrum involving only the HH level of the WW are shown in Fig. 2.39. For flat-band condition, the absorption spectrum is dominated by the spatially direct absorption (HH_1 to E_1, labeled "1" in the right panel), with a small peak at *higher* energy due to the spatially indirect absorption (HH_1 to E'_1, labeled "1′"). At $F = F_R$, the cw absorption spectrum consists of two nearly equal peaks separated by ΔE. With stronger electric field, the absorption is once again dominated by a strong peak due to the spatially direct absorption within the WW, with a weak peak at *lower* energy due to spatially indirect absorption.

When such a system is excited with an ultrashort laser-pulse with a spectral width larger than the energy spacing between any two electronic levels, the laser creates a non-stationary superposition of the two eigenstates. If the laser photon energy is low enough that it can only excite transitions from the HH_1 level of the WW, then the non-stationary wavepacket is initially created in the WW. The time evolution of this non-stationary state is determined time-dependent Schrödinger equation. Since the eigenenergies of the two eigenstates are

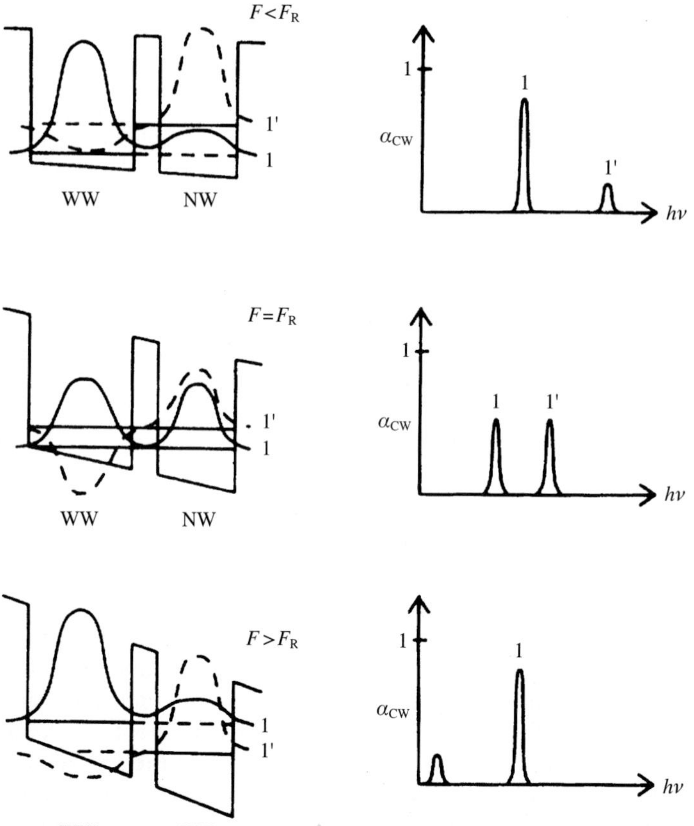

Fig. 2.39. Schematic of the electron wavefunctions for the two lowest electron levels (*solid curve* for level 1, *dashed curve* for level 1′ in the *left panel*) and the absorption spectra (*the right panel*) involving the HH_1 level of the WW and level 1 and 1′ in an *a*-DQWS. The resonance between the two electron levels occurs at the field F_R. For $F \ll F_R$ and $F \gg F_R$ the wavefunction for level 1 (1′) is localized primarily in the WW (NW). The wavefunctions are completely delocalized at resonance [2.236]

slightly different, their phases will evolve at a slightly different rate so that the wavepacket will be in the NW at time $h/(2\Delta E)$ after the excitation. The fraction of the wavepacket in the NW at time $h/(2\Delta E)$ is determined by the applied field F; at $F = F_R$, the wavepacket will be completely in the NW at time $h/(2\Delta E)$. As time progresses, the wavepacket will return to the WW and the oscillations will continue in the absence of dephasing. The period of the oscillation is given by

$$\tau_{osc} = h/\Delta E \ . \tag{2.81}$$

These ideas can be described in a more quantitative form by constructing an electron wavepacket from the two electronic eigenstates

$$\Psi(t) = \Psi_1 \exp(-\mathrm{i}E_1 t/\hbar) + \Psi_2 \exp(\mathrm{i}E_2 t/\hbar) \tag{2.82}$$

so that

$$|\Psi(t)|^2 = |\Psi_1|^2 + |\Psi_2|^2 + 2\Psi_1\Psi_2 \cos(2\pi t/\tau_{osc}) \ . \tag{2.83}$$

Thus the electron wavepacket undergoes coherent oscillations between the WW and the NW. This coherent oscillation is expected to lead to a periodic variation in the TI-FWM and pump-probe transmission signals, just as in the case of the HH-LH quantum beats. Such oscillations are also expected to lead to coherent THz emission. We discuss the former here and the latter in Sect. 2.12.1.

Such periodic oscillations have been observed by *Leo* et al. [2.228] in FWM and pump-probe transmission experiments. A high-quality 150/25/120 *a*-DQWS was grown in the i-region of an i-n$^+$ sample and a reverse bias was applied through a transparent Schottky contact on the top. Figure 2.40 displays the cw absorption spectrum of the sample at 10 K for various electric fields. It clearly shows the anti-crossing behavior for the HH as well as the LH transitions for fields close to F_R.

The TI-FWM and differential pump-probe transmission signals from this sample were investigated [2.228] by using a tandem synchronously-pumped IR dye laser with 500 fs pulsewidth. The spectral width of the laser was about 4.2 meV, about a factor of two larger than the transform-limited width and wide enough to excite both the eigenstates at F_R. Figure 2.41 depicts the differen-

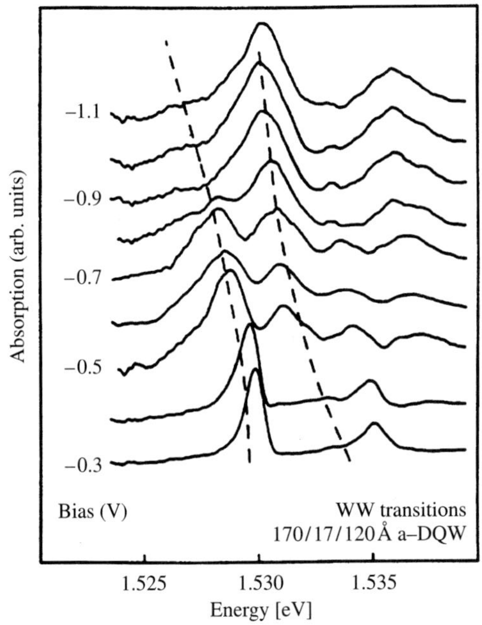

Fig. 2.40. cw Absorption spectra of a 170/17/120 GaAs/AlGaAs/GaAs *a*-DQWS as a function of the applied negative bias which tunes the electron levels through a resonance [2.228]

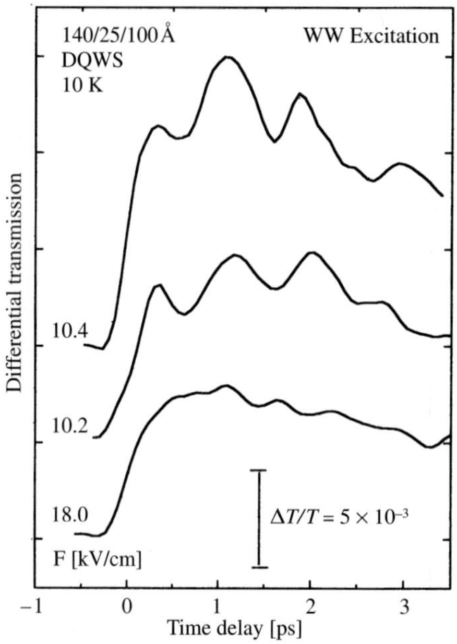

Fig. 2.41. Pump-probe differential transmission spectra showing coherent oscillations of the excitonic wavepacket in a 140/25/100 Å *a*-DQWS as the electron levels are tuned through a resonance by an applied electric field [2.228, 236]

tial pump-probe transmission as a function of the time delay τ_d between the pulses. The results clearly show periodic oscillations with $\tau_{osc} = 1.3$ ps. This corresponds to an energy splitting $\Delta E = 3.2$ meV, close to the calculated value of ≈ 3 meV for this structure and the splitting observed in the cw absorption spectra. In contrast to the simple model described above, the oscillations are damped because dephasing destroys the coherence of the wavepacket. The differential pump-probe transmission signal has a non-zero value after the oscillations are damped out because of the incoherent component of the signal arising from the real population (the diagonal component of the density matrix) which decays with a much longer recombination time T_1.

Figure 2.42 exhibits the TI-FWM signal obtained from the same sample under the same conditions. This signal also shows periodic oscillations with the same τ_{osc}. However, the TI-FWM signal is plotted on a semi-log scale and decays strongly as the oscillations damp out. This is because there is no incoherent component to the TI-FWM signal and the decay of the TI-FWM signal is related to the dephasing time T_2 as discussed earlier. Similar oscillations in the TI-FWM signal have also been observed when the laser is tuned close to the lowest interband-transition energy (HH_1' to E_1') in the NW. In this case, the wavepacket is initially created in the NW and the excitation also creates free electron-hole pairs in the WW. The free carriers lead to increased dephasing of the coherent wavepacket which leads to faster damping of the oscillations and a more rapid decay of the TI-FWM signal. Another difference is that the field at which the resonance is observed is different in this case compared to

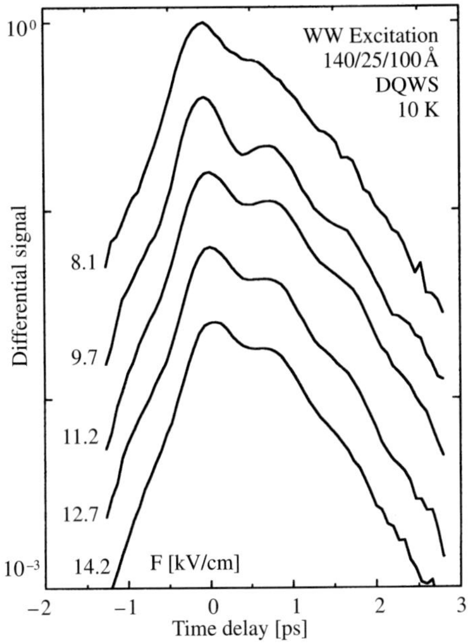

Fig. 2.42. TI-FWM signals as a function of time delay, showing coherent oscillations of the excitonic wavepacket in the *a*-DQWS of Fig. 2.41 as the electron levels are tuned through a resonance by an applied electric field [2.228, 236].

the case of WW excitation only. The period of the oscillations is the same in both cases.

These observations can be understood by incorporating the heretofore neglected Coulomb interaction between the localized hole wavepacket and the oscillating electron wavepacket. Coulomb interaction between electrons and holes in different wells has been analyzed theoretically [2.229, 230] and experimentally [2.231, 232] for the case of the double quantum wells. Since the binding energies of spatially direct and indirect excitons are different, it was shown that the field at which the resonance occurs is slightly larger (smaller) for photon energy such that the wavepacket is excited in the WW (NW). However, the energy splitting at resonance ΔE is the same in the two cases. Thus, excitonic effects explain both the constant τ_{osc} and different resonant fields observed experimentally. Excitonic and free-carrier effects in coherent interband and intraband dynamics in double quantum wells has been analyzed recently by *Binder* et al. [2.233] by a numerical solution of the multisubband semiconductor Bloch equations. These will be discussed briefly in Sect. 2.9.5.

These experimental results can be qualitatively accounted for by considering a simplified three-level model [2.228]. If $|0\rangle$ is the ground state, and $|1\rangle$ and $|2\rangle$ are optically coupled excited states with energies E_1 and E_2 and spectral weights w_1 and w_2, then for δ-function pulses and homogeneously broadened levels, the diffracted signal is given by

$$I_b(\tau_d) \propto \Theta(\tau_d)\left(\frac{w_1^2}{2\gamma_1}+\frac{w_2^2}{2\gamma_2}+\frac{2w_1w_2(\gamma_1+\gamma_2)}{(\gamma_1+\gamma_2)^2+(\Delta E)^2}\right)$$
$$\times[w_1^2 e^{-2\gamma_1\tau_d}+w_2^2 e^{-2\gamma_2\tau_d}+w_1w_2\cos(\Delta E\tau_d)e^{-(\gamma_1+\gamma_2)\tau_d}] \quad (2.84)$$

and the transmitted probe intensity is given by

$$I_c(\tau_d) \propto \Theta(-\tau_d)\{w_1^2 + w_2^2 + w_1 w_2 [1 + \cos(\Delta E \tau_d) e^{\gamma_3 \tau_d}]\} , \tag{2.85}$$

where $\Theta(\tau_d)$ is the Heaviside step function, $\gamma_1(\gamma_2)$ are dephasing rates for 1-0 (2-0) transitions, and γ_3 is that of the 1-2 transition. It was found that a good agreement with the experimental results may be obtained with $\gamma_1(\gamma_2) = 0.9$ (1.3) meV, and $\gamma_3 = 0.7$ meV.

It should be noted from these equations that the dephasing rate entering into the expression for the FWM intensity is $\gamma_1 + \gamma_2$, the sum of the "interband" dephasing rates for 1-0 and 2-0. In contrast, the dephasing rate entering into the pump-probe transmission intensity is γ_3, the "intraband" dephasing rate 2-1. These results show that different dephasing rates enter into different experiments. This distinction is important to keep in mind as we discuss other phenomena associated with coherent charge oscillations. We will see in Sect. 2.12 that charge oscillations in *a*-DQWS (discussed here) and superlattices (discussed in Sect. 2.9) lead to coherent emission at THz frequencies. The dephasing rate entering in the THz signals is the intraband dephasing rates. It should also be noted that scattering events that affect the excited states 1 and 2 in a similar manner may not disturb their relative phase. This will be important in understanding the data femtosecond quantum beats in exciton emission discussed in Chap. 6 on exciton dynamics.

It might be mentioned that the coherent oscillations in *a*-DQWS are analogous to the transitions between the two states of the ammonia molecule, and the above analysis is applicable to both systems. The analysis given above assumes that the dephasing rates are smaller than the oscillation frequency. The problem of tunneling in the presence of dissipative processes has been discussed by *Leggett* and coworkers [2.234] and will be discussed in connection with the incoherent resonant tunneling dynamics in Chap. 7. One of the consequences of dissipation is to slow down the frequency of tunneling [2.234–236]. *a*-DQWS may provide an ideal system for studying this because variable dissipation rates may be obtained by varying parameters such as temperature and carrier density. Although an initial attempt [2.237] to observe this behavior was not successful, further investigation of this phenomenon may be interesting.

Other wavepacket oscillations have been observed in semiconductor nanostructures since the first observation discussed above. *Feldmann* et al. [2.238] have recorded the coherent dynamics of excitonic wavepackets in strained InGaAs/GaAs quantum well structures. The wavepacket consisted of bound states of the excitons as well as the continuum states, so the observation was similar to the wavepackets in high Rydberg states of atoms. A completely unrelated phenomenon, Rabi flopping in semiconductor, has been reported recently by *Cundiff* et al. [2.126]. These Rabi oscillations are related to AC or dynamic Stark effect, as discussed in Sect. 2.3.

2.9 Bloch Oscillations in a Semiconductor Superlattice

In 1928, *Bloch* [2.239] demonstrated theoretically that an electron wavepacket composed of a superposition of states from a single band and peaked at a given quasi-momentum **k** undergoes periodic oscillations in the momentum and real space under an applied electric field, if interband transitions are neglected. The oscillation period T_B is inversely proportional to the applied field F and the periodicity of the lattice d in the field direction. This proposal has generated considerable controversy and discussion over the years [2.240–246], centering around (1) the correct theoretical approach to describe the motion of an electron in an infinite solid under an applied electric field (e.g., what are the correct boundary conditions), (2) the existence of a *discrete* Wannier-Stark ladder in a solid, and (3) whether interband tunneling is indeed negligible for times shorter than a few Bloch oscillation periods. From a theoretical point of view, the current consensus appears to be that the original picture of discrete Wannier-Stark ladder is correct provided that these states are considered as metastable states and if delocalization of states due to resonance between states of different bands is properly taken into account [2.247]. However, neither the Wannier-Stark ladder nor Bloch oscillations have been demonstrated in bulk solids.

One of the conditions for the observation of Bloch oscillations is that the oscillation period must be smaller than the dephasing time T_2. One mechanism governing T_2 is interband tunneling which becomes important at high fields. Since d can be much larger in semiconductor superlattices, T_B in superlattices can be much smaller than the corresponding T_B in bulk for a given electric field. This makes superlattices attractive candidates for the observation of the Wannier-Stark ladder and Bloch oscillations. There have, indeed, been demonstrations of a discrete Wannier-Stark ladder, and of periodic oscillations in optically-excited superlattices in applied electric field.

In this section, we will first present a semiclassical picture of the Bloch oscillations, then discuss a tight-binding picture of Bloch oscillations, present experimental investigation of the Wannier-Stark ladder and Bloch oscillations, and finally discuss implications of excitonic effects for these oscillations.

2.9.1 Semiclassical Picture

According to the semiclassical picture, the rate of change of quasi-momentum in an applied field F is given by

$$\hbar \dot{k} = eF \tag{2.86}$$

so that

$$k = k_0 + eFt/\hbar \ . \tag{2.87}$$

The time to go from $-\pi/d$ to π/d is T_B, the period for the Bloch oscillations. If the dispersion relation for the miniband is given by

$$E = E_0 - 2\Delta \cos(kd) , \tag{2.88}$$

then the group velocity, $(\partial E/\partial k)/\hbar$, and the position z of the wavepacket are given by

$$v(t) = (2\Delta d/\hbar) \sin(2\pi t/T_B) \tag{2.89}$$

and

$$z(t) = z_0 + (2\Delta/eFd) \cos(2\pi t/T_B) . \tag{2.90}$$

Therefore, the electron undergoes periodic motion in the momentum as well as real space with a temporal period T_B and spatial period L given by

$$L/d = 4\Delta/eFd , \tag{2.91}$$

where 4Δ is the miniband width, and eFd is the Wannier-Stark splitting. The miniband then breaks up into the Wannier-Stark ladder

$$E_p = E_0 + eFpd , \tag{2.92}$$

where the index $p = 0, \pm 1, \pm 2 \ldots$ denotes different Wannier-Stark states. There is a similar expression for the Wannier-Stark states for the heavy and light holes. In this simple picture, interband optical transitions show a series of peaks at photon energies

$$h\nu_p = h\nu_0 + eFpd , \tag{2.93}$$

where $h\nu_0$ is the energy for the vertical transition, and negative (positive) p corresponds to Stokes (anti-Stokes) transitions.

2.9.2 Tight-Binding Picture

One can also consider a superlattice as made up of an infinite number of quantum wells separated by barriers with a period d. The Wannier-Stark ladder eigenstates in the tight-binding scheme are well known [2.248]

$$\chi_p(z) = \sum_n \mathrm{J}_{n-p}(L/2d)\phi(z-nd) , \tag{2.94}$$

where $\phi(z)$ is the wavefunction resulting from a single site potential, and J_p is a Bessel function of the first kind of order p. One can then form an initial wavepacket using a superposition of these eigenstates [2.249] and calculate the time evolution of such a wavepacket. *Dignam* et al. [2.249] showed that the expectation value of z is given by an equation of the form (2.90), with the difference that z_0 is determined by the amplitudes of various eigenstates forming the wavepacket, and the amplitude of the oscillations is given by A_z determined by L and the amplitudes of the eigenstates. A_z has an upper limit given by L. However, it is possible to choose an initial condition such that $A_z \ll L$. In particular, one can create a wavepacket in a specific well, so that $A_z = 0$ [2.249]. In this case the electrons experience a breathing motion, but there is no change in the expectation value of z with time, i.e., there are no Bloch oscillations in the semiclassical sense. This is the very special case examined by *Bastard and Ferreira* [2.247]. While this is strictly true only for the simple energy-band dispersion discussed above, one expects that, for this special case, A_z would be $\ll L$ even for more complicated structure. For the general case, the electron wavepacket undergoes oscillatory motion whose spatial period can be calculated for any given initial conditions [2.249]. The breathing mode and the semiclassical oscillations are simply the two extremes in the range of possible motions.

2.9.3 Qualitative Quantum-Mechanical Picture

For a finite superlattice with N periods and miniband width 4Δ, the simple energy-dispersion relation given by (2.88) shows that, in the absence of an applied electric field, the energy levels are unequally spaced. For F such that $NeFd > 4\Delta$, one obtains the equally-spaced Wannicr-Stark ladder [2.241, 250–252]. A number of observations of the Wannier-Stark ladder in superlattices have been reported in the last few years [2.241, 250–252]. These represent the first experimental verification of the concept of the Wannier-Stark ladder in a solid.

If one excites this Wannier-Stark ladder by a pulsed laser whose spectrum encompasses more than one energy level, then one excites a wavepacket made up of a superposition of various eigenstates. Such a wavepacket would be expected to undergo periodic oscillations [2.249, 253]. This is an extension, to a more complex system, of the concepts developed above for the case of the coherent oscillations of an electronic wavepacket in the asymmetric double quantum well system. The periodic motion of the electronic wavepacket in this case is, in general, the much sought-after Bloch oscillations. This oscillatory motion should then be detectable by FWM experiments and other experiments, just as in the case of *a*-DQWS. In fact, the use of the photon-echo technique was proposed by *Manykin* et al. [2.254] and *von Plessen* et al. [2.255] for just such a purpose for bulk and superlattice cases, respectively. In the next section we describe FWM experiments which demonstrate such oscillations.

2.9.4 Observation of Bloch Oscillations in Semiconductor Superlattices

The first observation of oscillations in FWM signals associated with Bloch oscillation in a semiconductor superlattice was reported by *Feldmann* et al. [2.256]. A superlattice under an applied field was optically excited with an ultrafast laser whose center energy was between the first and the second Stokes peaks of the Wannier-Stark ladder, and whose spectral width was larger than the Wannier-Stark splitting. The time-integrated FWM signal showed two or three peaks in the FWM signal at various applied biases. The period of oscillations varied linearly with the applied field, and agreed well with the Wannier-Stark ladder splitting observed in a photocurrent measurement on the same sample. Calculations based on the theory of photon echo [2.255] agreed with the experimental results. These results show that the observed oscillations are manifestations of the Bloch oscillations of the optically created wavepacket in the superlattice.

We present here the results later reported by *Leo* et al. [2.257] and *Bolivar* et al. [2.258] who investigated a 40 period $GaAs/Al_{0.3}Ga_{0.7}As$ superlattice with 100 Å quantum wells and 17 Å barriers, with calculated miniband width of 19 meV for electrons and 2 meV for holes. Figure 2.43 displays the time-integrated FWM signal at 10 K obtained using 100 fs (17 meV spectral width) pulses from a mode-locked Ti:Sapphire laser centered at 1.537 eV, on the Stokes side of the Wannier-Stark ladder (lower than the energy for the vertical

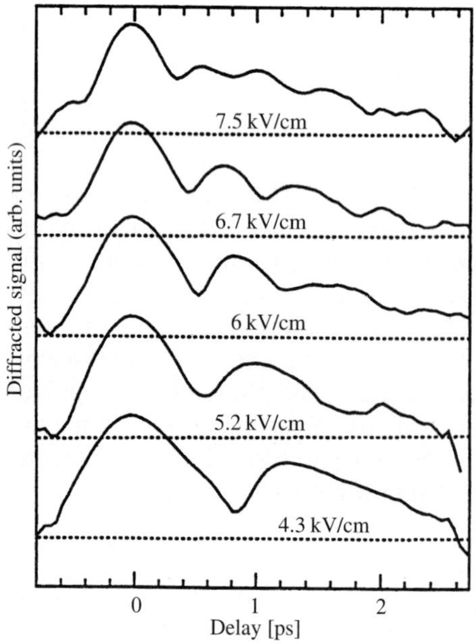

Fig. 2.43. TI-FWM signals from a 40 period 100/17 Å $GaAs/Al_{0.3}Ga_{0.7}As$ superlattice excited by a 100 fs laser pulse tuned slightly below the center of the superlattice transitions for several different electric fields on the superlattice. The oscillations correspond to optically excited Bloch oscillations in the superlattice [2.257]

interband transition). The FWM signal shows oscillations with a period which increases with increasing electric field. The Wannier-Stark splitting deduced from the measured period varies linearly with the applied electric field, as expected for the Bloch oscillations. The fact that the excitation was chosen to create a wavepacket centered on the Stokes side of the vertical interband transition implies that the observed oscillations correspond to Bloch oscillations rather than a breathing motion, as discussed above. The calculated spatial amplitude of the oscillations for the smallest applied field is approximately 10 periods or more than 1,100 Å. It would be interesting to correlate the amplitude of the oscillations with the excitation energy and compare with the tight-binding calculations [2.249].

Leisching et al. [2.259] have reported a detailed study of Bloch oscillations in GaAs/AlGaAs superlattices as a function of temperature, applied electric field, miniband width, and excitation conditions. They observed Bloch oscillations in TI-FWM signals in samples with an electron miniband width ranging from 13 to 46 meV (larger than the optical phonon energy of ≈ 37 meV in bulk GaAs), and lattice temperature up to 200 K. They see several well defined oscillations under a variety of conditions, and the period of the oscillations varies inversely with the applied electric field. The Wannier-Stark splitting calculated from the oscillation period is plotted in Fig. 2.44 as a function of the applied electric field for lattice temperatures of 10 and 77 K. Data for two different $GaAs/Al_{0.3}Ga_{0.7}As$ superlattices are displayed, a 67/17 superlattice (period $d = 84$ Å, and electron miniband width of 37 meV) and a 111/17 superlattice ($d = 128$ Å and electron miniband width of 13 meV). The dashed lines have

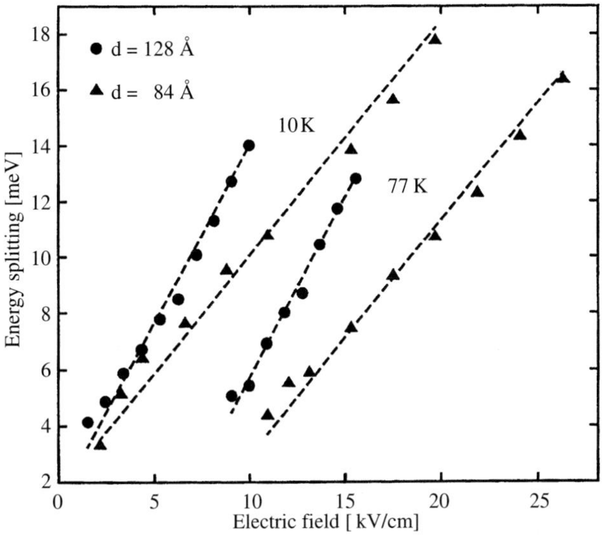

Fig. 2.44. Wannier-Stark splitting as a function of the applied electric field, determined from the period of optically excited Bloch oscillations in two GaAs/AlGaAs superlattice samples at two temperatures; *d* is the superlattice period [2.259]

the appropriately calculated slope. Possible causes for the offset in electric field were discussed by *Leisching* et al. [2.259]. The maximum oscillation frequency in Fig. 2.44, corresponding to a splitting of ≈ 17 meV, is ≈ 4 THz. The wider superlattice has a lower upper limit, as might be expected. The largest Wannier-Stark splitting deduced from the data corresponds to a frequency of 5 THz. Note, however, that the *amplitude* of the oscillation at the highest field becomes quite small, indicating a reduced overlap of the wavefunctions in the neighboring quantum wells, or a strong Wannier-Stark localization. Note also that the electron miniband width for the narrower superlattice is approximately equal to the optical phonon energy. The role of optical phonons in Bloch oscillations in semiconductor superlattices has also been discussed by *von Plessen* et al. [2.260].

FWM signals from Bloch oscillations discussed above result from the modulation of *interband* polarization at the Bloch frequency. An *intraband* polarization is also generated in this case, similar to the polarization between the excited states $|1\rangle$ and $|2\rangle$ generated in the case of *a*-DQWS discussed in Sect. 2.8. The intraband polarization induces anisotropic changes in the refractive index which can be detected by the technique of Transmittive Electro-Optic Sampling (TEOS) discussed briefly in Chap. 1. *Dekorsy* et al. [2.261] have reported observation of Bloch oscillations by this technique. Initial excitation conditions were shown to play an important role in determining the oscillation amplitude, and frequency, amplitude, and dephasing of Bloch oscillations were investigated over a wide range of electric fields. Dephasing times of 2.8 ps was deduced from these measurements for the intraband polarization, about twice the value of 1.4 ps deduced for the interband dephasing from the FWM measurements under identical conditions. The larger intraband dephasing time makes it possible to observe more periods in the TEOS measurements compared to the FWM measurements. *Dekorsy* et al. [2.262] have also recently reported observation of Bloch oscillations at room temperature, and shown that the oscillation frequency can be tuned from 4.5 to 8 THz by applying an external bias.

2.9.5 Influence of Excitons

We have discussed the experimental observations as if electrons are undergoing Bloch oscillations and the holes play no role in the experiments. In the GaAs/AlGaAs superlattice system, the hole wavefunctions are strongly localized so that the holes are not expected to undergo Bloch oscillations. However, the holes play an important role in the process through their Coulomb interaction with the electrons; i.e., excitonic effects must be considered. The concentration of oscillator strength at the discrete exciton energy and the long dephasing times of excitons would also play a role. Also, excitonic effects introduce changes in the optically-excited initial wavepacket which has significant influence on the Bloch oscillations, as discussed below.

The problem of excitonic effects in superlattices under electric field has been treated theoretically by *Dignam* and *Sipe* [2.252, 263], and has been applied to the case of Bloch oscillations by *Dignam* et al. [2.249]. For a superlattice under an applied electric field, they have calculated the maximum value of $|z_e - z_k|$, the absolute difference between the centers of the electron and hole wavefunctions during an oscillation. From this they calculated $D_\alpha(\Omega, \bar{\omega}_c)$, the oscillating part of the total dipole. Here, α is an index which takes on two values, ni for non-interacting electrons and holes, and ex when excitonic effects are taken into account. $\bar{\omega}_c \equiv (\bar{\omega} - E_0/\hbar)$ measures the deviation of the central laser frequency $\bar{\omega}$ from the frequency, $E_0/\hbar = 2\pi\nu_0$ of (2.93), of the vertical, $p = 0$ in (2.93), interband transition. Ω is a measure of the spectral width of the exciting optical electric field. These frequencies are expressed in terms of the angular Bloch frequency $\omega = 2\pi/T_B$.

Figure 2.45 a displays the oscillating portion of the total dipole $D_{ni}(\Omega, \bar{\omega}_c)$ for non-interacting electrons and holes as a function of the spectral bandwidth and the detuning from vertical transition. The darker areas indicate stronger oscillating dipole, as shown by the scale at the right. The calculations were performed for a 95/15 GaAs/$Al_{0.3}Ga_{0.7}As$ superlattice which roughly corresponds to the samples investigated in the first studies [2.256–258]. Figure 2.45 b presents similar results for the case when excitonic effects are included. One can make a number of observations from these calculations. First of all, the non-interacting case is symmetric with respect to detuning from the vertical

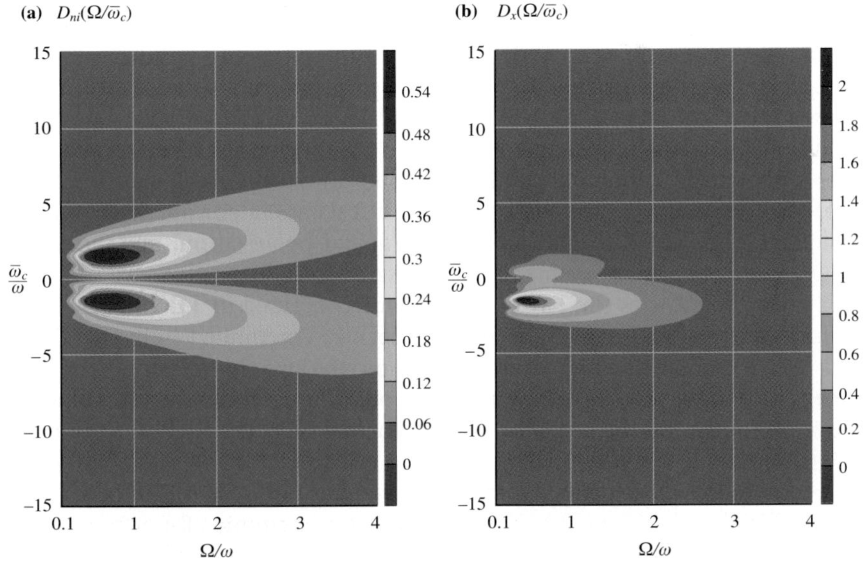

Fig. 2.45. (**a**) Calculation of the oscillating part of the dipole moment for non-interacting system, and (**b**) when excitonic interaction is considered. Note the strong modifications introduced by excitonic interactions; various angular frequencies are defined in the text [2.249]

transition, whereas the exciton case strongly favors excitation on the Stokes side of the vertical transition. Second, for zero detuning, the non-interacting case shows a zero-oscillating dipole. This corresponds to the special case of breathing motion already discussed above. The oscillating dipole does not vanish at zero detuning when excitonic effects are included, and the ratio of the oscillating dipole at optimum detuning and zero detuning is a factor of $2-3$ for spectral width of the order of the Bloch frequency. The calculations for the excitonic case predict that the maximum oscillating dipole would occur for spectral width of the order of the Bloch frequency and for the laser tuned one to two Bloch frequency below the vertical transition energy. Detailed experimental results are not yet available for a comparison.

Excitons introduce another subtle effect: the energy spacing between various Wannier-Stark levels does not remain constant when excitonic effects are included [2.252, 263]. Therefore, when the laser spectrum is wide enough to encompass several Wannier-Stark levels (especially true at low fields), one expects interference effects leading to additional (slower) oscillations superposed on the amplitude of the Bloch oscillations [2.249]. This effect remains to be investigated experimentally.

Thus, excitonic effects are clearly important in understanding the energy-level scheme in the superlattice and the interband polarization created by photoexcitation. The long dephasing times of interband excitonic polarization compared to interband free carrier polarization (Sects. 2.5 and 2.10) are also important in the observation of the Bloch oscillations in FWM experiments which probe the interband polarization. However, TEOS measurements (discussed in Sect. 2.9.4) and the THz emission measurements (to be discussed in Sect. 2.12) probe intraband polarization. *Binder* et al. [2.233] have shown that excitons as well as free carriers contribute to the intraband polarization, and hence THz emission, from *a*-DQWS, with their relative importance determined by excitation and dephasing conditions. *Meier* et al. [2.264] have calculated both the FWM and THz signals resulting from Bloch oscillations in superlattices with narrow minibands and found substantial differences between the two. They have shown that while excitonic effects are important for FWM experiments, they are not important for intraband polarization, and hence for TEOS and THz measurements of Bloch oscillations. They predicted different oscillation frequencies the FWM experiments compared to TEOS or THz experiments. *Leisching* et al. [2.265] have recently determined both the TEOS and FWM signals arising from Bloch oscillations in a $GaAs/Al_xGa_{1-x}$ superlattice under identical experimental conditions. Their results do, indeed, show a difference in the oscillations frequencies in FWM and TEOS measurements.

From TEOS measurements, intraband dephasing times of 2.8 ps [2.261], 2.4 ps [2.265] and 0.37 ps [2.262] have been deduced in three different samples at 10 K. The dephasing time in the last sample is reduced to 0.13 ps at 300 K. *Leisching* et al. [2.265] have argued that intraband polarization of continuum states rather than exciton states contribute to their TEOS measurements. These values are considerably longer than those for interband free-carrier polarization discussed in Sect. 2.10, and not completely understood at the present time. Monte Carlo simulations have been performed recently by *Rossi* et al. [2.266] to address this issue. Even if the oscillations in the TEOS signals are due to intraband polarization of the continuum states, excitons and exciton-exciton interaction may continue to play an important role, depending on the relative values of various parameters. The situation may be similar to that discussed for *a*-DQWS by *Binder* et al. [2.233] who found that both excitons and free carriers contribute to the intraband polarization, and the dominant contribution is determined by excitation and dephasing conditions (Sect. 2.8). This discussion shows that the influence of Coulomb interaction must be evaluated carefully for each experimental situation.

2.9.6 Concluding Remarks

It is clear that the observed oscillations have the same physical origin as the Bloch oscillations predicted 65 years ago. Excitonic effects play an essential role in these observations. However, as discussed above, the relative importance of excitons and free carriers may depend on the nature of the experiment as well as on the physical parameters of the system. This is a rapidly advancing field and there will undoubtedly be more investigations of this interesting phenomenon. It is clear that ultrafast spectroscopy has provided new insights into one of the most intriguing phenomena in solid-state physics.

We conclude this section with the observation that the Bloch oscillations have an oscillating dipole associated with them, just like the dipole associated with the coherent oscillations of the optically excited wavepacket in the case of an asymmetric double-quantum well. This oscillating dipole should lead to coherent sub-millimeter radiation, just as in the case of the double-quantum well system. *Waschke* et al. [2.267] were the first to report such THz radiation from superlattices. These related observations on Bloch oscillations will be discussed in Sect. 2.12.3.

2.10 Coherent Spectroscopy of Free Electron-Hole Pairs

Our discussion so far in this chapter has concentrated on coherent spectroscopy of excitons. A number of interesting observations have also been made for the case when the laser photon energy is such that it excites interband polarization of the continuum states. Some of these investigations are discussed in this section.

2.10.1 Transient Oscillations

Transient oscillations in the vicinity of the pump-photon energy in pump-probe differential transmission spectra have been observed for negative time delays [2.127, 131] when the continuum states of the band are excited. These transient oscillations are the precursor to the spectral hole-burning observed at positive time delays, as discussed in Sect. 2.4. Spectral hole-burning in semiconductors is discussed in Chap. 3.

2.10.2 Free-Carrier Dephasing in Bulk GaAs

Dephasing of free electrons and holes in bulk GaAs has been investigated using coherent [2.49] as well as incoherent [2.268] spectroscopy. We have seen in Chap. 1 that the shortest pulses generated to date are of 6 fs duration. Using such pulses, *Becker* et al. [2.49] have investigated TI-FWM signal from 0.1 μm thick bulk GaAs at room temperature. The continuum states in a semiconductor can be considered as being inhomogeneously broadened in momentum space under certain conditions as discussed earlier (Sect. 2.1.2). In this case the measured signal corresponds to a photon echo, and T_2 corresponds to $4\tau_{decay}$, as discussed earlier. τ_{decay} was found to vary from 3.5 to 11 fs as the density changed from 1.5×10^{17} to 7×10^{18} cm^{-3}. These correspond to some of the fastest transients ever measured. They interpreted these fast transients as due to dephasing of interband polarization by screened carrier-carrier scattering in GaAs. Note that these times are several hundred times shorter than the dephasing times for *excitons* discussed earlier. The times are also considerably shorter than the 190 fs momentum scattering times measured close to the band edge in bulk GaAs at 77 K by *Oudar* et al. [2.268] using a pump-probe transmission technique. *Portella* et al. [2.269], using linearly polarized 9 fs infrared pulses, found that the redistribution of carrier momentum occurs in the first tens of femtoseconds after excitation. The relaxation times are found to vary with carrier density, from 25 to 60 fs for a carrier density range from 8×10^{17} to 8×10^{16} cm^{-3}. They attributed the results to a rapid redistribution of the carrier momentum via a screened Coulomb interaction between the carriers.

2.10.3 Free-Carrier Dephasing in Intrinsic Quantum Wells

The degenerate FWM technique was used to investigate free carrier dephasing in intrinsic quantum wells at room temperature by *Bigot* et al. [2.270]. They measured the decay of the photon-echo signal as a function of time delay between the two incident pulses. The sample consisted of several periods of 96 Å GaAs quantum well and 98 Å $Al_{0.3}Ga_{0.7}As$ barriers. The 8 – 10 fs infrared pulses used for these measurements corresponded to a very wide spectrum (from 730 to 900 nm) encompassing not only the continuum states of interest, but also the $n = 1$ and $n = 2$ excitons as well as a small part of the 3D-like continuum states above the barrier. The measured τ_{decay} range from 16 to 50 fs as the carrier density is varied (Fig. 2.46). These represent some of the fastest dephasing processes measured in quantum wells and show the power of ultrafast spectroscopy with femtosecond pulses. The dependence of τ_{decay} on

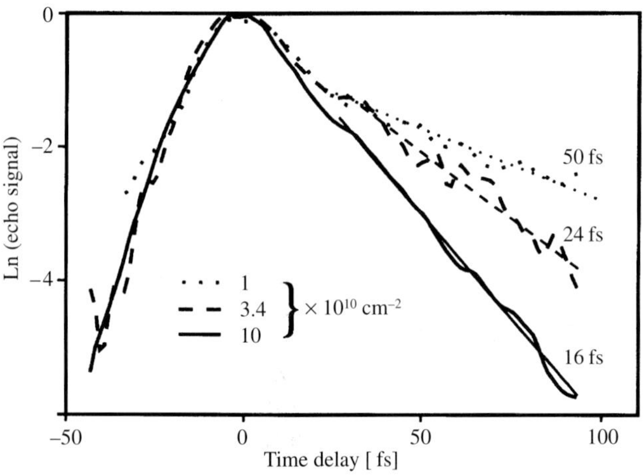

Fig. 2.46. Photon-echo signal from a 96/98 Å GaAs/$Al_{0.3}Ga_{0.7}As$ MQW structure at 300 K for three excitation densities obtained using 10 fs pulses [2.270]

density is different for bulk GaAs and GaAs quantum wells. This has been explained in terms of the dependence of screening on dimensionality [2.270]. A further analysis of the data may be interesting in view of more recent data [2.91] showing the importance of excitonic effects (Sect. 2.5).

2.10.4 Free-Carrier Dephasing in Modulation-Doped Quantum Wells

If the barriers of a quantum well sample are doped with donors, the electrons from the barriers transfer into the quantum wells leaving behind charged impurities. By using undoped spacer layers in the barriers, the carriers and impurities can be physically separated by a significant distance so that ionized impurity scattering, one of the main sources of scattering at low temperatures, is reduced considerably. This provides a sources of high-mobility 2D carriers that have remarkable transport properties (e.g., integer and fractional quantum Hall effect [2.271]). Such modulation-doped quantum wells [2.272] also provide an excellent system for investigating free-carrier effects in semiconductors without the interference of excitons because the presence of free carriers destabilizes (or screens out) the excitons. We discuss in this section some recent coherent experiments on such a system. In later chapters, we will discuss the use of modulation-doped quantum wells for investigating electron-hole scattering rates and for determining the different carrier-phonon scattering rates for electrons and holes.

The electron gas in n-modulation-doped nanostructures has received the most attention, and we concentrate here on this system, although many of the properties of the hole gas in p-modulation-doped nanostructures are similar. At low temperatures, the electron gas in the n-modulation-doped quantum

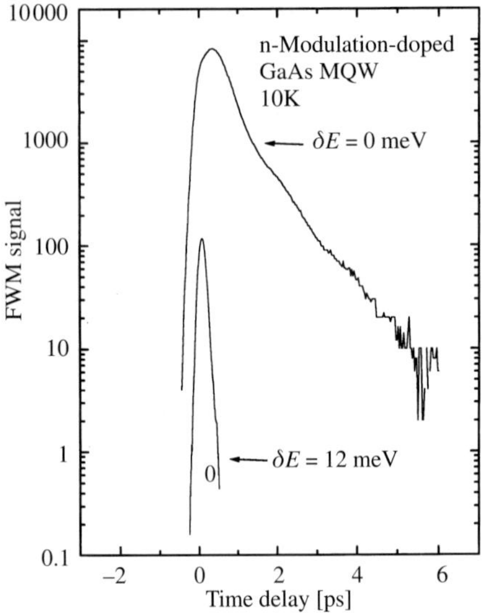

Fig. 2.47. TI-FWM from n-modulation-doped sample B at 10 K for two different excitation photon energies. The curves are labeled by electron excess energy δE as defined in the text [2.275]

wells becomes degenerate and develops a sharp Fermi edge. This Fermi liquid has unique properties described by the Landau Fermi Liquid theory [2.273], and exhibits Fermi edge singularity in interband optical spectra [2.274]. The validity of the Fermi liquid theory has been established by many transport measurements, and the Fermi edge singularity has been investigated by optical spectroscopy [2.274]. We discuss in this section the results on intraband scattering characteristics of electrons in the presence of a Fermi liquid, obtained by using coherent spectroscopy, and compare the results with the predictions of the Fermi liquid theory.

Kim et al. [2.275] measured TI-FWM signals from two modulation-doped $GaAs/Al_{0.3}Ga_{0.7}As$ quantum wells as a function of excitation photon energy, excitation density and temperature. Sample A (B) had 60 (100) periods, 80 Å (60 Å) wells, 100 Å barriers, and a nominal electron concentration of 2×10^{11} cm^{-2} (6×10^{11} cm^{-2}) corresponding to a Fermi energy of 7 meV (20 meV). Figure 2.47 displays the TI-FWM signal as a function of τ_d for sample B for two excess energies δE, the energy separation of the center of the laser spectrum from the Fermi edge singularity peak in photoluminescence excitation spectrum. The data show a striking decrease in the decay time τ_{decay} of the TI-FWM signal for a relatively small increase in the excess energy.

For a given δE, τ_{decay} was measured as a function of excitation density, and extrapolated to obtain the decay time at zero density, $\tau_{decay}(0)$. This quantity is thus not influenced by the photoexcitation, but is related to the dephasing of the interband polarization by the unperturbed crystal. *Kim* et al. [2.275] found that $\tau_{decay}(0)$ varied from about 10 ps when $\delta E<0$, to 100 fs (limited by their time resolution) for δE larger than the Fermi energy. Since this is an inhomogeneously broadened system, this corresponds to T_2 as long as 40 ps.

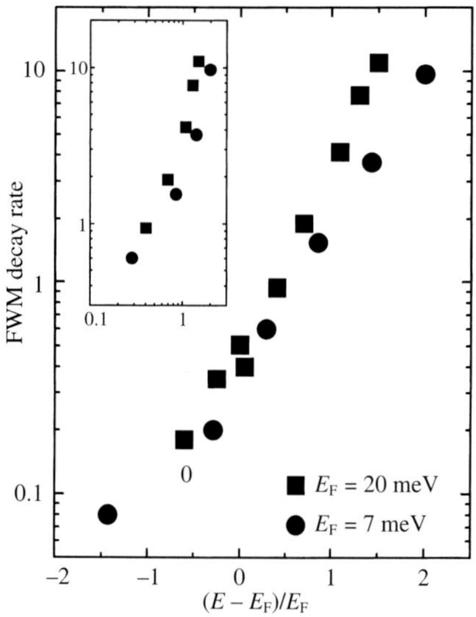

Fig. 2.48.
Zero-density decay rate of TI-FWM, $1/\tau_{decay}(0)$ as a function of $(E-E_F)/E_F$ for two samples of n-modulation-doped GaAs quantum wells; *triangles:* sample A, *solid circles:* sample B. E is the electron energy and E_F is the electron Fermi energy. The log-log plot for positive values (*inset*) shows a quadratic dependence [2.275]

Figure 2.48 shows a plot of $1/\tau_{decay}(0)$ vs $\delta E/E_F$ for both samples. A log-log scale for positive $\delta E/E_F$ in the inset shows that the decay rate varies quadratically with $\delta E/E_F$.

The dephasing of the interband polarization is in general related to many different scattering processes. However, in this case it was argued that electron-electron scattering is the dominant dephasing process and that the measured dephasing time represents the scattering time of photoexcited electrons with the electrons in the Fermi sea. The Landau Fermi liquid theory predicts that, for bulk semiconductors, this scattering rate increases quadratically as a function of the excess electron energy above the Fermi energy. Furthermore, the scattering rate for a given excess electron energy is expected to vary inversely with the Fermi energy. For 2D system this is multiplied by a logarithmic term [2.276] that can be ignored for the present case. The measurements provide a confirmation of both these predictions. While this simplified analysis of the data has provided useful information, the measured decay rate is averaged over a distribution of excess electron energies. A more rigorous analysis is required to quantitatively analyze the data. Also, it has been argued recently [2.277] that scattering of electrons with heavy and light holes, and collective effects such as plasmon emission and interaction with coupled phonon-plasmon modes should also be considered. The effect of averaging over the excess electron distribution must be analyzed before a meaningful comparison with theory can be attempted.

2.10.5 Time-Resolved FWM from Modulation-Doped Quantum Wells

TR-FWM provides further information about the nature of the FWM signal. Both samples A and B were investigated at 10 K using the self-diffracted two-beam technique. Figure 2.49 displays the time-resolved FWM signal from sample B in the $2q_2 - q_1$ direction as a function of time t following the arrival of the *second* pulse on the sample for co-circularly polarized beams. The results for various τ_d show a single peak that moves linearly with τ_d. This photon-echo like behavior indicates that the system behaves like an inhomogeneously broadened system. This is in agreement with our discussion that the continuum states are expected to behave as intrinsically inhomogeneously broadened in the momentum space.

The low-density sample A reveals a different behavior. TR-FWM shows the photon-echo signal plus a weak prompt signal close to $t = 0$ for co-circularly polarized beams. Also, in contrast to the results on sample B, the echo broadens as the time delay increases [2.278]. This behavior was explained by noting that with increasing τ_d only electrons close to the Fermi energy (i.e., electrons with sufficiently long dephasing times) contribute to the signal. Therefore, the FWM spectrum is expected to become narrower so that the echo broadens. Measurement of SR-FWM under identical conditions would test this hypothesis. Also, the TR-FWM signal for cross-circularly polarized beams from sample A shows only a prompt signal and no echo (inset of Fig. 2.49). The position of the prompt signal remains constant but its strength decreases as τ_d increases. The observation of the prompt signal was explained by noting that for sample A the Fermi energy is comparable to the exciton binding energy so that the exciton is not completely destabilized. Therefore, residual excitonic effects persist and lead to the prompt signal.

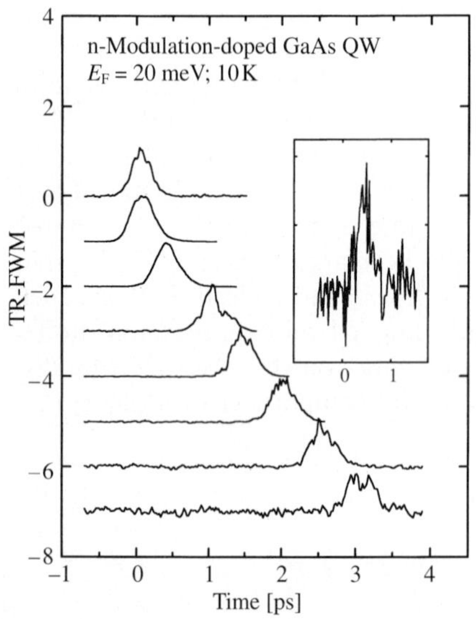

Fig. 2.49. TR-FWM signal from sample B at 10 K for co-circularly polarized beams. The curves are normalized to the same height and displaced vertically for clarity. The second pulse is delayed by τ_d with respect to the first pulse and arrives at the sample at time $t = 0$. Time delays τ_d, from top to bottom, are −0.2, 0, 0.5, 1.0, 1.5, 2.0, 2.5 and 3.0 ps. Inset shows the TR-FWM signal from sample A at 10 K for cross-circularly polarized beams for $\tau_d = 1$ ps [2.275]

2.11 Coherent Phonons

Coherent generation of phonons in condensed media by nonlinear frequency mixing of two lasers separated by the phonon frequency is a well known technique. Lasers used in such studies are in the transparency or weakly absorbing region of the material. Coherent Stokes Raman Scattering (CSRS) and Coherent Anti-Stokes Raman Scattering (CARS) are well known techniques that have been a part of the arsenal of experimentalists in the field of nonlinear optics. The CARS studies are also usually carried out using picosecond or subpicosecond pulses in order to preserve sufficient spectral resolution. They provide excellent dynamic range but no information on the phase of the Raman modes. A recent review of its applications to semiconductors is given by *Bron* [2.279, 280].

Femtosecond pulses have been used recently to investigate time-resolved Impulsive Stimulated Raman Scattering (ISRS) in dye molecules and molecular crystals [2.281–283]. Generation of coherent optical phonons in semiconductors [2.284–289], metals [2.290–292] and superconductors [2.293] by femtosecond excitation of these solids in the strongly absorbing spectral region has also received considerable attention in recent years. The coherent phonon oscillations are generally investigated using pump-probe reflectivity or transmission techniques or pump-probe Reflective Electro-Optic Sampling technique (REOS) [2.294, 295], discussed in Sect. 1.3.1. We discuss here the results obtained on coherent generation of phonons by impulsive excitation of the bulk semiconductor GaAs [2.284, 285, 287–289]. The results on CARS experiments on GaAs and GaP are discussed in Chap. 5 on phonon dynamics.

Cho et al. [2.284] were the first to observe the generation of coherent optical phonons in III–V semiconductors by impulsive excitation in the absorptive regime. The (100) surface of bulk GaAs was excited by 50 fs pulses from a linearly polarized CPM dye laser at 2 eV. The laser photons are strongly absorbed in GaAs and create electron-hole pairs with holes from all three valence bands. The change in the reflectivity of the sample was measured by an orthogonally polarized weak probe beam derived from the same laser, as a function of time delay between the pump and the probe pulses. Measurements were performed at various angles θ between the direction of polarization of the probe beam and the [110] axis of the GaAs crystal. Measurements were made from 10 to 300 K at various excitation densities ranging from 10^{17} to 4×10^{18} cm^{-3}.

Figure 2.50 displays the normalized differential reflectivity ($\Delta R/R_0$) as a function of the delay between the pump and the probe pulses for three different values of θ for GaAs at 300 K excited to a density of 10^{18} cm^{-3}. The autocorrelation trace of the laser indicates that the time resolution of the system is ≈ 100 fs. For $\theta = 90°$, the transient reflectivity rises during the excitation pulse, decays rapidly within 200 fs, and then increase gradually to a quasi-stationary value on picosecond time scale. For $\theta = 45°$ and $135°$, this last quasi-stationary component shows minute oscillations with the frequency of 8.8 ± 0.15 THz, matching exactly the frequency of the LO phonon in GaAs at

(a)

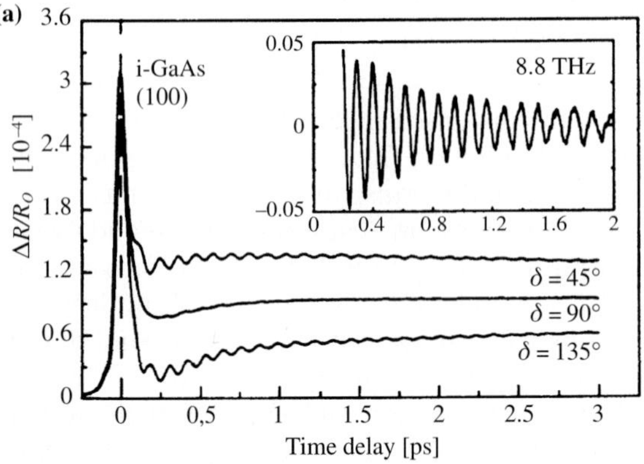

(b)

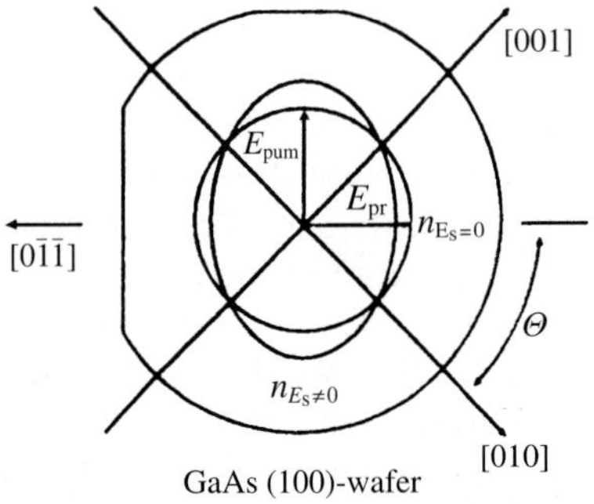

Fig. 2.50. **(a)** Time-resolved reflectivity changes of (100)-oriented intrinsic GaAs at an excitation density of 1×10^{18} cm^{-3} for orthogonal polarization of pump and probe beams. The parameter θ is the angle between probe polarization and the [010] crystal axis **(b)**. The magnified view of the oscillations in the inset shows that the frequency of oscillations corresponds to the LO phonon frequency in GaAs [2.284]

the Brillouin zone center. The oscillations at $\theta = 45°$ and $135°$ are out-of-phase with each other, and the quasi-stationary value for $\theta = 45°$ $(135°)$ is higher (lower) than that for $\theta = 90°$.

From the four-fold symmetry of the oscillations, *Cho* et al. [2.284] have shown that the measured oscillations in the reflectivity are related to the electro-optic effect in the zinc blend semiconductor GaAs. The intense photoexcitation generates LO phonon vibrations in the semiconductor as discussed below. The electric field associated with the LO phonon vibrations, modulates the linear susceptibility through the first-order electro-optic effect. This, in turn, leads to the modulation of the reflectivity. It was shown [2.284] that the measured differential reflectivity is given by

$$\Delta R \cong 0.5 \left(\frac{\partial R}{\partial n} \right) n^3 r_{41} \Delta E \ , \tag{2.95}$$

where n is the index of refraction, r_{41} is the electro-optic coefficient, and ΔE is the change in the longitudinal surface electric field.

In addition to the electro-optic contribution to the differential reflectivity, the measured differential reflectivity is influenced by isotropic changes resulting from photoexcited carriers which alter the surface electric field. The electro-optic contribution can be directly measured by a technique known as Reflective Electro-Optic Sampling (REOS) [2.294, 295], where the probe beam is polarized along the (001) axis ($\theta = 90°$), and the difference in the differential reflectivity along $\theta = 45°$ and $\theta = 135°$ is measured using a polarization beam splitter, $\Delta R_{eo} = \Delta R(\theta = 45°) - \Delta R(\theta = 135°)$ (Sect. 1.3.1). REOS from bulk GaAs has been investigated in detail [2.284, 285, 287] as a function of excitation and doping densities and temperature. From these studies and detailed consideration of the results it was concluded that an ultrafast longitudinal depolarization of the crystal lattice by the screening of the surface space-charge field by photoexcited carriers is responsible for launching the coherent LO phonons in the surface space-charge region of GaAs. Note that this mechanism is different from the displacive excitation of coherent phonons observed in metals such as Bi and Sb [2.290–292]. Both these models are based on a phenomenological description of the coherent phonon amplitude in terms of the oscillation frequency, damping and the driving force. A microscopic model for Ge has been developed by *Scholz* et al. [2.289] in terms of representation in real space. *Kuznetsov* and *Stanton* [2.296] have recently developed a microscopic model in terms of momentum space. These researchers have argued that, for deformation potential coupling, the coherent oscillations are caused by a macroscopic occupation of just one zero wavevector phonon mode, rather than by a synchronous motion of many modes with locked phases.

The observation of coherent phonon oscillations allows one to investigate the influence of carrier density on the phonon modes and the dephasing mechanisms of phonons [2.284, 285, 287]. At low carrier densities (10^{17} cm^{-3}), the oscillations decay with a single exponential from which the dephasing time was deduced to be ≈ 4 ps. The temperature dependence of the dephasing time in the temperature range 10–300 K was in accord with the results of CARS experiments [2.297] (Sect. 5.1.5). At intermediate densities, $(5-10)\times 10^{17}$ cm^{-3}, a second fast dephasing component with a time constant of $\approx$ 400 fs appears. At still higher densities, mode beating resulting from LO and TO phonons is observed. Strong damping and inhomogeneous densities may have prevented observation of coupled plasmon-phonon modes in these experiments.

These experiments clearly provide a powerful means of investigating phonons and their interactions with carriers in semiconductors directly in the time domain. We conclude this section by noting that thermalization, relaxation and cooling of photoexcited carriers in semiconductors (Chaps. 3, 4) lead to the generation of *incoherent* phonons whose dynamics has been investigated by pump-probe Raman spectroscopy (Chap. 5) and which have profound influence on the carrier cooling rates as discussed in Chap. 4.

2.12 Terahertz Spectroscopy of Semiconductor Nanostructures

Femtosecond lasers provide an ideal means of generating ultrafast electrical transients with THz bandwidth. Cerenkov radiation in electro-optic crystals [2.298, 299] provides one means of generating such transients. However, the large dielectric constants of electro-optic crystals makes it difficult to extract these fast electrical signals. In recent years femtosecond excitation of photoconducting switches [2.300–302], unbiased [2.303] and biased [2.304] semiconductor surfaces, and strained layer heterostructures [2.305] have been used to generate THz electromagnetic transients. Physical mechanisms for generating these transients and the application of this radiation for spectroscopy in the THz frequency range have been investigated. Recent conferences on Ultrafast Electronics and Optoelectronics [2.306, 307] and on Ultrafast Phenomena [2.308] contain a discussion of recent results.

We have discussed in Sects. 2.8 and 2.9 the investigation of the coherent oscillations of optically excited excitonic wavepackets by means of nonlinear optical techniques such as pump-probe spectroscopy and FWM spectroscopy. The oscillation frequency of these wavepackets is determined by the relevant energy splitting which is typically in the 1–20 meV range. Since 1 THz corresponds to a splitting of 4.14 meV, these wavepackets are oscillating at THz frequencies. In the a-DQWS case discussed in Sect. 2.8, the hole remains in the WW whereas the electron wavepacket oscillates between the WW and the NW. This oscillation corresponds to a time-dependent dipole moment which should lead to emission of coherent electromagnetic radiation in the THz frequency range. Such radiation from the motion of charge in semiconductor nanostructures has, indeed, been observed from a-DQWS [2.309], semiconductor superlattices [2.267] and single-quantum wells [2.310], and has provided direct evidence for optically excited charge oscillations in semiconductor nanostructures. Some aspects of these observations have been reviewed recently [2.311] and are discussed in this section.

2.12.1 Coherent Oscillations in a-DQWS

The first observation of THz radiation from an oscillating charge packet in a semiconductor nanostructure was reported by *Roskos* et al. [2.309]. They investigated the same a-DQWS samples in which *Leo* et al. [2.228] demonstrated coherent oscillations in nonlinear optical measurements such as pump-probe transmission and FWM spectroscopy. We discuss here the results on the 150/25/100 GaAs/$Al_{0.2}Ga_{0.8}As$/GaAs a-DQWS with 10 periods isolated by 200 Å $Al_{0.2}Ga_{0.8}As$ barriers. The a-DQWS was in the i region of an i-n^+ structure to which a reverse bias was applied through a transparent Schottky gate on the top. The sample was mounted on the cold finger of a cryostat and maintained at 10 K. A Ti:Sapphire laser with 160 fs pulsewidth, 76 MHz repetition rate and 14 meV spectral width was used for the experiments. The

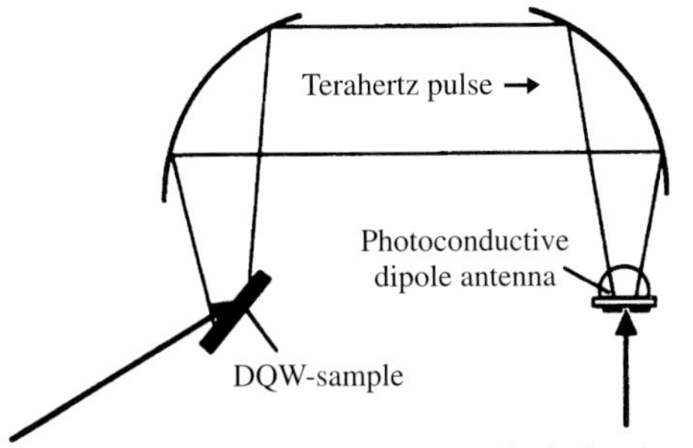

Fig. 2.51. Schematic of the experimental arrangement for the measurement of the coherent THz radiation emitted by an optically-excited semiconductor nanostructure. Paraboloid mirrors collect and focus the radiation on a photoconductive antenna (Fig. 2.52) gated by a delayed laser beam [2.309]

laser beam was divided into two beams, with one used to excite the sample after suitable attenuation and the other used as a gating beam on the photoconducting dipole antenna (see below) to detect the emitted radiation. The sample was weakly excited at an angle of $\approx 45°$ by an unfocused beam of 1 mm diameter; the estimated excitation density was $< 5\times 10^9$ cm^{-2}. The THz radiation emitted along the reflected optical beam was collected and focused on a 50 μm subpicosecond photoconducting dipole antenna [2.301, 312] by a pair of off-axis paraboloid mirrors. The photoconducting antenna is also excited by an appropriately delayed pulse train from the Ti : Sapphire laser. This arrangement is schematically shown in Fig. 2.51.

Radiation damaged Silicon-On-Sapphire (SOS) is widely used as a material for photoconducting antenna. A typical THz antenna consists of a 50 μm long planar dipole with a 5 μm gap, loaded with a coplanar transmission line ending in contact pads [2.301, 312], as shown schematically in Fig. 2.52a. The THz field acts as a bias on the antenna and the charge excited by suitably delayed laser gating pulse is collected as a function of the delay between the optical pulse train exciting the semiconductor nanostructures and that exciting the detecting antenna. The amplitude and the phase of the THz field is determined by the measured signal. Figure 2.52b illustrates how one such structure can be used as a transmitter and another as a receiver. For THz emission from a semiconductor nanostructure, the sample under investigation is the emitter so only one photoconducting antenna structure is needed as a detector.

The coherent electromagnetic transients emitted by the *a*-DQWS sample were measured using this technique by *Roskos* et al. [2.309], and are presented in Fig. 2.53 for several different DC bias fields on the *a*-DQWS. The Fourier transforms of the waveforms are shown in Fig. 2.54. By analyzing these results, it was demonstrated that the emission originated from *two distinct physical mechanisms*: there is an initial THz transient, lasting approximately for the duration of the laser pulse, resulting from the polarization created in the biased quantum wells in which the electron and hole wavefunctions are asymmetric

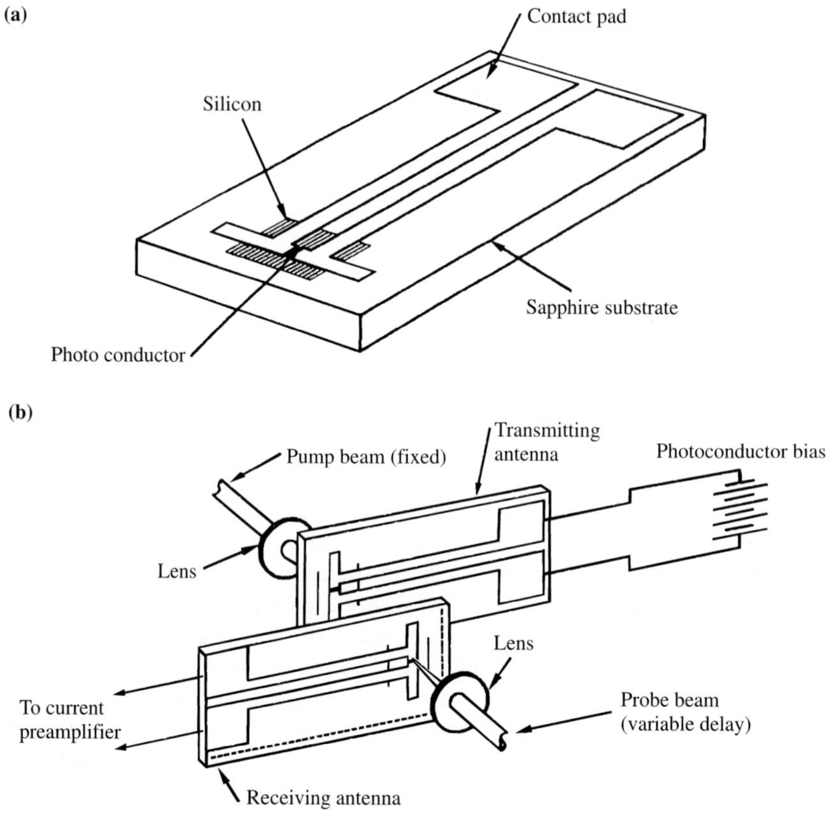

Fig. 2.52. **(a)** Schematic of a photoconducting antenna structure consisting of the dipole, photoconductor (radiation-damaged Si on sapphire), contact pads and coplanar strip transmission lines between photoconductor and contact pads; **(b)** a schematic of a typical transmitter-receiver set-up [2.301]

because of the bias. This radiation has a broad spectrum extending from DC to the inverse of the pulse width of the optical pulse, and its strength diminishes as the bias is lowered and the flat-band conditions are approached. Such radiation would also be expected from a biased single quantum well or a biased bulk semiconductor, or a typical semiconductor surface with surface space charge field.

The second source of THz transients is the charge oscillation of the optically excited wavepacket. Just as in the case of the nonlinear optical measurements discussed in Sect. 2.8, the amplitude of this radiation is maximum when the two electronic levels are in exact resonance. This radiation has a narrow spectrum, ideally determined by the dephasing time of the transitions. The temporal extent of the transient is determined by the dephasing processes and is much longer than the laser pulsewidth. The measurements show that the coherent THz transients persist for several picoseconds, showing that the dephas-

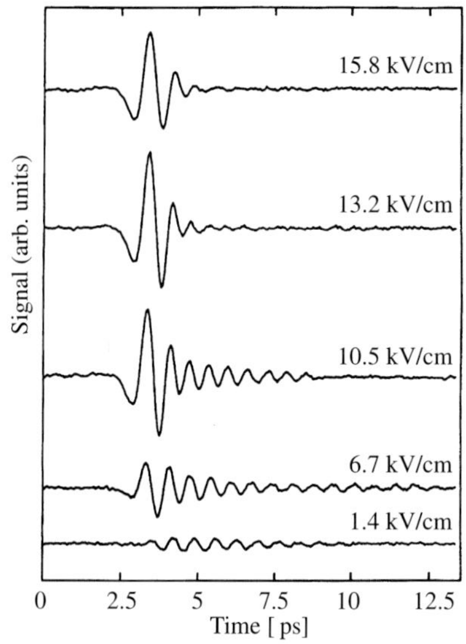

Fig. 2.53. Coherent THz transients from a GaAs/AlGaAs *a*-DQWS at 10 K excited by a 100 fs laser pulse, at various applied electric fields. There is an instantaneous and an oscillatory component to the coherent transients. [2.302, 309]

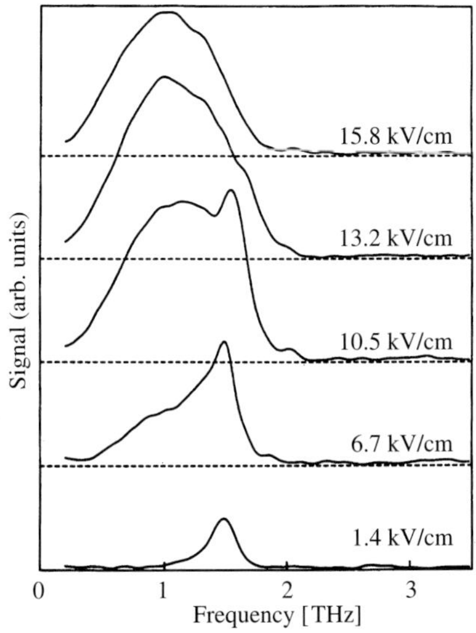

Fig. 2.54. Fourier transform of the THz transients in Fig. 2.53, showing the spectra of the THz radiation emitted by the optically excited *a*-DQWS. The spectra are not corrected for the detector response; it is estimated that the sharp feature close to 1.5 THz is a factor of 3 stronger than the peak of the broad spectrum [2.309]

ing times for the polarization between the two excited states $|1\rangle$ and $|2\rangle$ are several picosecond. Since the dephasing times are long, the Fourier transforms of the temporal traces show a sharp peak at ≈ 1.5 THz in Fig. 2.54. Note that the spectra are not corrected for the frequency response of the antenna and it is estimated that the resonant emission peak at ≈ 1.5 THz is three times as intense as the broadband emission due to the initial population transient. Finally, the center frequency of this radiation is expected to be the lowest at the resonance and increase with either increasing or decreasing bias as the splitting between the electronic level increases away from resonance. Experimentally, it was observed that there was only a small change in the frequency of the emission; and this was explained by the influence of the WW and the NW resonances [2.309].

The emission of THz transients from the semiconductor surfaces was first discussed in terms of a photoinduced current model [2.303]. However, this approach can not explain the orientation dependence of the observed signal. Subsequently, an optical rectification theory which explains all the major observations, including the crystal-orientation dependence, was proposed [2.313]. From a nonlinear optical point of view, the THz transients from semiconductors and their nanostructures can be explained in terms of the second-order nonlinear susceptibility for difference-frequency generation, $\chi^{(2)}(\omega_1 - \omega_2; \omega_1, \omega_2)$. In this picture, the initial signal following the optical pulse results from the diagonal (population) terms of the density matrix, whereas the signal corresponding to the oscillating charge wavepacket results from the off-diagonal (coherent) terms of the density matrix. It can be shown [2.313–315] that the polarization responsible for the emission of the THz transients from charge oscillations is proportional to the dipole between the upper two levels and to the coherence between the upper two levels ($|1\rangle$ and $|2\rangle$ in the notation of (2.84)). The observation of THz emission from the wavepacket oscillations in an optically-excited *a*-DQWS can be explained in terms of these quantities using the same three-level model (Fig. 2.14) that was used to explain the observation of quantum beats and wavepacket oscillations in this system (Sect. 2.8).

This first measurement of the THz oscillations from an optically excited semiconductor nanostructure clearly demonstrated the presence of optically excited charge oscillations, and was followed by a number of interesting experiments. Some of these are discussed in the following subsections.

2.12.2 HH-LH Oscillations

Nonlinear optical measurements have shown the existence of HH-LH quantum beats, as discussed in Sect. 2.6.2. This forms a three-level system in a single quantum well, and it is natural to ask if a coherent optical excitation of this three-level system leads to the emission of coherent THz transients. This was investigated by *Planken* et al. [2.310], and analyzed theoretically by *Luo* et al. [2.314].

Planken et al. [2.310] investigated the THz emission from a 15-period MQW structure consisting of 175 Å GaAs quantum wells separated by 150 Å $Al_{0.3}Ga_{0.7}As$ barriers. The barriers are sufficiently thick and high so that coupling between the wells can be neglected, and each quantum well can be treated as independent. The quantum wells were in the i-region of a i-n$^+$ structure so that a transparent Schottky gate was used to apply bias perpendicular to the quantum-well planes as before. The sample at 10 K was excited by

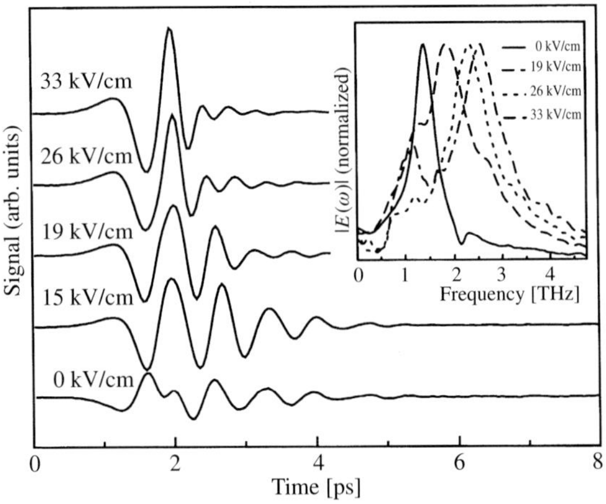

Fig. 2.55. Coherent THz transients from a 15-period 175/150 Å $GaAs/Al_{0.3}Ga_{0.7}As$ MQW sample at 10 K excited by a 100 fs laser pulse tuned between the HH and LH excitons, for several values of electric field applied perpendicular to the interfaces. The Fourier spectra of the transients (*inset*) show that the frequency of the THz emission can be tuned by the electric field [2.310]

pulses from a Ti:Sapphire laser at an appropriate wavelength so that a coherent superposition of HH and LH states was excited, and the THz emission was detected with a photoconducting dipole antenna as discussed above.

The observed coherent THz transients are evident in Fig. 2.55 for several values of the perpendicular electric fields on the sample. The signal once again consists of the initial transient and an oscillatory part. The oscillatory signal is present only when both the HH and the LH are excited, and hence corresponds to the HH-LH quantum beats [2.150, 151], discussed in Sect. 2.6.2. The period of the oscillation decreases as the field increases. The apparent decrease in the amplitude of the oscillatory part with increasing field may be a result of the decrease in the response of the photoconducting antenna at higher frequencies. The spectra of the oscillatory part of the signal (inset of Fig. 2.55) show that the center frequency of the oscillatory signal changes from 1.4 THz at flat-band to 2.6 THz at the highest field. This semiconductor MQW system thus provides a source of tunable THz radiation. The change in the frequency arises from the change in the HH-LH splitting with applied electric field and the measured changes are in good quantitative agreement with the calculations of the HH-LH splitting.

The observation of THz radiation may be surprising at first because the transition-dipole moment vanishes for the inter-subband transition between the HH and the LH at the in-plane wavevector $K_{||} = 0$ for all electric fields. Since the polarization responsible for the THz radiation is proportional to the transition dipole moment, no radiation is expected for HH and LH at $K_{||} = 0$. However, the wavefunctions of $K_{||} = 0$ excitons involve electron and hole wavefunctions away from $K_{||} = 0$ where transition dipole moment for HH-LH inter-subband transition is finite due to valence band mixing effects. Therefore, coherent excitation of HH and LH excitons is expected to emit coherent THz transients. *Luo* et al. [2.314] have evaluated the inter-subband dipole moment between the HH and the LH excitons taking full account of the valence-band mixing using the Luttinger-Kohn Hamiltonian. They obtain good quantitative agreement between the calculated HH-LH splitting and the observed THz frequency at various electric fields. They also found that the dipole moment at all $K_{||}$ decreases as the electric field is increased, in contrast to the expectation for the envelopes of the HH and the LH wavefunctions [2.311]. Using optical Bloch equations for the three-level system, they have also calculated the THz transients at various electric fields. A quantitative comparison of the calculated and observed amplitude of the radiated THz electric field would be interesting. It is also interesting that THz emission from both the *a*-DQWS and the HH-LH system is explained in terms of the optical Bloch equations for a three-level system and so far there is no need to invoke the more elaborate semiconductor Bloch equations (Sect. 2.1.2) which include interactions between the excitons.

2.12.3 Bloch Oscillations

As discussed in Sect. 2.9.4, FWM spectroscopy of semiconductor superlattices under applied electric field has shown the existence of Bloch oscillations of the optically-excited wavepacket under appropriate conditions. Initial considerations showed that the oscillation of the optically excited wavepacket corresponds to a breathing motion [2.316] and would not radiate THz signals because it does not correspond to an oscillating dipole. Later considerations by *Dignam* et al. [2.249] indicated, however, that the above conclusion is valid only when the excitonic effects are ignored and when the exciting laser spectrum is centered at the vertical interband transition. In the general case when the laser spectrum is not centered at the vertical transition, or when the excitonic effects are properly treated, *Dignam* et al. [2.249] showed that there is indeed an oscillating dipole which should radiate at the Bloch oscillation frequency.

Direct coherent THz transients from a semiconductor superlattice system were first observed by *Waschke* et al. [2.267, 317–319]. They investigated a superlattice structure consisting of 35 periods of undoped 97/17 GaAs/$Al_{0.3}Ga_{0.7}As$ superlattice grown on n-doped GaAs substrate with 2500 Å thick undoped $Al_{0.3}Ga_{0.7}As$ buffer layer, and with a 3500 Å undoped $Al_{0.3}Ga_{0.7}As$ cap layer. The calculated miniband widths for this structure are 19 and 2 meV for the electrons and the heavy holes, respectively. A semitransparent Schottky gate on the top of the sample allowed the application of reverse bias to bring the superlattice from the miniband regime to the Wannier-Stark regime. The sample was characterized by extensive photocurrent and differential reflectivity measurements.

The experimental arrangement was similar to those in the other THz transient experiments described above. The sample at 15 K was excited with 100 fs (22 meV spectral width) pulses from a Ti : Sapphire laser tuned to the vertical ($p = 0$, (2.93)) HH excitonic transition. The laser was incident at an angle of 45° and excitation density $< 10^9$ cm^{-2} per quantum well. The coherent THz transient radiation along the reflected optical direction was collected and fo-

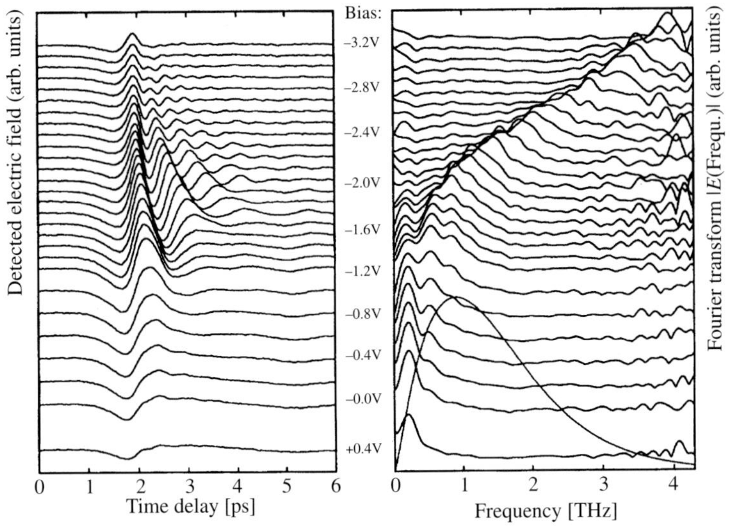

Fig. 2.56. *Left panel:* Coherent THz transients from a 97/17 Å $GaAs/Al_{0.3}Ga_{0.7}As$ superlattice at 10 K excited with a 100 fs laser pulse, for different bias voltages. The period of the oscillations decreases linearly with the increasing field between −1.6 and −3.2 V, as expected for Bloch oscillations; *right panel:* Fourier transform of the THz transients in the left panel [2.267]

cused on a 50 μm photoconducting antenna as before. Figure 2.56a exhibits the coherent THz transients measured from the superlattice for bias voltages ranging from +0.4 to −3.2 V. The THz transients are essentially structure-less in the low-bias region (0.4 to −1.0 V). For reverse biases larger than −1 V, oscillatory structure develops and up to six oscillation periods can be seen. The period of the oscillation decreases with increasing reverse bias. The maximum amplitude for the THz transient is observed at −1.8 V. The Fourier transform of the transients is depicted in Fig. 2.56b. For reverse biases larger than −1.8 V, the dominant spectral peak shifts nearly linearly with bias to a maximum of about 4 THz in these measurements. By a comparison with the photocurrent measurements showing the Wannier-Stark peaks, it was shown that the frequency corresponds to the Wannier-Stark splitting. Thus, the oscillations occur at the frequency expected for the Bloch oscillations.

An interesting aspect of this measurement is that the transient THz radiation is observed when the laser is tuned to the center of the direct vertical interband transition. From our discussion in Sects. 2.9.5 it is clear that only breathing motion of charges is expected under these conditions if Coulomb interaction between electrons and holes is ignored [2.249, 320]. Therefore, the observation of THz transients, and hence an oscillating dipole, for vertical excitation shows the importance of excitonic effects.

These initial measurements of THz transients from optically excited Bloch oscillations have been extended to other samples, other excitation conditions (including excitation into the continuum states) and to temperatures up to 200 K [2.321 – 323]. For excitation in the continuum, the FWM signal from an optically excited superlattice disappears within 100 fs as might be expected from the short dephasing times of free carriers discussed in Sect. 2.10.2 and 3. However, the oscillatory THz signals associated with Bloch oscillations persist for excitation energies as much as 80 meV above the bandgap [2.322]. The oscillatory signal is the strongest at the lowest bandgap, becomes weaker with increasing excitation photon energy, and shows a second maximum at the second minibandgap energy before going to zero. As already discussed in Sect. 2.9.5, Bloch oscillations at 300 K have been observed recently [2.262] using the TEOS technique. However, THz emission has not been observed at room temperature. The long dephasing times for intraband polarization make these observations possible, as discussed in Sects. 2.9.4 and 5. Finally, Bloch oscillations under a DC electric field and no optical excitation would be of immense fundamental and practical interest. Negative differential resistance under DC bias has been reported by *Sibille* et al. [2.324 – 327] in superlattices.

2.12.4 Coherent Control of Charge Oscillations

Considerable effort has been devoted in the past 30 years to control and manipulate dynamics of molecules and chemical reactions, using lasers that match specific transitions in molecules [2.328 – 333]. One possibility that was explored extensively was to excite a selected vibrational mode of the system so strongly that a selected molecular bond is broken. Rapid transfer of energy from the excited mode to the rest of the molecule frustrated efforts to achieve this goal. It is now recognized, and is amply clear from our discussion of coherent wavepacket oscillations, that constructive and destructive interference between various quantum mechanical states of the system play a key role in coherent dynamics and hence in coherent control. Efforts are now beginning to control these interference phenomena in an effective way.

In most of the multi-pulse experiments discussed above, the relative phase of the pulses did not play an important role, and the time delay between the center of the pulse envelopes was the relevant parameter. One must, however, never lose sight of the fact that the optical field within the pulse envelope oscillates rapidly at the optical frequency, and that the relative phases between the pulses may play an important role. The linear (Sect. 2.6.5a) and nonlinear (Sect. 2.7.1c) interferometric measurements discussed earlier clearly show the importance of the phase of the optical wave. The relative optical phase between two pulses can be controlled very accurately by precisely controlling the delay between the two pulses. Also, considerable progress has been made in recent years in generating controllable complex pulse shapes by using spatial manipulation of the spectrum of the ultrashort pulse by using spatial modulators in the Fourier plane [2.334 – 345].

Such phase-locked pulse sequences or complex pulse shapes can be used to control the interference between the polarizations and wavepackets, and to code and decode information [2.339], provided the delay between the two pulses or the overall duration of the complex waveform is shorter than the dephasing time of the induced polarization. Long dephasing times in atoms and molecules make them the ideal media for such manipulation, and considerable effort has indeed been devoted to exploring such systems with phase-locked pulses [2.332, 333]. In semiconductors, the dephasing times of excitons are in the few picosecond range, so it should be possible to explore the influence of phase-locked pulses and complex waveforms on wavepacket dynamics with currently available techniques. The first such measurements in semiconductors were reported by *Brener* et al. [2.346] using complex pulse shapes and by *Planken* et al. [2.347] using phase-controlled pulse sequences. We discuss here the experiments of *Planken* et al. [2.347] and the analysis by *Luo* et al. [2.315].

Planken et al. [2.347] investigated the same *a*-DQWS sample that was used to demonstrate the coherent wavepacket oscillations using FWM spectroscopy by *Leo* et al. [2.228] (Sect. 2.8) and using THz transients by *Roskos* et al. [2.309] (Sect. 2.12.1). It consists of 150/25/100 GaAs/$Al_{0.2}Ga_{0.8}As$/GaAs *a*-DQWS with 10 periods isolated by 200 Å $Al_{0.2}Ga_{0.8}As$ barriers. No bias was applied on the sample but the built-in electric field was such that the spatially direct exciton involving the HH and the electron in the NW was in resonance with the spatially indirect exciton involving the HH in the NW and the electron in the WW. The sample was excited to a density $< 10^{10}$ cm^{-2} using 80 fs pulses from a Ti:Sapphire laser with center wavelength tuned to match this resonance. The experimental arrangement was similar to that in Sect. 2.12.1 except that the sample was excited not by a single pulse but by a pair of pulses generated in a Michelson interferometer with different path lengths. One of the end mirrors of the interferometer was controlled by a piezoelectric transducer to actively stabilize the time delay within a fraction of the optical cycle so that the relative phase Φ between the two electric fields can be controlled precisely. The THz transients excited by the laser pulses were detected by a 50 μm photoconducting antenna as before.

Since the spectrum of the pulse exceeds the splitting of the two coupled levels, each photoexcitation pulse acting independently generates an oscillating wavepacket which emits coherent THz transient as before. The top two traces in Fig. 2.57 show the coherent THz transients measured when only the pulse 2 or pulse 1 is present. The oscillatory THz signal corresponds to the oscillating wavepacket excited by the optical pulse. The initial THz transient unrelated to the oscillating dipole is small compared to the earlier results (Sect. 2.12.1) because the total field is small. The lower two traces represent the THz transients when both pulses are present, when the overall delay between the two pulses is adjusted to be twice the oscillation period of the wavepacket, and the phase Φ between the two pulses is adjusted to be either 0 or π. For $\Phi = 0$, the emitted radiation is enhanced by nearly a factor of 3 showing that the wavepackets interfere constructively, whereas for $\Phi = \pi$ the emitted radiation is nearly quenched showing that the interference is destructive.

Similar measurements were performed when the delay between the two pulses was not an integer multiple of the oscillation period. For an overall

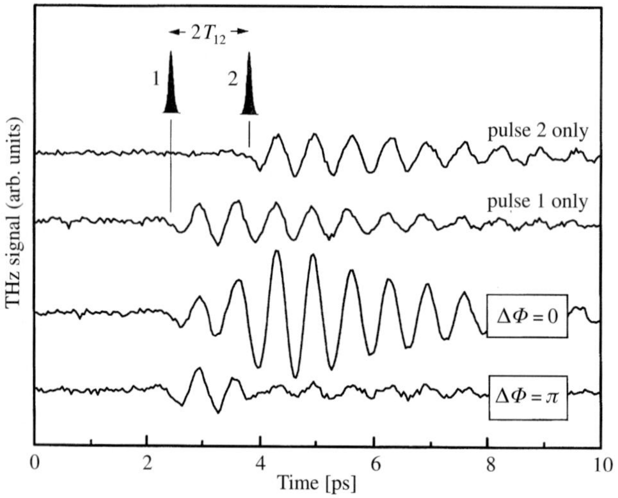

Fig. 2.57. Coherent THz transients from a 10-period 145/25/100 Å GaAs/$Al_{0.2}Ga_{0.8}As$/GaAs *a*-DQWS at 10 K, in a phase-locked double-pulse-excitation experiment with time-delay set at twice the oscillation period ($2T_{12}$); the curves, from top to bottom, are: pulse 2 only, pulse 1 only, pulse 1 and 2 with $\Delta\Phi = 0$, and pulse 1 and 2 with $\Delta\Phi = \pi$ [2.347]

delay of 2.5 periods, the THz transients were slightly stronger for $\Phi = 0$ or π but much weaker for $\Phi = \pi/2$. More interestingly, the *phase* of the THz transients was either retarded or advanced after the arrival of the second pulse. These measurements thus show that it is possible to control the amplitude and the phase of the emitted THz transients by an appropriate two-pulse sequence.

These experiments were analyzed in terms of the three-level optical Bloch equations by *Luo* et al. [2.315]. They analyzed the case of δ-function pulses and infinite relaxation times to obtain physical insights into the behavior and also presented results of numerical integration of the optical Bloch equations for finite laser pulse widths and relaxation times. They found that the evolution of the charge oscillations induced by a two-pulse sequence depends on the detuning energy, overall pulse delay, and the relative phase between the two optical fields. Thus, for an overall delay equal to one oscillation period, the condition for constructive interference changes from $\Phi = 0$ when the laser is centered at the lower level to $\Phi = \pi$ when the laser is centered in the middle of the two levels. On the other hand, for the laser centered in the middle of the two levels, the condition of the constructive interference changes from $\Phi = \pi$ to $\Phi = 0$ when the overall delay is changed from one to two oscillation periods. The numerical solutions of the three-level Bloch equations can provide detailed predictions of the THz transient under different values of these parameters. It is also important to realize that the arrival of the second pulse in the two-pulse sequence also influences the *population* (the diagonal components of the density matrix) of the upper levels in a constructive or destructive manner depending on the relative phase of the two optical fields.

2.12.5 Summary

The observation and investigation of THz transients have provided new insights into the physics of wavepacket oscillations in semiconductor nanostructures. The coherent THz transients provide not only a tunable source for linear spectroscopy in the THz region, but also provide a source synchronized with the optical pulse for nonlinear spectroscopy. THz emission from wavepacket oscillations has been demonstrated up to 200 K. The primary limitation to room temperature operation is the decrease in the dephasing times with increasing temperature. Another area of interest is the efficiency of the THz generation which is rather low. Currently the emitted THz power is in the range of nanowatts for optical excitation with milliwatts of power. Part of the low power efficiency is inherent in the optical excitation because of the large ratio (≈ 1000) of the photon energies in the near and far infrared. The work discussed here deals with optically-excited THz transients where the energy for THz emission comes from the optical fields. Electrical excitation of THz radiation is clearly of immense practical interest. One can expect to see considerable activities in these different areas in the near future.

2.13 Conclusions

The ability to investigate the coherent regime in semiconductors, with their ultrafast relaxation times, is a relatively recent development. As a result, there has been a flurry of activities in this field in recent years. These investigations have led to new insights into some very interesting fundamental phenomena in the coherent regime. These investigations have also led to a better understanding of some of the fastest scattering processes in semiconductors through investigations of processes that destroy coherence. Interesting phenomena of coherent oscillations in semiconductor nanostructures have been demonstrated and led to the observation of coherent THz emission from semiconductor nanostructures. Coherent phenomena in semiconductor nanostructures are likely to remain an active area of research in the coming years.

3. Initial Relaxation of Photoexcited Carriers

The interband polarization produced by an ultrafast laser is dephased by a variety of scattering processes, as discussed in the previous chapter. For real excitations, carriers and/or excitons remain in the crystal even after the coherence is destroyed. In most cases, the carrier or exciton distribution function after the loss of coherence is non-thermal; i.e., it can not be characterized by a temperature. We discuss in this chapter various experiments in GaAs and related semiconductors which probe these non-thermal distribution functions of free carriers and their relaxation to thermalized, hot distribution functions (non-thermal regime of Chap. 1). A discussion of non-thermal excitons is postponed to Chap. 6.

Collisions within a given type of electronic system (e.g., electrons in the Γ valley) are primarily responsible for thermalizing a non-thermal distribution function, although other scattering processes are occurring simultaneously and also influence thermalization to some extent. In general, electron-electron scattering is the most efficient process for thermalizing the electrons. However, since carrier-carrier scattering rates are density dependent, the relative importance of various scattering processes depend strongly on the density, and each case must be examined separately. A study of the non-thermal distribution functions and their approach to a thermal distribution, therefore, provides direct information about these carrier-carrier scattering processes. The generally slower carrier-phonon interaction determines the carrier-cooling dynamics (hot carrier regime of Chap. 1), a topic for the next chapter.

Since the carrier-optical-phonon scattering rate is of the order of $10^{13}\,s^{-1}$ for a carrier with sufficient energy to emit an optical phonon (Fig. 1.6), direct measurement of non-thermal distribution function requires a time resolution of about 100 fs, and this is the range in which most of the non-thermal distribution function measurements have been made. If the excited carriers have energies less than the optical phonon energy, then the time scale can be considerably longer. This would also be the case if the photoexcited carriers emit successive optical phonons and relax to less than an optical phonon energy before many carrier-carrier collisions occur; this would clearly require a low photoexcited density. Most of the measurements discussed in this chapter were performed using pump-probe transmission spectroscopy because of the ease of obtaining high temporal resolution with this technique. Also, the results primarily probe the electron dynamics, although we will discuss some experiments specifically aimed at understanding the more complicated behavior of holes.

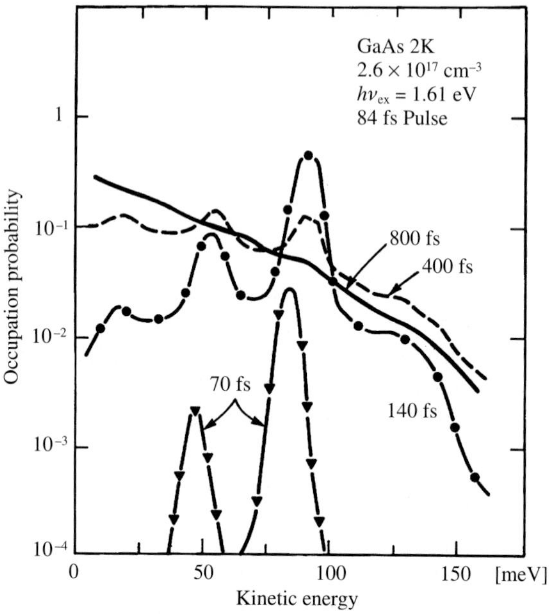

Fig. 3.1. Calculated electron-distribution function in bulk GaAs photoexcited at an excess energy of 80 meV and a density of 1×10^{16} cm^{-3} with an 84 fs pulse with a 6 meV spectral width at the photon energy 1.61 eV. Time delay 70 fs, 140 fs, 400 fs and 800 fs for inverted *triangles*, *filled circles*, *dashed line* and *solid line*, respectively [3.1]

The conduction band in GaAs has many subsidiary valleys and if the photoexcited electron has sufficiently high energy, transitions between the valleys may also play an important role in the initial relaxation of photoexcited electrons. Varying the photon energy of excitation provides the opportunity of studying the intravalley (within the Γ valley) as well as the intervalley dynamics.

The general approaches to the calculation of distribution functions discussed in Chap. 1 are well suited for calculating the non-thermal distribution functions and their evolution to thermal distributions. Figure 3.1 illustrates an early calculation by *Collet* and *Amand* [3.1] for bulk GaAs excited about 100 meV above its bandgap. The calculation clearly shows the non-thermal electron distribution functions including optical phonon replicas of the photoexcited distribution function, at early times. Quantitative comparisons with experiments are now possible using the balance equation and the ensemble Monte Carlo approaches discussed in Chap. 1.

Although we have separated the discussion of coherent and incoherent dynamics, in Chap. 2 and Chaps. 3–6, respectively, it should be emphasized once again that coherent phenomena influence the incoherent dynamics. In Sect. 1.4.5 we mentioned that most pump-probe and luminescence experiments are analyzed in terms of the semiclassical Boltzmann equation. It is clear, however, that coherent polarization influences the initial non-thermal carrier distribution. Efforts to include such effects in calculations [3.2–8] and to explore them experimentally [3.9, 10] are already under way. These effects will not be considered in detail in this book.

3.1 Non-thermal Distributions in GaAs

The first evidence for non-thermal distribution functions in photoexcited semiconductors was obtained using pump-probe spectroscopy which provides the simplest technique for obtaining time resolution limited by the laser pulsewidth. In this section, we discuss some experiments investigating non-thermal distribution functions in bulk GaAs, first by using pump-probe spectroscopy and then by luminescence spectroscopy.

3.1.1 Pump-Probe Spectroscopy

Oudar et al. [3.11] were the first to observe non-thermal distribution of optically-excited carriers in bulk GaAs. Using an amplified CPM dye laser, they generated a white light continuum in a water cell and amplified a part of the continuum in a Styryl 9 amplifier. The amplified pulse was spectrally filtered to generate a pulsewidth of 500 fs and was used to excite a 0.75 μ thick sample of GaAs at 15 K. The pump-photon energy was 1.538 eV, only 19 meV above the bandgap of GaAs. The change in the transmission induced by this pump was measured with a weaker 100 fs continuum pulse. The results are presented in Fig. 3.2. The transmittance shows a peak at the pump photon energy, corre-

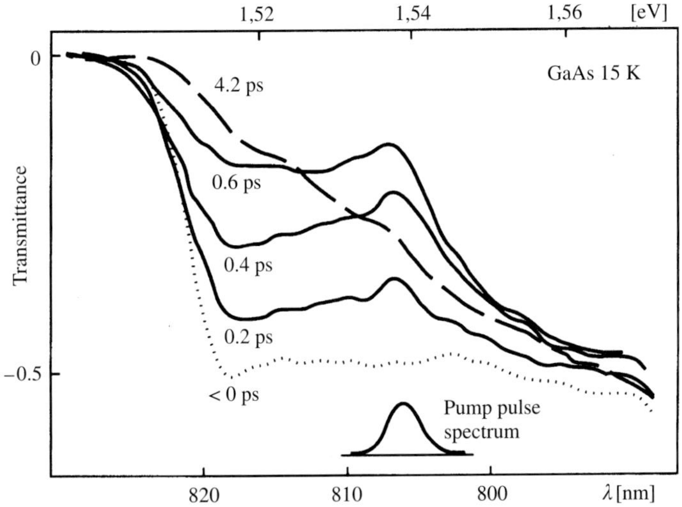

Fig. 3.2. Transmittance of bulk GaAs at 15 K at various time delays following excitation by a 0.5 ps pump pulse with the spectrum shown by the *lowest curve*. The *dotted transmittance curve* was taken before the arrival of the pump pulse (negative time delay) while the *long-dashed curve* was taken 4.2 ps after the pump pulse. The spectra clearly show a peak in the transmittance caused by the non-thermal carrier population which evolves into a thermalized (Maxwell-Boltzmann-like) tail in less than 4.2 ps [3.11]

sponding to the bleaching of absorption and the non-thermal occupation of the electron and hole states coupled to those photons. Note that the transmittance peak is quite large, indicating a strong bleaching of the absorption. The peak disappears slowly and becomes a Maxwellian tail at 4.2 ps, corresponding to thermalization of carriers. A dephasing time of 0.3 ps was deduced from the width of the hole in the absorption spectrum.

A systematic study of transient hole burning in the absorption spectrum of GaAs and several compositions of AlGaAs at 300 K was reported by *Lin* et al. for degenerate pump and probe wavelengths [3.12, 13], and also for continuum probe pulses [3.14]. The samples were excited using 35 fs pulses from a CPM dye laser (spectral width ≈ 50 meV) tuned to ≈ 1.97 eV, and probed with either the same laser or a white light continuum generated using high (kHz) repetition rate amplified pulses. Carrier densities in the range 1 to 10×10^{17} cm^{-3} were investigated. Transient holes in the absorption spectra, corresponding to the photoexcited non-thermal distribution functions were indeed observed close to the pump photon energy of 1.97 eV (Fig. 3.3). In some cases these researchers observed spectral holes at several probe wavelengths because excitations from the three different valence bands create electron distribution functions peaked at three different kinetic energies. The initial bleaching of absorption was found to be extremely rapid at all photon energies, indicating that the photoexcited carriers are redistributed over a wide range of energies in tens of femtoseconds. Luminescence measurements complementing these studies will be discussed in the next section.

In contrast to the measurements of *Oudar* et al. discussed above, the spectral holes in the experiments of *Lin* et al. [3.12–14] were extremely short-lived, and disappeared in *tens of femtoseconds*. Note that *Lin* et al. used much higher excitation photon energies. Contributions from intervalley scattering and emission of phonons may account for the difference, although no systematic study has been reported. Relaxation times in the range of *tens of femtoseconds* were also deduced by *Tang* and collaborators [3.15–18] using the equal-pulse-optical-correlation technique, which measures the combined transmitted flux of two equal intensity pulses as a function of the delay between the pulses. The focus of this technique is in determining various relaxation rates for the optically excited carriers, and not the investigation of the spectral hole. We have also seen in Chap. 2 that photon-echo signals in GaAs obtained with pulses shorter than 10 fs have extremely short decay times (4 to 15 fs) and have been interpreted in terms of extremely rapid carrier-carrier scattering times ranging from 15 to 60 fs. The spectral hole in GaAs quantum wells for low excitation energies was also found to evolve into a thermal tail in less than 200 fs (Sect. 3.3.1b).

Note that the spectral hole at the excitation photon energy is not universally observed and there appears to be some controversy in this regard. *Bradley* et al. [3.19] and *Taylor* et al. [3.20] have reported spectral hole burning in AlGaAs, whereas *Nunnenkamp* et al. [3.21] report bleaching near the bandgap but no spectral hole at the excitation photon energy in AlGaAs. These differences may be caused by different experimental conditions. It is worth mention-

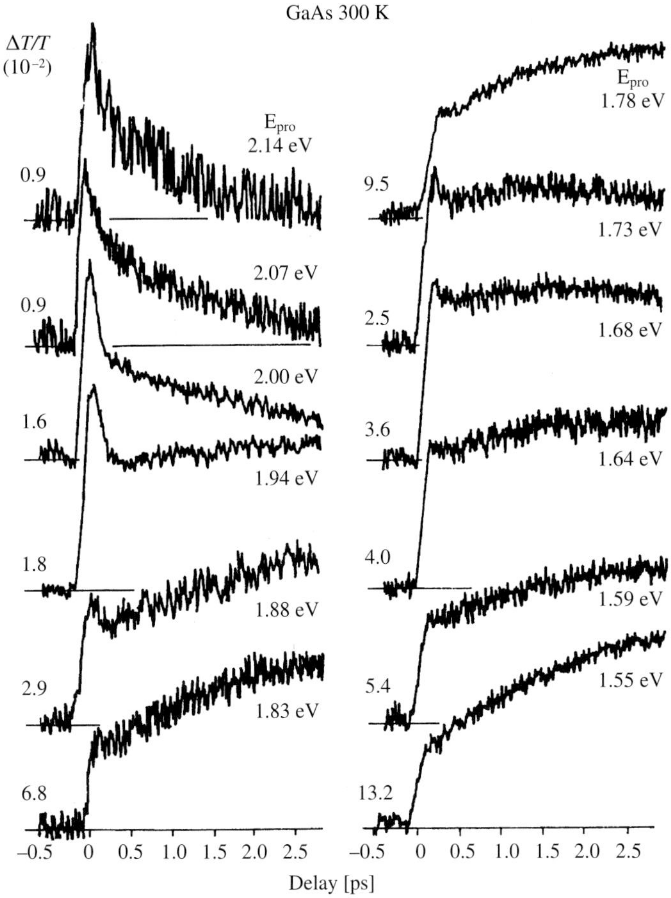

Fig. 3.3. Differential transmission in bulk GaAs at 300 K excited with 35 fs pulses at 1.97 eV at a carrier density of 1×10^{18} cm^{-3} at different probe energies as a function of time delay between the pump and the probe pulses. The initial peak in the curves observed for probe energies between 1.88 and 2.14 eV indicates the presence of spectral hole near the incident pump photon energy, whereas the initial peak in the curves observed for probe energies between 1.68 and 1.73 eV indicates the presence of spectral hole near the electrons generated from the spin split-off band [3.14]

ing here that *Nunnenkamp* et al. [3.21] pointed out the importance of the changes in the interband absorption coefficient induced by the pump as a result of modification in the Coulomb enhancement effects, a topic discussed in the section on interpretation of results in Sect. 1.4. *Bradley* et al. [3.19] also found that different electron and hole temperatures are required to fit the differential transmission data between 1 ps (when the spectral hole has disappeared) and 10 ps. *Taylor* et al. [3.20] have compared their results to static and dynamic screening of the carrier-phonon interaction, and found that the latter gives better agreement with the experimental results. A discussion of the screening of the electron-phonon and electron-electron interactions is given in Chap. 4.

It is also interesting to note here that the coherent spectral oscillations discussed in Chap. 2 are a precursor to the spectral hole burning discussed here. As already noted in Chap. 2, the pump-probe signals are influenced by both the coherent and incoherent effects, and this must be considered in the interpretation of the results. It should be emphasized that we have separated discussion of coherent effects, non-thermal behavior, cooling behavior, exciton dynamics in different chapters as a convenient way of organizing the discussion. In a real experiment, many of these effects are present simultaneously and it is the task of the researchers to design and carry out experiments which isolate one or two of these effects and provide a clear understanding of these effects.

Finally, we discuss in this section a recent experiment which sheds light on a different aspect of the photoexcited, non-thermal distribution function. *Foing* et al. [3.22, 23] investigated a 0.75 μm thick sample of GaAs at 20 K using 120 fs pulses to excite the sample and 30 fs continuum to probe the changes in the transmission through the sample. Special care was taken to assure that the pulses were not chirped. The excitation photon energy was ≈ 1.60 eV, approximately 80 meV above the low-temperature bandgap of GaAs, and the excitation density was varied between 5 and 25×10^{16} cm^{-3}. The unperturbed absorption spectrum of the sample and the laser spectrum are shown in the inset of Fig. 3.4. The transmission spectrum exhibits residual Fabry-Perot oscillations which need to be treated with care, as discussed in Sect. 1.3.1. The main signal in the pump-probe transmission spectrum at short times Fig. 3.4 is a peak in the differential transmission spectrum close to the pump-photon energy, corresponding to an increase in the transmission due to the non-thermal occupation of the photoexcited states. However, the spectral hole peaks at an energy slightly lower than the pump-photon energy (an effect also seen in quantum wells by *Knox* et al. [3.24] and discussed in Sect. 3.3.1), and is not symmetric. The spectral hole has a relatively long positive tail (bleaching) at lower energies, and there is a *negative* peak, corresponding to *increased absorption* at higher energies as can be seen for both excitation densities in Fig. 3.4. All these features were attributed by *Foing* et al. to many-body edge singularities resulting from the non-equilibrium distribution functions generated by photoexcitation. These edge singularities are similar to the strong enhancement observed close to the Fermi edge in modulation-doped quantum wells [3.25–27], which are analogous to such effects in the X-ray absorption edge [3.28]. The main difference here is that the distribution function is non-thermal so that there are *two edges* to the distribution function. Therefore, singularities appear near both edges and modify the simple spectral hole to the shape observed. These effects were first analyzed by *Asnin* et al. [3.29] and *Zimmermann* [3.30] using the non-linear Bethe-Salpeter equation in a study independent of these experiments. The experimental results are analyzed in detail by *Tanguy* and *Combescot* [3.31].

We conclude this section on non-thermal effects in bulk GaAs by noting that a number of different investigations on the initial relaxation of photo-

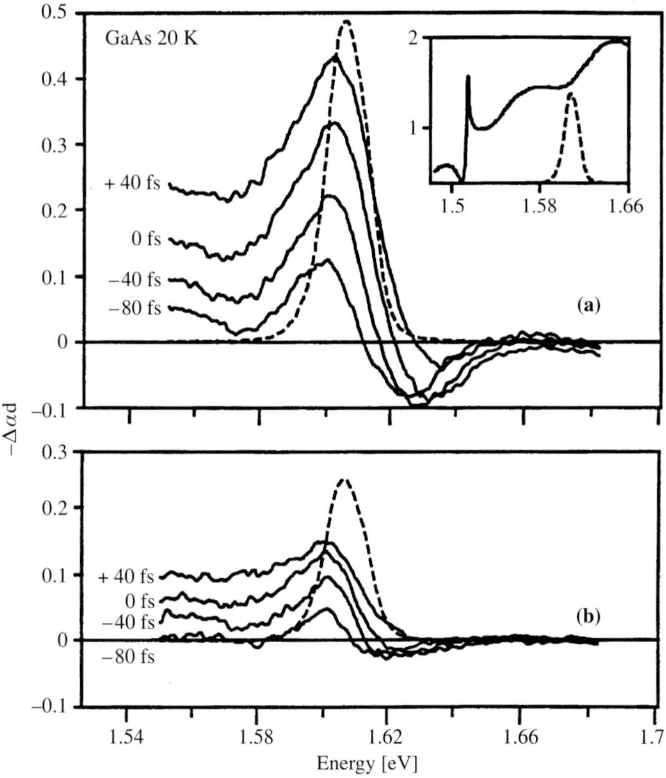

Fig. 3.4 a, b. Differential transmission spectra ($-\Delta\alpha d = \Delta T/T_0$) from a 0.75 μm thick GaAs sample at 20 K using *unchirped* 120 fs pulses. The results present a spectral hole in the absorption occurring at an energy slightly below the center of the laser spectrum, and an induced absorption occurring at higher energies. **(a)** 50 μJ/cm^2 (2.5×10^{17} cm^{-3}), and **(b)** 10 μJ/cm^2. The *inset* shows the unperturbed absorption spectrum and the laser spectrum [3.22, 23]

excited carriers, both experimental and theoretical, have been reported in the past several years. The reader is referred to the proceedings of the last several International Conferences on Hot Carriers in Semiconductors [3.32–35] for many contributions in this field. Some studies on quantum wells will be discussed in Sect. 3.1.

3.1.2 Luminescence Spectroscopy

Since the interband luminescence intensity at a given photon energy is the product of the electron and the hole distribution functions, occupancy of both the electron and the hole states involved in the transition is required to observe the corresponding luminescence signal. If it so happens that, following ultrafast photoexcitation, the electrons remain non-thermal for some time but the holes leave the photoexcited states very quickly, it may become very difficult to measure the luminescence signal at the excitation photon energy. In contrast, bleaching of absorption at the excitation photon energy persists so long as either electrons or the holes remain where they were excited. Also, ≈ 100 fs time resolution desirable for observing non-thermal distribution function in most cases is difficult to obtain in luminescence studies. Therefore, a well-defined spectral feature corresponding to a non-thermal distribution function (a spectral hole in the absorption spectrum or a spectral peak in the luminescence spectrum) is much easier to observe in pump-probe spectroscopy compared to luminescence spectroscopy.

While luminescence spectroscopy has this difficulty, it also has certain advantages for studying non-thermal distribution functions [3.36]. For high excitation energies the photoexcitation can excite electrons from the heavy-hole, light-hole and the split-off valence bands with corresponding electrons at different energies in the conduction band. Similarly, the measured transmission changes at a given photon energy monitor electron and hole populations from different regions of the conduction and valence bands simultaneously. This may make it difficult to obtain unambiguous information about carrier relaxation from pump-probe spectroscopy. The complications arising from this were present in the data by *Lin* et al. [3.14] discussed above. In contrast, the luminescence spectrum below 1.7 eV in GaAs at 300 K is dominated by the heavy holes so that it provides direct information about electrons and holes at a given wavevector.

We discuss in this section and in Sect. 3.3.2 some luminescence studies which have provided useful information about non-thermal carriers in GaAs.

(a) Femtosecond Luminescence Experiments. *Elsaesser* et al. [3.36] measured femtosecond dynamics of luminescence from GaAs using the technique of luminescence upconversion with a time resolution of ≈ 100 fs. The measurements were performed at room temperature so that the thermal tail of the hole-distribution function extended to high energies, and allowed measurement of the luminescence over a large spectral range. The sample, consisting of a 0.6 μm thick GaAs layer cladded with AlGaAs layers, was excited with 120 fs pulses from a hybridly modelocked dye laser operating at 1.93 eV so that only the GaAs layer was excited. A high-sensitivity photon-counting technique was utilized so that signal counts of less than 1 count/s could be measured. Several typical time-resolved luminescence spectra during the first 200 fs are displayed in Fig. 3.5. The excitation density, estimated for the center of the Gaussian-type focused spot, was varied from 1.7×10^{17} and 7×10^{17} cm^{-3}. Note that the spectra extend from the band-gap energy of 1.4 to 1.7 eV. The luminescence signal was undetectable for ≈ 200 meV below the excitation energy, so that the electron and hole states close to those directly photoexcited by the laser from the HH valence band to the conduction band were not accessible in this experiment.

In spite of this difficulty, *Elsaesser* et al. [3.36] were able to extract considerable information from this investigation. Figure 3.5 shows that even at the

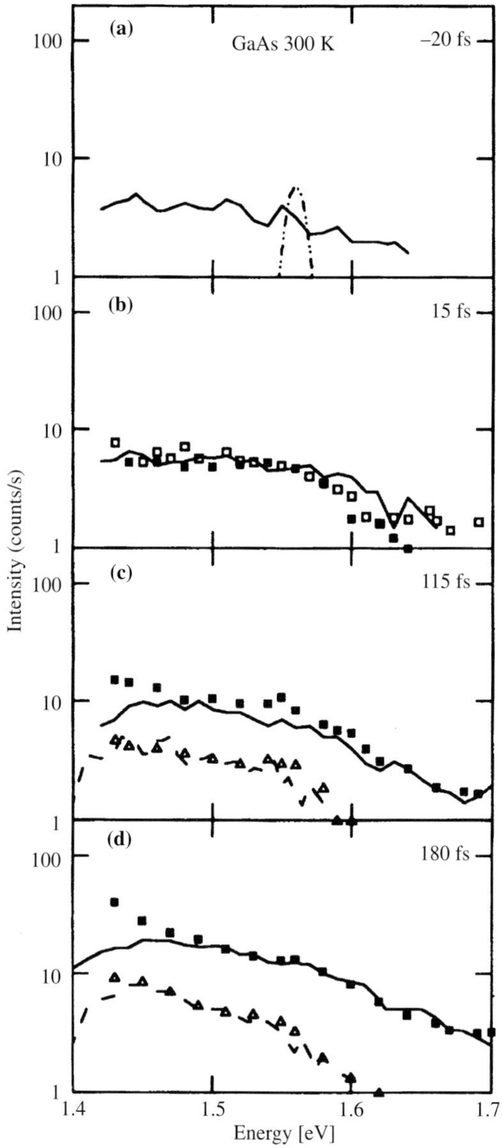

Fig. 3.5a–d. Time-resolved luminescence spectra of bulk GaAs at 300 K at various delays following excitation by a 120 fs pulse at 1.93 eV at an excitation density of 7×10^{17} cm^{-3} (*solid lines*) and 1.7×10^{17} cm^{-3} (*dashed lines*). The *solid squares* and *triangles* represent calculated spectra using Monte-Carlo simulations with static screening. The *open squares* in **(b)** are calculated from molecular-dynamics model. The *dash-dotted curve* in **(a)** centered at 1.57 eV represents the luminescence spectrum expected for the initial electron distribution excited from the spin-split-off valence band [3.36]

earliest time, the spectrum is extremely broad, indicating an extremely rapid relaxation of the photoexcited carriers. The dot-dashed peak in Fig. 3.5a is the expected peak position for the electrons excited from the split-off valence band recombining with the heavy holes. Comparison with the luminescence spectrum reveals that this peak is completely washed out at the earliest time measured. The rapid initial rise of luminescence intensities at all energies investigated is in agreement with the data of *Lin* et al. [3.14] discussed above. From the absence of any narrow structure and the rapid initial rise of the luminescence intensities at several energies, *Elsaesser* et al. [3.36] concluded that electrons in the Γ valley and holes in the HH band scatter rapidly and acquire a broad distribution function in <100 fs even at 1.7×10^{17} cm^{-3}. Furthermore, they found that only about 15% of the final luminescence intensity is reached after 180 fs, from which they concluded that most of the electrons are at energies not accessible in this study. These electrons are possibly in other conduction-band valleys, a subject of discussion in Sect. 3.2.

These luminescence spectra and the temporal profiles of luminescence intensities at several photon energies were compared with ensemble Monte-Carlo simulations using several different approaches to the screening of carrier-carrier interactions (see Chap. 4 for a more detailed discussion of screening). Using static screening, the simulations show that the holes thermalize within 100 fs whereas the electrons equilibrate much slower, leading to structured luminescence spectra. The absence of structured spectra show that this approach to screening is inadequate. The use of time-dependent static screening improves the fit, but is still inadequate in that it predicts a much sharper fall-off of luminescence intensity above 1.57 eV than observed. Finally, dynamic screening was included using a molecular dynamic approach [3.37–39], and provided a better agreement with the data. Note that the influence of dynamic screening on carrier relaxation has been calculated using other theoretical approaches and applied to experiments [3.40, 41].

These results are consistent with the results of *Lin* et al. [3.14] discussed above. Unfortunately, the data of *Lin* et al. [3.14] did not extend to sufficiently low energy to see if there is a spectral hole at photon energy of 1.55 eV corresponding to the HH-electron transition for electrons excited from the split-off valence band. *Elsaesser* et al. [3.36] did not see a peak in luminescence at this energy, and it would be interesting to make a direct comparison of luminescence spectra with differential transmission spectra in this energy range. Note also that the results of *Elsaesser* et al. [3.36] are rather different from those by *Oudar* et al. [3.11] discussed above for comparable densities. The experiments by *Oudar* et al. were performed at 15 K using 0.5 ps excitation pulses with an excitation photon energy only 19 meV above the bandgap. This suggests that it would be interesting to perform luminescence measurements with excitation energy close to the bandgap energy. Such experiments are described in the next section.

(b) Picosecond Luminescence Experiments. The band-to-band luminescence spectra correspond to the product of the electron and hole distribution functions, leading to certain limitations in the spectral range as discussed above. Another approach to investigating electron distribution functions is by monitoring the electron-to-acceptor luminescence which does not depend on the free hole distribution function, and can provide a large spectral range because the wavefunctions of spatially localized holes contain significant contributions from large wavevector states. This approach has been used by *Ulbrich* [3.42] to investigate cooling of photoexcited electrons in the range where acoustic phonons dominate, and will be discussed in Chap. 4. CW electron-acceptor luminescence has also been used to observe electron relaxation following photoexcitation [3.43–49] and for investigating intervalley scattering processes [3.47, 50], and will be discussed briefly in Sect. 3.2. *Snoke* et al. [3.51, 52] have recently used this process to investigate non-thermal electron distribution function in GaAs on picosecond time scale at very low excitation density.

Snoke et al. [3.51, 52] photoexcited Ge-doped GaAs (acceptor concentration: $(4-6)\times 10^{16}$ cm^{-3}; acceptor-binding energy: 40.5 meV) at 10 K with 5 ps pulses from a synchronously pulsed dye laser operating near the GaAs band-gap energy. The excitation density was varied between 4×10^{13} and 4×10^{14} cm^{-3}. Time-resolved spectra were measured using a streak camera with the temporal and spectral resolutions of 15 ps and 1 meV, respectively. The luminescence spectra are dominated by donor-acceptor pair recombination which, at short times, is independent of time because of its long decay times. Figure 3.6 exhibits the time-resolved spectra (after subtracting the time-independent donor-acceptor luminescence) obtained at two different excitation densities at several time delays. At $t = 0$, corresponding to the center of the excitation pulse, the higher intensity spectrum can be approximately fit to a Maxwellian as shown. However, the low-intensity spectrum is much broader and can not be fit to a Maxwellian. This is an indication of a non-thermal electron distribution function at low intensities and short times. At longer times, the low-intensity spectrum becomes narrower and can be characterized by a Maxwellian with 50 K at 55 ps.

Several other aspects of the data are interesting. The electron distribution function observed at the earliest times ($t = 0$) is already quite broad, much broader than expected on the basis of the valence-band parameters and the spectral width of the laser. It is likely that some redistribution of energy takes place during the pulse. Other possible sources of broadening are broadening due to dephasing during the coherent generation process [3.5, 10] and broadening of the acceptor level due to interactions between impurity levels. A more complete analysis of results should consider these and other possible sources of broadening at short times. More interesting is the observation that the thermalization time becomes independent at densities below 5×10^{13} cm^{-3}. Initially this independence was attributed to scattering of electrons with spatially correlated holes [3.52]. But more recent work [3.53] attributes it to a combined result of two competing effects: the scattering rate decreases with decreasing density because there are fewer particles to hit, but the scattering cross-section increases because of a reduction in screening. The independence of thermalization rate from density variation is an interesting phenomenon which will likely receive further attention in coming years. We conclude by remarking that the screening in 2D is expected to be density-independent, as we will discuss in Chap. 4.

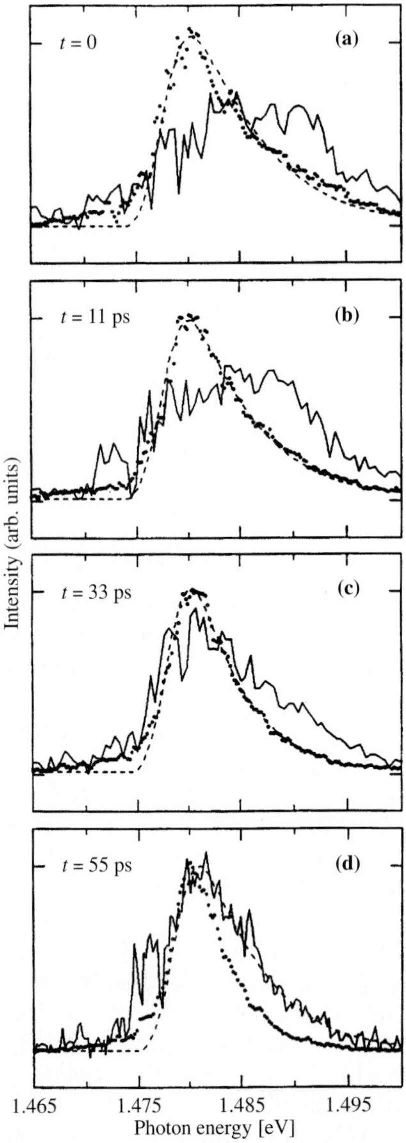

Fig. 3.6a–d. Time-resolved energy distribution of the conduction electrons in p-type GaAs:Ge at 10 K as seen in the band-to-acceptor luminescence following excitation by a 5 ps laser pulse at 1.536 eV for various time delays. *Dots* and *solid lines* correspond to 4×10^{14} cm^{-3} and 4×10^{13} cm^{-3}, respectively. The *dashed lines* correspond to the best fit Maxwell-Boltzmann distribution functions at different temperatures (**a**) 60 K, (**b**) 53 K, (**c**) 42 K, for the high-density excitation, and (**d**) 50 K for the low-density excitation [3.51]

3.2 Intervalley Scattering in GaAs

If the photoexcitation energy is sufficiently high, the photoexcited electrons may have sufficient kinetic energy to scatter into subsidiary conduction-band valleys, most likely by emitting phonons. While this introduces a complication in interpreting the results of such experiments, it also provides an opportunity to investigate the intervalley scattering processes. It is well known that electron scattering into subsidiary conduction-band valleys with heavier masses and hence lower mobilities forms the underlying physical mechanism for Gunn oscillations in semiconductors like GaAs. Measurement of velocity-field characteristics provides a means of determining the optical and acoustic deformation potentials through which intervalley scattering occurs. However, as discussed in Chap. 1, transport provides information averaged over the distribution function of the carriers. While this information is valuable, it is difficult to extract information about microscopic scattering rates from transport measurements. Thus values for the Γ-L deformation potential ranging from 2 to 20 eV/Å have been obtained by different researchers. A number of different optical techniques have provided a much more accurate determination of these important constants.

Kash et al. [3.54] applied non-linear techniques with a CO_2 laser to determine the effective intervalley deformation potential. *Collins* and *Yu* [3.55] measured pump-probe Raman scattering from phonons in GaAs and analyzed their data in terms of various intervalley scattering processes (Chap. 5). These early measurements were followed by time-resolved luminescence and pump-probe measurements. Before discussing some of these measurements, we note that the electron-acceptor luminescence, discussed in Sect. 3.1.2b in connection with non-thermal electron distribution function, has also been used to investigate energy relaxation of photoexcited electrons in cw experiments [3.43–49]. In these experiments, the electron relaxation by LO phonons in the Γ valley results in a series of luminescence peaks separated by the LO phonon energy. The nature of the luminescence spectrum changes dramatically for the photoexcitation energy large enough that the electron kinetic energy in the Γ valley exceeds the threshold for transfer to the subsidiary L and X valleys. These can be analyzed in terms of the intervalley transfer times which can then be converted into an effective deformation potential. These cw studies, which have been applied to a study of a number of physical problems, have been reviewed recently [3.49, 56, 57], and will not be discussed.

If electrons are excited with sufficiently high photon energy, one of the mechanisms that contributes to the scattering of electrons out of this initial region is the scattering to subsidiary valleys. Therefore, effects of intervalley scattering are always present in any pump-probe transmission experiment under these conditions. For example, such effects are present in the experiments by *Lin* et al. [3.14], and *Tang* and collaborators [3.15–18] discussed above. *Taylor* et al. [3.17] included the effects of intervalley scattering in their analysis, and *Lin* et al. [3.14] attempted to isolate specific transitions by varying the bandgap of AlGaAs by changing its composition. *Becker* et al. [3.58] investigated intervalley-scattering rates in GaAs with 6 fs pulses, and differentiated between various processes by varying the sample temperature. *Bigot* et al. [3.59] interpreted their interesting results in terms of resonant coupling between different valleys under certain conditions. The topic of intervalley scattering has also attracted much theoretical attention [3.60–62], partly in an effort to understand the ultrafast optical experiments and partly because of its relation to the high-field transport problem. Proceedings of the recent conferences on Hot Carrier in Semiconductors [3.32–35] and SPIE conferences on ultrafast phenomena in semiconductors [3.63] contain a number of other original contributions to the subject.

Luminescence spectroscopy provides another means of investigating intervalley scattering processes. We discuss here the work reported by *Shah* et al. [3.64] who excited GaAs and InP at 300 K with <500 fs pulses from a synchronously pumped dye laser at 2.04 eV. At this energy, the highest-energy electrons excited in GaAs can transfer to the subsidiary valleys whereas those in InP do not have sufficient energy for an intervalley transfer. Figure 3.7a displays the band structure of GaAs schematically on the left, and the time evolution of the near-bandgap (1.45 eV) luminescence intensity on the right. Similar curves were obtained for excitation densities between 5×10^{16} and 1×10^{18} cm^{-3}. At the end of the pulse, the intensity is less than 10% of the peak intensity. The early time evolution of the luminescence is plotted in Fig. 3.7b and compared with the time evolution in InP in Fig. 3.7c. Figure 3.7d presents the calculated temporal profile of the luminescence intensity in GaAs under various approximations [3.64].

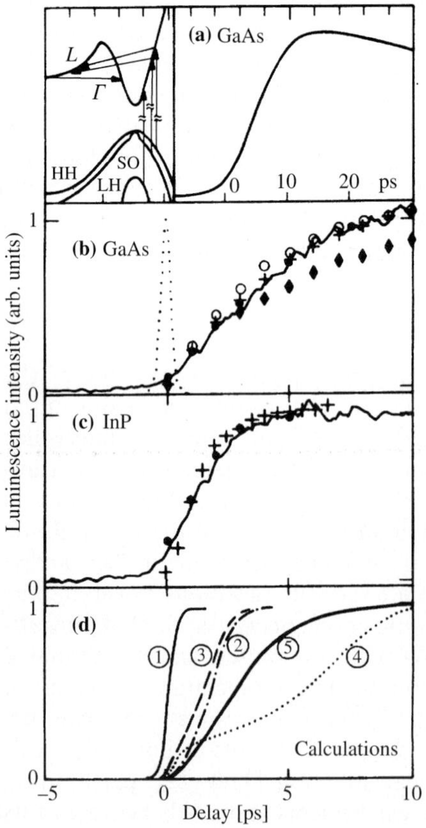

Fig. 3.7a–d. Luminescence intensity at 300 K excited by <500 fs pulses at 2.04 eV as a function of time delay. (**a**) GaAs luminescence at 1.45 eV; (**b**) same as (**a**) on an expanded scale; (**c**) InP band gap luminescence; (**d**) different calculations. The slow rise of GaAs luminescence was attributed to intervalley scattering and the value of the deformation potential $D_{\Gamma L}$ for intervalley scattering was determined from best fit between the data and Monte Carlo simulations. *Diamonds*, *crosses* and *open circles* show the results of Monte-Carlo simulations for $D_{\Gamma L}$ = (4, 6, and 8)$\times10^8$eV/cm, respectively [3.64]

Comparison of Fig. 3.7 b and c clearly shows that the near-bandgap luminescence in InP rises much more rapidly than in GaAs. These results were analyzed by ensemble Monte-Carlo simulations using the effective intervalley deformation potential as a parameter. The symbols in Fig. 3.7 b, c represent the results of these simulations for various values of the deformation potential. From these comparisons, the value for the intervalley deformation potential was deduced to be $(6.5 \pm 1.5) \times 10^8$ eV/cm for GaAs at 300 K. It should be emphasized that this is an effective deformation potential that takes into account all processes that bring about transfer of electrons between the different valleys. From the cw luminescence technique discussed above, *Ulbrich* et al. [3.47] deduced a somewhat smaller value at low temperatures. *Zollner* et al. [3.60, 61, 65] have performed detailed calculations for the effective deformation potential and found that phonon-absorption processes become important at 300 K and account for the difference in the values obtained by the two groups.

Oberli et al. [3.66] repeated the time-resolved luminescence measurements with a laser tuned above and below the threshold for creating electrons that transfer to the L valley in GaAs at 300 K. These results are depicted in Fig. 3.8. For the lower excitation-photon energy, the photoexcited electrons do not have sufficient energy to transfer to the L valley, so that the luminescence intensity near the bandgap energy rises quickly. For the higher photon energy, the electrons quickly transfer to the L valley and return slowly to the Γ valley because the density of final states for the return to the Γ valley is small. This leads to a relatively slow rise of the near-bandgap luminescence intensity for the higher photon energy. The comparison between the two curves shown in Fig. 3.8 clearly indicates the importance of intervalley scattering processes in GaAs under appropriate conditions.

This concludes our discussion of initial relaxation of photoexcited carriers in bulk semiconductors, with GaAs as the primary example. In the next section, we discuss measurements in semiconductor quantum wells.

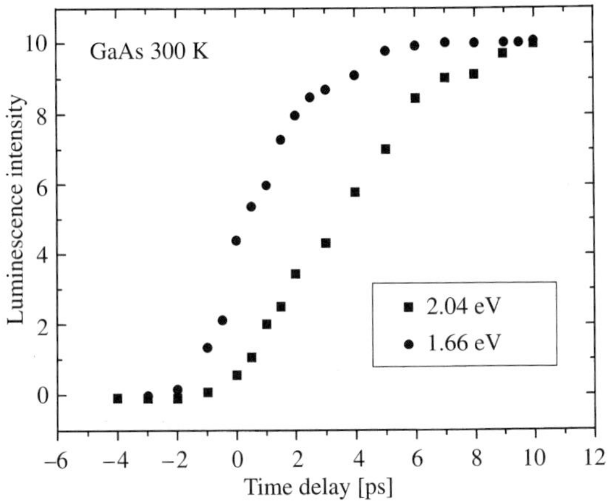

Fig. 3.8. Time evolution of the near-band gap luminescence intensity in GaAs at 300 K for two different excitation energies [3.66]

3.3 Initial Carrier Relaxation in Quantum Wells

Quantum confinement leads to formation of several subbands in the conduction and the valence bands, and also leads to confined phonon modes for the acoustic and optical phonons. These two changes introduce different considerations in the initial relaxation of photoexcited carriers.

The initial electron and hole distribution functions created by an ultrashort pulse can be rather different in quantum wells and bulk semiconductors. In a semiconductor like GaAs, photoexcitation at sufficiently high photon energies creates three electron and hole distribution functions, corresponding to interband transitions involving the heavy-hole, light-hole and split-off valence bands. This can lead to complications in the interpretation of pump-probe spectroscopy data, as discussed in Sect. 3.1.1. This point is illustrated by the calculations reported by *Stanton* et al. [3.67]. Assuming a laser pulse with a spectral width of 40 meV and assuming that no relaxation takes place during the laser pulse, they calculated energy distribution of the normalized electron density for electrons in GaAs at 300 K at the end of the laser pulse. Their results (Fig. 3.9 a) show three well-defined peaks corresponding to the three valence bands, as discussed above. Figure 3.9 b presents the results of a similar calculation for a 150 Å GaAs-AlGaAs quantum well at 300 K. A comparison of Fig. 3.9 a and b clearly shows that the optically-excited region is considerably broader in quantum wells than in the bulk. Such a broad initial electron-distribution function enormously complicates the analysis of the results in quantum wells at high excitation energies. While it is possible to obtain useful information from high-excitation-energy experiments in quantum wells [3.16, 18, 68], we will discuss here only the case of relatively low-energy excitations where the excitation is close to the fundamental bandgap so that only one or two subbands are excited in the conduction and valence bands.

We will discuss three different kinds of experiments: pump-and-probe transmission spectroscopy, time-resolved luminescence spectroscopy, and pump-and-probe Raman spectroscopy. The first two deal with the non-thermal carrier distribution functions in quantum wells, whereas the last concentrates on the inter-subband scattering of electrons in quantum wells.

3.3.1 Spectral Hole Burning in GaAs Quantum Wells

Pump-probe transmission experiments on undoped quantum wells have demonstrated the existence of non-thermal distribution functions and their relaxation to a hot, thermalized distribution function, whereas those in modulation-doped quantum wells have investigated the influence of a high density of carriers on this initial relaxation. We discuss these experiments in this section.

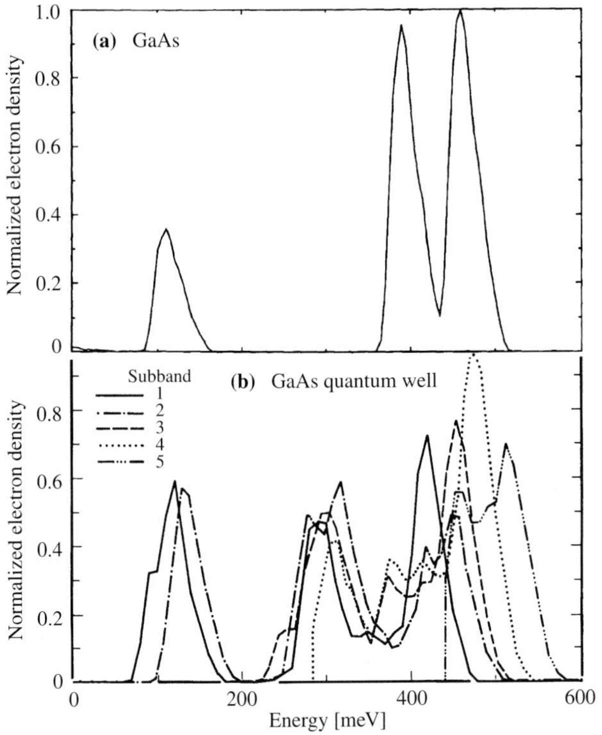

Fig. 3.9. (**a**) The initial normalized electron density as a function of electron energy for bulk GaAs excited at 2 eV, with three peaks corresponding to electrons excited from the heavy-hole, light-hole and spin split-off valence bands; (**b**) same as (**a**) but for a 150 Å GaAs quantum well with $x = 0.4$ AlGaAs barriers. The initial distribution has a large width because of the many subbands involved in the photoexcitation [3.67]

(a) Pump-and-Probe Transmission Spectroscopy in Undoped Quantum Wells. We first discuss the experiment on undoped quantum wells by *Knox* et al. [3.24] in which a multiple-quantum-well sample with 96 Å GaAs wells and 98 Å AlGaAs ($x = 0.3$) barriers was excited just above the lowest interband transition with a 100 fs laser pulse with a repetition rate of 8 kHz. A 50 fs broadband continuum was used to probe the transmission changes induced by this pump pulse. The sample was at 300 K and the excitation density was estimated to be 2×10^{10} cm^{-2}.

Figure 3.10 exhibits the differential transmission spectra $(T - T_0)/T_0$, where T and T_0 are the transmission of the sample with and without the pump pulse, for three different pump-photon energies, and zero delay between the pump and the probe pulses. The excitation is below the $n = 2$ transitions, but excites transitions from both the HH and the LH $n = 1$ valence subbands to the $n = 1$ electron subbands. The spectra show oscillatory signals close to the $n = 1$ HH and LH excitons resulting from bleaching, broadening and the shift

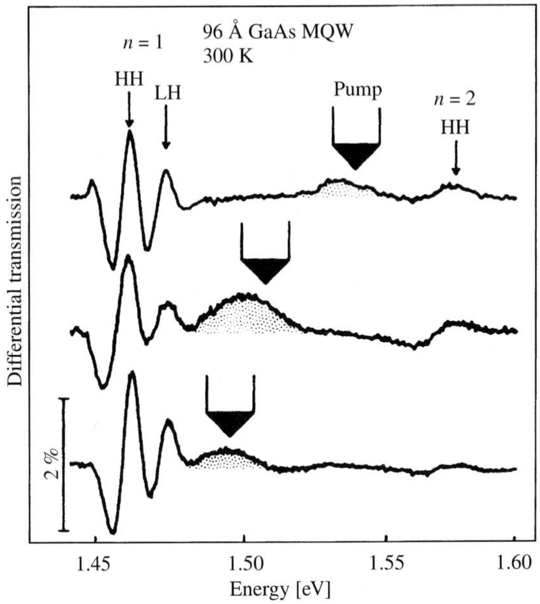

Fig. 3.10. Differential transmission spectra at nominally zero delay for three different excitation energies, showing that the spectral hole tracks the excitation photon energy. Data for a 65 period, 96 Å GaAs quantum well sample at 300 K at an excitation density of 2×10^{10} cm^{-2} [3.24]

of the exciton resonances. We concentrate here on the increased transmission peak observed in the continuum states. Figure 3.10 reveals that this peak moves with the laser-photon energy, demonstrating that this corresponds to a spectral hole in the absorption spectrum corresponding to the initial non-thermal electron and/or hole distribution functions. Note that the excess energy for all three cases shown in Fig. 3.10 is less than an optical phonon energy. This result reveals that there is a peak in the *sum* of the electron and the hole distribution functions, i.e., the distribution function for at least one carrier type (electron or hole) is peaked. As mentioned already, the non-thermal peaked distribution function is more likely for electrons than for holes because of the higher scattering rates for the holes. Therefore, these results are a clear demonstration that at least one carrier type has a non-thermal distribution function peaking close to the energy at which it is photoexcited. Finally, note also that the peak in the differential transmission spectrum occurs slightly below the center of the laser spectrum indicated by the arrows. This is similar to the observation by *Foing* et al. [3.22, 23] discussed above, but increased absorption above the laser energy was not observed in the quantum well case.

Figure. 3.11 exhibits that, with increasing time delay between the pump and the probe, the peak in the differential transmission spectrum due to hole-burning broadens and eventually develops into a Maxwell-Boltzmann-like tail peaking at the bandgap energy and decaying at higher energies. The transformation from a peaked non-thermal distribution function to a Maxwell-Boltzmann-like distribution function occurs in about 200 fs. This time is a measure of the carrier-carrier scattering rate responsible for the thermalization of the carrier-dis-

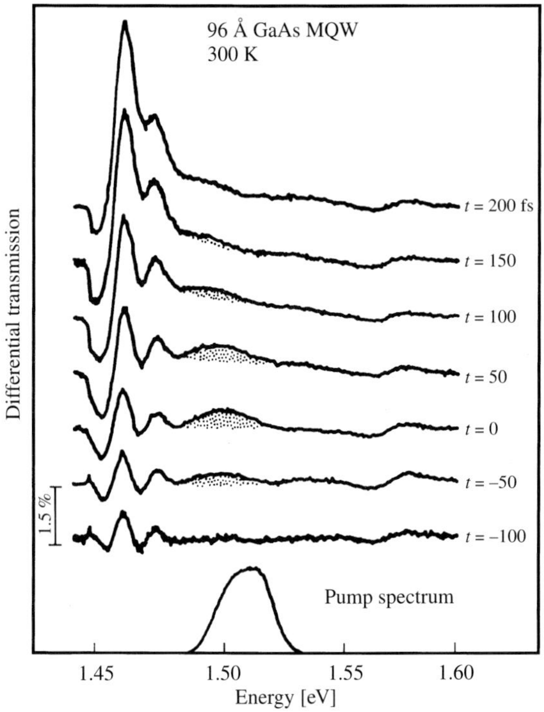

Fig. 3.11. Differential transmission spectra at various time delays for the GaAs quantum-well sample investigated in Fig. 3.10 for the laser pump at 1.507 eV as shown. The spectral hole evolves into a thermal tail in less than 200 fs [3.24]

tribution function. Figure 3.12a shows that the spectral hole disappears in less than 100 fs at an excitation density of 5×10^{11} cm^{-2}. The approach to a Maxwell-Boltzmann-like distribution function is much faster in this case since the density is about a factor of 25 higher than the density at which Fig. 3.11 was obtained.

(b) Pump-and-Probe Transmission Spectroscopy in Modulation-Doped Quantum Wells. Comparison of undoped and modulation-doped quantum wells provide a means of isolating the effects of electron-electron, electron-hole and hole-hole scattering mechanisms. Both n- and p-modulation-doped quantum wells were investigated by *Knox* et al. [3.69, 70] at room temperature. Figure 3.12b shows the results of pump-probe differential transmission measurements for a 120 Å GaAs quantum well sample with n-modulation-doping density of 3×10^{11} cm^{-2}, room-temperature electron mobility of 2000 cm^2/V s, and for a photoexcited carrier density also equal to 3×10^{11} cm^{-2}. The results reveal no indication of a spectral hole even at the earliest time. *Knox* et al. estimated that the relaxation time for the spectral hole was <10 fs in this case. Figure 3.12c displays similar results for a 110 Å GaAs quantum-well sample with p-modulation-doping density of 2×10^{11} cm^{-2} with a room-temperature hole mobility of 200 cm^2/V s. The differential transmission spectra are quite complex, showing both positive and negative values. The re-

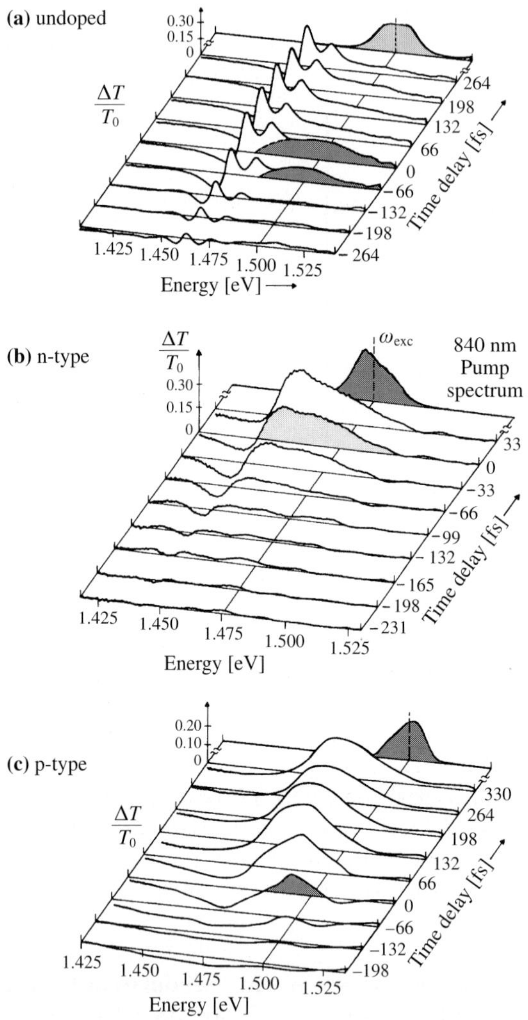

Fig. 3.12. Differential transmission spectra at various time delays for **(a)** undoped, **(b)** n- and **(c)** p-modulation-doped GaAs quantum wells at 300 K for sample parameters and excitation densities discussed in the text [3.69]

sults were attributed to a broadening and a shift of the Fermi edge whose origin was not determined. The relaxation to a thermalized (Maxwell-Boltzmann-like) distribution function occurred in about 60 fs.

These results indicate that the presence of a cold-electron or hole gas has a profound influence on the thermalization of photoexcited carriers, and leads to different thermalization rates under different conditions. It is reasonable to expect that cold plasma will influence the thermalization rates, but a quantitative understanding of the behavior is not yet available. It should be emphasized that, at 300 K, the electron and hole plasmas investigated above were non-degenerate, and an entirely different behavior may be expected at low temperatures (see, for example, the FWM results on modulation-doped quantum wells discussed in Sects. 2.10.4, 5). Detailed temperature-dependent measurements may be interesting. Note also that this problem of relaxation in the presence of cold plasma has been investigated by using cw electron-acceptor luminescence spectroscopy [3.71, 72]. Finally, it is interesting to note that electron-hole momentum scattering times have also been estimated from the observation and analysis of negative absolute mobilities in GaAs quantum wells [3.73–75], and carrier-carrier scattering has been investigated using numerical techniques such as Monte-Carlo simulations [3.76, 77].

3.3.2 Non-thermal Holes in n-Modulation-Doped Quantum Wells

The pump-probe transmission measurements in modulation-doped quantum wells discussed above were performed at 300 K. Under these conditions, both the n- and p-modulation-doped samples are non-degenerate so that there are empty states available even at the bottom of the conduction band or the top of the valence band. A different physical situation exists when the temperature is lowered to less than 10 K. In this case the carriers are degenerate so that the states between the band extremum and the Fermi energy are essentially filled. This puts severe restrictions on the available final states for certain scattering events. One consequence of this is the long dephasing time [3.78] for electrons observed in the four-wave-mixing experiments discussed in Chap. 2. Such degenerate plasma also provides an opportunity to investigate the initial relaxation dynamics of a specific type of carrier (electron or hole) using the complementary technique of ultrafast luminescence. The idea of using modulation-doped semiconductors to investigate relaxation processes of electrons and holes separately was first used for studying carrier-phonon interactions, and is discussed in Chap. 4.

As discussed in Chap. 1, pump-probe transmission spectroscopy is generally dominated by electrons and it is difficult to obtain information about the initial relaxation of holes in semiconductors. Luminescence spectroscopy, on the other hand, is not well-suited to investigate non-thermal dynamics at the excitation photon energy. *Tomita* et al. [3.79] have reported an experiment designed specifically to investigate the initial hole dynamics using luminescence spectroscopy. The idea of the experiment is to use an n-modulation-doped quantum well sample at low temperatures so that all states below the Fermi energy are filled, and excite it with an ultrashort laser with the following restrictions: (1) the excitation density is small compared to the doping density, and (2) the excitation photon energy is close to the Fermi energy. These conditions assure that the average excess energy imparted to the electron gas by the photoexcitation is small compared to the Fermi energy. Under these conditions, the change in the electron distribution function below the Fermi energy caused by photoexcitation may be sufficiently small that luminescence dynamics provides direct information about holes.

Tomita et al. [3.79] investigated an n-modulation-doped 60 period multiple-quantum-well sample with an electron density of 5×10^{11} cm^{-2} corresponding to a Fermi energy of 18 meV, and with a quantum-well width of 50 Å. The sample was weakly excited with <200 fs pulses from a Ti:Sapphire laser at 1.78 eV at a density of 1×10^{10} cm^{-2} or lower. Even assuming that all the excess energy was dissipated in the electron gas and there was no transfer to the lattice, the electron temperature was estimated to increase to <90 K. Since kT at this temperature is much less than the Fermi energy, this results in a <5% change in the electron distribution function near the bottom of the conduction band. Thus interband luminescence involving these electrons directly reflect the distribution function of holes at the corresponding wavevectors.

The time-resolved luminescence spectra measured under these conditions could be fitted to a thermalized distribution function of holes at times as early as 100 fs after the pulse. However, the measured spectrum corresponded to only a small range of hole energies and could not determine the hole distribution function at higher energies. In order to obtain further information about the hole distribution function, *Tomita* et al. also measured the time evolution of the luminescence intensity at the band gap energy, which is directly proportional to the number of holes near the top of the valence band. These results are shown in Fig. 3.13a for the first 1000 fs and in Fig. 3.13b for the

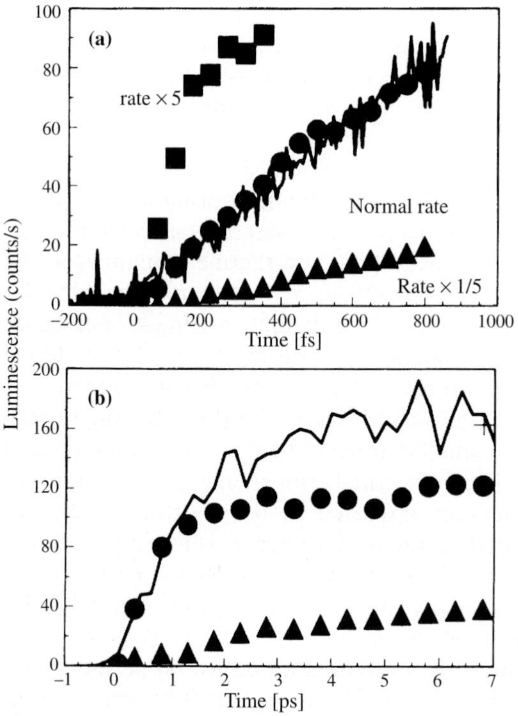

Fig. 3.13. (**a**) Femtosecond and (**b**) picosecond time evolution of the luminescence intensity at the bandgap of n-modulation-doped 50 Å GaAs quantum well at 10 K. Solid curves are experimental results; Monte Carlo results for normal, five times larger and five times smaller scattering rates are represented by circles, squares and triangles in (**a**). The triangles in (**b**) show the effect of turning off electron-hole scattering in Monte-Carlo simulations [3.79]

first 7 ps. These results indicate that there is a large (nearly a factor of 7) increase in the intensity between the end of the laser pulse and 1000 fs thereafter. It was argued that such a large increase could not be accounted for simply by the cooling of a thermalized hole distribution function, and that a non-thermal hole distribution function, extending to energies higher than those accessible by the spectral measurements, was essential to explain the results. This was supported by extensive Monte-Carlo simulations. Figure 3.13 also presents comparison of the data with Monte-Carlo simulations using different parameters. The comparison in (a) shows that the normal carrier-carrier scattering rates provide the best fit to the results, and the comparison in (b) indicates that turning the electron-hole scattering is the dominant scattering and thermalization process for the photoexcited holes. If it is assumed that all the holes are thermalized at 1000 fs, then detailed analysis yields that 33%, 67% and 80% of the holes are thermalized at 100, 300 and 800 fs from the center of the laser pulse.

While these first results on the thermalization of holes in GaAs quantum wells provide important information, the inability to measure the spectrum corresponding to high hole energies precludes a direct determination of the non-thermal hole distribution function. It will be worthwhile to explore different approaches to obtain direct information on the non-thermal hole distribution function. In closing it should be mentioned that hole thermalization has been investigated in bulk GaAs [3.80, 81] and Ge [3.82].

3.3.3 Intersubband Scattering in Quantum Wells

Electron-scattering processes between subbands have been investigated by a number of ultrafast spectroscopy techniques, but scattering of holes between subbands has remained largely unexplored because of the complexity of the valence subbands. The basic idea is to probe the distribution function of carriers in different subbands. While interband spectroscopy can provide such information, it is complicated by the fact that interband transitions involve both electrons and holes. A better approach would be to monitor the dynamics of inter-subband absorption or inter-subband electronic Raman scattering following either interband or inter-subband excitation by a pump pulse. The probe pulse for inter-subband absorption measurements needs to be in the mid- to far-infrared region of the spectrum because typical subband spacings are in the 10 – 100 meV range. Visible or near infrared lasers can be used for the Raman measurements.

(a) Pump-Probe Absorption Spectroscopy. Conflicting values for the inter-subband scattering rates for electrons in GaAs quantum wells have been reported in the literature. *Seilmeier* et al. [3.83] investigated an n-modulation-doped quantum-well sample (well width: 50 Å; doping density; $\approx 5 \times 10^{11}$ cm^{-2}). They excited the inter-subband transition from subband 1 to 2 (with an energy separation of 150 meV) by a picosecond infrared pulse and monitored the return of the excited carriers to subband 1 by measuring the absorption of a weak probe pulse at the same photon energy as a function of the delay between the pump and the probe. They found relaxation times between 10 and 15 ps, much larger than the calculated values of 0.5 ps [3.84, 85]. The high doping leads to band bending which facilitates transfer of electrons excited to the upper band into the barriers. It was later suggested [3.86–88] that the return of the carriers from the barrier to the quantum well may have contributed to the long times obtained in these measurements. More recent time-resolved absorption measurements [3.89] found a $2 \rightarrow 1$ relaxation time of less than 12 ps for 120 Å wells. Cooling of photoexcited hot carriers (to be discussed in Chap. 4) clearly play an important role in the dynamics of interband and inter-subband absorption and must be considered in analyzing such results. For example, results by *Elsaesser* et al. [3.90] show a time-dependent broadening of inter-subband absorption spectrum in InGaAs quantum wells caused by hot carrier effects. *Bäuerle* et al. [3.91] have investigated inter-subband electron relaxation in n-modulation-doped 82 Å $In_{0.53}Ga_{0.47}As/In_{0.52}Al_{0.48}As$ quantum wells which have large confinement barriers. Using four different pump-probe schemes, they deduced that the inter-subband relaxation time at 10 K is less than 3 ps, more in agreement with theoretical expectations. The inter-subband electron relaxation rates have also been investigated using pump-probe Raman spectroscopy, discussed in the next section.

(b) Pump-Probe Raman Spectroscopy. Raman scattering provides a powerful technique for investigating ultrafast dynamics of phonons and inter-subband transitions. We recall from Chap. 1 that such measurements are a form of pump-probe spectroscopy: the modes of interest are excited by a strong pump pulse, and probed by measuring the Raman signal from a weak probe pulse which is variably delayed with respect to the pump pulse. Chapter 4 discusses how relaxation and cooling of photoexcited plasma produces a large population of nonequilibrium phonons (hot phonons) and Chap. 5 explains how time-resolved Raman scattering is used to obtain considerable information about hot phonon dynamics in semiconductors. In this section we discuss how time-resolved *electronic* Raman scattering is employed to obtain information about inter-subband scattering rates.

Electronic Raman scattering in semiconductors can probe both single-particle excitations and the collective oscillations of a plasma [3.92, 93]. In quantum wells, these excitations have both inter-subband and intra-subband components. Experimentally one can measure both the polarized spectra $z(x', x')\bar{z}$ and depolarized spectra $z(x', y')\bar{z}$. Here z is the direction of the incident beam along the $z = (001)$ growth direction, $\bar{z} = (00\bar{1})$ is the direction of the back-scattered beam, and the symbols in the bracket refer to the polarization of the incident and the scattered light respectively with $x' = (110)$ and $y' = (1\bar{1}0)$. Single-particle excitations are observed in the depolarized spectra induced by the spin-density fluctuations whereas the collective oscillations are observed in the polarized spectra induced by the charge density fluctuations [3.92, 93]. The pump and probe energies are usually identical and chosen so as to make use of the resonant enhancement of the Raman signal.

Oberli et al. [3.94] investigated *single-particle Stokes* scattering in undoped quantum wells at low temperatures for quantum-well thickness of 116 and 215 Å, corresponding to $E_{21} > \hbar\omega_{LO}$ and $E_{21} < \hbar\omega_{LO}$, where E_{21} is the energy difference between electron subbands 1 and 2, and $\hbar\omega_{LO}$ is the optical phonon energy. A strong pump pulse near 1.9 eV excited a relatively high density ($\approx 2 \times 10^{11}$ cm^{-2}) of electrons from the various HH and LH subbands as well as from the spin split-off subbands into various conduction subbands. A suitably delayed probe pulse of equal intensity measured the Stokes Raman signal corresponding to either E_{12} or E_{23} single particle excitation. The former monitors the population difference between subbands 1 and 2 whereas the latter monitors the population difference between subbands 2 and 3. For the 116 Å well, the results showed scattering time τ_{21} (for transition from subband 2 to 1) of less than 10 ps, the experimental time resolution. However, for the 215 Å well, the population in the second subband persisted for several hundred picoseconds. This was deduced from the long rise time of the E_{12} single-particle Stokes Raman signal (where the rise time was attributed to the emptying of the second subband), and also from the decay time of the E_{23} single particle Stokes Raman signal (where the decay was attributed to the decay of the population in the second subband). The time evolution of the single-particle E_{23} scattering is depicted in Fig. 3.14. The least square fit to the data results in a lifetime of 325 ± 25 ps for the population of the second subband. This was explained by invoking the fact that $E_{21} < \hbar\omega_{LO}$ for the 215 Å quantum well so that optical phonon emission is not allowed and the transitions must proceed by scattering with acoustic phonons and other electrons or impurities. In contrast, $E_{12} > \hbar\omega_{LO}$ for the 116 Å quantum well so that emission of the

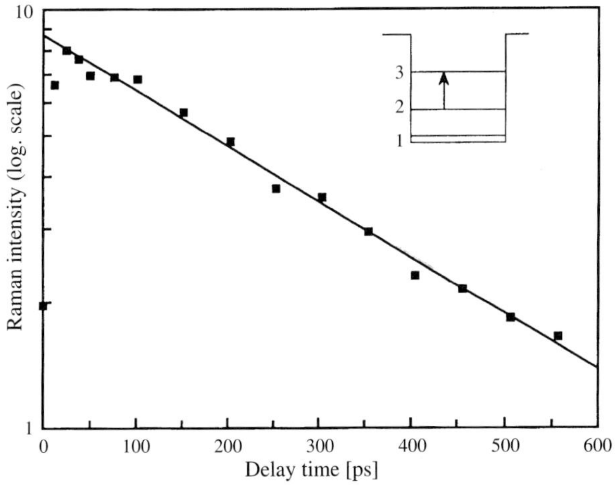

Fig. 3.14. Stokes Raman intensity of the single particle inter-subband excitation as a function of time delay between the pump and the probe beams, for the 215 Å GaAs quantum well sample at 5 K [3.94]

LO phonons makes the population lifetime in the second subband <10 ps, limited by the time resolution of the experiment.

An alternative interpretation of the results was proposed later by *Goodnick* and *Lugli* [3.86–88] based on their Monte-Carlo simulations. As noted above, the pump pulse excites electrons in various subbands. As we have seen above, electron-electron collisions thermalize these electrons to a hot electron gas in times less than 1 ps at these high densities. This hot-electron gas cools slowly because of hot-phonon effects (Chap. 4) and maintains a high population of electrons in the second subband for a long time. According to this interpretation, the single-particle Stokes Raman scattering probes this thermalized population, and the observed long decay time (325 ps) for E_{23} is due to the long cooling time of the hot-electron gas rather than due to long inter-subband scattering time. Further experiments will be necessary to resolve this satisfactorily.

Tatham et al. [3.95] have investigated time-resolved *anti-Stokes scattering from collective modes* in undoped GaAs/AlGaAs quantum wells (well thickness of 146 Å) at 10 K. The excitation photon energy was in resonance with the $n = 4$ interband optical excitation. The excitation density was $\approx 10^{10}$ cm^{-2}, yet very good signal-to-noise ratio was obtained with the use of a multichannel detector. The low excitation density is essential in eliminating various complicating effects that appear at high densities. Figure 3.15 exhibits the depolarized $(z(x', y')\bar{z})$ anti-Stokes Raman spectra in the energy range 25–70 meV for five different time delays. The strong peak close to 37 meV is due to anti-Stokes scattering from the GaAs-like LO phonon, and there is also a broad luminescence background. The second feature at 51 meV in the anti-Stokes spectrum in Fig. 3.15 (marked C_2-C_1) corresponds to the anti-Stokes

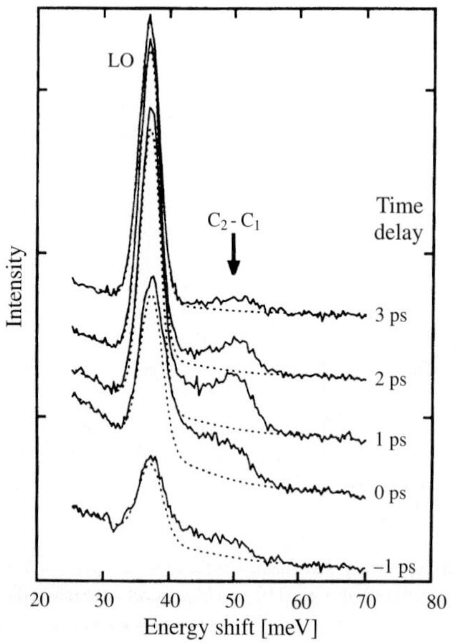

Fig. 3.15. Time-resolved anti-Stokes Raman spectra from a 146 Å GaAs quantum well at the $n = 4$ resonance, showing the LO phonon peak at 37 meV and the collective oscillation peak at 51 meV. The *dotted curves* are fits to the LO phonon and the broad luminescence background [3.95]

scattering from the collective modes corresponding to the transition from subband 2 to 1. *Tatham* et al. have argued that at low densities the single particle and the collective modes are coupled by Landau damping of the collective plasmon modes, so that the measured dynamics of the mode at 51 meV reflects the dynamics of the population of the $n = 2$ electron subband.

Figure 3.16 displays the anti-Stokes intensities of the nonequilibrium LO phonons and the inter-subband $2 \to 1$ transition as a function of time delay between the pump and the probe pulse. Also shown is the autocorrelation profile of the laser pulse which gives the instrumental time resolution. The inter-subband intensity reaches a peak at about 0.5 ps after the center of the excitation pulse and then decreases with a time constant of 1 ps. Taking into account the instrumental resolution, the decay constant for the population of subband 2 was deduced to be somewhat smaller than 1 ps. The LO phonon population decays with a time constant of 5.3 ps, somewhat smaller than in the bulk.

These results have been analyzed in terms of slab and confined phonon modes, with potentials given by $\phi_s = \sin(k_z z)$ and $\phi_c = \cos(k_z z)$, respectively (Chap. 5). From the expressions for the inter-subband scattering rates for these potentials have been given by different researchers [3.96–98], it can be concluded that the two rates differ by not more than a factor of 2. Therefore, these experiments can not distinguish between the different phonon models. However, *Tatham* et al. [3.95] used a rate equation analysis and found the characteristic exponential decay time to be less than 1 ps. This improves the limit of < 10 ps found by *Oberli* et al. [3.94]. The results clearly indicate that inter-subband relaxation rate, for subband energy spacing larger than the optical phonon energy, is faster than 1/ps. Unfortunately, these measurements have

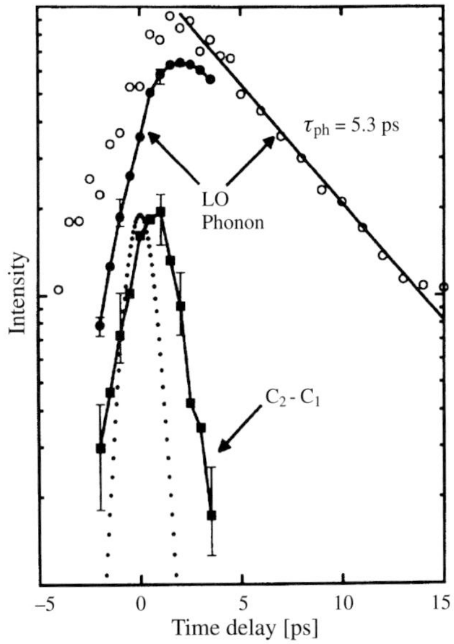

Fig. 3.16. Anti-Stokes Raman intensities of non-equilibrium LO phonons and C_2-C_1 inter-subband excitations in a 146 Å GaAs quantum well at the $n = 4$ resonance. Also shown is the laser autocorrelation trace which gives the system time resolution [3.95]

not yet been repeated for thicker wells where the inter-subband spacing is smaller than an optical phonon energy.

3.4 Summary and Conclusions

We have discussed some of the processes that contribute to the relaxation of photoexcited carriers after the initial coherence is lost. These include hole-burning in bulk as well as undoped and modulation-doped quantum wells, intervalley scattering, inter-subband relaxation, non-thermal holes in n-modulation-doped quantum wells, and the extremely rapid redistribution of electrons in the Γ valley of GaAs. These are interesting dynamical processes occurring on femtosecond time scale in semiconductors and their nanostructures. We conclude this chapter by noting that there are other equally interesting phenomena that we have not discussed; e.g., intervalley scattering in type-II superlattices, and spectral hole burning in semiconductor lasers.

4. Cooling of Hot Carriers

As discussed in Chap. 3, the non-thermal, photoexcited electrons and holes interact among themselves and other carriers present in the semiconductor. They achieve a distribution function characterized by a temperature in times of the order of 100 fs for the typical densities present in most femtosecond experiments. Initially, the temperatures of the electrons and the holes may be different, but a common temperature T_c for the carrier system is typically achieved in times of the order of a picosecond. The thermalized carriers are hot, i.e., their distribution function is characterized by a temperature T_c higher than the lattice temperature T_L. Although some of the energy of the electronic system is lost to the lattice via carrier-phonon interactions during this thermalization process, most of the energy typically remains within the electronic system so that the non-thermal regime provides information primarily about carrier-carrier interactions. The next phase of the relaxation occurs as the thermalized, hot electron-hole distributions cool and approach the lattice temperature. Much of our information about the carrier energy loss processes to the lattice comes from a study of this cooling process.

Carrier-phonon interaction is one of the most fundamental scattering processes in semiconductors and plays a role in many different aspects of semiconductor physics. High-field transport is one such area, and investigation of high-field transport has provided considerable information about carrier-phonon scattering processes and rates [4.1–6]. Similarly, cw photoexcitation has been shown to create a steady-state hot-carrier distribution [4.7–11]. A study of carrier temperature as a function of excitation intensity provides some information about the important scattering processes but is unable to provide quantitative information on scattering rates. We discuss in this chapter how ultrafast spectroscopy on picosecond time scales has been used to explore this aspect of semiconductor physics. We note that, although the process by which carriers gain energy is different for applied electric fields and photoexcitation, the process by which they lose energy to the lattice (i.e., by the emission of phonons) is the same in both cases. Therefore, information gained by optical studies is useful in understanding high-field transport, and vice versa. The primary advantage of the optical techniques is that they not only provide a means of generating a non-equilibrium carrier distribution, they also provide the best technique for measuring these distribution functions. In contrast, the transport measurements provide information that is averaged over the distribution function.

The basic concept behind studies involving cw photoexcitation is that the rate of transfer of energy from photons to the electronic system must be equal to the rate of transfer of energy from the electronic system to the lattice under steady-state condition. The first rate is not well determined because the initial excess energy is distributed among various excitations in different ways depending on the experimental conditions. One technique to avoid this is to use an electric field to heat the carrier system and optical techniques to measure the carrier-distribution function at various electric fields to quantitatively deduce energy-loss rates. This technique has been successfully applied to the case of quantum wells [4.12]. Another technique is to photoexcite with a short laser pulse and measure the *rate of change* of the carrier temperature. In contrast to the case of cw photoexcitation, this approach does not require the knowledge of the fraction of the initial excess energy given to the electronic system.

We discuss in Sect. 4.1 a simple theory of carrier-energy-loss rate and how *cooling curves* (T_c as a function of time following pulsed excitation) can provide quantitative information about carrier energy-loss rate to the lattice. We then outline some early experimental investigations (Sect. 4.2) and how other factors such as degeneracy, screening and hot-phonon effects influence the cooling curves (Sect. 4.3). More recent experimental results are discussed in Sect. 4.4, and effects in bulk and quasi-2D semiconductors are compared in Sect. 4.4.2.

4.1 Simple Model of Carrier Energy-Loss Rates and Cooling Curves

Expressions for the energy-loss rate $\mathrm{d}E/\mathrm{d}t$ of an electron or a hole of a given energy E by various phonons can easily be derived using the Fermi golden rule and are well-known for GaAs-like semiconductors and their nanostructures [4.1–6, 10, 13, 14]. The average energy-loss rate to the lattice per carrier, $\langle \mathrm{d}E/\mathrm{d}t \rangle$, is given by

$$\left\langle \frac{\mathrm{d}E}{\mathrm{d}t} \right\rangle = \frac{\int_0^\infty \frac{\mathrm{d}E}{\mathrm{d}t} f(E) g(E) \mathrm{d}E}{\int_0^\infty f(E) g(E) \mathrm{d}E} , \tag{4.1}$$

where $f(E)$ is the carrier-distribution function, and $g(E)$ is the density of states.

Let us consider a simple model which assumes that (1) the energy bands are parabolic, (2) the carriers have a Maxwell-Boltzmann distribution function characterized by the temperature $T_c > T_L$ (the lattice temperature), (3) the carrier density is high enough to establish the Maxwell-Boltzmann carrier distribution, but low enough that screening is not important, (4) the carrier population is sufficiently low so that the effect of Pauli exclusion principle is negligible; i.e., the occupation probability is $\ll 1$ for all states, and (5) the phonon generation and decay rates are such that phonon occupation remains low, and the phonon occupation number is given by the Bose-Einstein occupation number for the lattice temperature T_L. This simple model of energy-loss rate is very useful for the understanding experimental results; we will refer to this simple model quite frequently in our discussion. Analytical expressions for the energy-loss rate can be obtained for this simple model and have been given in the literature [4.1, 2, 10, 14, 15].

These expressions show that, at low temperatures, the energy-loss rate is small and determined by the deformation potential and piezoelectric interaction with acoustic phonons. There is a strong mass dependence to the deformation potential scattering so that the heavier hole dominates for the case of GaAs. For temperatures above about 40 K, optical phonon scattering becomes dominant. Only polar optical (Fröhlich) interaction is important for the s-like electrons in the Γ but both polar and non-polar optical phonon scattering is important for holes and electrons in other valleys. Considering all these factors, and the reduction in the hole-phonon scattering rate by a factor of 2 resulting from complicated valence band structure in GaAs [4.16, 17], it can be shown that hole-phonon interaction is about a factor of 2.5 stronger than the electron-phonon interaction in GaAs. The calculated energy-loss rate to the lattice per electron-hole pair for a thermalized distribution of electrons and holes at temperature T_c for $T_L = 2$ K [4.10, 18, 19] is depicted in Fig. 4.1 for GaAs for the simple model discussed above.

Energy loss to the lattice by interaction with polar optical phonons has received considerable attention since this mechanism dominates for electrons at $T_c > 40$ K in GaAs. Many early studies were analyzed in terms of only this mechanism. If $T_L \approx 2$ K and $T_c \gg T_L$, the average energy-loss rate to the lattice (per electron-hole pair) by polar optical scattering mechanism can be obtained by averaging (4.1) over the carrier-distribution function [4.1]:

$$\left\langle \frac{\mathrm{d}E}{\mathrm{d}t} \right\rangle = (\hbar\omega_{\mathrm{LO}}/\tau_{\mathrm{LO}}) \exp(-\hbar\omega_{\mathrm{LO}}/kT_c) \ , \tag{4.2}$$

where $\hbar\omega_{\mathrm{LO}}$ is the LO phonon energy. $1/\tau_{\mathrm{LO}}$ is a characteristic rate for polar optical phonon scattering and is given by (1.2)

$$(\tau_{\mathrm{LO}}^{3\mathrm{D}})^{-1} = \frac{e^2\sqrt{2m\hbar\omega_{\mathrm{LO}}}}{4\pi\hbar^2}\left(\frac{1}{\varepsilon_\infty} - \frac{1}{\varepsilon_\mathrm{s}}\right) \equiv W_0 \tag{4.3}$$

for bulk semiconductors, and

$$\tau_{\mathrm{LO}}^{2\mathrm{D}} = (2/\pi)\,\tau_{\mathrm{LO}}^{3\mathrm{D}} \tag{4.4}$$

for 2D semiconductors [4.20]. Here m is the carrier mass, ε_∞ and ε_s are the optical and static dielectric permittivities, related to the optical and static di-

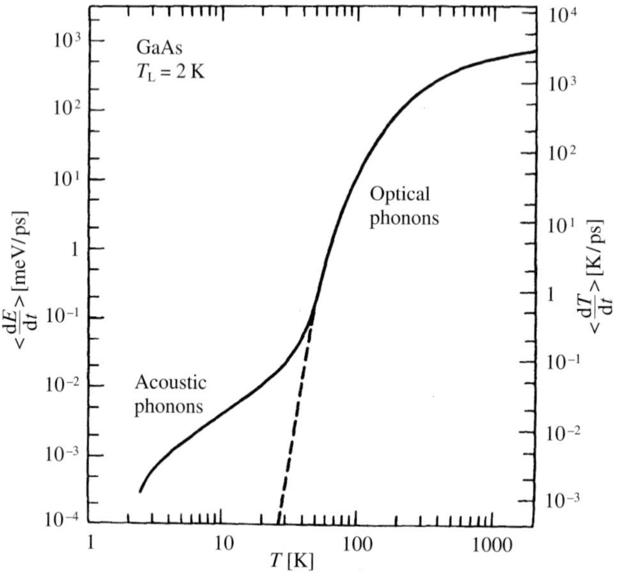

Fig. 4.1. Calculated average energy-loss rate per electron-hole pair for a Maxwellian plasma in GaAs at $T_L = 2$ K. Acoustic deformation-potential scattering (primarily for holes) dominates at low temperatures, whereas optical-phonon scattering dominates at higher carrier temperatures. The *dashed curve* shows the optical-phonon contribution to the energy-loss rate at low temperatures [4.10]

electric constants (K_∞ and K_s, respectively) by the free-space permittivity ε_0 ($\varepsilon_\infty = \varepsilon_0 K_\infty$ and $\varepsilon_s = \varepsilon_0 K_s$). The frequencies of the LO and the TO phonons are related by the Lyddane-Sachs-Teller relationship

$$(\omega_{LO}/\omega_{TO})^2 = \varepsilon_s/\varepsilon_\infty \ . \tag{4.5}$$

For electrons in GaAs, $\tau_{LO}^{3D} = 130$ fs and $\tau_{LO}^{2D} = 80$ fs [4.20, 21]. W_0 is not equal to the emission rate/absorption for a phonon at any carrier energy, but is related to it by a $\sinh^{-1}$ function of the carrier energy and phonon occupation number. $W_0 = 2\alpha\omega_{LO}$, where α is the polaron coefficient for the semiconductor (= 0.088 for GaAs, and 0.127 for InP). Note that (4.2) applies only under the very restrictive conditions of the simple model described above. One must consider more complex expressions or resort to a numerical calculation of the energy-loss rate when the assumptions of the model are violated.

The first measurements to explore the energy-loss rate of photoexcited carriers [4.7] were performed using cw excitation. These measurements established that electrons and holes generated by cw photoexcitation can be characterized by a temperature higher than the lattice temperature. Since the power input from photoexcitation must equal the power loss to the lattice in steady-state conditions, these measurements also allowed a plot of the inverse of the carrier temperature as a function of the photoexcitation intensity. The slope of such a plot was found to be equal to $\hbar\omega_{LO}$, thus showing by comparison with (4.2) that the LO phonons play an important role in the energy relaxation process. While this provided important qualitative insights into the relaxation processes of photoexcited carriers in semiconductors, it was difficult to obtain *quantitative* information about energy-loss rates because it was difficult to determine the optical power input to the electronic system.

Time-resolved spectroscopy provides a means of obtaining quantitative information about interaction of carriers with the lattice by measuring the "cooling curves" (the temperature of the carriers as a function of time following photoexcitation), and comparing them with calculated cooling curves. Cooling curves can be calculated from the expressions for the energy-loss rate in a straightforward manner simply by equating the average rate of loss of energy of an electron-hole pair to the energy-loss rate per electron-hole pair to the lattice,

$$\mathrm{d}\langle E\rangle/\mathrm{d}t = \langle \mathrm{d}E/\mathrm{d}t\rangle \tag{4.6}$$

so that

$$\int_0^t \mathrm{d}t = \int_{T_0}^{T_c} \mathrm{d}T \left(\frac{\mathrm{d}\langle E\rangle/\mathrm{d}T}{\langle \mathrm{d}E/\mathrm{d}t\rangle} \right), \tag{4.7}$$

where T_0 and T_c are the carrier temperatures at time $t = 0$ and $t = t$, respectively. Both the energy-loss rate and the cooling curve have to be calculated by numerical integration in the general case. Figure 4.2 presents the calculated cooling curves for two different T_0 for the simple model discussed above for which the energy-loss rate are shown in Fig. 4.1 and for which $\langle E\rangle = 3kT$. T_0 is determined by the excess energy per electron-hole pair given to the carrier system by the photoexcitation. The cooling curve is initially faster for higher T_0, but the temperatures and the cooling curve become nearly identical for different T_0 at longer times. C_0 is an empirical parameter that multiplies the calculated energy-loss rate (Fig. 4.1); e.g., $C_0 = 0.1$ means that the calculated energy-loss rate is reduced by a factor of 10. Figure 4.2 shows that a change in C_0 is equivalent to shifting the curve along the time axis on a semilog plot. A simple way to determine the energy-loss rate is to compare experimental cooling curves with calculations using C_0 as an empirically adjustable parameter. However, a better approach would be to calculate the energy-loss rate for the physical situation relevant to the experiment, rather than using the simplified energy-loss rate as in Fig. 4.2. Note that one can make an absolute comparison of data with the calculated cooling curves (with T_0 determined as discussed above), or one can use an experimentally determined T_0 at any given delay t_0 and compare data at other delays with calculations using appropriate modifications of (4.2). The latter approach eliminates the need to know the fraction of the excess photoexcited energy given to the electronic system.

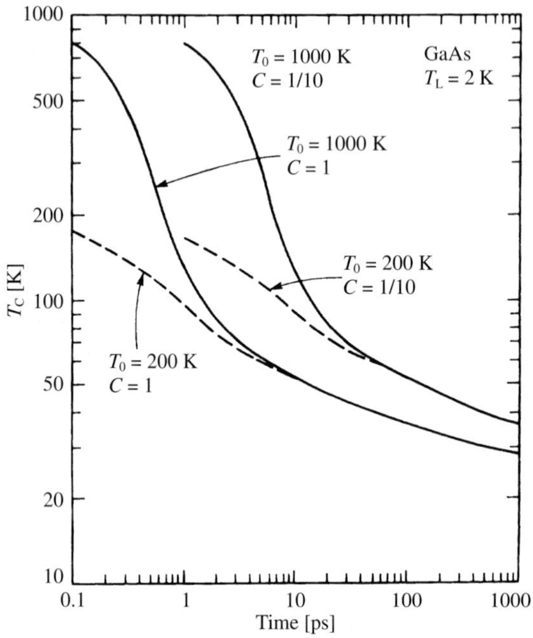

Fig. 4.2. Calculated cooling curves, variation of the carrier temperature T_c as a function of time, for photoexcited electron-hole plasma in GaAs at 2 K with the initial temperatures $T_0 = 1000$ K and 200 K, and for two values of the carrier-phonon coupling parameter C [4.10]

4.2 Cooling Curves: Early Measurements and Analysis

This section presents some early experimental results and analyzes them using the simple model discussed above. This discussion allows us to draw some important conclusions and naturally leads into a discussion of a more realistic model in the subsequent sections.

The first demonstration that photoexcitation led to a hot-carrier distribution characterized by a temperature higher than the lattice temperature was performed employing a cw argon laser to excite bulk GaAs at 2 K [4.7]. These measurements also showed that optical-phonon scattering is the dominant energy-loss process for photoexcited plasma in the temperature range investigated. Following this initial study, such cw photoexcitation studies were extended to higher excitation densities [4.22], other semiconductors [4.7, 23–25], and ternary semiconductors [4.25] with complicated phonon behavior [4.26]. Single-particle electronic Raman scattering was also used for studying photoexcited and electric field-heated carriers in GaAs [4.27, 28]. Electric field-induced changes in the luminescence spectra were also explored [4.29, 30]. The energy-loss rate for acoustic phonon scattering was investigated by using free-electron acceptor luminescence [4.29, 30] and by using Raman scattering [4.31]. The dependence of hot-carrier temperature on excitation photon energy was also determined [4.32]. Heating in semiconductor laser diodes was investigated by analyzing the spontaneous emission spectrum for various current injections [4.33]. Considerable information about the exchange of energy between photoexcited electrons and holes, and the background carriers, as well as about the processes responsible for the exchange of energy between carriers and the lattice was obtained from such studies [4.34]. However, quantitative information about the energy-loss rate was available only after the first time-resolved measurements were performed nearly 10 years after the initial cw photoexcitation studies [4.35, 36].

Shank et al. [4.36] and *Leheny* et al. [4.35] investigated a 1.5 μm thick GaAs crystal sandwiched between two AlGaAs layers. The sample was maintained at 10 K and excited with 0.5 ps pulses from a 6150 Å amplified CPM laser Raman shifted to 7500 Å. The excitation wavelength (7500 Å) was chosen so that the AlGaAs-cladding layers were not excited. Pump-probe transmission spectroscopy was utilized to measure the Differential Transmission Spectrum (DTS) at various time delays following the excitation pulse. The probe was a filtered (8000–8200 Å) output from the 0.5 ps white light continuum

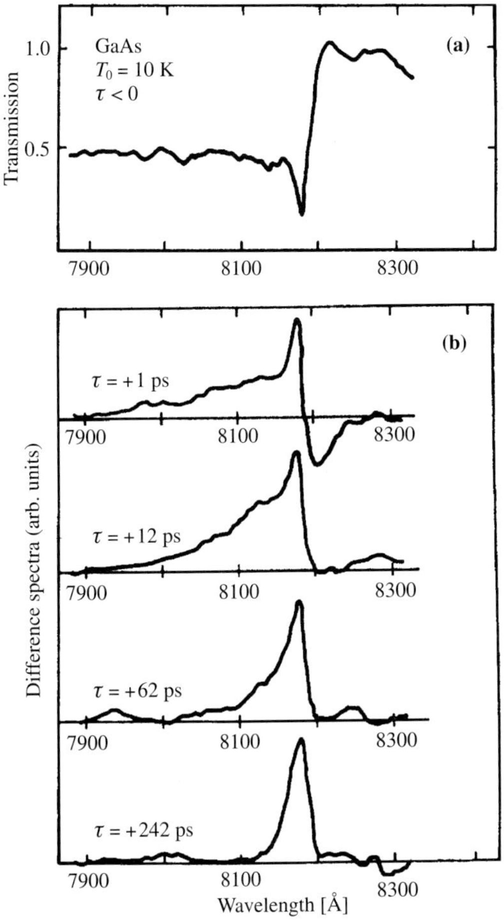

Fig. 4.3. (**a**) The transmission spectrum of unperturbed GaAs at 10 K. (**b**) Differential transmission spectra obtained with 0.5 ps pulses at various time delays. The high-energy tail becomes steeper with increasing time delay, showing the cooling of the plasma with time [4.35]

generated by passing the amplified CPM laser pulses through a liquid. Differential transmission spectra for several delays are shown in Fig. 4.3. The increased transmission at higher energies at short delays results from hot photoexcited carriers. The increasing sharpness of the high-energy tail with increasing delay is a result of carrier cooling. The DTS can be calculated using the carrier temperature and density as parameters [4.35, 36]. If the electron and the hole temperatures are different, then the number of parameters in the calculation increases. From a fit of the observed DTS with calculated DTS, one can determine T_c as a function of time delay, or the cooling curve. T_c is an effective temperature if electrons and holes are at different temperatures.

The experimental cooling curve obtained from such DTS is displayed in Fig. 4.4. *Leheny* et al. [4.35] also determined T_c at two other excitation density for a time delay of 10 ps. These two data points as well as the results obtained by *von der Linde* and *Lambrich* [4.37] using pump-probe spectroscopy, and by *Tanaka* et al. [4.38] by means of luminescence spectroscopy are also included in Fig. 4.4. In these analyses, electrons and holes were assumed to be at the same temperature, not an unreasonable assumption for the time scales investigated here. For times shorter than a couple of picoseconds, the electron and hole temperatures may be different and a detailed lineshape analysis may be necessary to deduce effective temperatures [4.39].

As discussed above, these results can be compared with the calculated cooling curves using C_0 as a parameter. The results of *Leheny* et al. [4.35] at the excitation density of 5×10^{16} cm^{-3} can be fit reasonably well with $C_0 = 0.5$,

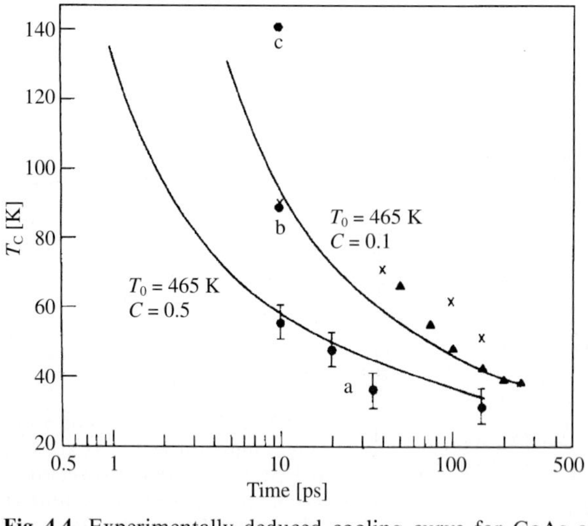

Fig. 4.4. Experimentally deduced cooling curve for GaAs at 10 K. *Circles: Leheny* et al. [4.35] at excitation desnity of of (a) 7×10^{16} cm^{-3}, (b) 3×10^{17} cm^{-3} and (c) 1×10^{18} cm^{-3}; *triangles: von der Linde* and *Lambrich* [4.37]; *crosses: Tanaka* et al. [4.38]. The *solid (dashed) curve* is the calculated cooling curve assuming $T_0 = 465$ K and $C = 0.5$ (0.1); from [4.10]

indicating that the energy-loss rate is reduced by a factor of 2 compared to the prediction of the simple model discussed above. The results by *von der Linde* et al. [4.37] and *Tanaka* et al. [4.38] were obtained at higher excitation density. They are shifted to the longer time on the semi-logarithmic plot of Fig. 4.4. $C_0 = 0.1$ is required to fit these results and similarly low values of C_0 are required to fit the high-density results of *Leheny* et al. [4.35]. These results lead to the conclusion that the energy-loss rate of photoexcited plasma in bulk GaAs is reduced compared to the predictions of the simple model discussed above, and the energy-loss rate decreases as the excitation density increases.

The signal-to-noise ratio in the DTS results of Fig. 4.3 is not good so the error bars in the deduced temperature are large. While more recent DTS measurements have much better signal-to-noise ratio, DTS measures small changes between two large signals and is not ideally suited for making measurements with the large dynamics range necessary for accurately determining the carrier temperature from the measured spectra. Luminescence spectroscopy provides a much better technique because there is no need to extract small differences between two large signals. The development of luminescence upconversion technique made it possible to obtain a time resolution limited by the laser pulsewidth (Sect. 1.3.3c). It has now become the preferred technique for measuring the temperatures of the photoexcited carriers.

The luminescence spectra reported by *Kash* et al. [4.40] on a 0.5 μm thick layer of $In_{0.53}Ga_{0.47}As$ grown lattice-matched on InP and maintained at 10 K are shown in Fig. 4.5. The sample was excited with 8 ps pulses from a Rh6G dye laser at 6100 Å at two different excitation densities. For the high-density case, the spectra at early times reveal a broad plateau, indicating that the carrier density is high enough for the chemical potential to be above the bandgap of the semiconductor. With increasing time delay, the spectra become narrower, indicating the decay of the photoexcited carriers, and the high-energy tail becomes steeper, indicating a decrease in the carrier temperature T_c. The spectra exhibit a dynamic range of more than 3 orders of magnitude and therefore allow a more accurate determination of the carrier temperature from the high energy slope of the spectra (Fig. 4.3). The cooling curves obtained from T_c at various time delays is plotted in Fig. 4.6. Comparison with the curves calculated from the simple model discussed above shows once again that the energy-loss rate in InGaAs is also much smaller than that predicted by the simple theory.

The experimental results and the simple analysis presented above clearly demonstrate that the carrier energy-loss rates to the lattice are reduced compared to those predicted by the simple theory. The above analysis, however, does not identify the physical process(es) responsible for this reduction in the energy-loss rates. Initially the reduced energy-loss rates were attributed [4.35] to a reduction in the electron-phonon interaction because of screening effects which were neglected in the simple model discussed above. Large reductions in the energy-loss rate were indeed predicted theoretically [4.41]. However, these reductions were predicted on the basis of static screening approximation which are not appropriate to the situation under investigation. Alternate possi-

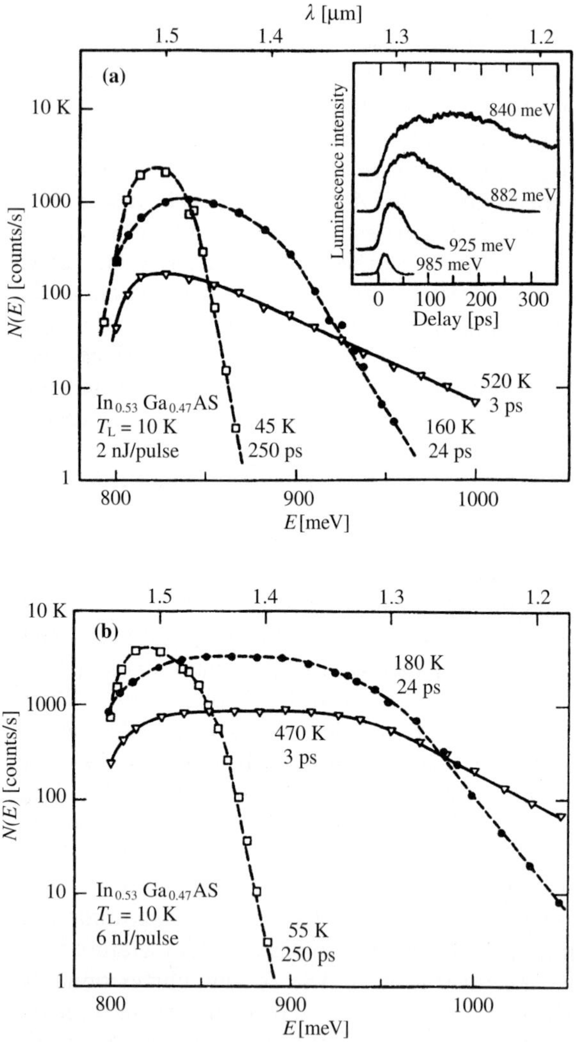

Fig. 4.5. Luminescence spectra of InGaAs at 10 K at three different time delays for an excitation density of **(a)** 7×10^{17} cm^{-3}, and **(b)** 2×10^{18} cm^{-3}. The *inset* shows the time evolution of the spectra at several emission photon energies [4.40]

bilities of a nonequilibrium population of optical phonons reducing the energy-loss rate were also raised soon after [4.42–46]. To resolve this question and to obtain further insights into the processes leading to a reduced energy-loss rate, extensive theoretical and experimental investigations of different aspects of this phenomenon were undertaken during the 1980s. The conclusion of these more recent investigations is that in a vast majority of cases effects of screening are small and the hot phonons play a dominant role in the cooling rates. Some important aspects of these investigations are discussed in the next two sections.

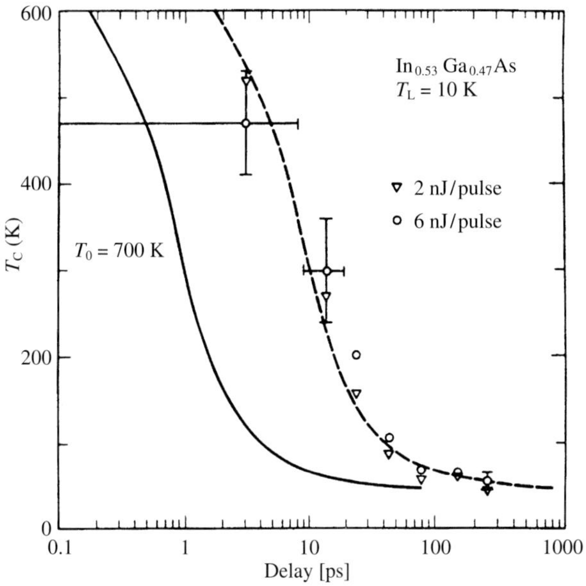

Fig. 4.6. Cooling curve for photoexcited InGaAs at $T_L = 10$ K for two different excitation densities. The solid (*dashed*) curve is calculated from the simple model discussed in the text with $T_0 = 700$ K and $C = 1$ (0.1) [4.40]

4.3 Other Factors Influencing the Cooling Curves

While introducing the simple model in Sect. 4.1, we discussed several assumptions underlying the model. It is clear that breakdown of any of these assumptions would lead to a modification of the energy-loss rate predicted by the model. Although non-parabolicity of bands (e.g., warping and anisotropy of the valence bands, presence of subsidiary valleys for the conduction band) may be important for specific cases, this effect will not be considered here. Some of the other effects are discussed below.

4.3.1 Pauli Exclusion Principle and Fermi-Dirac Statistics

If the photoexcited or the background carrier density is high or the temperature is sufficiently low, the carriers can no longer be described by the Maxwell-Boltzmann distribution function, and the correct carrier-distribution function must be utilized in the calculation. This modifies the energy-loss rate because the final state after interaction with a phonon may be occupied (Pauli exclusion principle), and also because the average energy of such a non-Maxwellian distribution function can be quite complicated. If the carrier density is sufficiently large so that the carriers can still be characterized by a temperature, one simply replaces the Maxwell-Boltzmann distribution function by the Fermi-Dirac distribution function. With this replacement, it may no longer be possible to obtain analytical expressions for the energy-loss rate and the cooling curves. But numerical calculation of these curves is straightforward. It is found that the effect of degeneracy is rather small at the densities below 1×10^{18} or 1×10^{12} cm^{-2} in quantum wells [4.20]. Therefore degeneracy effects are not expected to explain the large reduction in the energy-loss rate observed experimentally at moderate densities.

4.3.2 Hot Phonons

One of the basic assumptions of the simple model discussed earlier is that the phonon occupation number remains unaffected by the photoexcitation and can be equated to the occupation number at the lattice temperature. When the density of photoexcitation is moderate to high, a large number of phonons may be generated as the carriers relax by the emission of phonons. The optical phonons near the zone center have a rather low dispersion so that their group velocities are small and the generated phonons are unlikely to leave the photoexcited volume. In this case, the phonon occupation number is determined by the generation rate and the anharmonic decay rate into acoustic phonons. This anharmonic decay rate is found to be several picoseconds by time-resolved Raman studies [4.47–51], and will be discussed in more detail in Chap. 5. This decay time is sufficiently long that a large non-equilibrium phonon population is created at moderate to high photoexcitation density. The first evidence of such non-equilibrium phonon population (hot phonons) was obtained in Raman spectra of cw photoexcited GaAs by *Shah* et al. [4.52], and will be discussed in Chap. 5 on phonon dynamics.

The non-equilibrium phonon population may lead to a reabsorption of phonons by carriers, and hence a reduction in the net energy-loss rate to the lattice. This mechanism for the reduction of energy-loss rate was first considered by several researchers [4.42, 44–46]. The effect of non-equilibrium phonons or the "hot-phonon effect" was calculated by simultaneously solving the Boltzmann equation for the carrier and the phonon system. Three different numerical techniques have been used for this purpose: (1) the direct numerical integration, using discretization in wavevector, energy and time [4.44, 53–55], (2) the ensemble Monte-Carlo technique [4.56–60], and (3) the method of moments [4.46]. Each of these techniques have their strengths and weaknesses. We will follow here the balance equation approach based on the method of moments, discussed in detail by *Pötz* and *Kocevar* [4.46], because it is conceptually and computationally the simplest. We note, however, that this simplicity is achieved by assuming a thermalized carrier-distribution function with a characteristic temperature T_c, and a simplified band structure. Therefore, the validity of this approach is limited to times beyond the internal thermalization times (Chap. 3).

Before undertaking a more detailed discussion of hot phonons, it is useful to note that the concept of hot phonons is not restricted to photoexcited systems. In fact, the idea of hot phonons was initially introduced for understanding the high-field transport in the elemental semiconductor Ge [4.1]. In this case, hot acoustic phonons were important, but the concept is similar to that used for the case of hot optical phonons. We note that most of the photoexcitation experiments, cw and time-resolved, have explored the carrier temperature and time range where optical phonon dominate the energy-loss rates. Therefore, hot optical phonons have received more attention in this field. However, it is straightforward to include non-equilibrium effects for different types of phonons in calculations, and such effects were in fact included in the first calculation of hot phonons related to energy-loss rates [4.45].

In the discussion of the simple model above, the energy-loss rate was calculated by averaging the rate of loss of energy of a carrier with a given energy E over the carrier-distribution function. An alternate approach to determining the energy-loss rate is to first calculate $\partial N_q/\partial t$, the rate of generation of phonons of wavevector q, and then calculate the average energy-loss rate as [4.1, 15]

$$\left\langle \frac{\mathrm{d}E}{\mathrm{d}t} \right\rangle = -\frac{1}{nV} \sum_q \hbar\omega_q \frac{\partial N_q}{\partial t} . \tag{4.8}$$

$\partial N_q/\partial t$ is the difference between the emission and absorption rates of phonons of wavevector q, and can be calculated using the Fermi golden rule in much the same way that $\mathrm{d}E/\mathrm{d}t$ is generally calculated. For a Fermi-Dirac distribution function with temperature T_c and Fermi energy $\eta k T_\mathrm{c}$, it can be shown that [4.14, 15]

$$\left(\frac{\partial N_q}{\partial t} \right) = \frac{m^2 k T_\mathrm{c} V |M_q|^2}{\pi \hbar^5 q} [N_q(T_\mathrm{c}) - N_q] \times \ln \left\{ \frac{1 + \exp\left[\eta - \frac{\hbar^2}{8mkT_\mathrm{c}} \left(q - \frac{2m\omega_q}{\hbar q} \right)^2 \right]}{1 + \exp\left[\eta - \frac{\hbar^2}{8mkT_\mathrm{c}} \left(q + \frac{2m\omega_q}{\hbar q} \right)^2 \right]} \right\} \tag{4.9}$$

so that

$$\left\langle \frac{\mathrm{d}E}{\mathrm{d}t} \right\rangle = -\frac{m^2 k T_c V}{2\pi^3 \hbar^5 n} \int_0^\infty \mathrm{d}q \left[q\,|M_q|^2 \hbar\omega_q (N_q(T_c) - N_q) \right.$$

$$\left. \times \ln \left\{ \frac{1 + \exp\left[\eta - \frac{\hbar^2}{8mkT_c} \left(q - \frac{2m\omega_q}{\hbar q} \right)^2 \right]}{1 + \exp\left[\eta - \frac{\hbar^2}{8mkT_c} \left(q + \frac{2m\omega_q}{\hbar q} \right)^2 \right]} \right\} \right] . \tag{4.10}$$

Here m is the carrier-effective mass, n is the carrier density (number in volume V), $|M_q|^2$ is the squared carrier-phonon interaction matrix element, $N_q(T_c)$ is the Bose-Einstein phonon occupation number at the carrier temperature, $N_q(T_L)$ is the phonon occupation number at the lattice temperature, and N_q is the non-equilibrium phonon occupation number for phonons of wavevector q. The decay of optical phonons into acoustic phonons by anharmonic decay process can be described by

$$\left(\frac{\partial N_q}{\partial t} \right)_{\mathrm{ph-ph}} = -\frac{N_q - N_q(T_L)}{\tau_{\mathrm{ph}}} , \tag{4.11}$$

where τ_{ph} is the lifetime of the phonon, assumed to be the same for all q. Under steady-state conditions, the sum of (4.9) and (4.11) is set to zero to determine the nonequilibrium phonon population N_q for a given density n and temperature T_c. The average energy-loss rate in the presence of nonequilibrium phonons is then calculated using (4.10). For time-resolved experiment, one needs to also consider the generation of photoexcited carriers with a laser pulse of a given shape and photon energy, and perform time-dependent calculations.

It is instructive to see the predictions of the simple model considered above. Figure 4.7 illustrates a steady-state calculation of average electron energy-loss rate by polar optical phonon scattering in GaAs for electron density 2×10^{17} cm^{-3}, $T_c = 50$ K and $T_L = 2$ K. The bottom panel shows the nonequilibrium phonon occupation number, normalized to $N_q(T_c)$, as a function of q. The top panel exhibits the energy-loss rate in interval 0.1 q_0, where $\hbar^2 q_0^2/2m = \hbar\omega_{\mathrm{LO}}$ with and without the effects of hot phonons. It is clear that there is a substantial reduction of carrier energy-loss rate due to the hot-phonon effect.

We emphasize once again that the above model is extremely simplified, and is presented here only for illustrating the effect. Many time-dependent calculations using more realistic models have been reported in the literature [4.46, 56, 58–60]. Results of ensemble Monte-Carlo simulations will be presented in

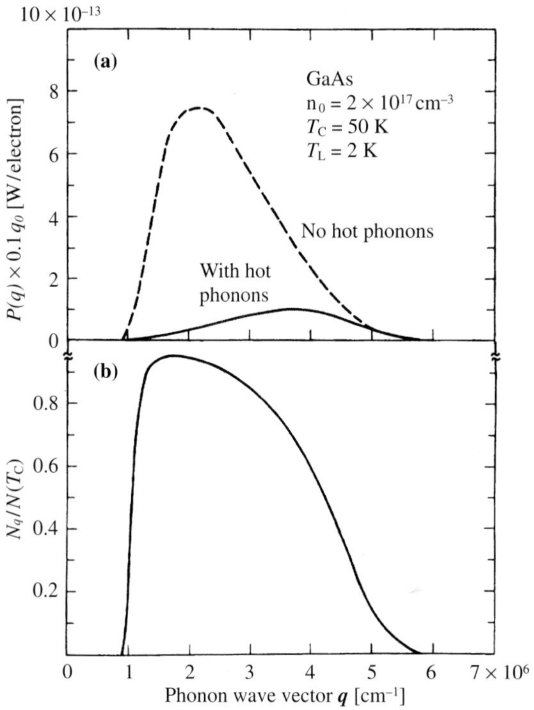

Fig. 4.7. (a) Electron energy-loss rate, and **(b)** relative phonon-occupation number as a function of phonon wavevector calculated from the simple hot-phonon model described in the text for bulk GaAs with $T_L = 2$ K and $T_c = 50$ K and electron density of 2×10^{17} cm^{-3} [4.133]

Chap. 5 on phonon dynamics. Extensive calculations via the method of moments have been reported by *Marchetti* and *Pötz* [4.61–63], and summarized by *Pötz* and *Kocevar* [4.46]. They have considered undoped GaAs/AlGaAs single quantum wells at 10 K in which a 0.25 ps laser pulse excites electron-hole pairs with excess energy of 150 meV per pair. They found that the results were insensitive to the quantum-well width, so we discuss here density-dependent results for 25 Å thick GaAs quantum wells.

Figure 4.8 displays the phonon-distribution function for a 25 Å GaAs quantum well for two excitation densities, each at three different time delays. The results show a marked dependence on the excitation density. The calculated cooling curves for electrons and holes in the 25 Å quantum well are plotted in Fig. 4.9 for three different densities on a linear scale. Note that it takes several picoseconds for the electron and hole temperatures to become equal. The results clearly show a strong reduction in the cooling as the excitation density, and hence the nonequilibrium phonon population, increases.

4.3.3 Screening and Many-Body Aspects

A large density of photoexcited or background carriers introduces several additional modifications to the simple model introduced in Sect. 4.1. These modifi-

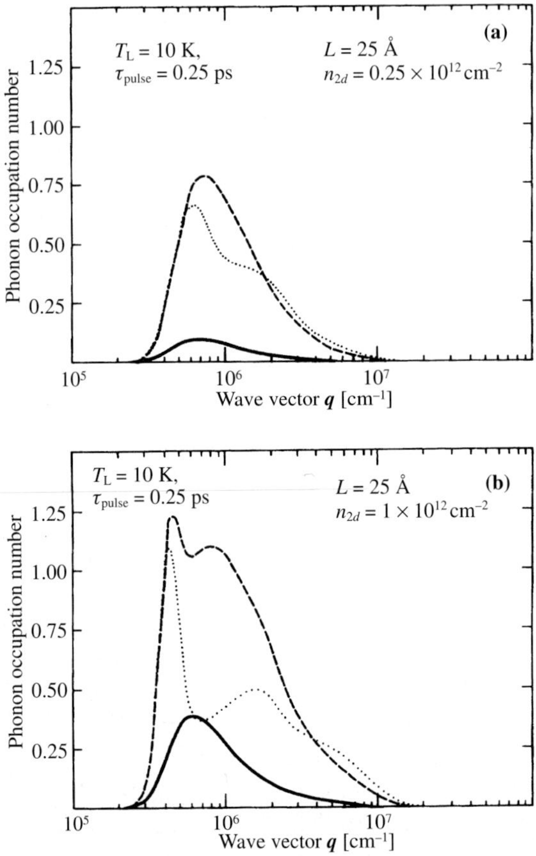

Fig. 4.8 a, b. Phonon-occupation number vs wavevector (phonon-distribution function) calculated using the method of moments for a 25 Å undoped GaAs quantum well at 10 K excited by a 0.25 ps laser pulse. **(a)** 0.25×10^{12} cm^{-2}, and **(b)** 1×10^{12} cm^{-2} for three different time delays: 0.1 ps (*solid line*), 1 ps (*dashed line*) and 5 ps (*dotted line*) [4.46]

cations result from the fact that, for large carrier densities, one must consider (i) the collective modes of the electronic system (plasmons), (ii) plasmon-phonon coupled modes of the system, and (iii) dynamical screening. These inter-related phenomena can be treated in a unified manner by considering the frequency and wavevector-dependent dielectric function $\varepsilon(q, \omega)$ of the semiconductor which can be thought of as the sum of three contributions:

$$\varepsilon(\boldsymbol{q},\omega) = \varepsilon_0 + \varepsilon_L(\boldsymbol{q},\omega) + \varepsilon_e(\boldsymbol{q},\omega) \ , \tag{4.12}$$

where ε_0 is the free space permittivity, ε_L is the lattice contribution to the dielectric function, ε_e is the electronic contribution to the dielectric function, and $\boldsymbol{q}$ is the plasma wavevector. ε_e includes the contribution of the valence electrons as well as the free electrons and holes.

Generally, ε_e is calculated under the Random Phase Approximation (RPA) [4.64] which assumes that density fluctuations whose phases vary in a coherent manner are dominant, and those with randomly fluctuating phases nearly can-

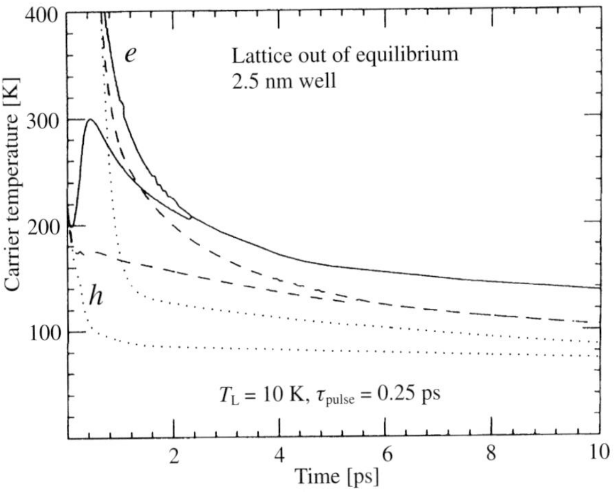

Fig. 4.9. Cooling curves for electrons and holes in a 25 Å GaAs quantum well excited by a 0.25 ps laser pulse, calculated using the method of moments including hot-phonon effects. 1×10^{12} cm^{-2} (*solid line*), 2.5×10^{11} cm^{-2} (*dashed line*) and 2.5×10^{10} cm^{-2} (*dotted line*). Note that the cooling curves show a strong dependence on the density, and it takes several picoseconds for the electrons and the holes to reach the same temperature [4.46]

cel each other and are negligible. One can separate the response of the valence electrons from those of the free carriers (electrons and holes) by including the former in the optical dielectric function ε_∞. *Lindhard* [4.65] derived the following form for the electronic dielectric function of a semiconductor within the one-band approximation (assuming $\varepsilon_L = 0$)

$$\varepsilon_e(q,\omega) = \varepsilon_\infty \left[1 - V_q \sum_k \left(\frac{f(k-q)-f(k)}{\hbar\omega + i\hbar\delta + E_{k-q} - E_k} \right) \right] , \tag{4.13}$$

where $f(\boldsymbol{k})$ is the Fermi distribution function, E_k is the energy of the carrier at $\boldsymbol{k}$, δ is an infinitesimally small quantity, and the summation over vector $\boldsymbol{k}$ is carried out for all band states. The Lindhard dielectric function is valid for 3D as well as 2D. In the above sum and other similar sums that follow, it is understood that $\boldsymbol{k}$ and $\boldsymbol{q}$ are 3- or 2-dimensional vectors according to the system dimensionality. V_q, the Fourier transform of the Coulomb potential $e^2/4\pi\varepsilon_\infty r$, is given by [4.6, 66]

$$V_q = \frac{e^2}{\varepsilon_\infty q^2 L^3} \quad 3\text{D} \tag{4.14}$$

and

$$V_q = \frac{e^2}{2\varepsilon_\infty q L^2} \quad 2\text{D} . \tag{4.15}$$

The dielectric function and the electronic polarizability $P(q, \omega)$ are related by

$$\varepsilon_{\mathrm{e}}(q,\omega) = \varepsilon_\infty [1 - V_q P(q,\omega)] \tag{4.16}$$

so that

$$P(q,\omega) = \sum_k \left(\frac{f(k-q) - f(k)}{\hbar\omega + \mathrm{i}\hbar\delta + E_{k-q} - E_k} \right) . \tag{4.17}$$

We note that if more than one type of carrier is present, both $\varepsilon(q, \omega)$ and $P(q, \omega)$ must include intraband as well as interband contributions with appropriate matrix elements. For example, the polarizability $P(q, \omega)$ can be written as the sum of various $P_{ij}(q, \omega)$:

$$P_{ij}(q,\omega) = \sum_k \left(\frac{M_{ij}^2(k,q)[f_j(k-q) - f_i(k)]}{\hbar\omega + \mathrm{i}\hbar\delta + E^j_{k-q} - E^i_k} \right) . \tag{4.18}$$

Different limits of the Lindhard functions allow one to extract the plasma frequencies as a function of q, and also the single particle excitation regime. When a plasma mode falls within the continuum of the single-particle excitations, the plasmon is Landau damped and decays into the pair excitations. It can be shown that, for small q, the (screened) plasma frequencies in 3D are given by

$$\omega_q^2 = \omega_\mathrm{p}^2 \left(1 + \frac{q^2}{\kappa^2} \right) , \tag{4.19}$$

where

$$\omega_\mathrm{p}^2 = \frac{ne^2}{\varepsilon_\infty m} \tag{4.20}$$

and those in 2D are expressed by

$$\omega_q^2 = \omega_\mathrm{p}^2 \left(1 + \frac{q}{\kappa} \right) , \tag{4.21}$$

where

$$\omega_\mathrm{p}^2 = \frac{ne^2 q}{\varepsilon_\infty m} . \tag{4.22}$$

For 3D, the screening wavevector κ, the inverse of the screening length, is given by the Fermi-Thomas expression in the degenerate case

$$\kappa = \sqrt{3e^2 n / 2\varepsilon_\infty E_\mathrm{F}} , \tag{4.23}$$

where n is the carrier density and E_F is the Fermi energy. For the non-degenerate case, the screening wavevector is given by the Debye-Hückel formula

$$\kappa = \sqrt{e^2 n k T / \varepsilon_\infty} \ . \tag{4.24}$$

The 2D screening wavevector is given by

$$\kappa = \frac{m e^2}{2 \pi \varepsilon_\infty \hbar^2} [1 - \exp(-\pi n \hbar^2 / m k T)] \ . \tag{4.25}$$

Note that the 2D screening wavevector becomes independent of n at low temperatures and high densities.

Using the Lindhard dielectric function, the screened interaction is given by

$$V_s(q) = V_q / \varepsilon_e(q, \omega) \ , \tag{4.26}$$

which provides a means of calculating interaction among photoexcited carriers and also between photoexcited or electrically-injected carriers, and a background of thermal carriers. The latter has been considered for different cases in the literature [4.20, 67–69] and will not be discussed here. Proper calculations of the screened interaction among the photoexcited carriers are essential for understanding the carrier-relaxation dynamics following photoexcitation such as those discussed in Chap. 3. The calculations for dynamically-screened carrier-carrier interactions in a two-component electron-hole plasma are quite complicated. Monte-Carlo simulations provide one of the best methods of calculating the dynamics of photoexcited plasma, but inclusion of dynamic screening in Monte-Carlo simulations is also difficult. One approach is to replace dynamic screening with time-dependent static screening [4.57, 60, 70–72] where $\varepsilon(q, \omega)$ is evaluated at $\omega = 0$, with a screening wavevector determined by the density and temperature of each kind of carrier (electron in Γ valley, electron in L valley, heavy holes, etc.) as a function of time. Another approach is to incorporate screened interaction in real space by using a molecular dynamics technique [4.6, 73]. Some numerical results obtained by both these approaches were discussed in Chap. 3. Another alternative for small-q scattering (polar optical phonon or electron-electron scattering), the plasmon-pole approximation [4.66], has also been used in recent years [4.74].

For the cooling of hot carriers, the main topic of this chapter, one is primarily interested in how the carrier-phonon interactions are affected by the presence of the high-density plasma. The carrier-phonon interaction is also screened by an equation similar to (4.26), with an appropriate V_q for the carrier-phonon interaction. The effect of screening and many-body effects on carrier-phonon interaction have been considered under many different assumptions. The simplest approximation is to neglect the collective modes and assume that the carrier-phonon interaction is screened by the static dielectric constant $\varepsilon(q, \omega = 0) \equiv \varepsilon(q, 0)$ such that the screened interaction potential is given

$$\varepsilon_e(q,0) = 1+(\kappa^2/q^2) \quad \text{in 3D} \tag{4.27}$$

and

$$\varepsilon_e(q,0) = 1+(\kappa/q) \quad \text{in 2D} \tag{4.28}$$

with the screening wavevector κ given by (4.23–25). The early calculations [4.41] and analysis of experimental results [4.35, 36] were based on this approach. However, it turns out that the static screening approach overestimates the screening of both electron-phonon and electron-electron interaction in most cases and is, therefore, not suitable for the analysis of the cooling curves. The correct approach to the average energy-loss rate in semiconductors is to use dynamic screening through the use of the frequency- and wavevector-dependent dielectric constant and the renormalized phonon frequencies because of plasmon-phonon coupling. One can then compare the results for screening and hot-phonon effects, and determine the dominant physical phenomenon contributing to the reduction in the energy-loss rates.

The origin of the coupled modes can be understood from the following simple picture. In the long-wavelength ($q = 0$) limit, it can be shown that the electronic dielectric function ($\varepsilon_L = 0$) approaches the Drude limit:

$$\varepsilon_e(0,\omega) = \varepsilon_\infty \left(1-\frac{\omega_p^2}{\omega^2}\right) . \tag{4.29}$$

The lattice contribution to the dielectric function is given by

$$\varepsilon_L(0,\omega) = \varepsilon_\infty \frac{\omega_{LO}^2-\omega_{TO}^2}{\omega_{TO}^2-\omega^2} . \tag{4.30}$$

The total dielectric function in the long-wavelength limit is then given by

$$\varepsilon(0,\omega) = \varepsilon_\infty \left(1-\frac{\omega_p^2}{\omega^2}\right) + \varepsilon_\infty \left(\frac{\omega_{LO}^2-\omega_{TO}^2}{\omega_{TO}^2-\omega^2}\right) . \tag{4.31}$$

The zero of the dielectric function gives the frequencies of the coupled modes. For low densities ($\omega_p \cong 0$), there is only one root at $\omega = \omega_{LO}$. For finite densities, there are two roots corresponding to the upper branch $\omega = \omega_+$ and the lower branch $\omega = \omega_-$. This shows that the coupled plasmon-phonon modes are renormalized compared to the bare plasmon and phonon frequencies. The nature of the upper and lower branches depends on the magnitude of the plasma frequency. If $\omega_p \ll \omega_{TO}$, then the lower branch is plasmon-like and the upper branch is phonon-like. For the opposite case of $\omega_p \gg \omega_{LO}$, the lower branch is phonon-like while the upper branch is plasmon-like. *Das Sarma* and coworkers [4.20, 21, 75–78] have given expressions for the average energy-loss rate of an electron plasma to the lattice for the general case of dynamic screening and renormalized phonon frequencies. Using these expressions, these researchers have calculated the average energy-loss rates for a number of different cases under steady-state conditions. For the case of noninteracting, nondegenerate electrons in 3D with dispersionless plasma frequency, and no Landau damping, *Das Sarma* et al. [4.21] have shown that

$$\left\langle \frac{\mathrm{d}E}{\mathrm{d}t} \right\rangle = \frac{\hbar\omega_+}{\tau_+} \exp\left(-\frac{\hbar\omega_+}{kT}\right) + \frac{\hbar\omega_-}{\tau_-} \exp\left(-\frac{\hbar\omega_-}{kT}\right) , \tag{4.32}$$

where $1/\tau_+$ and $1/\tau_-$ are the effective scattering rates which can be evaluated using the formulae given in [4.21]. These expressions have utilized the plasmon-pole approximation to evaluate the renormalization of the phonon frequencies, but not for the polarizability of the electron gas. We emphasize that this simple expression considers only a single-component plasma consisting of electrons and assumes simple spherical band structure. Note also that screening can depend on the carrier-distribution function [4.79] so that these results can not be applied directly to time-resolved studies where the distribution function changes with time.

In spite of these simplifications, the above expression contains the essential physics of the modifications introduced by dynamic screening and plasmon-phonon coupling [4.75]. First, we note [4.21] that the effective scattering rate corresponding to the plasma-like mode is very small so that the prefactor for the phonon-like mode is much larger than that for the plasmon-like mode in (4.32). On the other hand, the exponentials always favor the lower-frequency ω_- mode. Therefore, one can draw the following conclusions: (1) at high densities, both the prefactor and the exponential are larger for the second term in (4.32), so that the second term (ω_- mode) dominates. This is the phonon-like mode, so that plasmon-phonon coupling is not important in this case. (2) At low densities, the ω_+ mode is phonon-like so that the prefactor for the first term in (4.32) is larger. Therefore (a) at high temperatures the first term dominates and the plasmon-phonon coupling is once again not important, but (b) at low temperatures, the second, plasmon-like term dominates because of the exponential. Therefore, it is for the case of low temperatures and low densities that the phonon renormalization introduced by the plasmon-phonon coupling leads to important modifications of the average energy-loss rate. These considerations apply to 3D case as mentioned above. Although the case of 2D electron gas is more complicated, the above considerations qualitatively hold also for the case of 2D electron gas [4.21].

Some insight into the importance of various approximations is gained from Fig. 4.10 which compares the average electron energy-loss rate due to LO phonons in bulk GaAs for various approximations [4.21, 77, 80]. These steady-state calculations are for two different electron densities and assume that the electrons are characterized by an electron temperature T and that the lattice is <10 K. The figure compares seven different approximations in order of

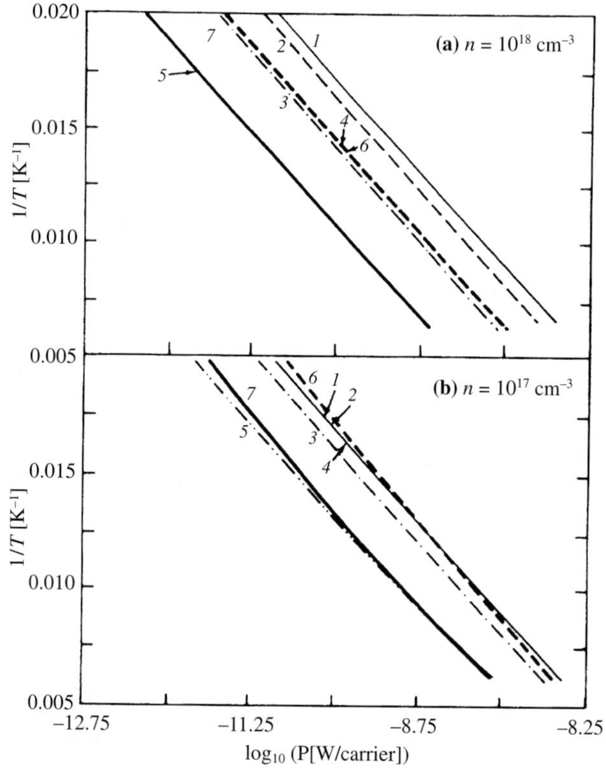

Fig. 4.10. Steady-state electron energy-loss rates as a function of the carrier temperature calculated for bulk GaAs at $T_L = 0$ K for two different electron densities (**a**) 1×10^{17} cm^{-3}, and (**b**) 1×10^{18} cm^{-3}. Different approximations are compared: (*1*) simple model (Sect. 4.1), (*2*) quantum statistics without screening, (*3*) static screening, (*4*) dynamical screening, (*5*) dynamical screening including hot-phonon effects, (*6*) dynamical screening including plasmon-phonon coupling, and (*7*) dynamical screening including both plasmon-phonon coupling and hot-phonon effects [4.21, 77, 80]

increasing sophistication, as indicated in the figure caption. Comparing curves *1* and *2*, we see that the Pauli exclusion (curve *2*) makes a noticeable reduction in the energy-loss rate compared to the simplified model discussed above (and represented by curve *1*) at 1×10^{18} cm^{-3} but not at 1×10^{17} cm^{-3}. Static screening (curve *3*) overestimates the screening compared to dynamical screening (curve *4*). Comparing curves *4* and *6* we see that the inclusion of phonon renormalization does not make a noticeable difference at these densities and temperatures.

Note that the most sophisticated calculation (curve *7*) gives nearly identical results with curve *5* which includes the hot-phonon effect but ignores screening. Furthermore, even static screening (curve *3*), which is known to overestimate the reduction in carrier-phonon coupling, gives a reduction in the energy-loss rate that is smaller than the hot-phonon effect (curve *5*). These calculations, therefore, show that the hot-phonon mechanism is the single most important cause for the reduction in the energy-loss rate compared to the simple model discussed above. One might also note that, as discussed by *Pötz* and *Kocevar* [4.45, 46], screening and hot-phonon effects work in tandem; i.e., if the screening is increased, fewer phonons are generated so the hot-phonon effect is reduced, and vice versa.

These calculations provide an excellent framework for understanding how the various mechanisms we have discussed above contribute to a reduction in the carrier energy-loss rate in semiconductors such as GaAs. However, these calculations do not provide a direct comparison with most photoexcitation experiments for several reasons. Most photoexcitation experiments are performed on undoped semiconductors so that an equal number of electrons and holes are present. The presence of two- or three- (if light holes are important) component plasma introduces enormous complications in the calculation of the electronic polarizability. Furthermore, coupling of the heavy holes to the lattice is stronger by a factor of 2 to 3 in semiconductors like GaAs, and less susceptible to screening because of the presence of nonpolar coupling to the lattice for the holes. Also, depending on the density and the temperature, the electrons may be degenerate while the holes may be non-degenerate. All these factors must be considered when making a detailed comparison with the results.

Time-resolved measurements present additional problems. The full theory of dynamical screening is very difficult to implement on femtosecond time scales where the distribution function may be non-thermal and evolving rapidly with time. Screening may depend on the carrier-distribution function and therefore may change as the distribution function changes [4.79]. Furthermore, even after a temperature is established for each component of the plasma, the temperature of the different components may be different for some time [4.81, 82]. Also, the electron and hole densities are time-dependent, bringing in additional complications. A judicious choice of approximations and experimental parameters may be essential in making a detailed quantitative comparison between experiments and calculations. We will discuss some of these in the next section.

4.4 Further Experimental Investigations of Energy-Loss Rates

The experimental investigations of carrier cooling curves proceeded along several different avenues. One point of interest was in determining which of the physical processes discussed above is the most important in reducing the energy-loss rate. Since the calculations incorporating realistic experimental conditions are extremely difficult, the best approach here seems to be to try to isolate various processes by a judicious choice of experimental parameters or techniques, so that the results can be analyzed in an unambiguous manner. We will begin with a discussion of some such experiments in this section. The second point of interest was in determining how the cooling rates compared in bulk semiconductors and their quantum wells, and if they depended on the thickness of the quantum well or the excitation density. This aspect has generated considerable controversy over the years and will also be discussed in this section. Finally, semiconductors other than GaAs have also received some attention. For example, carrier cooling studies in bulk and quantum-well InGaAs [4.83 – 86] as well as in bulk CdSe [4.87], as well as carrier heating in semiconductor LEDs and lasers have been reported, but will not discussed here.

4.4.1 Direct Measurement of the Energy-Loss Rates

(a) Electric-Field Measurements. We recall that the first measurements of the carrier heating by photoexcitation were performed under cw photoexcitation. These results could not be analyzed quantitatively because both electrons and holes are present, and because the fraction of the excess energy given to the electron and the hole system is not known. Both of these problems could be removed by using doped semiconductors, a DC electric field to heat such carriers, and the photoluminescence spectrum to measure the carrier temperature. The photoexcitation used for measuring photoluminescence is kept sufficiently low that it does not heat the carriers. Such measurements were first performed in n-doped bulk GaAs [4.29, 30], but were not analyzed in terms of the electron energy-loss rates to the lattice. We discuss here some experiments by *Shah* et al. [4.12] on n- and p-modulation-doped quantum wells. Modulation-doping [4.88, 89] provides high carrier densities with good mobilities and good sample quality, and thus is an ideal system for studying carrier energy-loss rate studies.

Two n-modulation-doped and one p-doped samples of GaAs/AlGaAs quantum wells were investigated. The samples were lithographically prepared with six contacts as for standard Hall measurements. A DC current was passed across the two opposite contacts, and the voltage drop across two adjacent contacts was measured to determine the electric field and the electrical power dissipation (and hence the energy-loss rate at that field). The region between the two voltage contacts was *weakly* excited by a cw laser close to the semiconductor bandgap, and PL spectra were measured, and used to determine the carrier temperature, at various electric fields. Typical PL spectra for the n-type sample 1 and the p-type sample are shown in Fig. 4.11. Combination of these electrical and optical measurements allows one to determine the electron or the hole energy-loss rate as a function of the electron or the hole temperature. Such results are displayed in Fig. 4.12 for all three samples. Note that the absolute value of the energy-loss rate in Watts/carrier is obtained from such experiments, in contrast to the photoexcitation experiments discussed above. The PL spectra (4.11) exhibit that the hole temperature is considerably lower than the electron temperature for the same applied electric field. While some of this difference is due to the difference in the mobility, it can be shown that most of the difference is due to a difference in the energy-loss rates of electrons and holes, as is evident from Fig. 4.12. For the simple model discussed above, the

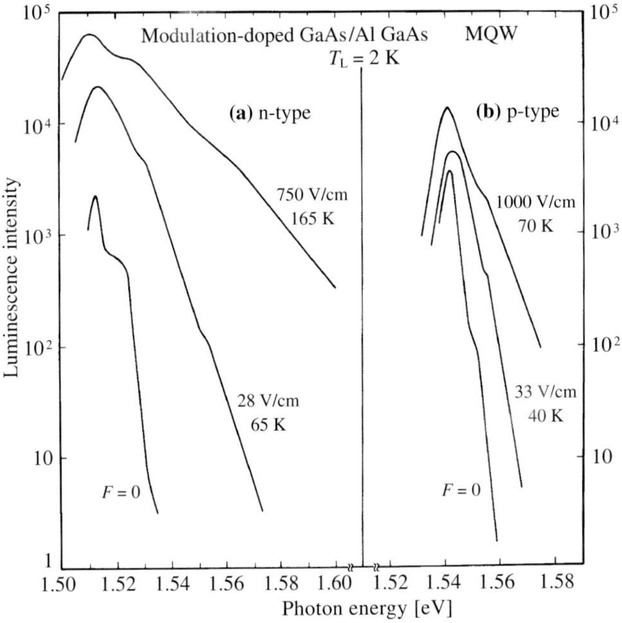

Fig. 4.11 a, b. Photoluminescence spectra of two modulation-doped GaAs/$Al_xGa_{1-x}As$ multiple-quantum-well samples at 1.8 K for several different electric fields applied parallel to the interface: (**a**) Sample 1: $n = 7\times10^{11}$ cm^{-2}, 262 Å quantum well, 20 periods, $x = 0.2$, $\mu = 63{,}000$ cm^2/Vs; (**b**) Sample 3: $p = 3.5\times10^{11}$ cm^{-2}, 94 Å quantum well, 15 periods, $x = 0.45$, $\mu = 36{,}000$ cm^2/Vs. Note that the slope of the high-energy tail for the p-type sample is steeper (lower T_c) for higher fields. The shoulders in the spectra correspond to contributions from higher subbands [4.12]

polar and nonpolar optical-phonon energy-loss rates for the holes is about a factor of 2.5 larger than for the electrons, but the measurements brought to light the then surprising result that the electron energy-loss rate is nearly 25 times smaller than that for the holes. Several arguments showed that this large difference is caused by hot-phonon effects, and in fact a comparison with calculations including hot-phonon effects gives a good agreement with the experimental results (4.12). The holes are less susceptible to hot-phonon effects under the conditions of the experiments because of the larger phase space available to holes. More detailed calculations [4.46] support these conclusions.

We note that many of these arguments are based on comparisons of the experimental results with calculations under different approximations. A *qualitative* argument against screening is that the difference between the experiments and the simple model decreases as the electron temperature increases. This is consistent with the hot-phonon model but not with the model of screening because screening is not expected to be strongly temperature dependent under these conditions.

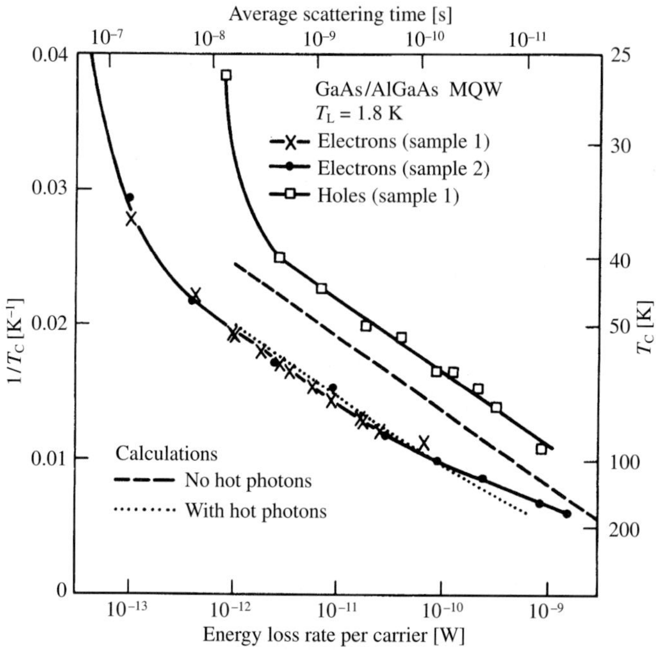

Fig. 4.12. Inverse of the carrier temperature ($1/T_c$) vs energy-loss rate per carrier to the lattice for three modulation-doped GaAs/$Al_xGa_{1-x}As$ multiple-quantum-well samples at $T_L = 1.8$ K. Samples 1 and 3 are described in the caption of Fig. 4.11, and Sample 2: $n = 3.9 \times 10^{11}$ cm^{-2}, 258 Å quantum well, 15 periods, $x = 0.23$, $\mu = 79{,}000$ cm^2/Vs. Points are experimental, the *solid curves* are to guide the eye, and the *dashed curve* is calculated for electrons in 3D using the simple model discussed in Sect. 4.1 and the *dotted curve* includes the hot-phonon effect in the model [4.11, 12]

A more rigorous comparison between theory and experiments was carried out by *Das Sarma* et al. [4.21]. The calculations include dynamic screening with plasmon-phonon coupling as well as hot-phonon effects. Figure 4.13 a compares the calculation with the results of *Shah* et al. [4.12]. The calculations are in good agreement and demonstrate that the energy-loss rate is dominated by the $\omega = \omega_+$ mode for the range of parameters applicable to this experiment. Fig. 4.13 b presents the calculations for a sample with slightly lower density and makes a comparison with the results of *Yang* et al. [4.90] using the same electric-field heating technique. The results show that for this lower density sample, the energy-loss rate is dominated by the lower coupled-mode branch ($\omega = \omega_-$ mode). These calculations provide one explanation for the apparently contradictory results on similar experiments by the two groups.

These results demonstrate that electric field-induced heating of carriers in doped semiconductors provide an excellent technique for quantitatively determining the energy-loss rate of the carriers to the lattice.

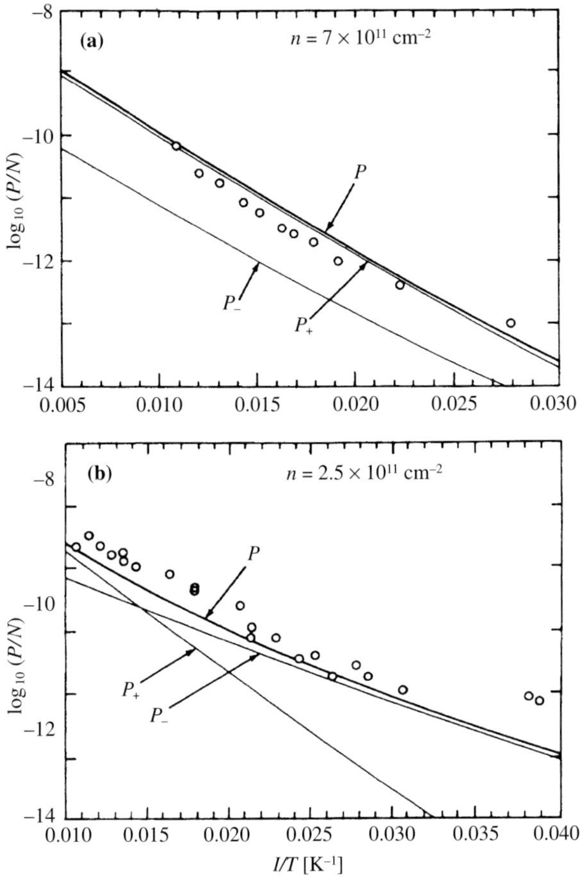

Fig. 4.13 a, b. Comparison of experiment and theory of electron energy-loss rates vs electron temperature. Experiments performed using DC electric field-induced heating of electrons and the theory includes coupled plasmon-phonon and hot-phonon effects. n-Modulation-doped samples with (**a**) 7×10^{11} cm^{-2} [4.12], and (**b**) 2.5×10^{11} cm^{-2} [4.90]. Theory and comparison from [4.21]

(b) Time-Resolved Measurements in Quantum Wells. The fact that holes have a higher energy-loss rate to the lattice can be used to advantage in time-resolved experiments on p-modulation-doped quantum wells. If the photoexcitation density is much less than the doping density, then the holes dominate the energy-loss rate, and problems of two-component plasma and time-dependent density discussed above can be largely removed. Also, the holes are non-degenerate during the time delay of interest, thus removing the additional complications arising out of quantum statistics. The first such measurements were reported by *Kash* et al. [4.91, 92], and many other measurements have been reported since [4.93–101].

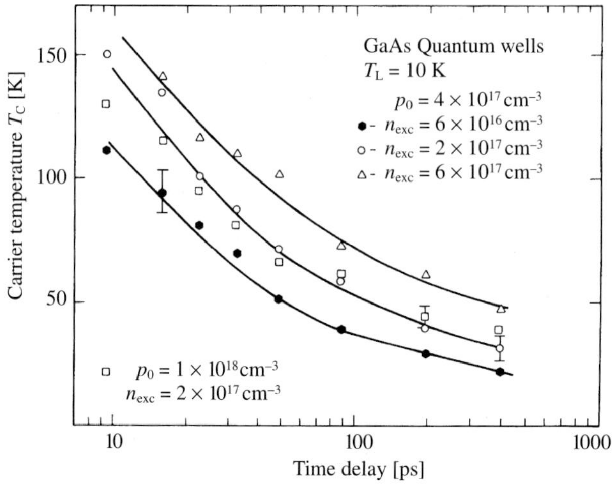

Fig. 4.14. Carrier temperature as a function of time delay (cooling curves) for two samples of p-modulation-doped GaAs quantum wells at T_L = 10 K at different excitation densities. Note that the curves on the semilog plot shift to the right, showing a reduction in the energy-loss rate, with increasing excitation density [4.91]

Kash et al. [4.91] investigated two p-modulation-doped quantum-well samples with well widths of $a = 90$ Å, and doping densities of $p_0/a = 4\times10^{17}$ cm^{-3} and $p_0/a = 1\times10^{18}$ cm^{-3}. The excitation density varied from $n_{exc}/a = 6\times10^{16}$ cm^{-3} to $n_{exc}/a = 6\times10^{17}$ cm^{-3}. For each sample and excitation density, luminescence spectra were recorded at various time delays using the up-conversion technique with a time resolution of about 8 ps. The carrier cooling curves determined from such measurements are depicted in Fig. 4.14. These curves show two important results: (1) For the same excitation density, the cooling curves do not depend on the doping density, and (2) the cooling curves shift to longer times as the photoexcitation density is increased, indicating a reduction in the energy-loss rate at a given temperature. The first observation was interpreted as a sign that screening is not important, while the second was interpreted as a sign that an excitation dependent hot phonon effect was becoming important.

Since these experiments deal with essentially a single-component, non-degenerate plasma of constant density, the numerically calculated slope of the cooling curve at each temperature can be used to determine the hole energy-loss rate at that temperature. The solid curve in Fig. 4.15 displays the energy-loss rates obtained from this time-resolved experiment. The dots show the results of electric-field heating experiment for p-modulation-doped sample which agree with the simple model calculation indicated by the dashed curve. The results of the time-resolved experiments reveal a much smaller energy-loss rate at high temperatures which correspond to early times. At low temperatures or long delay times, the measured energy-loss rate comes close to that predicted

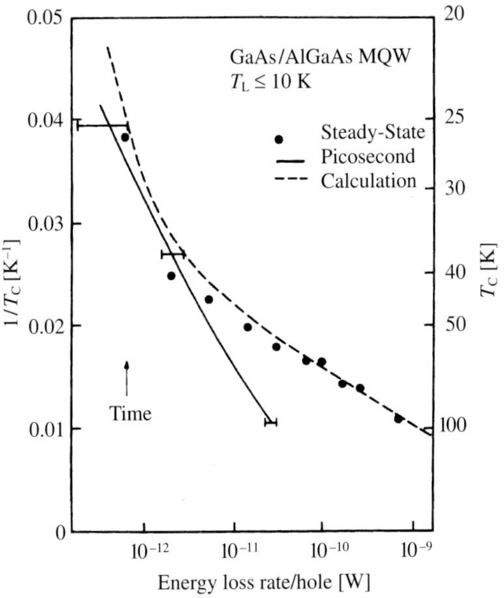

Fig. 4.15. Hole energy-loss rate as a function of the hole temperature, determined from the results of Fig. 4.14 as discussed in the text. Also shown are the results of the DC electric field-induced heating results (Fig. 4.12) for the same sample (*points*) and the results of a calculation based on the simple model of Sect. 4.1 [4.91]

by the calculation of the simple model. Since screening is not expected to vary significantly with temperature (or time), these results were used to conclude that hot-phonon effects were responsible for the reduction in the energy-loss rate at early times. Note that the semiconductor is excited much more strongly in pulsed photoexcitation than in the case of field-induced heating discussed in the previous section. This leads to hot-phonon effects for holes in time-resolved experiments whereas none was observed for holes in the field-induced case.

Rühle and *Polland* [4.102] performed a different experiment in which they photoexcited a cold plasma and measured how it heated the bath temperature. They found that the heating rates were independent of density, from which they deduced that screening of polar interaction is not important and that the reduction in cooling rates observed in other experiments must be caused by hot-phonon effects. However, the same researchers have shown [4.103] that the analysis is complicated by the effect of phonon underpopulation (cold phonons) and different electron and hole temperatures, so that this experiment, while consistent with the hot-phonon model, does not directly express the dominance of the hot-phonon effects. Their experiments on bulk AlGaAs [4.103] near the direct-indirect cross-over do show that hot phonons are the dominant cause for the reduction of energy-loss rate in this case.

While these experiments provide a direct experimental evidence against screening and for hot phonons, a more quantitative comparison with theory is desirable. Monte-Carlo simulations might be suitable for this purpose. It would also be desirable to improve the sensitivity of the system so that measurements can be performed at lower excitation density. We conclude this section by noting that many measurements on doped system have been reported since this initial study and have obtained results in agreement with this initial study [4.93–101].

4.4.2 Bulk vs Quasi-2D Semiconductors

The initial studies of energy-loss rates using time-resolved spectroscopy were performed on bulk GaAs using the pump-probe transmission technique [4.35–37] or luminescence [4.38, 104]. These studies showed that there is a substantial reduction in the energy-loss rate at high excitation density, as discussed above. A number of studies on energy-loss rate in quasi-2D systems were performed in early 1980s and found a substantial reduction in energy-loss rate at high densities. There was, however, considerable controversy in the early days because some studies found that the reduction in the energy-loss rate in quasi-2D and bulk semiconductors was comparable [4.105] whereas others found that the reduction was much larger in quasi-2D semiconductors [4.106, 107]. These initial experimental studies were complemented by calculations of the energy-loss rate in quasi-2D semiconductors [4.108–114], as reviewed by *Ridley* [4.115].

These early investigations were followed by a number of extensive investigations [4.12, 84–86, 93–95, 97, 101–103, 106, 107, 116–132]. The dependence of the energy-loss rate on the dimensionality as well as on the thickness of the quantum wells has been investigated. The controversy on the magnitude of the reduction in energy-loss rate in bulk versus quasi-2D semiconductors continued for quite some time. The controversy arises because it is difficult to keep all the parameters that influence the energy-loss rate the same in different experiments, and this leads to apparently contradictory results in some cases. For example, some experiments excited not only the quantum well but also the barriers. In this case, the relaxation of the carriers from the barriers to the well also plays a role in determining the cooling rate. Excitation at very high photon energies may lead to transfer of some electrons to the subsidiary valleys (Sect. 3.2) which may also affect the cooling rate. These issues have been discussed in detail by *Shah* [4.11], and *Pötz* and *Kocevar* [4.46] in recent reviews.

Leo et al. [4.94,120–122] investigated the dependence of the energy-loss rate on the thickness of undoped GaAs MQW at various densities, using a streak camera to measure time-resolved luminescence spectra. For every density and quantum-well thickness, they determined the factor $\alpha (= 1/C_0$ of Sect. 4.1) by which the measured energy-loss rate was reduced compared to the simple model (Sect. 4.1). Some of their results are displayed in Fig. 4.16. These results indicate that at low densities, the energy-loss rate is close to that expected for the simple model, but that there is a sharp reduction in the energy-loss rate (by as much as a factor of 100) at high excitation densities. Also, the results show that the measured energy-loss rate is independent of the quantum-well thickness. *Leo* et al. have also compared quantum wells and bulk GaAs,

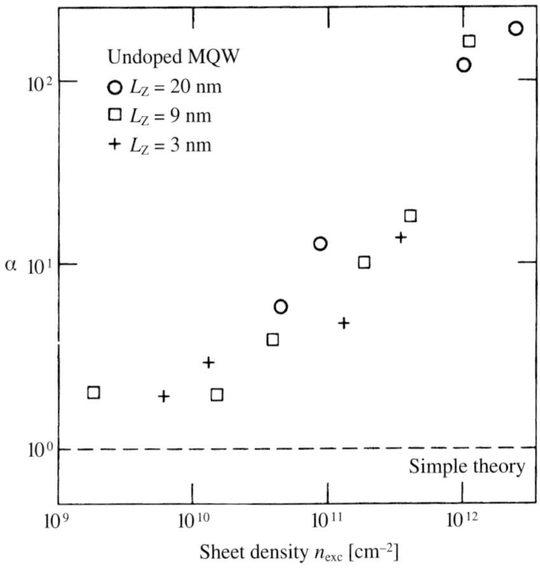

Fig. 4.16. Experimentally determined reduction in the energy-loss rate of photoexcited carriers in GaAs quantum wells as a function of excitation density, for quantum wells of three different thickness [4.121]

and concluded that the reduction in the energy-loss rate for equivalent densities is the same in the two cases.

Ryan and coworkers [4.95, 97, 127] have also carried out extensive investigations of the energy-loss rates in quantum wells. Figure 4.17 exhibits the measured cooling curves for n- and p-modulation-doped GaAs quantum wells. The cooling curves for all n-modulation-doped samples are the same within experimental error, and the same holds for the two p-modulation-doped quantum wells. Note that the p-modulation-doped samples have a lower temperature, consistent with the higher energy-loss rate for the holes as already discussed above.

These studies show that there is no dependence of the energy-loss rate on the well thickness and the reduction in the energy-loss rate is similar in magnitude for bulk and quantum-well GaAs. These conclusions are also in agreement with the theoretical predictions [4.46, 61–63]. The general consensus at this time appears to be that for the parameter range generally encountered in such experiments, bulk and quantum-well semiconductors are very similar as far as the energy-loss rates and hot-phonon effects are concerned, provided proper care is taken to make valid comparisons. The results for extremely thin wells and for very high densities may show some apparent differences, but extreme care in the interpretation of results is required because a large number of physical processes contribute in such regimes.

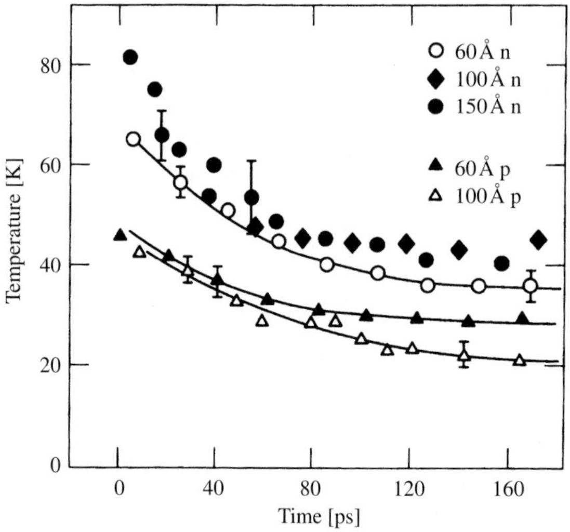

Fig. 4.17. Cooling curves determined by time-resolved luminescence spectroscopy on n- and p-modulation-doped GaAs quantum wells of different thickness, showing that the energy-loss rate is nearly independent of the quantum well thickness [4.97]

4.5 Conclusions

In this chapter we have discussed how the hot thermalized plasma cools to the lattice temperature by interacting with phonons. Ultrafast spectroscopy provides a powerful technique for investigating this phenomenon and has provided much insight into the processes responsible for the cooling of the hot plasma. It is clear that the simple model of energy-loss rate discussed in Sect. 4.1 is inadequate in many circumstances, and complications arising from quantum statistics, screening, coupled plasmon-phonon modes and nonequilibrium phonons must be considered in interpreting the experimental results. Nonequilibrium or hot phonons make a particularly strong contribution and lead to substantial reductions in the energy-loss rate under many conditions. In the next chapter we discuss how one can use ultrafast spectroscopy to investigate the dynamics of these nonequilibrium phonons.

5. Phonon Dynamics

We have discussed in earlier chapters that one of the initial relaxation processes for the photoexcited carriers is successive emission of optical phonons until the energy of the carrier is lower than the threshold for the emission of optical phonons. This is known as the phonon-cascade model and is appropriate mostly at low photoexcitation densities. Evidence for the cascade model was obtained in the form of peaks in the photocurrent and photoluminescence excitation spectra in GaAs [5.1 – 3] and in a number of different experiments in II-VI semiconductors [5.4 – 8] many years ago. If the photoexcitation density is sufficiently high, carrier-carrier collisions dominate and thermalize the carriers to a hot plasma before the emission of a large number of phonons. However, this hot plasma cools to the lattice temperature primarily by emitting optical phonons for carrier temperature $> \approx 40$ K. For temperature below 40 K and for carrier energies less than the optical-phonon emission, acoustic-phonon emission may become important (Fig. 4.1). The difference in the energy of the photoexcited carriers and the average energy of the carrier distribution is given to the lattice primarily in the form of optical phonons. This is true whether the cascade process dominates or the photoexcited carriers thermalize to a hot plasma which then cools towards the lattice temperature. Therefore, these relaxation and cooling processes, discussed in previous chapters, generate a large number of optical phonons in most photoexcitation experiments. Large population of nonequilibrium phonons can also be generated using coherent mixing techniques.

The dynamics of phonons is governed by the relaxation dynamics of the photoexcited carriers and the decay of phonons by anharmonic processes into acoustic phonons. The phonons themselves influence the dynamics of carrier cooling, as discussed in Chap. 4. Therefore, investigation of phonon dynamics may provide additional complementary information about dynamical processes in semiconductors and their nanostructures.

This chapter discusses the information obtained on the dynamics of photoexcited phonons by incoherent and coherent time-resolved light-scattering techniques treated in Chap. 1. Section 5.1 begins with a discussion of phonon generation and detection in bulk semiconductors like GaAs, and then reviews the results obtained in GaAs. Section 5.2 begins with a discussion of phonons in quantum wells, phonon generation and detection in quantum wells, and then deals with results obtained in GaAs quantum wells. In bulk GaAs, the primary information obtained is about the phonon-decay rates and their variation with lattice temperature. Since the phonon modes in quantum wells are much more complicated, the analysis becomes more complicated. However, the complexity also provides an opportunity to obtain additional information about carrier-phonon interactions for different modes and possibly use this information to distinguish between different models for phonon modes in quantum wells. We will conclude in Sect. 5.3 with a brief summary and conclusions.

Investigation of the phonon dynamics in semiconductors and their nanostructures, using both incoherent and coherent generation of phonons, has been a very active field of research [5.9–49] and a number of excellent review articles are available [5.28, 33, 39, 41, 50]. Finally, the nonequilibrium acoustic phonons generated by the decay of the optical phonons or by some other means can be investigated by using completely different techniques [5.16, 51–53], but will not be discussed here.

5.1 Phonon Dynamics in Bulk Semiconductors

Sections 5.1.1–1.4 discuss incoherent generation of optical phonons by photoexcited carriers in bulk semiconductors, with emphasis on GaAs, and their detection by incoherent Raman scattering using time-resolved pump and probe technique. Section 5.1.5 treats the coherent generation of optical phonons using two ultrashort lasers with the difference frequency tuned to the optical-phonon energy, and the detection of these phonons by Coherent Anti- Stokes Raman Scattering (CARS).

5.1.1 Phonon Generation in Bulk Semiconductors

A schematic of the generation of optical phonons of energy E_{phonon} by an electron initially at an energy E_i in the conduction band is depicted in Fig. 5.1. For simple parabolic bands with an effective mass m, the minimum and maximum phonon wavevectors generated by this process are given by energy and momentum conservation

$$q^{\min} = \sqrt{2m/\hbar^2}\left(\sqrt{E_i} - \sqrt{E_i - E_{\text{phonon}}}\right) , \tag{5.1}$$

$$q^{\max} = \sqrt{2m/\hbar^2}\left(\sqrt{E_i} + \sqrt{E_i - E_{\text{phonon}}}\right) , \tag{5.2}$$

where E_{phonon} is the phonon energy. For an electron in the Γ valley of GaAs, with an energy of 200 meV, $q^{\min} \approx 5\times10^5\ \text{cm}^{-1}$ and $q^{\max} \approx 1.1\times10^7\ \text{cm}^{-1}$. Similar equations apply for phonon absorption. In a typical III–V semiconductor, the conduction band is mostly s-like so that it interacts with optical phonons through a Fröhlich interaction for which the interaction rate is in-

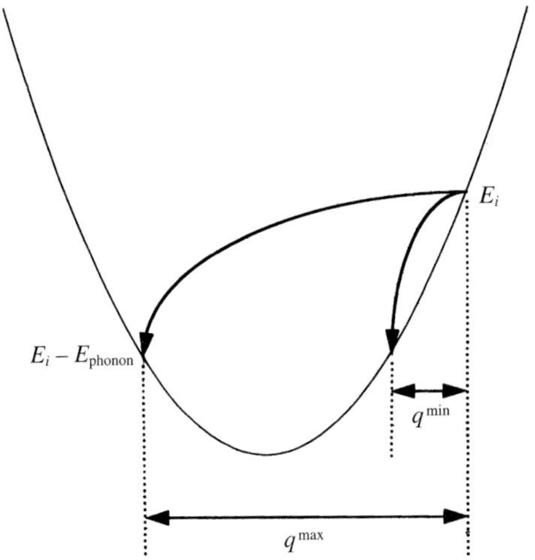

Fig. 5.1. Schematic diagram for generation of phonons by an electron of energy E_i in a parabolic band with the effective mass m. The band structure imposes limits on the minimum and the maximum wavevectors of emitted phonon

versely proportional to the square of the phonon wavevector. As a result the small wavevector phonons are preferentially produced, and the distribution is strongly peaked near $q^{\min}$ [5.12].

These phonons can scatter to other wavevectors within the optical phonon branch, be reabsorbed by the carriers or decay into two acoustic phonons through anharmonic interactions (see the phonon-dispersion curves in GaAs in Chap. 1). It is generally believed that the scattering within the phonon branch is not important and that the decay into acoustic phonon is the dominant decay process for optical phonons at all wavevectors. If the decay times of the optical phonons into acoustic phonons is sufficiently long, the phonon population becomes much larger than the equilibrium population at the lattice temperature. This non-equilibrium phonon population is a function of the phonon wavevector, and it is generally not possible to describe the entire phonon population by a single temperature. Nonetheless, these nonequilibrium phonons are generally referred to as hot phonons, and we will also use this terminology. It is, however, clearly understood that these are nonequilibrium phonons with non-thermal distribution functions; the phonon distribution function in this case may be described by a *wavevector-dependent* "temperature" if desired.

Methods of calculating the phonon-distribution function were considered in Chap. 4. A representative calculation for the steady-state non-equilibrium phonon-distribution function in GaAs under certain specific conditions is displayed in Fig. 4.7 as a function of the phonon wavevector. The calculation shows that the phonon-distribution function peaks near $10^6\ \text{cm}^{-1}$, and has certain minimum and maximum wavevectors as discussed above. Since there is a distribution of electron energies (corresponding to the Maxwell-Boltzmann distribution function at 50 K), the minimum and maximum are not abrupt.

5.1.2 Phonon Detection by Raman Scattering

Raman scattering provides an ideal technique for probing the phonon distribution in bulk semiconductors [5.27]. In such experiments one generally uses the back-scattering geometry so that the phonon wavevector probed by the Raman spectroscopy is approximately given by twice the incident wavevector of the photon

$$q^{\text{Raman}} \approx 4\pi n(\lambda)/\lambda \ , \tag{5.3}$$

where $n(\lambda)$ is the refractive index of the semiconductor at the wavelength λ of the incident laser. For excitation near 2 eV, $q^{\text{Raman}} \approx 8\times10^5\ \text{cm}^{-1}$, a value that is close to the peak of the generation rate of optical phonons by photoexcited electrons in GaAs as discussed above. q^{Raman} can be varied slightly by changing the angle of backscattering, and by a large amount by measuring 90° or near-forward Raman scattering. For the latter two cases, one would need to use either a very thin sample or a laser in the transparent region of the sample. The technique of time-resolved Raman scattering as a function of time delay between the pump and probe pulses was discussed in Chap. 1.

5.1.3 Steady-State Results in GaAs

The first evidence for the photoexcited non-equilibrium optical-phonon population was obtained by cw Raman scattering experiments in GaAs [5.54]. GaAs has longitudinal optical (LO) and transverse optical (TO) phonon branches. The LO-TO splitting at the zone center is ≈ 3 meV or 24 cm^{-1}. A crystal of GaAs at 300 K was excited with a cw argon laser operating in the blue-green spectral region, and the same laser was also used for measuring the Raman scattering spectra. The scattered light was collected in the back-scattering geometry and the polarizations of the incident and the scattered beams were such that both the TO and LO phonons were allowed. Figure 5.2 shows the Stokes and anti-Stokes Raman spectrum for two different intensities of the laser. It can be deduced from the figure that, with increasing excitation power, the ratio of the anti-Stokes to Stokes intensity for the LO phonon increases but the same ratio for the TO phonon remains essentially unchanged. Since the ratio of the anti-Stokes to the Stokes intensity is a direct measure of the phonon occupation number at the phonon wavevector probed by the Raman geometry, these results indicate that the LO phonon population increases while the TO phonons (and hence the lattice temperature) remain essentially unperturbed.

Figure 5.3 shows that the LO phonon occupation number deduced from the Raman spectra increases nearly linearly as a function of the photoexcitation intensity, except at the highest intensity where the lattice temperature, as deduced from the TO phonon occupation number, also increases. The highest photoexcitation intensity was $\approx 1\times10^4$ W/cm^2. A phonon lifetime was de-

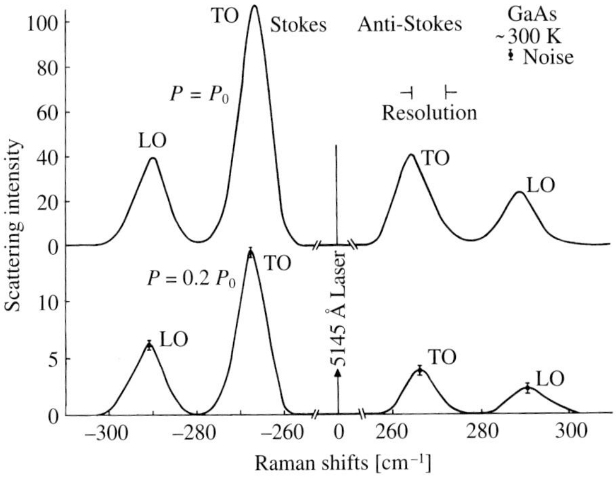

Fig. 5.2. Raman spectrum of GaAs at 300 K obtained with a cw argon-ion laser for two different intensities of excitation [5.54]

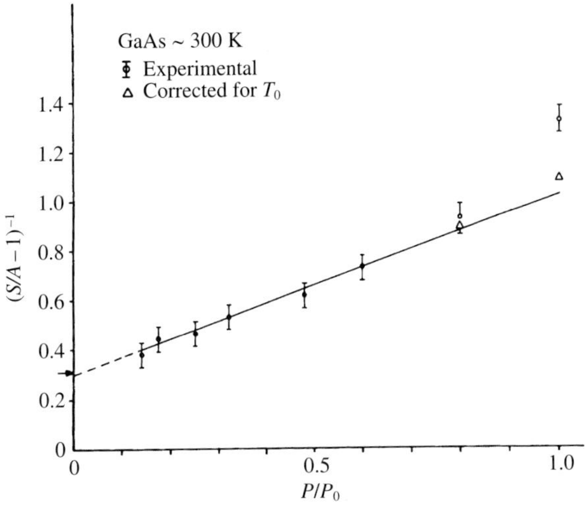

Fig. 5.3. LO-phonon occupation number deduced from the ratio of the anti-Stokes to Stokes Raman intensity as a function of cw argon-laser power used to excite the crystal of GaAs at $T_L = 300$ K [5.54]

duced from these results using a simple model. This approach was examined in more detail in later investigations [5.50, 55].

These early results clearly indicate that hot phonon are indeed generated by the photoexcitation and that the interaction with LO phonon is stronger than that with the TO phonons. This is what one expects because electrons, which are created with larger kinetic energy and are the primary source of optical-phonon generation, interact primarily with the LO phonons. Many other investigations have been reported subsequently [5.33, 39]. For example, *Kash* et al. [5.24] compared back scattering to forward scattering and showed that there are no hot phonons near $q = 0$, as expected from (5.1).

Apart from providing important insights into physics, the single-beam Raman scattering experiment described above demonstrates that light scattering provides a direct probe for the phonon population. Therefore, by combining it with ultrashort pulses, it should be possible to directly probe the dynamics of the phonons. We concentrate on this dynamical behavior in the remainder of the chapter.

5.1.4 Phonon Dynamics in GaAs

The first measurement on the dynamics of hot phonons in semiconductors was performed by *von der Linde* et al. [5.9] in GaAs at 77 K. They used a variant of the excite and probe technique (Chap. 1) utilizing a 2.5 ps dye laser at 2.16 eV for both pump and probe beams. The pump, polarized along [010] direction, excited the (100) face of GaAs. The probe was of equal intensity but polarized along the [001] direction and the scattered light with [010] polarization was measured. This geometry blocked the collection of LO phonon Raman signal generated by the pump beam, and allowed the measurement of the LO phonon Raman signal generated by the probe beam. The probe was delayed with respect to the pump by a variable delay τ_d. Since the equilibrium LO phonon population is very small at 77 K, the anti-Stokes LO signal provides a direct measure of the LO phonon population. The anti-Stokes signal consists of two parts: that induced by the phonons generated by the pump and that induced by the phonons generated by the probe pulse. The latter is independent of the delay τ_d, and provides a constant background which can either be subtracted or eliminated by using appropriate chopper and phase-locked detection.

Von der Linde et al. [5.9] demonstrated that the anti-Stokes signal is not due to lattice heating and that it must be attributed to hot phonons. Figure 5.4 shows, on a logarithmic time scale, the temporal evolution of the anti-Stokes LO phonon signal after subtracting the constant background, with the zero of time corresponding to the peak of the pump pulse. It reveals a 10–90% rise time of 6 ps, consistent with the pulse widths and a fast generation of optical phonons. At longer times, the signal decays exponentially with a time constant of 7 ± 1 ps. The inverse of this time constant determines the net rate at which phonon leave the k-space region probed by the anti-Stokes Raman scattering. This decay rate is determined by the anharmonic LO phonon decay into acoustic phonons, any scattering process within the phonon branch which removes the phonon from the k-space being probed (7.6 to 7.8×10^5 cm^{-1}), and reabsorption effects related to the presence of the high-density electron-hole plasma. These researchers also measured the Raman line width carefully, and found that it corresponded to 6.3 ps. The Raman line width is determined by all scattering processes which dephase phonons, including the anharmonic decay of optical phonons into acoustic phonons. The fact that the two numbers are the same within the experimental error was interpreted to indicate that the anharmonic decay of phonons was the dominant factor in the phonon dynam-

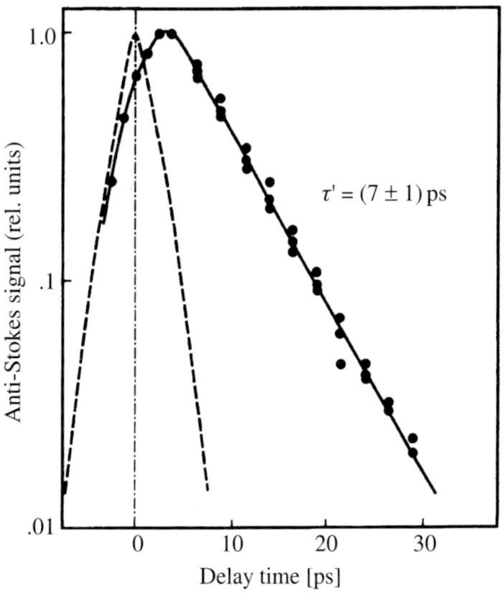

Fig. 5.4. The intensity of the anti-Stokes Raman scattering signal generated by the probe pulse from GaAs at $T_L = 77$ K photoexcited by a pump pulse, as a function of time delay between the pump and the probe. The *dashed curve* is the autocorrelation function of the laser pulse [5.9]

ics experiment. It might be mentioned that one calculation [5.56] indicated that the anharmonic decay time in GaAs is ≈ 30 ps, considerably longer than the experimental values.

Additional information concerning the temperature dependence of the phonon-decay rate, and concerning the rate of *creation* of optical phonons was obtained by *Kash* et al. [5.17, 23, 39, 57], who investigated the phonon generation and decay in GaAs by studying time-resolved Raman scattering from the (100) surface of GaAs using 600 fs compressed pulses from a dye laser. The principle of the measurement is the same as above, but the signal levels are extremely low as a result of the wide spectral width of the pulse, low excitation density, $(1-10)\times 10^{16}$ cm^{-3}, and probe intensity only 10–20% of the pump intensity. Furthermore, a subtraction procedure was needed to remove the background thermal-phonon population at room temperature, and obtain the anti-Stokes intensity attributable to the photoexcited phonons. Their measured anti-Stokes signal attributable to photoexcitation was less than 0.1 counts/s, but they were able to obtain reasonable data on the rise of the spectrally-integrated photoexcited anti-Stokes LO phonon signal. Initial data were obtained for 5×10^{16} cm^{-3} [5.17] but later results were obtained for a range of intensities.

Results for the anti-Stokes Raman scattering intensity obtained by *Kash and Tsang* [5.39, 57] are shown in Fig. 5.5 as a function of time delay for three different temperatures at an excitation density of 2×10^{16} cm^{-3}. The results indicate that the phonon decay-time constant increases with decreasing temperature, from about 3.5 ps at 300 K to 6.4 ps at 80 K to 8.8 ps at 8 K [5.23]. These values are in agreement with the results of *von der Linde* et al. [5.9] discussed above. On the basis of the measured temperature dependence of the decay rate, it was concluded that the LO phonons decay into two LA phonons

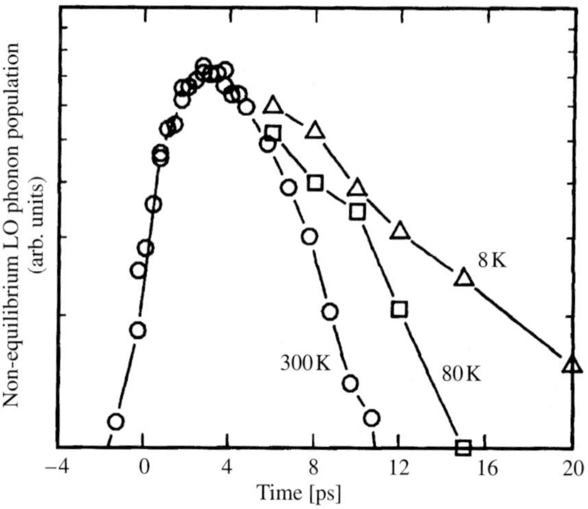

Fig. 5.5. The intensity of the anti-Stokes Raman scattering from GaAs in the region of the optical-phonon energy as a function of time delay for three different lattice temperatures [5.39]

of equal energy and opposite wavevectors. These results should be compared with the results of coherent techniques [5.58–60] discussed in Sects. 2.11 and 5.1.5. More recent measurements using CARS [5.60], to be discussed in Sect. 5.1.5, have reached a different conclusion.

The results of *Kash* and *Tsang* [5.39] for the anti-Stokes intensity as a function of time delay for GaAs at 300 K for two different excitation-photon energies [5.39] are exhibited in Fig. 5.6. Also shown are the calculated curves for

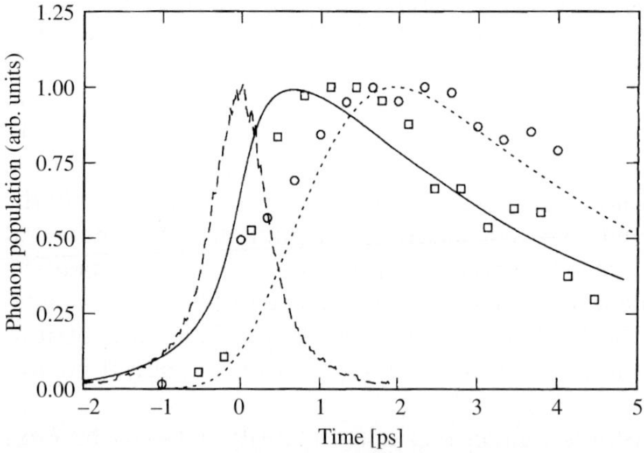

Fig. 5.6. The intensity of the anti-Stokes Raman scattering from GaAs at 300 K in the region of the optical phonon energy as a function of time delay. *Squares* (*circles*): excitation-photon energy 2.09 eV (1.91 eV) at an excitation density of 1×10^{16} cm^{-3} (8×10^{16} cm^{-3}). The *dashed curve* shows the autocorrelation of the 1.91 eV laser pulses. The *dotted curve* represents the predicted response if the phonons were generated instantly and decayed with a time constant of 3.5 ps. The *solid curve* gives the predicted response for the 2.09 eV excitation using the cascade model with a 190 fs phonon emission time and a 3.5 ps phonon decay-time constant [5.39]

two different conditions as described in the figure caption. The results indicate that the peak of the anti-Stokes signal occurs at times much longer than expected if the phonons are generated instantly. *Kash* et al. [5.17] have argued that most of the phonons detected in this Raman scattering experiment are emitted by electrons with kinetic energy between 100 and 200 meV. By assuming the cascade model, the average time for the emission of LO phonons was deduced to be 165 fs [5.17] and 190 fs [5.39]. *Kash* and *Tsang* [5.39] have argued that for the low-density results presented in Fig. 5.6 electron-acceptor luminescence results support the cascade model. However, carrier-carrier scattering dominates at higher densities, as shown recently for densities as low as 1×10^{17} cm^{-3} [5.61, 62], and also supported by ensemble Monte-Carlo simulations at densities of 5×10^{16} cm^{-3} discussed in Sect. 5.1.6. In this case, the electrons thermalize quickly and the peak in the anti-Stokes signal occurs when the electron-distribution function is such that the density of the electrons is maximum between 100 and 200 meV. In this case, the cooling of the photo-excited plasma discussed in Chap. 4 determines the phonon creation dynamics, and must be considered in analyzing the Raman data on phonon dynamics.

Another interesting application of the generation of nonequilibrium phonons by photoexcitation and their detection by Raman scattering was provided by *Collins* and *Yu* [5.12, 13]. They excited a crystal of bulk GaAs at 10 K with 4 ps pulses from dye lasers which were tuned from about 1.75 to 2.17 eV. They found that the nonequilibrium phonon-occupation number depended strongly on the excitation photon energy, as shown in Fig. 5.7 for zero time delay between the pump and the probe. The increase in the phonon population between 1.75 and 1.95 eV can be accounted for by the increase in the phonon generation either in the cascade model, or in the hot-plasma cooling model. *Collins* and *Yu* [5.12, 13] carried out a detailed analysis including non-parabol-

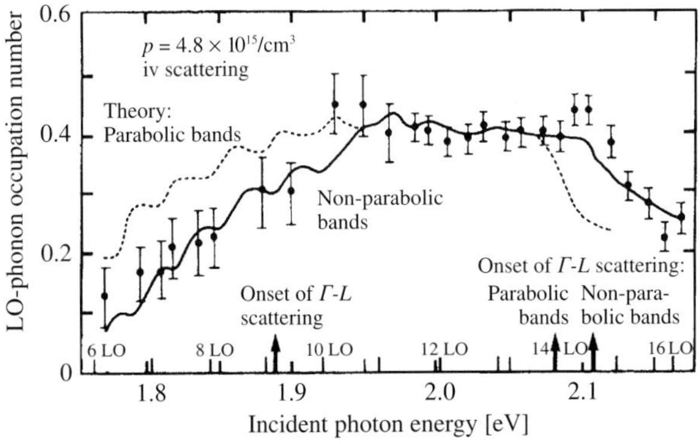

Fig. 5.7. LO-phonon occupation number as a function of the pump-photon energy deduced from the ratio of anti-Stokes and Stokes Raman intensities from GaAs at 10 K excited with 4 ps dye laser pulses, for zero time delay between the pump and the probe pulses. *Solid circles* are the experimental data and the *dashed* and *solid curves* are calculations for parabolic and non-parabolic bands [5.12, 13]

icity effects and the influence of intervalley scattering. Intervalley scattering rates were determined from this analysis. It turns out that their measurement was not very sensitive to the value of the intervalley scattering rate [5.63] so more accurate value for this parameter was later obtained using different techniques discussed in Chap. 3. However, these results indicate that considerable amount of complementary information about relaxation processes in photoexcited semiconductors can be obtained by using the technique of time-resolved Raman scattering from phonons.

5.1.5 Coherent Generation and Detection of Phonons

We have so far considered the phonons generated by photoexcited electrons and holes as they relax to equilibrium. Such phonons are generated incoherently, i.e., they do not have any well defined phase relationship with each other. Phonons in semiconductors and other solids, and vibrational modes in molecules may also be generated by completely different methods. For example, infrared photons in the Restrahlen region of the semiconductor can be absorbed directly by lattice vibrations and generate appropriate optical phonons [5.64]. Phonons and phonon polaritons can also be generated by mixing two phase-coherent, temporally overlapping lasers with photon energies differing by a phonon energy. The coherent phonons so generated may be probed by a third, delayed laser by measuring the Coherent Anti-Stokes Raman Scattering (CARS) as a function of the time delay between the third laser and the first two [5.65]. This is very similar to the four-wave-mixing experiments discussed in Chap. 2, and also requires energy and momentum conservation conditions (phase matching) for strong signals. Such CARS measurements provide a powerful technique for measuring the dephasing of the phonons, which is controlled not only by the decay of the optical phonons into acoustic phonons, but also by scattering within the optical branch and any other phase-destroying collision processes. This technique has been applied to a number of different solids and also to some semiconductors and is reviewed by *von der Linde* [5.28] and *Bron* [5.44, 49]. We discuss here some results on GaAs and GaP.

CARS experiments for studying LO-phonon dynamics require at least two ultrafast pulse sources at photon energies in the transparency region of the sample, and photon energies ω_l and ω_s such that $\omega_l - \omega_s = \omega_{LO}$, the LO-phonon energy at zone center. Excitation of the sample with the laser beam at ω_s and a part of the laser beam at ω_l coherently excited LO phonons in the crystal. The second part of the laser beam at ω_l is variably delayed, and used to measure CARS at $\omega_l + \omega_{LO}$ in the phase-matched direction. Reasonably high pulse energies are required to obtain good signal-to-noise ratio, but the excitation density can be kept low by exciting a large volume.

Phonon dynamics in GaAs has been investigated by *Vallee* and *Bogani* [5.60] using CARS. 5 ps Pulses from a Nd:YAG laser operating at 1.054 µm were split into three beams, the first and second were used as pump and probe beams (30 µJ and 10 µJ, respectively) at ω_l, and the third beam was used to

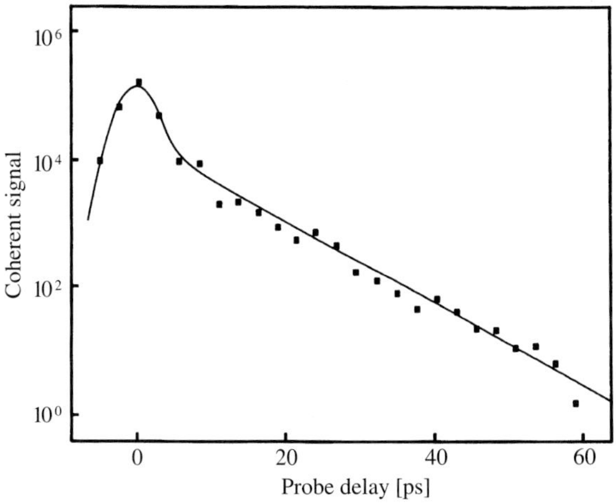

Fig. 5.8. CARS signal from semi-insulating GaAs at 77 K as a function of the time delay between the pump and probe pulses. The long time signal is due to phonons and decays exponentially with a time constant of 6.4 ps [5.60]

generate the ω_s by stimulated scattering in $CHBr_2Cl$ (10 μJ) such that the frequency detuning ($\omega_l - \omega_s$) matched the LO-phonon frequency in GaAs. A 625 μm thick high purity crystal of semi-insulating GaAs with (100) surfaces was used in these experiments. The excitation density was sufficiently low (much less than 10^{17} cm^{-3}) so that plasma-phonon interaction effects were not important, as verified by checking that the measured decay time was independent of intensity. Figure 5.8 shows the intensity of the anti-Stokes-Raman scattering from GaAs at 77 K. The fast initial response is limited by the pulse width in this case. This very fast response is in general agreement with the results of coherent spectroscopy of electronic states of semiconductors discussed in Chap. 2, and is indeed attributed to the electronic response. The long-time signal is from phonons. Note that the measurements have a dynamic range of about 10^3, considerably larger than the incoherent Raman scattering measurements discussed above. The measured decay time constant $\tau_{decay} = 6.4 \pm 0.4$ ps corresponds to a dephasing time of 12.8 ps.

The measured dephasing rate of the CARS signal ($\Gamma = 2/T_2$) includes contributions from all dephasing phenomena, such as elastic scattering within the phonon branch (pure dephasing with rate $2/T_2'$), as well as from the anharmonic decay of the optical into two or more acoustic phonons (decay rate $1/T_1$) i.e., $\Gamma = 2/T_2 = (2/T_2') + 1/T_1$. Therefore, the anharmonic decay time constant of the optical phonon can not be shorter than the measured $T_2/2$. The measured decay time constant can be equated with the anharmonic population decay time constant if the anharmonic decay is the only dephasing process. A maximum value of 9.2 ± 0.6 ps was obtained for $T_2/2$ at 6 K.

The phonon decay rate ($\Gamma = 2/T_2$) was deduced from measurements between 6 and 300 K with a dynamic range of at least three orders of magnitude. Figure 5.9 exhibits the measured decay rates as a function of the lattice temperature. Also shown for comparison is the expected temperature dependence of the anharmonic decay rates for three different models of the anharmonic decay

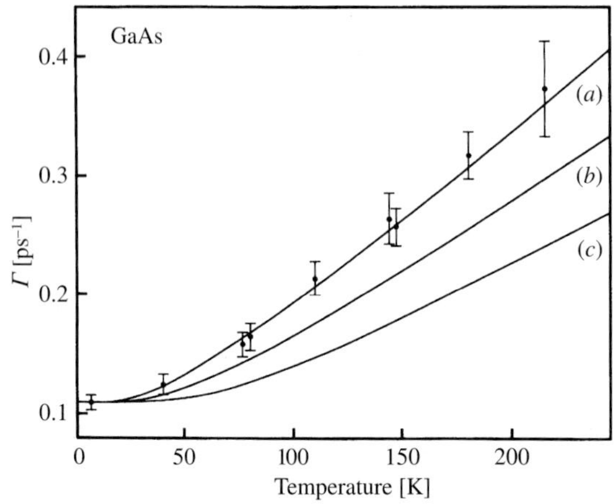

Fig. 5.9. The reciprocal of the measured decay rate of the CARS signal as a function of the lattice temperature, deduced from curves such as shown in Fig. 5.8. Also shown are the calculated decay time constants for the anharmonic decay of the LO phonon into (*a*) a TA(L) and a LO(L) phonons, (*b*) a TA(X,K) and a LA(X,K) phonon, and (*c*) two LA phonons of half the LO energy [5.60]

of the optical phonons in GaAs. Such a comparison is meaningful if the dephasing is substantially due to the anharmonic decay. The data reveal significant deviation from the model of anharmonic decay of the LO phonon into two LA phonon of equal energy and opposite wavevectors considered earlier [5.39, 57]. The results in Fig. 5.9 indicate that the best fit with the data is obtained for the model which considers the zone center LO phonons decaying into TA(L) and LO(L) acoustic phonons.

The phonons in GaP and the influence of electron-hole plasma on the phonon dephasing have been investigated extensively using CARS [5.14, 15, 29, 42, 44, 49, 66, 67]. The CARS signal measured from a high-purity crystal of GaP at 5 K excited to a density of 10^{16} cm^{-3} is plotted as open circles in Fig. 5.10 as a function of the time delay between the first two pulses and the third probe pulse [5.14, 15, 66]. Note that the data are obtained over a dynamic range of 10^5 with excellent signal-to-noise ratio. The initial response represents the electronic response and the slow response is due to the dephasing of phonons [5.14, 15, 66], as confirmed by the disappearance of the slow component when the detuning $\omega_l - \omega_s$ is different from the optical-phonon frequency. The decay time constant of $\tau_{decay} = 26$ ps is much longer than that measured in earlier measurements, presumably because the high purity of this sample eliminated some of the dephasing processes contributing to the earlier results [5.10, 28]. The measured decay time shows that the dephasing time T_2 of the LO phonons in high-purity GaP is about 50 ps. For the results presented in Fig. 5.10, it was shown that the primary process contributing to the dephasing of phonons is the anharmonic decay into two acoustic phonons, so that the anharmonic decay time in GaP is ≈ 26 ps, considerably longer than the

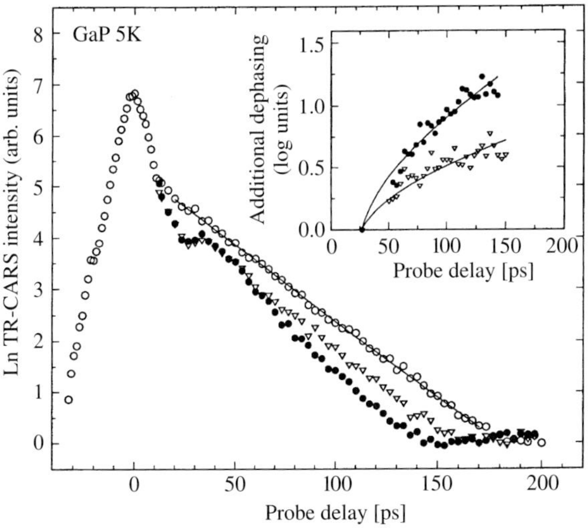

Fig. 5.10. CARS signals from a high-purity crystal of GaP at 5 K as a function of time delay. *Open circles:* no plasma present, *closed circles* and *open triangles:* in the presence of an electron-hole plasma at two different intensities. The data in the absence of a plasma show an exponential decay with a time constant $\tau_{\text{decay}} = 26 \pm 2.5$ ps. The additional dephasing induced by the plasma is displayed in the inset [5.42]

values discussed above and in Sects. 5.1.4 and 2.11 for the phonon decay time in GaAs.

Bron and collaborators [5.42, 44, 49] also investigated the influence of an optically excited two-component plasma on the dephasing of phonons in GaP. In addition to the three low-intensity beams used for the CARS experiment, they used a fourth intense beam at ω_s, at a fixed time after the coherent generation of LO phonons, to excite a two-component electron-hole plasma via two-photon absorption. The results for two different densities are also depicted in Fig. 5.10. The dip at 26 ps is consistent with the arrival of the pulse injecting the plasma, but its origin is not yet understood. The results show that the dephasing rate of the phonons increases following the injection of the electron-hole plasma. The additional dephasing induced by the plasma is plotted in the inset and compared with a theoretical model which incorporates interaction between the phonons and the plasma thorough a polarization field. These results illustrate that variations of the CARS technique provides a means of obtaining additional information about the interaction of phonons with other elementary excitations in the semiconductor.

The discussion is this section indicates that coherent generation and detection of phonons provides considerable information about phonon dynamics in semiconductors, often with much better dynamic range and signal-to-noise ratio. We conclude this section by noting that the two lasers differing in photon energy by a phonon frequency may, in fact, be derived from a single femtosecond laser with spectral width larger than the phonon frequency. For materials like GaAs, with the optical-phonon energy of ≈ 37 meV, one would require a pulse shorter than 50 fs to achieve this. The coherent excitation of phonons observed in Sect. 2.11 provided an excellent example of this. It might also be remarked here that various pulse-shaping techniques have been developed in recent years which allow great flexibility in modifying the pulse shapes and have been applied for impulsive excitation of molecules and semiconductors as discussed in Sect. 2.12.

5.1.6 Monte-Carlo Simulation of Phonon Dynamics in GaAs

One way to obtain a deeper understanding of the physics involved is to compare the experiments to an appropriate Monte-Carlo simulation. We discuss here the results of an ensemble Monte-Carlo calculation by *Lugli* et al. [5.63] who considered a realistic band structure of GaAs and InP. For excitation of GaAs near 2 eV, transitions from the three valence bands create electrons with different kinetic energies in the conduction band. Some of these electrons are at sufficiently high energies that they can transfer to subsidiary conduction band valleys as discussed in Sect. 3.2 dealing with intervalley scattering. *Lugli* et al. [5.63] calculated the electron-distribution function as a function of time delay following excitation of bulk GaAs at 300 K by a 400 fs pulse for an excitation density of 5×10^{16} cm^{-3}. Figure 5.11 exhibits the calculated results for three different time delays measured from the center of the laser pulse. At early times the distribution function is clearly non-Maxwellian with a weak shoulder for $E = 0.24$ eV and a rapid drop beyond $E = 0.3$ eV because of the transfer

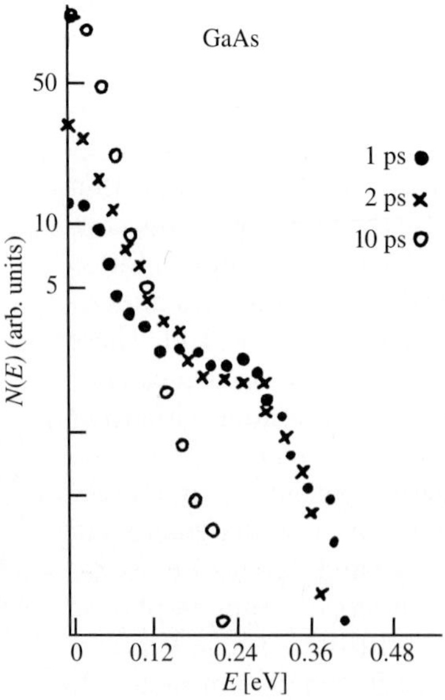

Fig. 5.11. Electron-distribution function in GaAs at 300 K following photoexcitation at 2 eV using 400 fs pulses at three different time delays, calculated using ensemble Monte-Carlo simulations with realistic band structure [5.63]

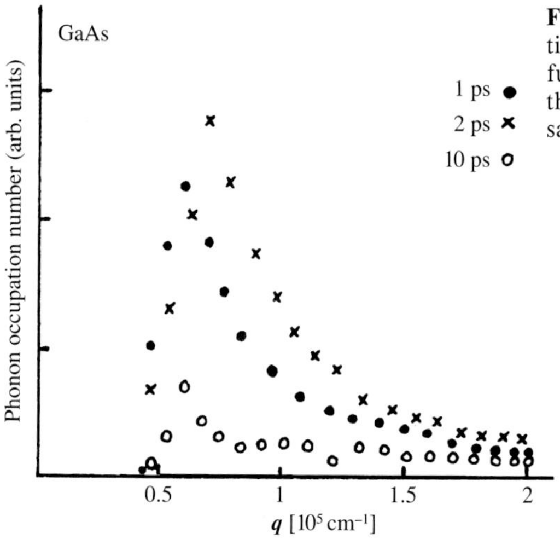

Fig. 5.12. Calculation phonon-occupation number in GaAs at 300 K as a function of the phonon wavevector for three different time delays, for the same parameters as in Fig. 5.11 [5.63]

of electrons to the subsidiary valleys. Note that the distribution function is not a series of peaks expected for the cascade model, demonstrating the inadequacy of the cascade model at even short times and moderately low densities.

Figure 5.12 plots the calculated phonon-occupation numbers under the same excitation conditions at the same three time delays. The calculations assume that the anharmonic decay rate of the phonons is independent of the phonon wavevector, and is taken to be 3.5 ps at 300 K and 7 ps at 77 K to match the experimental values [5.9, 17, 39]. The results clearly show a peak in the phonon-occupation number near 8×10^5 cm^{-1}, close to the wavevector probed by Raman back-scattering technique. These results also indicate that the phonon-occupation number decays much more rapidly at smaller wavevectors than at larger wavevectors. This is primarily because the phonons at smaller wavevectors are much more likely to be reabsorbed by the hot-electron gas. The calculated time evolution of the anti-Stokes Raman-scattering signal are in good agreement with the data of *von der Linde* et al. [5.9] and *Kash* et al. [5.17, 39] discussed above. One can conclude from these results that, while the cascade model may well be valid in low-density experiments, the experimental results can also be explained when the electron-distribution function does not show (Fig. 5.11) the characteristic peaks expected for the cascade model. Also, while the anharmonic decay term is assumed to be independent of phonon wavevector, reabsorption of phonons very much depends on the phonon wavevectors. This is because the electron-distribution function is evolving during the time that the Raman scattering is measuring the phonon evolution. As a result of the strong role played by phonon reabsorption, the phonon decay time depends on the phonon wavevector so that the measured decay time does not necessarily reflect the phonon decay rate [5.63].

5.2 Phonon Dynamics in Quantum Wells

In this section, our goal is to discuss the dynamics of hot phonons in quantum wells. Quantum confinement brings about changes not only in the nature of phonons, but also in the momentum distribution of the phonons generated by relaxation of carriers. The match between the wavevectors of the phonons generated by the carriers and the wavevector probed by Raman scattering is also different. We first consider these issues and then discuss the information that time-resolved Raman scattering has provided about phonons in quantum wells. No experiments investigating the dynamics of coherently generated phonons have been reported in quantum wells.

5.2.1 Phonons in Quantum Wells

The nature of phonons in bulk semiconductors has been extensively investigated by neutron scattering as well as by light scattering techniques, and is well understood for elemental semiconductors like Si and Ge, compound semiconductors like GaAs, and their ternary (e.g., InGaAs) and quaternary alloys (e.g., InGaAsP). Quantum confinement brings about many additional complexities to the phonon modes, just as it does to the electronic band structure. There still appears to be some discussion in the literature about the relative merits of different approaches used to calculate phonon modes in quantum wells. Since this subject has been discussed at length in the literature [5.41, 46, 68 – 87], we restrict the discussion here to some elementary remarks concerning the optical-phonon modes. Zone-folding of acoustic phonons leads to many interesting effects [5.88, 89], but will not be discussed here.

Microscopic lattice dynamical models which consider displacement of individual atoms in the lattice in a proper way have been discussed in the literature [5.79, 81, 82, 85, 90]. While the microscopic approach is clearly desirable, calculations based on the microscopic model are computation intensive, and become extremely time-consuming and difficult when the barrier material is a ternary semiconductor such as AlGaAs rather than a binary semiconductor such as AlAs. It is therefore desirable to introduce simplified models which can provide reasonable estimates for various rates and provide physical insights into the phonon dynamics.

Three such models have been introduced and discussed in the literature. These models assume a dielectric continuum with appropriate boundary conditions at the interfaces. The slab mode or the dielectric continuum model [5.68] assumes electrostatic boundary conditions, which require that the potential vanish at the interface, but does not put any boundary conditions on the atomic displacement at the interface. This model predicts the existence of both confined phonon modes whose displacements have anti-nodes at the interface and whose potentials vanish at the interface, and interface modes whose potential and frequency strongly depend on the in-plane wavevector. Both types of modes have indeed been observed in Raman scattering [5.90, 91]. Some aspects of the slab mode model have been criticized, notably the electrostatic boundary conditions which lead to discontinuities in the ionic displacement at the interface, and the omission of dispersion in the bulk optical-phonon mode branches.

In order to remove some of these objections, *Huang* and *Zhu* [5.74] proposed a modified-dielectric continuum model that satisfies, besides the electrostatic boundary condition, the continuity of the displacements at the interface, and includes the bulk-phonon dispersion. If the dispersion is neglected, then the scalar potentials of the confined modes are similar to those in the slab-mode model, but the ionic displacements in the Huang-Zhu model have nodes at the interface, in contrast to the slab-mode model. There is a corresponding change in the mode-numbering scheme in the two models, with the $n = 1$ mode in the slab-mode model corresponding to the $n = 2$ mode in the Huang-Zhu model. When the dispersion of the bulk-phonon modes is considered, the Huang-Zhu model predicts a qualitatively new feature, namely a mixing between the confined and interface modes. Such a mixing makes the classification into interface and confined modes questionable, especially for thin quantum wells.

A completely different continuum model was proposed by *Babiker* [5.70, 71]. In this mechanical model, hydrodynamic boundary conditions are applied to the continuum model and both retardation effects and the phonon dispersion are included in the model. This is known as the guided mode model and produces displacements similar to the Huang-Zhu model, so that mode numbering in these two models is similar. However, the scalar potentials are quite different, with the guided-mode model predicting anti-nodes and hence discontinuities at the interface.

The vibrational amplitudes (U_n) and potentials (V_n) for the confined longitudinal optical-phonon modes in a quantum well at position $-L/2 \leq z \leq L/2$ in these three continuum models are given by:

Slab modes (or dielectric continuum model) [5.68]:

$$U_{n^+} \propto \sin(n\pi z/L) \qquad n = 1, 3, 5 \ldots , \tag{5.4}$$

$$U_{n^-} \propto \cos(n\pi z/L) \qquad n = 2, 4, 6 \ldots , \tag{5.5}$$

$$V_{n^+} \propto \cos(n\pi z/L) \qquad n = 1, 3, 5 \ldots , \tag{5.6}$$

$$V_{n^-} \propto \sin(n\pi z/L) \qquad n = 2, 4, 6 \ldots ; \tag{5.7}$$

Guided modes (or mechanical model) [5.70, 71]:

$$U_{n^+} \propto \cos(n\pi z/L) \qquad n = 1, 3, 5 \ldots , \tag{5.8}$$

$$U_{n^-} \propto \sin(n\pi z/L) \qquad n = 2, 4, 6 \ldots , \tag{5.9}$$

$$V_{n^+} \propto \sin(n\pi z/L) \qquad n = 1, 3, 5 \ldots , \tag{5.10}$$

$$V_{n^-} \propto \cos(n\pi z/L) \qquad n = 2, 4, 6 \ldots , \tag{5.11}$$

Huang-Zhu model [5.74]:

$$V_{n^+} = \sin[\mu_n \pi z/(ma+a)] - 2[z/(ma+a)] \sin(\mu_n \pi/2) \qquad n = 3, 5, 7 \ldots , \tag{5.12}$$

$$V_{n^-} = \cos[n\pi z/(ma+a)] - (-1)^{n/2} \qquad n = 2, 4, 6 \ldots , \tag{5.13}$$

where $a/2$ is the lattice constant (2.83 Å for GaAs), $2m$ is the number of atomic layers in the quantum well (so that the quantum-well thickness is $L = ma$) and μ_n are the successive solutions of the equation

$$\tan(\mu\pi/2) = \mu\pi/2 \tag{5.14}$$

so that μ_n are close to n. These equations take into account the conclusion of the microscopic models that the potential and the displacement for the symmetric mode and their derivatives for the anti-symmetric modes vanish not exactly at the interface but some distance into the barrier, at $z = \pm(m+1)a/2$. The $n = 1$ mode is absent because the corresponding bulk mode becomes an interface mode in the Huang-Zhu model.

The guided-mode model does not allow the existence of the interface modes because of the use of rigid mechanical boundary conditions. The slab-mode model and the Huang-Zhu model predict the same properties of the interface modes. The expressions for the potential and the displacements for the interface modes are given in the literature [5.74, 90–92].

Rücker et al. [5.85] have compared the potentials and the displacements predicted by the three continuum models with the calculations for a microscopic model in Fig. 5.13 for the several lowest (i.e., highest frequency) GaAs-like optical-phonon modes for a 56 Å GaAs/AlAs quantum well ($2m = 20$ periods) . The $n = 1$ mode of the mechanical model has been shifted up to allow comparisons of modes with the same number of nodes. Specific values for the wavevectors are assumed for the calculation of the microscopic model, as indicated in the caption. Several interesting observations can be made immediately from Fig. 5.13. First of all, the mechanical or the guided modes model does not allow any interface modes as already mentioned. Secondly, while the potentials for the dielectric continuum model agree with those of the microscopic model, the displacements do not. Conversely, the displacements calculated from the mechanical model agree with those of the microscopic model, but the potentials do not. The modified continuum model proposed by *Huang* and *Zhu* [5.74] predicts displacements and potentials that agree with the microscopic model. Comparisons are made for specific wavevectors because the displacements and the potentials depend on the phonon wavevectors [5.85].

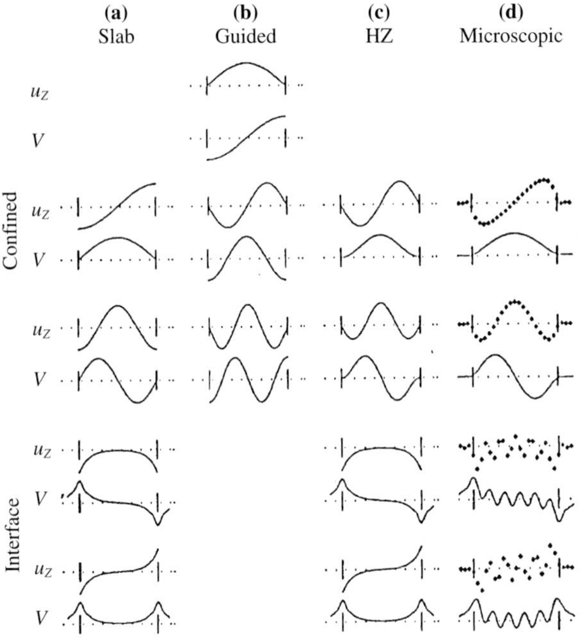

Fig. 5.13 a – d. Comparison of the z component of the displacement (u_z) and the corresponding potentials for different models for GaAs-like optical phonon modes in a (001)-oriented 56 Å (20 layers) GaAs quantum well surrounded by two AlAs barriers. The quantum-well interfaces are marked by the short vertical bars. The confined modes for the highest frequency (lowest mode number n) and two interface modes are displayed from top to bottom in order of decreasing frequency. (**a**) Dielectric continuum (slab-mode) model, (**b**) mechanical model, (**c**) Huang-Zhu model, and (**d**) microscopic model, with diamonds indicating the displacements of the anion. The *plots* are for $q_z = 0$ and $q_{\parallel} = 0.15$ Å^{-1}. The $n = 1$ modes of the mechanical model are shifted up to allow comparisons of the modes with the same number of nodes [5.85]

When the quantum well consists of a binary well and a binary barrier, there are two pairs of interface mode branches, with symmetric and anti-symmetric potentials. Most of the theoretical attention has been paid to such binary-binary quantum wells. When the quantum well, or the barrier or both, are made of ternary semiconductors, the situation becomes very complicated because a ternary semiconductor such as AlGaAs has two sets of optical-phonon modes, one a GaAs-like and another an AlAs-like mode. Therefore in a GaAs/AlGaAs quantum well, there is an overlap between the GaAs phonon in the quantum well and GaAs-like phonon branch in the barrier. This leads to new propagating optic modes and fewer confined modes, and complicates the carrier-phonon interactions.

5.2.2 Phonon Generation in Quantum Wells

We consider here the case of generation of phonons by electrons. The valence subbands are considerably more complicated but qualitatively similar considerations apply to the generation of phonons during the relaxation of holes.

The energy band diagram as a function of the in-plane electron wavevector, $k_{\|}$, is shown in Fig. 5.14 for two electron subbands with an inter-subband separation of Δ. Phonons can be generated either by inter-subband relaxation (process indicated by 1) or intra-subband relaxation of electrons (indicated by 2). In either case, the generated phonon has a well defined wavevector in the plane ($q_{\|}$), but not in the direction perpendicular to the plane. Energy and momentum conservation determine the wavevector of the emitted phonon. The magnitude of the phonon wavevector, $q_{\|}$, can be calculated in terms of the initial and final electron wavevectors. If the carrier is initially in the second subband with kinetic energy E_i, and if it relaxes to the first subband by emitting a phonon of energy E_{phonon}, then the minimum $q_{\|}$ is given by

$$q_{\|}^{\min} = \sqrt{2m/\hbar^2}\left|\sqrt{E_i+\Delta-E_{\text{phonon}}}-\sqrt{E_i}\right| , \tag{5.15}$$

where m is the carrier effective mass, and $E_i+\Delta-E_{\text{phonon}}$ must be positive. For a 100 Å GaAs quantum well with $Al_{0.3}Ga_{0.7}As$ barriers, $\Delta = 90$ meV so that for a phonon energy of 36 meV, $q_{\|}^{\min} = 3.1\times10^6\ \text{cm}^{-1}$ when $E_i = 0$. For higher initial energy, the minimum phonon wavevector is somewhat smaller.

From similar considerations it can be shown that the minimum in-plane phonon wavevector for intra-band relaxation of electrons is given by

$$q_{\|}^{\min} = \left(\sqrt{2m/\hbar^2}\right)\left(\sqrt{E_i}-\sqrt{E_i-E_{\text{phonon}}}\right) , \tag{5.16}$$

where E_i is the initial energy of the electron. This is similar to (5.1) except that it is only valid for the wavevector parallel to the interface planes. When the electron is exactly a phonon energy above the bottom of the subband, $q_{\|}^{\min} = 2.5\times10^6\ \text{cm}^{-1}$. Once again, the minimum wavevector decreases somewhat as the initial energy of the electron in the subband increases.

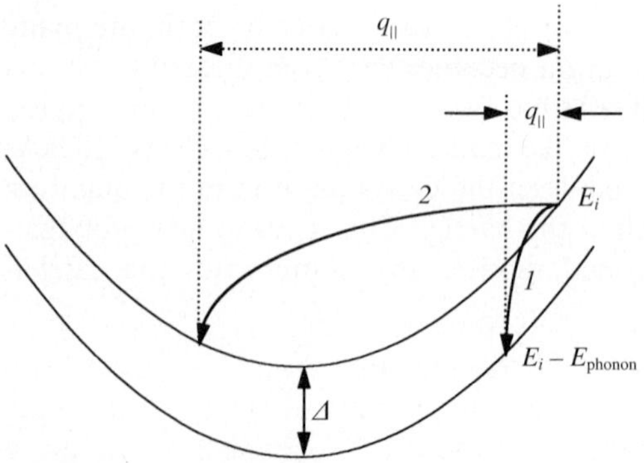

Fig. 5.14. Schematic of the phonon generation process in 2D

5.2.3 Phonon Detection by Raman Scattering

Most of the light-scattering experiments in semiconductors are performed in the nearly back-scattering geometry. The incident laser beam is typically incident at the Brewster angle (about 75° in GaAs) to the surface normal but the large refractive index (≈ 3.6 for GaAs) makes the internal angle rather small for the incident and scattered light in the back-scattering geometry. For GaAs quantum wells using a laser at 1.6 eV, close to the fundamental band gap energy of the quantum well, the *in-plane* phonon wavevector probed in Raman scattering experiments is $\approx 7 \times 10^4$ cm^{-1}, about one order of magnitude smaller than the $q_{\|}^{\min}$ generated by the relaxing electrons. The discrepancy is even larger for the phonons generated during the relaxation of holes. Note that this is in contrast to the situation in bulk semiconductors where the third dimension provides additional flexibility and makes the phonon wavevector probed in back scattering close to the wavevector for which the rate of generation of phonons is maximum.

These numbers indicate that there should be no contribution to the Raman scattering from the phonons generated during the relaxation of carriers photoexcited within the quantum well, unless there is efficient momentum scattering of phonons within the optical-phonon branches, or unless the selection rules are not strictly obeyed. In bulk semiconductors it is generally believed that such scattering within the optical-phonon branches is not strong.

In semiconductor heterostructures, Raman scattering from non-equilibrium phonons is indeed observed but only when either the incident or the scattered light is in resonance with an interband transition. It is believed that the selection rules are not strictly obeyed in resonant Raman scattering, and impurities and interface roughness may also play a role in the breakdown of momentum selection rules [5.46, 93]. While these processes make it possible to observe photoexcited hot phonons in Raman scattering experiments, the nature of these processes is not well understood. It is therefore essential to exercise sufficient caution in interpreting the results of these experiments.

One consequence of these considerations is shown in Fig. 5.15 which compares the phonon-occupation factors (assumed to be related to the anti-Stokes and Stokes intensities by the usual relation, see Sect. 5.2.6) of photoexcited bulk GaAs and GaAs quantum wells [5.41] as a function of photoexcited carrier density. For excitation with 1.3 ps pulses with photon energy of 1.945 eV, the phonon-occupation number in bulk GaAs at 38 K increases monotonically with density and then saturates, as shown in Fig. 5.15 a. Phonon-occupancy numbers close to 1 are achieved. Figure 5.15 b, c illustrate the results for two different quantum wells, which were excited with 5 ps pulses at photon energy 1.82 eV which is below the absorption edge in the barrier. For comparison with the bulk, the carrier sheet density per well is divided by the quantum well thickness to obtain a volume density. The most remarkable part of the comparison is that the phonon-occupation numbers in quantum wells are much smaller than those in the bulk. However, this does not imply that phonon generation in quantum wells is reduced. These results rather reflect the fact that the match

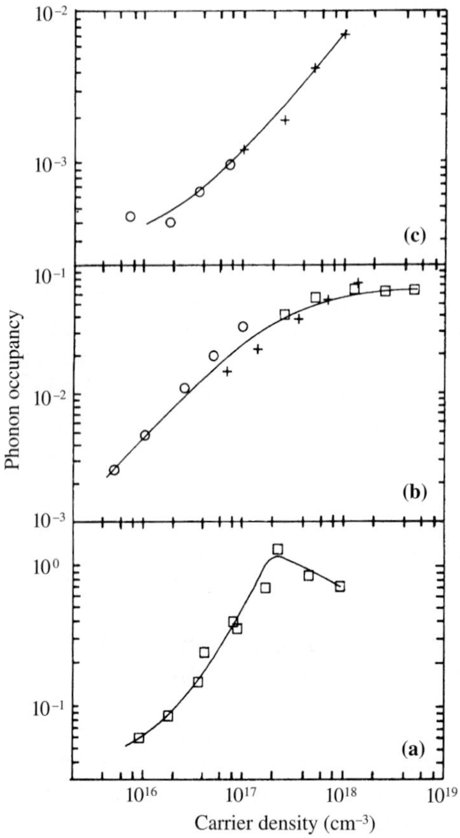

Fig. 5.15a–c. Hot-phonon occupation number, determined from the Stokes-anti-Stokes Raman scattering intensity ratios in the back-scattering geometry, as a function of photoexcited carrier density: **(a)** bulk GaAs, **(b)** 146 Å GaAs quantum well, and **(c)** 112 Å GaAs quantum well. The carrier densities for quantum wells are determined by dividing the areal density in the well by the well width [5.41]

between the phonons generated by the relaxation of photoexcited carriers and the phonons probed in the back scattering geometry is not very good.

In spite of this, light scattering has been successful in obtaining considerable information about phonon dynamics and the nature of electron-phonon interaction in quantum wells. We discuss some of these results in Sects. 5.2.5–2.7. Electronic Raman scattering measurements to investigate inter-subband scattering in quantum wells were discussed in Sect. 3.3.3b.

5.2.4 Monte-Carlo Simulation of Phonon Dynamics in GaAs Quantum Wells

Considerable progress has been made in recent years in modeling the phonon dynamics in quantum wells following ultrafast photoexcitation. The early work assumed the phonons to be bulk-like, but it was soon realized that this represented an unacceptable approximation. Considerable effort has now been devoted to understanding phonon modes, as discussed above. Monte-Carlo simulations which include these different phonon modes are now available and comparisons with experiments are possible in certain cases.

As an illustration, we discuss the Monte-Carlo simulations of phonon dynamics in a 51 Å GaAs/AlAs quantum well excited with 500 fs pulses at 2.153 eV [5.40]. The simulation used a microscopic model for the phonon modes in quantum wells [5.82]. The top three panels in Fig. 5.16 show the occupation number of the GaAs and AlAs interface modes, and of the slab modes in GaAs, as a function of the wavevector of the phonons for three different time delays following the photoexcitation. The bottom panel ilustrates the time evolution of the phonons at a specific wavevector. Several interesting observations can be made based on these results. First of all, the AlAs interface mode has the largest occupation number, indicating that this mode makes a strong contribution to the relaxation of photoexcited carriers in thin wells. The ratio between the occupation numbers of the AlAs and GaAs interface-phonon modes is about 3, in agreement with the time-integrated Raman data (Sect. 5.25). The Raman signal from the sample with the binary (AlAs) barrier was too weak to allow measurement of the phonon dynamics, so no direct comparison with experimental results was possible. The second point that the researchers made was that there is no evidence of the phonon-reabsorption effects of the kind that were discussed above for the bulk. Thus, the phonon population decays exponentially (Fig. 5.16) and the phonon-decay rate can be directly obtained from the measured decay rate of the anti-Stokes Raman signals.

5.2.5 Hot-Phonon Dynamics in GaAs Quantum Wells

We first discuss some results obtained by *Ryan* and coworkers [5.21, 31, 37, 40, 41, 43, 94]. Figure 5.17 shows the time-integrated Stokes and anti-Stokes Raman spectra from a multiple quantum well sample with 56 Å GaAs wells and 157 Å $Al_{0.36}Ga_{0.64}As$ barriers at 30 K in the $z(x, x)\bar{z}$ geometry where z is the direction of the incident beam parallel to the growth axis and $x = (110)$ is the direction of the incident and scattered light polarizations. The Stokes spectrum, excited by 4 ps pulses (spectral resolution of ≈ 0.1 meV) at 1.836 eV, in resonance with the $n = 2$ (C_2-HH_2) excitonic resonance, displays the even-order GaAs-confined phonon modes with $m = 2$ at 36.6 meV. It also shows the interface mode marked IF predicted by the slab mode as well by the Huang-Zhu models, and the TO-phonon mode. The time-integrated anti-Stokes spectrum, excited at 1.789 eV so that the outgoing (scattered) anti-Stokes photon is in resonance with the $n = 2$ (C_2-HH_2) excitonic resonance, is displayed in Fig. 5.17b. It reveals strong IF and $m = 2$ modes, with the IF mode approximately two times stronger than the $m = 2$ mode. While this does not provide an accurate quantitative comparison of the carrier-phonon scattering rates for the two phonons, it provides a qualitative idea of their relative strengths.

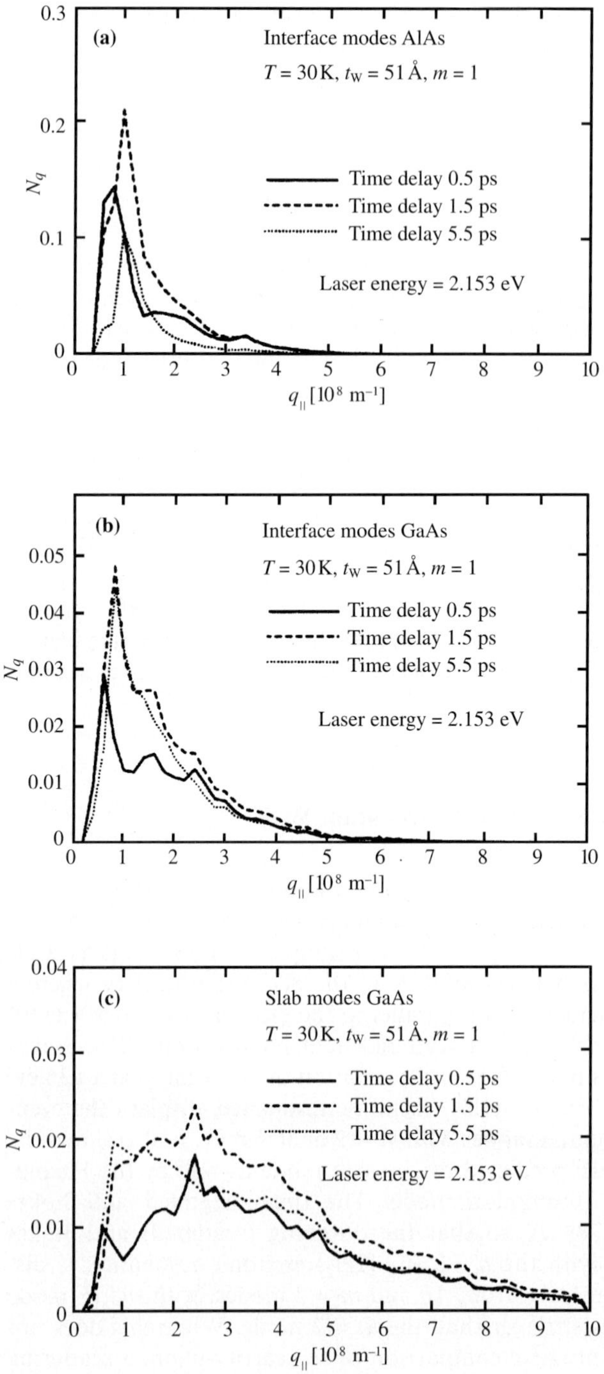

Fig. 5.16 a – d. Monte-Carlo simulation of phonon dynamics in a 51 Å GaAs quantum well structure [5.40]. (Figure 5.16d see opposite page.)

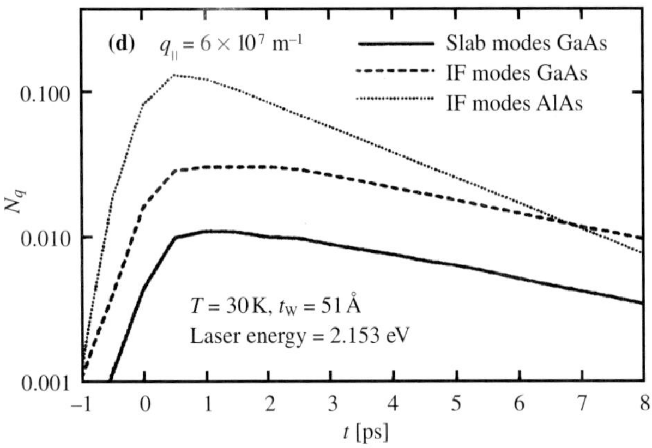

Fig. 5.16 d

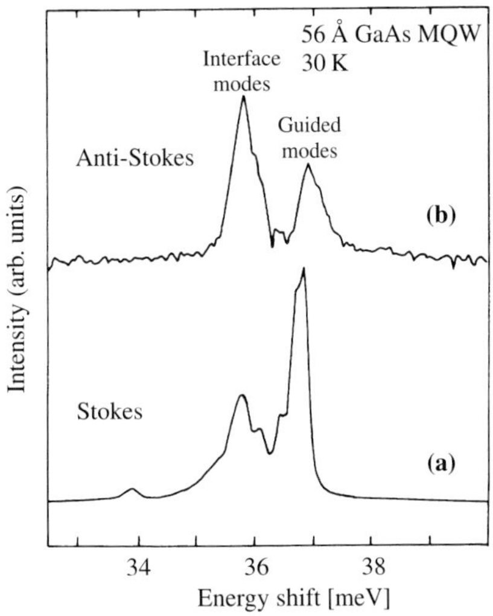

Fig. 5.17. (**a**) Time-integrated Stokes and (**b**) anti-Stokes Raman spectra from a GaAs MQW sample with 56 Å well width. The Stokes spectrum shows resolved even-m GaAs confined modes and interface modes, and weak structure near the TO mode of GaAs (at 33.5 meV) and the GaAs-like LO mode of the AlGaAs barriers. The anti-Stokes spectrum presents nonequilibrium $m = 2$ confined and interface modes [5.41]

The time-resolved anti-Stokes spectra at various delays obtained from the same sample under the same conditions as above are shown in Fig. 5.18. The IF and the $m = 2$ signals rise within 2 ps and then begin to decay. The time evolution of the anti-Stokes intensities of the two modes is depicted in Fig. 5.19. The rise time is limited by the instrumental resolution, but the underlying rise times are 1 – 2 ps. The details of the rise times are related to the phonon generation during the relaxation of photoexcited carriers. Since the predicted relaxation time is $\approx$ 50 fs for interface phonons and $\approx$ 3 ps for the confined modes [5.78, 95], the rise in the signal within 1 – 2 ps seems to rule out the latter as the dominant scattering mechanisms for electrons.

Results with sub-picosecond pulses, when neither peak is spectrally resolved, show a decay time of 4.5 ps [5.41], significantly shorter than the bulk GaAs value of about 9 ps obtained from incoherent and coherent phonon

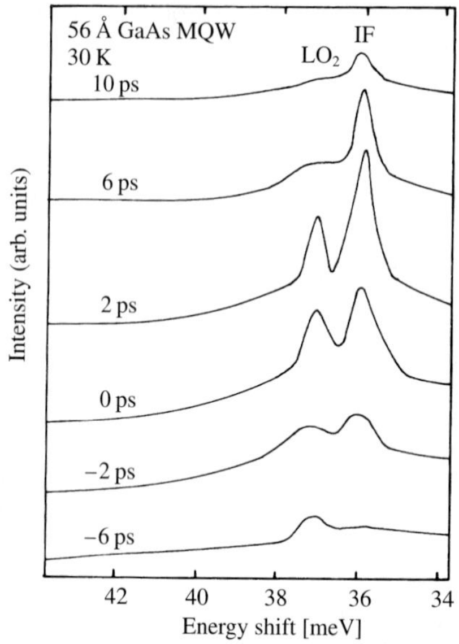

Fig. 5.18. Anti-Stokes Raman spectra of the 56 Å GaAs quantum-well sample under the same conditions as in Fig. 5.17 for various time delays, obtained with 4 ps laser pulses [5.41]

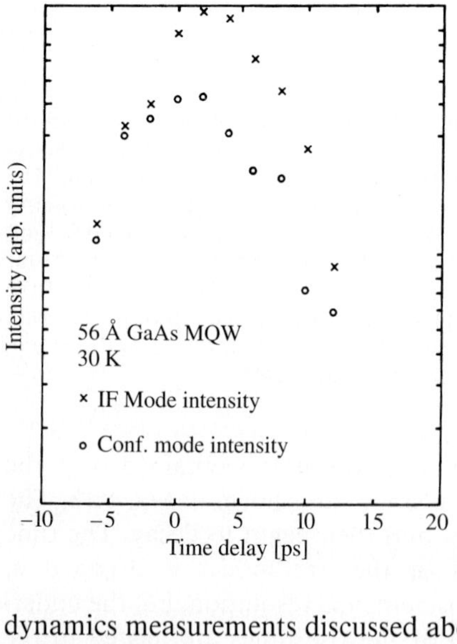

Fig. 5.19. Time evolution of the anti-Stokes Raman intensities of the confined and the interface modes in the 56 Å GaAs quantum well under the same conditions as in Figs. 5.17 and 18 [5.41]

dynamics measurements discussed above. It would be interesting to investigate if the anharmonic decay of optical phonons in quantum wells has a larger rate than in the bulk.

5.2.6 Determination of Phonon Occupation Number

Time-resolved Raman scattering has thus provided valuable information about the relative scattering strengths and the decay rates of phonons in quantum wells. However, it was argued correctly by *Ryan* and *Tatham* [5.41] that the usual technique of obtaining the phonon-occupation number by taking the ratio of the anti-Stokes to the Stokes intensity does not work in these cases because of complications related to the asymmetric shape of the Raman cross-sections close to a resonance. We discuss here a method of obtaining quantitative information about the occupation number of various phonons proposed by *Tsen* et al. [5.35, 45, 46].

For a phonon mode with occupation number n_q, the Stokes and anti-Stokes intensities are given by

$$I_{\mathrm{S}}(h\nu_i) = I_i \sigma_{\mathrm{S}}(h\nu_i)(n_q+1) \ , \tag{5.17}$$

$$I_{\mathrm{AS}}(h\nu_i) = I_i \sigma_{\mathrm{AS}}(h\nu_i)(n_q) \ , \tag{5.18}$$

where σ_{S} and σ_{AS} are the Stokes and anti-Stokes Raman cross-sections for the incident laser with intensity I_i and photon energy $\hbar\nu_i$. If the cross-sections are independent of photon energy, then the phonon-occupation number can be simply determined by taking the ratio of the Stokes and anti-Stokes intensity at the same photon energy:

$$n_q = \left(\frac{I_{\mathrm{S}}(h\nu_i)}{I_{\mathrm{AS}}(h\nu_i)} - 1\right)^{-1} . \tag{5.19}$$

However, the Raman cross-sections for optical phonons are known to be strong functions of the incident photon energy for bulk semiconductors as well as their quantum wells, so that the assumption of equal Stokes and anti-Stokes cross-sections for a given incident photon energy may not be valid. Hence, hot phonon-occupation number can not be determined from (5.19) especially in quantum wells where resonance may be essential for observing photoexcited phonons in Raman scattering as discussed in Sect. 5.2.3.

However, neglecting the fact that the cross-section is a tensor, time-reversal symmetry requires [5.96] that the two cross-sections be related by

$$\sigma_{\mathrm{S}}(h\nu_i) = \sigma_{\mathrm{AS}}(h\nu_i - \hbar\omega_{\mathrm{LO}}) \ , \tag{5.20}$$

where $\hbar\omega_{\mathrm{LO}}$ is the optical-phonon energy. This relationship has been experimentally tested and found to be valid over a wide range of conditions [5.35, 45] provided one measures polarized scattering and no perturbations (such as a magnetic field) are present which destroys the time-reversal symmetry. The phonon-occupation number is then given by

$$n_q = \left(\frac{I_{\mathrm{S}}(h\nu_i)}{I_{\mathrm{AS}}(h\nu_i - \hbar\omega_{\mathrm{LO}})} - 1\right)^{-1} . \tag{5.21}$$

In other words, one needs to obtain the Stokes and anti-Stokes intensity at two different incident laser frequencies separated by the phonon frequency in order to determine the hot-phonon occupation number. We discuss in the next section experimental determination of the hot-phonon occupation number using this method.

5.2.7 Experimental Determination of the Hot-Phonon Occupation Number and Dynamics

It is clear that the technique of using the same laser to generate and detect the phonons is not acceptable if one needs to compare the intensities at two different photon energies. This is because the generation rates may be different at the two photon energies. In order to overcome this difficulty, *Tsen* et al. [5.35, 45, 46] employed two synchronized dye lasers operating at two different photon energies. The pulse width of the two lasers was 3 ps, the jitter between the pulses was 2 ps, and the pulse repetition frequency was 76 MHz. We discuss here some investigations of thin GaAs quantum wells using this technique.

Tsen et al. [5.35, 45, 46] investigated five GaAs/AlAs quantum-well samples with quantum well thickness ranging from 2 to 6 nm. The pump laser was tuned such that it created electrons in the first subband with an excess energy of 200 meV in each sample. The probe laser was tuned close to the lowest exciton resonance for maximum Raman intensity. The intensity of the pump laser was adjusted to photoexcite a carrier density of $(2 \pm 0.4) \times 10^{10}$ cm^{-2}. The Stokes and the anti-Stokes intensities were measured for the probe laser tuned to $h\nu_i$ and $h\nu_i - \hbar\omega_{LO}$, respectively, so that (5.21) can be utilized to deduce the phonon-occupation number. Probe laser intensity was sufficiently low so that no anti-Stokes signal was observed in the absence of the pump. All the Raman spectra were measured in the polarized geometry $z(x', x)z'$ where $x' = [110]$ and $x = [001]$.

The Stokes and anti-Stokes spectra for a 20 Å GaAs/AlAs quantum-well sample at 10 K are displayed in Fig. 5.20. The results were obtained with the probe laser at 1.9 eV which were delayed by 1.5 ps from the pump pulses at 2.1 eV. The various features in the spectra are marked using the notation of *Sood* et al. [5.91, 92]. The Stokes-Raman spectrum (Fig. 5.20a) shows the $m = 2, 4, 6$ confined LO modes, the GaAs-like interface mode IF1, the AlAs-like interface mode IF2, and the TO phonon. The phonons in the anti-Stokes spectrum (Fig. 5.20b) are generated by the intra-subband relaxation of electron and holes, and the anti-Stokes spectrum is dominated by the two interface modes and the LO_2 mode.

These experiments were repeated for four other samples with quantum well width of 30, 40, 50 and 60 Å. The phonon-occupation numbers were obtained using (5.21) for each of the phonon modes for each sample. The phonon-occupation numbers determined in this manner are plotted for various phonon modes as a function of quantum-well width in Fig. 5.21. Some general trends are evident from these results. First of all, the phonon-occupation numbers are

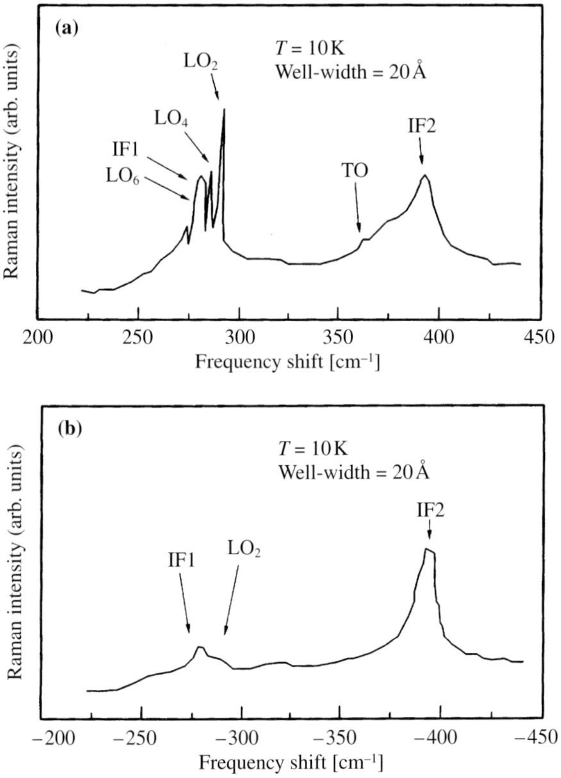

Fig. 5.20. (**a**) Stokes and (**b**) anti-Stokes Raman spectra of a 20 Å GaAs/AlAs multiple-quantum-well structure at 10 K with 3 ps pump pulses at 2.1 eV and 3 ps probe pulses at 1.9 eV at a time delay $\tau_d = 1.5$ ps [5.35]

rather small compared to bulk because of the mismatch (Sect. 5.2.3) between the phonon wavevector detected in Raman scattering and the phonon wavevector at which the phonon-generation rate peaks. Second, the AlAs-like interface phonon dominates at small quantum well widths but decreases in importance as the quantum-well width increases. LO_2 dominates at larger quantum-well widths.

Tsen et al. [5.35, 45, 46] compared these results to the predictions of the various phonon models discussed above. The guided-mode model is unable to account for the presence of the interface modes. The slab-mode model and the Huang-Zhu model both predict the importance of the confined-phonon modes and the interface modes. However, the slab-mode model predicts the presence of confined modes with odd integer number in contrast to the results, whereas the Huang-Zhu model predicts the presence of even-integer confined modes as observed. Therefore, of the three models, only the Huang-Zhu model is able to explain all the observed results qualitatively. *Tsen* et al. [5.45, 46] calculated the phonon-occupation number for various phonon modes predicted by the Huang-Zhu model by assuming that the carrier plasma can be described by a Maxwell-Boltzmann distribution function characterized by a temperature.

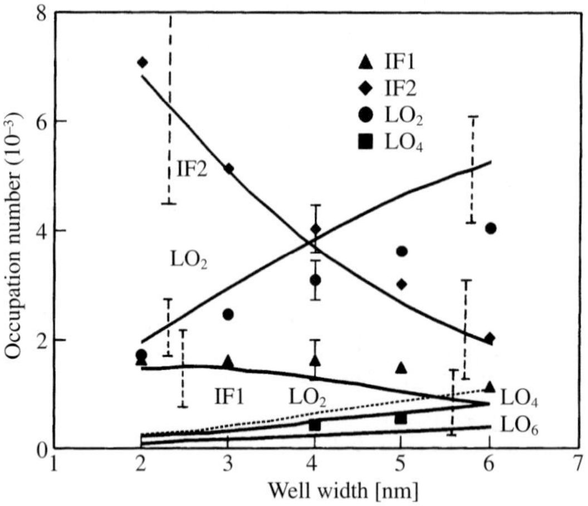

Fig. 5.21. Hot-phonon occupation numbers of the various confined and interface-phonon modes (Fig. 5.20) in GaAs/AlAs quantum wells as a function of the quantum-well width. The points are experimental with error bars indicated by *solid vertical lines*. The *solid curves* are calculated with the Huang-Zhu model for $q = 4.6 \times 10^6$ cm^{-1}. The *dashed vertical bars* give the variations in the calculated phonon populations when q is varied between 4.3×10^6 and 4.9×10^6 cm^{-1}. The *dotted curve* was calculated with the "mechanical model" discussed in Sect. 5.2.1 [5.35]. Recent studies of resonant Raman scattering indicate that the phonon modes LO_4, LO_6, LO_8 ... are related to peaks in the density of states of *interface* modes [5.102]

These calculated results (solid curves in Fig. 5.21) give good agreement with the experimental results.

Using this two-laser time-resolved Raman back scattering technique, *Tsen* et al. [5.45, 46] have also measured the *dynamics* of various phonon modes in GaAs/AlAs quantum wells of varying well widths. Figure 5.22 shows their results on the dynamics of AlAs-like interface mode (IF2 in Fig. 5.21) in a 60 Å GaAs/AlAs quantum-well at 10 K. The points are experimental data, and the solid curve is a fit to the data assuming a cascade model for the generation of the phonons (with a phenomenological generation rate Γ_{ph}) and a phenomenological phonon lifetime τ_{ph}. A fit to the observed time evolution by using Γ_{ph} and τ_{ph} as adjustable parameters gives $\Gamma_{ph} = (1.25 \pm 0.1) \times 10^{12}$ s^{-1} and $\tau_{ph} = 6 \pm 1$ ps. Repeating these measurements for quantum well samples with different thicknesses, *Tsen* et al. [5.35] found that Γ_{ph} for the AlAs interface mode decreases with quantum-well width between 20 and 60 Å. The results for Γ_{ph} are depicted in the inset of Fig. 5.22, and compared with the Huang-Zhu model (dashed curve) [5.46]. It was argued that since the theory considers contributions from all allowed phonon wavevectors, but only a limited range of phonon wavevectors participate in resonant Raman scattering, the theoretical scattering rates will always be larger than experimental rates. For this reason they used the range of phonon participating in the resonant Raman scattering as an adjustable parameter and obtained a good fit (solid curve in the inset of Fig. 5.22) for $\Delta q = 1 \times 10^6$ cm^{-1} around $q = 4 \times 10^6$ cm^{-1}.

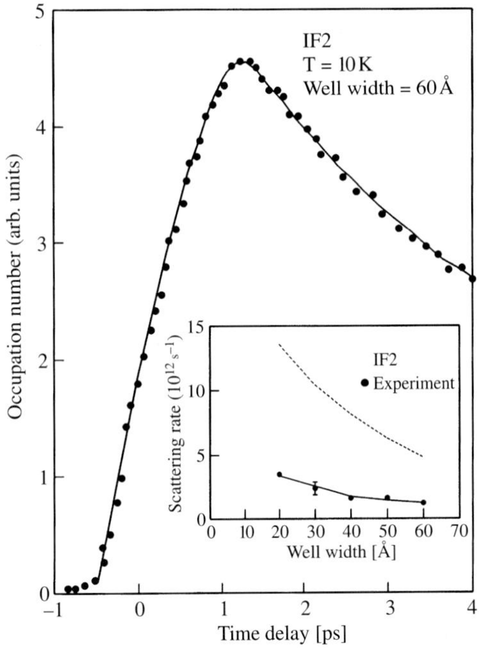

Fig. 5.22. Phonon dynamics of the AlAs-like interface mode in a 60 Å GaAs/AlAs quantum-well structure at 10 K excited to a density of 5×10^9 cm^{-2}. The *solid curve* is a fit to the data by the model discussed in the text. The *inset* shows the variation of the electron-phonon scattering rate for this phonon and a comparison with two theoretical models [5.46]

They also determined the phonon lifetime τ_{ph} for quantum well widths ranging from 20 to 100 Å, and found that τ_{ph} remains constant (≈ 6 ps) for the LO_2-confined phonon mode, increases from 5 to 8 ps for the GaAs-like interface mode with increasing well width, and increases from 5 to 9.5 ps for the AlAs-like interface mode as the well width is increased from 20 to 100 Å [5.46].

It is important to ask how general is this method of deducing the phonon-occupation numbers from Raman back scattering spectra at different photon energies. *Ruf* et al. [5.45] have further investigated this question, including the behavior at higher resonances. They have concluded that this procedure works for all cases they have investigated. Since (5.21) is based on time-reversal symmetry, anything that breaks the symmetry (such as a magnetic field) modifies it. The equation as stated is valid for polarized scattering. For depolarized scattering, one has to interchange the polarizations of the incident and the scattered light between anti-Stokes and Stokes scattering. The technique will also work for magnetic field, if one reverses the magnetic field when changing from Stokes to anti-Stokes. These considerations can thus get quite complicated.

5.3 Conclusions

It is clear from the results presented in this chapter that time-resolved Raman scattering on picosecond and subpicosecond time scales have provided valuable information about phonons which participate in the relaxation of photo-excited carriers in both bulk and quantum-well semiconductors. This information complements the information about carrier dynamics determined from time-resolved pump-probe and luminescence spectroscopies. In the case of quantum wells, Raman back scattering has provided very valuable information not only about the dynamics of phonons, but also about the phonon-occupation numbers of various phonon modes which made it possible to distinguish between various models of phonons in quantum wells.

In spite of this considerable success of time-resolved Raman back scattering, it is important to reiterate that this technique provides information about only a small range of phonon-wavevector space. In some sense this is equivalent to performing time-resolved pump-probe transmission measurements at a single optical wavelength, corresponding to the dynamics of the electrons and holes at one kinetic energy. The challenge of obtaining information about a range of phonon wavevectors is still unmet. Two-phonon scattering [5.97, 98] and phonon-assisted luminescence side bands [5.16, 51, 99] may provide a means for achieving this goal. Finally, impulsive Raman scattering from molecular vibrational modes has provided considerable information [5.100, 101], and can lead to potentially interesting results in semiconductors.

6. Exciton Dynamics

One major consequence of the Coulomb interaction between electrons and holes in semiconductors is the formation of excitons with a hydrogenic series of bound states converging to the bandgap energy. It is now well-known that excitons profoundly influence optical properties of a large number of insulators, molecular crystals and semiconductors. Theory of excitons was first developed by *Frenkel* [6.1], *Peierls* [6.2], and *Wannier* [6.3] in 1930s in an attempt to understand the basic absorption mechanism in solids. Excitons are classified into two types: *Frenkel* excitons for which the electron-hole pair is very tightly bound, and *Wannier* excitons where the electron orbits around a hole with a large radius. Excitons in semiconductors are generally Wannier excitons with large radii [6.4].

The dynamics of excitons is determined by their interactions with their surroundings. These interactions include scattering with acoustic and optical phonons, scattering from impurities, defects, interfaces and surfaces, and scattering with other excitons and free carriers present in the crystal. Investigation of exciton dynamics can therefore provide information about these scattering processes which are of fundamental interest in semiconductor physics. The goal of this chapter is to review some of the more recent experimental work on the dynamics of excitons, primarily in III–V semiconductors and their quantum wells. The literature on excitons in semiconductors is very extensive and we will only discuss some selected topics, related to the dynamics of exciton formation, exciton scattering into and out of its radiative state, exciton-spin relaxation and recombination of excitons.

We begin with a brief discussion of the basic concepts in the next section, where we review the nature of excitons and exciton-polaritons and their interactions with other elementary excitations such as phonons. We will see that resonant and non-resonant excitations of excitons lead to very different exciton dynamics and access very different physics. The bulk of the chapter is devoted to a discussion of exciton dynamics under non-resonant excitation (Sect. 6.2) and resonant excitations (Sect. 6.3).

6.1 Basic Concepts

There are many excellent texts and reviews in the literature about various properties of excitons in semiconductors and their heterostructures [6.5–11]. This section provides a brief overview of some of the fundamental properties of excitons.

6.1.1 Exciton States

In an exciton, an electron orbits around a hole in much the same way that an electron orbits around the nucleus in a hydrogen atom. Many of the properties of excitons can therefore be derived from the known properties of the hydrogen atom when the effective masses and dielectric constants are properly included. Like the hydrogen atom, the exciton has a series of states $n = 1, 2, 3$, etc. The binding energies of these excitons are given by

$$E_{\mathrm{b}n} = E_0/n^2 \tag{6.1}$$

for bulk (3D) and by

$$E_{\mathrm{b}n} = E_0/(n-1/2)^2 \tag{6.2}$$

for the ideal 2D case. The exciton-transition energies are given by

$$E_{\mathrm{x}n} = E_{\mathrm{g}} - E_{\mathrm{b}n} \ . \tag{6.3}$$

The exciton Rydberg E_{R} is expressed by

$$E_{\mathrm{R}} = \hbar^2/2m_{\mathrm{r}}a_0^2 = \mathrm{e}^2/4\pi\varepsilon a_0 \ , \tag{6.4}$$

where a_0 is the exciton Bohr radius. In the ground state ($n = 1$), the exciton radius, defined as the radial distance at which the wavefunction is 1/e of its maximum, is a_0 in 3D and $a_0/2$ in 2D. The reduced effective mass m_{r} is given by

$$1/m_{\mathrm{r}} = 1/m_{\mathrm{e}} + 1/m_{\mathrm{h}} \tag{6.5}$$

and ε is the permittivity of the semiconductor in the static limit. The exciton oscillator strength f_n for the nth excited state ($n = 1, 2, 3, \ldots$) is expressed by

$$f_n \propto 1/n^3 \quad \text{for 3D excitons} \tag{6.6}$$

and

$$f_n \propto 1/(n-1/2)^3 \quad \text{for 2D excitons} \ . \tag{6.7}$$

The near-bandgap absorption spectrum, therefore, consists of a series of sharp excitonic transitions of diminishing strength and decreasing spacing, merging into the continuum band-to-band absorption which is also modified by the Coulomb interaction between electrons and holes, as first discussed by *Elliott*

[6.12, 13]. The ratio of the Coulomb-enhanced absorption coefficient to that without Coulomb enhancement, known as the Sommerfeld or Coulomb enhancement factor, is given by

$$S(h\nu) = \frac{\left(\pi/\sqrt{\Delta}\right)\exp\left(\pi/\sqrt{\Delta}\right)}{\sinh\left(\pi/\sqrt{\Delta}\right)} \quad \text{for 3D} , \tag{6.8}$$

$$S(h\nu) = \frac{\exp\left(\pi/\sqrt{\Delta}\right)}{\cosh\left(\pi/\sqrt{\Delta}\right)} \quad \text{for 2D} , \tag{6.9}$$

where $\Delta = (h\nu - E_g)/E_R$ with E_g the bandgap energy and E_R the exciton Rydberg in 3D. As $\Delta \to 0$, the Sommerfeld enhancement factor approaches $2\pi/\sqrt{\Delta}$ (so that the absorption coefficient approaches a finite value) in 3D, and 2 in 2D.

In addition to the relative motion of electrons and holes that we have discussed above, the exciton also has a center-of-mass motion. If $\boldsymbol{r}$ and $\boldsymbol{R}$ denote the relative and center-of-mass coordinates of the exciton in real space, there are corresponding relative and center-of-mass momenta $\boldsymbol{k}$ and $\boldsymbol{K}$ in momentum space. In addition to the internal energy discussed above, the exciton also has a kinetic energy given by $\hbar^2 K^2/2M$, where $M(=m_e+m_h)$ is the exciton mass. Therefore, the total energy of the exciton can be expressed by

$$E_{\text{total},n} = E_g - E_{bn} + \hbar^2 K^2/2M . \tag{6.10}$$

This energy spectrum is schematically illustrated in Fig. 6.1 where the zero in energy corresponds to the unexcited semiconductor. Also shown is a schematic of the absorption spectrum.

The exciton wavefunction at any K is made up of a linear combination of a large number of electron and hole wavefunctions at different k values. Typically, one needs to include states up to $k \sim 1/a_0$. Excitons behave like bosons when the closest-approach distance between their centers of mass is larger than the exciton radius. However, it often happens in exciton dynamics for Wannier excitons that the reverse is true, in which case the Pauli exclusion principle which applies to the constituent particles of the excitons comes into play [6.14]. This has important consequences for excitons at high densities.

6.1.2 Exciton-Polaritons

If we plot the photon dispersion $\hbar\nu_K = \hbar c K/n_0$ on the same plot as the exciton dispersion, it will intersect the dispersion of 1s exciton at K_0 given by

$$(\hbar^2 K_0^2/2M) - \hbar c K_0/n_0 + (E_g - E_{b1}) = 0 , \tag{6.11}$$

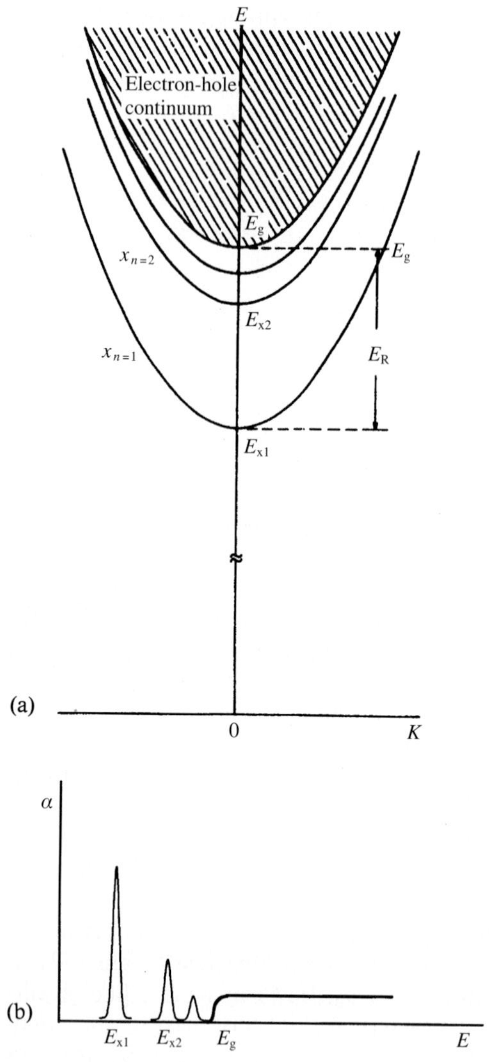

Fig. 6.1. Exciton-energy levels as a function of total wavevector $\boldsymbol{K}$ and the exciton principal quantum number n. Also shown is a schematic of the absorption spectrum including the Sommerfeld enhancement of the continuum absorption

where c is the velocity of light in vacuum, and n_0 is the refractive index of the semiconductor. The exciton-photon interaction removes this degeneracy and introduces a modified joint dispersion for the new quasi-particles known as exciton-polaritons. We will refer to these as simply polaritons in this chapter. Polaritons in bulk semiconductors have been investigated extensively beginning with the early work by *Hopfield* [6.15] and *Toyozawa* [6.16] in late 1950s. We refer the reader to the excellent reviews available in the literature [6.17–23]. In the absence of spatial dispersion ($M \rightarrow \infty$) and damping, a Longitudinal-Transverse (LT) splitting of Δ_{LT} develops between the two branches and there are no allowed modes between $E_g - E_{b1}$ and $E_g - E_{b1} + \Delta_{LT}$. The LT splitting Δ_{LT} is given by

$$\Delta_{LT} = 8\pi\, |\mu_{cv}|^2\, |\psi_1(r = 0)|^2 \;, \tag{6.12}$$

where $\psi_1(r = 0)$ is the wavefunction of the $n = 1$ exciton state at the center of the exciton. Since $|\psi_1(r = 0)|^2$ is inversely proportional to the Bohr radius of the exciton, the LT splitting increases as the Bohr radius decreases.

Spatial dispersion and damping introduce complicated dispersions in the region of overlap [6.9, 11, 24, 25] and will not be discussed here. Polaritons in bulk semiconductors are stationary states of the system so that interaction with surfaces, defects or phonons is required for radiative recombination of polaritons. Additional Boundary Conditions (ABCs) are required to fully specify the polariton problem in the bulk [6.15, 26, 27].

Recently, there has been considerable interest in polaritons in quantum wells. Interaction of quantum-well excitons with electromagnetic field leads to two kinds of modes [6.28 – 32]: *radiative polariton modes* for which the electric field has the plane-wave form outside the well, and *surface polariton modes* which decay exponentially outside the well. Conventional optical experiments couple to the radiative modes which, unlike bulk polaritons, are not stationary states of the system and have a finite radiative lifetime. This is because excitons with in-plane momentum $K_{||}$ smaller than the crossing of the photon line with the exciton-dispersion couple to a continuum photon mode [6.33, 34]. Surface polaritons can only be excited with special techniques, and are stationary states (like bulk polaritons) which can decay only by coupling to defects and phonons. We will be primarily concerned with the radiative polariton modes. The fact that these modes have a finite intrinsic radiative recombination rate Γ_r contrasts with the stationary nature of bulk polaritons and has important implications for the dynamics of excitons, as we will discuss in Sect. 6.3.

In quantum wells with growth axis along the z direction, there are three possibilities for photon polarization [6.35]: two modes polarized in the xy-plane, with the transverse mode T being perpendicular to the $K_{||}$ of the exciton, and the longitudinal mode L being parallel to $K_{||}$. The third mode Z is polarized along the z-axis. For HH excitons GaAs in quantum wells, the oscillator strength vanishes for the Z mode.

6.1.3 Exciton Fine Structure

The discussion so far has ignored the fact that an exciton is made of electrons and holes with definite symmetry properties. In bulk GaAs with its cubic symmetry, the conduction band at the Brillouin-zone center has s orbital character with two-fold degeneracy due to the electron's spin ($S = 1/2$). The valence band has p-orbital character and the top valence band is four-fold degenerate ($J = 3/2$) due to spin-orbit interaction. In axial symmetry (e.g., II – VI semiconductors, or III–V semiconductor under uniaxial stress or III–V heterostructures), the valence band is further split into Heavy Hole (HH; $m_j = \pm 3/2$) and Light Hole (LH; $m_j = 1/2$), each two-fold degenerate. The fine structure of excitons resulting from these symmetry considerations and from the exchange interaction (a residual part of the Coulomb interaction) has been discussed at length in the literature [6.17, 36]. For the case of reasonably large HH-LH splitting, so that interaction between them can be neglected, these considerations lead to a two-fold degenerate (dipole)-allowed HH exciton separated from a nearly two-fold degenerate dipole-forbidden HH exciton state by $3\Delta_z/2$ where Δ_z is the z component of the exchange energy. The dipole-forbidden states are split by the transverse component of the exchange. This is shown schematically in Fig. 6.2a and the optical transitions between the conduction band and the HH and LH valence bands are depicted in Fig. 6.2b. This fine structure and experimental determination of various splitting have been discussed recently by *Harley* [6.36].

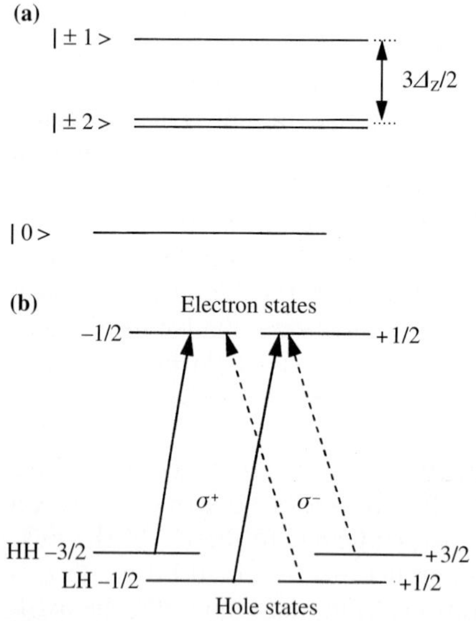

Fig. 6.2. **(a)** Fine structure for HH-exciton states in a quantum well grown on (100) substrate; **(b)** allowed optical transitions for HH and LH excitons

6.1.4 Dynamical Processes of Excitons

In this section we present a brief qualitative discussion of various dynamical processes related to excitons. For quantum wells, each of these processes very likely depends on parameters such as well-width, and barrier heights.

(a) Momentum and Energy Relaxation. Scattering with impurities, defects and interfaces or surfaces lead to relaxation of the exciton momentum but a negligible change in the kinetic energy of the excitons. Exciton-exciton scattering as well as exciton-free carrier scattering also leads to a change in the exciton energy and momentum. These processes contribute to the dephasing of excitons as measured by the coherent techniques discussed in Chap. 2. Since only the momentum parallel to the interface ($K_{\|}$) is a good quantum number in quasi-2D systems, the absorption of incident photons in a given direction creates an exciton with well defined $K_{\|}$. In the absence of any scattering, these excitons can only radiate in the reflection or transmission direction. Momentum scattering of excitons by the processes mentioned above changes $K_{\|}$ and allows these scattered excitons to radiate in other directions. Femtosecond luminescence studies under resonant excitation provides information about such events and will be discussed in Sect. 6.3.3.

Exciton-phonon scattering leads to a change in the exciton energy as well as momentum, except that the scattering can be quasi-elastic for low-energy acoustic-phonon scattering. At low temperatures, the exciton-phonon interaction is primarily via acoustic phonons because there are too few optical phonons present to be absorbed and the exciton energy is not large enough to emit an optical phonon. The influence of exciton-phonon interaction on the exciton-homogeneous line width was discussed in Chap. 2. At room temperature, optical-phonon scattering plays a significant role in determining the exciton line width in materials like GaAs quantum-wells. Exciton ionization by optical-phonon absorption is also important at room temperature as we will see below.

Exciton-phonon interaction has been treated extensively in the literature for bulk excitons as well as for quantum-well excitons [6.37–40]. One new feature that enters the quantum-well case is the fact that phonons themselves are subject to quantization effects (Chap. 5) and many different phonon modes may need to be considered for a proper treatment of exciton-phonon interaction in quantum wells.

(b) Spin Relaxation. Time-resolved studies using polarized light have the potential of providing information about spin relaxation processes, in addition to the momentum and energy relaxation processes. Spin relaxation of free carriers and excitons in bulk semiconductors has been investigated extensively, and a number of basic mechanisms of spin relaxation have been considered [6.41]. Confinement introduced by the quantum wells leads to several important changes. For example, the lifting of the degeneracy of the HH and LH valence bands at $k = 0$ results in slowing of the hole-spin relaxation in quantum wells. Further, the increased overlap between electron and hole wavefunctions leads to an increase in the exciton exchange interaction, and hence in the spin-relaxation rate.

As discussed above, the lowest exciton consists of two optically-allowed states ($|+1\rangle$ and $|-1\rangle$) and two optically-forbidden states ($|+2\rangle$ and $|-2\rangle$). If only one of the constituent particles in an exciton (i.e., electron or hole) flips its spin, then the exciton spin state changes between an allowed state and a forbidden state. If m_c and m_v are the indices representing the conduction and valence electron-spin states, we characterize the electron spin with an index $m_e(=m_c)$ and the hole spin with an index $m_h(=-m_v)$. Thus, a $|+1\rangle$ exciton is formed from a $+3/2$ hole and a $-1/2$ electron, and would be transformed into a $|-2\rangle$ exciton if the hole flipped its spin from $+3/2$ to $-3/2$. An exciton can make a transition between the two optically-allowed states by either flipping the spin of the exciton as a whole, or by successive flips of the electron and the hole spins within the exciton. We will designate the former as a *direct* exciton spin-flip, whereas the latter as an *indirect* exciton spin-flip. We will characterize the exciton transitions between an allowed and a forbidden state as either an *electron* or a *hole* spin-flip, with the clear understanding that this implies electron or hole spin-flip *within an exciton* rather than free electron or hole spin-flip.

The electron and the hole spin-flip can proceed by several mechanisms which have been discussed in the case of bulk semiconductors quite extensively [6.41, 42]. The direct exciton spin-flip proceeds via the exchange interaction [6.43, 44]. The increased exciton exchange interaction discussed above leads to a more important role for the direct exciton spin-flip of excitons compared to the indirect spin-flip, especially for the narrower quantum wells [6.36]. The exchange interaction can be divided into a short-range part and a long-range part [6.36, 43–49]. The short-range interaction does not directly flip the HH exciton spin but requires a second-order process involving LH excitons. Therefore, the long-range exchange interaction dominates the direct spin flip of HH excitons. Exciton spin-flip via exchange interaction has been recently investigated theoretically by *Maialle* et al. [6.43]. In Sect. 6.3 we will discuss recent experimental results and compare them with exciton-spin relaxation theories.

(c) Recombination Dynamics. Bulk polaritons have been investigated by light scattering, luminescence and absorption spectroscopies [6.19, 21, 22, 50]. From a dynamical point of view, polariton bottleneck [6.50] as well as recombination dynamics of bulk excitons [6.51, 52] have been analyzed extensively. In this chapter, we will concentrate mostly on the dynamics of excitons in quantum wells, with GaAs as a representative material system.

As already discussed, radiative recombination of bulk polaritons is qualitatively different from that of quantum-well polaritons. Bulk excitons require the assistance of surfaces, defects or phonons to radiate, whereas polaritons in quantum wells with in-plane wavevectors $K_{\|} \leq K_0$ are metastable states with well-defined intrinsic radiative rates that have been calculated by a number of groups [6.34, 53–55]. The radiative decay rate of the exciton amplitude vanishes for $K_{\|} > K_0$ and is a function of $K_{\|}$ for $K_{\|} < K_0$. The magnitude of the radiative decay rate of the exciton amplitude is characterized by a constant Γ_0 which is determined by the excitonic oscillator strength per unit area and the index of refraction. The radiative decay rate for the exciton *population* is then characterized by a $2\Gamma_0$. The oscillator strength should be calculated for finite barrier heights [6.56–58] to obtain a realistic estimate of Γ_0. For a 100 Å GaAs quantum well with 30% AlGaAs barriers, $2\Gamma_0$ is estimated [6.56] to be 1/(25 ps) for the L and T modes.

Physically observable quantities related to the radiative recombination of excitons can be derived by appropriate integration and averages from Γ_0 and the functional dependence of the recombination rate on the exciton energy. One may define two additional physical quantities: W_R, the average radiative decay rate of exciton population (averaged over optically-allowed and optically-forbidden or dark excitons) within the homogeneous line width Γ_h of the exciton, and W_D, the radiative decay rate of the exciton population which is thermalized at a temperature T. Since the radiative decay rate depends, in general, on $K_{\|}$, an appropriate integration or average must be performed in order to calculate these decay rates. One simplifying assumption is to assume that all optically-allowed excitons with $K_{\|} < K_0$ recombine with the same finite rate but optically-allowed excitons with $K_{\|} > K_0$ and optically-forbidden excitons have zero radiative rates. Under these simplifying assumptions, and using $E_0' = \hbar^2 K_0^2/2M$ the radiative decay rates W_R and W_D for exciton populations have been calculated [6.54, 55]. For $\Gamma_h \ll E_0'$:

$$W_R = 2(2\Gamma_0)/3 \tag{6.13}$$

and

$$W_D = 2E_0'(2\Gamma_0)/3kT \qquad kT \gg E_0' \,. \tag{6.14}$$

For $\Gamma_h \gg E_0'$

$$W_R = 2E_0'(2\Gamma_0)/3\Gamma_h \tag{6.15}$$

and

$$W_D = 2E_0'(2\Gamma_0)(1 - e^{-\Gamma_h/kT})/3kT \,, \tag{6.16}$$

which reduces to the same expression as (6.14) for $kT \gg \Gamma_h$. Note that in time-resolved spectroscopies at short times, exciton distribution may well be non-thermal so that these equations must be used with caution. For quantum wells it may be necessary to consider the fact that the angular distribution of the emitted radiation also depends on the distribution of in-plane momenta ($K_{\|}$) of the excitons. Finally, the above discussion does not include consideration of superradiance effects which have recently been considered theoretically for multiple quantum wells [6.59–61].

(d) Formation Dynamics following Non-Resonant Excitation. If the excitation-photon energy corresponds to the continuum states, this excitation must relax to $K_{||} = 0$ exciton before exciton can radiate. For continuum excitation, the thermalization and cooling of the photoexcited plasma (Chaps. 3 and 4), and formation of excitons by optical-phonon emission are the initial relaxation processes. Excitons can be formed over a wide range of energy and $K_{||}$. Once the carrier temperature is low enough that exciton formation by optical-phonon emission becomes unlikely, exciton formation proceeds via emission of acoustic phonons. Energy and momentum conservation requires that excitons be formed at large $K_{||}$, corresponding to kinetic energy typically equal to half the exciton-binding energy. The formation of these non-thermal excitons is the first stage of the formation dynamics. In the second stage, these non-thermal excitons interact among themselves and with acoustic phonons to form a thermalized hot-exciton gas whose temperature may be higher than the lattice temperature. The hot excitons then cool to the lattice temperature by emitting acoustic phonons. These processes parallel the relaxation processes of photoexcited electron-hole pairs (Chaps. 3 and 4) except that optical phonons play a relatively minor role at low lattice temperatures because of low kinetic energies of excitons. During this process of relaxation and thermalization, the exciton spins also undergo relaxation and thermalization.

The dynamics at moderate and high temperatures can be quite different because exciton ionization by absorption of acoustic and optical phonons becomes important. If the exciton ionization and formation rates are high enough that an equilibrium is established then the exciton and free carrier populations are coupled by the well-known Saha equation.

The excited states ($n > 1$) of excitons do not play a strong role in the absorption or emission of light because their oscillator strengths are small, as discussed above. However, they can act as intermediate states in the relaxation of photoexcited electron-hole pairs to $K_{||} = 0$ excitons. The excited states can also play a role in determining the thermal equilibrium distributions at higher temperatures. It may be difficult to determine the role of excited exciton states in these relaxation and thermalization processes experimentally.

The formation dynamics can be monitored by measuring time-resolved exciton radiation. In materials like GaAs where phonon-assisted exciton recombination is weak, one monitors the direct recombination of excitons which provides only indirect information about the exciton distribution. In this case, only excitons within the homogeneous linewidth Γ_h of $K_{||} = 0$ can radiate as discussed above. One is then left with the problem of determining the complex formation, relaxation and thermalization dynamics of excitons from the measured luminescence of $K_{||} = 0$ excitons. Such studies have been performed in GaAs and have provided valuable information (Sect. 6.2.1). In semiconductors like Cu_2O phonon-assisted recombination dominates for excitons. In such cases, more direct information about exciton dynamics can be obtained by measuring the time-resolved phonon-assisted exciton recombination spectrum (Sect. 6.2.2).

(e) Relaxation Dynamics Following Resonant Excitation. Resonant excitation by an ultrashort laser pulse leads to a completely different exciton dynamics.

We have seen in Chap. 2 that exciton-dephasing times in high-quality samples at low temperatures are several picoseconds, corresponding to a homogeneous linewidth Γ_h of a fraction of a meV. The spectral line width of a 100 fs laser is approximately 20 meV, considerably larger than Γ_h. The process of generation of excitons by such a short laser pulse is quite complex and interesting. The laser creates polarization and populations over a wide spectral range; however, the polarization and population decay quickly everywhere except within the exciton linewidth. The dynamics of excitons is the dynamics of the photo-excited polarization and population, and can be divided into two regimes which may overlap in time: the first is the coherent regime which is destroyed by various dephasing processes discussed in Chap. 2. If exciton dephasing is primarily due to elastic scattering processes, as expected at very low temperatures and excitation densities, then the exciton population is isotropically distributed in $K_{||}$-space and is within Γ_h of the exciton resonance when the coherence is destroyed. Information obtained on dephasing and momentum scattering rates from femtosecond resonant luminescence studies is discussed in Sect. 6.3.3. The second, incoherent regime of exciton dynamics involves radiative recombination, as well as relaxation of energy, momentum and spin of excitons, which transform the initial cold non-thermal exciton distribution into a thermal distribution. This relaxation takes the opposite course from that for the non-resonantly-created excitons, and its dynamics is investigated using picosecond resonant luminescence, as discussed in Sect. 6.3.2.

The radiative recombination properties of resonantly-excited excitons in quantum wells are very interesting and have come under increasing scrutiny both theoretically and experimentally. From a theoretical point-of-view, we have already noted that the excitons in quantum wells are metastable states with finite radiative lifetimes. Another interesting aspect is the angular dependence of the radiation. In the absence of any dephasing and in a perfect crystal, the radiation is emitted in the reflection direction and can be thought of as resonant Rayleigh scattering [6.62, 63]. In a structure with rough interfaces, the resonant Rayleigh scattering may develop an angular spread in a time of the order of the inverse of the inhomogeneous linewidth of the exciton [6.64]. In a perfect crystal, the excitons distribution becomes more isotropic as the excitons dephase. The redistribution of exciton allows the radiation to be emitted over a larger external angle, eventually making the emission nearly isotropic. As the excitons thermalize and occupy states outside of the homogeneous linewidth of excitons, only the excitons within Γ_h can radiate. All these factors have to be considered in analyzing experimental results at early times. Various aspects of the dynamics of resonantly-excited excitons are discussed in Sect. 6.4.

6.2 Experimental Results: Non-Resonant Excitation

As discussed in Sect. 6.1, the exciton dynamics following non-resonant excitation is dominated by the formation of $K_{||} = 0$ excitons and recombination of excitons. Since the binding energy and oscillator strengths of excitons are considerably larger in quantum wells, the excitons play a more prominent role in the near-band-edge optical properties of quantum wells. For III–V semiconductors such as GaAs, this makes it easier to study dynamics of excitons in quantum wells rather than in bulk. In this section we discuss several experiments performed under non-resonant excitation. We begin with a discussion of the formation dynamics of excitons in GaAs. We will see that considerable information about energy and momentum relaxation of excitons is obtained by monitoring the luminescence of $K_{||} = 0$ excitons. However, the inability to directly monitor the *distribution function* of excitons makes it difficult to obtain *quantitative* fits to the data. Although most of this book deals with III–V semiconductors, in Sect. 6.2.2 we will digress a little and discuss some results for Cu_2O which do permit a determination of the distribution function of the excitons, and hence allow a more quantitative comparison with the theory. We then return to GaAs quantum wells and discuss the spin-relaxation dynamics of excitons during the formation of $K_{||} = 0$ excitons (Sect. 6.2.3). Finally, we treat recombination dynamics of thermalized excitons in GaAs quantum wells (Sect. 6.2.4).

6.2.1 Exciton-Formation Dynamics in GaAs Quantum Wells

The dynamics of exciton formation following excitation in the continuum can be followed by monitoring the time-resolved luminescence spectra. The results depend on the nature of samples being investigated. If the quantum-well samples exhibit large Stokes shift between absorption and emission spectra, this is indicative of strong inhomogeneous broadening most likely resulting from large fluctuations in the well-widths. For such samples, relaxation of excitons from regions of narrow well-width to regions of wide well-width proceeds simultaneously with the relaxation of large $K_{||}$ excitons to $K_{||} = 0$ excitons. Therefore, the spectral shape and position as well as luminescence intensity varies as a function of time following photoexcitation. Investigation of such samples [6.65] reveals that energy relaxation to regions of wide well-widths dominates the time-resolved behavior, and has provided information about these processes.

It is clear, however, that investigations of such samples provide information that depend on the nature and quality of the sample, and can not provide information about the intrinsic processes that lead to relaxation of large $K_{||}$ excitons to $K_{||} = 0$ excitons. In order to obtain information about these intrinsic processes, one needs to investigate samples which have a negligible Stokes shift and are essentially homogeneously broadened.

We discuss here the results on such a high-quality GaAs quantum-well sample with 80 Å well-width and barriers with 30% Al reported by *Damen* et al. [6.66]. The exciton linewidth at 2 K was about 0.65 meV, and the sample exhibited a Stokes shift of less than 0.1 meV. The sample was excited by an LD 750 dye laser tunable between 720 and 800 nm and with pulse widths <0.7 ps. Figure 6.3 displays, for four different lattice temperatures, the time evolution of the HH exciton luminescence at 1.557 eV after the sample was excited at 1.605 eV at an excitation density of 1×10^{10} cm^{-2} per well. The data indicate that it takes *several hundred picosecond* for the exciton luminescence to reach its peak at 5 K. With increasing temperature, the luminescence peaks at earlier times as shown in the figure.

As discussed in Sect. 6.2, the dynamics of exciton luminescence intensity is determined by (1) formation time for large $K_{\|}$ excitons, and (2) relaxation and thermalization of these non-thermal excitons to $K_{\|}=0$. The results presented in Fig. 6.3 cannot determine which one of these two processes is responsible for the long-observed rise time of the luminescence signal. The ideal way to determine this would be to measure the dynamics of band-to-band luminescence. However, this signal was too weak to be observed in these experiments at low temperatures. Therefore these authors used another technique to determine the formation of large $K_{\|}$ excitons. As mentioned in Chap. 2, excitons interact much more strongly with free carriers than with other excitons. If the photoexcited carrier density is high, exciton-free carrier scattering determines

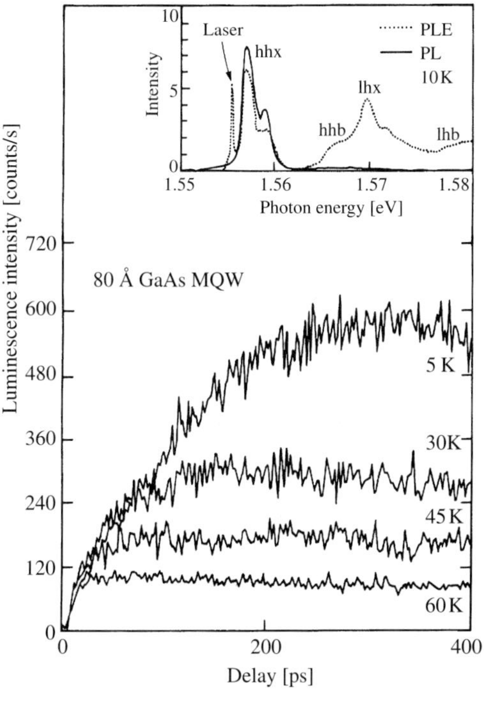

Fig. 6.3. The temporal evolution of exciton luminescence in GaAs quantum wells at four different temperatures. At low temperatures and excitation densities, it can take several hundred picoseconds for the luminescence to reach its peak. The *inset* shows the cw photoluminescence and photoluminescence-excitation spectra at 10 K; hhb and lhb indicate the onsets of band-to-band absorption involving the hh and lh valence bands [6.66]

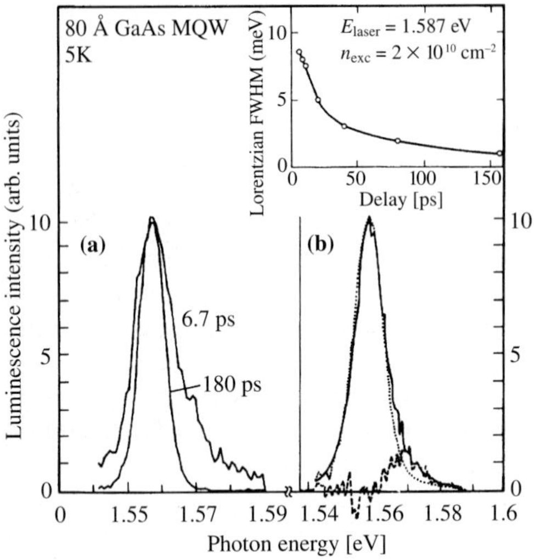

Fig. 6.4. (**a**) Luminescence spectra at two different time delays, showing the reduction in the width with increasing delay, (**b**) spectral shape analysis to determine the free carrier decay time and the formation time for large K excitons. The *inset* displays the Lorentzian width as a function of time delay determined from such analysis [6.66]

the linewidth of excitons. The linewidth would be expected to be large initially, but would then decrease as the free carriers form excitons. Therefore, if one monitors the dynamics of the exciton spectral linewidth following excitation by a short pulse into the continuum, then one can obtain a reasonably good measurement of the decay time of the free carriers.

Damen et al. [6.66] analyzed the line shape of excitons in the time-resolved luminescence experiments as a convolution of a Gaussian of fixed width and a Lorentzian of variable width. By fitting the luminescence line shape at various delays, the Lorentzian width was determined at each delay. Figure 6.4 shows time-resolved spectra at two time delays, a typical fit to the spectral shape, and the variation of the Lorentzian width as a function of time delay in the inset. Since the Lorentzian width depends linearly on the free carrier (primarily free electron) density, the results lead to the estimate that the carrier decay time is <20 ps. In a high-purity sample with a low defect density, this decay can be associated with the formation of large $K_{||}$ excitons. Combining these results with the long rise time of the luminescence signal in Fig. 6.3, one concludes that the long rise time is primarily determined by the thermalization and relaxation of large $K_{||}$ excitons to $K_{||} = 0$.

The thermalization and relaxation time of excitons is determined primarily by exciton-acoustic phonon and exciton-exciton interactions. At low densities, the peak of the luminescence is reached several hundred picoseconds after the excitation. Since the exciton decay time is comparable to this rise time, it is difficult to obtain a completely reliable determination of the relaxation time. The issue is further complicated by the fact that one does not know if the distribution of excitons can be described by a temperature, i.e., if the excitons are thermal or non-thermal. From a theoretical point of view, it should be possible to calculate the exciton distribution as a function of time. While exciton-phonon interaction in quantum wells has been investigated as discussed above, the author is not aware of any detailed calculations of the exciton-distribution function.

Experimental results for different excitation densities [6.66] show that the luminescence intensity increases faster at higher excitation density. This can be qualitatively understood as follows. Increasing the density of excitons increases exciton-exciton interaction and hence the rate of thermalization of excitons. Excitons therefore achieve a thermalized, hot (i.e., exciton temperature higher than the lattice temperature) distribution quickly. At higher densities, exciton-exciton collisions become important, so that excitons are likely to reach $K_{||} = 0$ faster than at low density where only scattering with acoustic phonons is important. This leads to a faster increase in the luminescence intensity.

The observations discussed above show no dependence of the exciton formation time on the photoexcitation energy. We discuss here some recent measurements by *Blom* et al. [6.67] which do exhibit such a dependence. These researchers argued that the long luminescence rise time (hundreds of picoseconds) compared to the short (<20 ps) formation time prevented the observation of oscillatory behavior. Therefore, they investigated thin wells with large acoustic-phonon scattering rate and moderately high exciton density both of which increased the relaxation rate of large $K_{||}$ excitons to $K_{||} = 0$ excitons and made the luminescence rise time comparable to the short formation time.

For these studies, *Blom* et al. [6.67] used two samples with a 26 Å GaAs quantum well, surrounded by 500 Å AlGaAs barriers ($x = 0.3$ or 0.4) on each side. This Separate Confinement Heterostructure Single Quantum Well (SCH-SQW) was repeated 10 times in a sample with a 100 Å AlAs barrier between each period. The samples were maintained at 8 K and were excited to a density of 5×10^{10} cm^{-2} using excitation-photon energy between 1.92 and 2 eV, below the barrier bandgap. Luminescence upconversion technique [6.68] was used with a spectral resolution of 2 meV. Figure 6.5 displays the temporal evolution

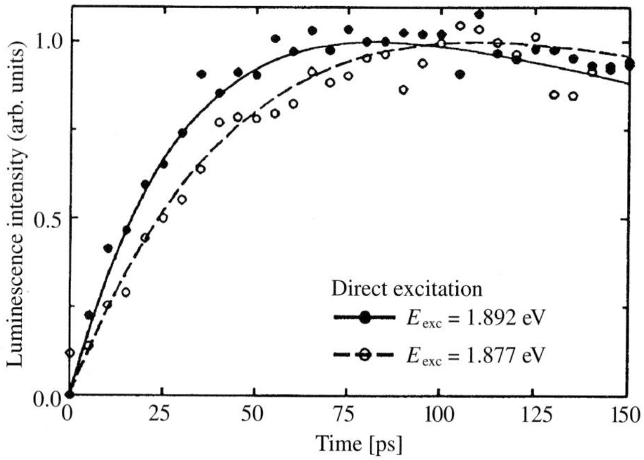

Fig. 6.5. Temporal evolution of the luminescence intensity from a 26 Å GaAs/$Al_{0.3}Ga_{0.7}As$ MQW sample at 8 K at an excitation density of 5×10^{10} cm^{-2} for two different excitation-photon energies [6.67]

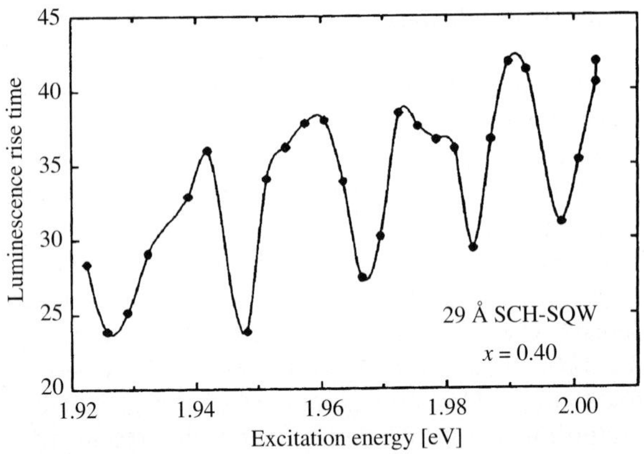

Fig. 6.6. Time constants for the rise of luminescence deduced from curves such as shown in Fig. 6.5 for a 26 Å GaAs/$Al_{0.4}Ga_{0.6}As$ MQW sample at 8 K at an excitation density of 5×10^{10} cm^{-2} as a function of excitation-photon energy showing oscillatory behavior. *Solid line* is a guide to the eye [6.67]

of the luminescence intensity at the HH exciton at 1.705 eV for two different excitation energies of 1.892 and 1.877 eV. By fitting the data points, they deduced time constants for exponential rise of 27 and 32 ps, respectively. Results of measurements at different excitation energies (Fig. 6.6) clearly show an oscillatory behavior, with a period of about 20 meV. The authors excluded hot-exciton relaxation and hot-electron effects as possible explanations because of the observed period. On the basis of Monte-Carlo simulations [6.69] they proposed that LO-phonon-assisted exciton formation becomes likely whenever (i) the sum of the electron and hole energies after multiple emission of LO phonons equals the LO phonon (36.8 meV) energy minus the exciton-binding energy ($\approx$ 10 meV), and (ii) either the electron or the hole has energy and momentum close to zero. On the basis of this model, they were able to obtain a reasonably good fit to the experimental results.

These experiments also point to a number of directions for future investigations. One obvious experimental question is how the results depend on the well thickness. Other questions include the role of carrier-carrier collisions and the validity of phonon-cascade model at 5×10^{10} cm^{-2} (an equivalent bulk density of 2×10^{8} cm^{-3} in a 26 Å well), the role of intervalley transfer for high excitation energies, and the role of confined and guided optical-phonon modes and folded acoustic-phonon modes in thin quantum wells.

There have been other investigations of the rise time of excitons as a function of excitation energy. For example, *Roussignol* et al. [6.70] found that, at low temperatures and excitation intensities, the rise time of the HH exciton luminescence increases considerably when the LH resonance is excited. This behavior disappears as the temperature or the excitation intensity is increased. *Deveaud* et al. [6.71], in their investigation of the fast radiative recombination of excitons in quantum wells (Sect. 6.3.2b), also reported dependence of the rise time on excitation energy very close to the HH exciton resonance.

The picture of non-thermal and hot excitons is very similar to the picture of non-thermal and hot free carriers discussed in Chaps. 3 and 4. However, the processes and their rates relevant for the case of excitons are not investigated as thoroughly and our understanding is only qualitative, with some empirical information about the relevant rates. A more systematic experimental investigation and detailed comparison with theory will be necessary for a quantitative understanding of these processes.

6.2.2 Exciton Relaxation Dynamics in Cu_2O

Cu_2O is a very interesting material whose excitonic properties have been investigated extensively [6.72, 73]. Direct transition between the ground state and the lowest exciton is dipole forbidden so that phonon-assisted processes dominate in absorption and emission. The lowest exciton is split into a para- and an ortho-exciton. The absorption spectrum is dominated by creation of the ortho exciton (energy = E_X^{ortho}) with the emission of a 13.8 meV phonon. Therefore, excitation at $E_X^{ortho} + 13.8$ meV creates the ortho exciton at $K = 0$. Increasing the excitation energy creates the ortho exciton with finite kinetic energy at higher K values. At low temperatures, the ortho excitons of kinetic energy E can recombine with the simultaneous emission of the 13.8 meV phonon and a photon at energy $h\nu = E_X^{ortho} + E - 13.8$ meV. If the recombination rate is independent of the exciton energy, the recombination spectrum reflects the *distribution function* of the ortho excitons. A schematic of the phonon-assisted ortho exciton absorption spectrum and phonon-assisted emission spectrum for thermalized ortho excitons is shown in Fig. 6.7.

This brief discussion makes it clear that by appropriately choosing the excitation-photon energy, one can create the ortho exciton with high kinetic energy (large K). The relaxation of these photoexcited excitons to $K = 0$ exciton can

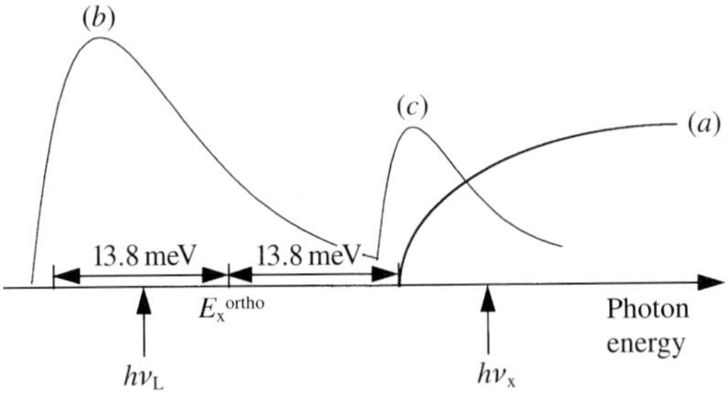

Fig. 6.7. Schematic of the absorption and emission of Cu_2O in the vicinity of the ortho exciton (*a*) absorption with emission of Γ_{12}^- phonons (13.8 meV), (*b*) luminescence from thermalized excitons with the emission of Γ_{12}^- phonons; (c) same as (*b*) but with the absorption of Γ_{12}^- phonons. In the absence of any thermalization, excitons created by photons at energy $h\upsilon_x$ luminesce at $h\upsilon_l = h\upsilon_x - 27.6$ meV at low temperatures (when phonon absorption in negligible)

then be monitored by measuring the luminescence spectrum as a function of time following the excitation pulse. This is precisely what was done by *Snoke* et al. [6.74].

A high-purity crystal of Cu_2O at a lattice temperature of 16 K was excited [6.74] by a synchronously-pumped dye laser with a pulsewidth of 5 − 7 ps and center energy of 2.054 eV, approximately 21 meV above $E_X^{\mathrm{ortho}} = 2.033$ eV, generating ortho excitons with kinetic energy of about 7 meV. The relaxation and thermalization of these non-thermal excitons to a thermal distribution was monitored by time-resolving the phonon-assisted luminescence of the ortho exciton using a streak camera with a time resolution of approximately 10 ps. The results at various time delays are depicted in Fig. 6.8. At small delays, the spectra clearly show the non-thermal exciton distribution. With increasing time delay, the excitons relax and eventually become thermalized at delays greater than 100 ps.

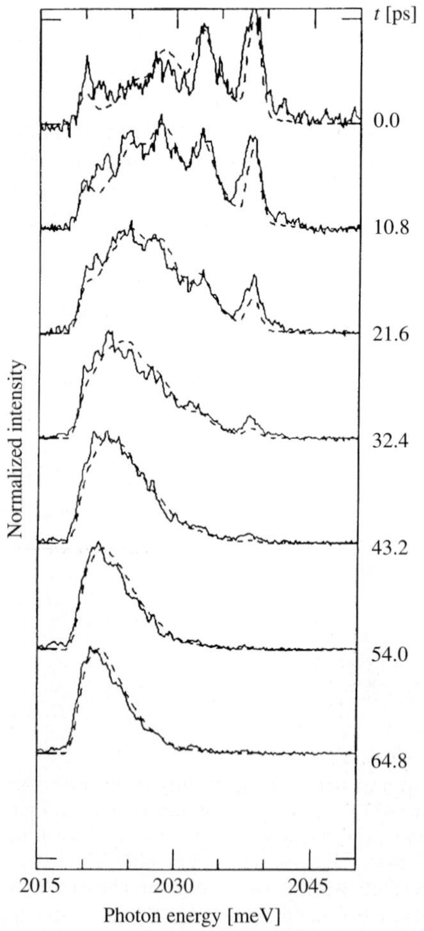

Fig. 6.8. The *solid curves* show the phonon-assisted luminescence of the ortho excitons in Cu_2O at 18 K, at various delays after excitation by a dye laser pulse at 2.067 eV generating ortho excitons with ≈20 meV kinetic energy. The *dashed curves* represent the predictions of a model [6.74]

A model incorporating numerical solution of the exact Boltzmann equation with absorption and emission of acoustic phonons via deformation potential interaction was developed. The dashed lines in Fig. 6. 8 show the exciton distributions at various time delays, calculated according to this model using an effective deformation potential of 1.8 ± 0.03 eV. The distribution corresponds to a Maxwell-Boltzmann distribution with a temperature of 23 K at 111.8 ps. With further increase in the time delay, the exciton temperature approaches the lattice temperature of 17 K.

Snoke et al. [6.74] studied the effect of exciton at higher photon energies so that optical phonons also play a role in the relaxation of excitons. They also investigated higher excitation density so that exciton-exciton scattering makes an important contribution. From these studies, they were able to determine the energy-loss rate per exciton in Cu_2O as a function of temperature of excitons under various approximations. This is the excitonic counterpart of the energy-loss rate of electron-hole pairs discussed in Chap. 4.

These measurements show that the ability to determine the *distribution function* of the excitons provide much more quantitative information compared to the case of GaAs in which only $K = 0$ excitons participate in recombination. We should note that Cu_2O is just one of the many materials exhibiting phonon-assisted luminescence. The use of the phonon sidebands of bound excitons in CdS and GaP for studying phonon dynamics has been discussed previously [6.75, 76]. II–VI semiconductors, in particular, exhibit strong LO-phonon-assisted free-exciton luminescence sidebands which may be useful in investigating exciton dynamics.

6.2.3 Spin Relaxation Dynamics in GaAs Quantum Wells

During the process of formation of $K_{\|} = 0$ excitons following excitation in the continuum, the electron-hole pairs and excitons also undergo spin relaxation. These processes can be investigated by a study of polarized time-resolved luminescence on appropriate time scales. Early studies of spin relaxation were reported by *Seymour* et al. [6.77, 78] and a number of other studies have been reported in the literature [6.79–86]. We discuss here some results reported by *Damen* et al. [6.87, 88], and *Viña* et al. [6.89].

The measurements were performed on the high-quality sample discussed in Sect. 6.2.1 on exciton formation dynamics, and also on n- and p-modulation-doped quantum wells. Polarized time-resolved luminescence measurements were performed with a streak camera for resonant and non-resonant excitation [6.88], and for non-resonant excitation [6.87, 89] with the luminescence upconversion setup [6.68]. The spin relaxation measurements on the intrinsic quantum-well sample using the upconversion technique show that the exciton-spin relaxation time decreases with decreasing excitation density, varying from 50 ps at 2×10^9 cm^{-2} to 150 ps at 1×10^{10} cm^{-2}. This trend of faster spin relaxation at lower densities is consistent with the streak-camera measurements which showed a spin relaxation time <20 ps at a somewhat lower density ($\approx 5\times10^8$ cm^{-2}). The relaxation of exciton spin was attributed to electron-hole exchange interaction in the exciton. With increasing density, it is well-known that the exciton-binding energy decreases and the exciton radius increases. This reduces the overlap between electron and hole wavefunctions, and hence leads to a reduction in the exchange interaction. This provides a qualitative explanation for the increase in the spin-relaxation time with increasing density.

Another observation was that the spin-relaxation time for resonant excitation was longer than that for non-resonant excitation at the same density [6.88]. This was qualitatively explained as follows [6.87]: For resonant excitation, the excitons are created close to $K_{\|} = 0$ and the range of hole wavevectors involved in excitons is small (of the order of the inverse exciton radius, i.e., $1/a_0$). For non-resonant excitation, the excitons are first formed at large $K_{\|}$ for which the hole wavevectors are large because of the larger mass of holes. The holes with larger k have a larger spin-orbit mixing which leads to a faster spin relaxation. This argument qualitatively explains the larger spin-relaxation rate of excitons for non-resonant excitation.

These studies provide a good qualitative understanding of spin relaxation dynamics in semiconductors, but a quantitative understanding would require further investigations.

6.2.4 Recombination Dynamics of Thermalized Excitons in GaAs Quantum Wells

Radiative recombination of excitons continues to be an area of interest in the field of semiconductors. Increasing purity of semiconductor materials has made it possible to explore the radiative recombination rates by reducing the importance of non-radiative processes, at least at low temperatures and moderately high densities so that the impurities are saturated. The enhanced oscillator strength of excitons and diminished role of impurity-related processes in quantum wells have made quantum wells an interesting system for investigating free exciton-radiative recombination dynamics.

Many investigations of recombination properties in various semiconductors have been reported for more than 30 years [6.90–92]. We note that radiative recombination dynamics in high-quality bulk GaAs has been extensively investigated (see, for example, [6.51, 52]. However, we concentrate in this subsection on investigation of radiative recombination dynamics in high-purity GaAs quantum wells.

The polariton effects play a different role in the two-dimensional excitons, and do not create a polariton bottleneck as in the bulk semiconductors. We have discussed these aspects of exciton-polaritons in quantum wells in Sect. 6.1, and will consider experimental results on resonantly-excited exciton-polaritons in Sect. 6.3.3b. We have also discussed above how non-resonantly excited electron-hole pairs relax to $K_{\|} = 0$ excitons. In this section, we investigate how these thermalized excitons decay radiatively.

The decay of thermalized excitons in GaAs quantum wells was first investigated by *Feldmann* et al. [6.93]. They measured the decay time of free-exciton luminescence in quantum-well samples which showed less than 5 meV Stokes shift between luminescence and PhotoLuminescence Excitation (PLE) spectra and which had exciton linewidths varying from 3 to 10 meV. These were state-of-the-art samples at the time, although not as good as those used for more recent studies on the dynamics of exciton formation and spin relaxation discussed above. These samples were excited with a synchronously-pumped dye laser of ≈ 5 ps pulsewidth. The excitation-photon energy was such that the barriers were not excited and the excitation density was approximately 5×10^{10} cm^{-2}, about an order of magnitude higher than those discussed above for the investigation of exciton formation dynamics. At these densities, the rise of exciton luminescence is considerably faster [6.66] than that shown in Fig. 6.3, and may be close to the 20 ps time resolution of the streak camera used for these studies. As we will see, the measured decay times are considerably longer and hence are not influenced by the rise time of the luminescence.

Figure 6.9 exhibits the decay times of exciton luminescence intensity as a function of temperature for several quantum-well samples of different well-width. The most interesting feature of the results is that the decay time *increases* with increasing temperature. This is opposite of the behavior in most semiconductors where the decay time decreases with increasing temperature because non-radiative processes become increasingly important at higher temperatures. The increase in the luminescence decay time with increasing temperature can be understood within the framework discussed above, without invoking nonradiative processes. According to (6.15), the radiative decay rate W_R is

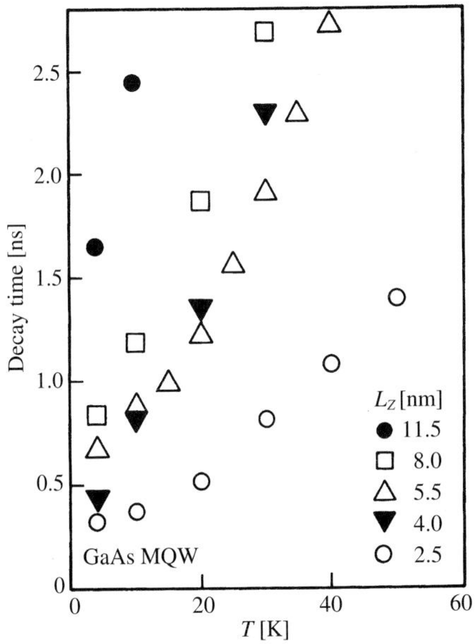

Fig. 6.9. Temperature dependence of the decay time of non-resonantly excited exciton luminescence for GaAs quantum wells of different thickness [6.93]

inversely proportional to kT when $kT \gg \Gamma_h$. This is because the fraction of the population within Γ_h varies inversely with the temperature whereas the radiative recombination rate for the exciton within Γ_h remains constant. We recall from Chap. 2 that the dephasing rate of excitons and hence Γ_h for excitons increases with temperature because of increased interaction with acoustic phonons. However, so long as $kT \gg \Gamma_h$, the linear dependence of W_R on $1/kT$ is expected to hold. The data also show that the decay time increases with increasing well width. This result can be qualitatively explained by noting that the exciton oscillator strength increases with decreasing well width because of the increased overlap of the electron and the hole wavefunctions [6.58]. A quantitative fit to theory was not attempted. The temperature dependence of the radiative decay rate has recently been reviewed by *Colocci* et al. [6.94]. Finally, we note that the dependence of W_R on temperature is expected to be different for quantum wires, and has been investigated recently by *Akiyama* et al. [6.95] and *Gershoni* et al. [6.96].

Eccleston et al. [6.97] investigated another prediction of (6.15), namely the dependence of W_R on Γ_h. Although this experiment utilized resonant excitation of excitons, we discuss this experiment in this section on non-resonant excitation because the experimental results of interest did not depend on this excitation condition. *Eccleston* et al. [6.97] recognized that, for typical high-quality samples under normal conditions, the inequality $kT \gg \Gamma_h$ generally holds. In Chap. 2 we discussed that the homogeneous linewidth of excitons, Γ_h, can be increased by increasing the density of excitons. Therefore, at a given T, Γ_h can be made comparable to kT if a sufficiently large density of excitons is excited in the system. As these excitons recombine, the density of excitons and hence Γ_h decreases so that W_R becomes a function of time. This would be expected to lead to a distinct non-exponential decay of luminescence intensity.

A sample with a narrow exciton linewidth shows a linear dependence on exciton density at a lower density and is preferable for such studies. Therefore, *Eccleston* et al. [6.97] used a very-high-quality sample with quantum well-width of 270 Å which had a (predominantly homogeneously-broadened) exciton linewidth (FWHM) of 0.3 meV. The sample at 1.8 K was excited by 5 ps pulses from a mode-locked Styryl 9 laser tuned to the HH exciton resonance. Cross-linearly polarized photoluminescence intensity was measured applying a streak camera with a system time resolution of 25 ps and spectral resolution of 0.5 meV. Figure 6.10 exhibits the measured time evolution for several excitation densities. The decay beyond 100 ps is exponential at the lowest density, but becomes distinctly non-exponential at higher densities as Γ_h becomes density-dependent (and hence time-dependent), in accord with the qualitative discussion given above. The data were fitted using the simplified equation (6.15) derived in Sect. 6.1.4c and the fits are also shown in the figure.

These experiments indicate that the picture of exciton recombination dynamics for thermalized excitons is consistent with the predictions of (6.15). The predicted dependence on both kT and Γ_h have been experimentally verified. It should be noted, however, that (6.15) uses the approximation that the

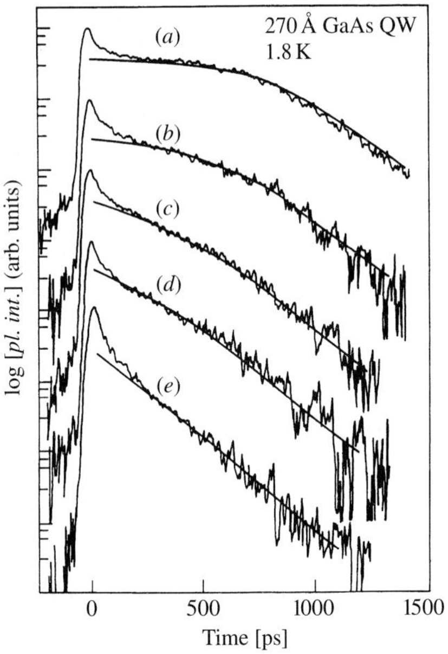

Fig. 6.10. Time evolution of cross-linearly polarized resonant luminescence from a 270 Å thick GaAs quantum well sample at 1.8 K for five different excitation densities: (*a*) 8×10^{10} cm^{-2}; (*b*) 5.2×10^{10} cm^{-2}; (*c*) 2.5×10^{10} cm^{-2}; (*d*) 1.5×10^{10} cm^{-2}; and (*e*) 0.18×10^{10} cm^{-2} [6.97]

radiative rate of excitons within Γ_h is constant and inversely proportional to Γ_h. This is only an approximation for a Lorentzian line shape of a homogeneously-broadened spectral line. Furthermore, the microscopic theory of radiative recombination of exciton-polaritons predict a dependence of recombination rate on $K_{\|}$ as well as on the nature of the mode (L or T), as discussed in Sect. 6.2. The experiments so far have not been sufficiently refined to investigate these finer features of the theory of exciton recombination dynamics.

6.3 Experimental Results: Resonant Excitation

As discussed in Sect. 6.2, the dynamics for resonantly-excited excitons is quite different from that of non-resonantly excited excitons. Pump-and-probe as well as luminescence spectroscopies have been used to investigate the dynamics of resonantly-excited excitons. The pump-and-probe technique generally provides better time resolution but is based on a nonlinear response of the semiconductor which may be difficult to interpret at times. Resonant luminescence, on the other hand, can be hampered by scattered light problems but may be easier to interpret because it is directly related to the density of the radiating state. Progress has been made on both fronts, and we will discuss some results obtained by both techniques.

6.3.1 Pump-and-Probe Studies

Pump-and-probe techniques have provided extensive information about nonlinear concepts at the exciton resonance resulting from excitation in the continuum as well as at the exciton resonance. A number of interesting phenomena such as the saturation, broadening and shift of the excitonic transition resulting from a number of effects such as screening, phase-space filling, exciton-exciton and exciton-electron (hole) collisions, and bandgap renormalization. Investigation of these nonlinear effects has been a very active area of research, and has provided considerable insight into the many-body effects in semiconductors. Since we are primarily interested in the dynamics of excitons, we do not discuss these very interesting phenomena here but refer the reader to a number of excellent reviews in the literature [6.14, 98–103]. We focus in this section on two pump-probe experiments which investigated dynamical behavior of excitons.

(a) Exciton Ionization Dynamics. We first discuss an early experiment in which ionization dynamics of resonantly-excited excitons was investigated by pump-and-probe transmission studies. We will see, however, that the experiment also provided considerable information about the nonlinear process of exciton bleaching.

In most semiconductors the exciton-binding energy is smaller than the optical-phonon energy. Therefore, at temperatures when the optical-phonon occupation number is not too small, a resonantly-created exciton can ionize into the continuum states by absorbing an optical phonon. One of the first experiments on femtosecond exciton dynamics was performed under such conditions by *Knox* et al. [6.104].

These studies were performed on a 65-period multiple-quantum-well sample with 96 Å GaAs quantum wells and a 98 Å $Al_{0.3}Ga_{0.7}As$ barriers. The sample was excited with 150 fs pump pulses centered either at the HH-exciton energy (resonant excitation) or about 70 meV above the HH exciton (non-resonant excitation). The differential transmission spectrum near the HH-exciton resonance was measured by a femtosecond continuum probe. Figure 6.11 exhibits the absorption at the HH resonance as a function of the time delay between the pump and the probe pulses for resonant excitation with the sample at 300 and 15 K, and for non-resonant excitation with the sample at 300 K. The results for resonant excitation at 300 K show a relatively rapid initial decrease in the HH-exciton absorption with the maximum bleaching occurring at a delay of 100 fs (near the end of the 150 fs excitation pulse), followed by a partial recovery occurring over several hundred femtoseconds. In contrast, the results for non-resonant excitation at 300 K display a relatively slow decrease in absorption which does not recover on the time scale of the experiment.

The initial decrease in absorption was attributed to a bleaching of the exciton absorption resulting primarily from phase-space filling due to the presence of a large number of cold excitons; the effects of screening are expected

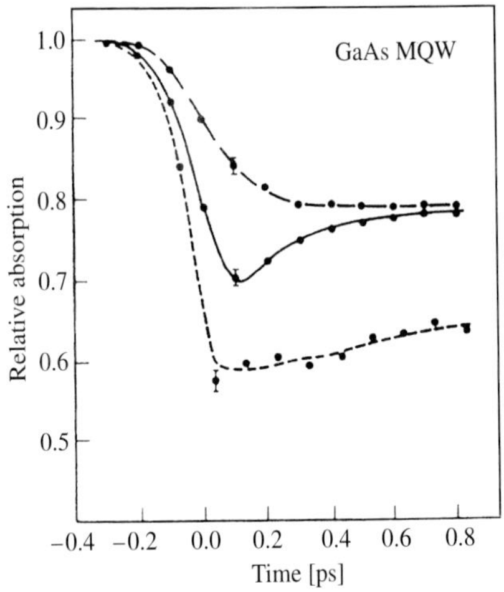

Fig. 6.11. Dynamics of the changes of the heavy-hole exciton peak absorption as a function of time delay between pump-and-probe pulses. *Long dashed lines:* non-resonant excitation, 300 K; *solid lines:* resonant excitation, 300 K, and *short dashed lines:* resonant excitation, 15 K [6.104]

to be small in 2D systems [6.100]. As the cold excitons created by the pulse are ionized into the continuum by the absorption of optical phonons, the phase-space filling effects are reduced, as discussed in the next paragraph. Therefore, the partial recovery of the bleaching was attributed to the ionization of cold excitons by LO-phonon absorption. From an analysis of the data, *Knox* et al. [6.104] deduced that the effective time constant for exciton ionization by the absorption of optical phonons at 300 K is approximately 300 fs. This value is in reasonable agreement with the value deduced from the exciton linewidth in the absorption spectrum [6.100] as well as with the exciton-dephasing rates obtained from four-wave-mixing measurements discussed in Chap. 2 [6.105]. The fast recovery of bleaching is absent at low temperature because there are fewer optical phonons and negligible exciton ionization at low temperatures. A systematic temperature-dependent measurement has the potential of providing information about other (not involving LO phonons) interesting dynamical processes affecting the resonantly-excited excitons. No such measurements have been reported.

The observation that the initial bleaching recovers as the excitons are ionized is very interesting and provides considerable information about the nature of phase-space filling and screening in quasi-2D systems. The exciton wavefunction is made up of the electron and hole wavefunctions extending from $|k| = 0$ to $|k| \approx 1/a_0$. At room temperature, kT is larger than the exciton-binding energy so that the free electrons and holes created by the exciton ionization are spread over a wavevector space larger than $1/a_0$. The phase space overlap between the high-energy free electrons and holes and the exciton is not very large. Therefore, these free carriers are not as effective in bleaching the exciton absorption as the cold excitons, and ionization of excitons into the continuum leads to a reduction in the bleaching and a partial recovery of the absorption. Once the carriers and excitons equilibrate at 300 K, the bleaching remains constant until the density begins to decrease because of recombination. The slow increase in the bleaching of exciton absorption for non-resonant excitation is a result of the cooling of the photoexcited electron-hole pairs to room temperature. It is clear from these arguments that carrier temperature plays an essential role in determining the effectiveness of the free carrier in exciton bleaching. This has been discussed in detail by *Schmitt-Rink* et al. [6.100].

Finally, recent studies have shown that the exciton-binding energies in II–VI quantum wells [6.106] can be larger than the optical-phonon energy. The role of optical-phonon absorption in these materials is thus reduced, even at room temperature. Pump-probe measurement on bleaching of excitons in some II–VI quantum wells have been reported recently [6.107, 108].

(b) Spin-Relaxation of Magneto-Excitons in GaAs Quantum Wells. Application of a magnetic field perpendicular to the quantum wells leads to quasi-zero-dimensional electrons, holes and magnetoexcitons. Spin and energy relaxation between these states can be investigated using time-resolved optical spectroscopies. Many different aspects of the nonlinear response of the electron-hole pairs and magnetoexcitons have been investigated in an interesting recent study by *Stark* et al. [6.109–112]. We discuss here the case of the resonant excitation of the 1s HH magnetoexciton at low temperatures. By using circularly polarized pump-and-probe beams, information about various relaxation processes was obtained.

Stark et al. [6.110] found that the nonlinear response can be divided into two temporal regimes. Exciton screening and phase-space filling dominate the response for times less than 1 ps. We will not discuss details of these results. We note, however, that a femtosecond nonlinear response was observed even when the pump and the probe had opposite circular polarizations. Significant contributions from phase-space filling are not expected in this case since the pump and probe couple to excitons which do not share any common states. The fact that a nonlinear response was observed may be attributed to other mechanisms such as dielectric screening and many-body effects, and underlines the complicated nature of the nonlinear response of excitons. In a recent study, *Maialle* and *Sham* [6.113] have shown that anti-parallel spin exciton-exciton correlation makes an important contribution to the third-order response. The complex nonlinear response of excitons presents a problem in the interpretation of the results, but it also provides an opportunity for investigating complex physical phenomena.

Stark et al. [6.110] found that the nonlinear response is independent of the pump polarizations after about 200 ps, showing that spin states have reached an equilibrium after this time. They deduced that the time constant for approach to equilibrium is 65 ± 5 ps at the magnetic field of 12 T for σ^- excitation, and attribute it to the spin relaxation of a hole in the magnetoexciton to the final optically-inactive $|+2\rangle$ state. Similarly for σ^+ excitation at 12 T, they deduce a time constant of 105 ± 10 ps and attribute it to the spin relaxation of an electron in the magnetoexciton to the final $|+2\rangle$ state. The magnetic field introduces a splitting of ≈ 1 meV between various 1s exciton states so that spin-relaxation is accompanied by energy relaxation in these cases. These relaxation rates are therefore expected to depend on the magnetic field. A systematic study of the dependence of these rates on magnetic field has not been reported. Note also that the values determined for a finite magnetic field can not be directly compared with the values for non magnetic field when the excitonic levels are nearly degenerate.

Exciton-spin relaxation has also been investigated at moderate magnetic fields in GaAs heterostructures [6.114]; however, we will not discuss these results because exciton localization may lead to different physics in these studies. We will also not discuss spin relaxation in shallow quantum wells that has been investigated recently [6.115]. We conclude this section with the remark that the interpretation pump-probe experiments have generally neglected radiative recombination dynamics. This is an excellent approximation in the femtosecond regime but may not be so in the picosecond regime, as we will see in the next section.

6.3.2 Picosecond Luminescence Studies

Time-resolved resonant luminescence spectroscopy provides another probe in the investigation of the dynamics of excitons. Resonant luminescence studies are difficult because the laser light scattered from surface imperfections interfere with the measurement of resonant luminescence. Resonant Rayleigh scattering yields another important contribution [6.62, 116–119] in the coherent regime and must be considered properly in the interpretation of the results. For luminescence studies using streak cameras, the time resolution is limited to ≈ 10 ps. Better time resolution can be obtained using luminescence upconversion systems, but resonant luminescence presents special problems [6.68] which can be overcome by using two-synchronized ultrafast lasers as shown recently [6.120, 121]. In this case the time resolution is limited by the laser pulse widths and jitter. Considerable progress has been made in recent years in investigating the dynamics of excitons using resonant luminescence spectroscopy. We discuss the highlights of these results in the following sections.

(a) Early Studies. *Deveaud* et al. [6.71] were the first to investigate the enhanced radiative recombination of exciton-polaritons in quantum wells expected from the theoretical considerations discussed above. They investigated a high-quality sample with a 45 Å wide single-quantum well of GaAs with AlAs barriers, with a HH-exciton linewidth of 2 meV (FWHM) and no Stokes shift. The sample was excited with a synchronously-pumped dye laser close to the HH-exciton resonance and the luminescence from the HH exciton was detected using a streak camera with 20 ps time resolution. The time evolution of luminescence for resonant and near-resonant excitations at 2.1 K for an excitation density of 3×10^9 cm^{-2} are shown in Fig. 6.12. The short-time transient ($t <$ 20 ps) was attributed to residual scattered light (from surface imperfections)

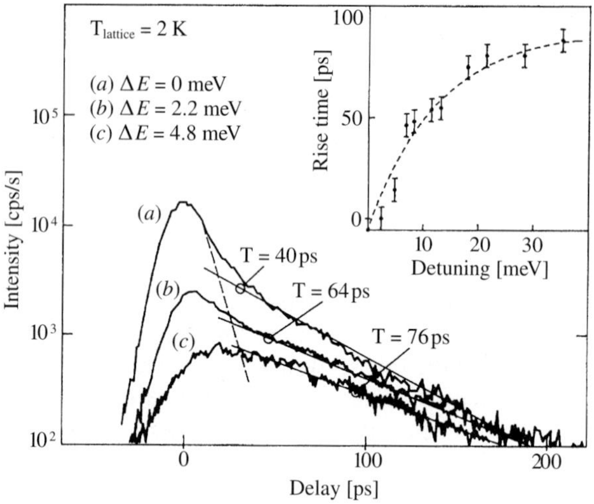

Fig. 6.12. Time evolution of HH exciton-luminescence intensity for three excitation energies at or close to the HH-exciton resonance. The initial drop in the resonant luminescence was attributed to residual stray light scattered from the surface. The *inset* shows the rise time as a function of detuning [6.71]

and the remaining transient was analyzed to yield a radiative recombination time of 10±4 ps. It was argued that this value is in good agreement with the value calculated by *Hanamura* [6.34] when corrected for the index of refraction of GaAs.

(b) Exciton Relaxation and Recombination Rates. While these early studies helped focus attention on this interesting problem, they were limited by the 20 ps time resolution and they ignored the full complexity of the exciton dynamics following resonant excitation. More recently, *Sermage* et al. [6.122, 123] have improved the time resolution for the streak-camera studies and provided a better analysis of the results. *Vinattieri* et al. [6.121] have developed a luminescence upconversion system using two synchronously-pumped dye lasers for investigating the picosecond dynamics of resonantly-excited excitons in GaAs quantum wells with 4 ps time resolution. Investigation of the polarization dependence of the luminescence dynamics allowed these researchers to investigate in detail spin relaxation, momentum relaxation and radiative recombination of resonantly-excited excitons. Various relaxation and recombination rates were obtained from an analysis of the data. It was concluded that many of these rates are comparable so that a *unified analysis* which considers the full complexity of the exciton dynamics is essential in obtaining a full understanding of the physics. We discuss the highlights of these studies in the following.

Vinattieri et al. [6.121, 124–126] investigated a multiple-quantum-well sample of GaAs with 150 Å wells, 150 Å $Al_xGa_{1-x}As$ barriers with $x = 0.3$, and 15 periods at temperatures between 10 and 40 K. The upper limit of tem-

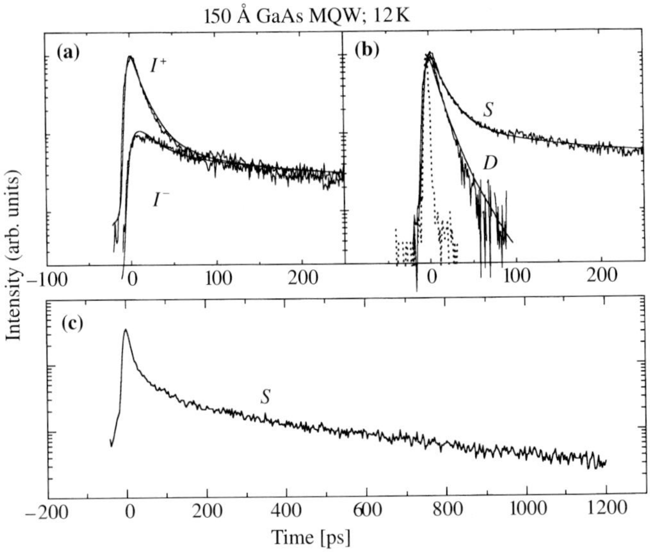

Fig. 6.13a–c. Time evolution of resonantly-excited HH-exciton luminescence in a 150 Å GaAs quantum well at 12 K: **(a)** I^+ and I^-, **(b)** S and D, **(c)** S on an extended time scale, showing the exponential decay at long times. The best fit to the data using the model discussed in the text are shown in **(a)** and **(b)** [6.121, 127]

perature was chosen to minimize the effects of light holes and free carriers. The sample was resonantly excited with either left- or right-circularly polarized photons from one laser, and the dynamics of the right-circularly polarized luminescence was measured using an upconversion system with a synchronized gating laser at a different wavelength. Figure 6.13 displays typical results at 12 K. Here I^+ and I^- denote the intensity of the $|+1\rangle$ exciton luminescence when the excitation laser is circularly polarized to excite the $|+1\rangle$ and $|-1\rangle$ excitons, respectively. S denotes the sum intensity $(I^+ + I^-)$ and the D denotes the difference $(I^+ - I^-)$. The total luminescence intensity S is also shown on an extended time scale.

A correct analysis of the results requires consideration of all the processes discussed in Sect. 6.1.4. Such an analysis would be quite complicated and has not been carried out. However, *Vinattieri* et al. [6.121] included all the essential physical processes in a simplified model shown in Fig. 6.14. The model divides all the $K_{||}$ states of the excitons into two manifolds with $K_{||} < K_0$ and $K_{||} \geq K_0$. Each manifold is divided into four states $|\pm 1\rangle$ and $|\pm 2\rangle$. Spin relaxation is allowed only within a manifold, with rates W_x for transitions between $|\pm 1\rangle$ states, and W_h and W_e for transitions between $|\pm 1\rangle$ and $|\pm 2\rangle$ states involving spin-flip of holes and electrons, respectively. Transitions between optically-allowed $|\pm 1\rangle$ states can occur in a single step (direct process with rate W_x) or by a two-step process involving both electron and hole spin-flips. Transitions between the two manifolds is characterized by a single rate W_k and occur only

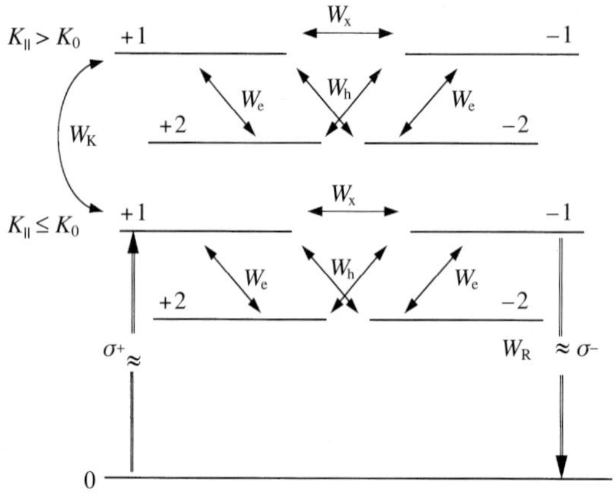

Fig. 6.14. Model for the exciton dynamics including direct and indirect exciton-spin relaxation, momentum relaxation and radiative recombination. This simplified model includes all features essential for a unified analysis of exciton dynamics [6.121, 127]

between like states ($|+1\rangle$ to $|+1\rangle$, etc.). The populations of various states are described by a set of eight coupled equations [6.127]:

$$\frac{dN}{dt}_i = F_{ij}N_j + G(t)\delta_{+1,i} \; , \tag{6.17}$$

where N_i is a column vector representing the populations of the various states, F_{ij} is an 8×8 matrix of different scattering rates, $G(t)$ is the generation rate by the optical pulse which is right-circularly polarized so that it generates only $|+1\rangle$ excitons. The five rates entering into F_{ij} include the four discussed above and W_R, the radiative rate of $|\pm 1\rangle$ (i.e., *optically-allowed*) exciton population with $K_\parallel < K_0$, given by [6.127]

$$W_R = \frac{4E_0}{3\Gamma_h}(2\Gamma_0) \; . \tag{6.18}$$

This equation assumes that all optically allowed excitons within the homogeneous linewidth Γ_h of K_0 recombine with a constant rate W_R which is proportional to the intrinsic rate Γ_0 and inversely proportional to Γ_h. This rate is larger than the rate in (6.15) by a factor of 2 because there the rate was averaged over allowed and forbidden excitons.

Best fits to the experimental results obtained using this model are also shown in Fig. 6.13a, b. It was argued that although there are five parameters, fits to different curves allowed determination of various parameters with reasonable errors. For example, the sum intensity S decays exponentially at long times when the excitons have reached thermal equilibrium with the lattice at temperature T_L. Under these conditions, the fraction f_0 of the exciton population in the radiative states can be expressed by

$$f_0 = \frac{1-\exp(-\Gamma_h/kT)}{1+\exp(\Delta/kT)} \tag{6.19}$$

and the decay rate W_D of S is given by

$$W_D = f_0 W_R \ . \tag{6.20}$$

The last two equations taken together are the same as (6.16) except that the effect of exchange splitting Δ between optically-allowed and forbidden excitons is explicitly considered here. Note that for $\Delta = 0$ the equations become identical. Once W_R is determined, it is found that the initial decay of S is primarily influenced by W_k and Γ_h. With Γ_h determined by four-wave-mixing measurements, a fit to the temporal variation of S allows a determination W_k with reasonable accuracy. Similarly, the early evolution of D is determined primarily by W_x and the long-time behavior of D allows a determination of the larger of W_h and W_e. *Vinattieri* et al. [6.121, 124] assumed that W_h is much larger than W_e and obtained the values of the various rates given in Table 6.1. Spin-flip dynamics under resonant two-photon excitation has been recently investigated by *Lohner* et al. [6.128].

One of the conclusions of this analysis was that the radiative recombination of excitons makes a significant contribution to the initial decay of the total luminescence intensity S. For the 150 Å quantum-well sample at 10 K, it was determined that about 50% of the decay is due to radiative recombination, with the other 50% resulting from scattering to the large $K_{\|}$ states and spin-flip into dark $|\pm 2\rangle$ exciton states. The relative contributions will vary with various parameters such as temperature and well thickness. However, this analysis shows that all the processes must be considered in a unified manner to extract quantitative information about these rates.

The value of the radiative rate deduced from these measurements is a factor of two smaller than the calculated value. This difference needs to be further examined.

(c) Dependence on Well Thickness. The measurements described above were repeated for quantum wells of different thickness and similar analysis was used to determine the relaxation and recombination rates. These rates are also listed in Table 6.1. The table shows that the radiative recombination rate increases with decreasing well width, a result that can be attributed to the increased overlap between electron and hole wavefunctions. This conclusion was also reached

Table 6.1. Scattering rates for spin relaxation, momentum relaxation and radiative recombination of excitons for three different GaAs quantum wells, determined from best fits to the data using a unified model for exciton relaxation

Well thickness (A)	W_D $10^9\,\mathrm{s}^{-1}$	Γ_0 $10^{10}\,\mathrm{s}^{-1}$	$\Gamma_{0\mathrm{th}}$ $10^{10}\,\mathrm{s}^{-1}$	W_k $10^{10}\,\mathrm{s}^{-1}$	W_x $10^{10}\,\mathrm{s}^{-1}$
80	2.8	4.5	4.6	10 ± 2	2.5 ± 0.5
130	1.8	2.6	3.1	8 ± 1.5	2.0 ± 0.5
175	1.54	2.2	2.5	5.5 ± 1.5	1.5 ± 0.5

in earlier studies [6.129]. Another conclusion of these studies is that the exciton momentum scattering rate W_k varies inversely with the well thickness, as expected for exciton-acoustic-phonon scattering. The exciton spin-flip rate W_x increased with decreasing well thickness, once again due to a larger exchange interaction resulting from the increased overlap between the electron and hole wavefunctions in narrower quantum wells. No dependence of well thickness was observed for the electron and hole spin-flip rates W_h and W_e.

(d) Electric-Field Dependence. Since the overlap of the electron and hole wavefunctions plays an important role in determining the radiative rate W_R and the exciton spin-flip rate W_x, a more controlled way of investigating these dependences is to apply an electric field along the growth direction. Such measurements have been reported by *Vinattieri* et al. [6.125, 127]. Figure 6.15 shows the measured dependence of various rates on the applied electric field for the 150 Å quantum well. W_x decreases by a factor of 5 whereas W_h and W_e remain unaffected by the electric field. For large fields, W_x becomes smaller than W_h leading to a larger role of the dark states in the dynamics, as evidenced by the increasing importance of long tails in the time evolution of the difference intensity D [6.125]. The measured [6.125] dependence of W_x on the field is compared in Fig. 6.16 with the dependence calculated by *Maialle* et al. [6.43, 44] for two quantum wells with infinite barriers and different thickness. The 185 Å well with infinite barriers give the same confinement energies as the 150 Å well experimentally investigated. The agreement is reasonably good. The fact that the calculation predicts a stronger dependence than measured was attributed to [6.125] certain approximations in the calculation.

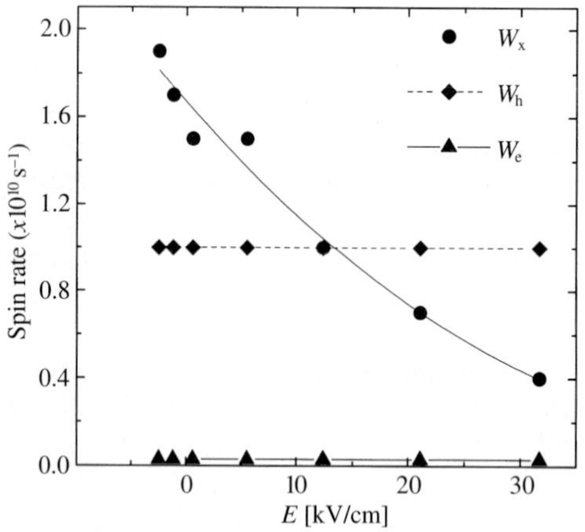

Fig. 6.15. Experimentally determined dependence of the direct (W_x) and indirect (W_h and W_e) exciton spin-relaxation rates on electric field [6.125, 127]

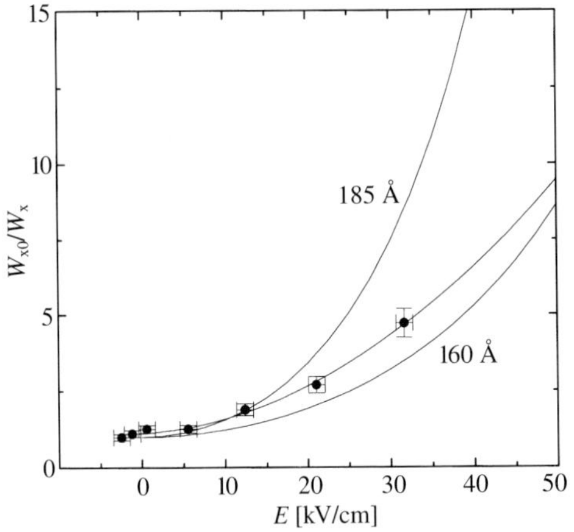

Fig. 6.16. Comparison of the experimentally determined spin-relaxation rate W_x with calculations [6.43, 44] for quantum wells of two different well widths with infinite barriers [6.125, 127]

6.3.3 Femtosecond Luminescence Studies

The picosecond studies of resonantly-excited excitons have provided valuable information on momentum and spin relaxation of excitons as well as on the intrinsic radiative recombination rate of excitons in quantum wells. These studies and their analysis assume that, for the time scale important in these studies, the excitons are already uniformly distributed within the homogeneous linewidth. This assumption is clearly not valid on shorter time scales.

As discussed in Sect. 6.1, only the momentum parallel to the interfaces ($K_{||}$) is a good quantum number in quantum wells. Consider the geometry defined in Fig. 6.17 such that the z is the growth direction, xz is the plane of incidence, and the radiation is also observed in the plane of incidence at an angle corresponding to the internal angle θ_l to the normal. Therefore, incident photons with momentum $\boldsymbol{q}$ (and magnitude q) can only create excitons with $K_{||} = q \sin\theta$ along the x-direction, as shown in Fig. 6.17. Initially the photoexcited excitonic ensemble is coherent so that radiation from the ensemble will interfere destructively in all directions except the reflection and transmission

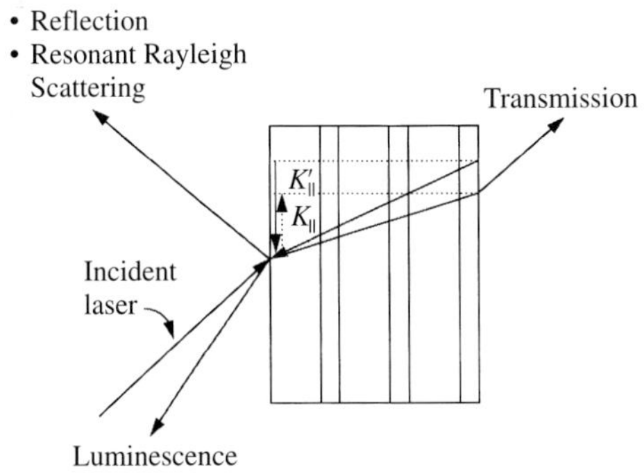

Fig. 6.17. Creation of excitons with definite momentum (schematic)

directions. Elastic scattering of these excitons by ionized impurities, interface roughness, or other excitons and quasi-elastic scattering by very-low-energy acoustic phonons nearly preserve $K_{\parallel}$ but change its direction. This leads to exciton radiation in other directions. Therefore, the dynamics of emitted radiation in a direction other than the reflected direction should provide new information about the underlying physical processes. Note that the excitons do not have to be uniformly distributed within the homogeneous linewidth of the excitons for emission to be observed in non-reflection or non-transmission directions. Thus, the investigation of the initial emission dynamics provides information about a subset of processes that lead to exciton dephasing.

For the geometry defined in Fig. 6.17, the change in the parallel momentum required to emit photons in the observation direction is given by

$$\Delta K_{\parallel} = q\,(\sin\theta + \sin\theta_1) \ . \tag{6.21}$$

Thus measuring the dynamics of the luminescence emitted in different directions is likely to provide further information about the initial elastic and quasi-elastic scattering dynamics of excitons. For all non-reflection and non-transmission directions, such dynamics should first show a build-up of population into the $K_{\parallel}$ states under study, followed by a decay of the population caused by all the processes discussed in Sect. 6.1.4.

Wang et al. [6.130–132] have recently investigated this initial dynamics of excitons on femtosecond time scales using the newly developed two-color upconversion technique (Chap. 1) which allows measurement of resonantly-excited luminescence with femtosecond time resolution. They investigated [6.131] two high-quality 15 period $GaAs/Al_{0.3}Ga_{0.7}As$ MQW samples, of well widths of 130 Å and 175 Å and barrier width of 150 Å. The PL line widths were 0.8 and 0.5 meV, respectively, and there was no Stokes shift. These measurements and three-beam FWM measurements [6.133] showed that the 175 Å sample was predominantly homogeneously broadened. The excitons were resonantly excited using a 100 fs Ti: Sapphire laser centered 6 meV below the excitons, thus minimizing the effects of LH excitons and the free electron-hole pairs. The resonant radiation was upconverted with 100 fs gating pulses at 1.5 μm from a synchronously-pumped optical parametric oscillator. The samples were at 10 K and the excitation density was $\approx 5\times10^{9}$ cm^{-2}. Different polarization geometries were investigated.

The temporal evolution of the co-polarized radiation from the 175 Å MQW sample for circularly-polarized excitation (Fig. 6.18) reveal several interesting features. The sharp spike at $t = 0$ is the laser light scattered from the surface imperfections (non-resonant Rayleigh scattering) and shows the time resolution ($\approx$ 100 fs) of the experiment. After this stray light has decayed to low levels, the signal rises slowly, reaching its peak at about 2.5 ps, and then decaying slowly. The signal also has an oscillatory component with a period of $\approx$ 1 ps, corresponding to the HH-LH exciton splitting of $\approx$ 4 meV in this sample.

We focus first on the initial rise of the luminescence. The data clearly show that there is a delayed rise of the luminescence. This represents the first

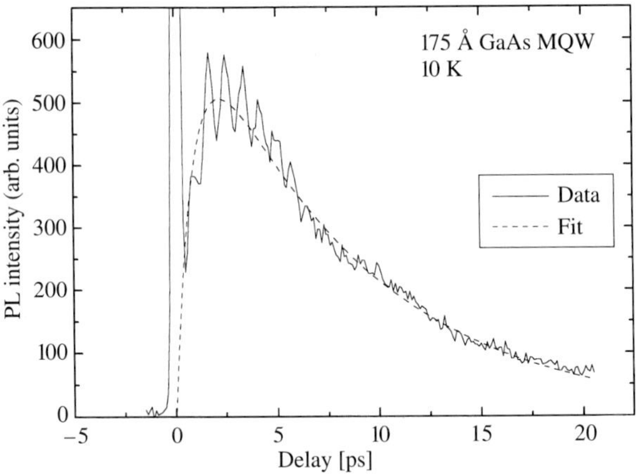

Fig. 6.18. Temporal evolution of co-circularly polarized resonant-exciton luminescence from a 175 Å GaAs quantum well at 10 K. The luminescence clearly shows a delayed rise and HH-LH quantum beats [6.131]

measurement of the elastic and quasi-elastic momentum scattering rates of excitons discussed above. A fit to the observed signal, ignoring the oscillatory component, gives a rise time of 1 ps. The initial decay time is 8 ps in the small range investigated here [6.134]. The 1 ps rise time corresponds to a rate of 10^{12} s^{-1} for scattering into the exciton states that radiate in the direction of observation. This rise time was shown to increase linearly as the excitation density was decreased. *Wang* et al. [6.130–132] found that, within the experimental error, the rise time did not depend on the luminescence angle θ_l, but the range of angles investigated was small because of experimental difficulties related to the scattered light.

An interesting question is how this momentum-scattering rate compares with the dephasing rate. The dephasing rate was determined by measuring the decay rate of the oscillatory component (quantum beat) of the signal. It was found to be considerably smaller than the momentum scattering rate under identical conditions. It was argued that this difference implies a breakdown in the impact approximation [6.131].

Quantum beats in resonant luminescence in atoms have been observed in the past [6.135], and recently in semiconductors [6.136–138] on longer time scales. The observation of beats on femtosecond time scale raises some interesting questions. As we have noted, radiation in a direction of observation can only occur if the excitons undergo momentum scattering. However, the presence of quantum beats in Fig. 6.18, persisting beyond the time when the luminescence peaks, implies that the scattering of excitons does *not* destroy the relative phase between the HH and LH excitons. Thus the scattering processes are such that they act similarly on HH and LH excitons. This is equivalent to the observation in atomic systems that the optically-induced coherence is retained while a scattering event changes the atomic velocity, provided the scattering interaction is state independent, i.e., it produces similar phase shifts for the states involved [6.139].

Non-resonant Rayleigh scattering, for example from the surface imperfections of a semiconductor, can be distinguished from resonant luminescence by the fact that non-resonant Rayleigh scattering is expected to follow the laser pulse. However, when one excites a semiconductor near one of its interband resonances, for example at the exciton resonance, the coherent polarization created by the excitation pulse persists until it is dephased, and the resonant Rayleigh scattering can have a temporal profile very different from the incident laser pulse. Resonant Rayleigh scattering has been considered in detail in recent years [6.62, 63, 116, 117, 119, 140–142]. For a perfect crystal, the resonant Rayleigh scattering is concentrated in the reflection and transmission direction. However, *Stolz* et al. [6.62] have shown that resonant Rayleigh scattering from excitons in GaAs quantum wells with interface roughness can be observed in other directions. As the interface roughness increases and its spatial coherence length becomes comparable and then smaller than the wavelength of light, the angular distribution of the resonant Rayleigh scattering broadens. *Zimmermann* [6.64] has argued that the broad angular distribution is not established instantaneously, but in time of the order of the inverse of the inhomogeneous linewidth of the excitons. If this is the case, then measurement of resonant emission in a non-reflection direction would be expected to increase in time, reach a peak in a time determined by the inhomogeneous linewidth and then decay with a time constant related to the dephasing rate. Thus, the temporal evolution of the resonant Rayleigh scattering could appear to be qualitatively similar to the temporal evolution of the luminescence expected on the basis of the model of momentum scattering discussed above.

Wang et al. [6.131, 143] have argued that the magnitude of the rise time, the observed dependence of the rise time on the intensity, and the fact that the measured decay time is longer than the dephasing time favor the explanation of the results in terms of resonant luminescence. However, one must await further theoretical developments (e.g., inclusion of dephasing other than radiative dephasing) and experimental investigations to resolve this question unambiguously.

Wang et al. [6.130–132] also investigated resonant luminescence in the cross-circular and co- and cross-linear geometries and were able to draw interesting conclusions on biexcitons in semiconductor quantum wells. We do not discuss these aspects here, but conclude with the comment that resonant luminescence has provided considerable information on atoms, and the ability to perform femtosecond measurements of similar phenomena in semiconductors should be very valuable.

6.4 Conclusions

Excitons play a very important role in the optical properties of semiconductors, and especially in semiconductor quantum wells. Ultrafast spectroscopy has been used extensively to study the dynamics of exciton formation, relaxation, scattering and recombination. Four-wave mixing (Chap. 2) has provided information concerning exciton-exciton, exciton-free carriers and exciton-phonon scattering rates. Pump-and-probe spectroscopy has led to a determination of exciton ionization rates and magneto-exciton spin-relaxation rates. Non-resonant luminescence spectroscopy has contributed to our understanding of exciton formation dynamics and radiative-recombination properties of a broad distribution of excitons. Resonant luminescence spectroscopy has furnished information about the influence of enhanced exciton recombination expected for 2D exciton-polaritons, and also about the spin-relaxation rates of excitons by exchange interaction and spin-relaxation rates of electrons and holes within excitons. A fairly comprehensive picture of exciton dynamics has been obtained through these studies.

One area that needs further studies is the dynamics of excitons in the coherent regime. Although nonlinear spectroscopy has provided information about the dephasing processes, the properties of emission from coherent excitons need to be further explored. For example, is the radiation pattern time-dependent and are there any superradiance effects associated with multiple quantum wells [6.33, 59–61]. We expect considerable activity in this area in the near future.

7. Carrier Tunneling in Semiconductor Nanostructures

As discussed in Chap. 1, ultrafast optical spectroscopy provides a powerful means of investigating relaxation as well as transport phenomena. The preceding chapters have discussed how ultrafast optical studies have provided invaluable information on the relaxation of interband polarization and non-thermal carriers, on the cooling of hot carriers, and on the dynamics of phonons and excitons in semiconductors. We now discuss the use of ultrafast spectroscopy to investigate dynamics of tunneling and transport in semiconductors and their nanostructures. Since transport over macroscopic distances in semiconductor nanostructures is controlled by many different physical processes, including tunneling, we begin with a discussion of tunneling in this chapter, and discuss carrier transport in Chap. 8.

Tunneling through a barrier is a quantum-mechanical phenomenon that has no classical analog. Tunneling phenomena manifest themselves in many different branches of physics. High-quality quantum nanostructures provide a versatile system for investigating the fundamental physics of tunneling because many parameters affecting tunneling can be controlled with a high degree of precision. Besides this fundamental interest, tunneling plays an important role in many devices based on superlattices in which perpendicular transport of carriers is important. Tunneling also plays a fundamental role in Resonant Tunneling Diodes (RTD) that have been shown to operate at frequencies exceeding several hundred GHz [7.1]. These factors have made investigation of tunneling in semiconductor nanostructures of interest to many researchers.

Considerable information about perpendicular transport in superlattices and tunneling in RTDs has been obtained by steady-state measurements of current-voltage characteristics, and by the investigation of dynamical properties using electrical techniques [7.2–7]. Investigation of high-frequency characteristics have provided valuable information about fundamental physics and device aspects of resonant tunneling diodes [7.1, 8]. Direct measurement in the time domain has also been performed using the electro-optic sampling techniques [7.9]. However, for tunneling as well as perpendicular transport, the best techniques for obtaining dynamical information on sub-picosecond and femtosecond time scales are *all-optical techniques*, such as excite-and-probe spectroscopy, transient four-wave-mixing spectroscopy or luminescence spectroscopy, whose time resolution is limited primarily by the laser pulsewidth, as discussed in Chap. 1. Time-resolved all-optical studies of tunneling began in late 1980s [7.10–12] and a large number of such studies have now been reported for different kinds of semiconductor nanostructures (for recent reviews, see [7.13, 14]).

The goal of this chapter is to present the basic concepts involved in tunneling and to discuss how all-optical techniques have led to new insights into this important physical phenomenon. The emphasis is on new concepts, and their illustration by various experiments. In order to investigate tunneling, one would like to design an experiment in which tunneling plays a dominant role and other physical processes have a negligible influence on the results. This is best accomplished by investigating *isolated structures* such as the Double-Barrier Structure (DBS) in which a quantum well is surrounded by two barriers, and the asymmetric Double Well Structure (a-DQWS) consisting of two quantum wells of unequal widths separated by a barrier. Such isolated structures are not influenced by transport over macroscopic distances, and hence provide quantitative information about tunneling rates. Our discussion on perpendicular transport in Chap. 8 will show that perpendicular transport is influenced by tunneling as well as other transport processes. The insights on tunneling obtained from measurements on isolated structures discussed in this chapter will prove to be important in understanding the interesting and complex phenomenon of perpendicular transport which plays an important role in many devices.

We note at the outset that the term tunneling time is used in two different contexts. In the first context, the term is defined as the time it takes for a wavepacket to travel from one side of the barrier to the other, and is also known as the traversal or dwell time. There is considerable theoretical discussion in the literature about the traversal time [7.15 – 19]. In the second context, the term tunneling time is defined simply as the reciprocal of the tunneling rate through a barrier. We will use the term in this second context. It should be noted that the two are distinct physical quantities.

The organization of the chapter is as follows. In Sects. 7.1 – 3 we discuss, for the two isolated structures mentioned above, some basic concepts related to tunneling, including various tunneling processes with and without the application of an external electric field. We will introduce the concept of optical markers, and show how optical spectroscopy allows a direct determination of the tunneling dynamics in such structures. The results on double-barrier structures are discussed in Sect. 7.4, those on a-DQWS for the non-resonant case are treated in Sect. 7.5, while those on a-DQWS for resonant case are delt with in Sect. 7.6. We conclude with a summary and some general observation in Sect. 7.7.

7.1 Basic Concepts: Optical Markers

Optical markers provide a powerful means of investigating tunneling and perpendicular transport in semiconductors. An optical marker is a thin region of the sample with unique optical properties (i.e., optical properties different from the rest of the sample). For example, an optical marker may have a luminescence or absorption spectrum that is different from those of other regions

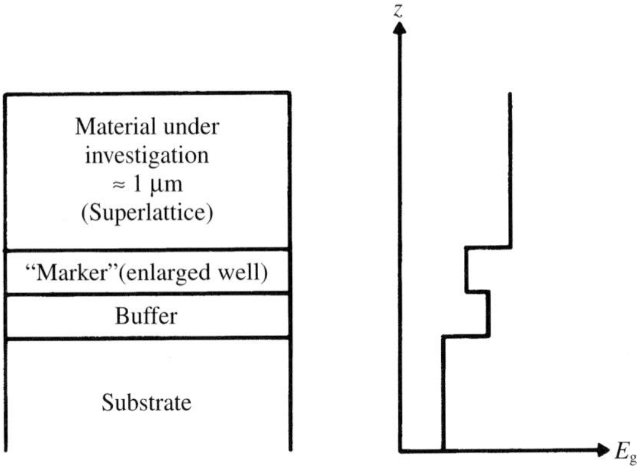

Fig. 7.1. Schematic illustration of the concept of a single optical marker which is a spatial region with unique spectral properties

of the sample. The marker can be a different semiconductor, the same semiconductor with different doping characteristics, or a more complicated nanostructure. The simplest case of a single optical marker is schematically illustrated in Fig. 7.1. The basic idea is that information about a specific *spatial* region, and hence about the dynamics of transport, can be directly obtained by studying some optical property (e.g., absorption or luminescence) in a specific *spectral* region. In the case illustrated here, the optical marker is an enlarged quantum well surrounded by a buffer layer and a superlattice of higher bandgap. If one measures, for example, the luminescence spectrum of such a structure, it will have four spectral features corresponding to the quantum well, the superlattice, the buffer layer and the substrate. By monitoring the relative intensities of the spectral features corresponding to the enlarged well and the superlattice as a function of time following pulsed photoexcitation, one can determine the dynamics of carriers in these regions.

In the double-barrier structure to be discussed in Sect. 7.2, the quantum well forms an optical marker and the spectral feature corresponding to the quantum well provides information about the carriers in the quantum well. In the double quantum well structure to be discussed in Sect. 7.3, the two quantum wells form the two optical markers. By monitoring the temporal evolution of the optical properties of the spectral features corresponding to the two quantum wells, one can obtain important information about the population in the two wells and hence the dynamics of tunneling. The extension of this concept to multiple optical markers, their application to investigating the dynamics of carrier transport, and some more examples of the optical technique are discussed in Chap. 8.

7.2 Basic Concepts: Double-Barrier Structures

A Double-Barrier Structure (DBS) consists of a quantum well of thickness a surrounded by a barrier (of thickness b and height V_0) and a bulk semiconductor (with bandgap energy lower than the lowest interband transition energy in the quantum well) on each side. A schematic of such a structure is shown on the left side of Fig. 7.2. The transmission coefficient $T(E)$ through the structure for an electron of energy E peaks strongly at the confinement energies of the bound states in the structure and is sketched on the right side of Fig. 7.2. Expressions for the transmission coefficient and the escape times as a function of the structure parameters are discussed below.

In a Double-Barrier Diode (DBD), the outer bulk semiconductors are doped to provide electrical contacts. A typical DBD with n-type contacts is shown schematically in Fig. 7.3 a under flat-band condition, and in Fig. 7.3 b for an applied voltage V. When no voltage is applied to the diode, the Fermi energy is below the electron confinement energy in the quantum well, and there is no tunneling and hence no current flowing through the diode (Fig. 7.3 c). As the applied voltage increases, the electron energy level in the quantum well and the Fermi level in the emitter cross each other so that the tunneling current goes through a maximum and then decreases showing a Negative Differential Resis-

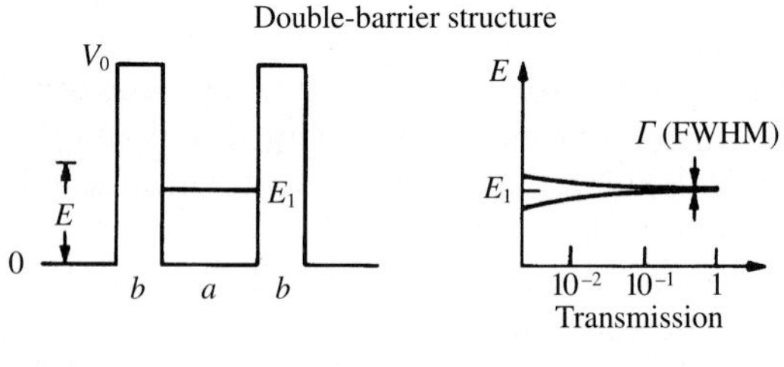

Fig. 7.2. Schematic of a Double-Barrier Structure (DBS) and the transmission through such a structure as a function of the electron energy

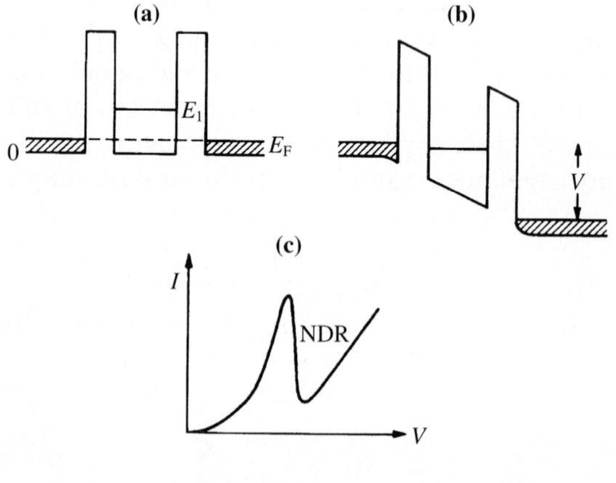

Fig. 7.3 a – c. Schematic of a Double-Barrier Diode (DBD) under two bias conditions and of the I – V curve for the DBD showing the Negative Differential Resistance (NDR)

tance (NDR). The electrical behavior of the double-barrier diode has been discussed extensively in the literature (see, for example, [7.1]).

For optical studies one generally uses a DBS where the important parameters are the transmission coefficient and the escape time τ_{esc} for an electron in the quantum well through the barriers. For an asymmetric DBS with an infinitely thick barrier on one side, the transmission coefficient of an electron in the well with energy E through the second barrier of width b and height V_0 is given by

$$T(E) = \frac{4E(V_0 - E)r}{[(r-1)E + V_0]^2 \sinh^2(2\pi b/\lambda) + 4E(V_0 - E)r} , \tag{7.1}$$

where λ is the de Broglie wavelength that describes the wavefunction penetration in the barrier and is given by

$$\lambda = h/\sqrt{2m_b(V_0 - E)} . \tag{7.2}$$

Here, $m_w(m_b)$ is the carrier-effective mass in the well (barrier) $r = m_b/m_w$. Since $\sinh(x)$ approaches $\exp(x)/2$, the transmission probability for $r = 1$ can be approximated for small λ as.

$$T(E) = \frac{16E(V_0 - E)}{V_0^2} \exp\left[-2b\sqrt{2m_b(V_0 - E)}/\hbar\right] . \tag{7.3}$$

For a DBS with two barriers of finite thickness, the transmission coefficient through the entire DBS for an electron of energy E incident from the left has several resonances at energies E_R corresponding to the resonant levels in the quantum well. The transmission coefficient exhibits a resonance (Fig. 7.2) of the form

$$T(E) = T(E_R) \frac{(\Gamma/2)^2}{(E - E_R)^2 + (\Gamma/2)^2} . \tag{7.4}$$

The Full Width at Half Maximum (FWHM) of the transmission resonance is related to the escape time by [7.1]

$$\Gamma = 2\hbar/\tau_{esc} , \tag{7.5}$$

where τ_{esc} is the escape time. If the two barriers in the DBS are asymmetric so that the transmission coefficients of the two barriers are different (T_l for the left barrier and T_r for the right barrier), then τ_{esc} is determined from the attempt frequency to escape (ν) multiplied by the average transmission coefficient $T_{av}(E)$ of the two barriers

$$\tau_{esc} = 1/(\nu T_{av}) , \tag{7.6}$$

where v is the frequency at which the electron attempts to escape one of the barriers (velocity of the electron/$2a$) given by

$$v = \frac{\sqrt{2E/m_{\mathrm{w}}}}{2a} . \tag{7.7}$$

For a symmetric DBS, the transmission coefficients of the left and the right barriers are equal ($T_{\mathrm{l}} = T_{\mathrm{r}}$). If the transmission coefficient of one of the barriers is reduced by increasing the barrier width or height, then the escape time increases. For a strongly asymmetric DBS, the escape time is twice as long as the escape time for the corresponding symmetric case, and the width of the transmission resonance decreases by a factor of two.

The calculated escape times for electrons and heavy holes for a strongly asymmetric GaAs/$Al_{0.3}Ga_{0.7}As$ DBS with 50 Å well width are depicted in Fig. 7.4 as a function of the barrier width. For thick barriers, the heavy hole-tunneling times can be orders of magnitude larger than those for electrons. Also shown in Fig. 7.4 are the calculated values of the coherent tunneling times (as defined in the next section) for two 50 Å quantum wells separated by a

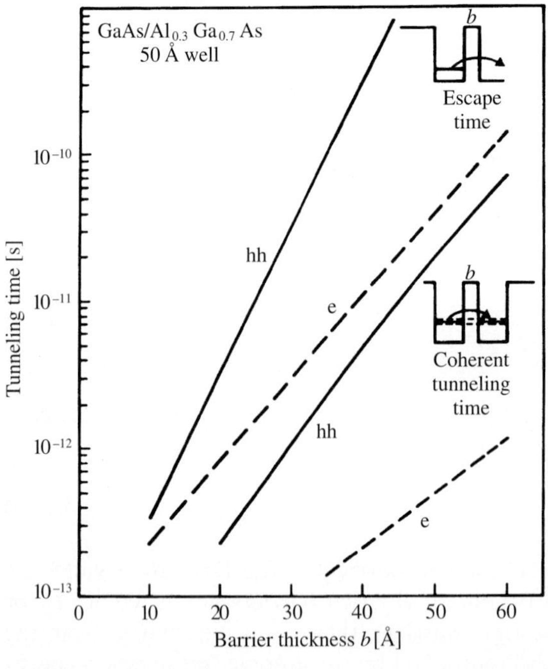

Fig. 7.4. The calculated escape times for electrons and heavy holes for a strongly asymmetric GaAs/$Al_{0.3}Ga_{0.7}As$ DBS with 50 Å well width as a function of the barrier width b; also shown are the calculated coherent tunneling times for electrons and holes in a symmetric double-quantum-well structure with 50 Å wells as a function of the barrier width b [7.27]

barrier of thickness *b*. For a given barrier width, the coherent tunneling time is much shorter than the escape time from a DBS.

We have discussed above the simplest case in which the conduction bands of the well and the barrier are both at the center of the Brillouin zone. More complicated situations in which the conduction band in the barrier occurs at a different point in the Brillouin zone have been considered in the literature [7.12, 20, 21]. Tunneling involving both the direct Γ valley as well as indirect L or X valleys in the barrier must be considered in such cases. Also the effective mass of the electron is not simply the mass at the bottom of the conduction band but must be calculated appropriately [7.12] using a two-band or more sophisticated model.

7.3 Basic Concepts: Asymmetric Double-Quantum-Well Structures

An asymmetric Double-Quantum Well-Structure (a-DQWS), consisting of a Wide Well (WW) and a Narrow Well (NW) separated by a barrier provides an ideal *isolated* structure for investigating tunneling. The conduction band for such a system under flat-band conditions is depicted in Fig. 7.5. We will describe such a structure as $a_L/b/a_R$ a-DQWS, where a_L is the thickness of the left well, b the thickness of the barrier, and a_R the thickness of the right well (all in Å). The parameters for the WW are denoted by unprimed quantities (a for WW thickness, E_i for the energy levels) while those for the NW are denoted by the primed quantities (a' for the NW thickness, and E'_i for the energy levels). For thick and high barriers, the two wells can be thought of as being uncoupled, and the energy levels of each well can be calculated indepen-

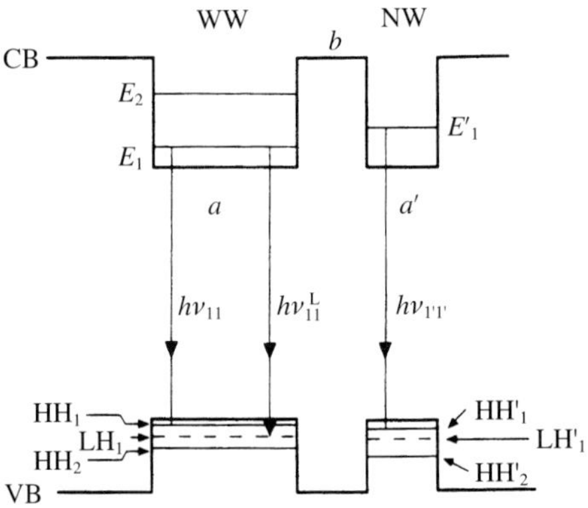

Fig. 7.5. Schematic of an asymmetric Double-Quantum-Well Structure (a-DQWS) under flat-band condition. The symbols define the notation used in the text

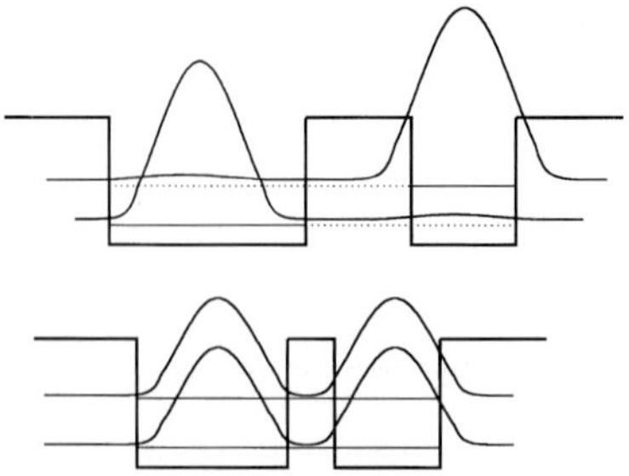

Fig. 7.6. Schematic illustration of the square of the wavefunctions in a weakly and strongly coupled a-DQWS

dently. There are a series of electron, Heavy Hole (HH) and Light Hole (LH) levels in each well corresponding to different quantized values of k_z [7.22]. Since these energies depend on the well thickness, no two levels at $k_{||} = 0$ are degenerate in general. There are, of course, higher energy states corresponding to $k_{||} \neq 0$, but there would be no coupling between states of different $k_{||}$ in the two wells even if the barriers were not thick and high.

This simplified picture of viewing the a-DQWS as two independent wells is only approximate. The correct way is to view the two quantum wells as a single system and calculate the eigenstates of the entire system. So long as different energy levels at $k_{||} = 0$ are far from resonance, this does not bring about any significant changes in the energy levels. But the important conceptual difference is that the eigenfunction of each level is an eigenfunction of the entire system and therefore can extend over the entire system. For thick and high barriers, the wavefunction for each level is localized primarily in one or the other well. As the barrier thickness and or height is reduced, the eigenfunction begins to acquire significant amplitude in both wells. This is illustrated schematically in Fig. 7.6.

Tunneling in a-DQWS is between two quasi-2D systems. The technique of optical markers in studying tunneling in a-DQWS, and concepts related to non-resonant and resonant tunneling are discussed in the next subsections.

7.3.1 Optical Markers in a-DQWS

The concept of optical markers was discussed in Sect. 7.1. The WW and the NW provide natural optical markers in the case of a-DQWS because the lowest interband optical transition (e.g., luminescence) in each well has a unique photon energy. If the barrier thickness and height are sufficiently large, the two wells can be treated as independent and the decay of luminescence at the photon energy unique to each well measures the recombination rate of that well. If the barrier thickness and heights are such that tunneling from one well (say the NW) to the other well (WW) occurs with a rate comparable or larger than

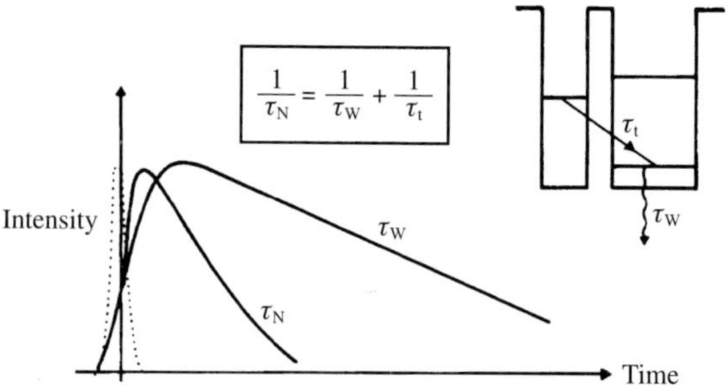

Fig. 7.7. Measurement scheme for tunneling rates in a-DQWS. The tunneling rate is obtained as the difference between the decay rates of the NW and the WW, assuming that the recombination rates in the two isolated wells are approximately the same

the recombination rate, then the measured decay rate of the NW would be faster than its normal recombination rate in the absence of tunneling. The normal recombination rates of the two wells should be approximately the same if the well thickness are not too different. In this case, the difference in the decay rates of the NW and the WW is a measure of the tunneling rate from the NW to the WW. This concept is schematically illustrated in Fig. 7.7, and has been utilized in a number of different experiments which will be discussed in Sects. 7.4–6 after discussing some basic concepts related to non-resonant and resonant tunneling in a-DQWS in the next two subsections.

7.3.2 Non-resonant Tunneling

When an electron in one eigenstate of the system, with its wavefunction localized primarily in one well, makes a transition to another eigenstate of the system, with its wavefunction localized primarily in the other well, this transition can be thought of as tunneling from the first well to the second. It should be clearly understood, however, that, strictly speaking, such a transition is an *inter-subband transition* between two eigenstates of the entire system. In Fig. 7.6 the tunneling indicated by the arrow between E_1' and E_1 is determined by the overlap of the two wavefunctions. Following the usual convention in this field, we will refer to such transitions as *non-resonant tunneling* since the carriers initially localized primarily in one well become localized primarily in the other well after the transition.

Such non-resonant tunneling must satisfy momentum- and energy-conservation requirements. This can be accomplished through elastic scattering such as impurity and interface scattering, or quasi-elastic scattering through the participation of low-energy acoustic phonons, or inelastic scattering through the emission of an optical phonon. Figure 7.8 shows the energy dispersion of the two lowest electron subbands in an a-DQWS. Tunneling from E_1' to E_1 mediated by elastic (impurity- or defect-assisted) process (I), Acoustic-Phonon

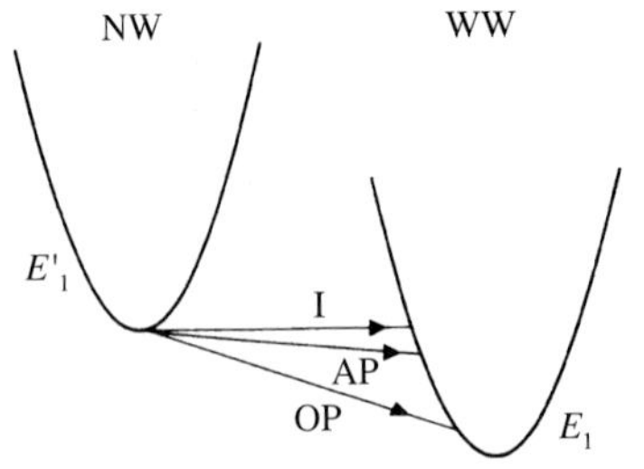

Fig. 7.8. The energy-band diagram for the lowest electron subband in the WW and the NW. Various assisted tunneling processes are indicated by the *arrows*

process (AP), and Optical-phonon assisted process (OP) are indicated by the arrows. Tunneling assisted by optical-phonon emission is possible only if the energy separation is large enough. It is also possible to *absorb* an optical phonon to satisfy the conservation conditions, but this process is generally negligible at low temperatures because the optical-phonon population is small.

7.3.3 Resonant Tunneling

When the electronic states in the two wells at the same in-plane wavevector are in resonance, tunneling interaction between them leads to a splitting of these levels. Such a resonance may occur, for example, between the lowest electron state of the Narrow Well (NW) and the first excited electron state of the Wide Well (WW), for certain specific values of the two well thickness and the barrier height. Such a resonance may also be induced in a more controlled fashion by applying an electric field along the growth direction. Figure 7.9 illustrates how various resonances can be attained by applying an appropriate positive or negative bias. In a practical system, it is often desirable to apply only reverse bias to avoid excessive dark current, non-uniformity of fields and other experimental problems. It may therefore be desirable, and at times necessary, to have different samples with different structures to investigate different resonances.

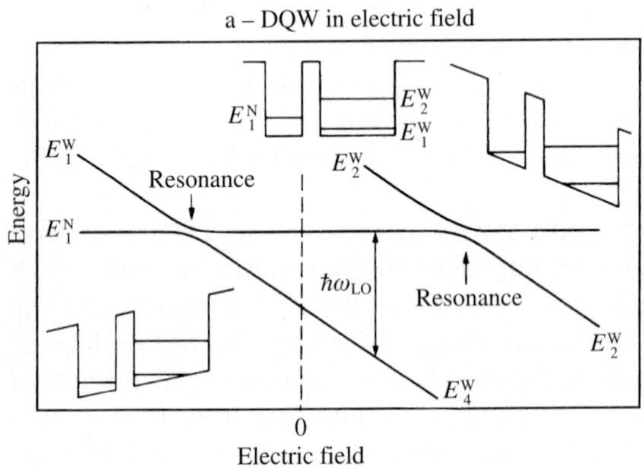

Fig. 7.9. Schematic of the variation of the electron-energy levels in an a-DQWS as a function of applied electric field. The two anti-crossing resonances and the threshold for LO-phonon-assisted tunneling are indicated

In Chap. 2 we discussed the situation in which an electric field induced a resonance between the two lowest electron states of each well (E_1 and E'_1). For this case, an electronic wavepacket resonantly excited in the WW by an appropriate short pulse undergoes coherent oscillations between the WW and the NW, provided that the dephasing rate is smaller than the oscillation frequency. *Kane* [7.23] has argued that half the period of oscillations provides the most direct measurement of the tunneling time. The frequency of the coherent oscillations is given simply by $\Delta E/h$, where ΔE is the energy difference between the levels E_1 and E'_1. The period of the coherent oscillations is given by

$$\tau_{\text{coh}} = h/\Delta E \ . \tag{7.8}$$

In general, ΔE and τ_{coh} have to be calculated numerically from the structure parameters; however a useful approximate expression for τ_{coh} in a symmetric DQWS is given by [7.24–27]

$$\tau_{\text{coh}} = \frac{\lambda b m_{\text{w}} \exp(2\pi b/\lambda)}{\hbar} , \tag{7.9}$$

where λ is a parameter that describes the wavefunction penetration in the barrier, it is given by (7.2). In Chap. 2 we discussed the observation of the coherent oscillations of excitons between the WW and the NW when the wavepacket was initially excited as a linear superposition of the two lowest eigenstates in the two wells.

We discuss here a different case in which an applied electric field induces a resonance between the ground electron state in the NW (E'_1) and the *first excited state* in the WW (E_2) (resonance R_1 shown schematically in Fig. 7.10). There is thus an electronic state (E_1) at energy lower than the energies of the

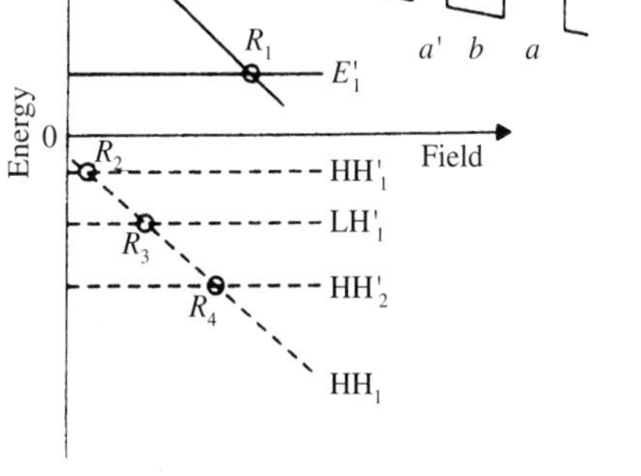

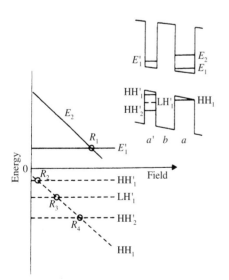

Fig. 7.10. Schematic of the electron and hole-energy levels of an a-DQWS under reverse bias which lowers the electron potential of electrons in the right well. Various resonances between the electron levels and between heavy-hole and light-hole levels are indicated by R_i

resonant states, so that relaxation to this lower state must also be considered. We consider here two different cases: (1) when the relaxation rate and the dephasing rate are sufficiently small compared to the oscillation frequency (Sect. 7.3.3), and (2) when the relaxation and/or dephasing rate is comparable to or larger than the oscillation frequency (Sect. 7.3.3b).

(a) Near-Ideal Case: Weak Collisions and Relaxation. Let us first consider the case of exact resonance R_1 shown in Fig. 7.10 when collision and relaxation rates are smaller than the coherent oscillation frequency. For excitation centered at HH'_1-E'_1 or HH_2-E_2 excitons with a spectral width wide compared to the splitting, a wavepacket created either in the NW or the WW would undergo coherent oscillations, like the case discussed in Chap. 2. But, in contrast to the case of the WW excitonic wavepacket discussed in Chap. 2, such an excitation would also excite continuum states in other subbands at $k_{||} \neq 0$. The presence of free carriers may increase the dephasing rate by introducing additional scattering mechanisms, such as exciton-free carrier scattering (Chap. 2). We recall that coherent oscillations for the case considered in Chap. 2 have been reported [7.28] when the initial wavepacket is created in the NW through excitation centered at HH'_1-E'_1. Since this case also creates additional free carriers, we expect that coherent oscillations would also be observed in the case shown in Fig. 7.10. If one probes the luminescence of the lowest exciton in the NW (E'_1-HH'_1), initially one would expect to see some oscillations in the luminescence intensity, corresponding to the coherent oscillations of the wavepacket. After the coherence is destroyed, the luminescence will decay with the sum of the recombination rate and the interwell (E'_1-E_1) tunneling rate. However, no observations of coherent oscillations in the case depicted in Fig. 7.10 have been reported, either in FWM or in luminescence.

If the excitation is in the continuum states rather than in resonance with the excitons, a very complicated situation develops. When the exciton states are in resonance, the continuum electron states in the E'_1 and E_2 subbands at the same $k_{||}$ are also in or close to resonance. There is therefore a splitting between these continuum states as well. However, unlike the case of excitons where all oscillator strength is concentrated at a given transition energy, the oscillator strengths for the continuum transitions are distributed in $k_{||}$-space. This case has not been treated theoretically. However, a simple qualitative picture would describe the luminescence dynamics if the dephasing rates are still small compared to the oscillation frequency. No oscillatory luminescence from excitons is expected since initially no excitons are created. The coherent electron-hole pairs would very likely dephase very quickly, and then relax to delocalized eigenstates with equal amplitudes in both the wells. Some of these pairs would then relax to form excitons (Chap. 6) in the NW and produce NW exciton luminescence. Thus, the initial dynamics of the exciton luminescence would be very different from the case of resonant excitation discussed in the previous paragraph. However, once the excitons are formed, their decay dynamics would be the same as before and would give information about the rate of interwell incoherent tunneling at resonance (which is formally the same as the rate of inter-subband scattering between E_1' and E_2 as discussed above).

It is interesting to inquire what happens to the tunneling rate as one tunes the levels E_1' and E_2 through each other with an applied electric field or by some other means, in this case of weak collisions and relaxation. When the electronic levels are far from resonance, one probes non-resonant tunneling as discussed in Sect. 7.3.2. When two electronic states are brought closer to resonance, one expects to see coherent oscillations, as discussed in Chap. 2. Away from exact resonance, the level separation is larger so that the coherent oscillation frequency is higher. However, the amplitude of the oscillation is smaller because the eigenfunction primarily associated with the NW (for example) has a smaller amplitude in the WW. The oscillation frequency is the smallest and the oscillation amplitude the largest at exact resonance. Beyond the exact resonance, the oscillation frequency begins to increase and the oscillation amplitude begins to decrease. The interwell tunneling rate depends on the overlap of the E_1' and E_2 eigenfunctions. These arguments show that this overlap is small away from resonance, is maximum at the resonance, and becomes small again as one tunes away from the resonance. Thus, assuming that the interwell tunneling rate is comparable or larger than the recombination rate in the NW, one expects the decay rate of the NW luminescence to show a non-monotonic behavior with a maximum at the exact resonance. The difference between the decay rate far from resonance and in the resonance region would then provide a direct measure of this interwell tunneling rate. Experiments which probe this dynamics will be discussed in Sect. 7.6.

(b) Tunneling in the Presence of Strong Collisions and Relaxation. The concept of coherent oscillations of an electronic wavepacket between two wells discussed above and in Chap. 2 breaks down when collisions or relaxation interrupt the phases at a rate comparable or higher than the expected coherent oscillation frequency. The transfer of the wavepacket from the well in which it is created to the other well slows down considerably as the collision/relaxation rates increase. When the collision/relaxation rates are much higher than the oscillation frequency, the coupling between the two states is destroyed. The two states then become independent, non-interacting states which can cross each other so that there is no delocalization of the wavefunction.

The general case of collisions/relaxation rates comparable to or larger than the ideal coherent oscillation frequency is of considerable interest in many branches of physics. *Leggett* and coworkers [7.29] have treated the general case of tunneling in the presence of dissipation with a density-matrix formalism. We consider here the special case in which relaxation is present but collisions are absent. In this case it was argued [7.30, 31] that the theory is particularly simple because $k_{\|}$ is a good quantum number and transitions occur only between states of the same $k_{\|}$. For this case one needs to consider only the effect of relaxation between E_2 or E_1' and E_1 and it can be shown that the equations of motion for the state amplitudes are the same as those for coupled damped harmonic oscillators. In this case, the effect of relaxation is to broaden the resonance by the relaxation rate ($1/\tau_R$) between E_2 or E_1' and E_1. If the relaxation rate is much larger than the coherent oscillation frequency ($1/\tau_R \gg 2\pi/\tau_{coh}$), then the tunneling rate at resonance ($1/\tau_t$) is given by [7.30, 31]

$$1/\tau_t \cong \frac{4\pi^2 \tau_R/\tau_{coh}}{\tau_{coh}} \ . \tag{7.10}$$

This equation indicates that in the limit of strong relaxation, the tunneling rate at resonance *decreases* as the inter-subband relaxation rate *increases*. This is because the rapid damping prevents the buildup of state amplitude in the WW. Therefore, one expects to see a non-monotonic behavior of the effective tunneling rate as a function of the relaxation rate. This result, which is *counter-intuitive* in a sequential picture of the coherent tunneling and relaxation processes, is a consequence of the loss of coherence due to relaxation. This non-monotonic behavior is schematically illustrated in Fig. 7.11 where the tunneling time goes through a minimum when the coherent tunneling time and the relaxation times are approximately equal. For relaxation times much longer than the coherent tunneling time, the tunneling time will be twice the relaxation time as discussed above. For relaxation time much shorter than the coherent tunneling time, the tunneling time will be larger than the coherent tunneling time by a factor proportional to the ratio of the relaxation time to the coherent tunneling time, as in (7.10). Tunneling in the presence of collisions and relaxation has also been discussed in [7.32, 33].

In a real physical system, there are scattering processes (e.g., carrier-carrier-scattering) or collisions destroying the phase coherence between the eigenstates of energies E_1' and E_2. From a full density matrix formalism required in this case it can be shown [7.30] that the width of tunneling resonance in energy increases from $\hbar/\tau_R$ to $(\hbar/\tau_R + 2\hbar/\tau_2')$ where $1/T_2'$ is the pure dephasing rate,

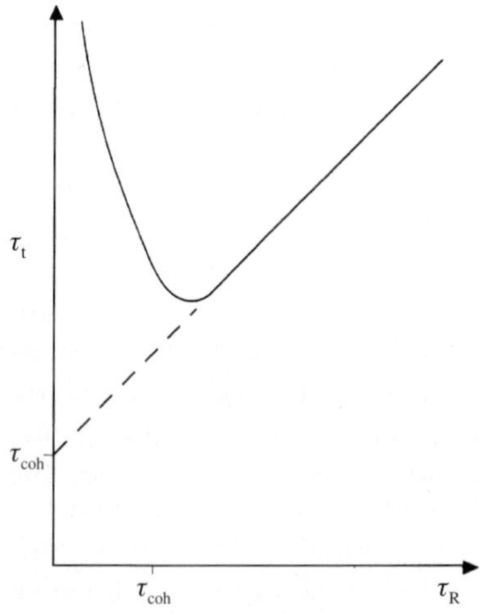

Fig. 7.11. Schematic illustration of the resonant tunneling times in the sequential picture (*dashed line*) and the unified picture of tunneling (*solid curve*). The unified model of resonant tunneling includes the effects of collisions and relaxation and predicts a non-monotonic behavior

or the additional decay rate of the phase coherence due to scattering processes (collisions). It is noteworthy that in either case, the integral of the tunneling decay rate with respect to the energy detuning is independent of the width, $\hbar/\tau_R$ or $(\hbar/\tau_r + 2\hbar/T_2')$, of the resonance and is given by

$$\text{Area} = h\pi^2/\tau_{\text{coh}}^2 \ . \tag{7.11}$$

The broadening of resonance can also be simply understood in terms of lifetime broadening of the level E_1' in the case of relaxation without any other collisions. Such a lifetime broadening reduces the available density of final states at any given detuning between the levels and hence reduces the tunneling rate. However, this does not change the area under the resonance curve. This is schematically illustrated in Fig. 7.12 for the near-ideal case and the case of strong collisions/relaxation. We note that the effect of collisions on the transmission through a double-barrier structure has been considered by *Stone and Lee* [7.34] who reached similar conclusions concerning the broadening of the transmission resonance but did not discuss the effect on tunneling rates.

One further point of interest concerns the effects of homogeneous versus inhomogeneous broadening. It should be obvious that the considerations of broadening of resonance discussed above applies only when the intrinsic homogeneous broadening due collision/relaxation is larger than any inhomogeneous broadening. If the sample quality is poor, the inhomogeneous broadening may dominate and broaden the resonance even if collision/relaxation effects are small. In a real system, both broadening mechanisms may contribute and may be difficult to separate.

To summarize this section, these considerations show that the sequential picture of tunneling followed by relaxation is not valid in the case of strong relaxation and collisions and that a *unified picture* must be used in this case.

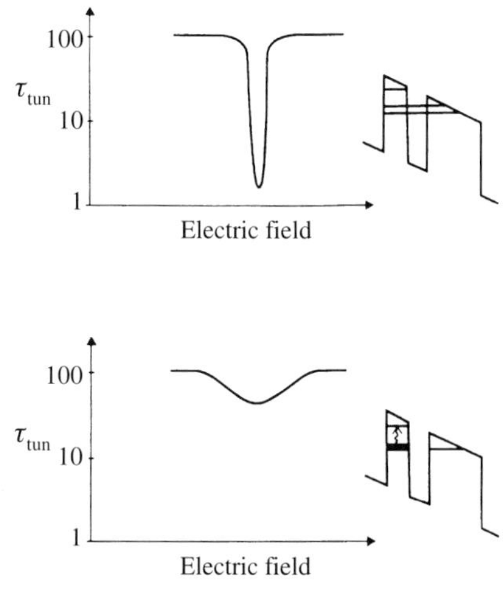

Fig. 7.12. Schematic diagram showing that the tunneling times are expected to exhibit a strong and narrow resonance for the near-ideal case of weak or no collisions (*top*) and a weak, broad resonance for the case of strong collisions/relaxation (*bottom*). The area of resonance is expected to be conserved

7.4 Tunneling in Double-Barrier Structures

Considerable information on tunneling in double-barrier diodes has been obtained by measuring the steady-state current-voltage characteristics [7.35, 36], and by studying the frequency response of the system [7.1, 8, 37–39]. We discuss in this section some results on the dynamics of tunneling in double-barrier structures obtained by using all-optical techniques.

7.4.1 Dependence on the Barrier Thickness

The first all-optical measurement on the dynamics of tunneling in DBS was performed by *Tsuchiya* et al. [7.10] using time-resolved luminescence spectroscopy. The sample consisted of a 62 Å quantum well surrounded by AlAs barriers of thickness varying from 28 to 62 Å. The quantum well in such GaAs/AlAs DBS was excited at the $n = 1$ light-hole exciton energy using 20 ps pulses from a dye laser, and the dynamics of the $n = 1$ heavy-hole exciton luminescence was investigated as a function of barrier thickness and temperature using a streak camera with a time resolution of 24 ps. Figure 7.13 displays the measured decay time constants as a function of barrier thickness for three different temperatures. The figure also shows the measured decay times for a 5-period Multiple Quantum Well (MQW) sample with a well width of 62 Å and AlAs barrier width of 51 Å, surrounded by 100 Å AlAs barrier on each side. The shortest decay time measured was about 60 ps.

The decay time increases rapidly with barrier thickness for small barrier thicknesses, and then asymptotically approaches the MQW values. The authors analyzed the results by equating the measured decay rate with the sum of the recombination rate and the escape rate out of the quantum well by tunneling. For thick barriers, the measured decay rate is essentially controlled by the recombination rate. However, for thin barriers there is a nearly exponential increase in the decay rate with decreasing barrier thickness. The decay rates in this range were attributed to the escape of electrons out of the quantum well. This escape rate can be calculated from the calculated width of the electron-transmission resonance through the DBS. These calculated rates are presented as dashed lines in Fig. 7.13 for two different barrier heights. The agreement between the calculated and measured rates for small barrier thickness supports the model of electron escape by tunneling, and provides a measurement of electron escape rates.

7.4.2 Dependence on Electric Field

Another technique for varying the escape rate out of a DBS is to apply an electric field perpendicular to the interface planes. The electric field increases the escape rate through one barrier and decreases the escape rate through the other

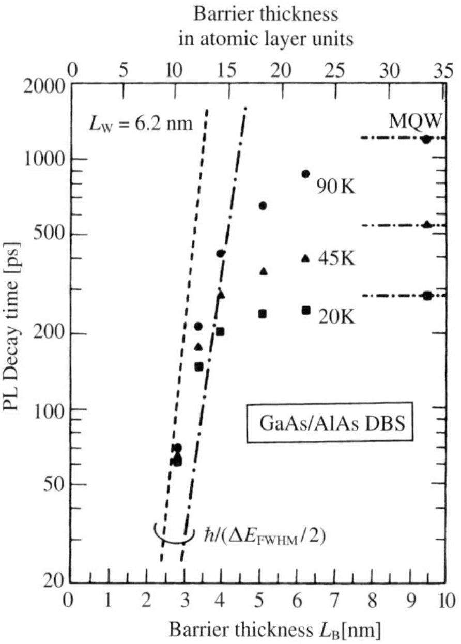

Fig. 7.13. Decay times of the luminescence from the 6.2 nm quantum well in a DBS plotted as a function of barrier thickness for three different temperatures. The tunneling escape process dominates for thin barriers (<4 nm) and recombination within the well dominates for thick barriers [7.10]

barrier by modifying the potential profile. Consider an asymmetric DBS in which one barrier has a very low transmission probability (escape rate) whereas the other barrier has a much higher transmission probability. In this case, the change in the measured decay rates can be attributed directly to the field-induced change in the transmission probability of the barrier with higher transmission probability. Such measurements were reported by *Norris* et al. [7.11].

Their results for a 30 Å GaAs quantum well with a 2000 Å $Al_{0.3}Ga_{0.7}As$ barrier on one side and an $Al_{0.3}Ga_{0.7}As$ barrier of much smaller thickness b on the other are shown in Fig. 7.14. The application of the electric field reduces the effective area of the thinner barrier and leads to a reduced decay time for all three samples. These results were fitted by approximating the tunneling escape time through a barrier as

$$\tau_{tun} = c \exp\left\{\frac{2}{\hbar}\left[\int_0^b dz\sqrt{2m(V_0-E-Fz)}\right]\right\}, \tag{7.12}$$

where F is the applied field along the z-direction, and c is a constant which was obtained from the transmission coefficient under flatband conditions (7.1–5). The calculated curves are also depicted in Fig. 7.14 and show reasonably good agreement with the experimental results.

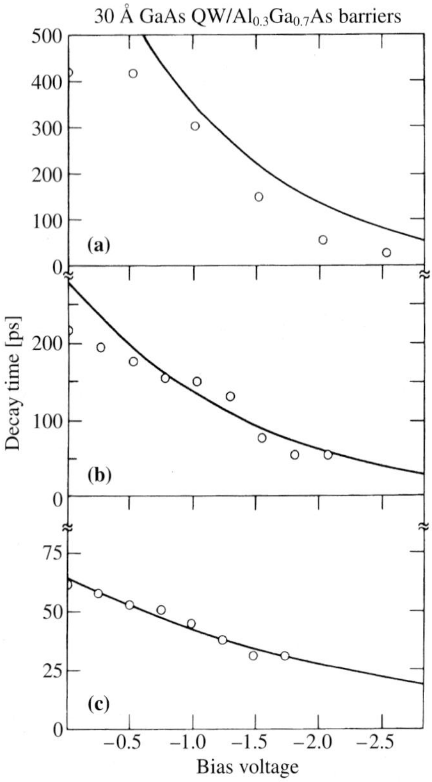

Fig. 7.14. Tunneling-escape rates as a function of applied electric field for DBS with three different barrier thickness **(a)** 121 Å, **(b)** 111 Å and **(c)** 85 Å at 6 K at an estimated injection density of 5×10^{16} cm^{-3}. Streak camera time resolution was 20 ps [7.11]

7.4.3 Summary

These results demonstrate that the dynamics of tunneling through symmetric or asymmetric DBS is well understood in terms of the escape of electrons out of the quantum well. One of the questions that arises is what is the role of excitons or electron-hole interaction in general. It may be argued that the exciton-binding energy is small compared to the tunneling-barrier height and it is reasonable to consider the process simply in terms of the electron-escape rate. A related question is the accumulation of charge in the quantum well during tunneling. This question has been investigated by steady-state electrical [7.40, 41] as well as optical studies [7.42–48], but will not be discussed here.

7.5 Non-resonant Tunneling in Asymmetric Double-Quantum-Well Structures

Tunneling in a-DQWS has been investigated by a number of different research groups for non-resonant (when no electronic levels are in resonance) as well as resonant (when two electronic levels in different wells are in resonance) conditions. We begin with a general historical background on both kinds of studies and then discuss some experimental results on non-resonant tunneling in this section, and on resonant tunneling in a-DQWS in Sect. 7.6.

Initial studies on isolated a-DQWS with 300 ps time resolution were reported by *Tada* et al. [7.49], and studies on non-isolated coupled quantum wells were reported by *Tsuchiya* et al. [7.50] and *Sakaki* et al. [7.51] with a time resolution of ≈ 50 ps. *Norris* et al. [7.52] reported on the dynamics of charge transfer luminescence (electrons and holes in separate wells) with 20 ps time resolution. Using time-resolved luminescence with sub-picosecond time resolution, *Oberli* et al. [7.53] reported a quantitative determination of the electron-tunneling rates and demonstrated the strong enhancement of tunneling rates at an electronic resonance. They also investigated the effect of barrier thickness on tunneling rates [7.54], and observed a threshold in the tunneling rate at the optical-phonon energy, demonstrating the importance of optical phonon-assisted tunneling processes [7.55].

Many other studies of tunneling in a-DQWS have since been reported, primarily using time-resolved luminescence spectroscopy. These studies include investigations of InGaAs/InP system [7.56], pressure dependence in GaAs [7.21, 56] by use of luminescence correlation spectroscopy to measure tunneling rates [7.12, 57], electron and hole tunneling in GaAs [7.58, 59], and resonant and non-resonant behavior as a function of well thickness [7.60] and electric field [7.14]. Tunneling between two quantum wells has also been used to demonstrate fast recovery of excitonic absorption [7.61]. In addition, resonant and non-resonant tunneling of holes has been investigated [7.30] and the results were used to demonstrate the importance of collisions and relaxation on tunneling. There have also been a number of studies of cw luminescence from coupled quantum wells (see, for example, [7.62, 63]) and study of spatially indirect excitons (see, for example, [7.64–67]). However, such studies provide only limited information about the dynamics of tunneling and will not be discussed here.

There have also been a number of theoretical studies related to tunneling in a-DQWS. One of the first was by *Weil* and *Vinter* [7.68] who investigated phonon-assisted transfer rates in a-DQWS. A discussion of coherent oscillations in a-DQWS has been given by *Luryi* [7.26, 69] and assisted relaxation processes have been treated by *Ferreira* and *Bastard* [7.70] and *Bastard* et al. [7.71]. Monte-Carlo simulations of tunneling in a-DQWS have been carried out by *Lary* et al. [7.72]. It should also be mentioned that general aspects of tunneling in presence of scattering have been discussed theoretically by *Stone and Lee* [7.34], *Leggett* et al. [7.29], *Wingreen* et al. [7.73], *Büttiker* [7.74], and *Gelfand* et al. [7.75].

Finally we recall that, as mentioned in Sect. 7.3, tunneling times are generally determined by measuring the decay rate of the lowest electron or hole level in one of the wells and properly accounting for the recombination times in the absence of tunneling.

7.5.1 Dependence on Barrier Thickness

Oberli et al. [7.53, 54] investigated the dependence of non-resonant tunneling rates on the barrier thickness for $63/b/100$ a-DQWS. The samples at 10 K were excited such that only the lowest electron subband in each well was excited. The rates for tunneling transition from the lowest electron level in the NW to the lowest electron level in the WW were investigated without any applied electric field on the samples, and were found to vary nearly exponentially with the barrier thickness b. The dependence of tunneling rate on barrier thickness has since been investigated by a number of different groups [7.14, 60, 76] in much greater detail. Figure 7.15 represents a compilation of data from various groups presented by *Rühle* [7.14]. In order to compare data from different structures with different potential barriers, the results are plotted as a function of d/λ where d is the barrier thickness and λ is given by (7.2). The general trend agrees with the exponential dependence expected from (7.3). However, the absolute values differ. It was suggested that breakdown of momentum conservation parallel to the interface by impurity or interface roughness might contribute to this variation. It should also be mentioned that an exponential dependence is expected only in the absence of resonance and under flat-band conditions. This may also contribute to the variation in values from different samples.

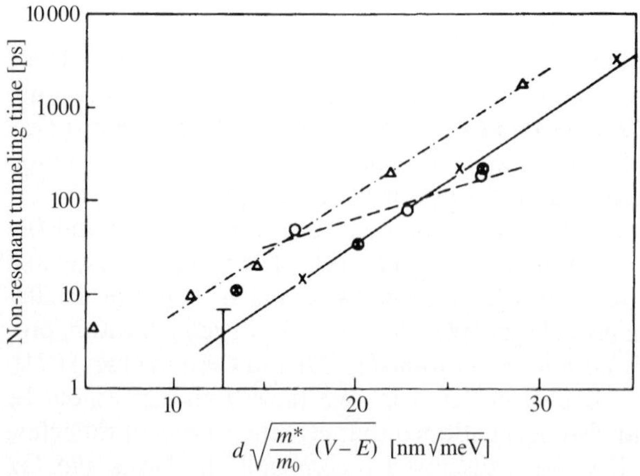

Fig. 7.15. Dependence of tunneling rates in a-DQWS for non-resonant condition as a function of a parameter proportional to d/λ as discussed in the text [7.76]

7.5.2 Resonant Phonon-Assisted Tunneling

The above results were obtained for samples with E'_1-E_1 larger than the optical-phonon energy so that the non-resonant tunneling times were between 10 and 100 ps. A question of considerable interest is the role of optical phonons in such tunneling or inter-subband transfer processes. This can be investigated by varying E'_1-E_1 applying an electric field perpendicular to the interfaces.

Such an investigation was first carried out by *Oberli* et al. [7.55], who investigated a 100/50/63 a-DQWS sample at low temperatures. The excitation-photon energy was such that electrons and holes were excited only in the WW. The measurements, using luminescence upconversion technique with sub-picosecond time resolution, concentrated on the decay time of the WW luminescence as a function of applied electric field which lowered the electron-energy levels of the NW with respect to those of the WW. Figure 7.16 exhibits the time evolution of the luminescence for three different applied electric field. The decay times of the luminescence determined from such curves at various electric fields are plotted in Fig. 7.17 as a function of the applied field. Then decay time under flat-band conditions is quite long, about 600 ps. With increasing field, the lowest electron levels E_1 and E'_1 cross each other and then E'_1 becomes lower in energy compared to E_1. As shown in Fig. 7.17, the WW decay time increases at small electric field, the decreases rapidly, goes through a minimum and then increases gradually. The initial increase in the decay time is not well understood. The minimum in the decay time is observed at ≈ 50 kV/cm. At this field, the energy difference E'_1-E_1 approximately equals the optical-phonon energy. These observations clearly demonstrate the importance of phonon-assisted resonance in the tunneling rate.

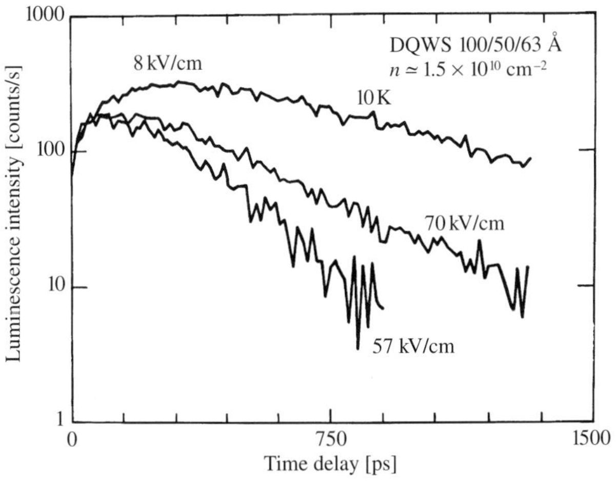

Fig. 7.16. Temporal evolution of the luminescence from the WW of a 100/50/63 a-DQWS at 10 K for three different electric fields. The luminescence decay time is determined from the long time exponential decay [7.55]

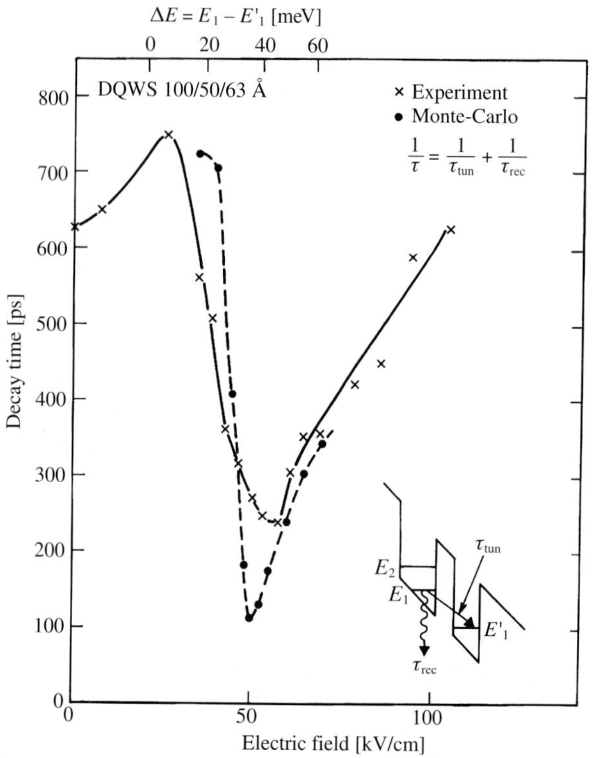

Fig. 7.17. The measured dependence of the decay time of the WW luminescence from a 100/50/63 a-DQWS sample at 10 K, as a function of the applied electric field. The separation between E_1 and E_1' are shown in the scale at the top. The *dashed line* is to guide the eye. The strong decrease in the decay time beginning at 35 kV/cm is attributed to LO-phonon-assisted electron tunneling from E_1 to E_1'. The solid points are the results of a Monte-Carlo simulation of the problem [7.55]

These results were compared in detail with ensemble Monte-Carlo simulations [7.55, 72, 77]. The decay of the WW luminescence was modeled as the sum of a constant recombination rate and the rate of transfer of electrons from E_1 in the WW to E_1' in the NW calculated by the Monte-Carlo simulation using wavefunction overlap and energy-and-momentum-conservation conditions. Initial calculations were performed using bulk-phonon modes [7.77]. The results of the calculation with $\tau_{rec} = 700$ ps are also shown in Fig. 7.17. The calculations reproduce the steep decline in the WW decay time, followed by a gradual increase with increasing electric field. The steep decline is a result of the onset of optical phonon-assisted tunneling of electrons between the WW and the NW. The gradual increase at higher electric field results from the decreasing matrix elements with increasing energy separation (larger phonon momentum) between E_1 and E_1', and a change in the wavefunction overlap. Thus, the calculations reproduce the important features in the data. However,

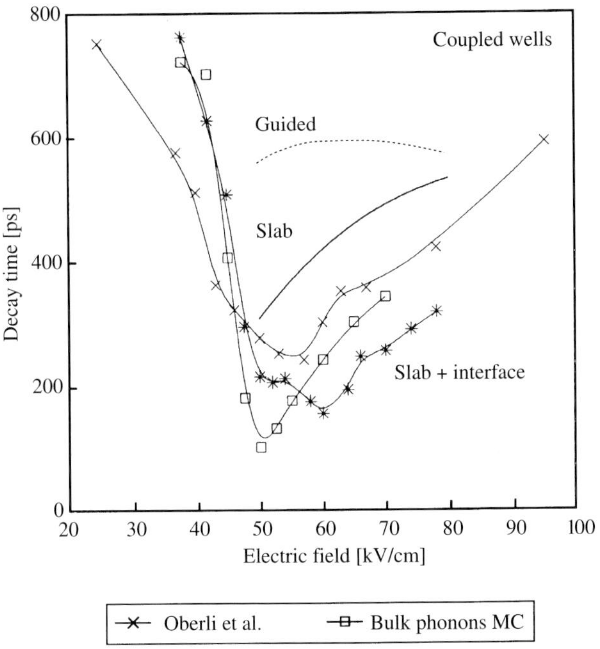

Fig. 7.18. Monte-Carlo simulations of phonon resonance in tunneling between the wells of an a-DQWS, using various phonon modes in the calculations [7.78, 79]. Also shown for comparison are the experimental results from [7.55]

there are also some disagreements, in particular in the absolute value of the maximum decay rate and the field at which the maximum occurs. It was suggested [7.55] that the complex phonon spectrum in quantum wells (Chap. 5) might contribute to these differences.

More recent Monte-Carlo simulations [7.78, 79] take into account these complexities in the phonon spectrum. Phonon-assisted tunneling rates between the two wells in an a-DQWS for different phonon modes are shown in Fig. 7.18. These results indicate that the interface and slab phonon modes dominate the tunneling process and that the guided mode contribution is negligible. The calculations reproduce the experimental shape remarkably well, and the shift of about 5 kV/cm between the experiments and the calculations is within the experimental error. The calculated curves still predict somewhat smaller tunneling times, but this depends to some extent on the value assumed for off-resonant decay time.

These results provide convincing evidence for the importance of the phonon resonance in electron tunneling. The phonon resonance has also been reported by *Roussignol* et al. [7.80]. It might be mentioned that the off-resonant decay time (due to tunneling or recombination) and impurity and disorder-induced tunneling transfer time must be sufficiently long in order to observe a strong phonon resonance. For large impurity concentration or interface disorder, elastic tunneling and/or non-radiative recombination process may make it difficult to fulfill this condition. *Heberle* et al. [7.14, 58] have reported that phonon-assisted tunneling plays a minor role compared to impurity- and interface roughness-assisted tunneling in the samples they investigated.

7.6 Resonant Tunneling in Asymmetric Double-Quantum-Well Structures

As discussed in Sect. 7.3, resonance between two electronic levels introduces qualitatively new features in the tunneling process. In the ideal case when collisions/relaxation are weak, coherent oscillations are expected. Experimental results on these oscillations using nonlinear time-resolved techniques were discussed in Chap. 2. In this section, we consider the case where the resonance occurs between a ground state and an excited state (for example, E_1 and E'_2 or H'_1 and H_2) so that relaxation to a lower state (E'_1 or H_1) is possible. We first treat the case of electrons and then of holes.

7.6.1 Resonant Tunneling of Electrons: Initial Studies

The first investigation of the effect of resonance on the electron tunneling rates in a-DQWS was performed by *Oberli et al.* [7.53] using luminescence upconversion spectroscopy with sub-picosecond time resolution. They investigated a 63/50/90 a-DQWS under an applied electric field which lowered the electron-energy levels of the WW with respect to those of the NW. Under flat-band conditions, E'_1 is between E_1 and E_2. Application of the electric field first brings E_2 into resonance with E'_1, and then brings E_2 below E'_1. The decay times for electrons in the NW, determined from the exponential decay of the NW luminescence at long times, are plotted as a function of the applied electric field in Fig. 7.19. Near flat-band conditions, the decay time is about 70 ps. This value is much smaller than the flat-band value for the WW luminescence in Fig. 7.17 because the energy separation allows optical phonon-assisted tunneling transitions in the present case, but not in the case investigated in Fig. 7.17.

As the electric field is increased, the first excited electron level E_2 approaches E'_1 and then becomes resonant with E'_1 at field F_R. It was shown [7.53] that the minimum in the decay time occurs at field F_R so that the maximum tunneling rate occurs when E_2 and E'_1 are in resonance. Further increase in the electric field brings the two levels out of resonance, with E_2 at lower energy than E'_1. This leads to an increase in the decay time of the WW, indicating a decrease in the tunneling rate. This *non-monotonic* behavior is the first demonstration of the strong influence of the electronic resonance on tunneling in a-DQWS.

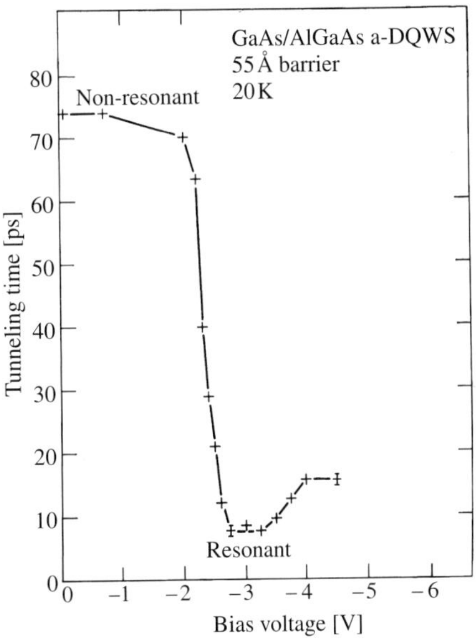

Fig. 7.19. The decay times of the luminescence from the NW in a 63/50/90 a-DQWS as a function of the electric field, as the electric field tunes E_2 through E'_1 (resonance R_1 of Fig. 7.10). The sharp drop in the decay time and the non-monotonic behavior of the decay time results from resonant-electron tunneling between E'_1 and E_2. The measured decay time at resonance (~7 ps) is considerably longer than the system resolution of 0.5 ps [7.53]

While these qualitative features of the data are well understood, the minimum value of the decay time is 7.5 ps, much longer than the experimental time resolution of ≈ 0.5 ps. The field at which the resonance occurred varied slightly with the intensity of the laser exciting the luminescence, indicating the presence of space charge effect at higher densities. However, the decay time at the minimum remained the same in the range of intensities investigated. This minimum decay time (7.5 ps) was considerably larger than the calculated half period of coherent oscillation (0.67 ps). A number of reasons were considered for this large difference [7.53], but the complete physical picture became clear only after the results and analysis of the resonant tunneling of holes. We postpone a detailed discussion of the discrepancy to Sect. 7.6.3.

There is another aspect of the problem that also needs some further attention. These measurements, as well as most others reported in the literature, have excited the levels under study in a non-resonant manner, i.e., the exciting photon energy is generally higher than the energy of the photon emitted by the transition under study. As discussed in Chap. 6, the photoexcited free electron-hole pairs form excitons with large in-plane momenta in about 20 ps, which then relax in a few hundred picoseconds to excitons with small in-plane momenta that couple to photons [7.81] at low excitation intensities and low lattice temperatures. This dynamics of exciton formation and relaxation is generally not considered in analyzing the data. While this is probably a good assumption when the tunneling times are long, a closer scrutiny of the assumption in the case of resonance when the tunneling times can be less than 10 ps may be useful.

7.6.2 Resonant Tunneling of Holes: Initial Studies

In the GaAs/AlGaAs quantum-well system, the conduction band offset is about 2/3 of the bandgap difference between the two semiconductors. Thus, the barrier for the holes is smaller but the heavy hole effective mass is much larger than the electron effective mass. These two factors have opposing influence on the tunneling rate but the net effect is to make the hole-tunneling rate smaller than the electron-tunneling rate. For this reason, the initial studies of tunneling in semiconductor nanostructures were performed on electrons. In fact, in the analysis of the results above, it is assumed that hole tunneling does not influence the results in a significant manner. As mentioned above, tunneling of electrons and holes in the same structures have been investigated [7.58, 59, 80]. It is also possible to design structures and experiments in which hole tunneling is the primary process.

One such experiment was performed by *Leo* et al. [7.30, 31] who investigated hole tunneling from the WW to the NW in a 63/50/90 GaAs/$Al_{0.3}Ga_{0.7}As$ a-DQWS as a function of electric field and temperature. There were eight periods of the asymmetric double-quantum wells separated by 150 Å $Al_{0.3}Ga_{0.7}As$ barriers. In the absence of any applied electric field, the built-in field in the structure makes HH_1 below HH'_1 but above LH'_1 and HH'_2 levels of the NW. With applied reverse bias, the HH_1 level is scanned through these two NW levels as shown in the schematic on the upper part of Fig. 7.20. The experiments concentrated on the field region encompassing the HH_1-LH'_1 and HH_1-HH'_2 resonances. The excitation-photon energy was chosen to excite only

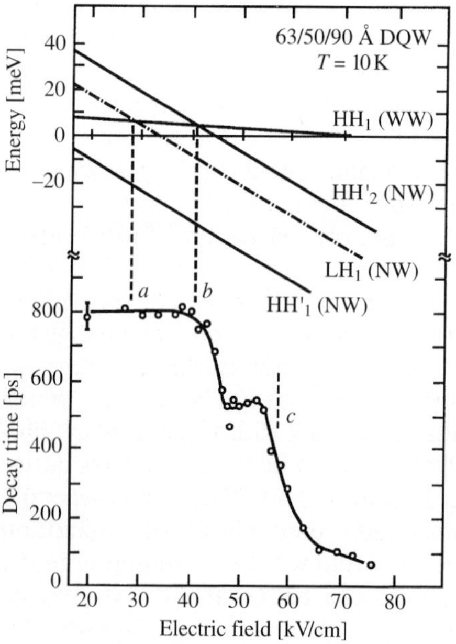

Fig. 7.20. Decay rate of the WW luminescence in a 63/50/90 a-DQWS at 10 K as a function of the applied electric field. The upper part of the figure shows a calculation of the various energy levels as a function of the applied electric field. Resonances occur at points *a* and *b* where the levels cross. The drop beginning at about 40 kV/cm corresponds to resonant tunneling of holes from HH_1 to HH'_2. The subsequent drop at fields >55 kV/cm is due to electrons [7.30]

the WW transition, and the decay of the WW luminescence intensity was measured as a function of the applied electric field using luminescence upconversion techniques with subpicosecond time-resolution. The decay rate of the WW luminescence is the sum of the recombination rate in the WW, and tunneling of holes from the WW to the NW. The electrons in E_1 can not tunnel to the NW and their tunneling rate through the 150 Å thick barrier is negligible except at very high electric fields. Therefore, hole-tunneling rate can be deduced from the experimental results.

The results are shown in Fig. 7.20. At low electric field the decay time is about 800 ps and corresponds to the recombination time constant in the WW because non-resonant tunneling of holes is negligible. The first important result is that the decay rate is unaffected by the resonance between HH_1 and LH'_1 at the field marked by "a" in Fig. 7.20. There is no change in the WW decay rate at this resonance even at 80 K and up to a density of 1×10^{11} cm^{-2}. Under these conditions, states with larger $k_{\|}$ are occupied and the mixing between the heavy and the light hole states is expected to be larger. The reason for the absence of this resonance is not clear at this time. HH-LH resonance is a subject of considerable interest in DBDs [7.82, 83] and a-DQWS, and has been invoked by several researchers to explain their cw [7.63, 70, 71] and time-resolved optical studies [7.84–86] on a-DQWS. One such optical investigation by *Heberle* et al. [7.14, 80, 87] will be discussed in Sect. 7.6.3.

The data in Fig. 7.20 do show a decrease in the WW decay time at the HH_1-HH'_2 resonance indicated by "b" in Fig. 7.20. The decay time decreases to about 500 ps, reveals a plateau and then shows a further decrease beyond about 55 kV/cm. This last decrease was demonstrated to be associated with the electron tunneling through the 150 Å thick barrier and was therefore not considered further.

The decrease in the decay time from about 800 to 500 ps at the HH_1-HH'_2 resonance implies that the resonant tunneling time for holes at this resonance is $(1/500 - 1/800 \text{ ps})^{-1} = 1330$ ps. Calculations yield that the expected splitting of the HH_1-HH'_2 level at resonance is 0.074 meV corresponding to $\tau_{coh}/2 = 28$ ps. Thus, the observed tunneling rate is about *a factor of 45 smaller* than expected theoretically for an ideal case. This behavior is qualitatively similar to that of electrons at the E'_1-E_2 resonance and will be discussed in detail in Sect. 7.6.4.

7.6.3 Resonant Tunneling of Electron and Holes: Further Studies

Rühle and coworkers [7.14, 21, 56, 76, 84, 85, 87–91] have reported extensive investigation of resonant tunneling of electrons and holes in a-DQWS. By monitoring the decay times of the lowest levels in the WW as well as the NW, and by applying positive as well as negative bias, they were able to investigate a number of different resonances in a single a-DQWS. Their results for a 100/60/50 sample are depicted in Fig. 7.21. The strongest resonance in the NW decay times (crosses) at 43 kV/cm corresponds to the resonance between

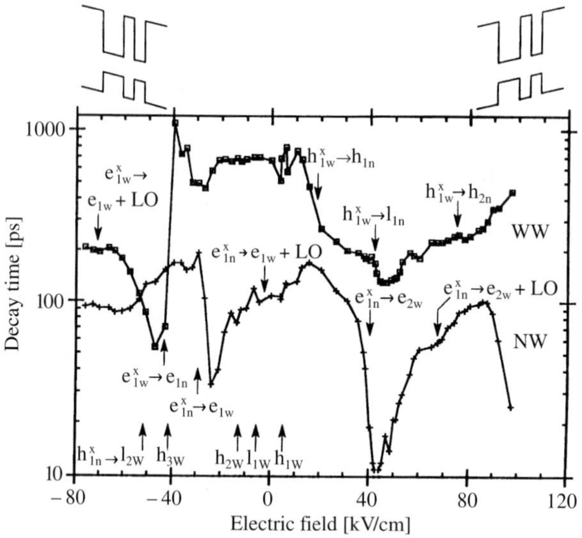

Fig. 7.21. Decay times of the interband luminescence from the WW and the NW in a 100/60/50 a-DQWS at various positive and negative applied electric fields. Various calculated resonance positions are indicated [7.14]

the first NW electron level with the second WW electron level. This corresponds to the case investigated by *Oberli* et al. [7.53] and discussed in Sect. 7.6.1. By a careful comparison of the measured and the calculated resonant fields, *Heberle* et al. [7.87] concluded that the resonance actually involves excitons. For this reason they have marked the resonance as $e^{x}_{1n} \rightarrow e_{2w}$.

The second strongest resonance in the decay time of the NW occurs at -21 kV/cm and corresponds to a resonance between the lowest electron (exciton) level in each well. This corresponds to elastic tunneling mediated by interface disorder, impurities or near-elastic tunneling mediated by low-energy acoustic phonons. This resonance was not observed by *Oberli* et al. [7.53]. Figure 7.21 also shows the fields at which phonon-assisted resonances are expected. Strong phonon resonances are not observed, presumably because the measured background decay times in Fig. 7.21 are already quite short (see the discussion at the end of Sect. 7.5.2). The decay times of the WW in Fig. 7.21 exhibit the strongest resonance at -43 kV/cm. This resonance was assigned to elastic tunneling between the two lowest electron (exciton) levels as indicated. In addition, the data also brings to light several broad and shallow resonances attributed to resonance between various heavy-hole and light-hole levels.

Heberle et al. [7.87] investigated the effective thickness dependence of the $e_{1n} \rightarrow e_{2w}$ resonant tunneling rates for different samples. They recognized that plotting the results as a function of $\int_0^b \sqrt{V(z)-E}\,dz$ where $V(z) = V_0 - Fz$ of 7.12) would allow a comparison between different samples with different bar-

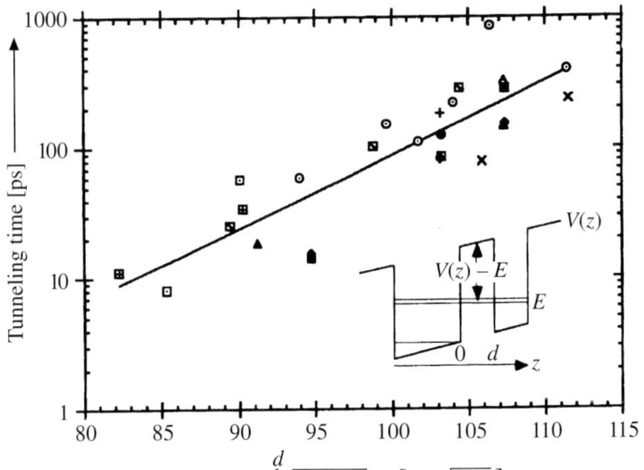

$\int_0^d \sqrt{V(z) - E}\,dz$ [nm $\sqrt{\text{meV}}$]

Fig. 7.22. Dependence of the tunneling times on the effective barrier thickness for various samples under the conditions of $e_{1n} \rightarrow e_{2w}$ resonance [7.14]

rier heights and different resonant electric fields. Figure 7.22 reproduces the extensive results of *Heberle* et al. [7.87] on 5 different samples of GaAs/AlGaAs a-DQWS with different well and barrier widths. The tunneling time does indeed increase exponentially with this parameter, but the slope is 3 times as large as predicted by the resonant tunneling model. They suggested that interface roughness or impurities at the interface may affect the transfer, and lead to this discrepancy.

Deveaud et al. [7.60] have taken a different approach to investigating resonant tunneling. Instead of applying electric field, they prepared several different a-DQWS ($a/b/a'$) using precisely controlled MBE techniques and extensively characterized the samples using X-ray techniques. For a given value of a' and b, they varied the WW thickness a to attain conditions of resonance. They found a shallow resonance (about a factor of 5 in the decay time) as a function of a. They also found that the on-resonance decay time is about a factor of 15 shorter than the off-resonance decay time for the same barrier thickness.

These measurements by no means exhaust all the measurements reported in the literature. However, they show the variety of information on tunneling rates available through time-resolved optical studies.

7.6.4 Unified Picture of Tunneling and Relaxation

The experimental results on resonant tunneling of electrons and holes discussed above show that the measured tunneling rates are considerably smaller than the expected coherent tunneling rates. In Sect. 7.3.3b we discussed how the presence of collisions and relaxation is expected to lead to a considerable reduction in the tunneling transfer rates between the two wells. We compare in this section some of the experimental results with the predictions of this unified model.

Leo et al. [7.30] first considered the influence of collisions and relaxation on resonant tunneling of holes. As discussed above, they found that the measured tunneling rate of (1/1330 ps) is a factor of 45 smaller than that expected on the basis of the calculated splitting of 0.074 meV. However, the full width at half maximum of the resonance curve in Fig. 7.20 corresponds to 6 meV. Therefore the area under the resonance curve is about a factor of 2.5 smaller than that predicted theoretically by (7.11). Conversely, by inserting the width of the resonance and the calculated coherent oscillation frequency, they estimated from (7.10 and 11) the peak tunneling rate to be (1/770 ps), about a factor of 2 larger than the measured rate of (1/1330 ps). Therefore, most of the factor of 45 disagreement between the results and the ideal model of tunneling is removed by considering the unified model with collisions and relaxation. Factors that might contribute the remaining factor of 2 discrepancy are experimental uncertainty in determining the area and the width of the resonance, the uncertainty in the barrier width. Other factors such as inhomogeneous broadening and the influence of interface disorder, nonuniform electric fields etc. must also be considered.

It is also interesting to consider how this model applies to the results on resonant tunneling of electrons. *Oberli* et al. [7.53] determined a resonant tunneling rate of (1/7.5 ps) which is about a factor of 11 smaller than that expected for ideal coherent tunneling. A comparison of the area of the measured resonance curve with (7.11) reduces the discrepancy to a factor of 3.5. While this improvement is substantial, additional effects such as inhomogeneous broadening as proposed by *Alexander* et al. [7.84] must be considered in order to explain the results on electron resonant tunneling properly. Application of the unified tunneling model with collisions and relaxation to other available data is discussed in detail by *Shah* et al. [7.13].

The problem of tunneling with relaxation has also been considered from a theoretical point of view by *Gurvitz* et al. [7.32]. They have argued that quantum-interference effects are important in the dynamics of a system in which tunneling and relaxation are simultaneously present. They found a non-monotonic behavior of the tunneling rate as the relaxation rate is increased, in agreement with the discussion of Sect. 7.3.3b. More interestingly, they have also considered the case when the system of double wells is coupled to a continuum through a barrier. The advantage of the system is that, unlike the a-DQWS, the relaxation rate into the continuum can be controlled by changing the width and/or height of the barrier. *Cohen* et al. [7.33] have recently investigated such a system experimentally and reported a non-monotonic decay rate of the coupled-well luminescence with increasing barrier width.

It is clear that the experimentally determined resonant-tunneling rates are considerably smaller than the coherent tunneling rates expected for an ideal system. Collisions and relaxation play an important role in slowing down the transfer between the wells by destroying the phase of the wavefunctions. Other extrinsic effects may also play a role in real systems depending on the quality of the samples. It is interesting to note that in the case of resonance between the lowest electron level in each of the two wells in an a-DQWS, there is no lower level to relax to, and one observes coherent oscillations of the electronic wavepacket, as discussed in Chap. 2. Here the oscillation frequency agrees well with the measured and expected splitting of the levels at resonance, and extrinsic sample-related effects are not important. Therefore, the two extreme cases have been investigated: the near-ideal case of coherent oscillations and the case of strong relaxation which drastically reduces the tunneling transfer rate. It would be interesting to investigate the transition between the two cases. For example, can one detect a change in the coherent oscillation frequency as the relaxation rate is increased from being much less than the oscillation frequency to being comparable and larger than the oscillation frequency. Although an initial attempt to observe this by varying the temperature was not successful [7.92], this would be an interesting direction to follow in the future.

7.6.5 Summary

Time-resolved luminescence studies have provided considerable information about the tunneling of electrons and holes under the conditions of various electronic resonances. The results show that tunneling rate does indeed increase at resonance. However, in the presence of relaxation, it is clear that one cannot use the simple sequential picture of tunneling followed by relaxation. A unified picture of tunneling, relaxation and scattering must be developed. Strong scattering or relaxation broadens the levels and therefore reduces the peak tunneling rate although the area of the tunneling curve (peak times the width) must remain constant.

7.7 Conclusions

We have reviewed in this chapter the use of ultrafast optical spectroscopy to investigate tunneling processes and rates in double-barrier structures and in asymmetric double-quantum-well structures. The former has been investigated primarily by measuring the decay of quantum well luminescence as a function of various parameters such as barrier thickness, temperature and electric field. a-DQWS has proven to be an excellent system for investigating a wide variety of tunneling phenomena and for determining the tunneling rates quantitatively. Dependence of the non-resonant tunneling rates on the barrier thickness has been measured. Clear demonstration of the resonance in the tunneling rate when the two participating levels are separated by an optical-phonon energy has been obtained. For resonance between two electron or two hole levels, the tunneling rate between the levels reveals a distinct maximum as expected. However, the value of the tunneling rate at resonance is found to be considerably smaller than expected for the ideal case. It is shown that relaxation and scattering have a profound influence on the tunneling rate at resonance, and that a unified model including tunneling, relaxation, and scattering is essential for a proper understanding and analysis of tunneling at resonance.

8. Carrier Transport in Semiconductor Nanostructures

We discuss in this chapter how ultrafast optical spectroscopy techniques can be used to investigate the dynamics of carrier transport in semiconductors over macroscopic distances. Traditionally, carrier transport has been investigated by electrical time-of-flight techniques [8.1], but all-optical techniques offer many advantages. Optical markers, successfully used in the study of tunneling, as discussed in Chap. 7, provide one means of investigating dynamics of transport. We will discuss an extension of this concept to multiple optical markers in this chapter. We will also discuss how screening of electric field in the sample caused by the motion of photoexcited electrons and holes in the opposite directions can be used to obtain information about carrier transport dynamics.

We will begin with a discussion of some basic concepts in Sect. 8.1, and then discuss several specific examples. These include dynamics of transport in graded-gap superlattices (Sect. 8.2), dynamics of vertical transport and carrier sweep-out in multiple quantum wells (Sect. 8.3), and capture of carriers in quantum wells (Sect. 8.4).

8.1 Basic Concepts

Electrical time-of-flight measurements is a powerful technique and has been applied with success to many different systems [8.2, 3] since the classic Haynes-Shockley experiment [8.1]. *Hybrid* techniques in which photocurrent response is measured following excitation by an ultrafast laser have also been applied to bulk semiconductors [8.2], silicon inversion layers [8.4], and quantum wells [8.5]. Such measurements have provided much useful information, but they can not match the time resolution of all-optical techniques. We concentrate in this chapter primarily on the use of all-optical techniques for investigating dynamics of transport in semiconductor nanostructures.

Two classes of transport phenomena have been investigated optically: *lateral* transport and *perpendicular* transport. In lateral transport, electrons, holes, electron-hole pairs or excitons created in one region of the sample move to another region of the sample in a motion parallel to the surface of the sample. In such cases, the transport can be studied by spatially-resolved optical studies which can be combined with time-resolved techniques to investigate dynamics of lateral transport. A simple technique is luminescence imaging, which is useful if both electrons and holes are transported, or if there is a background of one type of carriers so that the transport of the other type of carrier can be monitored by spatial imaging of luminescence. More sophisticated optical techniques such as pump-probe absorption or reflection can also be used in conjunction with spatial resolution to monitor lateral transport and its dynamics. We will discuss some examples of these concepts which are simple extensions of studies without any spatial resolution.

The spatial resolution obtained by these conventional imaging techniques is approximately 0.5 μm, determined by diffraction limits. Dramatic improvement in the spatial resolution in optical microscopy has been made possible with the development of near-field scanning optical microscopy (NSOM) [8.6, 7], and other techniques such as Scanning Tunneling Microscopy (STM) [8.8] and Atomic Force Microscopy (AFM) [8.9]. Considerable effort is currently under way to combine pico- and femtosecond time resolution with the techniques of NSOM [8.10], and STM [8.11–13] or Scanning Force Microscopy (SFM) [8.14–16]. Such techniques [8.17] may provide a powerful means for investigating lateral transport on a spatial scale much smaller than currently possible.

While lateral spatial resolution can be obtained in a straightforward manner, optical studies of perpendicular transport require development of new techniques because optical studies, in general, are not sensitive to the distance of the carriers from the surface in a homogeneous sample. One exception to this is the use of photons of very high energy so that the penetration depth is extremely small. However, this is not a generally applicable technique for the investigation of perpendicular transport, and two different approaches have been developed for investigating perpendicular transport by all-optical techniques.

The first approach involves a measurement of the change in the perpendicular electric field in the sample caused by the motion of electrons and holes. If the sample has a built-in electric field or is subjected to an applied electric field in the direction perpendicular to the sample surface, photoexcited electrons and holes travel in the opposite directions under the influence of the field. The separation of electrons and holes produces a space-charge field which screens the built-in or applied field, and thus alters the electric field on the sample. This change in the field can be investigated by monitoring any optical property that is sensitive to the field. For example, the optical absorption near the band gap is altered by the field, via the Franz-Keldysh effect [8.18] in bulk semiconductors, and the Quantum-Confined Stark Effect (QCSE) in quantum wells [8.19]. Thus information about the dynamics of electron and hole transport can be obtained by studying the absorption spectrum as a function of time delay. Similarly, the total power of the terahertz radiation emitted by a photoexcited semiconductor (Sect. 2.12) depends on the electric field. Therefore, the dynamics of the total electric field, and hence the motion of electrons and holes photoexcited by a pump pulse, can be determined by monitoring the THz power emitted by the sample in response to a variably-delayed probe pulse. This has been used recently to measure the dynamics of transport on 10 fs time scale [8.20].

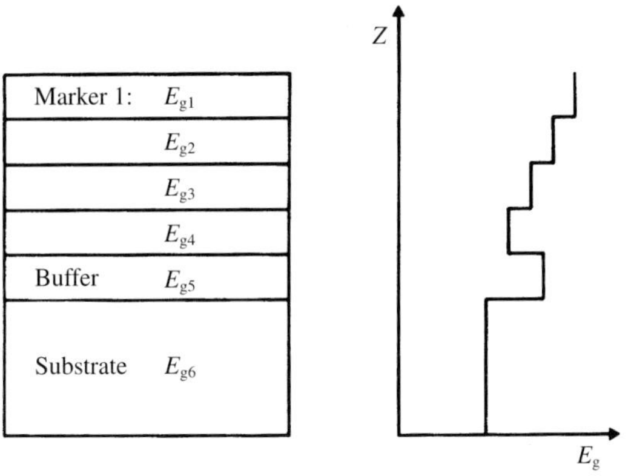

Fig. 8.1. Schematic diagram illustrating the concepts of multiple optical markers

The second approach to investigating perpendicular transport involves the utilization of optical markers, already discussed in connection with the investigation of the dynamics of tunneling in Chap. 7. The basic idea is that information about carrier density in a specific *spatial* region, and hence about the dynamics of carrier transport, can be directly obtained by studying optical properties in a specific *spectral* region. The concept of a single optical marker was discussed in Chap. 7 (Fig. 7.1). A structure with multiple optical markers is schematically illustrated in Fig. 8.1. If carriers are created near the surface of this structure by photoexcitation, their transport to the interior of the sample can be investigated by monitoring the modification in the optical property of the marker regions within the structure caused by the arrival of the carriers. *Hillmer* et al. [8.21] first proposed the use of two quantum wells with different widths to monitor ambipolar transport in AlGaAs over tens of micrometers. The utilization of an enlarged quantum well as a marker in a superlattice was first proposed and demonstrated by *Chomette* et al. [8.22]. The concept and application of the concept of optical markers for investigating the dynamics of perpendicular transport has been discussed in recent reviews [8.23–27]. We will discuss some applications of the optical marker concept in the next section.

8.1.1 Some Examples

There are many reports in the literature on the use of electrical, hybrid and all-optical time-of-flight measurements to determine various aspects of transport in semiconductors. Some of the examples of the all-optical measurements are briefly mentioned in this section. Perpendicular transport in superlattices, carrier sweep-out in MQW structures, and carrier capture in quantum wells, the main topics of this chapter, are discussed in the following sections.

Lateral transport using imaging techniques has been investigated in the case of modulation-doped quantum wells by *Höpfel* et al. [8.28, 29]. These studies found that, under certain circumstances, the minority carriers travel in the direction opposite to that expected for the applied field, i.e., the minority carriers have *negative absolute mobility*. For example, under certain conditions, photoexcited electrons in p-modulation-doped GaAs quantum wells have been shown to travel towards the *negative* electrode, rather than the positive electrode, as expected in the absence of the holes. It was shown that momentum scattering between the photoexcited minority carriers and the majority carriers was responsible for this unusual effect. Although these studies measured time-integrated luminescence, it should be possible to extend these measurements to short times and to obtain more information about the dynamics of lateral transport.

Lateral transport of excitons [8.30] and electron-hole pairs [8.31–33] has been investigated by *Hillmer* et al. via lithographically prepared apertures to obtain spatial resolution of the order of 100 nm. The idea is to deposit an opaque NiCr mask on the sample surface with a small (of the order of 1 μm) opening at the center. This central circular region of the sample is excited with a picosecond laser at sufficiently low intensity so that the carrier recombination time is independent of the excitation intensity. Only the luminescence emitted from the central circular region is collected, and its time decay is determined by the carrier recombination time and the carrier escape to regions under the opaque mask. By comparing the decay dynamics under various experimental conditions, it is possible to determine information about the exciton and carrier transport. Considerable activity is continuing in the field of lateral transport applying various techniques. As already mentioned in Sect. 8.1, the combination of NSOM, STM, AFM or other such techniques with ultrafast techniques promise dramatic improvement in the spatial resolution.

Perpendicular transport in Si was investigated by monitoring the luminescence from a thin doped region in the interior of a Si sample photoexcited near the surface [8.34], and time-of-flight in samples incorporating two quantum wells of different thickness was used to measure ambipolar transport of carriers over tens of micrometers with sub-nanosecond time resolution. All-optical time-of-flight measurements have also been reported in graded-gap alloy semiconductors [8.35], in quantum-well systems [8.21, 36, 37] and superlattices [8.38].

One of the first application of the concept of the screening of the applied field by the motion of photoexcited electrons and holes was reported by *Shank* et al. [8.39] who investigated the velocity overshoot in a 2 μm thick undoped GaAs in the intrinsic region of a p-i-n structure with AlGaAs p and n regions. The structure was reverse biased and maintained at 77 K. 0.5 ps pulses from an amplified, passively mode-locked dye laser were Raman shifted to 805 nm and excited only the intrinsic GaAs region. Differential absorption spectra were measured as a function of time delay and applied field strength. Figure 8.2 illustrates such spectra obtained at a delay of 20 ps for three different applied electric fields. The amplitude of the absorbance change was determined by adding the area under the positive and negative portions of the differential spectrum. They found that velocity overshoot at early times was nec-

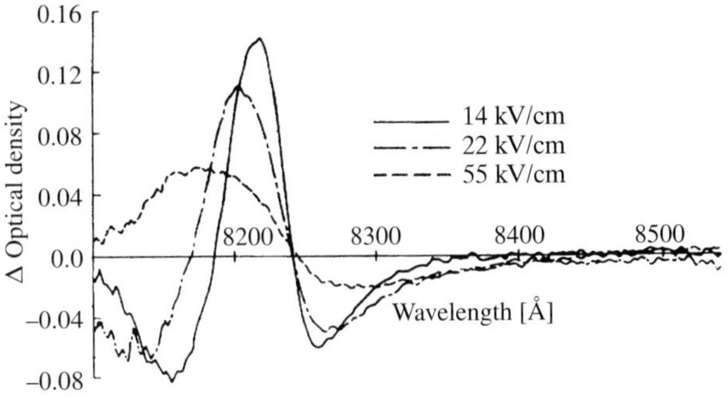

Fig. 8.2. Differential absorption spectrum (induced optical density change) in GaAs at 77 K for three different applied electric fields at a delay of 20 ps following excitation by a 0.5 ps pulse at 803 nm. The period of the oscillations changes with the change in the electric field [8.39]

essary to explain the results at certain electric fields, in agreement with the theoretical expectations [8.40]. More recent measurements of velocity overshoot, ballistic transport and carrier heating in GaAs under electric field using pump-probe measurements and terahertz measurements have been reported by *Norris* and coworkers [8.41 – 43]. As already mentioned, *Hu* et al. [8.20] have recently investigated velocity overshoot in GaAs by measuring the change in the THz power emitted by photoexcited GaAs as a function of time delay. Screening of the electric field resulting from the transport of carriers has also been used for investigating the dynamics of transport along the growth direction of multiple-quantum-well structures, and will be discussed in the next section.

This brief discussion shows that all-optical time-of-flight techniques provide a powerful means for investigating both lateral and perpendicular transport in semiconductors.

8.2 Perpendicular Transport in Graded-Gap Superlattices

In this section, we discuss investigations of carrier transport dynamics using the concept of multiple optical markers. In a multiple quantum-well sample, the barrier between the wells is sufficiently wide and high so that there is a negligible overlap of wavefunctions of different wells, and the wells can be treated as independent of each other. As the height or the width of the barrier is reduced, the overlap between the wavefunctions in different wells increases and the N independent and degenerate energy levels in a N period structure broaden into a band with N levels at different energies. We refer to such a structure as a superlattice, and the energy bands are referred to as minibands. In Chap. 2, we discussed how the lowest electron states in a symmetric double well structure split into an asymmetric and symmetric states with a splitting

$2\Delta_0$ determined by the tunneling matrix elements between the wells. In a superlattice with $2N+1$ periods, and with the same well thickness and barrier thickness and height as the symmetric double well structure, the $2N+1$ energy levels are given by

$$\varepsilon_k = E_1 - 2\Delta_0 \cos(kd) \ , \tag{8.1}$$

where $kd = i\pi/2(N+1)$ and $1 \leq i \leq 2N+1$. Therefore, for large N, the width of the superlattice miniband is $4\Delta_0$, twice the splitting in a symmetric double-quantum-well structure.

There is a miniband corresponding to each confined electronic level in an isolated quantum well, with the miniband width determined by various parameters such as the confinement energy, the barrier width and height, and the effective mass of the carrier. The lowest miniband is narrower than the others, and the heavy-hole minibands are much narrower than the electron minibands. As the number of periods in the superlattice are increased, the total width of a given miniband remains approximately constant but the number of levels in the miniband increases (8.1). Schematic behavior of the three lowest minibands in GaAs/AlGaAs superlattice as a function of the superlattice period is exhibited in Fig. 8.3. The miniband width increases as the superlattice period decreases, and for small periodicities, the miniband states can extend above the barrier energy. The effective mass plays an important role in determining the miniband width so that the HH miniband is much narrower than the electron miniband in the GaAs/AlGaAs superlattice. Figure 8.4 displays the electron

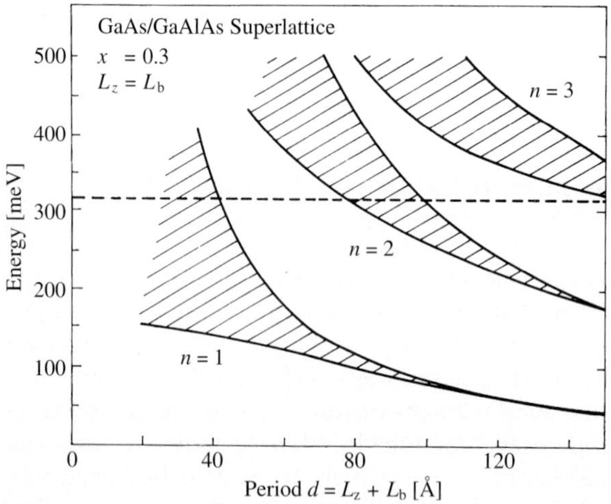

Fig. 8.3. Schematic behavior of minibands in a GaAs/AlGaAs superlattice as a function of the superlattice period d. Quantum levels in isolated wells (large d) broaden into minibands of increasing width for short-period superlattices. The *dashed line* shows the position of the bottom of the barriers [8.23]

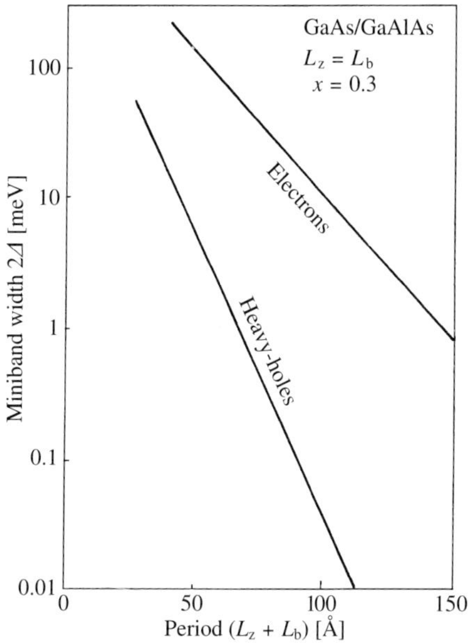

Fig. 8.4. Electron and heavy hole miniband widths as a function of the superlattice period calculated on the basis of a Kronig-Penney model [8.23]

and hole miniband widths as a function of the superlattice period, calculated using a Kronig-Penney model. Electron (hole) miniband widths in excess of 100 (10) meV are predicted by this simple model.

The effective mass and the mobility of the carrier in the perpendicular (z) direction can be approximately expressed as [8.44]:

$$m_z^* = \frac{\hbar^2}{\partial^2\varepsilon/\partial^2 k} = \frac{\hbar^2}{2d^2\Delta_0} , \tag{8.2}$$

$$\mu_z = e\tau/m_z^* , \tag{8.3}$$

where d is the superlattice period and τ is the carrier-scattering time.

The above picture is valid if the carrier scattering time is sufficiently long so that the wavefunction extends over several superlattice periods. As the scattering time decreases, the mean-free path of the carrier may become comparable or smaller than the superlattice period d. In this case, the concept of superlattice miniband breaks down; the superlattice states are no longer extended Bloch states but are localized in the z direction. In addition to localization by scattering, wavefunctions can also be localized by interface disorder or fluctuations in the thickness of the layers. Such fluctuations lead to a broadening of energy levels in an isolated well. When such broadening exceeds the halfwidth of the miniband ($2\Delta_0$), localization occurs. Disorder-induced localization was first discussed by *Anderson* [8.45], and evidence of Anderson localization in superlattices has been reported [8.46]. Regardless of the cause of

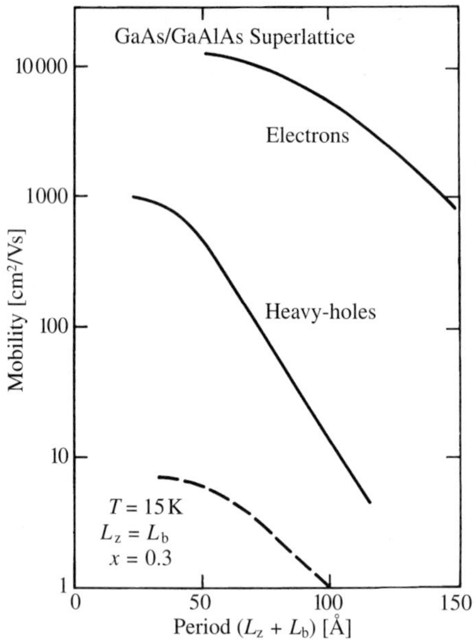

Fig. 8.5. Calculated electron and holes mobilities for Bloch transport in GaAs/AlGaAs superlattice as a function of superlattice period (*Deveaud* et al. [8.23]). The *dashed curve* shows a calculation of the electron mobility for hopping transport [8.48]

the disorder, the character of transport of carriers is expected to be completely different depending on whether the wavefunction is extended or localized in the z direction. Bloch transport is expected to dominate for the extended states whereas phonon-assisted hopping may dominate the transport for localized states [8.47, 48]. Figure 8.5 presents calculated mobilities of electrons and holes for Bloch transport, assuming that they approach the typical bulk mobilities 10,000 and 1000 cm^2/Vs (for electrons and holes, respectively) for small superlattice periods. Calculation of the hopping mobility for electrons by *Palmier* [8.47, 48] is shown as a dashed curve in the figure. More rigorous calculation of Bloch and hopping mobilities are reported in the literature [8.47–50].

We discuss next how the concept of optical markers was applied to investigate carrier transport in superlattices. Figure 8.6a depicts a superlattice with an Enlarged Well (EW) embedded a distance L from the surface. If photoexcitation creates electron-hole pairs close to the surface, then initially they radiate at the superlattice-interband transition energy. As the carriers are transported to the interior of the sample, e.g., by diffusion, they are captured into the EW. The luminescence spectrum would then be expected to show two peaks, one corresponding to the superlattice transition (by carriers not yet captured in the EW), and the other corresponding to the EW transition (by carriers captured in the EW). The relative intensities of the two transitions provide information about the spatial position of the carriers. Figure 8.6b sketches a specially designed superlattice with multiple optical markers. The alloy composition in the barrier is changed in a stepwise manner so that the interband

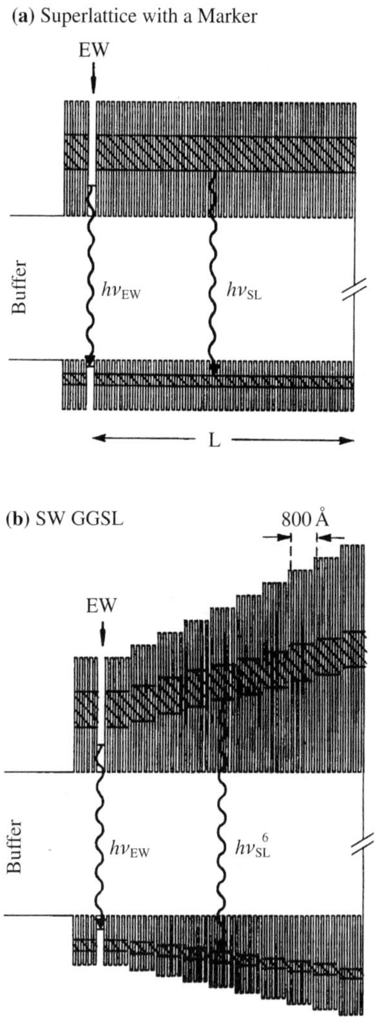

Fig. 8.6. (**a**) Schematic of a superlattice with an Enlarged Well (EW) as an optical marker; (**b**) a StepWise Graded-Gap Superlattice (SW GGSL) in which each step has a width of 800 Å and N periods of thickness d, but the alloy composition of the barrier is reduced in a stepwise manner (2%/step) away from the surface. An EW is embedded after the tenth step. The superlattice miniband also shows a stepwise decrease of the interband transition energy; the EW is at a still lower energy. The SW GGSL thus has eleven optical markers with distinct photon energy, ten for ten steps and one for the EW [8.23]

transition energy of each step of the superlattice is different. The parameters for such a Step-Wise Graded-Gap Superlattice (SW GGSL) structure are defined in the figure caption. There are eleven optical markers in this structure, and time-resolved luminescence spectroscopy should provide information about the transport of carriers in such a structure.

Figure 8.7 exhibits the cw spectrum of a SW GGSL sample with 20 Å wells and 20 Å barriers so that each step of 800 Å has 20 periods [8.23, 51]. The spectrum distinctly shows the luminescence transitions corresponding to the eleven optical markers. Although the electron-hole pairs are created close to the surface, the strongest luminescence corresponds to the EW which is 8000 Å away from the surface. This clearly brings to light the strong transport of

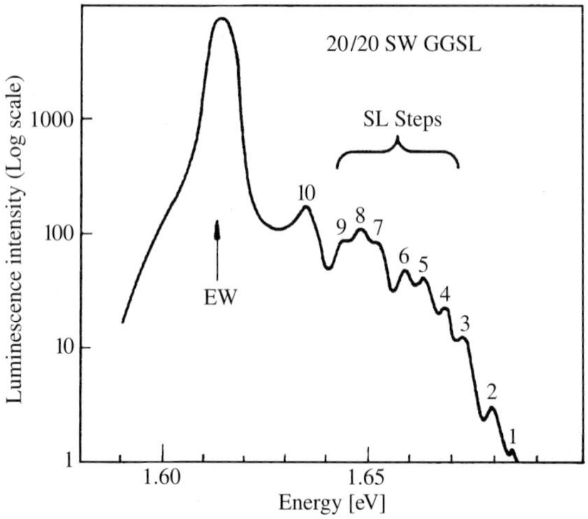

Fig. 8.7. cw Luminescence from a SW GGSL sample with 20 Å well- and barrier-widths, showing the 11 spectrally distinct transitions corresponding to the 11 optical markers. The strong EW luminescence intensity indicates a very efficient carrier transfer into the EW [8.51]

photoexcited carriers during the lifetime of carriers and points the way to investigating the dynamics of this transport by measuring the temporal evolution of the spectrum.

Such measurements were carried out by *Deveaud* et al. [8.51] for SW GGSL with three different periods of 20, 40 and 60 Å. The samples were excited by 300 fs pulses from a dye laser and the spectra were measured at various time delays following the excitation pulse. A typical series of time-resolved luminescence spectra of the 20/20 SW GGSL are depicted in Fig. 8.8 for four different time delays after the photoexcitation. 5 ps after the excitation pulse, the luminescence peaks at 1.68 eV, close to the second superlattice step. With increasing time delay, the luminescence shifts to lower energy indicating the transport of the carriers to the interior of the sample. At 100 ps, the luminescence peaks close to 1.62 eV corresponding to the EW transition. The fact that the total intensity of the luminescence has not decreased much during this time shows that most of the carriers have transferred to the EW within 100 ps. Similar measurements were performed on the other two samples and a reference alloy sample in which there were no quantum wells, but the Al concentration in the sample was varied in a stepwise manner (1% per 1000 Å step) to approximately reproduce the potential gradients of the SW GGSL samples. The results for all four samples are summarized in Fig. 8.9 in the form of distance traveled by the center of the carrier packet as a function time delay. The figure reveals that the center of the carrier packet travels with a constant velocity in the three SW GGSL and a reference AlGaAs alloy sample. The photoexcited carriers in the 10/10 and 20/20 SW GGSL, and the reference alloy sample have approxi-

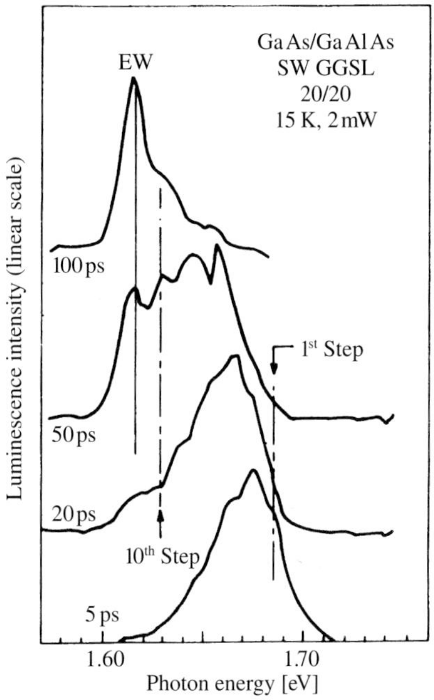

Fig. 8.8. Time-resolved luminescence spectra of a 20/20 SW GGSL sample at different delays after a subpicosecond excitation pulse. Note that the center of the luminescence band moves to lower energy with increasing time delay. The photon energies corresponding to the first SL step, the tenth SL step and the EW are shown [8.51]

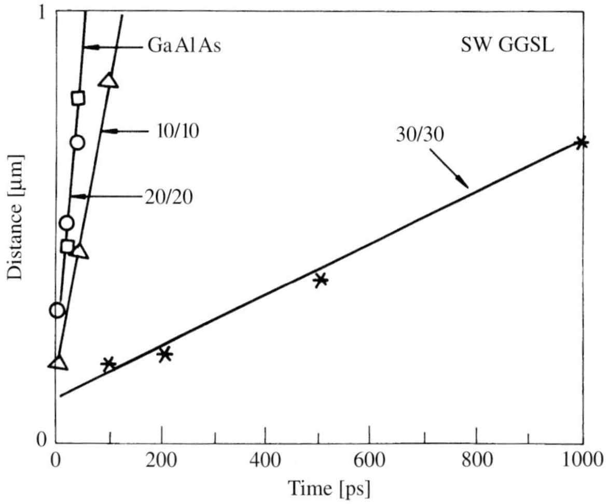

Fig. 8.9. Position of the center of the carrier wavepacket in the sample as a function of the time delay after the excitation pulse for four different samples. The movement is linear in time in all samples investigated. The effective velocities in the 10/10, 20/20 and the reference alloy sample are approximately the same, showing that the perpendicular transport in these superlattices proceeds via Bloch-like extended states. The drastic reduction in the effective velocity for the 30/30 sample shows the importance of localization effects for this sample [8.51]

mately the same velocity, but the 30/30 sample has a distinctly lower velocity. The non-zero intercept on the distance axis results from the effect of surface recombination velocity and the finite absorption length in the sample.

These results were interpreted as follows. The fact that the luminescence intensity remained approximately constant implies that both electrons and holes traveled together to the interior of the samples. The schematic minibands in Fig. 8.6 show that the SW GGSL structure has an effective electric field which exerts a force in the direction of the interior of the sample for both electrons and holes. However, the correct analysis considers the ambipolar diffusion of the electrons and holes in the presence of steps in the potential. The usual diffusion in a homogeneous sample also brings the carriers to the interior of the sample but the peak of the carrier distribution always occurs at the surface in the absence of strong surface recombination velocity. The difference here is that the ambipolar diffusion is driven by the steps in the potential. If the potential step is large compared to kT, the carriers can not diffuse back to higher potential energy and there is a motion of the center of the carrier packet to the interior of the sample, even in the absence of surface recombination.

Interesting physical information is obtained through an analysis of the distance vs. time curves in Fig. 8.9. The observation that the velocity of the carriers in the 10/10 and the 20/20 sample is approximately the same as the velocity of carriers in a reference alloy sample immediately shows that the *transport occurs via extended Bloch states of the minibands* in these two superlattice samples. The fact that the 10/10 sample has a lower velocity than the 20/20 and the reference samples is explained by arguing that increased interface roughness may lead to significant broadening of the energy levels in the 10/10 sample so that there is some transport through localized states in the 10/10 sample. Increasing the temperature to about 77 K where kT exceeds the localization energy increases the velocity of the transport in the 10/10 sample, in agreement with this model. Analyzing the results in terms of a step-driven diffusion model [8.51] reveals that the hole mobility in the 20/20 sample is 900 cm^2/Vs which corresponds to a hole mean-free path of 100 Å, larger than the superlattice period of 40 Å. This confirms that hole transport, and hence also the electron transport, occurs through extended Bloch states of the miniband.

The fact that the transport velocity decreased by nearly a factor of 20 in the 30/30 sample compared to the other three samples was attributed to a change in the nature of the hole transport from Bloch-like to hopping transport through localized states. The electron mobility not expected to change much as the superlattice period is increased from 40 to 60 Å (Fig. 8.5). In contrast, the hole mobility for Bloch transport begins to decrease rapidly (Fig. 8.5) as the superlattice period increases beyond 40 Å. Furthermore, the hole miniband width in an ideal 30/30 superlattice is only 2.9 meV, smaller than the lifetime broadening and the fluctuations in the energy levels due to interface roughness. Also, the potential steps in these SW GGSL are larger than the miniband width of 2.9 meV in the ideal case. As a result the hole transport is much slower in the 30/30 superlattice and the velocity of the transport of the carrier packet is much smaller than that in the other two SW GGSL samples.

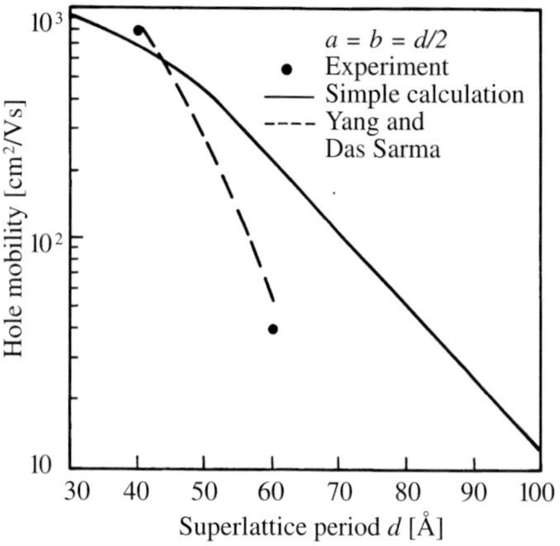

Fig. 8.10. Hole mobilities for perpendicular transport (determined from the all-optical time-of-flight measurements) as a function of the superlattice period. Also shown is a comparison with a simple model and the model presented by *Yang* and *Das Sarma* [8.52]; from [8.23]

It is therefore not surprising that there is a large reduction in the measured mobility, as the superlattice period is increased from 40 to 60 Å.

The transition from Bloch-like transport to hopping transport for holes is one of the more interesting aspects of these results. Figure 8.10 suggests that the experimental results show a much stronger decrease in the mobility than expected on the basis of a simple model of mobilities. This discrepancy has been examined more closely by *Yang* and *Das Sarma* [8.52] who argued that the case of superlattice minibands is quite different from that of the normal 3D solids because the miniband widths, Fermi energy, phonon energy and the temperature are all comparable. Under these conditions, they have argued that the quasiparticle approximation that makes the wavevector of electrons a unique function of electron energy breaks down. They calculated the mobility by replacing the usual δ function by a Gaussian. The experimental results are in reasonable agreement with the theory (Fig. 8.10) but more experiments are clearly required. We note that such studies might provide interesting information concerning intermediate disorder regime.

As shown in Fig. 8.5, the electron mobilities are expected to decrease rapidly with the superlattice period only when the period exceeds about 100 Å. In the measurements discussed above, the transport is ambipolar and dominated by holes because the carrier density is quite high. Another set of measurements was carried out on a different set of samples of the type (Fig. 8.6a) at very low excitation intensities so that the transport is dominated by electrons [8.23]. The time-resolved luminescence showed two peaks, one corresponding to the superlattice and another at lower energy due to the EW. The superlattice luminescence decayed in less than 20 ps in the 30/30 superlattice, but persisted for more than 200 ps for the 50/50 superlattice. These results thus suggest a large reduction in the electron mobility when the superlattice period is increased from 60 to 100 Å, in qualitative agreement with the results for the holes at much smaller superlattice periods. *Lambert* et al. [8.53] have reported measurements of electron mobility in superlattices as a function of temperature.

Deveaud et al. [8.54] have investigated electron mobility by using linearly-graded, p-doped superlattices and performed a detailed analysis using drift-diffusion model along with transfer to higher conduction band valleys and carrier capture in quantum wells (Sect. 8.4).

These results demonstrate that the technique of optical markers, coupled with time-resolved optical studies, provide an excellent means of obtaining fundamental information about the nature of transport in semiconductors and their nanostructures.

8.3 Carrier Sweep-Out in Multiple Quantum Wells

Another interesting transport phenomenon is the sweep-out of carriers optically excited in Multiple Quantum Wells (MQW). Apart from the interest in the fundamental physics of this phenomenon, there is also interest from a device point of view because such transport plays an important role in determining the response speed of devices based on multiple quantum wells, notably the SEED (Self Electro-optic Effect Device) device [8.55]. The carriers photoexcited in the quantum wells either decay by recombination, or are swept away to the contacts following thermionic emission or tunneling into the continuum states above the barriers. An understanding of these processes and their dependence on the structure parameters (e.g., barrier thickness and height) and physical parameters (e.g., temperature and electric field) is important for optimizing the device performance. It is also important to understand the behavior of electrons and holes, and determine which carrier determines the device response in a given set of circumstances [8.56].

The first approach to understanding this behavior was to measure the I–V characteristics of a MQW structure under cw photoexcitation. The results of one of the first measurements, performed by *Capasso* et al. [8.44, 57] on InGaAs/InP MQW at low temperatures, are presented in Fig. 8.11. The current shows two distinct peaks, the first corresponding to the field at which the electron levels E_1 and E_2 are in alignment and the second corresponding to the alignment of electron levels E_1 and E_3. These results directly indicate that the alignment of electron levels results in an increase in the photocurrent, and lead to the conclusion that electron tunneling plays a major role in the current transport in these MQW structures. Similar results showing structure in the I–V characteristics have been reported by others in other material systems [8.58].

Dynamics of transport and carrier sweep-out in MQW structures have been investigated either by hybrid techniques, in which the carriers are excited by a short optical pulse and the transport is monitored by measuring temporal evolution of the photocurrent, or by all optical techniques. These are discussed in the next two subsections.

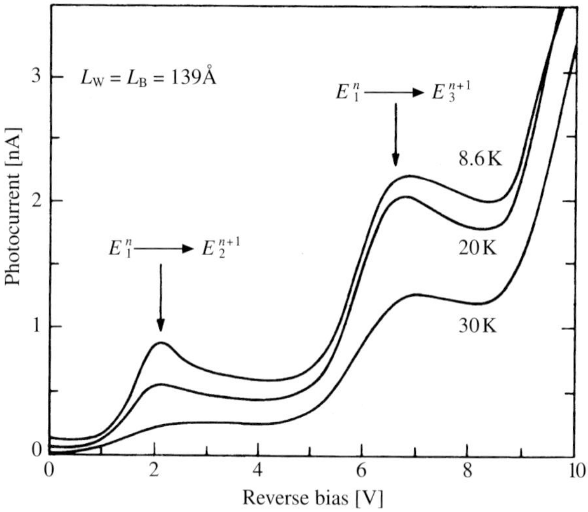

Fig. 8.11. DC photocurrent as a function of the applied voltage perpendicular to the interface plane for InGaAs/InP superlattice; note the two distinct peaks in the current [8.44, 57]

8.3.1 Hybrid Technique

Time-resolved current-voltage measurements, following photoexcitation of the MQW structures with nanosecond pulses, have provided considerable information on this subject. *Schneider* et al. [8.58] excited a sample of GaAs/AlAs MQW, sandwiched between $Al_{0.5}Ga_{0.5}As$ layers, with 530 nm pulses from a dye laser pumped by a nitrogen laser. The penetration depth of the exciting photons was about 125 nm in the MQW region and 300 nm in the AlGaAs windows. This helps generate electron-hole pairs close to the negative contact so that the holes are swept away quickly and the transport is determined mostly by electrons. Figure 8.12 displays the temporal evolution of photocurrent from a 50 period 123/21 GaAs/AlAs MQW structure at 100 K for different applied voltages. The temporal shape of the photocurrent depends *non-monotonically* on the applied voltage, showing that quantum effects (i.e., alignment of the confined electron levels in adjacent wells) play an important role. For example, the transients become shorter as the applied voltage changes from -1 to -3.6 V, but become longer with further change in the voltage to -8 V. A plot of the peak photocurrent at 80 K as a function of applied voltage (Fig. 8.13) exhibits a strong peak at -3.6 V (corresponding to the E_1 to E_2 tunneling transition), and a weak peak at -11 V (corresponding to E_1 to E_3 tunneling transition).

The time resolution of these electrical measurements can be improved to tens of picoseconds with faster detectors and measurement techniques. However, electrical techniques can not match the time resolution of all-optical techniques discussed in the next section.

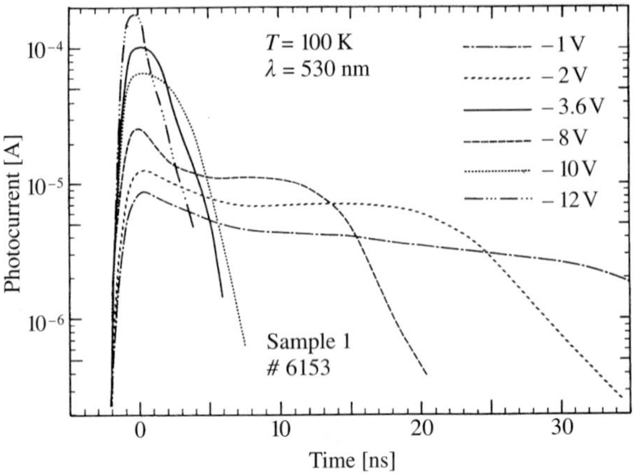

Fig. 8.12. Time evolution of the photocurrent in a 50 period 123/21 GaAs/AlAs MQW at 100 K for different applied voltages. The excitation was by 500 ps pulses at 530 nm [8.58]

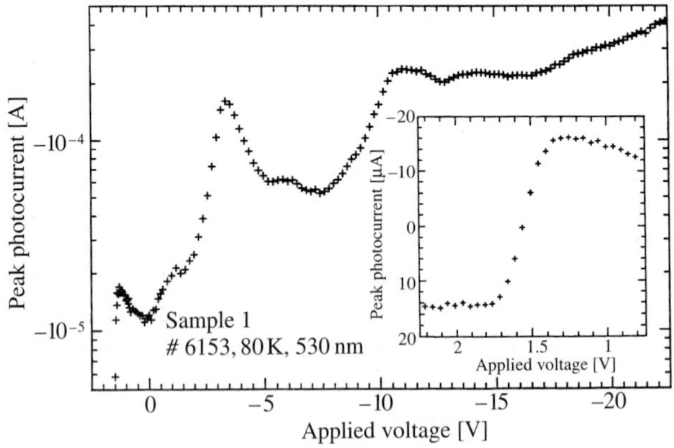

Fig. 8.13. Peak photocurrent on a logarithmic scale as a function of the applied voltage for a 123/21 GaAs/AlAs MQW at 80 K [8.58]

8.3.2 All-Optical Studies of Carrier Sweep-Out

Application of an electric field to a semiconductor leads to changes in the absorption spectrum of the semiconductor near its bandgap. In bulk semiconductors, this phenomenon is known as the Franz-Keldysh effect [8.17, 18] and has been extensively investigated theoretically and experimentally [8.59]. In quantum wells, the analog of this is known as Quantum-Confined Stark Effect (QCSE) and has also been investigated very extensively [8.19, 60]. The interest here is not only in the fundamental physics but also in applications to devices such as quantum well modulators and SEED devices, already mentioned above.

Electron-hole pairs photoexcited in a semiconductor under electric field travel in the opposite directions and produce a space-charge field which opposes the field initially present in the semiconductor. The reduction in the field produces a change in the optical absorption near the bandgap. In the limit that the space-charge field is small compared to the initial field, the change in the optical absorption is approximately proportional to the space-charge field. Thus, a measurement of the change in the absorption as a function of time following photoexcitation provides a means of determining the dynamics of carrier transport. As discussed in Sect. 8.1.1, such studies were utilized to investigate velocity-overshoot phenomenon in bulk GaAs [8.39].

Similar measurements have been reported in the case of multiple quantum wells embedded in the i-region of a p-i-n structure under reverse bias. *Livescu* et al. [8.61] excited a 65/57 $GaAs/Al_{0.31}Ga_{0.69}As$ MQW embedded in the i-region with 6–10 ps pulses from a synchronously-pumped dye laser. The differential absorption of a delayed probe beam derived from the same laser was measured as a function of time delay τ_d between the pump and probe pulses. The differential absorption spectrum displayed a typical derivative shape of the HH and the LH exciton peaks, and the wavelength of the laser was selected to give maximum differential absorption at each applied voltage. The temporal evolution of the differential absorption for four different applied voltages is shown in Fig. 8.14. It was argued that the rise time of the differential absorption reflects the build-up of the space charge due to the vertical motion of electrons and holes, whereas the decay of the differential absorption reflects the decay of the space charge as the charge spreads laterally due to rapid electromagnetic diffusion or dielectric relaxation in the doped contact layers. They

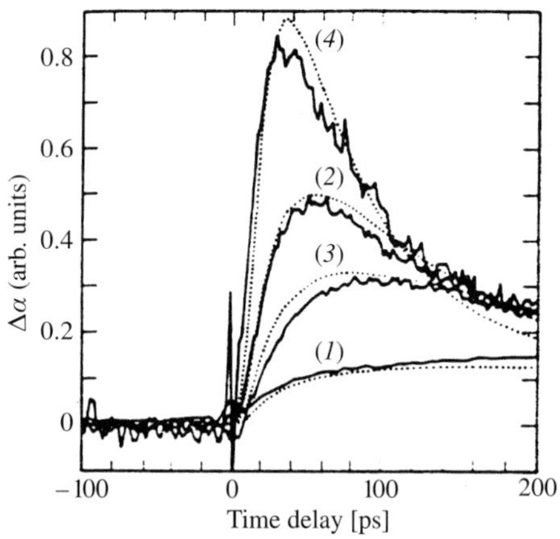

Fig. 8.14. Temporal evolution of room-temperature differential absorption (at the wavelength corresponding to the largest positive differential absorption) for different applied electric fields. Full lines are experimental and the dotted curves are calculated. The applied electric field and the rise time correspond to (*1*) 100 kV/cm, 375 ps; (*2*) 120 kV/cm, 60 ps; (*3*) 160 kV/cm, 120 ps; and (*4*) 240 kV/cm, 25 ps [8.61]

fitted their data to a rising exponential $1-\exp(-\tau_d/\tau)$ and the decaying function $1/(1+\tau_d/\tau_c)$, τ_c being the diffusive time constant. They calculated τ_c and obtained the rise time τ_d by fitting the experimental results. The non-monotonic dependence of τ on the field is obvious from Fig. 8.14 even without such an analysis. But a quantitative analysis allowed them to demonstrate that the rise time shows a local minimum value of about 50 ps at a field at which E_1 and E_2 and are expected to be aligned. Thus resonant tunneling makes a strong contribution to vertical transport in MQW. The value of 50 ps contains contribution from both tunneling and transport, and was much larger than expected for reasons that were not clear from these measurements. The results on tunneling in isolated structures discussed in Chap. 7 suggest that scattering and relaxation can reduce the tunneling rate considerably, an effect that may also be important here.

These measurements were extended by *Fox* et al. [8.62, 63] to quantum-well structures with different well and barrier widths, and different barrier heights using sub-picosecond pulses. Figure 8.15 depicts the measured escape times, defined as 10–90% rise of the differential electro-absorption signal, for three different samples at two different temperatures. The low temperature results show that with increasing reverse bias, the escape times decrease rapidly, go through a minimum and a maximum, and then decrease gradually. Room temperature results agree with those at low temperatures in the range where both results could be obtained. The non-monotonic behavior in the escape time is reminiscent of the behavior already seen in the cw and transient electrical measurements discussed above. The position of the minimum in each case agrees very well with the value of the field at which E_1 is expected to be in resonance with E_2 of the next well (indicated by the arrows in the figure), showing that resonant electron tunneling makes a significant contribution to the transport at the minimum. At these high fields the electron escapes into the states above the barrier after tunneling through only one or two barriers. Once an electron is in these states, it accelerates very quickly to its saturation velocity and reaches the contact without falling back into the wells.

The escape times at high fields were nearly independent of temperature. In contrast, the escape times at low fields showed a strong dependence on the temperature. In fact, the escape times at low fields showed a thermally activated behavior, with an activation energy of 49 ± 15 meV. It was argued that at high fields, the escape time is determined by tunneling through one or two barriers followed by transport. On the other hand, a pure tunneling mechanism is unlikely at low fields because it would require multiple steps of tunneling followed by relaxation within a well, so that a large number of barriers must be crossed [8.64]. Therefore, thermionic emission dominates at low fields. *Fox* et al. [8.62] found that the large activation energy would be consistent with either electron or hole escape from the quantum wells, and also with phonon-assisted tunneling.

These all-optical measurements have thus clearly demonstrated the importance of resonant tunneling in the perpendicular transport of carriers in MQW at resonant electric fields. Tunneling escape dominates transport at high fields,

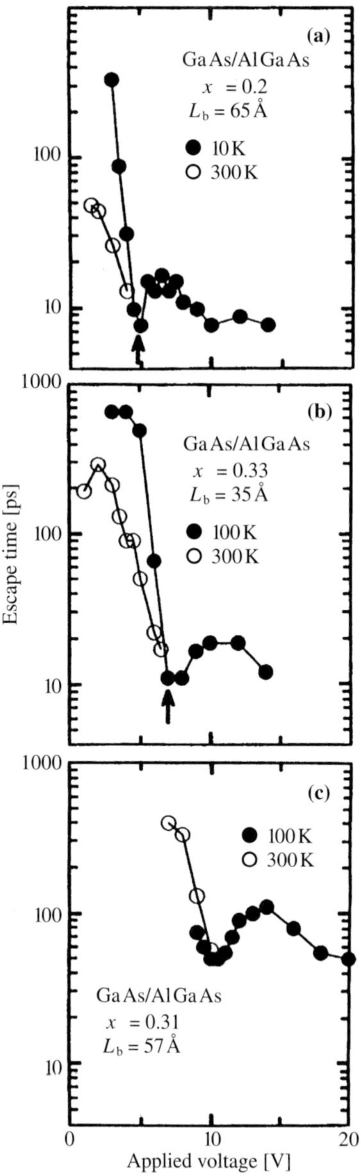

Fig. 8.15a–c. Escape times, defined as 10–90% rise time of the differential absorption signal, as a function of the applied voltage for different MQW samples and different temperatures [8.62]

whereas thermally activated transport dominates at low fields. These measurements showed that sweep-out of optically excited carriers can be accomplished in times less than 10 ps under appropriate conditions, a results of considerable interest in determining the speed of the modulators and SEED devices based on quantum wells. It is also interesting to compare the two approaches discussed in Chaps. 7 and 8. The strength of the techniques discussed in this chapter is that it provides information about transport dynamics which is of fundamental importance and is relevant to devices. On the other hand, the approach of using *isolated* structures in Chap. 7 allows one to eliminate the effects of transport and obtain unambiguous quantitative information about the tunneling rates, also of fundamental and device interest.

8.4 Carrier Capture in Quantum Wells

Carriers injected into the barrier region of a high-quality semiconductor quantum well structure are eventually captured into the quantum wells because they represent the lowest energy states of the system. The dynamics of this carrier capture process has received considerable attention, partly because of fundamental interest and partly because of its relevance to the dynamics of quantum well lasers. We conclude this chapter with a discussion of how the optical marker technique has been used successfully to obtain information about the capture dynamics.

8.4.1 Theoretical Predictions

From a fundamental point of view the interest in this process was generated at least partly by the prediction [8.65–68] that the capture rate into a quantum well depends on the well thickness in an oscillatory manner. This behavior is expected on the basis of the following simple quantum-mechanical picture: for very thin quantum wells, there is only one confined electron level in the well. As the well thickness increases, the lowest level (band) in the barrier begins to approach the top of the quantum well (bottom of the barrier), and becomes a confined level in the well with further increase in the well thickness. This process is repeated periodically as more and more levels are confined in the well. Figure 8.16 presents some of the energy levels (bands) of a $GaAs/Al_{0.3}Ga_{0.7}As$ MQW with 100 Å barrier as the well width is varied from 10 to 200 Å [8.67]. It shows that as the well thickness increases, the bottom of the lowest subband above the barrier enters the well at well widths which are multiples of 56.4 Å. This periodic variation is the basis of the oscillations in the capture rates as a function of the well width.

The capture from the barrier into the well is mediated primarily by the emission of optical phonons. For our discussion, we concentrate on this phonon-emission process, although capture mediated by carrier-impurity and carrier-carrier scattering have been considered in the literature [8.69–71]. The capture rates are determined by two main factors: (1) the overlap of wavefunctions between the initial and the final states, and (2) the energy separation between the initial and the final states. The energy separation must exceed the phonon energy so that a threshold behavior is expected at very low carrier temperatures when only the lowest barrier state is occupied. As the energy separa-

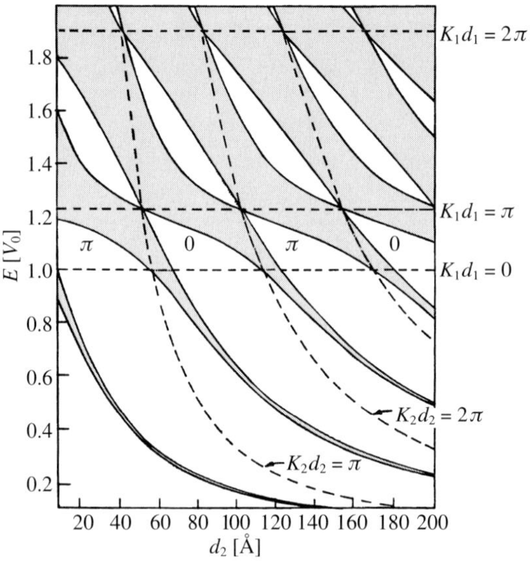

Fig. 8.16. Calculated band structure of $GaAs/Al_{0.3}Ga_{0.7}As$ multiple-quantum-well structure as a function of the well width [8.67]

tion increases above this threshold, the wavevector of the emitted phonon increases leading to a decrease in the capture rate because of the inverse wavevector dependence of the Fröhlich interaction. The form of the periodic variation in the capture rates is determined by a competition between these effects. If the carrier temperature is high, then an average over the initial distribution function must be performed. Another factor that has been considered is the periodic increase in the number of confined phonon modes with increasing well thickness [8.72]. However, the importance of this effect is not yet clear.

Numerical calculations of the electron (and hole) capture rates into quantum wells in multiple-quantum-well structures as well as in laser structures have been performed by a number of groups [8.65 – 70, 73 – 79]. These calculations do indeed show a periodic variation in the electron capture rates with the well thickness as expected from the discussion above. However, the precise form of the periodic variation is not the same in all calculations. We present in Fig. 8.17 the calculated results [8.80] for the electron capture times for a $GaAs/Al_{0.25}Ga_{0.75}As$ multiple-quantum-well structure with 200 Å barriers. These results are calculated using the model discussed in detail by *Deveaud* et al. [8.69]. The dotted curve display the LO-phonon-assisted electron capture rates for electrons at the bottom of the lowest barrier state whereas the solid curve depict the results for a thermal distribution of electrons at 295 K. The dotted curve shows a large anti-resonance in the capture rate (i.e., long capture times) resulting from factors discussed above. However, averaging over a thermalized initial electron distribution removes this anti-resonance, as indicated by the solid curve. These curves are representative of the calculations on capture rates reported in recent years. These recent results differ from those reported earlier by *Babiker* and *Ridley* [8.67] for reasons that are not entirely clear.

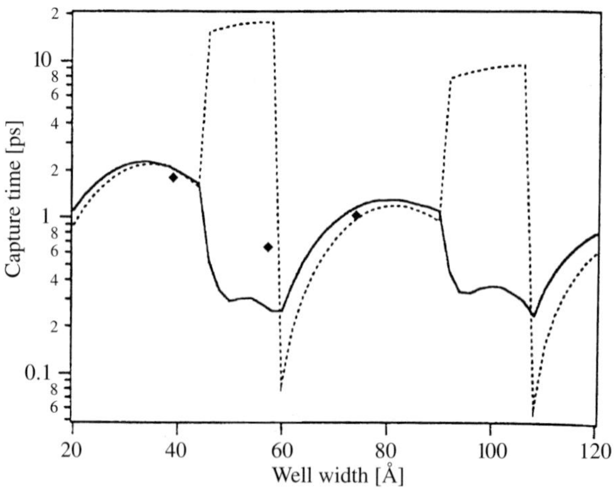

Fig. 8.17. Calculated LO-phonon-assisted capture rates for electrons in a GaAs/$Al_{0.25}Ga_{0.75}As$ multiple-quantum-well structure with 200 Å barriers, as a function of the well width. The *dotted curve* corresponds to the capture of electrons from the bottom of the lowest barrier level and the *solid curve* is calculated for an electron distribution corresponding to 295 K. Note that the removal of the anti-resonance for the latter case leads to a much smaller variation in the capture rate. The *points* correspond to experimental results obtained by the pump-probe technique [8.80]

Experimental investigation of carrier-capture rates is discussed in the next section.

8.4.2 Experimental Studies

In addition to the fundamental interest in observing such oscillations, it was argued even in the early days of quantum-well lasers that carrier capture is an important physical process in quantum-well lasers [8.65]. Therefore a measurement of capture rates and an understanding of its dependence of various parameters is obviously of considerable device interest. The early estimates of the capture rates were obtained by using cw PhotoLuminescence (PL) and PhotoLuminescence Excitation (PLE) spectroscopy. These values ranged from 0.1 to 50 ps [8.81 – 83], partly reflecting the uncertainty in determining the capture rates from cw measurements.

The application of the concept of optical markers to this system is clear. There are two markers, the barrier and the well, each with a distinct bandgap. The time evolution of these two spectral features should provide information about the capture of both types of carriers by the well. The decay of the barrier luminescence, for example, is determined by the capture of the carrier with the higher capture rate, assuming that recombination rates are slow compared to this capture rate. On the other hand, the rise of the quantum-well luminescence is determined by the capture of the carrier with slower capture rate, thermalization (Chap. 3) and cooling (Chap. 4) of the carriers in the quantum well, and processes such as exciton-formation dynamics (Chap. 6). With a careful choice of parameters, it may be possible to extract the capture rates for both types of carriers from such experiments.

One of the first studies of carrier capture in quantum wells was by *Westland* et al. [8.84] who measured carrier diffusion and trapping in quantum wells using all-optical techniques. We discuss here the results reported by *Deveaud* et al. [8.85] who investigated lattice-matched InGaAs/InP MQW with well thickness ranging from 9 to 150 Å, and barrier thickness of 200 and 500 Å as a function of temperature between 70 and 300 K. The samples were excited by a 300 fs visible dye laser and time-resolved luminescence spectra were obtained using the upconversion technique with 300 fs time resolution. The incident photons were absorbed by both the wells and the barriers. The total thickness of the MQW was about 1 μm so that no luminescence from the InP substrate is expected. Figure 8.18 displays the time-resolved luminescence spectra of a 150/500 MQW sample at 150 K for three different time delays. The high energy feature near 1.4 eV is the InP barrier luminescence whereas the low energy feature near 0.85 eV is the InGaAs quantum well luminescence. At 1 ps, the InP peak is about three times as strong as the InGaAs peak, approximately the same as the ratio of the two thickness. With increasing delay, the barrier luminescence decreases and the well luminescence increases, showing the transfer of photoexcited carriers from the barriers to the wells. In contrast to these results, samples with 200 Å barriers exhibit no barrier luminescence at low temperatures and low excitation intensities, regardless of the well width, even at the earliest times. Even when the barrier luminescence is observed at high excitation intensity or high temperature, the intensity of the barrier luminescence is much weaker than the well luminescence at all times.

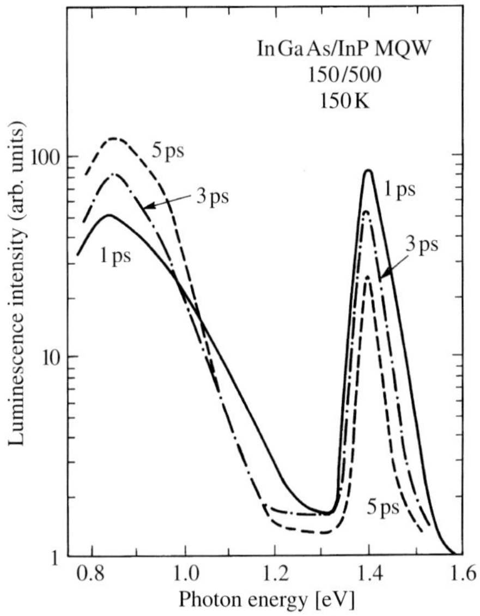

Fig. 8.18. Time-resolved luminescence spectra of InGaAs/InP 150/500 spectra at 150 K for three different time delays [8.85]

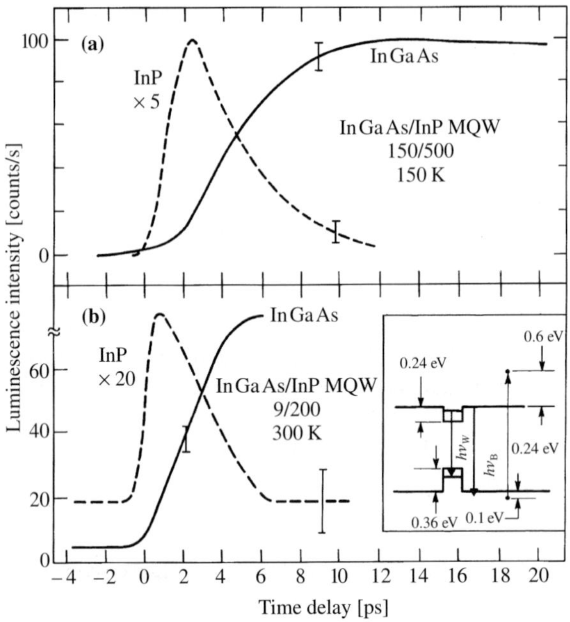

Fig. 8.19. **(a)** Temporal evolution of well- and barrier-luminescence intensities for 150/500; **(b)** 9/200 InGaAs/InP multiple-quantum-well samples. The *inset* shows a schematic of the conduction and valence bands in the well and the barrier [8.85]

The sharp contrast in the results between thin and thick barriers was interpreted to mean that qualitatively different processes dominate for thin and thick barriers. For thin barriers, an analysis of the temporal evolution of the well and barrier luminescence intensities (Fig. 8.19) led to the conclusion that the faster of the two capture times (presumably for holes) was <0.3 ps and that the slower of the two capture times was <1 ps. It was argued that for these thin barriers, quantum-mechanical effects of wavefunction overlap dominates, i.e., these are *quantum capture times.* On the other hand, for the 500 Å barrier sample, the deduced capture times were $2-3$ ps for both carriers. These times were not determined by quantum-mechanical processes, but by *transport of carriers in the barrier.* The second important conclusion from these studies was that the capture times were much shorter than calculated theoretically, and that no oscillatory behavior in the capture times was observed as a function of well thickness for reasons not completely understood at the time.

These studies have provided important insight into the carrier capture processes in terms of transport and quantum capture. However, two aspects of the quantum capture process need further discussion. The first is the question of the oscillatory behavior and the second is the absolute magnitude of the capture rates. A number of studies have been devoted to the question of the oscillation of the capture rates with well thickness, some finding no oscillations

[8.86] whereas others finding some oscillations [8.87]. We discuss here two recent studies by *Blom* et al. [8.76] and *Barros* et al. [8.80].

Blom et al. [8.76] investigated the problem of carrier capture in GaAs/$Al_{0.3}Ga_{0.7}As$ structures consisting of ten periods of a single GaAs quantum well surrounded by 500 Å $Al_{0.3}Ga_{0.7}As$ followed by a 100 Å AlAs cladding layer on each side. Four such SCH (Separate Confinement Heterostructure)-SQW (Single Quantum Well) samples were investigated with quantum-well widths of 30, 50, 70 and 90 Å. They eliminated the effect of the oscillatory behavior in the exciton formation times (Chap. 6) and the effect of carrier cooling on the capture times by measuring the rise time of luminescence for excitation both below and above the barrier, with the two excitation energies differing by an LO-phonon energy. Their measurements were performed at 8 K with an excitation density of the order of 5×10^{10} cm^{-2}. Most of their measurements were performed using the technique of luminescence upconversion but the results were also compared with those from luminescence correlation measurements for the barrier and buffer GaAs layer luminescence.

The capture times deduced from the analysis of these time-resolved studies are shown in Fig. 8.20 as a function of the quantum-well width. They have argued that the coherence length of holes is small compared to the SCH dimension so that holes behave classically. On the other hand, the coherence length of electrons is longer than the SCH dimension so that quantum capture picture applies to electrons. The results of their calculations of the ambipolar capture times of electrons and holes based on this model are also given in Fig. 8.20. The calculations show a reasonable agreement with the experimental results, both in terms of the absolute magnitude of the capture times and their dependence on the well widths.

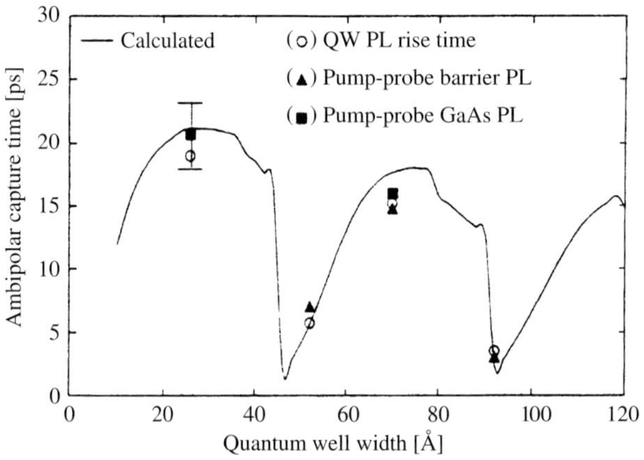

Fig. 8.20. Carrier capture time in a GaAs quantum well of thickness *a* surrounded by 500 Å thick $Al_{0.3}Ga_{0.7}As$ barrier and 100 Å AlAs cladding layer on each side, as a function of well thickness *a*. The sample temperature was 8 K and the results were independent of density in the range 3×10^{15} and 2×10^{17} cm^{-3}. Capture times obtained by three different techniques are displayed and compared with a calculation (*Blom* et al. [8.76])

Since these studies are influenced by both the classical diffusion process and the quantum capture process, they do not provide direct information about the quantum capture times. *Blom* et al. [8.76] calculated values of "local capture time" ranging from 0.1 to 1.8 ps on the basis of a model. A more direct investigation of the quantum capture process was reported by *Barros* et al. [8.80] using pump-probe transmission spectroscopy. They investigated three samples of 20-period GaAs/$Al_{0.25}Ga_{0.75}As$ MQW structures, with well widths of 39, 58 and 78 Å and barrier width of 200 Å. The samples were maintained at 300 K and excited at a photon energy corresponding to the $Al_{0.25}Ga_{0.75}As$ absorption edge at a density of 1×10^{11} cm^{-2}. The excitation at the absorption edge eliminated most of the carrier relaxation processes in the barrier, except for carrier thermalization which was assumed to be very fast. *Barros* et al. [8.80] deduced electron capture times to be 1.8, 0.65 and 1.2 ps for the 39, 58 and 78 Å quantum wells, arguing that the differential transmission signal was primarily due to electrons because the initial narrow energy distribution of holes quickly broadens to the broad thermal distribution at 300 K. These values are also plotted in Fig. 8.17.

While these measurements show a minimum in the capture time for the 58 Å well as expected theoretically, the range of capture times was much smaller than predicted by the theory. *Deveaud* et al. [8.69] considered the effects of LO-phonon-assisted as well as impurity-assisted capture. Impurity-assisted capture, and a high carrier temperature leading to a wide distribution function of electrons in the barrier, counteract the effect of anti-resonance in phonon-assisted electron capture from the bottom of the barrier state, as discussed in Sect. 8.41. This leads to a smaller variation in the capture times with well thickness compared to the predictions of the simple models which ignore the electron distribution and impurity-assisted capture. Note that even at low lattice temperatures, carrier temperature approximately corresponding to the width of the laser spectrum would be expected even for excitation at the bottom of the barrier. For a 100 fs laser pulse, this would correspond to a carrier temperature of the order of 200 K. Thus an ideal experiment to investigate the well-width dependence of the capture rates would be a low temperature experiment resonantly exciting the bottom of the barrier with a relatively long pulse (e.g., 1 ps corresponding to ≈ 1.5 meV bandwidth).

8.4.3 Implications for Lasers

Quantum-well lasers are expected to have a larger differential gain compared to the bulk lasers, and this is expected to lead to enhanced modulation bandwidth for the quantum-well lasers (see, for example, [8.88, 89]). However, it is only recently that the modulation response of quantum-well lasers have matched or surpassed that of bulk lasers [8.90]. It has been proposed recently [8.89, 90] that carrier transport in the SCH region is a major factor influencing the modulation response of the lasers. This provides an important reason for investigating the transport in the confinement region and capture by quantum wells [8.91].

The conclusion [8.85], discussed in Sect. 8.4.2, that transport in thick barrier is the rate-limiting capture process may thus have important implications for semiconductor quantum-well lasers. Therefore different types of laser structures have been investigated with the idea of determining the efficiency of capture for them. The most common type of laser structure provides separate confinement for carriers and the photons. A typical structure for such a SCH laser is shown in Fig. 8.21. A variation of these structures is the GRINSCH (Graded Refractive INdex Separate Confinement Heterostructure) structure in which the refractive index of the confinement region is graded in a Linear (L-GRINSCH) or Parabolic (P-GRINSCH) manner (Fig. 8.21). It is clear that ambipolar diffusion would dominate the transport in the SCH structure so that the capture rate in a SCH structure would be expected to vary as the square of the thickness of the confinement region, in agreement with the results of *Feldmann* et al. [8.81]. However, the transport in both the L- and the P-GRINSCH structures are expected to be faster because of the potential gradient. *Morin* et al. [8.75] have investigated different structures in detail by measuring the decay time of the barrier luminescence in various structures as a function of temperature (Fig. 8.22). They have concluded that the carrier transport in the confining layer is the rate-limiting process for the SCH structure. In contrast, the field-assisted transport in the GRINSCH structures is fast so that quantum capture is likely to be the rate-limiting process in GRINSCH structures. While these results clearly show that the transport and capture in GRINSCH structures are faster, another factor that must be considered in the comparison is the degree of confinement that each structure provides. Other studies on laser structures include a report by *Weiss* et al. [8.92] who investigated the carrier capture times in InGaAs quantum-well laser amplifier structures at high plasma densities. They found three different time constants, with the intermediate time constant being determined by transport in the confining layers. Implications of carrier capture effect for the quantum-well laser perfor-

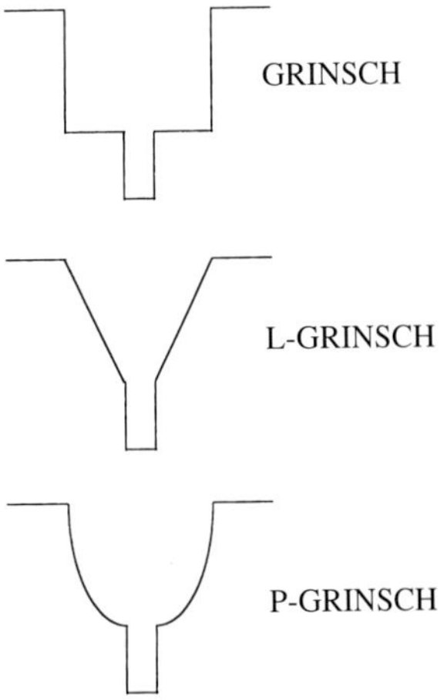

Fig. 8.21. Schematic of a quantum-well-laser structure with separate confinement layer for optical confinement and a cladding layer on each side. Three types of Separate Confinement Heterostructures (SCH) are shown: SCH with a constant confinement potential (*solid lines*), L-GRINSCH with a linearly graded refractive index SCH (*dashed lines*), and P-GRINSCH with a parabolically graded refractive index SCH (*dotted curve*)

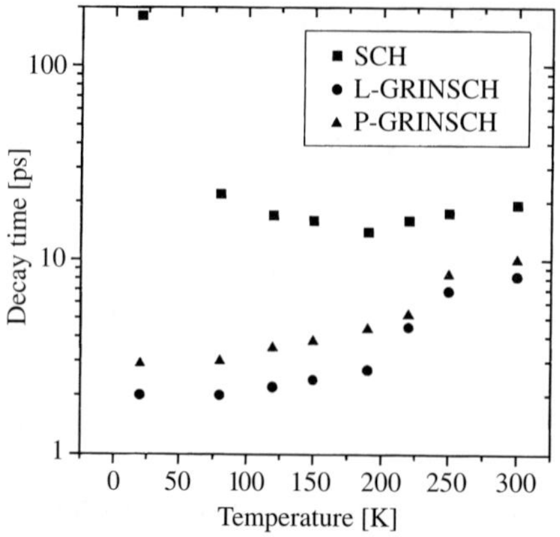

Fig. 8.22. Decay-time constants for the barrier luminescence decay for SCH, L-GRINSCH and P-GRINSCH GaAs/AlGaAs structures. For each of the structures, the GaAs quantum-well thickness was 50 Å, and the confinement barrier was 2000 Å wide on each side of the well. The confinement barriers ended in outer barriers of $x = 0.6$ AlGaAs. For SCH, the confinement barrier had $x = 0.3$ AlGaAs, whereas for the GRINSCH structures, the x value in the confinement region varied from $x = 0.3$ near the well to $x = 0.6$ at the outer barrier [8.75]

mance have also been discussed by others [8.70, 77, 93]. Application of the general principles discussed above should allow one to determine the relative importance of transport and quantum capture in carrier capture by quantum wells, and optimize the capture efficiency in the wells.

It should be remarked that electrical injection introduces different physical considerations because the electrons and holes are injected from opposite ends. Therefore, ambipolar transport is not important but the holes, being the heavier species, are still likely to determine the speed of the device. The relative importance of transport and quantum capture was discussed in [8.79].

8.4.4 Summary

Important insight into the nature of the carrier capture processes by quantum wells has been obtained by all-optical techniques as discussed in this section. These results and discussion show that the capture is determined by a combination of classical transport such as diffusion and drift, and by quantum capture. The relative importance of the two processes depend on the particular structure under investigation. If the coherence of the carrier wavefunction extends over the region of interest, then quantum-capture process dominates. In this case one finds capture times in the subpicosecond and picosecond range. The large variation in the capture times as a function of well width predicted by early simple models are not observed although recent measurements do show some evidence for an oscillatory behavior. It is proposed that the hot carrier distribution functions and effects of impurities and carrier-carrier collisions may explain the absence of very small capture rates at certain well widths expected from simple theories. While there are still some outstanding questions, such as the disagreement between the most recent and earlier calculations, and the influence of the confined phonon modes in the quasi-2D structures, it is reasonable to state that considerable insight into the carrier capture process in quantum wells has been gained by these time-resolved studies using optical markers.

8.5 Conclusions

The results presented in this chapter show clearly that ultrafast optical spectroscopy provides a powerful means of investigating the transport dynamics of photoexcited species such as electrons, holes, electron-hole pairs, and excitons in semiconductor nanostructures. Perpendicular transport has been investigated using either optical measurement of the screening of an applied or built-in electric field caused by charge transport, or by using the technique of optical markers in which unique spectral properties are associated with different spatial regions. Such studies have provided information about Bloch transport in superlattices and its transition to hopping transport with decreasing miniband width, about transport of carriers in multiple-quantum-well structures, and about carrier capture in quantum wells. These studies have provided fundamental insight and useful information about devices such as modulators and lasers. Lateral transport has been investigated using imaging and other techniques, and has provided information about diffusion and drift of carriers and excitons. Possibility of dramatic improvement in the spatial resolution exists with the refinement of techniques involving ultrafast lasers and high-resolution microscopy such as near-field scanning microscopy, scanning tunneling microscopy and atomic force microscopy.

9. Recent Developments

Many exciting new developments have occurred in the field of ultrafast spectroscopy of semiconductors since the completion of the first edition of this book. Ultrafast spectroscopy of semiconductors has remained a vibrant field of research. The goal of this chapter is to present an overview of these developments and discuss some selected recent developments in this field.

The first edition focused on ultrafast spectroscopy of III–V semiconductors and their quantum wells. Considerable progress has been made in the fabrication and investigation of lower-dimensional systems, quantum wires and quantum dots in recent years. Another type of nanostructure that has received considerable attention during the past few years is the semiconductor microcavity. Section 9.1 begins with a review of these nanostructures, and then discusses recent advances in ultrafast lasers and techniques. Much of the recent research in the field of ultrafast spectroscopy has focused on understanding the coherent response of semiconductors. Recent advances in this area are discussed in Sect. 9.2. This is followed by an analysis of results on resonant secondary emission from excitons in quantum wells (Sect. 9.3). It will be obvious from the discussion in Sect. 9.3 that it is not possible to divide the treatment on the basis of coherent and incoherent response. Some very interesting coherent properties of resonant secondary emission from quantum wells will be described. This will be followed in Sect. 9.4 by a discussion of relaxation and transport dynamics of carriers and excitons, and of high density effects.

A number of excellent books and review articles have appeared in this field in the last few years [9.1–3]. The reader is encouraged to refer to them as a complement to this monograph.

9.1 Nanostructures, Lasers and Techniques

9.1.1 Semiconductor Nanostructures

The band structure of bulk semiconductors and quantum wells is discussed briefly in Sect. 1.1.1, and a passing reference to lower-dimensional structures like quantum wires and dots is made in Sect. 1.1.1 c. In recent years considerable progress has been achieved in fabricating and characterizing semiconductor quantum wires, quantum dots and microcavities, and also in investigating their novel and unique properties. We will describe some of the new developments in ultrafast spectroscopy of these nanostructures in different parts of this chapter, once again restricting the discussion to III–V semiconductors. This section is intended as a brief introduction to quantum wires, quantum dots and microcavities.

(a) Quantum Wires and Quantum Dots. Quantum wells are planar structures in which the energy levels are confined in one direction; i.e., they are quasi-two-dimensional (quasi-2D) nanostructures. There is considerable interest in fabricating and investigating properties of semiconductor nanostructures with a confinement in two or three dimensions (quasi-1D and quasi-0D nanostructures, respectively). The interest in these lower-dimensional systems stems from the possibility of new physics and applications, and improved devices. The fabrication process for these lower-dimensional nanostructures can be classified in two broad categories: epitaxial growth techniques, and lithography or post-growth techniques.

Epitaxial growth techniques are generally used to grow on specially prepared substrates. Examples include wires grown on vicinal surfaces [9.4, 5], on high-index surfaces (e.g., (311) GaAs) [9.6], on specially patterned substrates (V-groove quantum wires) [9.7] or on cleaved edges (T-wires) [9.8]. Quantum dots have been prepared by self-organized growth process [9.9–15]. Some of these have been reviewed recently [9.16, 17]. While most of the work has concentrated on III–V semiconductors, self-organized growth of SiGe has also been reported [9.18]. Another technique, following the method of T-wires, is to grow quantum dots by twofold-cleaved edge overgrowth [9.19–22].

The second technique for preparing lower-dimensional nanostructures is to process the sample after the completion of the growth of quantum wells. The most common technique is to use high-resolution lithography combined with etching techniques. These have been reviewed recently [9.23]. Yet another technique is to utilize strain-induced confinement [9.24, 25]. Finally, local fluctuations in the potential due to disorder in quantum wells lead to extremely sharp spectral features in both photoluminescence and photoluminescence excitation spectra under certain conditions. These have been attributed to quantum dots [9.26–28].

The interest in quantum wires stems partially from the square-root singularity in the density-of-states at the bottom of the one-dimensional band. The possibility of reduced threshold for lasing [9.29] was one of the original motivations for introducing quantum wires. The nature of electronic states and interband matrix elements have been investigated extensively. It is interesting to note that Coulomb correlation effects in 1D systems reduces the matrix elements for interband transition near the bandgap [9.30, 31], at least partially counteracting the effects of the singularity in the density-of-states. Other areas of interest in quantum wires include the nature of the exciton and its binding, capture of carriers in quantum wires, formation and relaxation dynamics of excitons and relaxation of carriers in quantum wires.

Interest in quantum dots is driven by the same factors: fundamental physics of a novel system, and potential for new applications and improved devices. Quantum-dot lasers have already been demonstrated [9.32–39]. The discrete nature of electronic states in quantum dots and the dynamics of relaxation of carriers between these discrete states are two of the areas of interest in the investigation of quantum dots. We will discuss experiments on coherent spectroscopy on quantum wires and quantum dots in Sect. 9.2.13, and capture and relaxation processes in the incoherent regime in Sect. 9.4.3 for quantum wires and Sect. 9.4.5 for quantum dots. High-density and many-body effects in quantum wires are discussed in Sect. 9.4.4.

(b) Microcavities. The cavity lengths of typical gas and solid-state lasers are tens or hundreds of centimeters. The edge-emitting semiconductor laser is tiny by these standards but still has a cavity length of hundreds of micrometers. The advances in epitaxial-growth techniques such as Molecular Beam Epitaxy (MBE) and Metal-Organic Chemical Vapor Deposition (MOCVD) allows one to make multiple layers of high-quality semiconductors with precise control of the thickness. Epitaxial growth of alternate layers of two semiconductors with different refractive indices (e.g., GaAs and AlAs) allows one to fabricate a Bragg reflector much like a multiple-layer dielectric mirror. If two such Bragg reflectors are separated by a spacer layer, one has a planar Fabry-Perot cavity whose length can be in the sub-micrometer regime. A semiconductor microcavity is schematically depicted in Fig. 9.1. Fundamental physics of linear and nonlinear phenomena in microcavities is very interesting (e.g., the strong coupling between photons and excitons) and many different applications of microcavities are being considered. Verti-

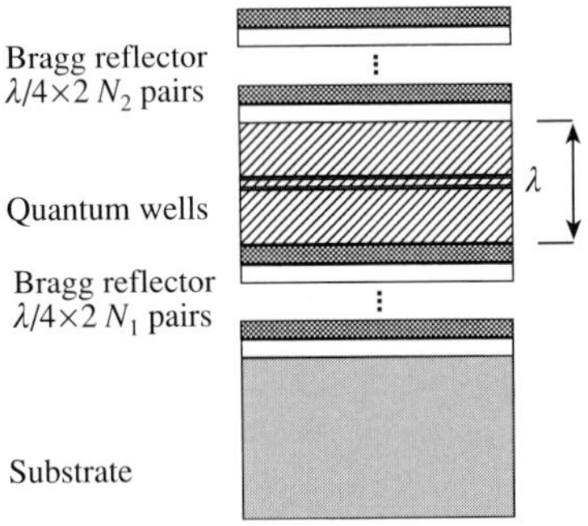

Fig. 9.1. Schematic of a semiconductor microcavity. The structure is grown epitaxially on a semiconductor (e.g., GaAs) substrate, and consists of: (*1*) the lower Bragg reflector (N_1 pairs of alternating $\lambda/4$ layers of low- and high-index semiconductors; e.g., AlAs and $Al_{0.11}Ga_{0.89}As$), (*2*) a λ spacer layer (e.g., $Al_{0.3}Ga_{0.7}As$) with two quantum wells at the antinode, and (*3*) the top Bragg reflector (N_2 pairs of alternating $\lambda/4$ layers of low- and high-index semiconductors; e.g., AlAs and $Al_{0.11}Ga_{0.89}As$)

cal Cavity Surface-Emitting Lasers (VCSELs) have many potential applications. This combination of interesting fundamental physics and applications has led to extensive research on microcavities in recent years. The field is vast [9.40–107] and we have chosen to discuss some selected results related to ultrafast spectroscopy in this chapter. This section serves as a brief introduction to the basic concepts of microcavity physics. The reader is referred to excellent books [9.62, 75, 108, 109] and recent review articles [9.53, 106, 107, 110] on this subject for an in-depth discussion.

It is now well known that spontaneous emission is not an immutable property of the emitter but is influenced by the surroundings of the emitter and the coupling of the emitter to the electromagnetic modes [9.111]. Microcavity controls the modes in which an emitter can radiate and hence controls properties such as the emission rate and the angular distribution of the emission. Early work on semiconductor microcavity focused on this aspect of controlling the emission properties by placing the emitter within a microcavity. The threshold current of a semiconductor laser is strongly affected by the number of electromagnetic modes to which the radiation can couple. The idea of a "threshold-less" laser, in which the semiconductor can radiate into only one mode, was considered at length in early studies on microcavities. This is in the so-called weak coupling regime of microcavities in which the coupling strength Ω_0 between the electronic and photon modes is weaker than the decay rate of the electronic system $\gamma = 2/T_2$ and the decay rate of the photon κ determined by the mirror reflectivities ($\Omega_0 < \gamma, \kappa$).

More recent studies have focused on the strong coupling regime in which $\Omega_0 \gg \gamma, \kappa$. Multiple reflections at the Bragg mirrors lead to the establishment of a stationary electric field in the microcavity. The layers are generally chosen such that the antinode of the electric field is at the center of the spacer layer of the microcavity. The strong coupling regime is most easily achieved by placing one or more (N) quantum wells at or close to the antinode(s) of the electric field in the microcavity and considering the coupling between the photon and the quantum-well exciton which has a large oscillator strength. If the quantum-well thickness is such that the exciton transition energy coincides with the cavity-mode energy, then the two modes couple strongly and lead to two normal modes that are separated by the so-called Rabi or normal-mode splitting given by

$$\hbar \Omega_0 = e \hbar \sqrt{\frac{N f_A}{2 \varepsilon m L_{eff}}} , \qquad (9.1)$$

where f_A is the exciton oscillator strength per unit area, ε is the dielectric constant of the cavity material, m is the free electron mass, and L_{eff} is the effective cavity length which is generally larger than the spacer layer thickness because the field penetrates in the Bragg mirrors. For a semiconductor microcavity with a few quantum wells ($N \sim 3$), the normal-mode splitting is of the order of 5 meV.

It is important to make a distinction between atomic systems and semiconductor. In ultrahigh-finesse cavities, coupling between the cavity mode and a single atom is sufficiently strong to give a measurable splitting. This is the true strong-coupling regime where the picture of a Jaynes-Cummings ladder with non-equal splitting in different rungs of the ladder [9.112] is valid. Such a ladder has indeed been observed in atomic systems [9.113]. The strong-coupling regime for a single- or a few-atom system must be treated quantum mechanically. In contrast, the splitting expected for a single exciton in a microcavity is ≈ 1 μeV, much smaller than the typical exciton linewidths and photon decay rates (κ) in a semiconductor microcavity. Therefore, a large number of excitons must interact with the cavity mode for an observable normal-mode splitting in semiconductor microcavities. This regime is far from the quantum statistical limit and can be treated semiclassically. This point has been discussed in the recent semiconductor literature [9.71, 106] where the term nonperturbative regime has been introduced for describing the (many-exciton) normal-mode coupling regime in semiconductors.

If the exciton resonance energy and the cavity mode resonance energy are close but do not match exactly, there is still interaction between the two modes. The two normal-mode energies are given by

$$E_{\pm} = \frac{E_{\rm X} + E_{\rm c}}{2} \pm \sqrt{(\hbar \Omega_0)^2 + (E_{\rm X} - E_{\rm c})^2}\,, \qquad (9.2)$$

where $E_{\rm X}$ ($E_{\rm c}$) is the exciton (cavity-resonance) energy. A typical semiconductor microcavity is prepared with a taper such that the cavity resonance energy varies in one direction of the sample. By varying the detuning between the exciton and the cavity, the normal-mode energies show a typical anti-crossing behavior (Fig. 9.2).

The coupling between the cavity and the exciton in a semiconductor microcavity is similar to the coupled exciton-photon mode (polariton) in semi-

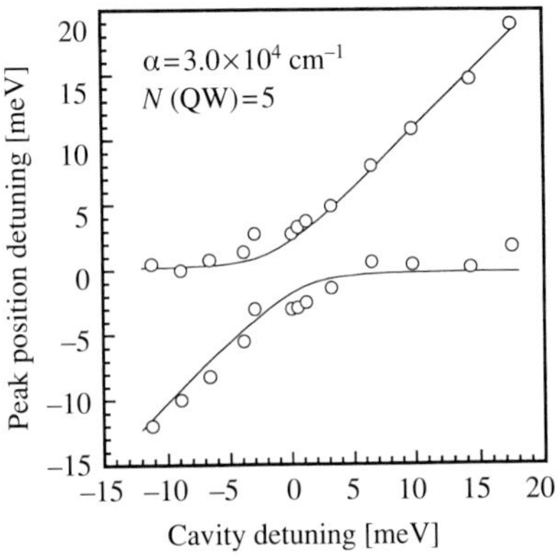

Fig. 9.2. Reflectivity-peak positions as a function of the cavity detuning for a 5-quantum-well sample at 5 K, showing the anti-crossing behavior of the normal modes [9.114]

conductors [9.115, 116] (Sect. 6.1.2) and semiconductor quantum wells (see, for example, [9.49, 117, 118]). The normal modes in a semiconductor microcavity are the polariton modes with a polariton splitting much larger ($\approx$ a few meV) than in bulk or quantum-well semiconductors (a small fraction of a meV in a typical III–V semiconductor). This discussion of the polariton in semiconductor microcavities so far has considered only the direction normal to the interfaces, and ignored the dispersion of the excitons and cavity resonance energies parallel to the interface planes. Modifying (9.2) to account for this in-plane dispersion gives

$$E_{\pm}\left(K_{\|}\right)=\frac{E_{\mathrm{x}}\left(K_{\|}\right)+E_{\mathrm{c}}\left(K_{\|}\right)}{2} \pm \sqrt{\left(\hbar \Omega_{0}\right)^{2}+\left[E_{\mathrm{x}}\left(K_{\|}\right)-E_{\mathrm{c}}\left(K_{\|}\right)\right]^{2}}, \quad (9.3)$$

where $E_{\mathrm{c}}\left(K_{\|}\right)=(\cos \theta) / E_{\mathrm{c}}(0) \equiv E_{\mathrm{c}}(0) \sqrt{1+\left(\left(2 \pi / L_{\mathrm{c}}\right) / K_{\|}\right)^{2}}$ and $E_{\mathrm{x}}\left(K_{\|}\right) = \hbar^{2} K_{\|}^{2} / 2 M$. M is the exciton mass, and L_{c} is the thickness of the cavity spacer layer. The dispersion curve for the coupled modes will depend on detuning as well as on the in-plane wave vector $K_{\|}$. Dispersion curves for cavity polaritons have been obtained from luminescence measurements [9.51]. A typical cavity-polariton dispersion curve is depicted in Fig. 9.3.

We have so far considered only *planar* microcavities. Three-dimensional microcavities, with lateral dimension of the order of 1 μm, have been fabricated using various techniques and have been shown to have many interesting properties [9.83, 97, 119–130]. The most direct technique for achieving such structures is to start with a planar microcavity, mask it with a circular or square mask whose dimension is ≈ 1 μm, and then etch away the unwanted semiconductor around the mask by chemical or reactive ion process. Such processing produces a micropost with a circular or square cross-section and supports only a few optical modes. We also mention that there has been considerable interest in recent years in photonic bandgap materials or structures [9.131–134]. These are structures that exclude photons within a certain range of frequencies, much like electronic bandgaps in semiconductors and insulators which exclude electronic states within a certain energy

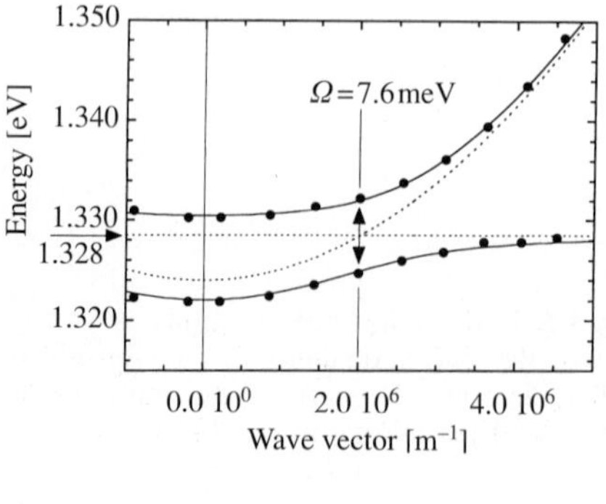

Fig. 9.3. Polariton dispersion curves deduced from angle-resolved luminescence measurements for a resonance occurring at an oblique angle of incidence (29°). The *points* are experimental, the *solid curves* are theoretical calculations and the *dashed curves* are the uncoupled exciton and cavity dispersion curves [9.51]

range. Semiconductor microcavities can be considered as photonic bandgap materials because only certain photon modes are allowed in semiconductor microcavities.

9.1.2 Ultrafast Lasers

Significant advances have been made in the field of generation of ultrashort laser pulses. The shortest pulse remained at 6 fs, or approximately 3 optical cycles at 6200 Å, for nearly a decade [9.135]. But the advent of femtosecond solid-state lasers (e.g., Ti:Sapphire laser) has heralded a new era in the field of ultrafast lasers. The fluorescence bandwidth of the lasing material allows for pulses as short as 4 fs. Indeed pulses as short as 6.5 fs have been obtained directly from a Ti:Sapphire oscillator [9.136]. Using the technique of self-phase modulation and pulse compression a new record of <5 fs, or less than 2 optical at 8200 Å, was established during the last two years [9.137–140]. Femtosecond pulses with intensities approaching 10^{21} W/cm^2 have been generated using the concept of chirped-pulse amplification [9.141, 142]. Femtosecond lasers with a variety of wavelengths from mid-infrared to UV have been developed using nonlinear techniques, and femtosecond pulses of X rays with 0.4 Å wavelength have been generated utilizing Thomson scattering of relativistic electrons with femtosecond pulses [9.143–145]. Compact and practical femtosecond lasers have been developed [9.146, 147] spurring real-world applications such as T-ray imaging [9.148–150] and chirped-pulse wavelength division multiplexing [9.151–153]. These developments in ultrafast lasers, along with new measurement techniques discussed in Sect. 9.1.3, are leading to exciting new results in the field of ultrafast studies of physical, chemical and biological systems.

9.1.3 Measurement Techniques

Techniques that provide information about the amplitude and the phase of the linear or nonlinear signal generated by the semiconductor in response to excitation by a single pulse or a sequence of pulses are discussed briefly in Sect. 1.3.4. Some results obtained by such techniques are described in Sect. 2.6.5 b, Sect. 2.7.1 c, and Sect. 2.12.4. These techniques have matured considerably in the last few years, and have found a widespread use in the investigation of semiconductors. Some of these measurement techniques are discussed in this section.

(a) Spectral Interferometry: Characterization of Weak Ultrashort Light Pulses. As discussed briefly in Sect. 2.7.1 c, characterization of both the amplitude and the phase of ultrashort light pulses is very complex. Many techniques have been developed to characterize the amplitude and the phase of a light pulse when the light pulses are of moderate or high intensity, such

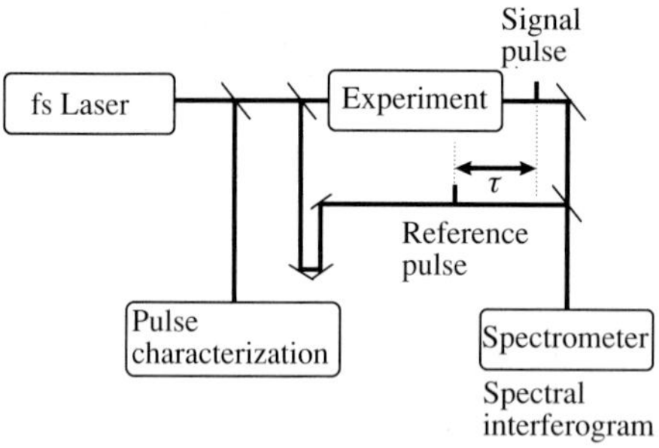

Fig. 9.4. Schematic of an experimental arrangement for spectral interferometry. The unknown signal from the experiment is collinearly combined with a fully characterized reference pulse to yield the spectral interferogram. The amplitude and the phase of the unknown signal from the experiment can be determined from the spectral interferogram, as discussed in the text

as from an ultrafast laser; e.g., the chronocyclic representation technique [9.154, 155] or the Frequency-Resolved Optical Gating (FROG) technique [9.156–160] or any of the other techniques [9.161–164]. In all these studies, nonlinear schemes requiring high intensities are used for a complete characterization of the light pulse. Therefore, such techniques are not well-suited for characterizing weak signals. In recent years, linear concepts for characterizing almost arbitrarily weak ultrashort light pulses have been developed [9.165–170]. These techniques combine a well-characterized light pulse with the unknown pulse to deduce information about the amplitude and the phase of the unknown pulse. We discuss some of these concepts in this section.

In spectral interferometry or TADPOLE, a femtosecond laser beam is divided into two beams. A schematic of the experimental arrangement for spectral interferometry is displayed in Fig. 9.4. The amplitude and phase of one beam is fully characterized using one of the nonlinear gating techniques discussed above. The second beam, after further division if necessary, excites the sample and generates a weak ultrashort light pulse as a result of linear or nonlinear processes in the sample. The spectrum of the reference pulse must be wider than the spectrum of the unknown pulse. The delay between the two pulses is set appropriately and the signal beam is collinearly combined with the fully characterized reference beam. The spectrum of the combined beam is recorded using a spectrometer and a (multichannel) detector. The intensity of this measured "spectral interferogram" is given by

$$I_{\mathrm{SI}}(\nu) = I_{\mathrm{REF}}(\nu) + I_{\mathrm{S}}(\nu) + 2\sqrt{I_{\mathrm{REF}}(\nu)\, I_{\mathrm{S}}(\nu)} \cos \times \left[\phi_{\mathrm{S}}(\nu) - \phi_{\mathrm{REF}}(\nu) - 2\pi\nu\tau\right] ,$$

where τ is the delay between the signal and the reference beams, $I_{\mathrm{SI}}(\nu)$ is the intensity of the spectral interferogram at the photon energy $h\nu$, $I_{\mathrm{REF}}(\nu)$

and $\phi_{\mathrm{REF}}(\nu)$ are the intensity and the phase of the reference pulse at ν, and $I_{\mathrm{S}}(\nu)$ and $\phi_{\mathrm{S}}(\nu)$ are the intensity and the phase of the unknown signal pulse at ν. Since $I_{\mathrm{REF}}(\nu)$ and $\phi_{\mathrm{REF}}(\nu)$ are known from the nonlinear gating characterization measurements, and $I_{\mathrm{S}}(\nu)$ can be found by blocking the reference beam, the only unknown $\phi_{\mathrm{S}}(\nu)$ can be determined from these measurements. The intensity $I_{\mathrm{S}}(t)$ and the phase $\phi_{\mathrm{S}}(t)$ of the emitted signal in the time domain can be obtained from the spectral interferograms using Fourier Transforms (FTSI) [9.165] or TADPOLE [9.166] or by means of the technique of Dual-Quadrature Spectral Interferometry (DQSI) [9.165]. It has been shown [9.165–170] that the amplitude and phase of extremely weak light pulses (10^{-21} J or only a small fraction of a photon per pulse) can be determined by this technique. Some results on phase and amplitude of the signal representing linear and nonlinear response of semiconductors to ultrashort pulses are discussed in Sect. 9.3.3 and Sect. 9.2.4, respectively.

(b) Excitation with Phase-Locked Pulses. Another technique that provides phase-sensitive information and is used extensively in coherent control experiments (Sects. 2.12.4, 9.2.3 and 9.3.2) is excitation by a pair of phase-locked pulses. For a wavelength of 8150 Å, close to the bandgap of GaAs, the period of the optical cycle is 2.72 fs which corresponds to a distance of 0.9 μm in air. Techniques have been developed to control the relative phase of the two phase-locked pulses to better than 1/100th of the optical cycle. Fine control of the delay can be obtained by a number of different techniques (see, for example, the references cited in Sects. 2.12.4, 9.2.3 and 9.3.2), and the whole system is typically stabilized by some feedback mechanism in conjunction with a reference He-Ne laser.

A schematic of a system that uses phase-locked pulses to investigate linear and nonlinear responses of a semiconductor is depicted in Fig. 9.5, along with the notation used for various parameters. The figure should be compared to the three-beam FWM setup shown in Fig. 1.11. The primary difference is that in Fig. 9.5 the pump pulse passes through a Michelson interferometer which divides the pulse into two pulses of equal intensity and separated by a precisely controlled phase delay. These phase-locked pulses 1 and 1′, with delay τ_{M} between them, then excite the sample. The FWM signals are along the same directions as without the Michelson interferometer. Another difference is that, in addition to measuring TI-FWM, one can also determine TR-FWM, transmission, reflection or emission with the phase delay τ_{M} and the gate delay τ_{G} as parameters.

An alternative technique is to rapidly scan the delay in the Michelson interferometer [9.171]. A fast data acquisition system that keeps track of the phase difference as a function of time during the rapid scan enables dependence of various measured quantities on the phase difference. Very low intensity measurements on semiconductors have been reported in this manner [9.171]. Various examples of the use of these techniques will be discussed in forthcoming sections (Sects. 9.2.3 and 9.3.2).

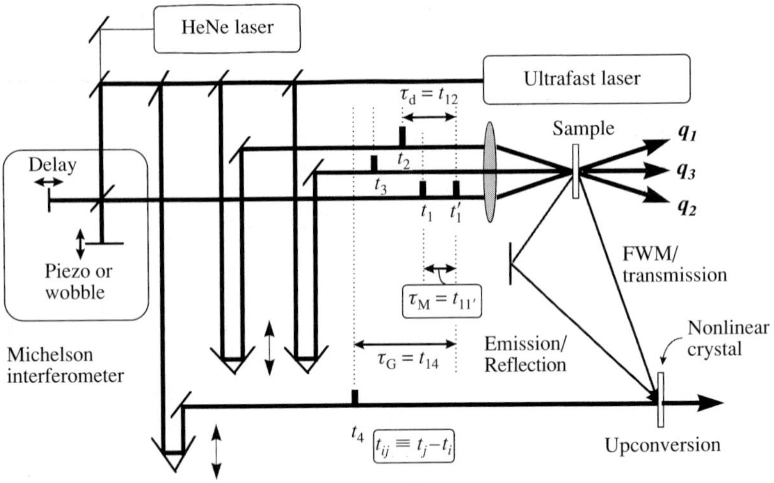

Fig. 9.5. Schematic experimental arrangement for using a pair of phase-locked pulses, generated in a stabilized Michelson interferometer, to excite the sample and then measure the linear or nonlinear response of the sample. The arrangement is similar to three-beam arrangement of Fig. 1.11, except that a pair of phase-locked pulses is used for excitation

(c) Ultrabroadband Electro-Optic Detection of THz Transients. Section 2.12.1 on THz spectroscopy discusses the use of photoconducting antenna for generating and detecting THz radiation. The bandwidth of THz detectors has recently been extended to >30 THz [9.172] (corresponding to a wavelength of ≈ 10 μm or a photon energy of ≈ 125 meV or 1000 cm^{-1}). This is achieved by using electro-optic crystals such as ZnTe. As discussed briefly in Sect. 2.11, an electric field applied to an electro-optic crystal modifies the polarization of the reflected or transmitted probe beam. The same principle is applied in the electro-optic detection of THz signals.

In Sect. 2.11 the pump pulse modified the electric field in the sample and resulted in an anisotropic change in the reflected/transmitted probe beam so that the sample itself acted as an electro-optic detector. For the detection of THz radiation, a linearly polarized femtosecond probe pulse and the THz pulse are incident on an appropriately oriented electro-optic crystal. If the two pulses overlap temporally, then the polarization of the transmitted probe pulse is modulated by the electro-optic effect in the crystal in proportion to the strength of the THz electric field. Therefore, the difference in the intensity of the two orthogonal polarizations of the transmitted probe pulse provides a measure of the amplitude of the THz electric field. Measurement as a function of the delay between the pump pulse exciting the THz radiation and the probe pulse allows one to determine the amplitude and the phase of the THz transient.

Several factors determine the bandwidth of the electro-optic THz detector. The thickness of the detector clearly plays an important role because of the variation in the velocities of the visible or IR probe pulse and the THz pulse. This is similar to the case of luminescence upconversion (Sect. 1.3.3 c) for which the group-velocity difference between the gating, signal and sum-frequency pulse plays an important role. There is an obvious trade-off between the signal strength and the bandwidth. Another factor that influences the bandwidth is the influence of Fresnel reflection/refraction in the region of strong variation in the dielectric constant in the vicinity of the phonon resonances. Typical optical phonon energies in III–V and II–VI semiconductors are in the 6–10 THz range, so that these effects are important in the frequency region of considerable interest for THz measurement. These effects have been considered in detail [9.172] for the case of ZnTe and GaP. These researchers have shown that bandwidth exceeding 50 THz are possible for a 6–10 μm thick ZnTe and GaP electro-optic crystal. Another factor that may significantly influence the frequency response of the detector is the frequency dependence of the electro-optic coefficient, which comprises of both electronic and lattice contributions (*Leitenstorfer* et al. [9.172]). The frequency dependence of the electro-optic coefficient in GaP has been investigated by *Faust* et al. [9.173].

These advances in the detection of THz radiation with wide bandwidth open up many interesting possibilities for using measurement of THz radiation as a probe for ultrafast processes in materials. An example of the use of such ultrabroadband THz detectors in studying ultrafast high-field transport in semiconductors will be discussed in Sect. 9.4.6.

9.2 Coherent Spectroscopy

Coherent spectroscopy of semiconductors and their nanostructures continues to be an active and important field of research. Many exciting new results have been obtained in the last few years. We begin the discussion with possibly the most exciting developments in this field that require going beyond the relaxation-time approximation of SBE (or the Boltzmann kinetics) to include quantum-kinetic effect (Sect. 9.2.1), and going beyond SBE to include four-particle correlation effects (Sect. 9.2.2). We follow these results with a discussion of results demonstrating coherent control of the population, spin and orientation of excitons, and coherent control of current in semiconductors (Sect. 9.2.3). Section 9.2.4 treats the progress made in making phase-sensitive measurements in coherent spectroscopy of semiconductors. The next few sections (Sect. 9.2.5–10) deal with interesting results on Bloch oscillations, coherent phonons, plasmon-phonon oscillations, HH-LH beats in the continuum and biexcitons. These results can be understood within the framework of the Semiconductor Bloch Equations (SBE) in the relaxation-time approximation. Finally, we treat novel systems such as semiconductor microcavities, quantum wires and quantum dots (Sect. 9.2.5–12). Femtosecond resonant emission from semiconductor nanostructures contains a coherent component. This aspect of coherent spectroscopy will be discussed in Sect. 9.2.11 on ultrafast emission dynamics.

9.2.1 Quantum Kinetics in Semiconductors

Nearly all the results presented in Chap. 2 have been interpreted in terms of the semiconductor Bloch equations (Sect. 2.1.2) with the relaxation-time approximation for the relaxation Hamiltonian (Sect. 2.1.1 d). A better approximation for the relaxation Hamiltonian can be obtained from the full semiclassical Boltzmann equation in which transition probabilities are calculated on the basis of the Fermi's golden rule. This Boltzmann kinetic approach has worked extremely well in numerous physical situations, but it must be remembered that it is based on a number of assumptions. It assumes that each collision strictly conserves energy and momentum. In any scattering process, energy conservation holds only if the carrier distribution function varies slowly on the scale of the collision time. Under these conditions, individual collision events become decoupled so that the memory of what has happened prior to the collision has disappeared, or all collisions can be considered independent. This is the Markovian approximation (Sect. 2.1.1 d) under which most of the results in coherent spectroscopy have been analyzed. The Markovian approximation can be considered valid if the average *interval* between collisions (the inverse of the scattering rate) is much longer than the collision *duration*. This is essentially the same as the condition for the validity of Fermi's golden rule. When these conditions are not satisfied, one must invoke quantum kinetics for the interpretation of results. Quantum kinetics in transport and optics of semiconductors has been discussed in detail in an excellent recent monograph by *Haug* and *Jauho* [9.1].

The duration of the collision time can be estimated as the oscillation period of the energy quantum exchanged during the collision. For example, the oscillation period of an optical phonon in GaAs is ≈ 115 fs and the oscillation period of an electron plasma is ≈ 150 fs for a carrier density of 5×10^{17} cm^{-3}. This can be compared with the inverse of the electron-phonon collision rate, i.e. the collision interval, of ≈ 200 fs (Sect. 4.1). The availability of reliable solid-state lasers with pulse widths ≈ 10 fs and recent improvements in experimental techniques and apparatus (Sect. 9.1) have made it possible to explore dynamics on the 10 fs time scale which can be much shorter than these characteristic times in semiconductors. For the interpretation of such experiments, one must, therefore, consider the possibility that the Boltzmann kinetic approach is not valid, and consider the experiments in terms of quantum kinetics, which removes the shortcomings of the Boltzmann approach described in the previous paragraph. Quantum kinetics has been analyzed from a theoretical point of view since the late 1980s (see the references cited in Sect. 2.1.1 d). However, it is only recently that experiments showing unambiguous quantum kinetic effects have been reported. We will discuss some of these experiments in this section.

During this discussion it would be useful to keep in mind that, as already mentioned in Sect. 1.4.2, the simplified picture presented in Sects. 1.3.1 and 1.4.2, for the analyses of differential transmission measurements, must be used with caution. Furthermore, the generation and the probing processes can not be viewed as independent processes on ultrashort time scales. This means that the simple picture (Sect. 1.1.5) of relaxation can no longer describe the experiments, and a more sophisticated formalism has to be developed to analyze the ultrashort time experiments. In particular, in Differential Transmission Spectroscopy (DTS) (Sect. 1.3.1), the change in the absorption at a given photon energy can not be simply related to the occupation of the states coupled by that photon energy.

To emphasize this point, we present here the results of a recent theoretical calculation [9.174] of the DTS of GaAs using the Keldysh non-equilibrium Green's function theory [9.1] in a realistic case. A 120 fs Gaussian pulse, with a central photon energy of 1.57 eV excites a sample of bulk GaAs at a low density of 8×10^{14} cm^{-3}. The calculations were performed to provide a realistic comparison with the results of *Fürst* et al. [9.175] that will be discussed in Sect. 9.2.1 a. Although the calculation considers the full effects of quantum kinetics, for our illustrative purposes of this introduction, we depict in Fig. 9.6 the results obtained for Boltzmann kinetics. The top part of the figure shows the sum of the calculated electron and hole distribution functions as a function of the photon energy for various delays after the excitation. The bottom part of the figure represents the calculated DTS for the same delays. The results clearly illustrate that the DTS at a given energy does not correspond to the sum of the distribution functions at the corresponding energy. Recent calculations of the lineshape of differential transmission in the coherent regime by *El Sayed* and *Stanton* [9.176], although not as rigorous, illustrate the same point by using a simplified model which allows approximate analytical solutions. These researchers pointed out that,

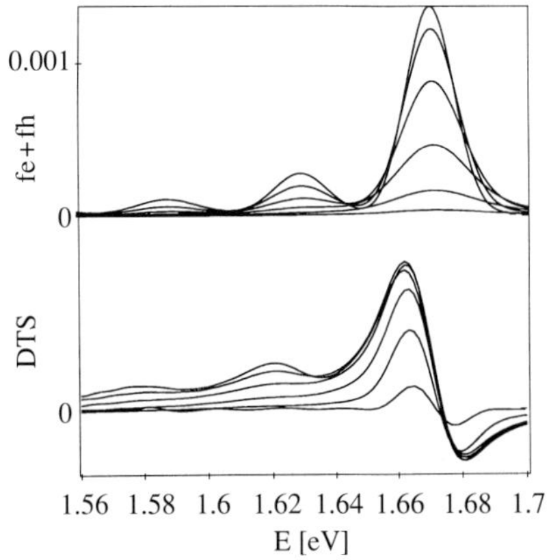

Fig. 9.6. The sum of the electron and hole distribution functions (*top*) and the Differential Transmission Spectra (DTS) (*bottom*) calculated using Boltzmann kinetics as discussed in the text. The six curves correspond to the delays of −100, −40, 0, 40, 80, and 120 fs (in order of increasing values at 1.64 eV); compare with the results of quantum kinetic calculations in Fig. 9.9 (courtesy of *L. Banyai*; see [9.174])

of the three contributions [Phase-Space Filling (PSF), Band-Gap Renormalization (BGR) and Local Field (LF)] that contribute to DTS, only PSF approximately follows the sum of the distribution functions. The LF contribution approximately resembles the first derivative of the distribution functions, whereas the BGR contribution resembles the negative first derivative. Therefore, DTS does not correspond to the sum of the distribution functions.

We first discuss phonon dynamics in the quantum kinetic regime (Sect. 9.2.1 a–c) and electron dynamics in the quantum kinetic regime in Sect. 9.2.1 d.

(a) Exciton-LO-Phonon Quantum Kinetics: Memory Effects in GaAs. We first consider the results on exciton-LO-phonon quantum kinetics reported by *Banyai* et al. [9.177]. A high-quality GaAs/$Al_{0.3}Ga_{0.7}As$ double heterostructure with 600 nm thick GaAs layer was investigated at 77 K. Two co-linearly polarized, equal intensity pulses from a Ti:Sapphire laser (FWHM = 14.2 fs, nearly transform limited) are incident on the sample along the directions $\boldsymbol{q}_1$ and $\boldsymbol{q}_2$. The self-diffracted TI-FWM signals along $2\,\boldsymbol{q}_2 - \boldsymbol{q}_1$, measured as a function of time delay τ_d between the pulses, is depicted in Fig. 9.7 for three different excitation densities. The FWM signal exhibits an oscillatory behavior similar to the quantum beat signals discussed in Sect. 2.6. The modulation becomes less distinct with increasing density, but the period of the oscillations is independent of density (100, 98 and 98 fs for the three densities investigated). The density-independent period excludes the interpretation of these oscillations as Rabi oscillations (Sects. 2.3 and 2.8) or plasma oscillations (Sect. 9.2.5). The oscillations period is also much too short for the oscillations to be ascribed to propagation beats (Sect. 2.6.5). Surprisingly, the observed oscillation period of ≈ 100 fs is smaller than the well-established LO phonon oscillation period of 115 fs.

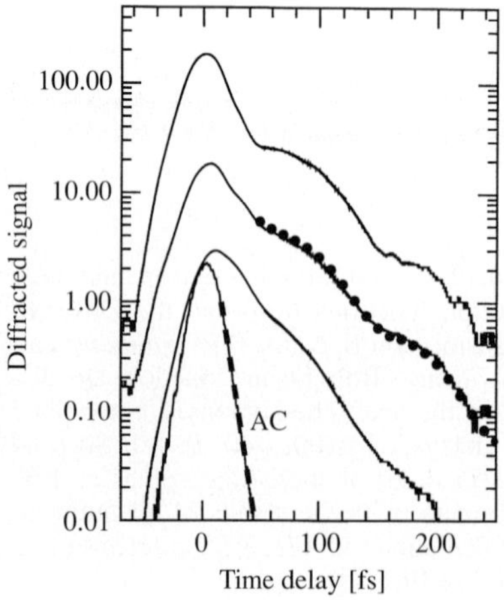

Fig. 9.7. TI-FWM from GaAs at 77 K using 14.2 fs pulses for three different carrier densities (from *top* to *bottom* 1.2, 1.9 and 6.3×10^{16} cm^{-3}). The curves are displaced vertically for clarity. The dots are calculated using quantum kinetic theory. The curve marked AC is the autocorrelation of the laser pulse [9.177]

Banyai et al. [9.177] argued that quantum coherence can not be separated from dephasing and relaxation on time scales shorter than or comparable to the inverse of the characteristic oscillation frequency, as discussed above. They used the SBE with retarded collision integrals for the LO-phonon scattering to properly account for the memory effect. The results of their quantum kinetic calculations, with the addition of a phenomenological dephasing time of 143 fs (most likely due to Coulomb scattering among carriers), are depicted in Fig. 9.7 as dots. The agreement with the experiments is remarkably good, including the fact that the theory predicts the correct oscillation period which deviates from the LO-phonon oscillation frequency, as outlined above. In fact, a decreased oscillation period was predicted by the theory before the experiments [9.178, 179]. The oscillations are interpreted as interference or beating between interband polarizations at wavevectors connected by LO-phonon scattering of an electron in the conduction band. These oscillations are therefore related to the band-to-band transitions, but strongly enhanced by excitonic effects. The oscillations in the theory disappear in the Markovian or Boltzmann kinetic limit.

(b) Quantum Kinetic Electron-Phonon Interaction in GaAs. *Fürst* et al. [9.175] have recently reported on pump-probe DTS measurements in bulk GaAs (5000 Å epitaxial layer grown on GaAs substrate by MBE) at 15 K that show the importance of energy nonconserving phonon scattering events and memory effects; i.e., the importance of quantum kinetic effects in the electron-phonon interaction on femtosecond time scales. They employed circularly polarized 120 fs Gaussian pulses to excite the sample at 1.67 eV, considerably above the bandgap energy of 1.52 eV, and co-circularly polarized 25 fs weak probe pulses centered at 1.64 eV to probe the dynamics. These pulses were selected for optimum temporal and energy resolution at the uncertainty limit, and co-circular polarization was used to minimize the influence of light hole transitions. The sensitivity of the system was improved so that measurements could be made at a density of 8×10^{14} cm^{-3} to minimize carrier-carrier scattering.

Differential transmission spectra at various time delays t_D between the pump and the probe are depicted in Fig. 9.8. For $t_D=0$ fs, DTS shows a red-shifted spectral hole with induced absorption at higher energy, qualitatively similar to the results of *Foing* et al. [9.180] mentioned earlier (Sect. 3.1.1, Fig. 3.4). The remarkable aspect of the results in Fig. 9.8 is that, for t_D between 40 and 80 fs, they observe a low-energy *shoulder* (not a peak) that is much broader than the peak due to bleaching near the pump energy at $t_D=0$. Furthermore, this broad feature develops into a well-defined peak, with a minimum between the no-phonon and one-phonon features, after ≈100 fs, the period of LO-phonon oscillations in GaAs. After ≈200 fs, the one-phonon replica has a width approximately equal to that of the no-phonon feature. *Fürst* et al. argued that for the case of energy-conserving Boltzmann kinetic picture, the one-phonon replica should replicate the width of the no-phonon peak and a minimum should exist between the

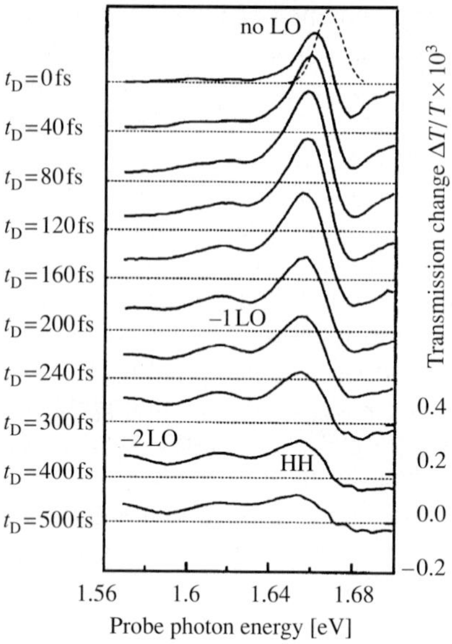

Fig. 9.8. Differential transmission spectra in GaAs at 15 K at a density of 8×10^{14} cm^{-3} for different time delays as indicated. The excitation spectrum is shown by the *dashed curve* [9.175]

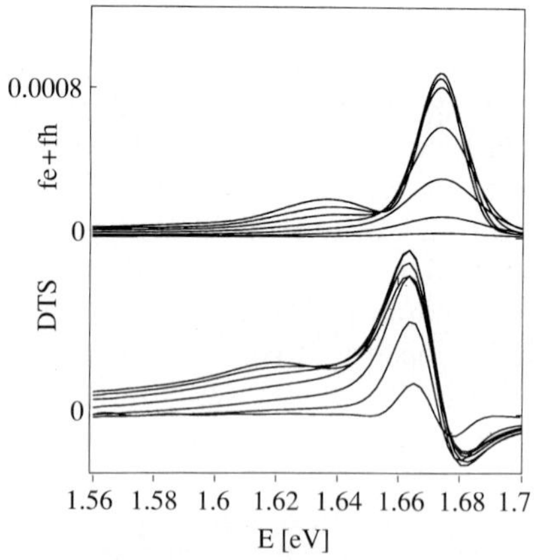

Fig. 9.9. The sum of the electron and hole distribution functions (*top*) and the Differential Transmission Spectra (DTS) (*bottom*) calculated using quantum kinetics, as discussed in the text. The seven curves correspond to the delays −100, −40, 0, 40, 80, 120, and 160 fs (in order of increasing values at 1.64 eV); compare with the results of Boltzmann kinetic caluclations in Fig. 9.6 (courtesy of *L. Banyai*; see [9.174])

two peaks at all times. The contrast of these observations with the predictions of the Boltzmann kinetic model was explained on the basis of quantum kinetics, and a model calculation for an analytical one-dimensional model provided qualitative agreement with the results. More recently, a numerical calculation for quantum kinetics in three dimensions has been performed [9.174]. Figure 9.9 displays the results for the sum of the electron and hole distribution functions (top) and DTS (bottom) as a function of photon energy. The difference in the predictions of the Boltzmann and quantum kinetic approaches can be obtained by comparing these results with those in Fig. 9.6. Note that the LO-phonon replica in DTS appears as a peak at t_D=0 for the Boltzmann kinetic calculation (Fig. 9.6). In contrast, the quantum kinetic calculation for DTS (Fig. 9.9) reveals that a broad low-energy shoulder develops first and the LO-phonon replica appears as a peak only after about 160 fs, in good qualitative agreement with the experimental results [9.175]. A detailed comparison can be found in [9.174].

(c) Exciton-LO-Phonon Quantum Kinetics: Reversal of Phonon Collision. A most interesting manifestation of memory effects and the control of carrier- or exciton-phonon scattering was recently demonstrated in bulk GaAs by *Wehner* et al. [9.181]. They argued that, since scattering processes are not instantaneous in time, they do not become irreversible for times short compared to the collision duration or the memory time. So they asked the question: Would it then be possible to reverse or enhance, after it has already started, an interaction process that would become irreversible if the system was left to itself? In order to answer this question, they investigated TI-FWM signal from bulk GaAs in the $2\boldsymbol{q}_2$-$\boldsymbol{q}_1$ direction when the sample was (same as in [9.177]) excited by two phase locked pulses (at times t_1 and $t_{1'}$) along $\boldsymbol{q}_1$ and a single pulse (at time t_2) along $\boldsymbol{q}_2$ (Fig. 9.5). TI-FWM signal was recorded as a function of the time delay $t_{21'}$ (measured from the first phase-locked pulse) for several delays $t_{11'}$ between the phase-

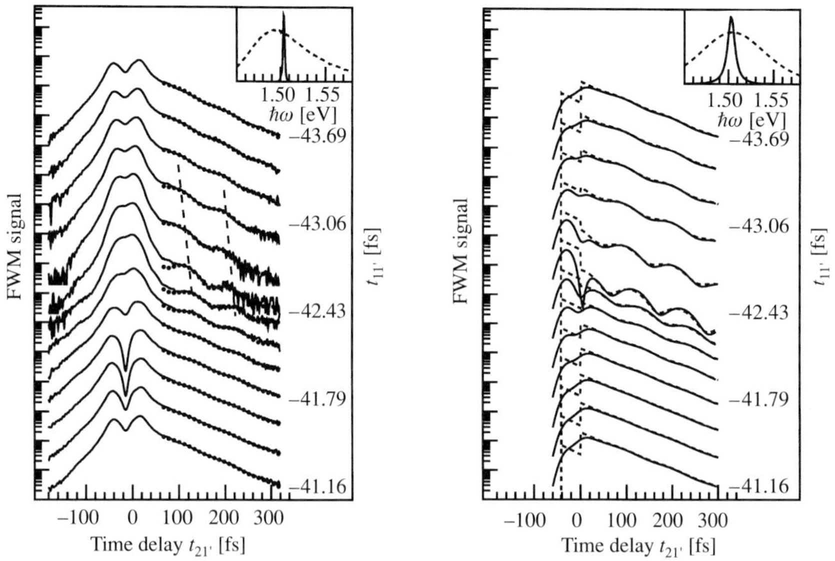

Fig. 9.10. *Left panel*: Experimental TI-FWM signal from GaAs at 77 K as a function of the time delay $t_{21'}$ for various fixed time delays $t_{11'}$ between the phase-locked pulses in steps of 0.21 fs. The curves are displaced vertically for clarity. The excited carrier density (incoherent sum) is 3.6×10^{15} cm^{-3}. The dots are fits from which parameters shown in Fig. 9.11 are determined. *Right panel*: Calculated TI-FWM signal with parameters corresponding to the experiment. The *dashed lines* are the linear response [9.181]

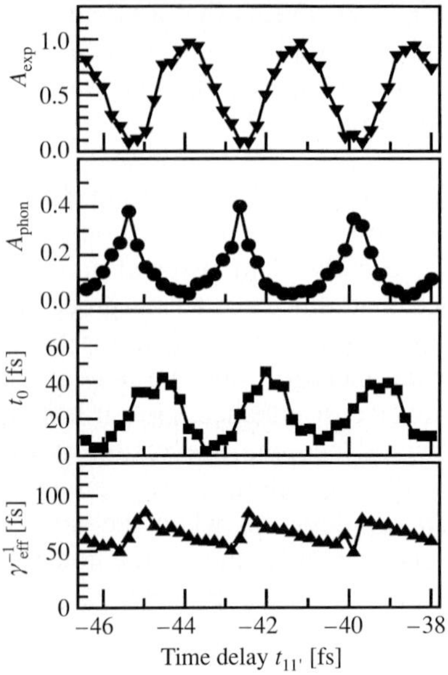

Fig. 9.11. Parameters of (9.4) extracted from the fit to the data in Fig. 9.10 as a function of the delay $t_{11'}$ between the phase-locked pulses [9.181]

locked pulses. The delay $t_{11'}$ could be controlled with attosecond (10^{-18} s) precision [9.182, 183].

Figure 9.10 (left panel) depicts the measured TI-FWM as a function of $t_{21'}$ for various fixed delays $t_{11'}$ in steps of 0.21 fs. They noted several remarkable features: (i) the non-Markovian phonon oscillations [9.177] discussed in Sect. 9.2.1 a almost disappear at certain $t_{11'}$ (e.g., –43.69 or –41.16 fs), (ii) pronounced phonon oscillations occur at certain $t_{11'}$ (e.g., –42.43 fs) at low signal levels with a more pronounced initial decay, and (iii) a more usual time profile with phonon oscillations can be seen at certain $t_{11'}$ (e.g., –43.06 and –41.79 fs). These features of the data were quantified by fitting the TI-FWM signal (temporally and spectrally integrated FWM signal) at time delay $t_{21'}$ to the expression, see (2.5),

$$I_{221}^{(3)} \cdot (t_{21'}) \propto A_{\text{exp}} \left\{1 + A_{\text{phon}} \cos\left[\omega_{\text{osc}} \left(t_{21'} - t_0\right)\right]\right\} \exp\left(-\gamma_{\text{eff}}\, t_{21'}\right), \quad (9.4)$$

where ω_{osc} was fixed at $2\pi/95$ fs^{-1}, close to the value deduced earlier [9.177]. Various parameters of this expression are deduced by fitting the data (shown as dots in the left panel of Fig. 9.10) and are plotted in Fig. 9.11. These results reveal clear oscillations as a function of the delay $t_{11'}$, or the phase between the two phase-locked pulses. The period of the oscillations equals the inverse of the optical field frequency ($h\nu = 1.503$ eV; $\nu = 3.63 \times 10^{14}$ s^{-1}; $1/\nu \approx 2.75$ fs). The remarkable (nonlinear) oscillations of the phonon amplitude A_{phon} in Fig. 9.11 shows that the amplitude of the phonon is indeed controlled by the relative phase between the two phase-locked pulses. The overall behavior was found to be independent of the carrier density between 8×10^{14} cm^{-3} to 3.6×10^{16} cm^{-3}.

Wehner et al. [9.181] have calculated the FWM response of GaAs using a model Hamiltonian with parameters appropriate to the experiments. The results of their calculations, depicted in the right panel of Fig. 9.10, indicate good agreement with the experimental results shown in the left panel.

(d) Electron Dynamics in the Quantum Kinetic Regime. The femtosecond response of a semiconductor at higher carrier densities when carrier-carrier scattering becomes important is more complicated. *Camescasse* et al. [9.184] have recently investigated ultrafast electron distribution through Coulomb scattering using non-degenerate pump and probe pulses derived from a regeneratively amplified Ti:Sapphire laser. The 130 fs pump pulses centered at 1.589 eV ($\approx$70 meV above the bandgap at 15 K) excited electrons from both the heavy-hole (HH) and the light-hole (LH) valence bands. The 30 fs probe pulses are selected such that they can probe the absorption saturation of the interband transition from the spin-orbit split-off (SO) valence band to the conduction band. Since there are no holes in the SO valence band, the differential absorption is only sensitive to electrons. Having pump and probe at different wavelengths also avoids any contribution from induced-grating coherence effects.

Differential absorption spectra from a 700 nm thick GaAs sample at 15 K are depicted in Fig. 9.12 for several delay times. The excitation density for these measurements was 6×10^{16} cm^{-3}, roughly two orders of magnitude larger than the lowest used in Sect. 9.2.1 a for investigating quantum ki-

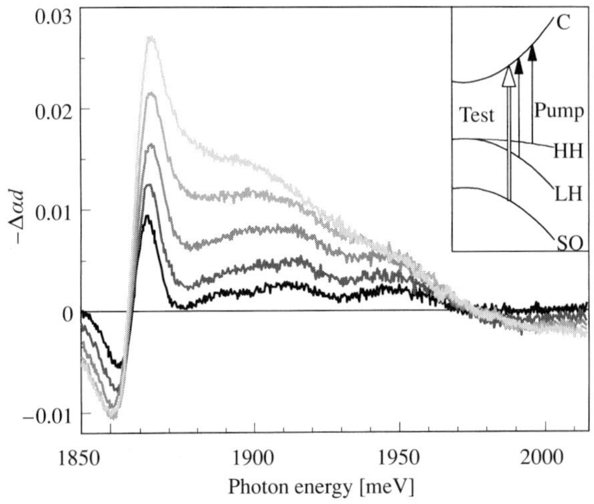

Fig. 9.12. Differential absorption spectra of GaAs at 15 K excited at 1.589 eV a density of 6×10^{16} cm^{-3} and probed near the spin split-off bandgap (see the *inset*). The pump-probe delays are –80, –40, 0, 40, and 80 fs from *bottom* to *top* [9.184]

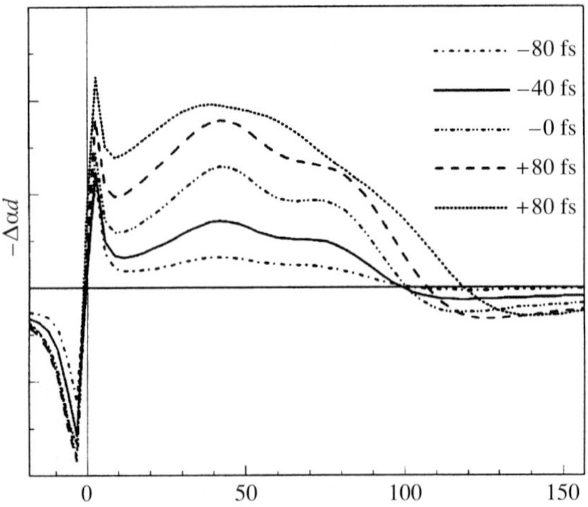

Fig. 9.13. Calculated differential absorption spectra corresponding to the experimental results in Fig. 9.12 [9.184]

netics related to electron-phonon scattering. The results indicate that at early times (<40 fs) the DTS show two peaks due to electrons excited from the HH and LH valence bands. The peaks begin to disappear at a time delay of 40 fs and have completely disappeared at 80 fs. In addition, there is induced absorption at high- and low-energy ends of the spectral range. Another remarkable feature of the data in Fig. 9.12 is that DTS is very broad at the earliest time delay reported.

DTS are affected by several factors other than phase-space filling resulting from excited electrons, as discussed above (Sect. 9.2.1). The large spectral width of the data at earliest times suggests that the data be analyzed in terms of energy non-conserving carrier-carrier scattering in the quantum kinetic model. *Camescasse* et al. have calculated the electron population and the resulting DTS from such a model and their results are depicted in Fig. 9.13. The agreement with the experimental results in Fig. 9.12 is good, especially in predicting that the pump-induced peak in DTS disappears in about 80 fs. This feature of the calculation is due entirely to quantum kinetic effects incorporated in the model. The calculations also reproduce the induced absorption at low and high energies. Although efforts to improve the calculations are continuing, the qualitative agreement shown in Fig. 9.12 clearly shows the influence of quantum kinetics in redistribution of photoexcited electrons.

Differential-transmission measurements in GaAs have also been reported by *Bar-Ad* et al. [9.185] for linearly polarized 30 or 100 fs pump pulses, and cross-linearly polarized 30 fs probe pulses, both centered at the same wavelength. They reported DTS as well as its time derivative DTS′ with respect to the time delay. They also found that the initial DTS is much broader than the photoexcited distribution functions. The DTS and DTS′ data can not be consistently explained in terms of the relaxation rate approximation of SBE. *Bar-Ad* et al. concluded that quantum kinetics must play an important role in carrier-carrier interaction at very short times [9.185].

We conclude this section on quantum kinetics by commenting that extremely broad luminescence spectra were reported by *Elsaesser* et al. [9.186] (Sect. 3.1.2 a; Fig. 3.5). The results were discussed in terms of various models of screening within the framework of Boltzmann kinetic approach. It would be interesting to see some calculations of luminescence spectra based on quantum kinetic approach.

9.2.2 Beyond the Semiconductor Bloch Equations

The Semiconductor Bloch Equations (SBE) [9.187] have provided a good framework for understanding most of the results presented in earlier chapters. We have seen in Sect. 9.2.1 that, although many recent experiments require going beyond the relaxation Hamiltonian based on Boltzmann kinetics and considering the effects of quantum kinetics, SBE still provided a good framework for understanding the results.

Coulomb interactions between electrons and holes play a central role in many physical systems. In most cases, the Coulomb interaction can not be treated exactly and various approximation schemes have to be devised. For example, for the case of an electron plasma response, the two-operator dynamics is coupled to four-operator terms [9.187]. To simplify the situation, one usually assumes the mean-field or Random-Phase Approximation (RPA). In RPA, one considers that sums over terms with unequal wavevectors oscillate rapidly and hence can be neglected. Therefore, the response is dominated by terms with equal wavevectors. This allows four-operator terms to be simplified to two-operator terms, and hence a closed-form solution for the two-operator dynamics. Similarly, an underlying assumption for SBE is that the Coulomb interaction between electrons and holes in a semiconductor can be represented by the Hartree-Fock approximation or RPA. This leads to a factorization of the four-operator expectation values into products of two-operator expectation values. These issues have been discussed in detail by, for example, *Haug* and *Koch* [9.187]. For the purposes of our treatment, any theoretical approach that removes the mean-field approximation will be called "going beyond the semiconductor Bloch equations".

Although SBE have been extremely successful in explaining many of the observed phenomena in the coherent response of semiconductors, it has been clear for some time that one needs to go beyond the mean-field or Hartree-Fock approximation of SBE to explain some experimental results. For example, the results on biexcitons in Sect. 2.7.2 were interpreted in terms of a *phenomenological* model with five levels. The SBE formalism was extended in an *ad hoc* manner to include these additional levels. This rather unsatisfactory state of the theory has now been considerably improved by more recent theories which go beyond the SBE formalism. *Muka-*

mel and co-workers [9.2, 188, 189] have developed these ideas in the context of molecular systems. Other extensions of the theory beyond SBE have used the diagrammatic approach [9.190], the correlation approach [9.191] or a refined truncation scheme [9.192]. The truncation scheme, called Dynamics Controlled Truncation (DCT), naturally extends the density matrix approach of SBE. *Kner* et al. [9.193] and *Chemla* [9.3] have developed an effective polarization model that is useful in providing an intuitive picture. We discuss here some recent experiments that show the need for going beyond the SBE formalism.

(a) Magnetically Enhanced Exciton-Exciton Correlations. We discuss in this subsection recent FWM results reported by *Kner* et al. [9.194] on GaAs at 1.6 K at high magnetic fields. A 250 nm thick high-quality sample of GaAs sandwiched between two AlGaAs layers was mounted on a c-axis sapphire substrate. Co-circularly polarized 100 fs pulses from a Ti:Sapphire laser were used in the standard self-diffracted FWM configuration, and TI-FWM, TR-FWM and SR-FWM were measured for magnetic fields up to 12 T. Strain in the thin sample lifted the degeneracy in the valence band so that the low-temperature absorption spectrum of the sample showed both the heavy-hole (HH) and the light-hole (LH) exciton peaks (Sect. 1.1.1 a). The use of co-circularly polarized configuration avoided excitation of the bound state of the biexciton (Sect. 2.7.2).

Figure 9.14 depicts the TI-FWM intensity on a logarithmic scale as a function of the time delay τ_d, for a number of applied magnetic fields perpendicular to the sample for a low excitation density of 5×10^{15} cm^{-3}. The laser center wavelength was tracked with the exciton resonance as it moved with the magnetic field. The curves are displaced vertically for clarity. At B=0, the data show a signal at negative time delays due to local fields and interaction-induced effects (Sect. 2.7.1 a), and HH-LH quantum beats (Sect. 2.6.2) at positive time delays. The peak signal strength increases by about a factor of 3 as the magnetic field is increased to 10 T because the number of free carriers produced by the laser pulse changes as the exciton binding energy changes with applied magnetic field. The most remarkable change,

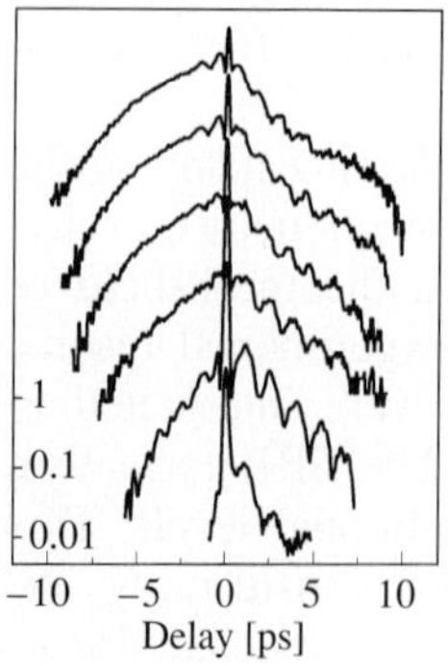

Fig. 9.14. TI-FWM signals from GaAs at 1.6 K for various magnetic fields (0, 2, 4, 6, 8 and 10 T from *bottom* to *top*) for an excitation density of 5×10^{15} cm^{-3}. The laser was tuned to maintain a constant exciton density. The curves are vertically displaced for clarity. The signal increased by a factor of 3 as the field increased from 0 to 10 T [9.194]

however, is the emergence of the negative time-delay signal even for a small magnetic field. This signal gains in strength as the field is increased, extends to $\tau_d=-9$ ps and completely dominates the nonlinear response at 10 T. The signal for $\tau_d>0$ behaves normally: it shows quantum beats with a period corresponding to the HH-LH splitting, and a field-independent decay time constant of 1.5 ps, given by $T_2/2$, where T_2 is the dephasing time of the excitons. In contrast, for low magnetic fields, the signal for $\tau_d<0$ rises nearly exponentially except for one oscillation near –0.5 ps. The time constant of the signal for $\tau_d<0$ is 0.75 ps ($=T_2/4$) for low fields, as expected from earlier discussion (Sect. 2.7.1 a), but steadily increases as B is increased, and the rise of the signal becomes strongly non-exponential at high magnetic fields. These results show that the FWM response is completely altered at low density and high magnetic fields even when bound states of biexcitons are not involved in the nonlinear response.

These results were qualitatively explained as follows: the nonlinear response of a semiconductor near its bandgap is determined by a number of processes such as Pauli blocking, static Coulomb interaction and Coulomb correlation (screening, collisional broadening, exciton-exciton interaction). At low densities, Pauli blocking dominates with some contribution from the Coulomb correlation giving rise to the signal at negative time delays, which increases as $T_2/4$. This response is in agreement with the predictions of the SBE, as discussed in Sect. 2.7.1. However, *Kner* et al. [9.194] argue that the strongly altered nonlinear response observed at low densities and high fields can not be explained on the basis of SBE, and increased Coulomb correlations due to exciton-exciton interactions are responsible for this behavior. Increasing the magnetic field enhances the exciton binding energy and leads to a slight increase in T_2 because fewer free carriers are generated. More importantly, the magnetic confinement increases the importance of Coulomb correlation to the nonlinear response of the semiconductor compared to Pauli blocking, and completely modifies the nonlinear response.

It is well known that increasing the carrier density increases screening, reduces the exciton binding energy and therefore should result in a reduction of the Coulomb correlation effects. Thus, at 10 T increasing the density would be expected to lead to a reversal of the influence of the magnetic field; i.e., the data at 10 T and a higher carrier density would be expected to resemble the data at 0 T and low densities if this model is correct. Density-dependent TI-FWM results [9.194] are depicted in Fig. 9.15. They show that the signal at negative time delays decreases in importance as the excitation density is increased to 1×10^{17} cm^{-3}. This provides further confirmation of the correctness of the model. They further demonstrated (Fig. 9.15) that the spectra at various time delays (i.e., SR-FWM signals) exhibit different relative intensities and different broadening for the HH and LH excitons.

These qualitative arguments were supported by theoretical calculations based on a model that included exciton-exciton correlations by taking into account terms beyond the Hartree-Fock terms in the coherent limit within the $\chi^{(3)}$ truncation scheme [9.192, 195, 196]. For negative time delays, the

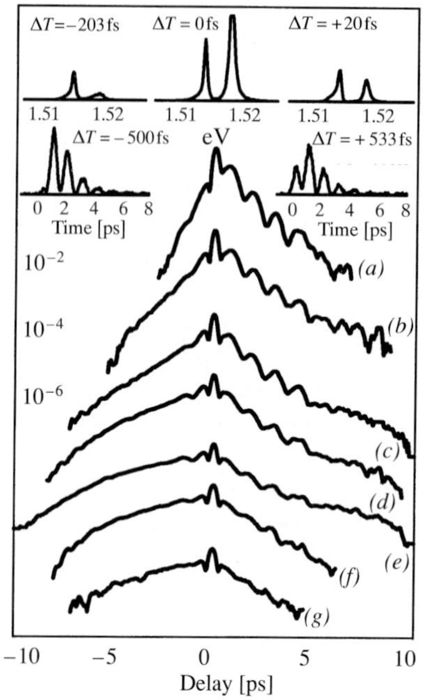

Fig. 9.15. TI-FWM from GaAs at $B=10$ T for different excitation densities *(a)* $N \approx 10^{17}$ cm^{-3}, *(b)* $N/2$, *(c)* $N/3$, *(d)* $N/6$, *(e)* $N/20$, *(f)* $N/63$, and *(g)* $N/200$. The times at the top show SR-FWM and TR-FWM signals [9.194]

calculations show the presence of a strong, slow-rising signal in qualitative agreement with the experimental results. This signal disappears when the calculations are performed in the Hartree-Fock limit. Furthermore, the calculations also predict changes in the spectral linewidths, as observed experimentally (Fig. 9.15). The calculations, however, predict strong beats for $\tau_d<0$ signal, in contrast to the observations. This discrepancy was attributed to the use of the coherent approximation in the calculations.

Kner et al. [9.193] and *Chemla* [9.3] have recently discussed an effective polarization model which provides an intuitive understanding of the additional processes that become important as one goes beyond the mean field approximation. Their approach leads to additional terms in the model developed to account for the local field within the semiconductor Bloch equations [9.197]. The additional terms are related to an effective four-particle correlation function which represents a two-photon bound biexciton transition. Using this model, one can evaluate different contributions to the FWM signals for the case when correlations between different pairs are important. Calculation of the TI-FWM response using the effective polarization model show (Fig. 9.16) the relative importance of different contributions at various positive and negative time delays. The total calculated signal is dominated by the exciton-exciton correlation (XXC) contribution and provides a qualitative agreement with the experimental results exhibited in Figs. 9.14 and 9.15. While this approach is useful in developing an intuitive picture of the physics, *Kner* et al. [9.193] and *Chemla* [9.3] emphasize that numerical sim-

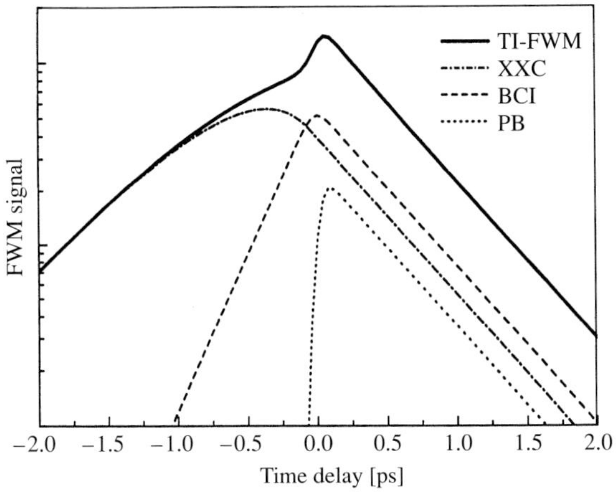

Fig. 9.16. Calculation of the TI-FWM signal in the effective polarization model. PB: Pauli blocking, BCI: bare Coulomb interaction, XXC: exciton-exciton correlation. Note that exciton-exciton correlations give the biggest contribution to the TI-FWM signal at large negative time delays [9.193]

ulations that account for the full complexity of the problem are essential in providing a quantitative understanding of the results.

(b) Importance of Excitonic Contributions to THz Radiation from a Superlattice. As mentioned above, *Axt* and *Stahl* [9.192] have recently proposed a microscopic model incorporating systematic truncation concepts which define the finite set of density matrices that have to be taken into account to describe the coherent response of the semiconductor exactly within a prescribed order of the electromagnetic field. This Dynamics-Controlled Truncation (DCT) scheme provides a systematic approach to going beyond the Hartree-Fock approximation, and hence SBE, in calculating the coherent nonlinear response of semiconductors.

Axt et al. [9.195] applied this approach to the nonlinear response (THz emission) of a superlattice with a narrow miniband (i.e., minibands whose energy width is comparable to the exciton-binding energy) under the influence of an applied electric field. Such superlattices have strong anti-crossing of energy levels due to excitonic effects [9.198, 199] and were investigated earlier using SBE by *Meier* et al. [9.200]. They found remarkable differences in the results predicted by the SBE and DCT. SBE predict that the source term for excitonic THz emission decays with 1/2 the interband dephasing time, whereas DCT predicts that the excitonic THz emission decays with intraband dephasing time. Therefore, for intraband dephasing time (T_2^{intra}) larger than half the interband dephasing time ($T_2^{\text{inter}}/2$), the two approaches give different predictions. SBE predicts that the response would be governed by free carriers whereas DCT predicts that there is a delicate cancellation of the free-carrier contribution, and the excitonic contribution

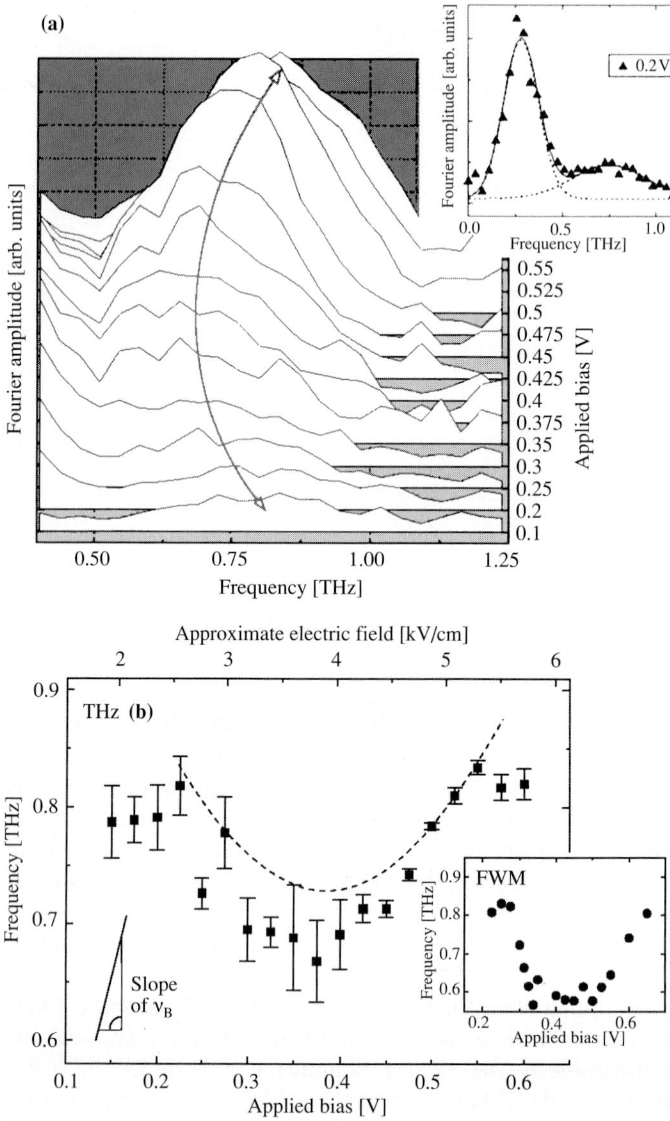

Fig. 9.17. (**a**) Corrected spectra of the high-energy peak of the THz emission (see the *inset*) for different applied bias, (**b**) the variation of the center frequency of the high-energy peak as a function of the applied electric field. The *points* are experimental and the *dashed curve* is a hyperbolic fit to the splitting between the Wannier Stark ladder states HH_{-2} and HH_{-1} [9.202]

dominates. If the excitonic contributions dominate (DCT model) then the strong anti-crossing behavior of exciton energies [9.198, 201] would predict that the THz emission frequency would be non-monotonic function of the electric field. SBE, on the other hand, predicts domination by free carriers so that no anti-crossing is expected and the THz emission frequency would depend linearly on the electric field, as expected for free-carrier Wannier Stark states.

Haring-Bolivar et al. [9.202] investigated linear and nonlinear optical properties of a 35 period GaAs/$Al_{0.35}Ga_{0.65}As$ superlattice with 135 Å wells and 17 Å barriers at 10 K. The electron miniband width of 7 meV is comparable to the exciton binding energy. Transmission spectra clearly showed the anti-crossing behavior expected for the excitons in a certain field range. In order to measure the interband dynamics and to analyze and control the excitation conditions, TI-FWM and linear transmission measurements were performed for different applied electric fields. This allowed them to adjust the excitation spectrum precisely in the anti-crossing region, to determine that $T_2^{\text{inter}}/2 = 0.4$ ps under the conditions of their THz experiments, and to calibrate the electric field in the sample. The THz transients have distinct low- and high-frequency components. From the bandwidth of the THz spectrum, T_2^{intra} was deduced to be 1.2 ps. Since this value is about a factor of three larger than $T_2^{\text{inter}}/2$ derived from the FWM measurements, it was argued that the requirements for expecting different behavior for SBE and DCT, as discussed in the previous paragraph, were satisfied.

Figure 9.17 a depicts the high-frequency part of the THz emission spectrum (see the inset), obtained by Fourier transforming the transient emission waveform, at various applied electric fields. It is seen that the center frequency varies non-monotonically, and goes through a minimum, as the applied field increases. These center frequencies are plotted in Fig. 9.17 b as a function of the applied electric field, and compared with the splitting derived from the linear transmission measurements (dotted curve). The inset in Fig. 9.17 b indicates the frequencies extracted from the FWM measurements. These results show that the THz emission frequencies do indeed vary as the excitonic frequencies in the anti-crossing region, as determined by linear transmission and FWM measurements. This result is in accord with the prediction of DCT and in contradiction to the prediction of SBE. These observations led *Haring-Bolivar* et al. to conclude that exciton correlation effects make an important contribution to the THz experiments and one needs to go beyond SBE to provide an adequate theoretical understanding of the results.

9.2.3 Coherent Control in Semiconductors

We have discussed in Sect. 2.12.4 the concept of coherent control of different properties of a system by using a pair of phase-locked pulses to excite the system. We have also discussed the first application of these concepts to semiconductors in which the THz emission from a system of coupled quantum wells was manipulated (enhanced or diminished) by varying the phase delay between the two phase-locked pulses exciting the coupled quantum-well system [9.203] (Sect. 2.12.4). More complex manipulation of the properties of a system can be achieved by using a more complex waveform for the femtosecond pulses generated by using the techniques [2.334–345] already mentioned in Sect. 2.12.4.

A number of new developments have occurred in this field since these early experiments. Recent experiments have demonstrated the beauty and the power of coherent control techniques in manipulating a number of important properties of semiconductors. This active field of research also includes a number of theoretical [9.68, 69, 204–206] investigations. We restrict our discussion in this section to some recent experimental results.

(a) Coherent Control of Exciton Density. The coherent control of exciton populations, polarization and other properties (e.g., spin) have been demonstrated in an elegant series of experiments by *Heberle* et al. [9.207, 208] and *Baumberg* et al. [9.209]. In the first of these experiments, *Heberle* et al. [9.208] investigated reflectivity and TI-FWM from a $GaAs/Al_{0.3}Ga_{0.7}As$ sample containing five 120 Å quantum wells separated by 120 Å $Al_{0.3}Ga_{0.7}As$ barriers. The sample at 4 K was resonantly excited by a pair of phase-locked 150 fs pulses from a Ti:Sapphire laser. An actively-stabilized Michelson interferometer allowed precise control of the time delay τ_{12} between the pulses to within a fraction of the optical period (the optical period T_{HH} corresponding to the HH exciton transition energy E_{HH} is $h/E_{HH}=2.68$ fs

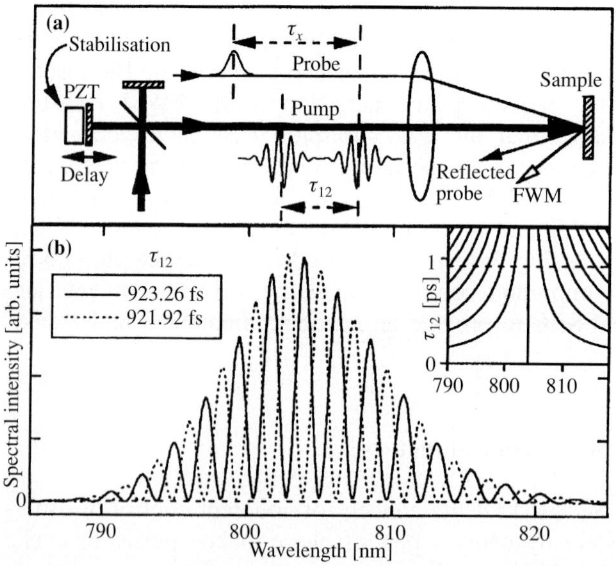

Fig. 9.18. (**a**) Experimental setup, (**b**) spectrum of a pair of phase-locked pulses separated by nT_L and $(n+1/2)T_L$ where n is an integer (344 in this case) and T_L is the period of the optical cycle at the center of the laser frequency. The *inset* shows calculated wavelengths of the peaks of the fringes as a function of the delay τ_{12} between the pulses [9.207]

for this sample). The differential reflectivity of the sample at the time delay τ_x following the first of the phase-locked pulses is monitored as a function of the τ_{12} at a fixed τ_x. A schematic of the experimental arrangement for reflectivity and FWM studies is exhibited in Fig. 9.18.

Reflectivity and PL data in Fig. 9.19 a show a linewidth of 0.7 meV for the lowest HH exciton. In the coherent control experiment depicted in Fig. 9.19 b, the reflectivity is measured at τ_x of 10 ps (a time delay long enough that the reflectivity monitors the population of excitons) as a function of τ_{12}. The two phase-locked pulses are co-linearly polarized and the probe is cross-linearly polarized. The measured reflectivity shows an oscillatory behavior as a function of τ_{12}, with the period of the oscillations corresponding to the optical cycle, as depicted in the insets of Fig. 9.19 b for three different τ_{12}. The maxima and minima of these oscillations are plotted in Fig. 9.19 b as a function of τ_{12}. The oscillations diminish in amplitude as τ_{12} increases. But clear oscillations persist for time delays much longer than

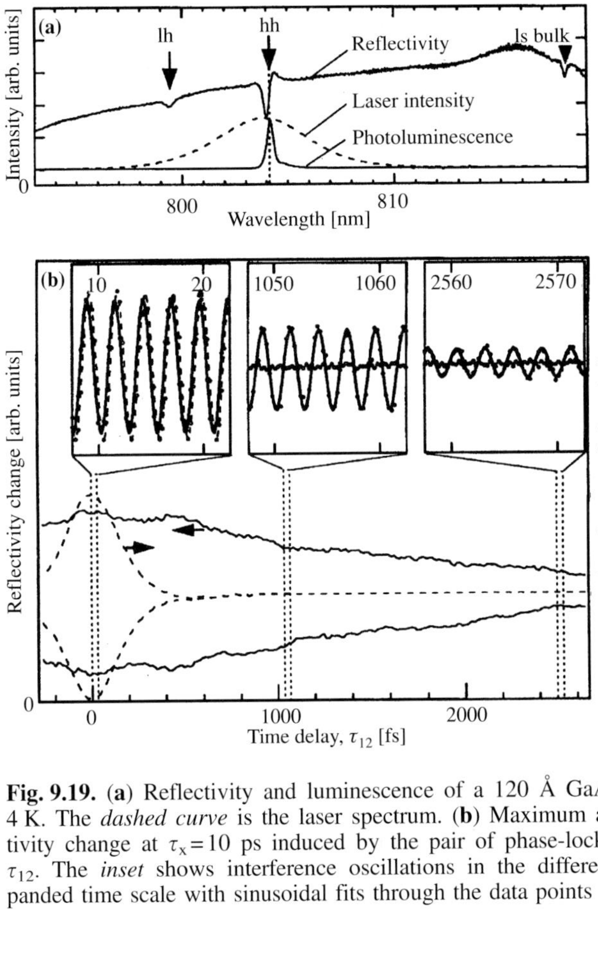

Fig. 9.19. (**a**) Reflectivity and luminescence of a 120 Å GaAs quantum well sample at 4 K. The *dashed curve* is the laser spectrum. (**b**) Maximum and minimum of the reflectivity change at $\tau_x = 10$ ps induced by the pair of phase-locked pulses as a function of τ_{12}. The *inset* shows interference oscillations in the differential reflectivity on an expanded time scale with sinusoidal fits through the data points [9.207]

the widths of the exciting pulses, as evident by comparison with the interferogram of the pump pulse (dashed curve in Fig. 9.19 b). The observation of these oscillations is clear evidence that the incoherent population of excitons is either enhanced or reduced by varying the phase between the two pulses. This coherent control is the result of either constructive or destructive interference between the second pulse and the polarization excited by the first pulse that remains at the time of the arrival of the second pulse. Therefore, the interference or the oscillation amplitude diminishes as the polarization created by the first pulse decays with increasing delay τ_{12}.

An alternative, instructive way of looking at this interference is in the spectral domain. Figure 9.18 b exhibits the spectra of the pair of phase-locked pulses for two time delays τ_{12} separated by half an optical cycle. The spectral fringes are out of phase as expected. The spacing between the spectral fringes decreases as the delay between the pulses increases, as shown in the inset of Fig. 9.18 b. Therefore, for small delays between the pulses, the period of the spectral fringes is larger than the linewidth of the HH exciton. For such delays, the peaks or the troughs of the spectrum are positioned on top of the exciton absorption peak and produces maximum constructive or destructive interference and gives the maximum contrast in the fringes. As the delay between the pulses increases, the fringe period will diminish so that more fringes are likely to excite the HH exciton. Conse-

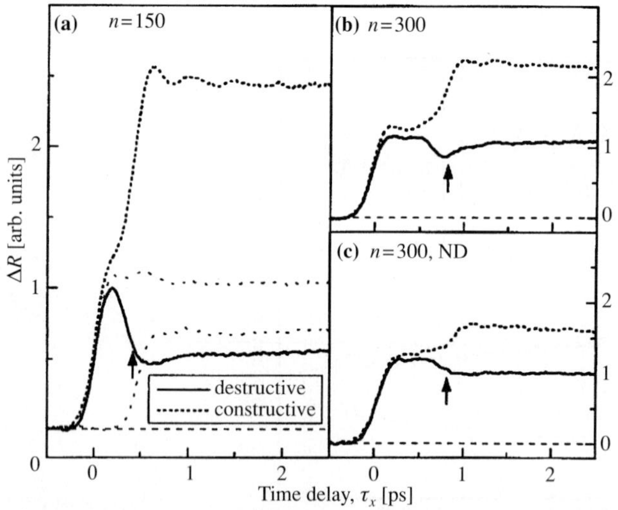

Fig. 9.20 a–c. Time-resolved enhancement (*dashed*, $\tau_{12}=nT_{HH}$) and destruction (*solid*, $\tau_{12}=(n+1/2)\,T_{HH}$ of the reflectivity induced by phase-locked pulses. Zero of time corresponds to the overlap of the gating pulse with the first of the phase-locked pulses. T_{HH} is the period of the optical cycle at the HH exciton transition energy. The *arrows* indicate the arrival of the second pulse. The *dotted lines* in (**a**) record the effects of each pump pulse alone. The second pump beam is attenuated by 60% using a Neutral Density (ND) filter in (**c**) [9.207]

quently, the contrast between constructive and destructive interference diminishes, leading to a reduction in the fringe contrast observed in Fig. 9.19 b. It is obvious from the spectral picture that it is possible to select the delay at which the spectrum peaks at the HH exciton but has a minimum at the LH exciton. Some interesting effects of this kind were reported by *Heberle* et al. [9.207] in the TI-FWM investigations.

Heberle et al. [9.208] also investigated the reflectivity as a function of τ_x for two fixed τ_{12} differing by half an optical cycle (1.34 fs). The results, shown in Fig. 9.20, are similar to the results for coherent control of the intensity of THz emission reported by *Planken* et al. [2.347] (Sect. 2.12.4). The dotted curves in Fig. 9.20 a display the change in reflectivity induced by each pulse individually. The solid curve shows the result when the sample is excited by two phase-locked pulses separated by $\tau_{12}=(n+1/2)_{\mathrm{THH}}$ where n is an integer. The reduction in the reflectivity at the second pulse exhibit that the arrival of the second pulse reduces the exciton population, a result of destructive interference. In contrast, at $\tau_{12}=nT_{\mathrm{HH}}$, the reflectivity increases when the second pulse arrives, indicating an increase in the exciton population due to constructive interference. Figures 9.20 b and c depict the results for $n=300$, corresponding to $\tau_{12}=804$ fs. The difference in the exciton populations for constructive and destructive interference diminishes compared to Fig. 9.20 a because the polarization has had more time to relax. The results in Fig. 9.20 b can be made more symmetric by decreasing the intensity of the second pulse to match the decreased polarization remaining at longer delays (Fig. 9.20 c).

Heberle et al. reported that they were unable to achieve a larger than 70% destruction in the population of the excitons in spite of extensive experimentation with various parameters such as density, temperature and polarization. Reasons for this incomplete destruction were not clear.

(b) Coherent Control of Exciton Spin Orientation: Faraday Rotation Studies. The experiments discussed in Sect. 9.2.3 a used co-linear polarization for the two phase-locked pulses to control the exciton density in semiconductors. Two co-circular phase-locked pulses gave similar results [9.207]. Using cross-linear or cross-circular polarization for the two phase-locked pulses allows coherent control of different properties of semiconductors. We recall (Sect. 6.1.2) that in a III–V semiconductor like GaAs, σ^+ light (right circular polarization or positive helicity) couples $m_{\mathrm{j}}=-3/2$ state of the heavy-hole (HH) to $m_{\mathrm{j}}=-1/2$ state of the electron to form $|1\rangle$ excitons. Similarly, σ^- light (left circular polarization or negative helicity) couples $m_{\mathrm{j}}=3/2$ state of the heavy-hole (HH) to $m_{\mathrm{j}}=1/2$ state of the electron to form $|-1\rangle$ excitons. These are the only exciton states that couple to light. Different combinations of these states are excited depending on the polarization state of the exciting light. This provides an opportunity to coherently control the fraction of exciton polarization in each state *without affecting the total density* of excitons. The dark states of the excitons ($|2\rangle$ and $|-2\rangle$) can be ignored for this discussion since the exciton spin relaxation times for conversion between the optically allowed states and the dark states is known to be much longer than the typical exciton dephasing time (see, for example, [9.210, 211]). We also recall here that optical orientation and alignment of exciton and free-carrier spin have been investigated extensively in the incoherent regime, and have provided a very effective means for studying various physical processes in semiconductors [9.212].

We consider here the case of two cross-linear phase-locked pulses considered by *Heberle* et al. [9.207]. Coherent control of exciton orientation using cross-circular polarization has been investigated by *Marie* et al. [9.213] and will be discussed in Sect. 9.2.3 c. Cross-linear pulses do not interfere with each other even if they overlap in time. However, the polarization resonantly excited in the sample by the first pulse oscillates with the period T_{HH} and can interfere with the electric field of the second pulse. In the notation of *Heberle* et al., the delay between the two phase-locked pulses $\tau_{11'}$ in the notation of Sect. 9.1.3 b is denoted by τ_{12}, and τ_2, the delay of the probe pulse from the first phase-locked pulse is denoted by τ_x. The interference between the electric field of the second pulse with the remaining polarization from the first pulse produces, in general, elliptical polarization whose ellipticity depends on the relative phase of the two pulses determined by τ_{12}. If the dephasing time is much longer than τ_{12} and n is an integer, the total polarization after the arrival of the second pulse is linear for $\tau_{12}=nT_{HH}$ and $(n+1/2)\,T_{HH}$, and circular for $\tau_{12}=(n\pm1/4)\,T_{HH}$. If dephasing occurs before the arrival of the second pulse, the unequal amplitudes of the two polarizations lead to elliptical polarizations with different ellipticities as a function of τ_{12}. This interference produces excitons of different angular momenta or different spin orientation, and can be probed by measuring the

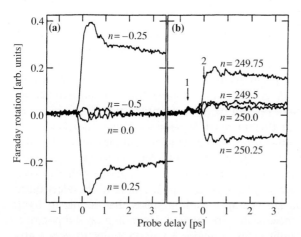

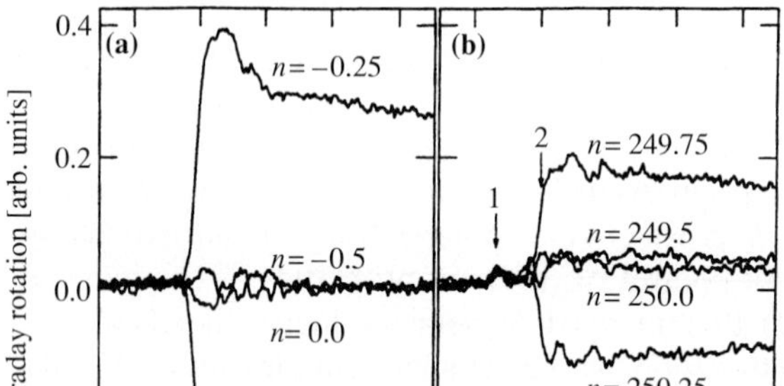

Fig. 9.21. Coherent control of the exciton spin, as evidenced by Faraday rotation as a function of the probe delay τ_x for cross-linearly polarized phase-locked pulses with different delays $\tau_{12}=nT_{HH}$. The pump pulses arrive simultaneously in (**a**) and are separated as indicated by the arrows in (**b**) [9.207]

Faraday rotation of a reflected probe pulse, which is sensitive to the different reflectivities of polarized light of different helicities (right or left circularly polarized light).

Measurement of the Faraday rotation of a probe pulse delayed by τ_x from the first pulse were made for various delays by *Heberle* et al. [9.207]. Their results are displayed in Fig. 9.21 for τ_{12} close to zero in (a) and for $\tau_{12} \approx 670$ fs in (b). In (a) the pulses overlap in time so that they produce linear or circular polarizations in the sample depending on their phase difference. Figure 9.21 b shows that, even when the two phase-locked pulses are delayed, $\tau_{12}=(n+1/4)\,T_{HH}$ produces a negative Faraday rotation ("destructive interference") whereas $\tau_{12}=(n+3/4)\,T_{HH}$ produces a positive Faraday rotation ("constructive interference") i.e., the interference can be controlled by controlling the phase between the two cross-linearly polarized pulses.

The coherent control of the exciton density and exciton spins can be modeled using the semiconductor Bloch equations. Such calculations have been shown to provide good agreement with the experimental data.

(c) Coherent Control of Exciton Spin Orientation: Resonant Emission Studies. Coherent control of the exciton spin orientation was recently demonstrated by *Marie* et al. [9.213] using the technique of time-resolved resonant emission [9.214]. These studies are discussed in this subsection. Earlier studies of femtosecond resonant emission dynamics of excitons are treated in Sect. 6.3.3, and other new developments in resonant secondary emission from quantum-well excitons will be described in Sect. 9.2.11.

Marie et al. [9.213] investigated a $GaAs/Al_{0.6}Ga_{0.4}As$ MQW sample with 30 periods, 100 Å wells and 200 Å barriers. The sample was maintained at 10 K and excited at an exciton density $\approx 10^9$ cm^{-2} by a pair of 1.4 ps pulses derived from a Ti:Sapphire laser tuned to the HH exciton resonance. The pulses were phase-locked using a Mach-Zender-type interferometer. The use of 1.5 ps pulses allowed *Marie* et al. to selectively excite the HH exciton and minimize excitation of LH excitons and free carriers. The resonant emission from the sample was upconverted in a $LiIO_3$ nonlinear crystal (Sect. 1.3.3 c) by using gating pulses from an OPO synchronously pumped by the same Ti:Sapphire laser that was used for exciting the sample. Different polarizations of the phase-locked pulses, and different time delays and phase differences between the phase-locked pulses were investigated.

Figure 9.22 displays the results for the case when the two phase-locked pulses have opposite helicities. In the notation of these researchers, the total delay between the phase-locked pulses is written as $\tau_{12}=t_1+t_2$ where t_1 is an integral multiple of the period T_{HH} ($t_1=n\,T_{HH}$). t_2 is a fine adjustment of the delay which can be controlled within $T_{HH}/20$. Figure 9.22 a displays the measured PL intensities I^X and I^Y for the X and Y linear polarizations as a function of gating delay τ_G measured from the first phase-locked pulse for the case of $t_2=m\,T_{HH}$ (m is zero or a small integer). Before the arrival of the second pulse, the PL excited by the σ^+ light has the same intensity ($I^X=I^Y$) for the X and the Y linear polarizations. Therefore, the linear polari-

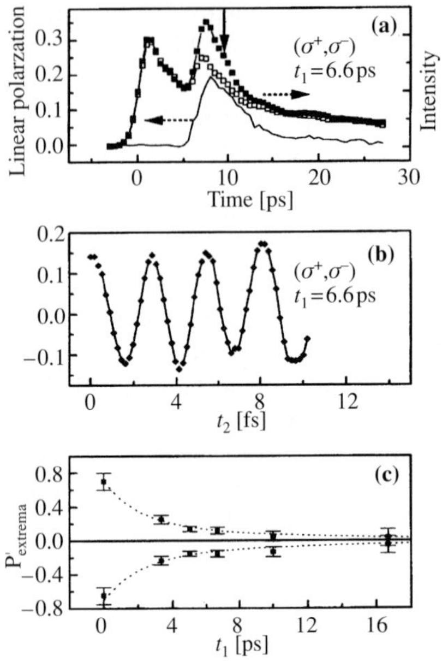

Fig. 9.22. Coherent control of the exciton orientation, as evidenced by linear polarization: (**a**) The temporal evolution of the intensity (I^X: *closed squares*; I^Y: *open squares*) and linear polarization P^l (*solid curve*) of excitonic emission resonantly excited by a pair of phase-locked pulses of opposite helicity for $t_1 = 6.6$ ps and $t_2 = mT_{HH}$ where $m = 0$ or a small integer. (**b**) Variation of the linear polarization as a function of t_2 for $t_1 = 6.6$ ps and τ_G set at 4 ps after the second pulse. (**c**) The maxima and the minima of the oscillations in the linear polarization P^l as a function of t_1 [9.213]

zation ($P^l = (I^X – I^Y/I^X + I^Y$) is zero. After the arrival of the second phase-locked pulse, the intensity of X polarized PL is enhanced whereas that of the Y polarized PL is reduced (X and Y orientation is immaterial in this case), leading to a non-zero P^l. These results can once again be understood in terms of the interference between the second phase-locked pulse and the polarization remaining in the sample after excitation by the first pulse. This is confirmed by varying the fine delay t_2 for a fixed t_1 (=6.6 ps), as depicted in Fig. 9.22 b. These results exhibit that the linear polarization P^l of the emitted PL intensity shows an oscillatory behavior with the period T_{HH}. This reflects the rotation of the orientation of "linear" excitons (i.e., $(|1\rangle \pm |-1\rangle)/\sqrt{2}$) in the plane of the quantum wells.

Figure 9.22 c depicts the variation of the extrema of the linear polarization of the PL intensity [i.e., the "amplitude" of the fringes in (b)] as a function of the time delay t_1. This amplitude is large at small t_1 and decreases at larger t_1, indicating that the coherent exciton polarization created by the first phase-locked pulse decays with time. The dephasing time T_2 of excitons (Sect. 2.2), as determined from the decay time constant in Fig. 9.22 c was 6±1 ps. This time was much longer than that corresponding to the inverse inhomogeneous linewidth of the exciton.

These experiments elegantly demonstrate the coherent control of exciton spin orientation in quantum wells. *Marie* et al. also investigated coherent control of excitons using phase-locked pulses with linear polarizations using time-resolved luminescence techniques. These results should be compared with those by *Heberle* et al. discussed in Sects. 9.2.3 a and b.

(d) Coherent Control of the Current. Understanding and controlling the magnitude and the direction of the electrical current in semiconductors is of considerable fundamental and potentially technological interest. We discuss in this section recent experiments which show that the magnitude and the direction of the current in semiconductors can be controlled by appropriate optical excitation [9.215–217]. The principle behind coherent current control in semiconductors is to excite the same final state in the continuum of a semiconductor by different optical routes. If the phase is preserved in each quantum pathway, strong interference in the final-state wavefunction is expected, as in the case of Fano resonances [9.218], see also Sect. 9.2.5 a. The initial state can be discrete [9.215] or a part of the continuum [9.217]. In both these experiments the same final states were excited by either a one-photon absorption at 2ω or two-photon absorption at ω. These transitions connect initial states to degenerate final states of different symmetry. Therefore, adjusting the relative phase of the two beams alters the combination of final states created, and leads to a current if the selected state does not have a fixed parity.

The first demonstration of coherent current control in a semiconductor was reported by *Dupont* et al. [9.215] who investigated a 55 Å δ-doped GaAs quantum-well sample at 82 K. They excited the donor state in the quantum well with two-photon absorption of 100 ns pulses from a 10.6 μm laser and with one-photon absorption using the frequency-doubled output at 5.3 μm. The two quantum mechanical pathways coupled to degenerate symmetric and anti-symmetric states in the continuum. The resulting photocurrent exhibited well-defined oscillations as a function of phase difference between the two beams. The current changed from a positive value at the maximum to a negative value at the minimum, demonstrating that the *direction* of the current was determined by the relative phase between the two beams.

In a more recent experiment, *Hache* et al. [9.217] excited interband transitions in bulk GaAs at 300 K using two-photon absorption of femtosecond pulses at 1.55 μm and one-photon absorption of its second harmonic at 0.755 μm. These lasers induce interband transitions in the semiconductor, i.e., they couple free-carrier continuum states in the valence and conduction bands of the semiconductor. In this case, the interference between one-photon and two-photon transitions coupling the same states results in an anisotropic distribution of carriers which results in a net current flow whose magnitude and direction can be controlled by the relative phase between the two beams.

Hache et al. investigated both low-temperature GaAs and normal GaAs. We discuss here the results on LT-GaAs because its carrier lifetime of 1 ps allowed measurements at lower densities using a high-repetition-rate Optical Parametric Oscillator (OPO). Figure 9.23 depicts the experimental schematic. The fundamental and frequency-doubled lasers were focused on the sample after a double-pass through a glass plate. The rotation of the glass plate introduced a relative phase delay between the two lasers. 200 μm×250 μm gold

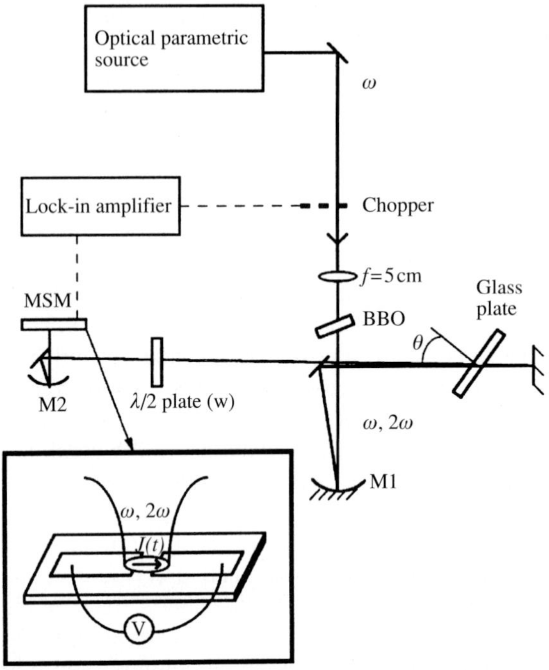

Fig. 9.23. Schematic diagram of the experimental setup. BBO is β-barium borate, the frequency-doubling crystal, M1 and M2 are curved gold mirrors with focal lengths of 10 cm and 2.5 cm, respectively [9.217]

electrodes of thickness 170 nm were deposited on the sample with gaps ranging from 5 to 50 μm. The lasers were focused at the center of the gap of these MSM structures using curved mirrors and the resulting photocurrent was measured using a lock-in amplifier. The OPO produced peak irradiance of 18×10^6 $\mathrm{W\cdot cm^{-2}}$ (3×10^6 $\mathrm{W\cdot cm^{-2}}$) and an excitation density $\approx 10^{13}$ $\mathrm{cm^{-3}}$ ($\approx 10^{14}$ $\mathrm{cm^{-3}}$) for ω (2 ω) lasers.

The total generation rate for the current density is a tensor, in general [9.216]. For the case when both lasers are linearly polarized parallel to the electrodes, the rate of generation of photocurrent is proportional to $\sin(2\phi_\omega - \phi_{2\omega})$, where ϕ_ω and $\phi_{2\omega}$ are the phases of the ω and 2ω lasers, respectively. The rate of dissipation is determined by the relaxation rates for electrons and holes (≈ 100 fs). Therefore, the measured current is expected to show oscillatory behavior as the glass plate in the path of the lasers is rotated. Knowing the indices of refraction and the thickness of the glass plate, the phase change can be calculated as a function of the angular rotation of the plate. Thus the measured current can be plotted as a function of the phase difference between the two frequencies. This is depicted in Fig. 9.24 for a 25 μm gap LT-GaAs MSM structure. The current oscillates periodically between positive and negative values showing that the direction of the current is determined by the relative phase. These results demonstrate that

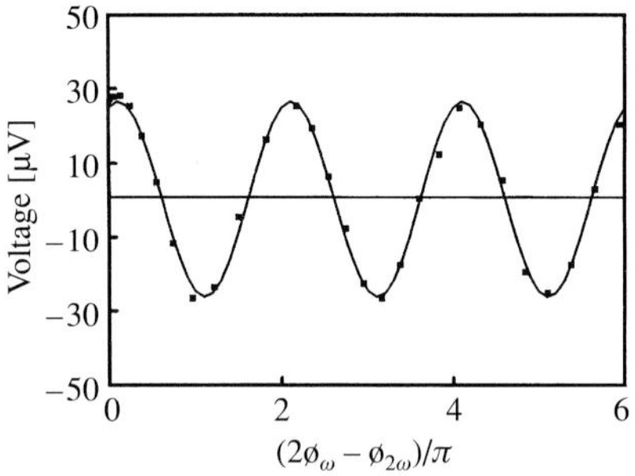

Fig. 9.24. Modulated component of the induced voltage on a Metal-Semiconductor-Metal (MSM) detector in the presence of ω and 2ω laser radiation as a function of $2\phi_\omega - \phi_{2\omega}$. The *solid curve* is a best fit for a sine function. The oscillations are a signature of the coherent control of the current [9.217]

the direction and the magnitude of the current can indeed be controlled coherently in a semiconductor at room temperature.

9.2.4 Phase Sensitive Measurements

Most of the measurements discussed in Chap. 2 on coherent spectroscopy investigated the *intensity* of the coherent nonlinear signals. Some notable exceptions were treated in Sect. 2.6.5 b and Sect. 2.7.1 c. Measurements of the *amplitude and phase* of the coherent nonlinear signals provide additional insights into the physical processes that influence the nonlinear response of the semiconductor. The use of phase-locked pulses to coherently control various processes is discussed in Sect. 9.2.3. We describe in this section some of the many investigations [9.219–222] of the electric field of the nonlinear signal emitted by a semiconductor when excited by an ultrashort pulse. Investigation of the amplitude and phase of femtosecond resonant emission from excitons in quantum wells is treated in Sect. 9.3.3, and of the THz electric field emitted by accelerating charge carriers in Sect. 9.4.6.

We recall that many of the amplitude and phase measurement techniques, such as those in Sect. 2.7.1 c, require the use of a nonlinear process, and hence are not sensitive enough to measure very weak signals. However, as discussed in Sect. 9.1.3 a, recent work has demonstrated that an almost arbitrarily weak signal can be determined indirectly through Fourier transform by the technique of femtosecond spectral interferometry utilizing a well-characterized reference pulse [9.160, 165–170, 223]. We illustrate the use of these techniques by discussing how they have been applied to obtain a deeper understanding of the polarization properties of FWM signals briefly outlined in Sect. 2.6.2 and Sect. 2.7.2 d.

Early in the study of HH-LH exciton quantum beats in quantum wells, *Schmitt-Rink* et al. [9.224] predicted that for the case of linearly polarized incident beams (angle θ_{12} between the two linear polarizations), the TI-FWM signal, for laser-pulse spectrum overlapping the HH and LH excitons, would be elliptically polarized, in general. The TI-FWM signal is linearly polarized only when θ_{12}=0 or 90°. They showed that while some of the predictions of this model were verified by experiments (Sect. 2.6.2, Fig. 2.20), the experiments revealed many unexpected results that were investigated later in considerable detail [9.225, 226] (see also Sect. 2.7.2 d). These later studies showed that the predictions of the optical Bloch equations were inadequate, and that semiconductor Bloch equations augmented with excitation-induced dephasing and biexcitonic effects were required to explain the complex set of results obtained as a function of various parameters.

Smirl and collaborators [9.219, 221, 222, 227–229] have extensively investigated this problem over the last few years using TR-FWM and spectral interferometry. They have argued that, in order to analyze the problem, one needs to know not just the intensity of one component of TI-FWM but a complete characterization of the temporal behavior of the degree and the state of polarization of the FWM signal. They have investigated the complete third-order nonlinear response of a GaAs quantum well employing the technique of ultrafast ellipsometry (in which TR-FWM is measured as a function of various combinations of incident and signal polarizations) and a dual-channel version of femtosecond spectral interferometry [9.221, 222] (Sect. 9.1.3 a). The former technique gives direct results, but it is time-consuming and requires relatively high signal levels, whereas the latter is extremely sensitive but requires considerable amount of data analysis [9.170].

We discuss here some recent results presented by *Buccafusca* et al. [9.222] who investigated the FWM signals from a multiple quantum well structure consisting of 10 periods of 140 Å GaAs quantum wells with 170 Å $Al_{0.3}Ga_{0.7}As$ barriers by means of the technique of dual-channel femtosecond spectral interferometry. The measurements were performed at 80 K where the HH exciton line is homogeneously broadened (FWHM ≈ 1.8 meV) and the HH-LH exciton splitting is ≈ 7 meV. The excitation pulse width was 150 fs, the spectral width was 15 meV, and the central energy was tuned 7 meV below the HH exciton. The excitation density was 2×10^{10} cm^{-2} and the delay between the two incident beams was fixed at 300 fs. Figure 9.25 exhibits the experimental setup for measuring the spectral interferograms for the X and Y polarizations. Information about the amplitude and the phase for each polarization could be obtained from such studies, thus completely characterizing the nonlinear response. Figure 9.26 depicts the spectra of the real and imaginary amplitudes of the electric fields in the x and y directions deduced from their measurements with Fourier transforms. These results were obtained for θ_{12}, the angle between the polarization vectors of the two incident pulses, set at 60°. Measurements at various θ_{12} were reported elsewhere [9.221]. These results contain a large amount of information. For example, the spectral width of the y-field is larger than that of the x-field, and the y-field is shifted to lower energy by about 1.5 meV, approximately the binding energy of the biexciton in this sample. The behavior of the imaginary parts of the x and y fields are totally different. *Buccafusca* et al. have demonstrated that all this information is important to obtain a correct picture of the nonlinear response of the system.

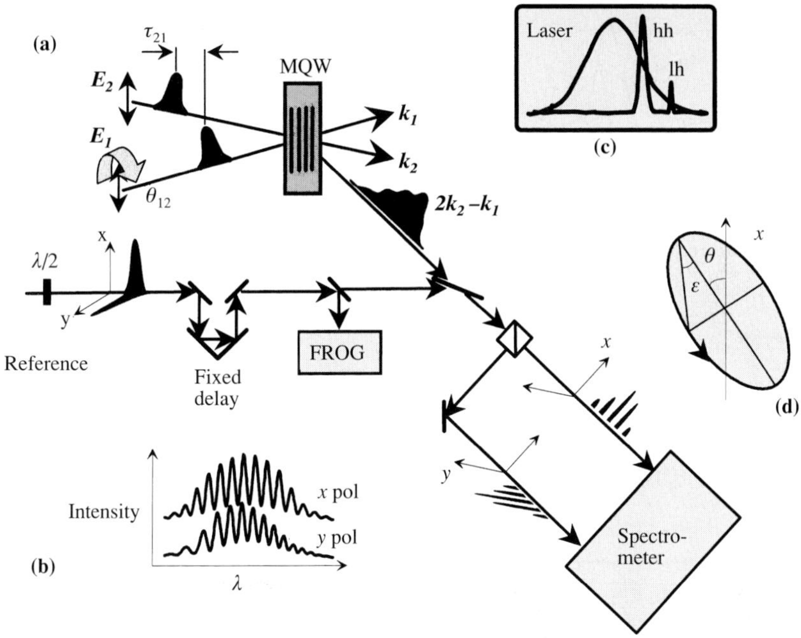

Fig. 9.25. (**a**) Schematic of the dual-channel spectral interferometry experimental setup for measuring the amplitude, phase and the polarization state of the FWM signal. (**b**) Typical spectral interferograms for the x and y polarizations. (**c**) Absorption spectrum of the sample showing the HH and LH exciton transitions and the laser spectrum. (**d**) Nomenclature used to describe the polarization state: θ denotes the orientation with respect to the x axis, and $\tan\varepsilon$ is the ratio of the minor and major axes of the polarization ellipse [9.221]

The results of Fig. 9.26 have been analyzed with inverse Fourier transforms to obtain the temporal dynamics of the FWM signal. From this analysis, they have deduced the temporal dynamics of the total intensity S_0, and the intensities and the phase difference of the x and y components of the FWM signal [9.222]. There are many interesting aspects of these results. The FWM signal has a delayed peak, shown earlier [9.230] to be related to many-body effects. The analysis also yields that the dynamics of the x and y intensities are different, consistent with the different spectral widths of the two components (Fig. 9.26), and the phase difference between the two components is a strong function of time. *Buccafusca* et al. also noted that the results are the same as found earlier by using the direct technique of ultrafast ellipsometry [9.229], but obtained much more simply without the tedious point-by-point measurements requiring high intensities.

The fact that the relative phase of the x and y components of the field varies strongly with time implies that the degree and the state of polarization of the FWM signal varies with time. Figure 9.27 a depicts the temporal evolution of the polarization and ellipticity of the signal deduced from the

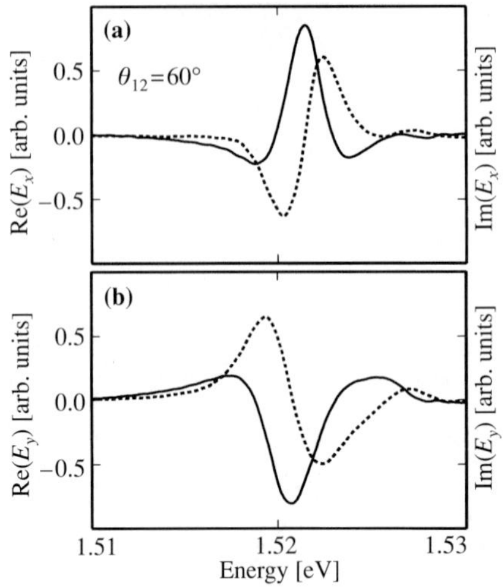

Fig. 9.26. Real (*solid*) and imaginary (*dotted*) parts of the complex spectral electric field for (**a**) x and (**b**) y polarizations of the FWM signal from GaAs at 80 K and $\approx 2\times10^{10}$ cm^{-2} [9.222]

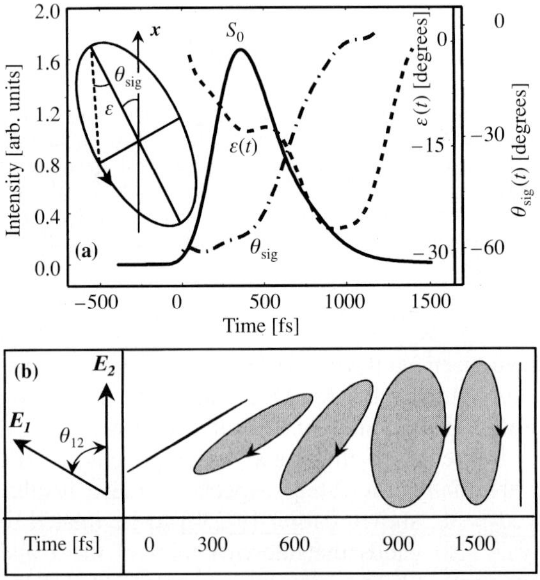

Fig. 9.27. (**a**) Total intensity $S_0(t)$ (*solid curve*), azimuthal angle $\theta_{sig}(t)$ of the polarization ellipse (*dot-dashed curve*), and ellipticity angle $\varepsilon(t)$ (*dashed curve*) that completely characterize the temporal evolution of the FWM signal. (**b**) Sketches of the polarization ellipse corresponding to the data in (**a**) for selected time delays [9.222]

data. The data show that the ellipticity and the orientation of the signal vary strongly with time. This implies that the polarization ellipse undergoes significant transformation as a function of time, as illustrated in Fig. 9.27 b. The signal is initially linearly polarized at θ_{signal} equal to $-\theta_{12}$, becomes elliptical and rotates in the clockwise direction and ends up as a vertical ($\theta_{12}=0$) linearly polarized signal.

These results have many implications for understanding the physics of nonlinear processes in quantum wells. First of all, a time-dependent polarization is not expected on the basis of optical Bloch equations, so that the many-body effects included in the semiconductor Bloch equations are essential to explain the results. *Paul* et al. [9.229] have considered a five-level model for the system (Sect. 2.7.2, Fig. 2.34) and included the effects of local field, excitation-induced dephasing and biexcitons in a phenomenological manner. By comparing the results with the calculation based on this model, they have concluded that leaving out any one of these three effects results in a significantly poorer fit with the data. Therefore, all three effects are important in the nonlinear response of quantum wells. Clearly, a quantitative comparison with calculations using a model with full many-body effects would be interesting.

We summarize this section with a few comments. These results have clearly demonstrated the power of the technique of femtosecond spectral interferometry. Although similar information can be obtained by using ultrafast ellipsometry, femtosecond spectral interferometry provides the potential of doing these measurements at a much lower density where it may be easier to isolate different physical effects. Application of femtosecond spectral interferometry for investigating resonant secondary emission from semiconductor quantum wells is discussed in Sect. 9.3.3. One can expect to see more applications of this technique in the future.

9.2.5 Exciton-Continuum Interaction

Coulomb interaction between various states leads to renormalization of their energy and the Rabi frequency (Sect. 2.1.2). Some important consequences of these can be phenomenologically understood in terms of a local field model (Sect. 2.1.2 d). Some experimental manifestations of these phenomena were discussed in Sect. 2.7. We have also noted how interaction between excitons modifies the dephasing rates of excitons (Sect. 2.2.1) as well as the coherent response of excitons. The latter was discussed in terms of excitation-induced dephasing in Sect. 2.7.1 d. Interaction between excitons and free carriers also increases the dephasing rates of excitons (Sect. 2.2.1), modifies the incoherent response of excitons [9.231], and accelerates the decay of excitonic wavepackets [9.232]. One of the interesting developments in recent years has been the realization that interaction between exciton and continuum states leads to important modifications in the coherent response of semiconductors. We discuss some of these results in this section.

(a) Dynamics of Fano Resonances in Semiconductors. It is well known that quantum mechanical coupling between a discrete state and an energetically degenerate continuum of states leads to new eigenstates that manifest themselves in an asymmetric lineshape, known as the Fano resonance, in linear spectroscopy [9.218]. Fano resonances are characterized by a pronounced minimum in the absorption spectrum at an energy where the transition amplitudes of the discrete and continuum states interfere destructively. Such resonances have been observed in a variety of atomic, molecular and semiconductor systems in frequency-domain experiments [9.233–235]. In bulk semiconductors, the application of a magnetic field leads to the formation of Landau levels in the conduction and valence bands. There is a one-dimensional continuum associated with each Landau level as well as a magnetoexciton associated with each pair of conduction and valence-band Landau levels with the same Landau quantum number n. Higher-order magnetoexcitons couple to energetically degenerate continuum states associated with the lower Landau levels via Coulomb interaction. The observation of Fano resonances in the magnetoabsorption spectrum of bulk GaAs resulting from this coupling has been reported recently [9.236]. We discuss here the results of recent experiments investigating the coherent response of semiconductors near Fano resonances in the time-domain using the technique of four-wave mixing [9.237].

Siegner et al. [9.237] investigated a high-quality GaAs bulk sample at 1.6 K. The strain introduced by the removal of the substrate for transmission measurements led to a splitting in the heavy-hole (HH) and light-hole (LH) exciton energies, with LH at lower energies. Application of the magnetic field thus resulted in two series of magneto-excitons corresponding to HH and LH excitons. Time-Integrated (TI), Time-Resolved (TR) and Spectrally-Resolved (SR) FWM experiments (Sect. 1.3.2) were performed with co-circularly polarized 100 fs pulses from a Ti:Sapphire laser tuned to $n=1$ HH magnetoexciton at a magnetic field of 10 T, which is energetically degenerate with the $n=0$ LH continuum. Figure 9.28 depicts the results of TI and TR-FWM measurements at a carrier density of about 10^{16} cm^{-3}. TI-FWM decays within the time resolution of the experiment whereas TR-FWM decays on a time scale of several hundred femtoseconds. SR-FWM has a narrow (FWHM ≈ 2 meV) spectrum (not shown). This spectral width is consistent with the decay constant of TR-FWM but *not* with the extremely fast decay of TI-FWM.

From an analysis of the data *Siegner* et al. found that the decay of the TR-FWM signal corresponds to the coupling strength between the discrete and continuum states deduced from the linear absorption spectrum at the Fano resonance. They concluded that the decay of TR-FWM is related to dephasing. On the other hand, the decay of TI-FWM signal is much faster and independent of the density of excitation over a range of 4×10^{15} to 4.4×10^{16} cm^{-3}. The atomic model of Fano resonance would predict the same decay time for TI-FWM and TR-FWM. *Siegner* et al. suggested that semiconductors behave differently because the Fano interference and nonlinearity that generates the coherent response both originate from Coulomb interaction, and they must be treated on an equal footing.

Another interesting aspect of the Fano resonance in semiconductors was demonstrated by *Bar-Ad* et al. [9.238]. With linear absorption spectroscopy,

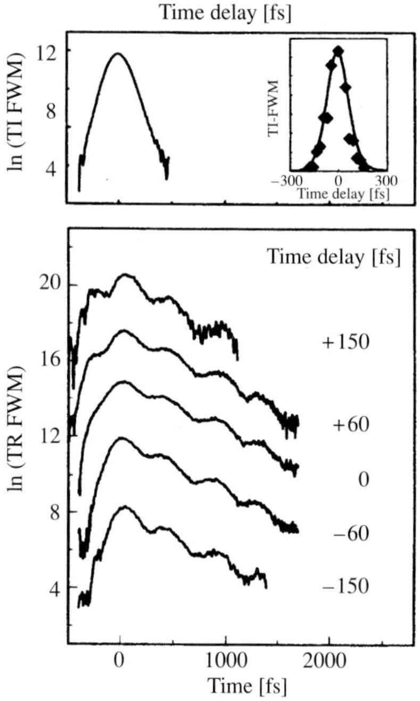

Fig. 9.28. Semi-logarithmic plot of the TI-FWM (*top*) and TR-FWM (*bottom*) from a Fano resonance in GaAs at 10 T. TR-FWM curves have been shifted for clarity. The inset in the top figure shows, on a linear scale, the numerically integrated intensity of the TR-FWM signal (*diamonds*) and the experimental TI-FWM (*solid curve*) [9.237]

they showed that Fano resonances in the magnetoabsorption spectrum disappear as the thickness of the bulk GaAs layer is reduced to 250 nm. This is because quantum confinement discretizes the continuum states. It is interesting that this occurs for a thickness that is an order of magnitude larger than the exciton Bohr radius.

Further studies have shown that the exciton-continuum interaction is much more general and makes an important contribution to the coherent response even in the case when the discrete exciton state is *not degenerate* with the continuum states. We discuss some examples of this in the next section.

(b) Non-Degenerate Interaction Between Excitons and Continuum. The clearest example of the *non-degenerate* interaction between excitons and continuum was reported by *Cundiff* et al. [9.239]. The experiments were performed on high-quality, symmetrically-strained, 50-period $In_{0.14}Ga_{0.86}As/GaAs_{0.71}P_{0.29}$ quantum well sample with 8.3 nm quantum wells and 8.0 nm barriers grown on GaAs. The HH exciton is nearly 70 meV below the LH exciton because of the strain. This allows the excitation of the HH exciton and/or continuum without any interference from the LH states. The samples were held at 8 K in a split-coil superconducting magnet and excited by two co-circularly polarized pulses with the wave vectors $\boldsymbol{q}_1$ and $\boldsymbol{q}_2$ separated by a time delay τ. TI-FWM and SR-FWM were measured in the direction $2\boldsymbol{q}_2 - \boldsymbol{q}_1$. The two pulses were derived from a Ti:Sapphire laser generating 40 fs pulses tuned $\approx$22 meV above the 1s HH exciton at 1.381 eV for no applied magnetic field. Two kinds of measurements were reported: standard, degenerate FWM with identical pulses, and partially nondegenerate FWM (PND-FWM) in which the pulse along $\boldsymbol{q}_1$ passed through a narrow-band filter, narrowing its bandwidth from $\approx$30 meV to 13 meV and stretching the pulse to 90 fs, but maintaining its coherence with the probe pulse.

Figure 9.29 depicts the results of both experiments for no applied magnetic field. Figure 9.29 a shows that the signal in the degenerate FWM experiment peaks at the 1s HH exciton energy even though the peak of the laser is in the continuum. This has been observed previously and attributed to the larger oscillator strength of excitons and more importantly the slow dephasing of excitons [9.240] (Sect. 2.5). What is more remarkable is the result that the excitons dominate the signal even in the PND-FWM experiments (Fig. 9.29 b); i.e., even when the spectrum of the first pulse along $\boldsymbol{q}_1$ does not overlap the excitons (see the front panel for the laser and the absorption spectra). The signal strengths in the two cases were approximately the same suggesting that this is not a small effect.

These results can not be understood in the independent transition picture which ignores Coulomb interactions between excitons and continuum states, and predicts a photon echo (Sect. 2.1.1 h) because the continuum states can be considered as inhomogeneously broadened in the momentum space (Sect. 2.1.2 a). These PND-FWM results clearly show that continuum states are interacting with the excitons to produce the observed strong coherent emission at the exciton energy. Such interaction can be introduced phenomenologically by postulating density-dependent material constants (dephasing rate, oscillator strength or transition energy) in SBE as discussed in the case of Excitation-Induced Dephasing (Sect. 2.7.1 d). However, a more intuitive picture is provided by the usual two-particle energy level picture of a semiconductor (Fig. 6.1) with a common ground state. It was argued by *Cundiff* et al. [9.239] that a pair of two-level systems can be transformed into a four-level system with a common ground state, and that interaction between the levels will remove the degeneracy between the levels and introduce additional terms in the coherent response of the system. These additional terms introduce coherent response at energies not excited by the first pulse, as observed in the experiments.

Another interesting aspect of the results in Fig. 9.29 is that the TI-FWM signal decays with essentially the autocorrelation width of the laser whereas the spectrum of FWM is much narrower than the laser spectrum; i.e., TI-FWM decays much more rapidly than expected from the FWM spectrum. This is similar to the observations (Sect. 9.2.5 a) that TI-FWM at a Fano resonance decays more rapidly than expected from the spectral width. In both cases, quantum interference masks the real dephasing rates and results in a pulse width-limited decay of TI-FWM if the continuum bandwidth is larger than the pulse spectral width. This feature of quantum interference is nicely illustrated in the case of PND-FWM experiments in the presence of a magnetic field (Fig. 9.30). For two identical pulses (Fig. 9.30 a), the pump

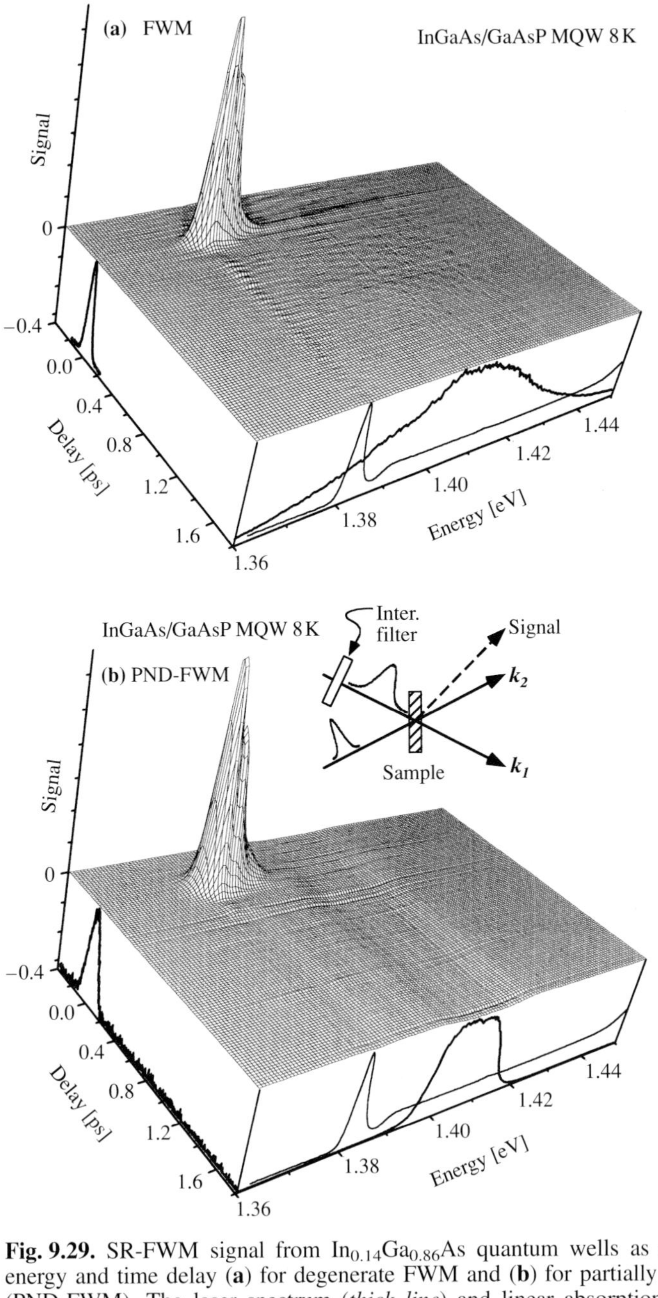

Fig. 9.29. SR-FWM signal from $In_{0.14}Ga_{0.86}As$ quantum wells as a function of photon energy and time delay (**a**) for degenerate FWM and (**b**) for partially nondegenerate FWM (PND-FWM). The laser spectrum (*thick line*) and linear absorption spectrum (*thin line*) are plotted on the front panel; the spectrally integrated signal is plotted on the side panel. Inset shows the experimental geometry [9.239]

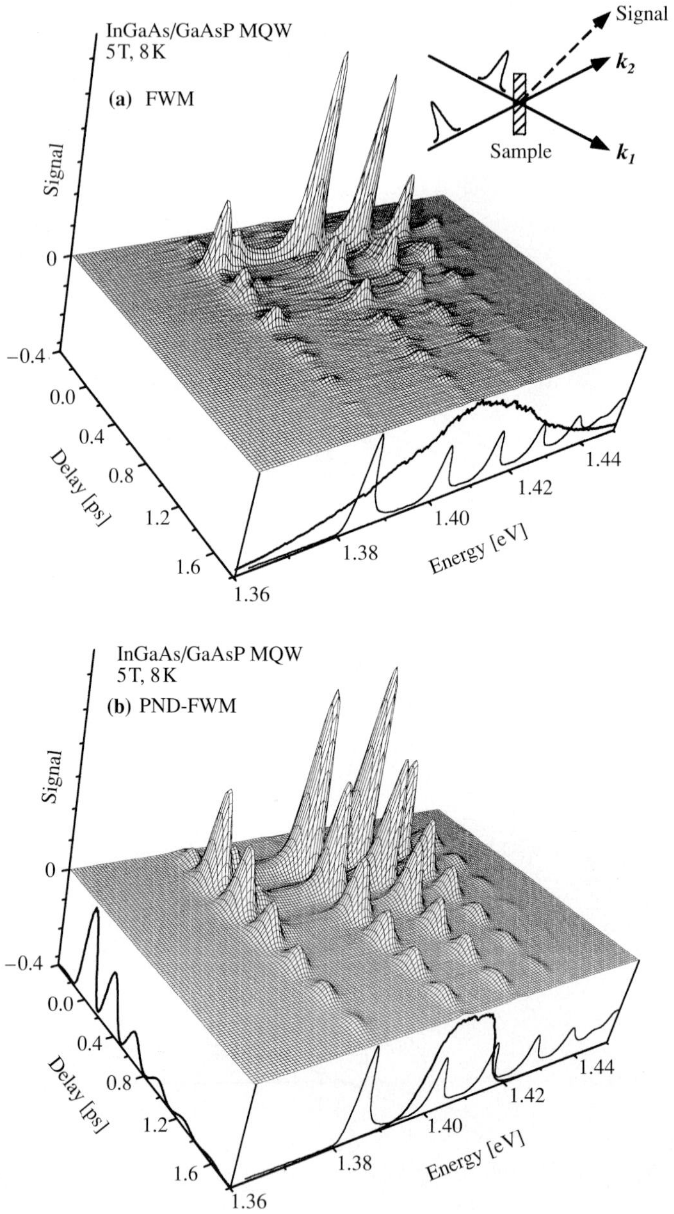

Fig. 9.30. Same as Fig. 9.29, but for $B=5$ T. Note the observation of quantum beats at the 1s magnetoexciton, even for PND-FWM in (**a**), with the period corresponding to the energy separation of the 2s and 3s magnetoexcitons. Simultaneous excitation of the 2s and 3s magnetoexcitons leads to quantum beats at the 1s magnetoexciton, even for PND-FWM in (**b**) [9.239]

pulse along $\boldsymbol{q}_1$ excites a superposition of several magneto-exciton states and FWM signals display complicated beating phenomena. For the PND-FWM case (Fig. 9.30 b), the pump pulse along $\boldsymbol{q}_1$ excites primarily the 2s and 3s magneto-excitons and so a well defined beat pattern with beat period corresponding to the energy difference between these two magneto-excitons is observed in TI-FWM. It is most interesting that the FWM signal at the 1s exciton, induced by the interactions between various magneto-excitons, also shows beating with *the same period* (corresponding to the energy separation of the 2s and 3s magneto-excitons). This is a clear manifestation of quantum interference between the exciton and continuum states.

(c) Further Two- and Three-Beam FWM Experiments on GaAs. The observation that the TI-FWM decays more rapidly than expected on the basis of the spectrum of FWM is a general phenomenon that has been reported for other systems [9.241–243]. *Wehner* et al. investigated TI-FWM and SR-FWM in bulk GaAs at 77 K using 14 fs pulses in spectrally-resolved two-beam FWM [9.241] and three-beam FWM [9.242] experiments (Sect. 1.3.2). They investigated a high-quality $GaAs/Al_{0.3}Ga_{0.7}As$ heterostructure grown by metal-organic vapor-phase epitaxy. The 600 nm thick GaAs layer resulted in an optical density of 0.3 for the continuum states and 1 for the exciton. All pulses were derived from a Ti:Sapphire laser with 15 fs sech^2 pulses tuned to the exciton resonance. Great care was taken to ensure that all pulses are nearly transform limited (time-bandwidth product of 0.31–0.33) at the sample. Figure 9.31 depicts their two-beam results for TI-FWM and SR-FWM in a three-dimensional plot. The FWM spectra exhibit a single peak that coin-

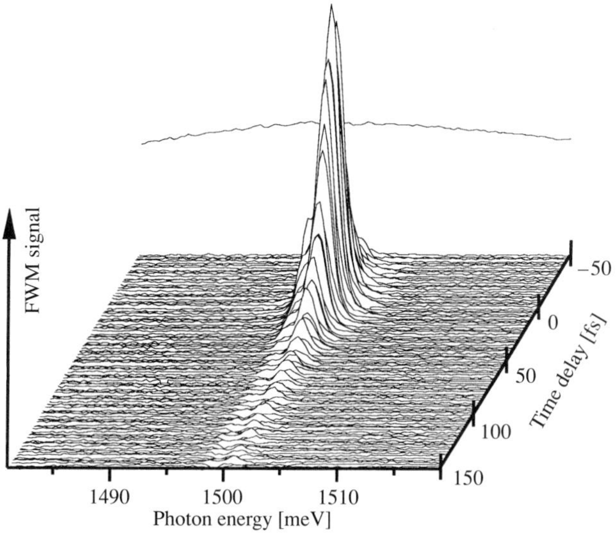

Fig. 9.31. SR-FWM at various time delays between the two beams in a two-beam FWM experiment on GaAs at 77 K for an excitation density of 4×10^{14} cm^{-3} [9.242]

cides with the spectral position of the exciton peak in the absorption spectrum and has a Half-Width at Half-Maximum (HWHM) of 1 meV. This narrow linewidth would correspond to a dephasing time of $T_2 = 660$ fs. On the other hand, the time constant for the initial decay of the TI-FWM signal in the time-delay domain was measured to be ≈ 12 fs (Fig. 9.31), much shorter than expected from the spectral width. At the low densities $\leq 2\times10^{15}$ cm^{-3} used in these experiments, the free electron-hole pairs excited in the continuum is not expected to reduce the exciton dephasing time so drastically. However, it was shown that exciton-continuum interaction leads to the dominant nonlinear optical response of the semiconductor. This is analogous to the idea of excitation-induced dephasing resulting from exciton-exciton interaction leading to the dominant nonlinear response in differential transmission measurements in bulk GaAs [9.244]. *Wehner* et al. [9.242] have noted that calculations based on simplified SBE which includes exciton-continuum interaction term provides a good agreement with their two- and three-beam data. They concluded that the response as a function of time delay is given by interference among polarization of continuum states.

Birkedal et al. [9.243] have also investigated this issue in GaAs quantum wells. A multiple quantum well structure containing 10 wells each of thickness 80 Å, 100 Å, 130 Å, and 160 Å, and separated by 150 Å $Al_{0.3}Ga_{0.7}As$ barriers was cooled to 5 K and excited by 100 fs pulses from a Ti:Sapphire laser. The spectral width of the laser was 15 meV and the center of the laser spectrum was located slightly above the HH exciton resonance of the 100 Å quantum wells; i.e., it excited different proportions of excitons and continuum states in each set of wells. The excitation density was estimated to be $\approx 1.5\times10^{9}$ cm^{-2} for the resonantly-excited 100 Å wells. They investigated TI-FWM and SR-FWM response of excitons in quantum wells of different thickness. The laser excited primarily the localized states associated with the HH excitons in the 80 Å wells, both the HH and LH excitons in the 100 Å wells, and excitons as well as the continuum states in the wider wells. As depicted in Fig. 9.32, the results are different for different wells, and consistent with the excitations of different states in different wells: photon echo-like response from the 80 Å quantum wells, strong HH-LH quantum beats from the 100 Å wells, a strong signal at HH excitons decaying extremely rapidly because of exciton-continuum interactions for the 130 Å and 160 Å quantum wells. Note that the excitons in the 100 Å wells are degenerate with the continuum states of the wider well but are not affected by this degeneracy. From a detailed analysis of these results, *Birkedal* et al. conclude that for the case when both the excitonic resonance and the continuum states are excited, there are two contributions to the FWM signal at the excitonic resonances: a dominant, prompt signal resulting from the interaction between excitons and the coherently-excited continuum states (interaction-induced nonlinearities), and a weak, slow component exhibiting the same features as the response of selectively-excited excitonic resonances, and resulting from the usual excitonic nonlinearities when only the excitonic states are excited. From a numerical simulation of SBE, it was concluded

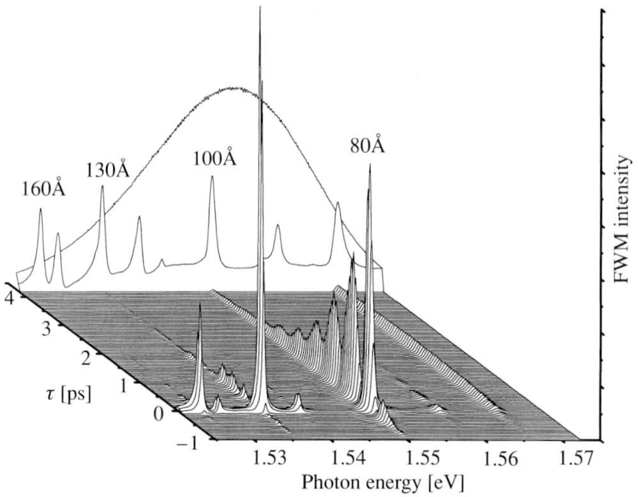

Fig. 9.32. SR-FWM signals from a quantum-well structure consisting of GaAs quantum wells of four different thickness, displayed as a function of the time delay τ in a three-dimensional plot. Quantum wells of different thickness behave differently, as discussed in the text. The absorption spectrum of the quantum wells and the laser spectrum are also displayed [9.243]

that interaction-induced nonlinearities are responsible for the exciton-continuum interactions.

The results of *Wehner* et al. and *Birkedal* et al. discussed in this subsection provide a concrete model for the intuitive picture of interaction-induced nonlinear response presented by *Cundiff* et al. [9.239] in Sect. 9.2.1 b. The results in this subsection clearly show that exciton-continuum interaction plays a dominant role in determining the nonlinear response of semiconductors on femtosecond time scales.

(d) Non-Degenerate FWM with Independently Tunable Femtosecond Lasers. Nearly all the FWM studies we have discussed are *degenerate* FWM investigations; i.e., the same laser is used in the two-beam and three-beam measurements. Two exceptions are the CARS (Coherent Anti-Stokes Raman Scattering) experiments discussed in Sect. 5.1.5, and partially-nondegenerate FWM discussed in Sect. 9.2.5 b. We conclude this section by mentioning recent results reported by *Kim* et al. [9.245] who used two independently-tunable, externally-synchronized femtosecond Ti:Sapphire lasers to perform a number of interesting two- and three-beam *non-degenerate* FWM experiments on a quantum well structure consisting of 30 periods of 100 Å GaAs quantum wells and 100 Å of $Al_{0.25}Ga_{0.75}As$ barriers at 10 K. One of the interesting results, related to the topics discussed in this subsection, was obtained with two beams derived from the same laser (frequency ω_1) and the third beam from the second laser (frequency ω_2). The delay between the

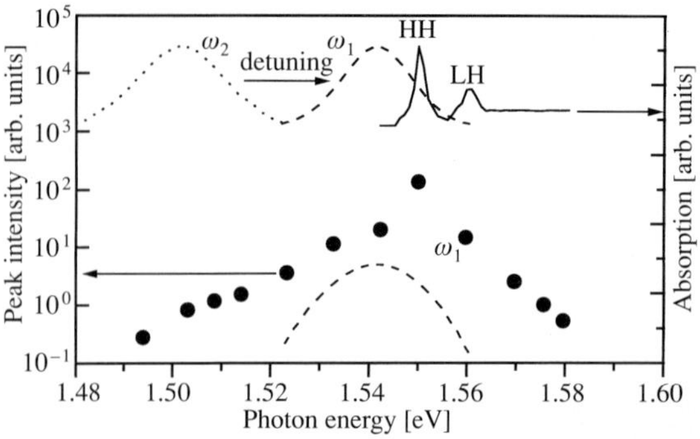

Fig. 9.33. Peak intensity of the non-degenerate FWM signal (at $\tau_1=150$ fs) for $\tau_{12}=10$ ps as a function of the frequency ω_2 for ω_1 tuned slightly below the HH exciton resonance, as shown by the *dashed line* (linear at the *top* of the figure, and logarithmic at the *bottom*). Note that the peak intensity varies by two orders of magnitude with detuning, with more rapid variation for $\omega_2>\omega_1$; i.e., ω_2 in the continuum. The *solid line* at the *top* shows the absorption curve [9.245]

pulses in the first two beams is τ_1, and that between the pulses in the third and the first beams τ_{12}. For ω_1 tuned to the HH exciton resonance, and ω_2 tuned considerably below the exciton resonance so there is no absorption, they observed beats in the FWM signal. With ω_1 set slightly below the HH exciton transition energy, TI-FWM signal was measured as a function of ω_2. In this experiment, the FWM signal as a function of τ_1 (for $\tau_{12}=2$ ps) exhibited beats at the HH-LH frequency. This was interpreted as the change in the strength of the population grating created by the two beams at ω_1. Figure 9.33 depicts the FWM signal as a function of ω_2 for $\tau_1=150$ fs and $\tau_{12}=10$ ps, and ω_1 tuned slightly below the HH exciton, as shown at the top of the figure. FWM signal is observed even when ω_2 is tuned completely below the exciton so that no absorption occurs, indicating that diffraction from the population grating occurs even when there is no absorption of the ω_2 beam. The signal is sensitive to detuning, decreasing by about two orders for a detuning of $\approx$–50 meV or $\approx$+30 meV. In other words, the TI-FWM signal for the continuum is much weaker than for below the band edge for the same detuning. They attributed this to destructive interference among the continuum states, and were able to explain some of the main characteristics of the off-resonant FWM signals by using the semiconductor Bloch equations.

9.2.6 Bloch Oscillations

The first observations of Bloch oscillations in semiconductor superlattices by pump-probe and FWM techniques (Sect. 2.9.4) as well as by THz emission technique (Sect. 2.12.3) are discussed in Chap. 2. Some of the recent developments in this field are outlined here.

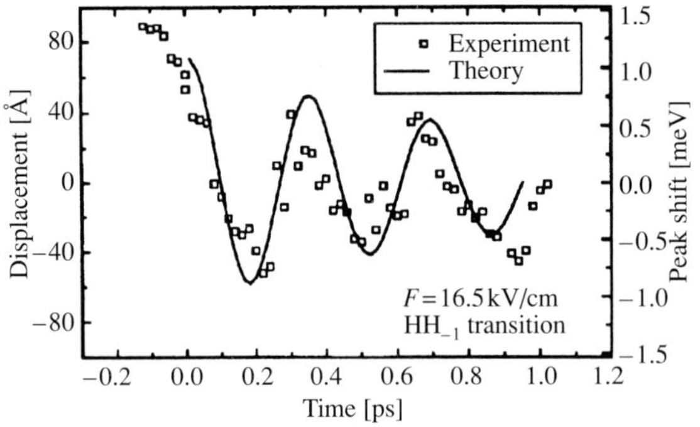

Fig. 9.34. Temporal evolution of Wannier-Stark ladder peak shifts (*right axis*) determined experimentally from the shifts in SR-FWM peaks, and Bloch electron displacement deduced from the data (*left axis*). Experiments were performed on a GaAs/$Al_{0.3}Ga_{0.7}As$ 67/17 Å superlattice at ≈ 10 K [9.246]

As mentioned previously (Sect. 2.9), a simple picture of Bloch oscillations in superlattices predicts that the amplitude of oscillations is given by Δ/eF, where Δ is the miniband width, e is the electron charge, and F is the electric field. However, as shown by theoretical calculations [9.200], excitonic effects and excitation conditions influence the nature of the optically-generated wavepacket and hence the amplitude of oscillations. We discuss here a measurement of the amplitude of Bloch oscillations by *Lyssenko* et al. [9.246].

The principle of this measurement is as follows: the interband transition energies of the Wannier-Stark ladder states are a sensitive function of the electric field. Therefore, any motion of the electron (or holes) that results in a screening of the applied electric field would produce a shift in the interband-transition energies, and a measurement of such shifts as a function of the delay after the excitation pulse provides a measure of the electric field at such delays from which the displacement of electrons can be deduced. *Lyssenko* et al. [9.246] investigated a GaAs/$Al_{0.3}Ga_{0.7}As$ superlattice with 67 Å GaAs well and 17 Å $Al_{0.3}Ga_{0.7}As$ barriers with a calculated electron miniband width of 38 meV. The hole miniband width is sufficiently small so that only electrons undergo oscillatory motion, and screen the field, to the first order. The superlattice, cooled to ≈ 10 K, is excited by 120 fs pulses from a Ti:Sapphire laser tuned slightly below the center of the Wannier-Stark ladder, and spectrally-resolved FWM as well as differential transmission spectra were recorded as a function of the delay between pulses. The FWM spectra exhibit peaks corresponding to various interband transitions, and the peak positions show oscillatory behavior as a function of the time delay. From these measurements, the researchers deduced the oscillation amplitude of the electron wavepacket. These results (Fig. 9.34) indicate that the center of the electron wavepacket oscillates as a function of time. The displacement has an initial value of ≈ 90 Å, and undergoes damped oscillations with the Bloch oscillation period. The damping is due to the dephasing of the wavepacket. A model theory that takes into account excitonic effects and a phenomenological dephasing time of 1.2 ps provides a good agreement with the data, as shown. The total oscillation amplitude increases to nearly 300 Å at lower fields ($\approx$13 kV/cm) and decreases to $\approx$100 Å at 23 kV/cm, in agreement with the predictions of a theory that includes excitons. These studies provide the first direct measurement of the amplitude of Bloch oscillations in semiconductor superlattices, and emphasize once again the importance of excitonic effects in Bloch oscillations [9.200].

A number of other interesting recent developments in the field of Bloch oscillations have been reviewed by *Kurz* et al. [9.247]. For example, Bloch oscillations are observed even when the laser is tuned more than one LO-phonon energy above the bandgap. Under these conditions, the photoexcited electrons emit LO phonons in times shorter than 200 fs (as discussed in Chap. 3). However, the Bloch oscillations decay with a time constant much longer than this [9.248, 249]. It was argued that this implies preservation of intraband coherence even after scattering by LO phonons. In some ways, this is similar to the observation that the HH-LH beats in emission persist for times longer than the momentum scattering time [9.214] (Sect. 6.3.3). *Martini* et al. have investigated the superradiant character of the THz radiation emitted by electrons undergoing Bloch oscillations and have deduced the spatial amplitude of the charge oscillations from the measured emission efficiency [9.250]. *Cho* et al. [9.251] have investigated Bloch oscillations in strain-balanced InGaAsP superlattices by means of time-resolved transmission experiments.

We end this section by noting that interesting observations of Bloch oscillations have been reported recently in systems that use atoms instead of electrons and a periodic light field instead of the periodic potential of a crystal [9.252–256].

9.2.7 Coherent Phonons

Section 2.11 discussed the generation and detection of coherent optical phonons in GaAs, Bi and Sb. Several new developments have taken place in this field. Emission of THz emission from such phonon oscillations has been observed in the small band-gap semiconductor Te [9.257, 258]. The emission shows a peak at about 3 THz, approximately corresponding to the two infrared active phonon modes. Reflectivity measurements, on the other hand, reveal several Raman active phonon modes. The mechanism of coherent phonon generation by short light pulses has been investigated in Sb [9.259]. These researchers found that the coherent driving force in absorbing materials like Sb can be described by a Raman process, as in transparent materials. They further argued that the Raman formalism provides a unifying approach for describing coherent phonon generation by impulsive as well as displacive forces. An overview of the experimental work and theory was presented by *Merlin* [9.260].

While most of the work has been concentrated on optical phonons, acoustic phonons have also received some attention. *Baumberg* et al. [9.261] used a femtosecond Ti:Sapphire laser to excite electron-hole plasma by near-bandgap absorption in a single GaAs quantum well located some distance from the sample surface. Deformation potential coupling with acoustic phonons resulted in the generation of an acoustic wavepacket that was well localized in space and time. The acoustic wavepacket travels in both directions away from the quantum well. The arrival of the acoustic wavepacket at the surface produces a geometric bowing of the surface as well as a change in the dielectric constant of the surface layer. These changes were detected by the change in the reflectivity of the second harmonic of the Ti:Sapphire laser (with very small penetration depth) in the usual pump-probe reflectivity measurement. The arrival time of the phonons at the surface was determined the acoustic velocity. It was argued that the oscillatory shape of the differential reflectivity signal indicated that the change in the dielectric constant of the surface layer was the dominant cause of the observed signal. Coherent acoustic phonons in PbS quantum dots were generated using femtosecond pulses from a frequency-doubled optical parametric oscillator pumped by a regeneratively amplified Ti:Sapphire laser [9.262].

9.2.8 Plasmon-Phonon Oscillations

As described in Sect. 4.3.3, the plasmon and phonon modes in semiconductors interact with each other to generate coupled plasmon-phonon modes. The plasma frequency for a single component plasma in bulk and two-dimensional semiconductors are given by (4.19–22). The coupling between plasmons and phonons become strong when the two frequencies become comparable. In bulk GaAs the electron plasma frequency becomes equal to the LO-phonon frequency at $\approx 7\times10^{17}$ cm^{-3}. We discuss here some recent femtosecond studies of coherent oscillations of the plasmons as well as coupled plasmon-phonon modes [9.263, 264].

(a) Pump-Probe Measurements. *Sha* et al. [9.263] excited a low-density plasma (2×10^{15} cm^{-3} to 1.5×10^{16} cm^{-3}) in the i-region of a GaAs p-i-n structure at 80 K by photoexciting with 100 fs pulses from a Ti:Sapphire laser tuned to GaAs band edge. The built-in field in the i-region was 21 kV/cm. Using a sensitive differential electroabsorption technique, they found that the time derivative of the electroabsorption as a function of the delay between the pump and the probe pulses showed oscillations whose frequency varied with the photoexcitation density (Fig. 9.35 a). The oscillation frequency showed a square root dependence on the excited carrier density (Fig. 9.35 b) as one expects for the plasma oscillations from (4.19). The plasma density was sufficiently low so that no coupling between the plas-

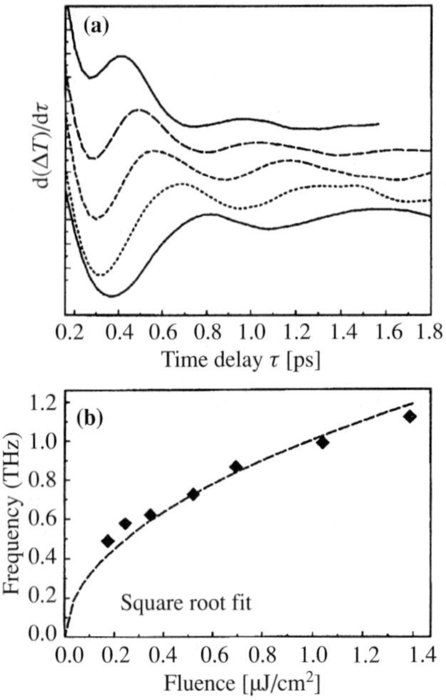

Fig. 9.35. (**a**) Time derivative of the change in the electroabsorption of a spectrally-filtered probe beam as a function of the time delay between the pump and probe for fluences of 1.0, 0.7, 0.5, 0.35, and 0.25 $\mu J/cm^2$. 1 $\mu J/cm^2$ corresponds to $\approx 1.1\times10^{16}\,cm^{-3}$. (**b**) Frequency of the oscillating field as a function of the incident fluence, extracted from the electro-absorption measurements in (**a**) [9.263]

mon and phonon modes was observed. *Sha* et al. were also able to deduce the maximum oscillation amplitude and the maximum kinetic energy and show that both these quantities vary inversely with the plasma density as expected.

These plasmon oscillations are excited by the initial ballistic motion of the electrons and holes under the influence of the built-in electric field. The separation of charge carriers induces a polarization that acts as a restoring force and leads to plasma oscillations. It is interesting to note that photoexcitation creates a two-component plasma (electrons and holes). A two-component plasma has two oscillation frequencies, corresponding to the optical plasmon mode and the acoustic plasmon mode. Optical as well as acoustic plasmon modes in photoexcited GaAs have been observed in Raman scattering experiments [9.265].

(b) Terahertz Spectroscopy. Plasma oscillations in a photoexcited GaAs p-i-n structure have also been investigated with THz correlation spectroscopy [9.264]. The p-i-n structure at 300 K was excited by a pair of 100 fs pulses from a Ti:Sapphire laser operating at 800 nm. The emitted THz power was measured as a function of the time delay between two femtosecond pulses. The measured correlation traces exhibited strongly damped oscillations whose frequency was determined from the Fourier spectrum of the correlation trace. The oscillation frequency varied as the square root of the photoexcitation density, as expected for the plasma frequency for densities up to $\approx 1.5\times10^{17}\,cm^{-3}$.

Cho et al. [9.266] and *Kersting* et al. [9.264] have also investigated n-doped GaAs samples. Photoexcited carriers produce a change in the surface electric field which results in a shift in the bandgap due to the Franz-Keldysh effect. *Cho* et al. [9.266] measured the change in the reflectivity of photoexcited n-GaAs at 300 K resulting from this effect using a probe beam, and observed well defined oscillations in the differential reflectivity as a function of time delay. Three samples with different n-doping (7.7×10^{16} cm^{-3}, 3×10^{17} cm^{-3}, and 2×10^{18} cm^{-3}) were analysed. Differential reflectivity signals exhibited beating phenomena for certain doping and excitation conditions, and the corresponding Fourier spectra showed two peaks corresponding to the lower and upper branches of the coupled plasmon-phonon modes (Sect. 4.3.3). *Cho* et al. [9.266] found that the measured oscillation frequencies agreed well with the calculated lower-branch frequencies provided the total density (i.e., the sum of the doping and photoexcitation densities) was used for the calculation of the plasma frequency. They presented data for the total density up to $\approx 2\times10^{18}$ cm^{-3}. At the high densities, the coupled mode approaches the TO phonon frequency and has primarily the phonon character. Therefore, the relaxation time of the oscillations increases as the density increases, i.e., as the coupled mode becomes more like a phonon mode.

Kersting et al. [9.264] investigated a sample with a doping density of 1.7×10^{16} cm^{-3} and investigated the photoexcitation density up to 1.8×10^{17} cm^{-3}. They found from their THz experiments that the plasma frequency in their doped samples was determined by the doping density and was not affected by the photoexcitation density or the excitation wavelength. From these observations, they concluded that cold plasma oscillations were excited in their experiments. According to their model, the cold electrons are confined by the surface depletion layer on one side and by the electric field at the interface between the epitaxial layer and the substrate on the other side. The observed plasma oscillations result from the response of the cold plasma to the screening of the surface field by the photoexcited carriers. This interpretation was supported by measurements of the intensities of the THz radiation as a function of the intensity of the second excitation pulse.

9.2.9 HH-LH Resonance in the Continuum

The observation of quantum beats in FWM spectroscopy of quantum wells is discussed in Sect. 2.6.2. All these observations correspond to the excitation of HH and LH *excitons* by a spectrally broad femtosecond pulse. We discuss in this subsection recent observations of quantum beats when the laser excites the continuum states of the semiconductor. These measurements provide insight into the quantum coherence of continuum states in semiconductors.

Dekorsy et al. [9.267] have investigated quantum coherence of hole continuum states in the valence subbands of GaAs quantum wells by using the

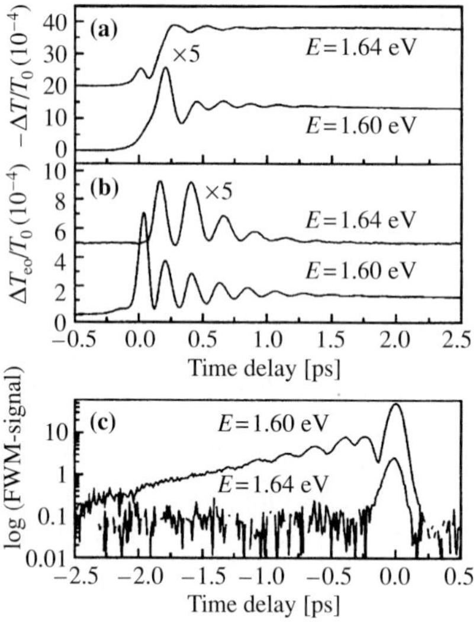

Fig. 9.36. (**a**) Differential Transmission (DT), (**b**) TEOS (Transmitted Electro-Optic Sampling) measured by subtracting the two polarization components of the transmitted beam, and (**c**) TI-FWM as a function of the delay between the pump and probe beams for a laser in resonance with the HH-LH resonance (1.60 eV) and the laser in the continuum (1.64 eV). Note the linear scales in (**a**) and (**b**) and the semi-logarithmic scale in (**c**) [9.267]

techniques of FWM, Differential Transmission (DT) and TEOS (Transmitted Electro-Optic Sampling) (Sects. 1.3.1, 2.11). They studied a sample with 40 periods of 65 Å GaAs quantum wells separated by 200 Å $Al_{0.35}Ga_{0.65}As$ barriers at 10 K. Measurements were performed with 50 fs pulses from a Ti:Sapphire laser tuned to either 1.60 eV (between the 1s HH and LH excitons; resonant excitation) or 1.64 eV (non-resonant excitation). In the latter case, the continuum states were primarily excited, and the excitation of the HH exciton decreased by more than 2 orders of magnitude compared to the excitation at 1.60 eV. The excitation density is $\approx 1\times10^{10}$ cm^{-2} (7×10^{9} cm^{-2}) for the resonant (non-resonant) excitation. Figure 9.36 depicts the results of the three types of experiments. The TI-FWM signal is measured in the $2\boldsymbol{q}_1-\boldsymbol{q}_2$ direction so that the time axis is inverted. For the differential transmission and TEOS signals, the beats are observed for non-resonant excitation for an excess energy as high as 75 meV where the laser intensity at excitons has decreased to 6×10^{-6} of the intensity at the center of the laser spectrum. Therefore, it was concluded that the beats are related to continuum states of the holes in the HH and LH valence bands.

For resonant excitation, TI-FWM at long times exhibits a slow decay (decay time constant $\tau_{decay}=620$ fs) with HH-LH excitonic beats superposed on it (Fig. 9.36 c). For non-resonant excitation, on the other hand, TI-FWM is extremely fast, following the autocorrelation of the exciting pulse. Since the decay of the TI-FWM signal is determined by dephasing of the interband polarization, this implies that interband electron-hole polarization decays extremely rapidly whereas the interband excitonic polarization decays slower. This is consistent with the results presented for dephasing of excitons and free carriers in Sects. 2.2 and 2.10.

The results for differential transmission and TEOS presented in Fig. 9.36 a and b are similar to each other, but strikingly different from the TI-FWM data discussed above. The TEOS results exhibit strong HH-LH beats for both resonant and non-resonant excitations. It is well-known [9.268–270] that, in contrast to TI-FWM signals, the differential transmission and TEOS signals contain both coherent and incoherent contributions, and that the decay of the beats is determined by the dephasing of the intraband polarization, in this case the polarization of the holes. For the resonant case, $\tau_{\rm decay}$ for the beats in the TEOS signal is $\approx$700 fs, not too different from the value of 620 fs for $\tau_{\rm decay}$ of the interband polarization. However, $\tau_{\rm decay}$ of the beats in TEOS for non-resonant excitation is $\approx$360 fs, much longer than the interband dephasing times determined from TI-FWM, but only a factor of 2 shorter than for the resonant excitation. Thus the dephasing of intraband polarization of free holes is surprisingly slow. *Dekorsy* et al. [9.267] also showed that the beat frequency increases with increasing excess energy of the free carriers. This is related to the fact that splitting between the HH and LH valence band states increases as the exciting laser imparting larger excess energy couples with hole states of larger wavevectors.

Joschko et al. [9.271] reported on differential transmission measurements on 0.5 μm thick undoped and n-doped GaAs sandwiched between $Al_{0.3}Ga_{0.7}As$ layers. The substrate is removed for transmission measurements, inducing small strain and a splitting of the HH and LH bands. The measurements were performed using 20 fs pulses from a Ti:Sapphire laser centered at 1.61 eV, about 90 meV above the bandgap at 8 K, the temperature of the sample. These researchers also observed temporal oscillations in the differential transmission signal. The period of the oscillations agreed with the splitting between the HH and LH states in the valence band, leading to the conclusion that the oscillations were due to quantum beats of HH and LH valence band states. The results are in good agreement with a calculation based on the semiconductor Bloch equations for a three-band model.

9.2.10 Biexcitons

The importance of biexcitons in the coherent response of semiconductor quantum wells is discussed in Sect. 2.7.2. Investigations of biexcitons, and their linear and nonlinear properties have continued during recent years from experimental [9.228, 229, 272–290] as well as theoretical [9.190, 194, 291–294] points of views. Coherent spectroscopy has provided invaluable information about nonlinear processes and interactions in semiconductors, as well as about other aspects of semiconductors. We discuss in this subsection an experiment that uses SR-FWM to study the binding energies of biexcitons in semiconductor quantum wells.

Birkedal et al. [9.282] investigated a multiple quantum well sample of GaAs with $Al_{0.3}Ga_{0.7}As$ barriers. The sample contained 10 wells each of 80, 100, 130 and 160 Å, separated by 150 Å $Al_{0.3}Ga_{0.7}As$ barriers. The

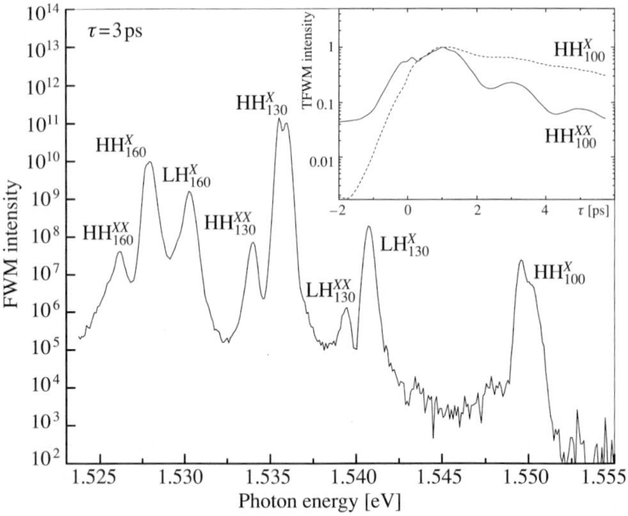

Fig. 9.37. SR-FWM at a time delay of 3 ps from a GaAs quantum-well sample consisting of quantum wells of thickness 80, 100, 130 and 160 Å; T=5 K. Note the large dynamic range of the data. The inset shows TI-FWM for the HH exciton and biexciton in the 100 Å wells [9.282]

sample was mounted on a sapphire disk after removing the substrate, maintained at 5 K in a cryostat, and excited by 100 fs pulses from a Ti:Sapphire laser. Two-beam self-diffracted SR-FWM measurements were performed using two co-linearly polarized beams. Since FWM experiments probe the exciton and biexciton states before they undergo scattering, they can provide information difficult to obtain by incoherent techniques such as photoluminescence or absorption.

The SR-FWM signal is depicted in Fig. 9.37 for a time delay τ_d=3 ps between the pulses. Note that the signal intensity covers a very wide dynamics range and shows rich structure. The data exhibit not only the HH excitons for the 160, 130 and 100 Å quantum wells, but also the HH biexcitons for the 160 and 130 Å quantum wells, and the LH biexciton for the 130 Å quantum well. These well-resolved structures allow an accurate determination of the binding energies of the biexcitons. From such data, it was deduced that the binding energy of biexcitons in quantum wells decreases as the well thickness increases, but the ratio of the binding energies of biexciton and exciton remains relatively constant $\approx$0.20. This ratio is larger than the value of $\approx$0.12 predicted by *Kleinman* [9.295]. The measured value of the ratio is in good agreement with calculations [9.282, 296] that predict a ratio of 0.228 for biexcitons in quantum wells. These results remove a long-standing uncertainty about the binding energy of biexcitons in quantum wells and demonstrate a useful application of coherent spectroscopy.

Note that, as discussed in Sect. 9.2.2, biexcitonic effects are not included in the semiconductor Bloch equations. But recent extensions of SBE that include four-particle correlations allow calculation of the nonlinear response of semiconductors including biexcitonic effects.

9.2.11 Coherent and Nonlinear Phenomena in Semiconductor Microcavities

Basic properties of semiconductor microcavities were discussed in Sect. 9.1.1 b. In this subsection we outline investigations of coherent properties of semiconductor microcavities using ultrafast techniques. Ultrafast dynamics of spontaneous and laser emission from semiconductor microcavities will be treated in Sect. 9.3.3.

As discussed in Sect. 9.1.1 b, the strong-coupling or nonperturbative regime in a microcavity leads to anticrossing behavior of the two normal (polariton) modes. Such anticrossing behavior has been observed in semiconductor microcavities [9.114]. The observation of anticrossing is an evidence for the coupling between the two states. The time domain manifestation of the coupling is the oscillation of the energy between the two coupled modes after excitation of a linear superposition mode by an ultrashort pulse. This is similar to the observation of quantum beats for the HH-LH system (Sects. 2.6.2 and 2.12.2) or the wavepacket oscillation for the asymmetric double quantum well system (Sects. 2.8 and 2.12.1). We emphasize that the observation of a doublet in the spectral domain does not prove that the two states are coupled [9.297]; demonstration of anticrossing is essential to show the coupling between the states. Similarly, observation of oscillations in the linear or nonlinear response of a system is simply equivalent to the existence of a doublet in the spectral domain. Therefore, it does not prove that the two states are coupled or have a common ground state. This was discussed in Sect. 2.6.4 in which methods of distinguishing between polarization beats (two independent two-level system or uncoupled states) and quantum beats (a three-level system or coupled states) were presented. It was shown in Sect. 2.6.4 that TR-FWM or SR-FWM provides a means of distinguishing between the two cases.

We analyse here the results of *Wang* et al. [9.61, 298] who investigated a semiconductor microcavity sample with SR-FWM in the reflection geometry. The microcavity sample consists of two 150 Å quantum wells separated by 100 Å $Al_{0.3}Ga_{0.7}As$ barriers at the center (antinode) of a λ cavity sandwiched between $AlAs/Al_{0.11}Ga_{0.89}As$ Bragg mirrors. The cold cavity linewidth was <1 meV and the exciton was inhomogeneously broadened with a linewidth estimated to be 2.4 meV from reflectivity studies. Two-beam self-diffracted SR-FWM measurements in the reflection geometry were performed on the sample at 10 K. The SR-FWM signal (FWM power spectrum) shows two peaks which correspond to the peaks in the reflectivity spectrum. The energy separation of the two peaks in the SR-FWM signal varies with the cavity-exciton detuning which is attained by probing different positions on the sample. *Wang* et al. measured the TI-FWM signal at the

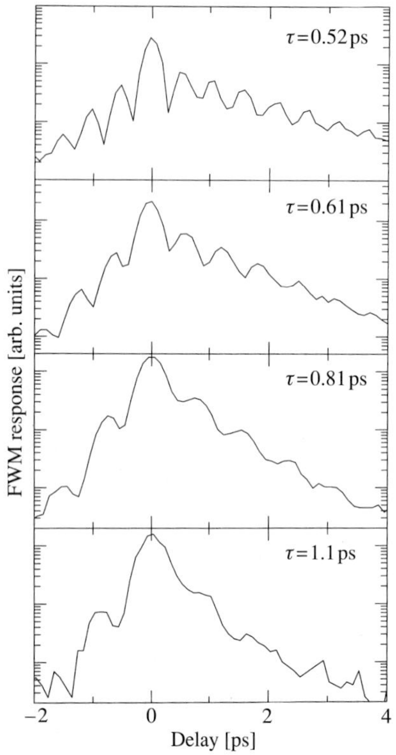

Fig. 9.38. TI-FWM signals from a GaAs microcavity at 10 K, at the peak of the cavity-like resonance as a function of time delay τ_d for various detunings between the cavity and the exciton modes. The results demonstrate that the oscillations are quantum beats and hence show the coherent exchange of energy between the modes. The oscillation frequency corresponds to the splitting in the SR-FWM spectra. The period of the oscillation τ for each detuning is indicated in the figure [9.61]

spectral peak of the cavity-like resonance as a function of time delay τ_d between the two pulses. The measured curves, depicted in Fig. 9.38 for various detunings, show well defined oscillations whose frequency corresponds to the energy separation of the two normal modes, as measured from the SR-FWM spectra. It was discussed in Sect. 2.6.4 b that spectrally-resolved transient FWM signal at the resonance frequency exhibits oscillations only if the two levels are coupled. Therefore, the results of Fig. 9.38 demonstrate that the two levels are coupled and that there is a coherent exchange of energy between the two normal modes. We note that *Kelkar* et al. have applied similar techniques and confirmed the existence of coherent oscillations in II–VI microcavities with very large normal-mode splitting [9.299].

As discussed in Sect. 2.6.4, quantum beats can also be distinguished from polarization beats by using the techniques of TR-FWM. *Koch* et al. [9.300] have reported on such an investigation on the same sample as discussed in the preceding paragraph. The TR-FWM signal is depicted in Fig. 9.39 a for time delays of 100 and 170 fs. In these traces, the (real) time t has been measured from the first pulse exciting the sample, and τ gives the delay between the two exciting pulses. The signal shows oscillations whose frequency corresponds to the splitting observed in the FWM spectrum. From such plots, one obtains the time t at which each minimum in the TR-FWM

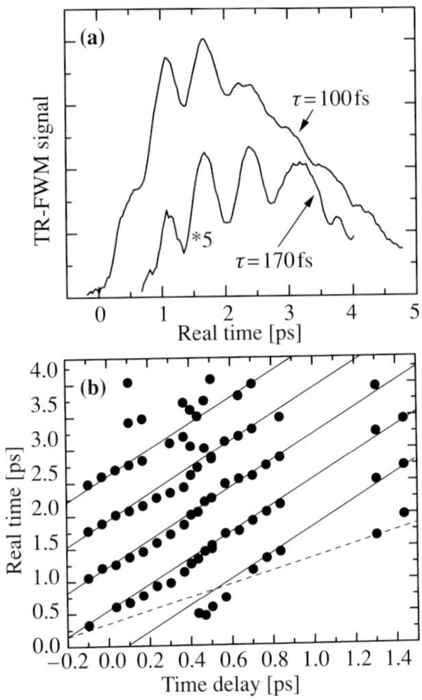

Fig. 9.39. (**a**) TR-FWM signals from a GaAs microcavity at 10 K for two delays τ. (**b**) Positions of the minima in TR-FWM in time as a function of the delay τ between the pulses. The *dashed line* represents a slope of 1 and the *solid lines* a slope of 2. Note the complex behavior of the data near τ=0.5 ps [9.300]

signal occurs as a function of time delay. Figure 9.39 b presents such t vs τ plots for several minima. These t vs τ plots are expected [9.301] to show a unity slope (i.e., a change in the peak position equals the change in the delay τ) for a system of coupled oscillators. In contrast to this expectation, the results in Fig. 9.39 b show a slope of 2. Furthermore, the plot in Fig. 9.39 b exhibits undulations around this slope as if there is an interaction between various $t=2\,\tau$ lines. *Koch* et al. [9.300] proposed a simplified model taking into account the fact that the emitted field in a microcavity is not simply the square of the third-order polarization (2.44 and 45). The third-order polarization acts as a source for the intracavity field in the observation direction, part of which escapes the microcavity and is detected. The simplified model gives an analytical solution for δ-function pulses, and shows that (i) the $t=2\,\tau$ slope is expected for the microcavity, and (ii) there are higher-order terms which lead to undulations similar to those observed. A more quantitative comparison must await a more refined theory.

Kuwata-Gonokami et al. [9.94] have investigated a degenerate TI-FWM signal using 1.9 ps Ti:Sapphire laser with a spectral width of 0.7 meV as the photon energy of the laser is tuned across the coupled-mode (polariton) energies. These investigations were carried out for various combinations of linear and circular polarizations for the three beams in the degenerate TI-FWM experiment in the phase-conjugate geometry. The experiments were performed on a 120 Å thick GaAs quantum well placed at the antinode of a $\lambda/2$ cavity with an AlAs spacer layer sandwiched between two GaAs/AlAs

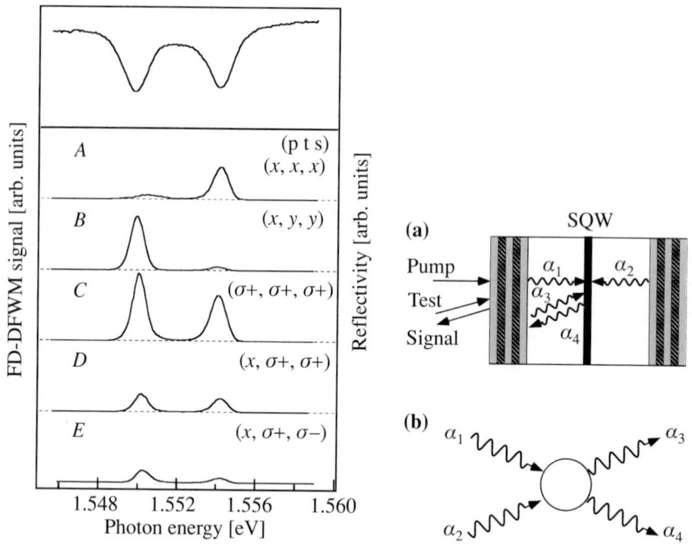

Fig. 9.40. *Left panel*: TI-FWM from a $\lambda/2$ planar microcavity in the reflection geometry as a function of the center frequency of the 1.5 ps laser (FWHM: 0.7 meV). Various polarization combinations discussed in the text are investigated. *Right panel*: (**a**) self-pumped phase conjugation geometry used in the experiments, (**b**) relevant processes for FWM based on the cavity-polariton picture [9.94]

Bragg mirrors (22 and 14.5 pairs for the bottom and the top, respectively). Figure 9.40 depicts the experimental configuration and results. The top of the left panel in Fig. 9.40 presents the reflectivity spectrum for no detuning between the exciton and the cavity modes. The spectrum shows that the cavity and the exciton linewidths are $\approx$1.5 meV and the normal-mode splitting is $\approx$4.3 meV. The FWM experiments were performed in a geometry depicted in Fig. 9.40 a, with the signal measured in the phase-conjugate direction. The TI-FWM signal for a zero time delay between the pump and the test beams is measured as the laser is tuned across the coupled mode energies. The results are depicted in the left panel of Fig. 9.40 for five different combinations of the polarizations of the three beams as indicated in the figure. The results reveal that the polarization of the three beams influence the FWM intensities as well as the relative strengths of the FWM signals at the two polariton energies. *Kuwata-Gonokami* et al. [9.300] proposed a model based on the cavity polariton concept to explain these results (Fig. 9.40 b). The process is described as stimulated polariton scattering of cavity polaritons. There are three terms contributing to anharmonicities: repulsion and attraction between excitons and phase-space filling. The last two effects dominate the excitonic nonlinearity in quantum wells. The theory provided a good agreement with the data of Fig. 9.40, as well as with the dependence of the signal on detuning.

Nonlinearity of excitons is responsible not only for the FWM signals but also for the collapse of the normal-mode splitting at high intensities. This is another aspect of semiconductor microcavities that is different from atoms in a microcavity. For independent two-level atoms, nonlinearity comes from the saturation of atomic transition. In contrast, the nonlinearity of excitons is quite complex [9.186] and results from many-body interactions in semiconductors. The nonlinearity of excitons in semiconductor microcavities has been investigated by a number of researchers [9.56, 71, 100, 302]. We discuss here the recent experimental and theoretical results on the saturation of polaritons in microcavities presented by *Jahnke* et al. [9.71, 106].

Jahnke et al. investigated a microcavity with two 80 Å $In_{0.04}Ga_{0.96}As$ quantum wells located at the antinodes of a 3/2 λ cavity formed with GaAs spacer layer and GaAs/AlAs Bragg mirrors. The small In concentration allowed transmission measurements without removing the GaAs substrate and also strain-shifted the LH exciton so that it does not interfere the coupling of the HH exciton with the cavity mode. The exciton linewidth was 1 meV and the resonant normal-mode splitting was 6.8 times the linewidth, one of the largest ratios reported. A 20 quantum well reference sample with nominally identical conditions was also grown for comparison. CW pump-probe transmission measurements were performed at 4 K using a Ti:Sapphire laser at 7870 Å (outside the stop band of the cavity) as the pump and an LED centered at 8500 Å as a broadband probe.

The results of the measurements for different pump intensities are depicted in the left panel of Fig. 9.41. (a) shows the saturation of the absorption at the HH exciton in the reference sample, whereas (b) displays the probe transmission spectra in the region of normal-mode splitting for several different intensities. Although the intensities of the two sets could not be compared directly because of the top Bragg mirror in the microcavity sample, calculations demonstrate that the unprimed and primed numbers in (a) and (b) correspond to each other. The remarkable result in Fig. 9.41 b is that the normal-mode splitting remains independent of intensity and then abruptly collapses (curve 4′). In other words, the normal-mode splitting does not gradually decrease and disappear, but collapses abruptly.

As we discussed in Sect. 9.1.1 b, normal-mode splitting is proportional to the square root of the exciton oscillator strength and independent of the exciton broadening whereas the width of the transmission peaks depend on exciton broadening. Thus, an investigation of the intensity dependence of the transmission spectra of a microcavity provides an opportunity for studying the two effects independently. The results in Fig. 9.41 b show that with increasing density of the excitons and the free carriers, the exciton first broadens and then loses oscillator strength at higher densities. *Jahnke* et al. [9.71] applied a kinetic approach to theoretically calculate the saturation of exciton states and the corresponding effects on the normal-mode transmission spectra of a microcavity. This approach describes coupled bound and continuum states of quantum-well excitons under the influence of phase-space filling, screening and dephasing due to additional unbound electron-hole pairs.

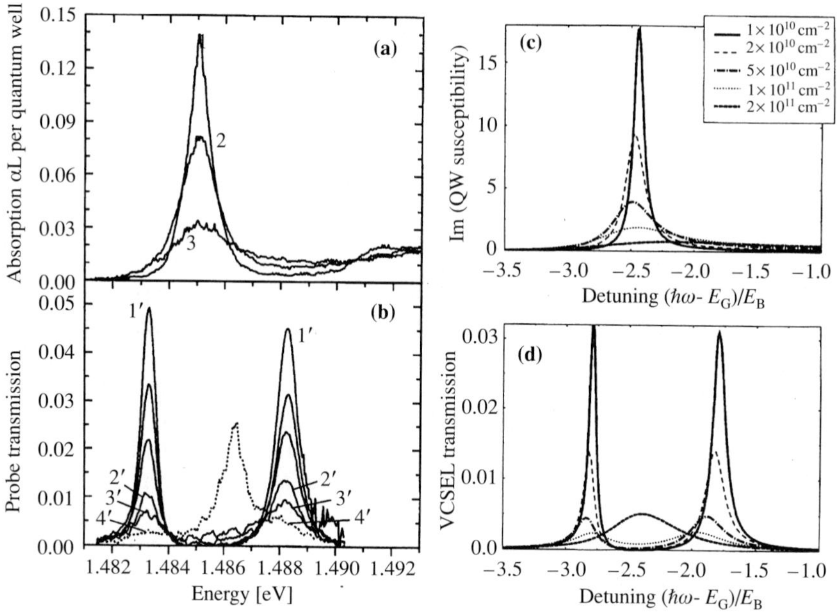

Fig. 9.41. *Left panel*: experimental probe transmission spectra with increasing excitation density for (**a**) a reference sample containing 20 quantum wells like the two in the microcavity sample, (**b**) the microcavity sample with zero detuning between the exciton and photon modes. *Right panel*: (**c**) imaginary part of the optical susceptibility for an 80 Å quantum well at 77 K at various densities as indicated, (**d**) calculated transmission of the microcavity for an increasing plasma density and bleaching of the exciton as per (**c**) [9.71]

Their results for the susceptibility of an 80 Å quantum well at various densities are depicted in Fig. 9.41 c, and the corresponding results for the transmission through a microcavity containing an 80 Å quantum well at the antinode are depicted in Fig. 9.41 d. There is excellent qualitative agreement between experiment and theory. Quantitative comparison is hindered by a lack of knowledge of the absolute densities in the microcavity.

Nonlinear emission dynamics from semiconductor microcavities in the nonperturbative regime has also been investigated recently with luminescence upconversion (Sect. 1.3.3 c) [9.96]. *Lyngnes* et al. observed normal-mode oscillations in emission and reflection. With increasing excitation density they found that the modulation depths of these oscillations is reduced without a change in the oscillation period. In other words, the coupled modes become weaker but their spacing is not reduced, in agreement with the cw pump-probe measurements discussed above.

Several further comments are appropriate at this point. The reason for the gradual decrease in normal-mode splitting reported in earlier studies [9.56, 100, 302] is the large inhomogeneous broadening of excitons. Under such conditions, high densities are required to induce broadening of the exciton comparable to the inhomogeneous broadening. At such high densities, the reduction of the exciton oscillator strength also becomes important. The effects of inhomogeneous broadening on the linear and nonlinear response of microcavities have been investigated [9.84, 302–308]. Interesting effects on the dependence of the polariton linewidths on the "excitonic" fraction of the polariton have also been attributed to inhomogeneous broadening [9.82, 95]. Secondly, the results reported here should be compared to those of *Wang* et al. [9.244] who reported (Sect. 2.7.1 d) that excitation-induced dephasing results in a broadening of the exciton absorption linewidth in bulk GaAs, and that this was the dominant nonlinear effect at low densities. The similarity between the two is clear. Finally, recently there has been some discussion of how luminescence characteristic of the strong-coupling regime, exhibiting polariton-like two-mode behavior, is transformed into a laser with a single emission line as the excitation intensity is increased. Bosonic approximation for bound excitons states has been considered, and the idea of "Boser action" related to Bose condensation has also been considered. However, more recent results have led to different conclusions [9.309]. It is well-known that, for a proper treatment of spontaneous emission, a semiclassical theory is inadequate and a fully quantum-mechanical treatment is required. *Kira* et al. [9.310] have recently developed a fully quantum-mechanical analysis of the interacting electron-hole-photon system and studied microcavity luminescence at various excitation densities in the framework of this theory. They found that the results on microcavity luminescence transforming into laser action at high excitation densities can be consistently explained within this framework in terms of fermionic electron-hole interaction. Finally, *Quochi* et al. [9.311] have recently reported a transition of the coupled exciton-photon modes at low intensities to a triplet at high intensities. One can expect continued interest in the coherent and nonlinear properties of semiconductor microcavities in the coming years.

We conclude this section on coherent and nonlinear effects in semiconductor microcavities by considering more applied aspects of four-wave mixing in microcavities [9.55, 67, 312–314]. In addition to Vertical Cavity Semiconductor Lasers (VCSELs), microcavities are attractive for many other optoelectronic devices because of the enhanced electromagnetic field at the antinode of the microcavity (see, for example, [9.315, 316]). *Jiang* et al. [9.312] reported on non-degenerate FWM in microcavities. They investigated a VCSEL emitting at the photon energy hf at room temperature. When a weak (1 μW) probe at frequency ($f \pm \Delta f$) is injected into the VCSEL, a phase-conjugate signal ($\approx$100 μW) is generated at $f \mp \Delta f$ for Δf close to 1 GHz. *Buhleier* et al. [9.67, 313] and *Collet* et al. [9.55] investigated degenerate FWM from a bulk GaAs asymmetric microcavity (with reflectivities of 0.9 and 0.8 for the back and the front mirror, respectively) using 170 fs and 900 fs pulses from a Ti:Sapphire laser. They found that the cavity-enhanced nonlinearity of the interband transition results in a diffraction efficiency of 0.5% at room temperature for 150 pJ, 900 fs pulses. This was a factor of 400 larger than the efficiency for a bare GaAs reference sample of the same thickness. Diffraction efficiency in microcavities has been theoretically analyzed in [9.55, 87]. *Tsuchiya* et al. [9.314] also considered nondegenerate FWM in microcavities, but with the goal of getting highly nondegenerate interaction. The phase-conjugate signals obtained by *Jiang* et al. [9.312] were restricted to GHz frequency shifts because normal-incidence geometry was used. *Tsuchiya* et al. [9.314] proposed and demonstrated a scheme whereby beams of different frequencies are incident on the cavity at different oblique angles chosen to minimize the phase mismatch between various beams. They found an external efficiency of 1.6% at 10 K for 1 ps pulses with 55 pJ pulse energy. These studies show that efficient nonlinear interaction is possible in semiconductor microcavities and may be potentially useful for optoelectronic and photonic devices.

9.2.12 Coherence in Multiple Quantum-Well Structures and Bragg/Anti-Bragg Structures

Coherent propagation and quantum beats of quadrupole polaritons in bulk Cu_2O [9.317] has been discussed in Sect. 2.6.5 a. It is known from bulk semiconductors that pulse break-up and polariton beating is important in optically thick samples but may be neglected for optically thin samples. Propagation in thick quantum wells was investigated by time-resolved and interferometric transmission techniques [9.318] and has been discussed in Sect. 2.6.5 b. In recent years it has become apparent [9.319–325] that there are important differences between the homogeneous case of the bulk semiconductor and the case of multiple quantum wells. There are several features that are important in this distinction. A multiple quantum-well structure is spatially inhomogeneous and is generally characterized by some degree of disorder (e.g., fluctuations in the quantum-well thickness). Furthermore, exciton-polariton are stationary states in bulk semiconductors but not in quantum wells because of the lack of translational symmetry in the growth direction, allowing intrinsic radiative decay of exciton-polaritons, as discussed in Sects. 6.1.2 and 6.1.4 c [9.326–329]. The emitted radiation travels in both forward and backward directions and leads to a radiative coupling between excitons in different quantum wells. This coupling is efficient because of the high radiative recombination rates of excitons in quantum wells (Sects. 6.1.2 and 6.1.4 c), and can lead to interesting collective effects.

Superradiant decay was predicted (*Citrin* [9.320]) for multiple quantum-well structures thinner than an optical wavelength in the material ($\lambda \approx$ 2500 Å for GaAs-like quantum wells). A typical periodicity in a multiple quantum-well structure is $\approx$250 Å, so that one needs to work with a structure with substantially less than 10 periods to satisfy this condition. *Stroucken* et al. [9.323] examined the coherent dynamics of radiatively coupled quantum-well excitons and found that the effect of light-induced coupling between different wells must be considered in high-quality multiple quantum-well structures. *Weber* et al. [9.330] have experimentally investigated coherent exciton dynamics as the number of quantum wells was varied (1, 2 and 5). In a pump-probe reflection study, they found that a fast initial decay component develops in the differential reflectivity as the number of wells is increased from 1 to 5. They attributed this result to collective effects in the multiple quantum-well structures. *Stroucken* et al. [9.321] considered the light propagation and disorder effects in multiple quantum wells. They reported observation of long tails in 110 fs pulse transmitted through a 20 period $In_{0.08}Ga_{0.92}As$/GaAs multiple quantum-well structure with 100 Å well width and 400 Å barrier width, an observation similar to that discussed in Sect. 2.6.5 b [9.331]. They also found that the transmitted pulse exhibits oscillations whose frequency varies with the detuning between the HH exciton energy and the energy of the center of the incident pulse. Calculations of the linear response with Maxwell's equations through the structure assuming weak disorder and infinite barriers provided qualitative agreement with the data. They conclude that a correct interpretation of the experimental results require an analysis that considers not only the dielectric inhomogeneities but also the disorder in the system. In the case of a nonlinear response of a multiple quantum-well structure, these factors along with the Coulomb interaction must be examined on an equal footing. This was done by *Stroucken* et al. [9.323] applying semiconductor Maxwell-Bloch equations.

The radiative coupling and the phase coherence between the quantum wells in a multiple quantum-well structure can be controlled by choosing the spacing between the quantum wells to be an odd integer multiple of $1/2\ \lambda$ (Bragg structure) or $1/4\ \lambda$ (anti-Bragg structure), where λ is the wavelength of light in the crystal at the exciton resonance. These structures have obvious similarities to microcavities because microcavities can be considered as Bragg or anti-Bragg structures that are folded in space.

From a theoretical analysis of such Bragg and anti-Bragg structures based on Maxwell-Bloch equations, *Stroucken* et al. [9.323, 324] have concluded that propagation-induced dipole-dipole interaction leads to a coupling of electronic excitations in different wells and results in a collective behavior of the electronic dynamics. Such light-induced coupling does not necessarily lead to a splitting of resonances but may result in a pure radiative broadening of the exciton; i.e., the radiative decay rate of the excitons increases because of this interwell coupling. *Hübner* et al. [9.332] were the first to investigate these ideas using degenerate TI-FWM from Bragg and anti-Bragg structures. They investigated one 10-period anti-Bragg structure and two 10-period Bragg structures (with slightly different Al content in the barriers), and compared the results with those in a single quantum well. The samples consisted of 200 Å GaAs quantum well(s) surrounded by $Al_{0.3}Ga_{0.7}As$ barriers of varying thickness. The TI-FWM signals from the samples were investigated at 8 K using 770 fs (3.3 meV) pulses from a Ti:Sapphire laser. Figure 9.42 b depicts the normalized TI-FWM signal as a function of the time delay between the pulses for the single quantum-well sample and the two Bragg structures. Bragg structure A, which comes close to satisfying the Bragg condition, shows a much faster initial decay of the TI-FWM signal

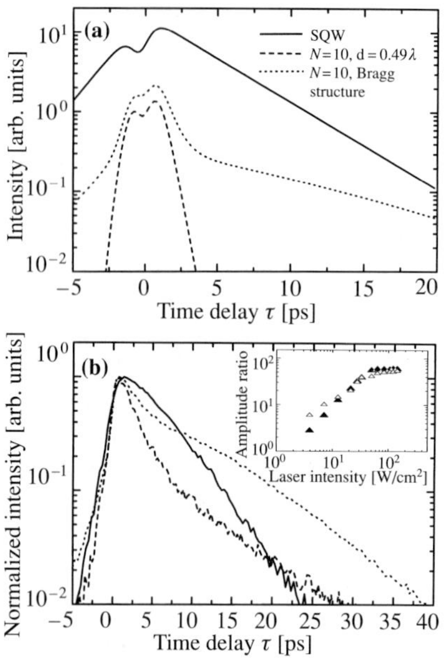

Fig. 9.42. (**a**) Calculated TI-FWM of a GaAs single quantum well (*solid line*), a 10-period Bragg structure (*dashed line*), and an almost-Bragg structure with a spacing equal to 0.49 λ (*dotted line*). (**b**) Normalized experimental TI-FWM signals of a single quantum well (*solid line*), sample A (*dashed line*), and sample B (*dotted line*) at an excitation intensity of 2 W/cm^2. The inset depicts the ratio of TI-FWM signal for the Bragg structure and the TI-FWM signal for the reference single quantum well; *solid triangles* for sample A and *open triangles* for B [9.332]

than the single quantum well. The fast initial component is much smaller in Bragg structure B. The fast initial component is followed by a slow decay in both samples (Fig. 9.42 b). Figure 9.42 a depicts the calculated response based on the semiconductor Maxwell-Bloch equations. Their calculations give a radiative decay time of 9.3 ps for excitons in a single quantum well and 0.96 ps for a 10-period Bragg structure. The experimental data for Bragg structure A resembles the calculation for a slightly imperfect Bragg structure (d=0.49 λ), and are in rough agreement with these values. *Stroucken* et al. also reported the intensity ratio of TI-FWM in different samples, variation of the signal with incident angle (change in the effective thickness of the Bragg structures), and TI-FWM from an anti-Bragg structure.

A number of different aspects of Bragg structures have been studied using a variety of techniques in recent years [9.324, 325, 333–336]. These studies were performed on a 10-period Bragg structure consisting of 200 Å GaAs quantum wells and 945 Å $Al_{0.3}Ga_{0.7}As$ barriers, giving a period close to half the wavelength of the HH exciton in the quantum well. A time-resolved reflected signal from the Bragg structure was investigated at 8 K using the upconversion technique [9.325, 333, 336]. The reflected signal exhibits a characteristic two-peak behavior as a function of time delay. The first peak corresponds to the reflection from the front surface of the sample, and has a width given by the cross-correlation between the exciting pulse and the gating pulse used for upconversion. The subsequent signal is the superradiant emission from the polarization excited in the Bragg structure. An interference between the signal reflected from the front surface and the signal radiated from the sample leads to a dip in the reflected signal and hence the two-peak behavior. The interference depends on the thickness of the top cap layer that separates the quantum wells from the top surface and the width of the exciting pulse. This shows that the effects of the structure must be properly analyzed in interpreting experimental results. *Hübner* [9.932] found that the radiation from the polarization induced in the sample is 100 times stronger for the Bragg structure compared to a single quantum well of the same thickness. They also deduced a radiative recombination time constant of 9.6 ps for the single quantum well and 2.5 ps for the Bragg structure. Theoretical studies have indicated that the radiative coupling between the wells is very sensitive to dephasing mechanisms as well as static disorder. This was investigated by measuring the time-resolved reflectivity of the Bragg structure at various excitation intensities [9.325, 333, 336]. They observed that the reflected signal after the pulse, resulting from the polarization created in the Bragg structure, decreased with increasing intensity. This was attributed to excitation-induced dephasing (Sect. 2.7.1 d) which destroys the coupling between the quantum wells responsible for the superradiant emission.

Another interesting aspect of the coupling between the quantum wells has been revealed by time-resolved four-wave mixing (TR-FWM) studies of *Kuhl* et al. [9.325] and *Hübner* et al. [9.334]. As discussed in Sect. 2.1.1 h (Figs. 2.3 and 2.4), TR-FWM signal (in a two-pulse FWM experiment) from a homogeneously broadened system (in the absence of many-body effects such as local fields and excitation-induced dephasing) is expected to show a rapid rise at the second pulse and then decay with $T_2/2$. In contrast, the TR-FWM signal from an inhomogeneously broadened system exhibits a broad photon-echo peak at a time (measured from the first pulse) equal to twice the time delay between the two pulses (τ_{12}). The results of TR-FWM from the Bragg structure described above are depicted in Fig. 9.43. These data indicate that for small τ_{12}, the signal has an asymmetric shape, with a sharp rise at the second pulse followed by a gradual decay, characteristic of free polarization decay expected from a homogeneously broadened system. However, for large τ_{12}, the signal has become more symmetric with the peak occurring at $2\tau_{12}$, a characteristic signature of the photon echo expected for an inhomogeneously broadened system. The plot of the temporal position of the peak as a function of τ_{12} in Fig. 9.43 reveals this clearly. This transition from behavior characteristic of a homogeneously broadened system to that of an inhomogeneously broadened system is explained briefly as follows: in an N-period Bragg structure, there are N modes with relative radiative coupling strengths a_n (with $n=1\ldots N$). If $n=1$ is the superradiant mode, then, in an ideal Bragg structure, $a_1=1$ and all others $a_n=0$. Because of inevitable imperfections, $a_1<1$ and other subradiant modes have non-zero a_n. The homogeneous linewidth of these modes is given by $\gamma_n=\gamma_0+a_n\gamma_{\rm rad}$ where $a_n\gamma_{\rm rad}$ is the radiative component (with $\gamma_{\rm rad}$ the radiative linewidth of the exciton in a single quantum well of the same width), and γ_0 is the nonradiative contribution to the linewidth. Disorder leads to a distribution in the homogeneous linewidths. Superradiant modes with large homogeneous line-

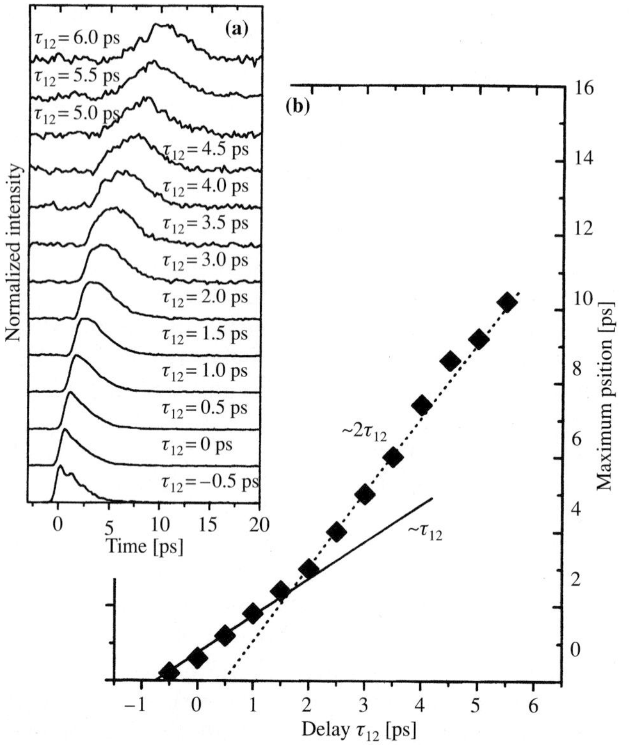

Fig. 9.43. (**a**) TR-FWM from the 10-period Bragg structure described in the text for various delays τ_{12} between the two pulses. α_{ext}, the angle between the normal to the sample surface and the bisector of the two incident laser beams is 45°. (**b**) The position of the maximum in TR-FWM as a function of τ_{12}, showing a slope of ≈ 1 (free polarization decay) and changing to larger slope at higher τ_{12} [9.334]

widths dominate at short delays while inhomogeneously-broadened subradiant modes dominate at long delays leading to a transformation in the nature of the TR-FWM signals as observed.

We note that *Baumberg* et al. [9.171, 337] have recently observed that temporal dynamics of the reflectivity of a femtosecond pulse from a multiple quantum-well structure also depends on the number of quantum wells in the structure and interpreted the results in terms of the temporal evolution of the polariton wavepacket composed of various allowed and forbidden (dark) states of polaritons in a multiple quantum-well structure. These ideas have been discussed recently by *Kavokin* and *Baumberg* who considered the phenomenon of exciton-photon coupling in quantum wells, including the effect of disorder [9.338].

We conclude by noting that the results discussed in this subsection demonstrate the strong influence of radiative coupling between the quantum wells in a multiple quantum-well structure on the ultrafast response of semiconductor multiple quantum wells. The radiative coupling depends strongly on the number of coupled wells, their spacing, and the propagation direction of the beams. The radiative coupling leads to substantial changes in the linear and nonlinear response of the structures, and in the dephasing and radiative recombination rates. The strong collective effects are observed in high-quality samples at low temperatures, and are destroyed by disorder, and intrawell dephasing caused by phonons or other excitons.

9.2.13 Coherent Properties of Quantum Wires and Dots

The first studies of quantum-well excitons by coherent spectroscopy were related to the measurement of the exciton dephasing rates from which the homogeneous linewidth and its dependence on temperature and density of excitons and free carriers were deduced [9.339–341]. The early studies on quantum wires have paralleled these studies in quantum wells. Although optical properties of quantum wires and dots have been investigated quite extensively, they have focused on the nature of electronic states, high-density effects and relaxation dynamics, and only a few studies on coherent properties have been reported.

Mayer et al. [9.342] reported on the first measurements of TI-FWM and TR-FWM on quantum wires. The quantum wires were prepared by etching a 25-period multiple quantum-well structure consisting of 106 Å GaAs wells separated by 153 Å $Al_{0.36}Ga_{0.64}As$ barriers. Holographic photolithography was used to etch rectangular grooves with a 2800 Å period. The active quantum well width was shown to be 600 Å. They observed a biexponential decay in the TI-FWM signals and attributed it to the presence of free and localized excitons. The homogeneous linewidth for free and localized excitons was measured as a function of temperature and excitation density. They found that the temperature coefficient, a in (2.73), for the free excitons in quantum wires was $\approx$6.2 μeV/K, about a factor of 2 higher compared to the free excitons in quantum wells. They were also able to deduce the dependence of the homogeneous linewidth of free and localized excitons on the density of excitons. They also observed quantum beats and attributed them to the two lowest laterally confined states of the $n=1$ HH exciton.

More recently, *Braun* et al. [9.343] have treated the dependence of the dephasing of excitons in quantum wires on the width of the wires. The quantum wires were fabricated starting from a multiple quantum-well structure consisting of 20 periods of 30 Å thick $In_{0.135}Ga_{0.865}As$ quantum wells with 600 Å thick GaAs barriers. Quantum wires were etched to a depth of 1.4 μm using high-resolution electron-beam lithography and etched by a reactive ion etching process. They used a spectrally-filtered (spectral width = 2.3 meV) Ti:Sapphire laser to measure TI-FWM signals for two different wire widths at several temperatures. They assumed that the decay time of the TI-FWM signal corresponds to $T_2/4$, i.e., the inhomogeneous limit (Sect. 2.1.1). The homogeneous linewidth deduced in this manner

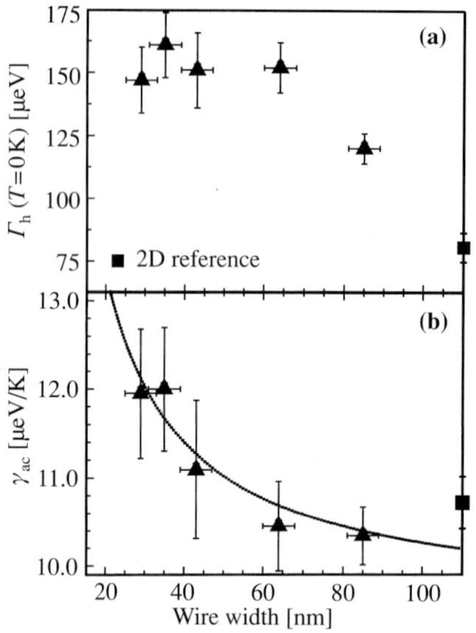

Fig. 9.44. Wire-width dependence of (**a**) the zero-temperature homogeneous linewidth, and (**b**) the temperature coefficient of the homogeneous linewidth. These values for excitons in quantum wires are deduced from TI-FWM studies [9.343]

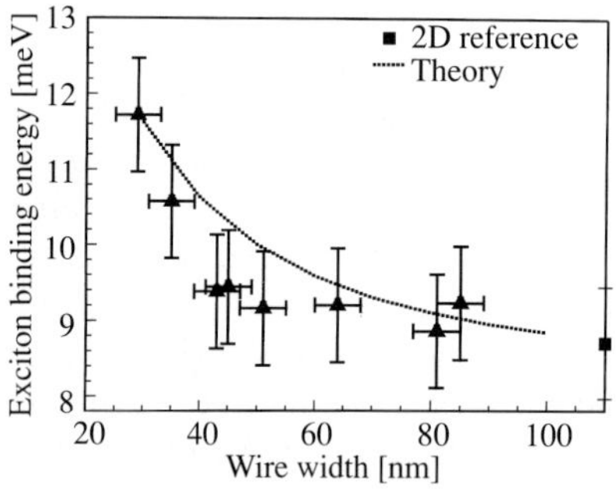

Fig. 9.45. Dependence of the exciton binding energy in quantum wires on the wire width deduced from the period of the beats observed in TI-FWM (*triangles*). The *solid square* indicates the value for the quantum-well reference. The *solid line* shows the results of a variational calculation [9.344]

were fit to an equation of the form $\Gamma_h(T) = \Gamma_h(0) + aT_s$, see (2.73), for each width, and the values of $\Gamma_h(0)$ and a (denoted by γ_{ac} by these researchers) were obtained as a function of wire width. These results are depicted in Fig. 9.44. They show that both $\Gamma_h(0)$ and a increase as the wire width is reduced. *Braun* et al. attributed the increase in $\Gamma_h(0)$ to enhanced fluctuations in the wire width as the width is reduced, and the increase in a to enhanced acoustic phonon scattering as the wire width is reduced. The experimental results are consistent with the theoretical expectation that the excess acoustic phonon scattering due to wires should increase inversely with the width of the wire.

In another investigation on the same set of samples, *Braun* et al. [9.344] studied TI-FWM using 65 fs pulses (spectral width 35 meV) from a Ti:Sapphire laser. Because of the increased spectral width of the laser, a wavepacket consisting of a linear superposition of exciton (and other) states was excited (Sects. 2.6–9). The binding energy of the excitons was deduced from the period of the quantum beats in the TI-FWM signal. Figure 9.45 depicts the exciton binding energy derived from these measurements as a function of the wire width. The results show that the binding energy increases as the wire width decreases. A theoretical calculation provided reasonable agreement with the data. We note also that *Bayer* et al. [9.345] have recently used low-intensity magneto-photoluminescence measurements of the diamagnetic shifts of modulated barrier $In_{0.1}Ga_{0.9}As/GaAs$ quantum wires and quantum dots of widely varying sizes to investigate the effects of confinement and of dimensionality on the binding energies of excitons. They found that the binding energies increase systematically with decreasing size and reducing dimensionality (going from wires to dots of the same size). The increase in the binding energy is found to begin for sizes as large as 10 times the Bohr radius and to reach 3.5 times the bulk value. The experimental results for the exciton binding energies and for their diamagnetic shifts were noted to be in good agreement with detailed calculations for these structures.

Recent studies have used photoluminescence and various microprobing techniques such as NSOM and sub-micrometer apertures to investigate narrow quantum wells. Under certain conditions (see the discussion in Sect. 9.4.5), such studies have demonstrated the existence of extremely sharp (tens of *micro* eV) spectral features [9.26, 346, 347] characteristic of three-dimensional confinement, i.e., 0D systems or quantum dots. We discuss here a recent *frequency-domain* experiment reported by *Bonadeo* et al. [9.348]. These researchers investigated a 42 Å single quantum well with 250 Å $Al_{0.3}Ga_{0.7}As$ barriers at 6 K using two independently-tunable single-frequency dye lasers with bandwidths less than 1 MHz. Apertures ranging in diameter from 0.2 μm to 25 μm were deposited on the sample surface and the substrate was removed to permit transmission measurements. Differential transmission measurements using homodyne detection were performed through a single aperture. Luminescence through a 0.5 μm aperture showed many sharp luminescence lines in atomic-like spectrum [9.28]. Degenerate differential transmission measurements also showed similar structure. Non-degenerate differential transmission measurements revealed a strongly resonant response resulting from energy transfer between different dots. *Bonadeo* et al. were able to obtain detailed information about relaxation and dephasing rates by studying the nonlinear response at several fixed detunings of the pump laser from the quantum-dot transition energy as the detuning of the probe laser was varied. They determined that the population decay time T_1 (due to spontaneous emission and energy transfer, etc.) for an exciton in a single quantum dot is typically ≈ 19 ps and T_2, the total decoherence time, is ≈ 32 ps. This implies that the pure dephasing rate ($= 1/T_2 - 1/2T_1$) is rather small. They concluded that the exciton in a single dot behaves like an atom, in that the nonlinearity arises primarily from a loss of oscillator strength rather than broadening and shift of the resonance frequency that are so important in higher-dimensional semiconductors.

9.3 Ultrafast Emission Dynamics

In the discussion of the hierarchy of time scales in semiconductors in Sect. 1.1.5, recombination was considered as the fourth and the last stage. It is generally true that the recombination that removes the photoexcited carriers and returns the sample to thermodynamic equilibrium is the slowest process. However, we have seen already in Sect. 6.3.2 that recombination from *resonantly excited* excitons in quantum wells can occur on picosecond time scale and hence competes with other relaxation processes in semiconductors. In fact, some of the early secondary emission is in the form of resonant Rayleigh scattering which is coherent in nature. We therefore discuss resonant emission dynamics immediately after the treatment of coherent spectroscopy in the previous section. Stimulated emission can also occur on picosecond time scale and is also discussed in this section. The dynamics of non-resonantly excited luminescence from semiconductors and their nanostructures is described in Sects. 9.4.3 and 9.3.4, as they relate primarily to carrier relaxation and transport.

This section is organized as follows: In Sect. 9.3.1 we outline recent results on the measurement of the *intensity* of the resonant secondary emission. This is followed by a discussion of *phase-sensitive* measurements of the resonant secondary emission; this includes measurement of intensity following excitation by phase-locked pulses (Sect. 9.3.2), as well as direct measurement of the amplitude and the phase of the emitted electric field using the technique of spectral interferometry (Sect. 9.3.3). Finally, Sect. 9.3.4 discusses resonant and non-resonant ultrafast emission and laser dynamics in semiconductor microcavities.

9.3.1 Resonant Secondary Emission from Quantum Wells: Intensity

Consider resonant excitation of a semiconductor near its bandgap with an ultrashort pulse. Since the laser excites an ensemble of dipoles, its radiation pattern is very different from the well-known pattern of a single dipole. Photoexcitation generally excites an area of the sample much larger than the wavelength. Therefore, the coherent radiation emitted by the dipoles or the macroscopic polarization created by photoexcitation interferes destructively in all directions other than the transmission and reflection directions, i.e., the primary directions, and coherent emission occurs only in the primary directions where the interference is constructive. If there are no phase-destroying collisions and only radiative dephasing is allowed, then only coherent emission is possible. The resonant coherent emission occurs only in the primary direction for a "perfect" sample, but may also be present in secondary directions (i.e., other than primary directions) if there are imperfections in the sample. The latter is the Resonant Rayleigh Scattering (RRS). If the interband polarization decays as a result of phase-destroying but energy-conserving scattering processes, resonant incoherent emission [i.e., Resonant

Photo-Luminescence (RPL)] in primary and secondary directions becomes possible.

Femtosecond Resonant Secondary Emission (RSE), i.e., coherent and incoherent resonant emission in directions other than the primary directions, has attracted much attention in recent years [9.213, 214, 349–358]. The initial studies on femtosecond resonant secondary emission from excitons in GaAs quantum wells by *Wang* et al. [9.214], discussed in Sect. 6.3.3, have been followed by a number of interesting studies [9.213, 353–358]. We analyse the measurement of the intensity of the femtosecond resonant secondary emission from GaAs quantum wells in this section. Phase sensitive measurements of the secondary emission using phase-locked pulses and the technique of spectral interferometry are treated in Sects. 9.3.2 and 9.3.3, respectively.

As discussed in Sect. 6.3.3 and above, femtosecond resonant secondary emission from excitons in quantum wells can originate from either RPL or RRS [9.349–352]. *Wang* et al. [9.214] argued that several factors favor the interpretation in terms of RPL under the experimental conditions investigated by them (Sect. 6.3.3). *Haacke* et al. [9.354] have extended these measurements to lower excitation density using the technique of two-color femtosecond upconversion [9.214] (Sect. 6.3.3) employing a CCD camera as a sensitive detector. They investigated a number of high-quality quantum wells at 10 K at excitation densities ranging from 5×10^{8} cm^{-2} to 5×10^{10} cm^{-2}. The samples were excited with 100 fs (18 meV) pulses from a Ti:Sapphire laser with a center energy of 6–8 meV below the HH exciton energy. The upconversion was accomplished by a synchronously pumped OPO operating at 1500 nm and provided temporal and spectral resolutions of 150 fs and 12 meV, respectively. Figure 9.46 shows the temporal evolution of the intensity of the RSE from the HH excitons in the 180 Å quantum wells for three different excitation densities. The thick curves are the experimental results, and the dotted curves are calculated from a theory of RRS [9.349, 350] discussed below. At higher densities, the curves shift to shorter times and are interpreted as RPL from excitons after collisions have destroyed coherence, in agreement with [9.214]. However, a different interpretation was given to the results at low intensities. Figure 9.47 a presents experimental data on the initial RSE after the oscillations in the data resulting from HH-LH quantum beats are removed by Fourier transforms. The shapes of the initial rise changes as the excitation density is lowered. It was argued that at the lowest intensities, the curves deduced from the experiments (Fig. 9.47 a) have the temporal shape expected for RRS resulting from interface roughness. Figure 9.47 b displays the calculated temporal behavior expected for a model of RRS from interface roughness [9.350]. The low intensity curves in Fig. 9.46 b clearly resemble the RRS model rather than a double exponential model also shown in Fig. 9.47 b. On the basis of this comparison it was concluded that RSE at low intensities results from RRS.

The theory [9.350] used for comparison with the data is based on semiclassical treatment and assumes that the disorder in the sample is spatially

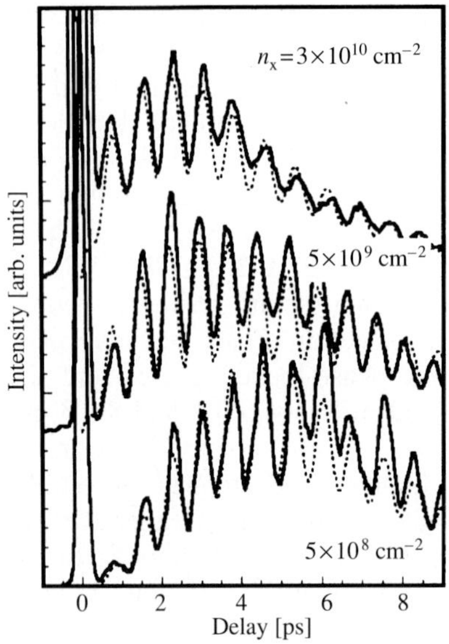

Fig. 9.46. Resonant secondary emission from a 180 Å GaAs quantum-well sample at 10 K for three different excitation densities. The oscillations result from HH-LH quantum beats. The curves are shifted vertically for clarity [9.354]

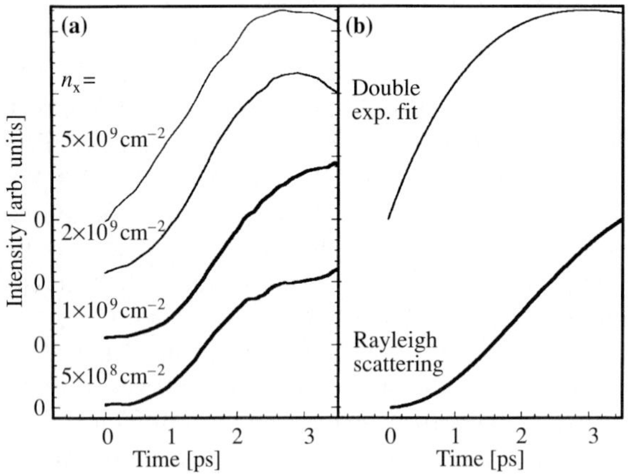

Fig. 9.47. (**a**) Resonant secondary emission after the beats in Fig. 9.46 have been removed. (**b**) Expected behavior of Rayleigh scattering and double exponential fit. The curves are shifted vertically for clarity [9.354]

correlated with a Gaussian correlation length. It then calculates the intensity of RRS assuming the validity of ensemble averaging, a concept discussed in more detail at the beginning of Sect. 9.3.2. Some recent results obtained using phase-sensitive techniques, to be discussed in Sect. 9.3.3, show that this assumption is not realistic because the experiments probe a specific realization of the disorder in the sample. Theory is also being extended to remove this assumption [9.359].

The measurements of the intensity of the resonant secondary emission have thus yielded important information about the dynamical processes in semiconductor quantum wells on femtosecond time scales. In the next subsection, we discuss phase-sensitive measurements of RSE which, by investigating the amplitude and phase of the emitted electric field, provide further insights into this subject.

9.3.2 Resonant Secondary Emission from Quantum Wells: Investigation by Phase-Locked Pulses

Most of the measurements of ultrafast dynamics determine the intensity of a signal as a function of time. This applies to pump-probe, FWM and emission experiments. Early experiments in semiconductors that probed both the amplitude and the phase of the signal have been discussed in Sects. 2.6.5 b and 2.7.1 c. Sophisticated techniques to obtain information about the amplitude and phase have been developed (Sect. 9.1.3) and applied to study coherent control (Sect. 9.2.3) and FWM signals (Sect. 9.2.4) in semiconductors. We describe in this subsection studies of the Resonant Secondary Emission (RSE) from quantum wells excited by two phase-locked femtosecond pulses. We begin with a general discussion of the nature of the RSE in quantum wells, then analyse two studies of RSE from quantum wells excited with a pair of femtosecond pulses: first studying time-integrated emission and thereafter time-resolved emission. Direct measurement of the amplitude and the phase of RSE from quantum wells using the technique of spectral interferometry is discussed in Sect. 9.3.3.

The nature of light scattered from an inhomogeneous medium has been a problem of interest in physics since the pioneering studies of Lord Rayleigh [9.360–363]. We first consider the simple example of Rayleigh scattering of a cw (monochromatic) laser from a rough semiconductor surface. We assume that the laser is far from any electronic resonance so that we are dealing only with non-resonant Rayleigh scattering. It is well-known that the scattered light forms a speckle pattern in this case. The speckles are stationary in space and time, and the size of the speckle is related to the spatial coherence which is determined by the diameter of the laser beam at the sample surface and geometrical optics [9.363]. The formation of speckles is a result of the constructive and destructive interference of light scattered from different points on the rough surface. Consider now the non-resonantly scattered light from this surface when illuminated by a femtosecond laser with a broad spectrum. The scattered light is present only during the laser pulse, but the speckle pattern will be the same as for the cw case because the spectral width of the laser is small compared to the center frequency of the laser.

The nature of scattered light is considerably more complex when the laser excites an electronic resonance in the system as, for example, the excitonic resonance in a quantum-well. In a real quantum-well structure, the interfaces are not atomically smooth [9.364] but exhibit roughness. This disorder localizes the excitons in space, and leads to a local variation (inhomogeneous broadening) of the exciton transition energy, and of the susceptibility of the system in the vicinity of the excitonic resonance. When a cw-laser beam in resonance with a subset of these localized excitons is incident on the quantum wells, the RRS shows a speckle pattern just as in the case of non-resonant Rayleigh scattering from a rough surface. If these localized excitons are resonantly excited by a femtosecond laser with a broad spectrum, the temporal response of the system is very complex. The coherent emission generally lasts longer than the pulse width, determined by the inhomogeneous broadening and the dephasing times in the system. Also, the speckle pattern is different from the case of scattering from a rough surface. A speckle pattern is expected if the scattered light is passed through a narrow spectral filter, just as in the case of scattering of a cw-laser beam from a rough surface. However, if the filter is tuned to a different frequency, the laser will excite localized excitons of a different energy, with different resulting speckle pattern. The temporal and spatial behavior of resonant Rayleigh scattering of a femtosecond pulse with a wide spectrum is very complex. The speckles will vary in space and time, and the variations will depend on the nature and degree of correlation of the disorder in the sample, i.e. on if and how the energies of the localized excitons are spatially correlated. For a totally uncorrelated disorder, i.e. localized excitons in close proximity having completely different energies, the temporal fluctuations in the speckle pattern are rapid. For a correlated disorder, they will be slower and may depend on the correlation function. Heterodyne detection of the emission using spectral interferometry, which provides information about the amplitude and the phase of the emitted electric field, is a promising technique for investigating the underlying physics. The first such measurements are discussed in Sect. 9.3.3.

It is possible to totally ignore the question of speckles and calculate the temporal evolution of the *intensity* of RRS in secondary directions. Such calculations assume the validity of the ergodic hypothesis and derive the intensity of the RRS by ensemble averaging. In this model, the generally accepted picture of RRS in secondary directions [9.349, 350] is as follows. The excitons are initially excited in phase by the femtosecond pulse. The initial coherent radiation is directed only in the primary (transmission and reflection) directions because of destructive interference in other directions. Since the phases of inhomogeneously broadened excitons at different energies evolve at different rates depending on their energy, the destructive interference in the secondary directions does not persist, with the result that the RRS intensity in secondary directions increases. The time evolution of RRS in secondary directions was calculated by *Zimmermann* [9.350] in the semiclassical formalism assuming ensemble averaging with a Gaussian correlation function (in space) for the disorder. The intensity is calculated as an average over an ensemble of different realizations of disorder with the same statistical properties. It is obvious that such an ensemble average can have

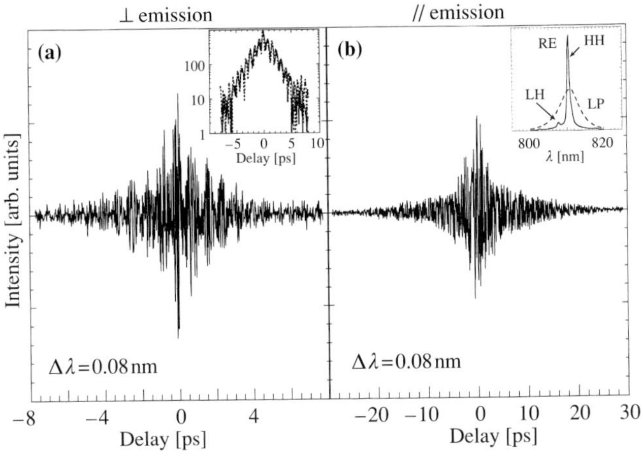

Fig. 9.48. Temporal interferograms of resonant secondary emission through a narrow band-pass filter of bandwidth 0.8 Å. The results are for a 180 Å GaAs quantum-well sample at 10 K excited by a pair of 100 fs phase-locked pulses; the delay refers to the delay between the phase-locked pulses. The emission is polarized (**a**) perpendicular and (**b**) parallel to the exciting laser polarization. Note the different scales for the delay. The inset shows the HH and LH features of resonant emission (RE), and the laserpulse (LP) [9.355]

non-zero intensity but, no speckle pattern is expected from such a calculation. The results of *Haacke* et al. [9.354] were compared to this theory, as already discussed in Sect. 9.3.1.

We will draw upon this qualitative discussion in analyzing the results of experiments treated below and in Sect. 9.3.3.

We first discuss the results of *Ceccherini* et al. [9.353] and *Gurioli* et al. [9.355] who investigated the coherent properties of RSE from GaAs quantum wells excited by a pair of phase-locked femtosecond pulses (Sect. 9.1.3 b). They investigated a single quantum-well structure consisting of a 180 Å GaAs quantum well with $Al_{0.3}Ga_{0.7}As$ barriers at 13 K and low excitation densities ($<10^9$ cm^{-2}). The excitons were inhomogeneously broadened, with a low-temperature PL linewidth of ≈ 1 meV. The HH excitons in the sample are resonantly excited with a pair of linearly polarized, phase-locked pulses with the delay τ_M. The intensity of the time-integrated RSE is detected as a function of τ_M (i.e., the "temporal interferogram" is measured) with spectrally broad or narrow detection. Figure 9.48 a displays the measured temporal interferogram [9.355] for RSE polarized perpendicular to the excitation beam for a spectral bandwidth of 0.8 Å (≈ 0.16 meV). The inset gives the envelope of the temporal interferogram on a semilog scale. The temporal interferogram measured in a broad spectral bandwidth (not shown) exhibits similar behavior. The fact that the two temporal interferograms are similar suggests that the RSE for cross-linear polarization is incoherent. The incoherent emission shows a temporal interferogram because of the interference in the sample between the polarization created by the first pulse with the electric field of the second pulse. This interference produces a population of excitons that depends on the relative phase between the two pulses, the same interference phenomenon that is responsible for the coherent control of excitons discussed in Sect. 9.2.3. Since the interference is produced in the sample, it is independent of the spectral width of detection. The width of the temporal interferogram is determined by the rate of loss of coherence of photoexcited polarization in the sample.

The temporal interferogram for co-linear polarizations measured in a broad spectral bandwidth (not shown) exhibits behavior similar to that for the cross-linear polarizations in Fig. 9.48 a. However, the temporal interferogram for co-linear polarizations determined with a narrow spectral width indicates a more complicated, two-component decay behavior, as depicted in Fig. 9.48 b (note the extended time scale). This behavior allows a discrimination between RRS and RPL. The rapidly decaying component represents RPL. It has the same temporal form as the incoherent emission detected with cross-linear polarization. This is the "coherent control" component resulting from interference of the excitation process in the sample. The fast decay constant of 2 ps results from the decay of the excitonic polarization due to loss of coherence. The slower component in Fig. 9.48 b decays as the inverse of the spectral-filter function and represents RRS.

The relative strengths of the two components in Fig. 9.48 b allows a determination of the *time-integrated* relative strengths of RPL and RRS. *Gurioli* et al. [9.355] deduced from an analysis of their results that the ratio of (time-integrated) RPL to RRS is $\approx$4 to 1 at the emission peak, and increases strongly in the wings where the luminescence dominates. These researchers further found that the femtosecond results can be simulated from a series of frequency-resolved cw measurements, of the kind first reported by *Hegarty* et al. [9.365], for a series of laser-excitation frequencies. Thus, *Gurioli* et al. concluded that these time-domain measurements and the earlier frequency-domain measurements [9.365] contain similar information about the relationship between coherent and incoherent resonant secondary emission.

These results suggest that a discrimination between RRS and RPL can be obtained in a time-domain experiment using phase-locked pulses. These experiments also demonstrate that the electric field in a narrow spectral band has a well-defined phase at a given spatial position so that it can interfere with the electric field of the emission excited by the second phase-locked pulse *at the same spatial position* on the detector. The measurements essentially give the autocorrelation of the emitted coherent field and provide no information on the phase of the emitted field, yielding the intensity spectrum. They also provide no information on the relation between electric fields at different spatial or spectral positions. In particular, the experiments do not show that the instantaneous electric field averaged over the detector area is non-zero. The experiments also provide no information about the *temporal dynamics* of RRS.

We next consider some recent results by *Woerner* and *Shah* [9.357] who also excited a multiple quantum-well structure with two phase-locked femtosecond pulses, but detected the RSE with femtosecond resolution using the two-color upconversion technique (Sects. 6.3.2, 6.3.3 and 9.3.1). *Marie* et al. [9.213] employed a similar technique with picosecond pulses to coherently control the optical orientation of excitons (Sect. 9.2.3 b). *Woerner* and *Shah* [9.357] investigated a multiple quantum-well structure with 10 periods of 170 Å GaAs quantum wells and 150 Å $Al_{0.3}Ga_{0.7}As$ barriers at 10 K, the same sample that was studied by *Wang* et al. [9.214] and discussed in Sect. 6.3.3. They investigated the RSE as a function of the excitation density, the detuning between the laser and the HH exciton energy, the Michelson delay τ_M [coarse delay and phase (or fine) delay] between the phase-locked pulses, and the gate delay τ_G at which the RSE is detected (τ_G is measured relative to the second pulse). The emission induced by each pulse consists of (Figs. 6.18 and 9.48) non-resonant scattering of the laser from the surface roughness (with the spectral and temporal characteristics of the incident laser pulse), and the RSE from the excitons whose temporal and spectral characteristics are under study. *Woerner* and *Shah* identified *three* different components of RSE, with strikingly different properties. For small τ_M, they found that the detected emission for $\tau_G=0$ (i.e., the detection coincident in time with the second pulse) shows oscillations or fringes (with period given by the optical cycle) as the phase between the two phase-locked pulses is varied. This demonstrates that a component of the RSE induced by the first pulse is in phase with the non-resonant surface scattering of the second laser pulse, i.e., with a delayed laser pulse. This is a fully coherent component that has not been considered in previous studies, and corresponds to the interference of the electric field of the laser and RRS at the detector. The time evolution of this component is determined from these studies by varying τ_M. This component was studied in detail by *Birkedal* and *Shah* [9.358] with the spectral interferometry technique (Sect. 9.1.3 a), and is discussed in Sect. 9.3.3. The second component of the RSE is apparent for large τ_G and small τ_M . Under these conditions, the emission shows fringes as the phase between the two phase-locked pulses is varied, but the fringe contrast is *independent* of τ_G and dependent on τ_M. This corresponds to incoherent resonant secondary emission or RPL which yields fringes because of the interference in the excitation process in the sample. The origin of this component is the same as the origin of the fast component in the temporal interferogram measured by *Gurioli* et al. [9.355] and the process underlying the coherent control experiments (Sects. 2.12.4 and 9.2.3).

The third component identified by *Woerner* and *Shah* [9.357] has been observed for intermediate gate delays and for Michelson delays τ_M extending to several picoseconds. For $\tau_M=0$ the RSE shows 100% fringe contrast at all τ_G, as expected. Figure 9.49 depicts, for three finite τ_M, the dependence of the RSE as a function of gate delay τ_G for the case when the two phase-locked pulses are in phase ($\Phi=0$, solid curves) and when they are out-of-phase ($\Phi=180^\circ$, dashed curves). The curves were obtained by taking

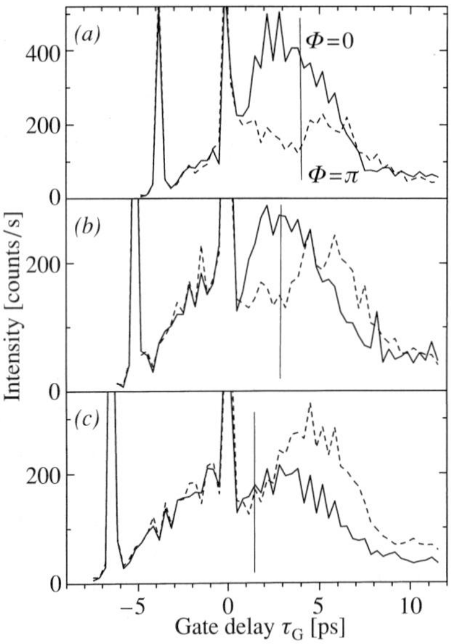

Fig. 9.49. Resonant secondary emission from a 170 Å GaAs quantum well at 10 K excited by a pair of 100 fs phase-locked pulses and detected by two-color upconversion at a gate delay τ_G from the second phase-locked pulse. The results show the maxima ($\Phi=0$: *solid line*) and minima ($\Phi=\pi$: *dashed line*) of the interference fringes observed when the delay between the phase-locked pulses (τ_M) is fine-tuned over several optical cycles for every gate delay. The two intense narrow peaks at $\tau_G=0$ and negative τ_G result from the non-resonant surface scattering of each pulse; their separation indicates τ_M for each data set (*a*), (*b*) and (*c*). Results indicate a complex dependence on τ_M, as discussed in the text [9.357]

an average over several fringe maxima and minima at each gate delay. The results reveal strong fringe contrast that depends on τ_M and τ_G. For a given τ_M, the fringe contrast increases gradually with τ_G. Interestingly, the phase of the fringes changes at certain time delays. *Woerner* and *Shah* argued that the origin of these fringes is neither the fully coherent component or the interference in the excitation process ("coherent control" component) discussed above. They believed that the interference occurs in the sample and not at the detector. However, the interference persists for $\tau_M > 1/\Gamma_{inh}$, where Γ_{inh} is the inhomogeneous linewidth. Therefore, if the second phase-locked pulse arrives after this time ($\tau_M > 1/\Gamma_{inh}$), interference in the excitation process is no longer possible. However, since Γ_h (the homogeneous linewidth) $< \Gamma_{inh}$, the microscopic polarization of individual excitons is not yet destroyed by collisions, and can interfere with the electric field of the second pulse. This interference can be quite complex because microscopic polarization of a large number of localized excitons has to be considered. A complete understanding of this complex interference behavior (Fig. 9.49) awaits further developments.

The study of resonant secondary emission from excitons in quantum wells excited with two phase-locked femtosecond pulses have thus yielded considerable insight into the nature of this emission. In the next subsection we discuss a direct measurement of the amplitude and phase of the electric field of the coherent part of resonant secondary emission from quantum wells.

9.3.3 Resonant Secondary Emission from Quantum Wells: Direct Measurement of Amplitude and Phase

Section 9.1.3 a presents a discussion of how the amplitude and phase of a weak, ultrafast signal can be determined applying the technique of spectral interferometry. The investigation of the polarization-dependent amplitude and phase of the FWM signal from multiple quantum wells is described in Sect. 9.2.4. We have also seen other examples of how the amplitude and phase of linear (Sect. 2.6.5 b) and nonlinear (Sect. 2.7.1 c) signals from semiconductors have provided useful information. We discuss in this subsection the first direct measurement of the amplitude and the phase of the femtosecond resonant secondary emission from semiconductor quantum wells.

Birkedal and *Shah* [9.358] investigated a multiple quantum-well structure consisting of 10 periods of 130 Å GaAs quantum wells with 150 Å $Al_{0.3}Ga_{0.7}As$ barriers as a function of a number of parameters such as temperature (up to 80 K) and excitation intensity. At 10 K and a low excitation density, the full width at half maximum of the HH exciton emission is 0.6 meV. The sample was excited with 100 fs pulses from a Ti:Sapphire laser operating at 82 MHz. RSE emitted by the sample was collected at an angle of 3° from the reflected beam and collinearly combined with a reference laser beam with an appropriate intensity. The combined beam was passed through an aperture to ensure complete overlap of the two beams, dispersed through a spectrometer and detected with a multichannel detector. The relative delay between the two beams was adjusted to give a sufficient number of fringes within the emission spectrum. Measurements were performed as a function of the excitation densities 1×10^7 cm^{-2} to 5×10^9 cm^{-2}. Three spectra were recorded for each set of parameters: the reference laser by itself, RSE by itself and the combined signal. The spectral interferogram is obtained from these measurements by subtracting the two individual spectra from the combined spectrum.

Figure 9.50 a depicts the spectral interferogram of the resonant secondary emission from this GaAs quantum well sample obtained at 10 K and an excitation density estimated to be $\approx 5\times10^8$ cm^{-2}. The spectral interferogram in Fig. 9.50 a clearly shows well-defined, high-contrast spectral fringes over the entire spectral range, and particularly over the emission spectrum which is plotted in Fig. 9.50 b for comparison. Following the discussion of the technique of spectral interferometry in Sect. 9.1.3 a, the observation of these

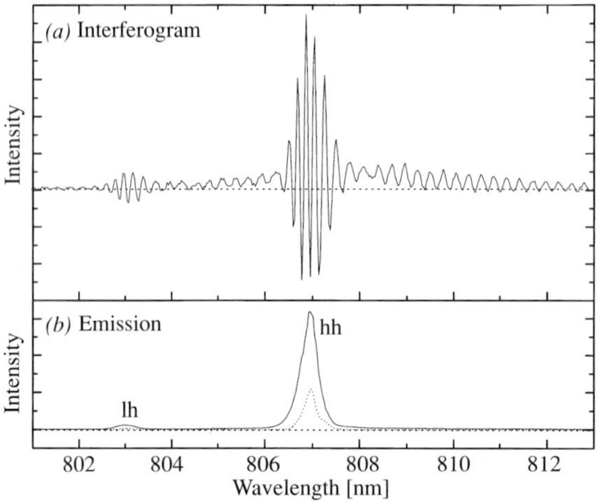

Fig. 9.50. *Upper panel*: Spectral interferogram of the resonant secondary emission from a 130 Å GaAs quantum-well sample at 10 K excited with a single train of 100 fs pulses obtained by the technique of spectral interferometry (Sect. 9.1.3 a). The excitation density is $\approx 4\times10^8$ cm^{-2}. *Lower panel*: time-integrated resonant secondary emission [9.358]

fringes provides a direct and clear demonstration that RSE contains a fully coherent component. The nature and implications of this coherent component are discussed below.

The spectral interferogram (Fig. 9.50 a) exhibits fringes over the entire spectral range. The fringes overlapping the emission spectrum of the excitons (both HH and LH) correspond to the coherent RSE or Resonant Rayleigh Scattering (RRS). The weak fringes over a broader spectral range result from an interference between the laser and the laser light scattered from surface roughness. These have a longer period, indicating that the non-resonant surface scattering occurs before RRS. Also, there is a noticeable phase change between the two sets of fringes.

The temporal evolution of the coherent component of RSE can be directly determined using the technique of Fourier transform spectral interferometry [9.165, 167] (Sect. 9.1.3 a). The results for the density of 5×10^8 cm^{-2} are shown in Fig. 9.51. Since the temporal evolution occurs on a time scale much longer than the pulse duration, de-convolution was not considered necessary. This is the first direct determination of the dynamics of the coherent resonant secondary emission from quantum wells. This emission has the same origin as the first of the three components identified by *Woerner* and *Shah* [9.357] and discussed in Sect. 9.3.2. The dynamics of the coherent component of resonant secondary emission was determined for various excitation densities [9.358]. The results show the relative insensitivity of the dynamics to excitation densities, except at densities above 5×10^9 cm^{-2}, and temperature. The observed dynamics is also compared with the theory based on ensemble averaging [9.350] in Fig. 9.51. In Sect. 9.3.1, it was discussed

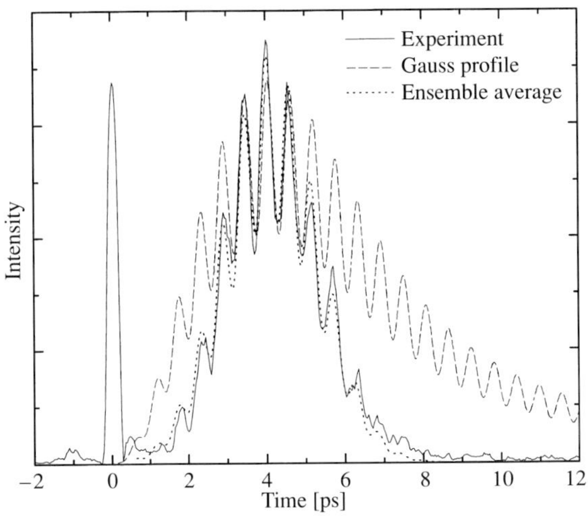

Fig. 9.51. The temporal evolution of the *coherent* component of the resonant secondary emission obtained by the technique of spectral interferometry from the data of Fig. 9.50 (*solid line*) compared with the predictions of a model for Resonant Rayleigh scattering based on ensemble averaging and with a Gaussian function superposed with beats [9.358]

that a comparison of the experimental data on the time evolution of the intensity of RSE with this theory was used to establish the importance of RRS in RSE at low excitation densities. The dynamics of RRS predicted by this model is compared with the dynamics determined by spectral interferometry in Fig. 9.51. The agreement is not good, showing that the ensemble averaging model, or the ergodic hypothesis on which it is based, does not provide a good description of coherent RSE (RRS) from quantum wells. Thus, it was concluded [9.358] that RRS in quantum wells occurs in the non-ergodic regime.

A striking aspect of the results is the observation of high-contrast spectral fringes in the interferogram in Fig. 9.50. As discussed in the introduction of Sect. 9.3.2, the spectral interferogram, representing the amplitude and the phase of the emitted field, provides information about the nature of disorder in the sample. The observation of high-contrast spectral fringes supports the idea of correlated disorder. Thus the assumption of spatial correlation in the ensemble-averaging model [9.349, 350] is justified. However, in contrast to the predictions of the ensemble-averaging model, the experimental results on the dynamics of RRS are well represented by a shifted Gaussian curve, as depicted in Fig. 9.51. The implications of the Gaussian dynamics are interesting. The ensemble-averaging model predicts [9.349] that the decay of the RRS signal can be used to determine the dephasing times of the system. The observation in Fig. 9.51 exhibits that the rise and the decay of RRS or the coherent component of the resonant secondary emission from high-quality GaAs quantum wells are determined by constructive and destructive interference effects, and can not be used to obtain information about dephasing times.

The experiments also suggest that the coherent component constitutes more than 25% of the resonant secondary emission at low densities. This result is in agreement with the frequency-domain results of *Hegarty* et al. [9.365] and the results of *Gurioli* et al. [9.355] using phase-locked femtosecond pulses.

This concludes our discussion of femtosecond resonant secondary emission from quantum wells using a variety of techniques. The next section (Sect. 9.3.4) outlines some results on ultrafast emission and nonlinear effects in semiconductor microcavities.

9.3.4 Microcavities

The dynamics of spontaneous and stimulated (laser) emission from semiconductor microcavities has been investigated by a number of researchers under varied conditions [9.57, 73, 85, 89, 96, 99, 366–378]. We discuss some of these results in this section.

Following the initial demonstration of normal-mode splitting in semiconductor microcavities (Sect. 9.1.1 b) by *Weisbuch* et al. [9.114], *Norris* et al. [9.57, 366] investigated the dynamics of primary emission (i.e., emission in the reflection direction) using the technique of luminescence upconversion (Sect. 1.3.3 c) and pump-probe reflectivity (Sect. 1.3.1 a). They found that for zero detuning (exciton transition energy E_x=cavity resonance energy E_c), they observed oscillations in the emitted intensity with a period that corresponds to the observed splitting in the reflectivity spectrum. The oscillations in the reflected intensity are a natural consequence of the splitting in the reflectivity spectrum, and does not show that the system is a coupled system (see the discussion in Sect. 9.2.11). In an interferometric pump-probe measurement, *Norris* et al. [9.57, 366] observed fringes extending to several picoseconds, much longer than the pulse duration. They attributed the fringes to interference between the reflected probe beam and scattered emission in the direction of the reflected probe, resulting from inhomogeneities in the quantum wells embedded in the microcavity. The presence of interference fringes shows that emission from quantum wells in a microcavity retains the phase memory for several picoseconds.

Yamamoto and collaborators [9.40, 47, 85, 304, 370, 373, 379–382] have investigated a number of different aspects, including temporal dynamics, of microcavities and VCSELs. We discuss here a recent experiment [9.373] in which *Huang* et al. investigated the *amplitude and phase* of the emission in the reflection direction ("primary emission") in the linear and nonlinear regimes with the technique of spectral interferometry (Sect. 9.1.3 a). They investigated the reflectivity from a $\lambda/2$ GaAs/$Al_{0.3}Ga_{0.7}As$ microcavity sample with a single 200 Å thick quantum well at the antinode. The sample was maintained at 4.7 K and excited resonantly with 200 fs pulses from a

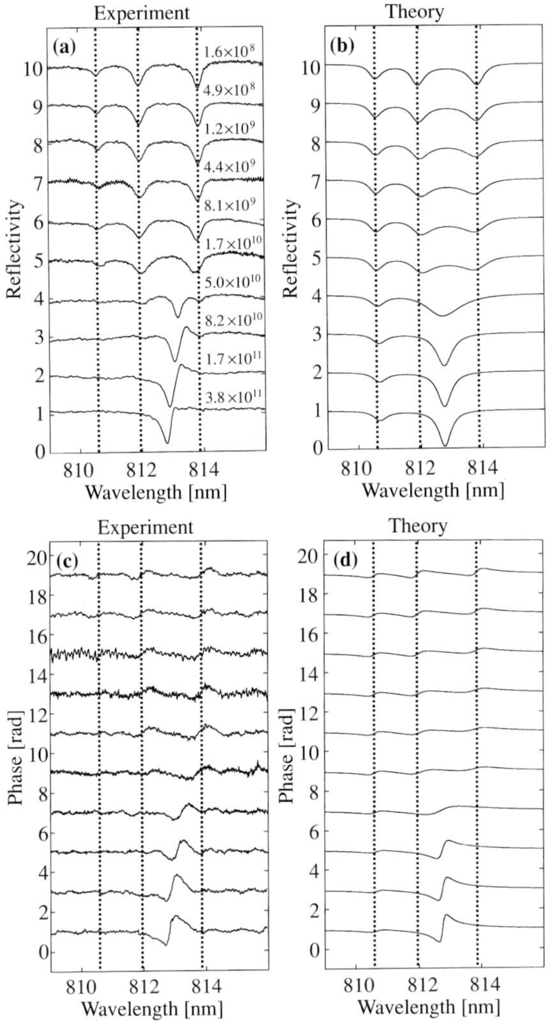

Fig. 9.52. Amplitude (**a**, **b**) and phase (**c**, **d**) of the reflected signal from a GaAs microcavity sample at 4.7 K for various excitation densities. The *left panels* (**a**, **c**) depict experimental results and the *right panels* depict the corresponding theoretical calculations [9.373]

Ti:Sapphire laser. The reflected beam was collinearly combined with a delayed reference beam derived from the same laser. Both HH and LH excitons couple to the cavity mode resulting in a three-mode cavity-polariton behavior. From these measurements, they determined the amplitude and the phase of the reflected beam as a function of the photon energy. They found that the reflectivity shows dips and the phase of the reflected signal exhibits changes at the cavity-polariton energies that depend on the amplitude of reflectivity at the resonance. With increasing excitation density (Fig. 9.52 a and c), the resonances broaden and the normal-mode splitting collapses as expected. A blue shift in the bare photon mode at high intensity is observed and attributed to a change in the refractive index due to free carriers. The researchers calculated amplitude and phase of the reflectivity spectrum based on a transfer matrix model in which the quantum well is treated as having poles in susceptibility at the HH and LH energies. The results of this calculation, depicted in Fig. 9.52 b and d for different intensities, provide a good qualitative agreement with the data, as can be seen by comparison with Fig. 9.52 a and c.

The influence of the magnetic field on the normal modes of the cavity-polariton in semiconductors has led to a number of interesting observations [9.89, 96, 103, 372, 383–386]. We discuss here a recent study which shows that spin-precession dynamics of electrons in a magnetic field leads to real-time modulations of the microcavity-laser emission. *Hallstein* et al. [9.89] investigated a 3/2 λ cavity with two 80 Å $In_{0.04}Ga_{0.96}As/GaAs$ quantum wells at the antinodes of a cavity formed by GaAs/AlAs-distributed Bragg mirrors of 99.6 % reflectivity. The microcavity was held at a temperature of 15K in a 16 T magnet. Choosing the z axis as the growth direction, the magnetic field was in the plane of the quantum well in the x direction (Voigt geometry). The microcavity was excited with circularly polarized, 2 ps pulses from a Ti:Sapphire laser at 7800 Å. The corresponding photon energy was above the microcavity stop band (i.e., both the quantum well and the barriers were excited). The (non-resonant) emission at 8350 Å was temporally resolved in a streak camera with a time resolution of 7 ps. Pronounced temporal oscillations in the emission intensity were observed, with a period of $\approx$100 ps at the field of 2T. The oscillation frequency and the modulation depth were found to be a function of the magnetic field and excitation intensity.

The oscillations were shown to result from “Larmor precession” of photo-excited electrons. Circularly polarized laser excitation leads to a preferential electron spin orientation [9.212]. The magnetic dipole interaction between the electron magnetic moment in the z direction with the magnetic field in the x direction leads to the Larmor precession of the electrons with the frequency $\nu_L = g_e \mu_B B/h$, where g_e is the electron Landé g factor. Consider the case when laser excitation with positive helicity creates a population of spin-up electrons. Thus, initially the gain is highest for emission of positive helicity, and the system lases for sufficiently high intensity. The laser emission is of positive helicity, and only (weak) spontaneous emission is present for the opposite helicity. As a result of Larmor precession, the population of the spin-up and spin-down electrons will oscillate with the Larmor frequency. After half of the Larmor period, electrons have transferred to the spin-down state so that lasing occurs for the opposite (negative) helicity and only (weak) spontaneous emission is observed for positive helicity. At a time delay corresponding to 1/4 of the Larmor period, there is no gain for either helicity and only spontaneous emission is observed for both helicities.

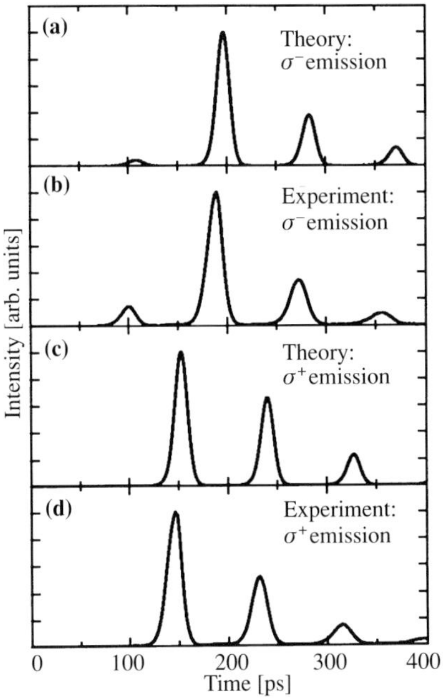

Fig. 9.53. Time-resolved stimulated emission of the quantum-well microcavity laser in a 2T magnetic field for opposite helicities of the emission; experiment and theory as indicated [9.89]

Therefore, the intensity of the total emission oscillates at twice the Larmor frequency. The intensity of emission with positive helicity oscillates with the Larmor frequency. The intensity of emission with a negative helicity also oscillates with the Larmor frequency but is 180 degrees out of phase. This is precisely what is observed. Figure 9.53 presents the measured temporal evolution of the intensity of emission for positive and negative helicities, along with a theory developed by *Hallstein* et al. [9.89]. They noted that a modulation depth greater than 99% and an oscillation frequency of 22 GHz was observed in circularly polarized emission at 2 T, with the oscillations persisting for several hundred picoseconds. Spin relaxation and recombination lead to a decay of the oscillations and laser emission. Increasing the magnetic field to 4 T doubles the oscillation frequency but decreases the modulation depth to 60%, indicating that complex emission dynamics in the microcavity imposes a high-frequency limit to the oscillations.

The frequency of the laser oscillations observed in these measurements was in the range of tens of GHz. *Tsuchiya* et al. [9.376] recently reported oscillations in the THz frequency range. They investigated two 150 Å GaAs/$Al_{0.3}Ga_{0.7}As$ quantum wells at the antinode of a λ cavity with $Al_{0.3}Ga_{0.7}As$ spacer layer. The sample was maintained at 10 K and excited at a position where the cavity resonance coincided with the HH exciton energy (1.543 eV). The sample was resonantly excited with 100 fs pulses from a Ti:Sapphire laser at an oblique angle θ to the normal. Emission was detected in the normal direction. The normal-mode splitting of the microcavity was 4.5 meV at low intensities, but was destroyed at the high inten-

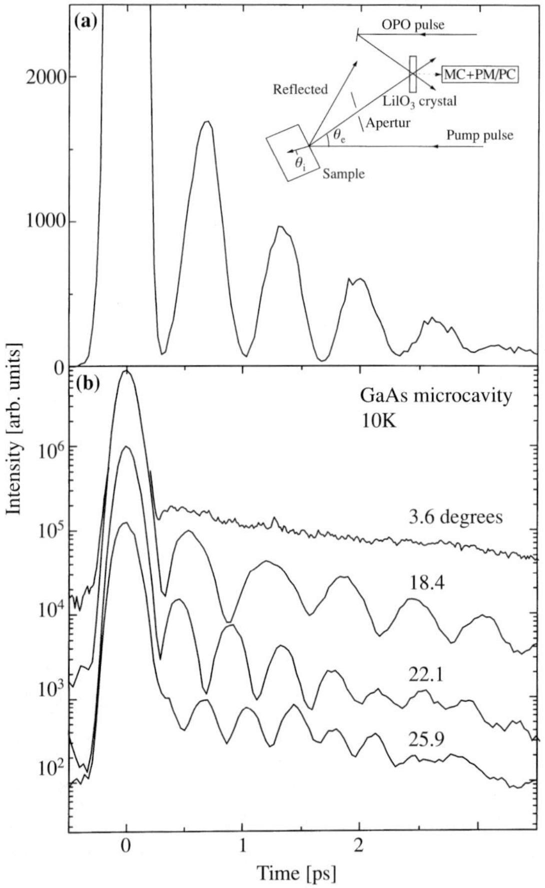

Fig. 9.54. THz oscillations in the emission normal to the surface from a GaAs microcavity at 10 K, resonantly excited by 100 fs pulses from a Ti:Sapphire laser with 20 mW average power. Excitation at an external oblique angle θ_{ext} normal to the microcavity surface: (**a**) linear scale, $\theta_{ext}=17°$; (**b**) semi-logarithmic scale; θ_{ext} as indicated. The frequency of the oscillations increases with increasing angle [9.376]

sity of excitation used in these experiments [9.57, 61, 71, 302, 366]. The emission in the normal direction was temporally resolved with the upconversion technique (Sect. 1.3.3 c). The temporal dynamics of the emitted intensity is displayed on a linear scale in Fig. 9.54 a for a pump power of 20 mW for $\theta=17$ degrees. The emission intensity displays strong oscillations with a period of ≈ 500 fs and a high modulation depth. The oscillations last for several picoseconds. Figure 9.54 b displays the emission intensity on a semilog scale and reveals that the oscillation frequency depends on the external angle of excitation θ. The oscillation frequency corresponded to the difference in the energy of the cavity modes in the normal and oblique directions. Two qualitative explanations based on scattering of the pump-laser light in the normal direction and many-body effects were proposed, but the physical nature of the interference phenomena is not clear.

Vertical Cavity Surface Emitting Lasers (VCSELs) have generated considerable interest in recent years because of their potential for practical applications. These are microcavities with p- and n-type contacts so that the application of forward electrical bias generates gain in the microcavity and leads to laser action normal to the surface. A number of measurements of laser dynamics on VCSELs have been reported in the literature. *Hasnain* et al. [9.387] investigated electrical and optical gain-switching in VCSELs with a streak camera and upconversion. They found that there was a delay of ≈ 50 ps in the turn-on of the laser for optical gain-switching. Electrical gain-switching was somewhat slower but approached this delay for a high amplitude of RF modulation. *Michler* et al. [9.367–369, 374] have extensively investigated the picosecond dynamical response of $In_{0.2}Ga_{0.8}As$/GaAs VCSELs at low temperatures as a function of a number of parameters such as the emission wavelength, the number of quantum wells in the active region of the VCSEL, and the cavity length (λ vs. 2λ). They found that the 2λ cavity produces shorter pulses (as short as 3.3 ps) and faster turn-on (as short as 8.2 ps) times. They have also investigated the dynamics of a coupled-cavity VCSEL in which the top cavity containing 100 Å GaAs quantum well is separated from the bottom cavity containing a 160 Å GaAs quantum well, by a Bragg mirror. They observed pulsewidths as short as 4.8 ps (10 ps) and peak delays as short as 13 ps (16 ps) for the shorter (longer) emission wavelength. Fast pulse fall times of 1 ps were noted for pulses of the shorter emission wavelength, which can be explained by the simultaneous interaction of the two photon modes with both gain regions of the two types of quantum wells. *Hense* and *Wegener* [9.378] have reported that a VCSEL that is already lasing can be turned off when it is optically excited by a short pulse with a photon energy higher than the laser photon energy. This apparently unexpected dynamics can be explained by the heating of the carriers (in the lasing medium) that reduces the gain below threshold in spite of the increase in the density of electron-hole pairs. Subsequent cooling of the carriers by interaction with the lattice (Chap. 4) leads to increased gain and turns the laser on again. However, the turn-on is slow because of the low gain of the VCSEL [9.387]. *Elsaesser* et al. [9.377] showed that the laser can be turned on faster by using a bulk semiconductor as the gain medium.

Sermage et al. [9.99, 371] have investigated relaxation dynamics of cavity-polaritons by measuring resonant and near-resonant secondary emission dynamics at low intensities. The emission was detected using a streak camera with a time resolution of 3 ps. They investigated a $\lambda/2$ microcavity consisting of two 120 Å strained quantum wells of $In_{0.14}Ga_{0.86}As$ in an AlAs spacer layer. The quantum wells were at the anti-node of the cavity formed by two GaAs/AlAs Bragg mirrors with 20 and 13.5 periods. A reference sample, similar to the microcavity sample but with the top mirror removed, was also investigated. The samples were maintained at 10 K and excited by transform-limited 1.5 ps pulses from a Ti:Sapphire laser operating at 82 MHz. The narrow spectral width (≈ 1 meV) of the laser permitted selective excitation of either the upper or lower branch of the cavity-polariton because the normal-mode splitting was 4.8 meV. The sample was excited at an external angle of 8 degrees, and luminescence with a maximum external angle of about 11 degrees was collected, dispersed by a spectrometer and detected by a streak camera. Excitation of the lower branch generated resonant luminescence at the lower-branch energy. Excitation of the upper polariton branch resulted in two emission peaks, one due to resonant luminescence from the upper branch and the other due to near-resonant luminescence from the lower branch. Detuning between the cavity and the exciton was varied by selecting different positions on the tapered microcavity sample.

Some of the results of *Sermage* et al. [9.371] are displayed in Fig. 9.55. (a–c) show the temporal dynamics of the emission under different conditions. The initial strong emission peak from non-resonant surface scattering is absent in these experiments. Figure 9.55 a presents the time evolution of the resonant secondary emission of the lower polariton branch for three different detunings between the cavity mode and the exciton. Figure 9.55 a also shows the luminescence dynamics of the reference sample. Negative detuning means that the cavity mode is at a lower energy than the exciton resonance at 1.395 eV. The lower branch is photon-like for large negative detuning and exciton-like for large positive detuning. Just the opposite is true for the upper polariton branch. Figure 9.55 b depicts the resonant secondary emission from the upper branch for three different detunings. Finally, Fig. 9.55 c exhibits, for three different detunings, the luminescence dynamics of the lower polariton branch when the upper branch is resonantly excited. Figure 9.55 a and b illustrate that the resonant emission has an intense, rapidly decaying component which is followed by a slowly decaying component for certain detunings. When the excitation energy is such that it excites a photon-like mode [large negative (positive) detuning for lower (upper) branch excitation], the initial strong decay totally dominates and there is no delayed emission [see curve (a) in Fig. 9.55 a and curve (c) in Fig. 9.55 b]. Under these conditions, the decay is controlled by the decay rate of the photon from the cavity, i.e., by the mirror reflectivities. As the detuning is varied such that the excitation begins to excite the normal mode with a larger exciton-like character, the emission develops a component which decays slower. This component of the luminescence has the characteristics of resonant secondary emission from quantum-well excitons (Sect. 6.3.2). For detuning such that the excitation excites a primarily exciton-like normal mode [large positive (negative) detuning for the lower (upper) polariton branch], the emission dynamics is totally controlled by excitonic processes (Sect. 6.3.2). Figure 9.55 d plots the initial decay time of emission as a function of detuning when the lower polariton branch is resonantly excited (corresponding to Fig. 9.55 a). The initial decay is extremely fast for large negative detuning (lower branch primarily photon-like), but becomes

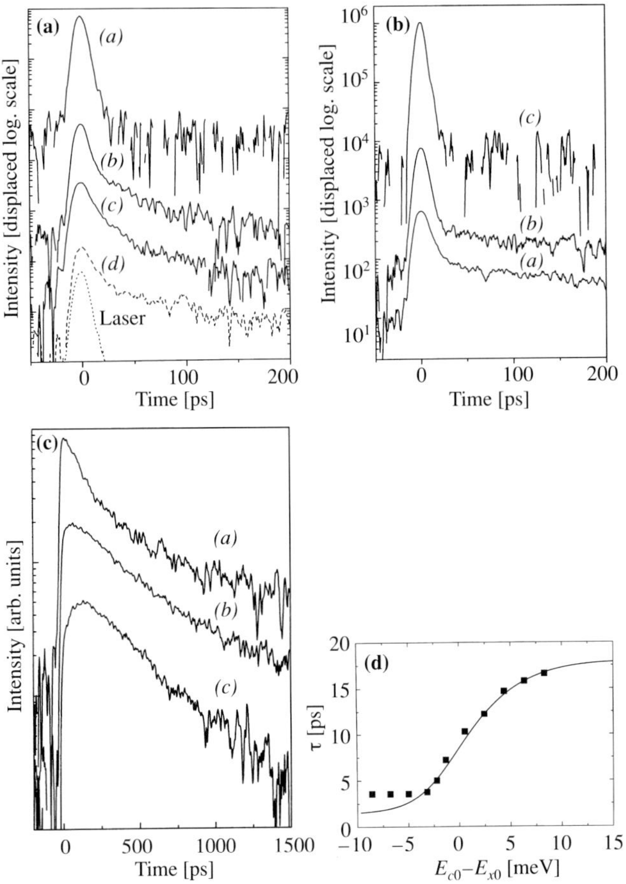

Fig. 9.55. Luminescence decay curves for an $In_{0.14}Ga_{0.86}As$ microcavity sample at 10 K at low excitation densities. (**a**) Resonant luminescence of the lower polariton branch for different detunings: (*a*) –8.5 meV, (*b*) –1.5 meV, (*c*) +3.5 meV. The decay curve of resonant luminescence from the reference sample is shown in (*d*), and the instrument response is indicated by the *dotted curve* marked "laser". (**b**) Resonant luminescence of the upper polariton branch for different detunings: (*a*) –4.8 meV, (*b*) –1.0 meV, (*c*) +4.7 meV. (**c**) Decay of the lower polariton luminescence when the upper branch is resonantly excited for different detunings: (*a*) –6.5 meV, (*b*) –1.5 meV, (*c*) +3.5 meV. (**d**) Initial resonant luminescence decay times as a function of exciton-cavity detuning; *points* are experimental, the *dashed curve* is a two-level model and the *full curve* is the polariton model [9.371]

slower and saturates near 17 ps as the detuning becomes positive (lower branch becomes exciton-like). The dotted and solid curves in Fig. 9.55 d represent calculations of the initial decay times for two different models [9.371]. A reasonable fit to the data is obtained. *Bloch* and *Marzin* [9.375] have recently considered this problem in more detail and have found an excellent quantitative agreement with the data discussed here.

An additional relaxation mechanism is active when the upper polariton branch is excited. The fact that luminescence at the energy of the lower polariton branch is observed when only the upper branch is resonantly excited demonstrates that relaxation from the upper to the lower branch is effective on the time scale of the decay rates of the upper branch. Figure 9.55 c shows the temporal dynamics of the lower polariton emission when the upper branch is resonantly excited. The data indicate that there is a finite rise time of the emission from the lower branch, and this rise time depends on detuning. The decay of the emission is relatively slow and independent of the detuning in this case. No quantitative analysis of this relaxation dynamics is available at this time.

Wainstain et al. [9.104] have also investigated the rise and decay times of the two polariton modes in a microcavity comprising of two groups of three 75 Å $In_{0.13}Ga_{0.87}As$ quantum wells separated by 100 Å GaAs barriers positioned at the antinodes of a 3 $\lambda/2$ cavity. They have performed the measurements as a function of temperature and varied the external angle of incidence to excite at different points of the upper and lower branches of the polariton dispersion curves. Using these and cw studies [9.388], they have investigated the effects of polariton relaxation and compared the results with a theoretical calculation.

We conclude this section by noting that most of the work on 3D microcavities is related to microcavity lasers where the goal is to achieve ultralow threshold currents. There are, however, some reports of optical studies of 3D microcavities. These include investigations of the size dependence of the confined optical modes [9.389], of the strong-coupling regime [9.83], and of the lifetimes of the optically confined modes using picosecond spectroscopy [9.97]. Ultrafast spectroscopy of 3D microcavities will no doubt receive increasing attention in the coming years.

9.4 The Incoherent Regime: Dynamics and High-Intensity Effects

Much of the discussion so far has focused on the coherent regime (Sect. 1.1.5) where the excitations in the system have maintained a well-defined phase relation with the electromagnetic field exciting them. In the remainder of this chapter, we consider ultrafast studies in semiconductors after the coherence is destroyed by various scattering processes. Ultrafast relaxation dynamics and nonlinear effects at high intensities in semiconductor nanostructures are discusssed in Sects. 9.4.1–5, and ultrafast carrier-transport in Sect. 9.4.6.

9.4.1 Carrier Dynamics

Relaxation dynamics of photoexcited carriers in semiconductors has been investigated extensively using ultrafast spectroscopy techniques (Chaps. 3 and 4). The emphasis in recent years has shifted from relaxation dynamics to the coherent response of semiconductors (Sect. 9.1). However, carrier relaxation dynamics continues to be investigated and we discuss in this subsection some recent reports.

(a) Electron Dynamics. We first describe an experiment by *Lutgen* et al. [9.390] in which they investigated femtosecond relaxation of electrons following intersubband excitation in an n-modulation-doped quantum-well structure consisting of 50 periods of 80 Å $In_{0.53}Ga_{0.7}As$ quantum wells separated by 140 Å $Al_{0.48}Ga_{0.52}As$ barriers which were δ doped at the center to provide an electron density of 5×10^{11} cm^{-2}. Samples were cooled to 8 K and electrons in the $n=1$ subband were excited to the $n=2$ subband with 130 fs infrared pulses with a photon energy of 200 meV (corresponding to 6.2 μm) resonant with the intersubband transition. The electron dynamics was probed with 100 fs near-infrared pulses by monitoring the change in the interband absorption corresponding to $n=1$ and $n=2$ electron subbands. The pump intensity was such that 10–20% of the electrons were excited from the $n=1$ to $n=2$ subband.

The differential transmission signal near 1.14 eV, corresponding to the interband transition to the $n=2$ electron subband, shows a rapid rise within the pump pulse and then a decay with a time constant of 1 ps. Comparison with an ensemble Monte Carlo simulation led to the conclusion that scattering of electrons with polar optical and interface phonons is responsible for this rapid depopulation of the $n=2$ subband. The experiments also monitored the Differential Transmission Spectrum (DTS) near the interband transition involving the $n=1$ subband. The DTS signal, depicted in Fig. 9.56, initially decreases because the depletion of electrons by the pump pulse reduces the occupation of the final state in the interband absorption. The decrease corresponds to a depletion of about 15% of the electrons, as determined from a comparison with a Monte Carlo simulation. Subsequently, scattering of the thermal electrons with the photoexcited electrons in the $n=1$ and $n=2$ subbands further depletes the population of thermal electrons. Although the electrons return to the $n=1$ subband with a time constant of 1 ps, they are not immediately observable in the DTS because they are outside the energy range probed. Since there is no change in the electron density in the sample, the spectrally integrated differential transmission signal would be expected to be zero for a thermalized distribution if the matrix elements are independent of energy. This has not been the case initially, indicating that a substantial fraction of the electrons are outside the energy range probed, and probably have a non-thermal distribution function. Subsequent electron-electron interactions thermalize the electron distribution. The electron-phonon interactions cool the electrons so that the negative and posi-

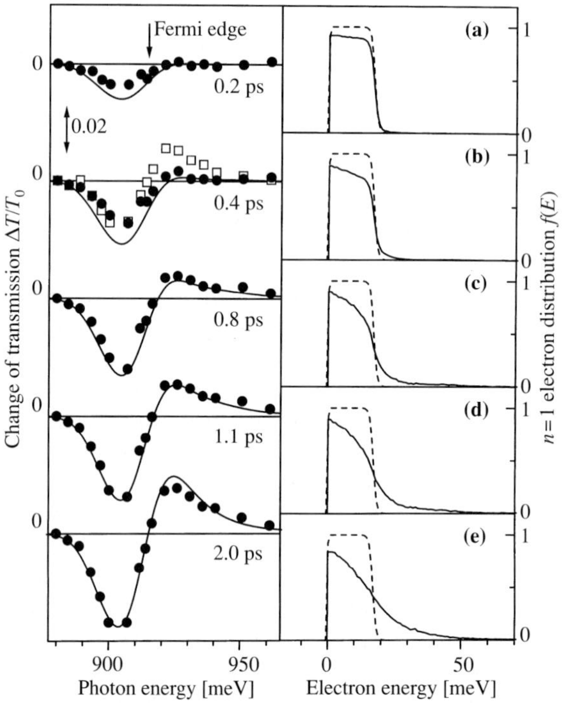

Fig. 9.56. *Left panel*: Transient Differential Transmission Spectra (DTS) of n-modulation-doped InGaAs quantum wells near $n=1$ interband transition after femtosecond inter-subband excitation. The delays between the pump and the probe are indicated in the figure; the *open squares* in the second plot from the top are for a delay of 11 ps. The probe pulses have FWHM of 13 meV. *Solid lines* are results of Monte Carlo simulation. *Right panel*: Transient distribution function of $n=1$ electrons calculated by the ensemble Monte Carlo simulation. The *dashed line* is the initial Fermi distribution [9.390]

tive part of DTS become more nearly equal (see the open squares representing the data for a delay of 11 ps). The right-hand panel of Fig. 9.56 illustrates the calculated electron distribution function at various time delay showing how the thermal tail in the distribution function develops with time. *Lutgen* et al. [9.390] concluded from these measurements that the electrons in the $n=1$ subband remain non-thermal for times as long as 2 ps. The slow thermalization was attributed to a strong reduction in electron-electron scattering due to Pauli blocking.

Inter-subband scattering of carriers in quantum wells has been a subject of interest for many years and the early work with pump-probe spectroscopy and Raman spectroscopy is discussed in Sect. 3.3.3. This subject continues to be investigated because of an interest in the fundamental physics and because of the interest in inter-subband lasers reactivated by the work on quantum cascade lasers by *Capasso* and co-workers [9.391]. Some recent work on inter-subband scattering was reported in [9.392–394]. For quantum wells of widths such that the inter-subband energy spacing E_{12} is larger than the Longitudinal Optical (LO) phonon energy $\hbar\omega_{LO}$, it is generally accepted that scattering of carriers by LO phonons dominates the carrier relaxation process. Since the Fröhlich interaction depends inversely on the wave vector of the phonon, this process is most efficient for $E_{12}=\hbar\omega_{LO}$ for carriers at the bottom of the second subband. Times in the sub-picosecond regime have been obtained for this case (Sect. 3.3.3). For $E_{12}<\hbar\omega_{LO}$ this process is not possible for carriers near the bottom of the $n=2$ subband and the expectation is that the inter-subband scattering rate of the carriers by acoustic phonons is rather small. Totally convincing evidence has not yet been presented for very long inter-subband scattering times for wide quantum wells for which $E_{12}<\hbar\omega_{LO}$.

Hartig et al. [9.395] have recently reported an investigation of inter-subband scattering of electrons in a p-doped asymmetric double quantum-well (a-DQWS) structure. (The reader is referred to a discussion of the basic concepts of a-DQWS in Sect. 7.3, various aspects of tunneling in a-DQWS in Chap. 7, and coherent oscillations of electronic wavepacket in a-DQWS in Sects. 2.8 and 2.12.1). The sample was designed such that the $n=2$ level in the 220 Å wide well was in (near-) resonance with the $n=1$ level of the 90 Å narrow well separated by a 30 Å barrier. The p-doping was quite high ($\approx 10^{11}$ cm^{-2}) to allow measurements of time-resolved luminescence at a low photoexcitation density. Tunneling interaction splits the $n=2$ level into symmetric and anti-symmetric states. Because of the hybridization of the wave functions, the $n=2$ states of the coupled structure at perfect resonance would have half its amplitude in the narrow well. Therefore, the inter-subband scattering rate from each of the two delocalized $n=2$ states to the $n=1$ state (localized primarily in the wide well) is reduced by a factor of 2, further facilitating measurement of the dynamics of the $n=2$ population. This structure was cooled to 20 K and excited at 1.532 eV, slightly below the (e-hh)$_2$ transition energy by a Ti:Sapphire laser (spectral width: 22 meV). The electron wavepacket, initially created in the wide well, undergoes coherent oscillations (Sects. 2.8 and 2.12.1) which are damped out quickly. The electrons also undergo scattering between the two $n=2$ levels and between the $n=2$ and $n=1$ levels. Information about the inter-subband scattering rates can be obtained by monitoring the population of the $n=2$ states by time-resolved luminescence. *Hartig* et al. [9.395] measured the time-resolved luminescence using the two-color upconversion scheme with 150 fs gating pulses from a synchronized optical parametric oscillator [9.214]. Figure 9.57 a depicts their results for an undoped 210 Å multiple quantum-well structure at 77 K for a delay of 600 fs. There is a hint of the emission from the $n=2$ subband at 1.548 eV, the expected step in intensity being broadened by the 11–14 meV spectral width of the gating pulse in the upconversion scheme. Figure 9.57 b presents the results at various delays for the a-DQWS sample described above. The emission from the $n=2$ subband is clearly observed for several picoseconds. From measurements such as these, *Hartig* et al. deduced an inter-subband scattering time of 2.9 ps in

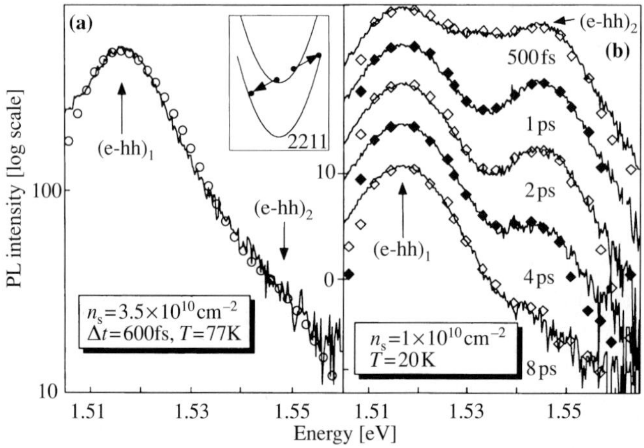

Fig. 9.57. (**a**) Spectrum of a 210 Å GaAs quantum-well sample at 77 K for a delay of 600 fs and 3.5×10^{10} cm^{-2}. Circles represent calculated PL spectrum under thermal equilibrium between $n=2$ and $n=1$ electrons. *Inset*: most important inter-subband carrier-carrier scattering process. (**b**) The PL spectra of a p-doped coupled quantum-well sample for various delays. The diamonds represent their results of Monte Carlo simulation [9.395]

these structures for photoexcited densities $\leq 5\times10^{10}$ cm^{-2}. Because of the hybridization of the wave function discussed above, this would correspond to a time of 1.5 ps in a quantum-well structure. They attributed this time to electron-hole scattering. At densities above 5×10^{10} cm^{-2}, electron-electron scattering begins to play a more important role. By comparing the results with an ensemble Monte Carlo simulation, *Hartig* et al. deduced that, at higher excitation densities, inter-subband electron-electron scattering increases the mean energy of the electron in the $n=2$ bands in the first picoseconds. They also deduced that about 75% of the carriers scatter from the $n=2$ to $n=1$ by carrier-carrier scattering and the remaining transfer via LO-phonon emission made it possible by the heating of the electrons by inter-subband carrier-carrier scattering. These studies have provided good insights into the physics of inter-subband scattering at high carrier densities. However, inter-subband scattering rates for acoustic phonons still remains undetermined.

We have so far discussed the relaxation of *population* from a higher subband to a lower subband in a quantum well. Recently, *Kaindl* et al. [9.396] have investigated *dephasing* of inter-subband polarization in a n-modulation-doped $In_{0.53}Ga_{0.47}As/InP$ multiple quantum-well structure with 60 Å wells and 50 periods. The inter-subband separation was 257 meV. Using TI-FWM measurement technique in the infrared, they measured the decay of the resonantly excited inter-subband polarization and deduced that the dephasing time is 320 and $\leq$200 fs for the lowest (1.5×10^{11} cm^{-2}) and the highest (1.5×10^{12} cm^{-2}) electron density, respectively. We also mention that a cw photoluminescence-measurement technique has been used recently

[9.392] to determine inter-subband scattering rates in an n-modulation-doped GaAs/AlGaAs single heterostructure in the presence of a magnetic field.

(b) Hole Dynamics. We have discussed inter-subband relaxation of electrons in Sect. 9.4.1 a. There is also considerable interest in the topic of hole relaxation in the valence bands of bulk and 2D semiconductors. Since the time-resolved pump-probe and luminescence signals are usually dominated by electron dynamics in an undoped semiconductor, one approach to investigate holes is to use n-doped semiconductors for which hole dynamics can be isolated [9.397–401]. Relaxation of non-thermal holes in n-modulation-doped quantum wells [9.399] has been treated in Sect. 3.3.2, and some early work on hole relaxation is referenced in that subsection. We discuss here some of the more recent work on hole relaxation.

A new technique, based on investigating heating of a cold photoexcited carrier distribution [9.402], was used recently [9.403–405] to investigate heating of holes in undoped GaAs as a function of temperature, carrier density and excitation photon energy. The basic concept behind the technique is to photo-excite a cold-carrier distribution (the excess carrier energy less than the thermal energy $3kT_L/2$, where T_L is the lattice temperature) and monitor the Differential Transmission (DT) at a higher probe photon energy. [The idea of probing at a photon energy higher than the pump energy was used by *Camescasse* et al. [9.184] to study *electron* dynamics (Sect. 9.2.1)]. Since the effective mass of the heavy holes is much larger than that of the electrons, the holes occupy states with a much larger wave vector as they equilibrate with the lattice by absorption of phonons. Simple estimates show that, for GaAs with the probe photon energy of 250 meV above the bandgap energy, the heavy hole contribution to DT is larger than that of electrons by a factor of 5, 50 and 10^4 at T_L of 300, 190 and 100 K, respectively. The DT signal is determined primarily by bandgap renormalization, modification of the Coulomb enhancement factor and occupation of the hole states coupled to the probe energy.

Femtosecond dynamics of DT was investigated for a 0.2 μm thick GaAs sample with AlGaAs cladding and anti-reflection coating as a function of temperature (100–300 K), density (3×10^{16} cm^{-3}–7×10^{17} cm^{-3}) and pump wavelength (8100–8400 Å) [9.403–405]. The pump and probe pulses were derived by spectral filtering the (spectrally broadened) output of a short fiber pumped by a high-power Ti:Sapphire laser operating at 76 MHz [9.406]. Transform-limited, independently tunable pump and probe pulses (90 fs) with a nearly Gaussian pulse shape were used to measure DT with a noise level of a few times 10^{-7}. The temporal evolution of DT at the probe energy of 1.68 eV for three different temperatures and a carrier density of 2×10^{17} cm^{-3} is depicted in Fig. 9.58. The pump-photon energy was varied at each temperature to keep the initial hole excess energy small, as indicated in the figure caption. All curves show a rapid initial induced absorption (decrease in differential transmission) due to bandgap renormalization

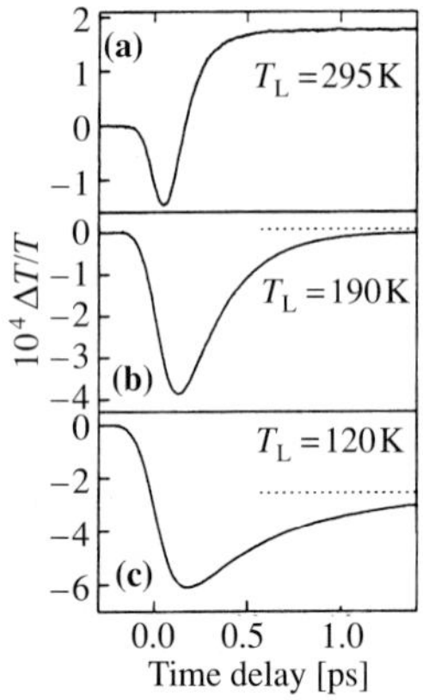

Fig. 9.58. Measured transient transmission change in GaAs for a probe photon energy of 1.68 eV and three lattice temperatures as indicated. The carrier density is 2×10^{17} cm^{-3} and the pump photon energy is (**a**) 1.48 eV, (**b**) 1.5 eV, and (**c**) 1.51 eV [9.405]

(BGR) resulting from the photoexcited carriers. BGR increases with increasing carrier density and decreasing temperature; it also depends, in principle, on the carrier distribution function [9.407]. Induced absorption due to BGR is reduced by the occupation of the hole states coupled to the probe beam (state-filling effects). The light-hole contribution is negligible so that the recovery of the induced absorption provides information about the heating of the initially cold holes by absorption of phonons. The recovery-time constant increases as the lattice temperature decreases because of the decrease in the number of phonons available for absorption. The level to which the signal recovers depends on the relative strength of BGR and state-filling effects. At 295 K, BGR is small so that state filling leads to an eventual bleaching of the signal; at lower temperatures, BGR is larger and state filling is not sufficiently strong so that induced absorption persists at all times shown. Eventually, as the carriers recombine both BGR and state filling vanish and so must the differential transmission signal.

Other data obtained [9.403–405] show that the hole thermalization time constant increases as the excess carrier energy is reduced by increasing the pump wavelength, decreases by about 25% as the carrier density was increased from 3×10^{16} cm^{-3} to 1×10^{17} cm^{-3}, but then remains nearly constant with further increase in density. An analysis of these data provides useful information about hole dynamics in GaAs. The hole thermalization time constant varies from 550 fs at $\approx$120 K to 120 fs at room temperature, and was found to be independent of the probe energy in the range in which holes dominate the DT signal. Using the calculated value of polar optical phonon scattering of holes, a value of 40 eV was deduced for the optical deformation potential for holes. These results show that useful information about the hole dynamics in intrinsic semiconductors can be extracted by this technique. A similar scheme to investigate the heating of resonantly excited excitons was used by *Kim* et al. [9.408] to determine exciton phonon interaction and is discussed in the next subsection.

9.4.2 Exciton Dynamics

Following the investigations on exciton formation and relaxation (Chap. 6), there have been a number of further studies [9.409–413]. The most direct information about exciton dynamics has been obtained when it is possible to monitor the *distribution* of excitons in the center-of-mass momentum space, as discussed in Sect. 6.2.2 for the case of Cu_2O. Phonon-assisted excitonic transitions in II–VI semiconductors provide such opportunities [9.414, 415], recently exploited for ZnSe quantum wells [9.413]. However, in keeping with the main theme of this book, we discuss here some recent results on III–V semiconductors.

We first consider the use of differential transmission at heavy-hole (HH) and light-hole (LH) excitons as a probe of relaxation processes in GaAs quantum wells following resonant excitation of HH and LH excitons. In contrast to the earlier differential transmission measurements at the HH exciton ([9.416], Sect. 6.3.1 a), *Kim* et al. [9.408] investigated the differential transmission spectra (DTS) so that information for both HH and LH (light-hole) excitons was obtained. The experiments were performed on a quantum-well structure consisting of 40 periods of 65 Å GaAs wells surrounded by 200 Å $Al_{0.35}Ga_{0.65}As$ barriers. The energy difference between the HH and LH excitons in these narrow quantum wells is $\approx$25 meV, larger than the HH exciton binding energy. Therefore, the LH exciton transition energy overlaps the continuum interband transitions involving heavy holes. The sample at 300 K was resonantly excited at the HH or LH exciton energy using spectrally-filtered $\approx$300 fs pulses from a Ti:Sapphire laser, generating a density of $\approx 1.1\times10^{10}$ cm^{-2}. They found that the excitation at HH or LH excitons creates differential transmission (DT) signals at both the excitons, but the dynamics of the signals is different for HH and LH excitons. For resonant excitation at the HH exciton, the dynamics of DT at the HH exciton is quite similar to the earlier results for resonant excitation of the HH exciton (Fig. 6.11: [9.416]). However, additional information was available because of pumping and probing at different combinations of the two excitons. For example, the DT at LH for resonant excitation at either LH or HH did not produce an initial peak in DT. The earlier results [9.416] were explained on the basis of exciton ionization. By analyzing the additional information obtained in their experiments, *Kim* et al. argued that their results can not be explained by exciton ionization alone, and that they also needed to invoke inter-valence band relaxation processes by interaction with LO phonons to explain their results. From a rate-equation analysis of the data at different temperatures, the time constants for a HH-LH transition were deduced to vary from 230 fs at 300 K to 900 fs at 150 K. No transfer was observed at 8 K as expected.

Exciton dynamics in quantum wires is also of interest. However, most of the studies on the dynamics in quantum wires (QWR) have concentrated on the carrier capture into the wires and the carrier relaxation in wires. These studies will be discussed in Sect. 9.4.3. The dynamics of exciton formation

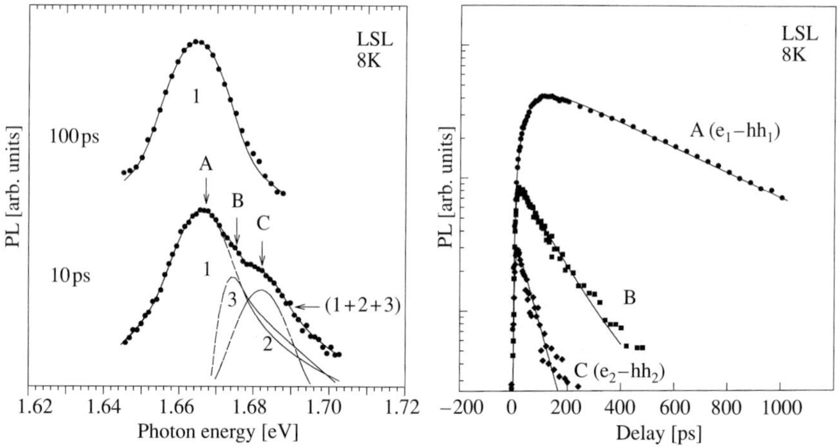

Fig. 9.59. (**a**) Time-resolved PL spectra obtained by upconversion on a GaAs quantum-wire array sample grown on a vicinal surface at two different delays after non-resonant excitation; sample temperature 10 K. (**b**) Temporal evolution of the signal at various spectral positions [9.417]

and relaxation in quantum wires is just beginning to be investigated. *Kumar* et al. [9.417] have analysed the dynamics of carrier capture, exciton formation and exciton relaxation in QWR arrays grown on vicinal substrates [9.418, 419]. The periodicity of the QWR array is 32 nm and the lateral potential is modulated with 5% modulation in the Al concentration. The sample was cooled to 5 K and excited with picosecond pulses from a dye laser at 6400 Å so that the electron-hole pairs were excited in the 2D continuum states above the confined QWR states but not in the barriers. The excitation density was $\approx 5\times10^4$ cm^{-1}, about an order of magnitude below the expected Mott density. The time and energy resolutions in the experiment were 3 ps and 4 meV, respectively. Figure 9.59 a shows the time-resolved luminescence spectra at two different delays. The lowest energy feature is identified as due to recombination of exciton associated with the lowest electron and hole states of the QWR. A lineshape analysis was used to identify the two higher-energy transitions marked B and C with excitons of higher-energy states of the quantum wires. Figure 9.59 b presents the time evolution of the emission at these three excitons. The emission at higher-energy excitons rises within the time resolution of the system and decays with a time constant of $\approx$50 ps. In contrast, the luminescence at the lowest exciton energy increases with a time constant of 55 ps and decays very slowly. *Kumar* et al. also investigated the dynamics of the linewidth of the excitons [9.420] and deduced that the exciton formation time is $\approx$35 ps. From an analysis of these results it was further concluded that the exciton cooling and relaxation for the lowest-energy exciton occur in 20 ps, and the exciton recombination time is $\approx$450 ps.

Finally, we note that radiative recombination dynamics of excitons in quantum wires has been investigated theoretically [9.421] and experimentally [9.422–424]. In contrast to the case of quantum wells where the exciton-radiative lifetime increases linearly with temperature [9.320, 328, 425], the radiative lifetime in quantum wires has been shown to increase as the square root of the temperature [9.422–424]. Thus, the radiative recombination lifetime of excitons in quantum wires is longer than that of excitons in quantum wells at low temperatures, but shorter at higher temperatures.

9.4.3 Quantum Wires: Capture and Relaxation Dynamics

We discuss in this subsection some studies of relaxation dynamics of carriers in quantum wires. These include the carrier capture into quantum wires, and hot electron relaxation in quantum wires. Similar treatments on quantum wells are described in Sect. 8.4 and Chap. 4, respectively.

Dynamics of carrier capture and carrier relaxation in quantum wires have been investigated experimentally as well as theoretically. Studies have been reported on rectangular-etched quantum wires, V-groove and T-shaped wires (Sect. 9.1.1 a). Electronic states in quantum wires have been considered theoretically by a number of groups [9.16, 31, 426–430]. Theoretical investigations of carrier capture and energy relaxation processes in quantum wires have also been reported in [9.431–441]. Experimental studies of capture and relaxation dynamics have been treated in [9.442–448]. We discuss here some representative results of these studies.

The dynamics of carrier capture in quantum wires was investigated by *Ryan* et al. [9.446, 449] for GaAs quantum wires grown by molecular beam epitaxy on (001) GaAs substrate. Very-high-quality V-groove wires were grown by this technique [9.16, 450]. An array of grooves, each 250 nm wide and 110 nm deep, parallel to the $(1\bar{1}0)$ axis, was produced by lithography. A 50 Å GaAs quantum well was grown with $(GaAs)_8(AlAs)_4$ superlattice barriers, growing in a distinct crescent shape, as shown by the TEM profiles in Fig. 9.60. The layer thickness at the bottom of the groove was measured by TEM to be 93 Å; it decreases rapidly with distance away from the center of the groove to a value of 22 Å where it merges with the superlattice barrier. The combined effects of layer bending and narrowing gives rise to a lateral confinement potential of 15 meV. *Ryan* et al. [9.446, 449] obtained a two-dimensional compositional map from high-resolution TEM images of the structure, and calculated a full two-dimensional potential profile. By solving the Schrödinger equation for this potential profile, they calculated the energy states for the structure. On the basis of these calculations it was concluded that there are three types of states: 1D quantum wire (QWR) states (Fig. 9.60 a), bulk-like Extended Superlattice States (ESL) (Fig. 9.60 b), and Localized Superlattice (LSL) states (Fig. 9.60 c). Figure 9.60 also depicts the wave functions of several representative states.

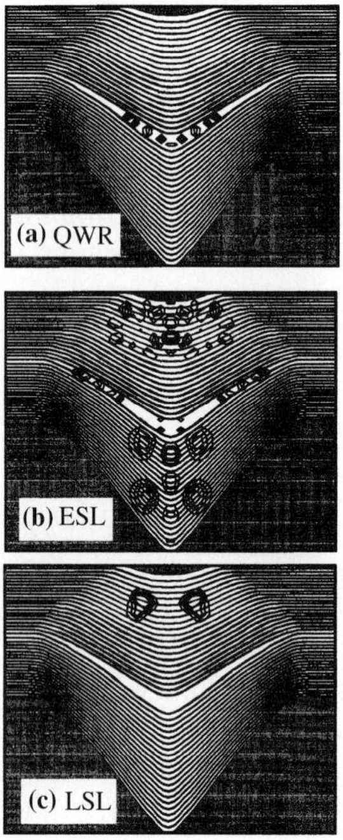

Fig. 9.60. Potential profiles of the GaAs V-groove structure with $(GaAs)_8(AlAs)_4$ superlattice barriers obtained from TEM data (AlAs layers are *shaded*), together with contour plots of the electronic probability density for (**a**) the $|1,1\rangle$ quantum wire (QWR) state, (**b**) a typical Extended SuperLattice (ESL) state, and (**c**) the lowest-energy Lateral SuperLattice (LSL) state [9.446]

The quantum wire sample was cooled to 10 K and excited with 5 ps pulses at 1.9 eV (high into ESL continuum states). The luminescence was detected with a streak camera with 30 ps time resolution. The experiments were performed at a relatively high excitation density of 5×10^6 cm^{-2}. The time-integrated luminescence spectrum consisted of a doublet at 1.563 and 1.580 eV, corresponding to the two lowest QWR states, a strong emission at 1.72 eV corresponding to the lowest LSL state, and a high energy shoulder at 1.78 eV corresponding to the ESL states. The 1.78 eV emission increases with increasing pump power and exhibits a high-energy tail. The time-resolved luminescence data are displayed in Fig. 9.61. The QWR emission at 1.563 eV shows (Fig. 9.61 a) a rise time of $\approx$150 ps and a decay time of $\approx$400 ps. From the observation that the QWR emission at higher energies (inset in Fig. 9.61 a) exhibits a much faster rise, the authors concluded that the capture into the QWR states occurs in <30 ps, and the long rise time seen at the lowest QWR energy corresponds to the dynamics of carrier relaxation within the quantum wires. Figures 9.61 b and c display the data for the LSL and ESL emission peaks, respectively. The fact that the decay times of the LSL state ($\approx$400 ps) is considerably longer than the rise time for the lowest QWR state shows that the photoexcited carriers do not pass through the LSL states before they are captured in the QWR. This conclusion is reinforced by the fact that the decay of the ESL emission is $\approx$30 ps, agreeing with the rise time of the higher-energy QWR states and confirming that the carriers are captured into the QWR states directly from the ESL

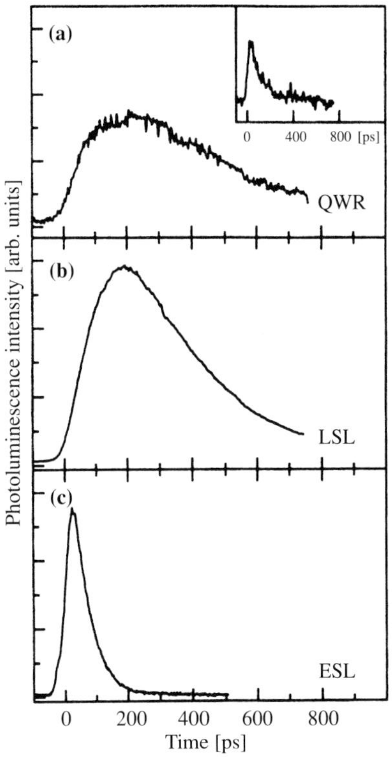

Fig. 9.61. Time-resolved luminescence from V-groove quantum wire at 10 K at a high photo-excited density of $\approx 1\times10^7$ cm^{-1} for (**a**) the $|1,1\rangle$ QWR state, (**b**) LSL states, and (**c**) ESL states. The inset shows the QWR signal measured at a much higher energy where intrawire relaxation effects are less pronounced [9.446]

states. Following the arguments presented for capture of carriers into quantum wells (Sect. 8.4), *Ryan* et al. [9.446, 449] argued that the capture of holes is fast, and hence determines the decay of ESL luminescence, whereas the capture of electrons gives the rise of QWR luminescence. The efficient capture of carriers by quantum wires is attributed to special nature of the growth process which results in a strong overlap between initial and final states.

Carrier capture in V-groove quantum wires was also investigated by *Christen* et al. [9.16, 451, 452] with the cathodoluminescence technique. They found that the carrier capture depends strongly on the structure of the wires investigated. For a vertical stack of quantum wires, they noted the capture time to be in the hundreds of picoseconds. In contrast, for a lateral submicrometer pitch quantum-wire array, the onset of quantum-wire luminescence was within their time resolution of 25 ps.

A number of studies of quantum wires using high spatial resolution have been reported [9.448, 453–460]. *Richter* et al. [9.448] have recently investigated the carrier capture into QWR states by measuring the steady-state and time-resolved (250 ps) photoluminescence (PL), and photoluminescence via near-field scanning microscopy (NSOM) [9.461]. A spatial resolution of better than 2000 Å was obtained by using a nanometer-sized aperture of the fi-

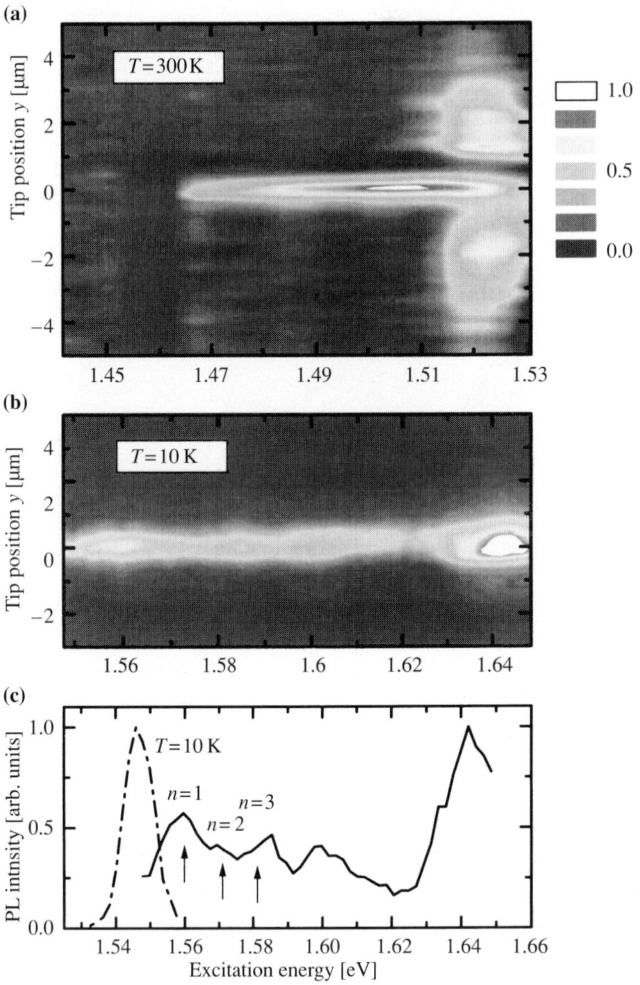

Fig. 9.62. Near-field PLE spectra of a quantum-wire structure at $y=0$ for local excitation through a fiber tip, as a function of energy (x-axis) and distance of the fiber tip from the wire (y-axis). (**a**) 300 K, detection at 1.459 eV, (**b**) 10 K, detection at 1.544 eV, (**c**) cross-section at $y=0$ for 10 K (*solid line*), PL spectrum (*dash-dotted line*). For 300 K and excitation above 1.515 eV, carriers excited in the surrounding quantum well are captured in the quantum wire and contribute to quantum wires PL; for other conditions depicted in the figure, only direct excitation of the wire (near $y=0$) leads to wire PL [9.448]

ber tip. A GaAs quantum wire grown on (311)A substrate by MBE was studied with an excitation density in the range 10^4–10^5 cm^{-1}. The luminescence was excited by means of tunable lasers and the fiber tip. Photoluminescence excitation (PLE) spectra of the QWR emission were measured as a function of the distance of the exciting tip from the quantum wire at $y=0$. Figure 9.62 depicts typical spatially-resolved PLE spectra at 300 K (a), at

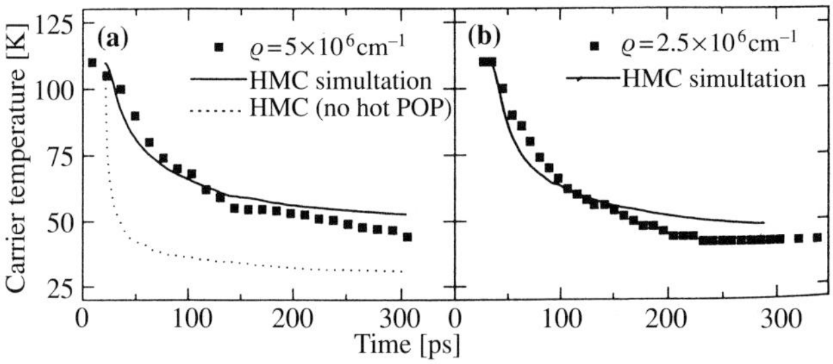

Fig. 9.63. Time evolution of the carrier temperature for two different quantum-wire structures at two different densities; squares are temperatures determined from the high-energy tail of the PL spectra at various delays, the *solid* (*dotted*) curve represents hybrid Monte Carlo simulation with (without) hot polar optical phonons [9.443]

10 K (b), and the cross section of (b) at $y=0$ in (c). For 10 K, the wire luminescence is excited only for the tip very close to the wire. In contrast, for 300 K the carriers excited in the quantum-well states contribute to the luminescence of the QWR even when the excitation tip is a few micrometers away from the wire. From studies such as these, they concluded that the capture of carriers into QWR states involves (i) real-space transfer over several micrometers, determined by diffusion of holes, and (ii) emission of optical phonons.

The dynamics of carrier relaxation in quantum wires has been experimentally investigated by a number of groups [9.422, 443, 444, 447, 462–465]. Early studies on etched quantum wires indicated that carrier relaxation is slower than that reported for similar quantum-well systems [9.462–464]. *Maciel* et al. [9.443, 444] have investigated carrier cooling in the same V-groove samples used in the carrier-trapping studies discussed above. They excited the sample at 10 K with sub-picosecond pulses from a Ti:Sapphire laser operating at 1.63 eV, well below the LSL states, and detected the luminescence with ≈ 2 meV spectral resolution using a streak camera with 20 ps time resolution. The excitation density was 2.5–5×10^6 cm^{-2}. Time-resolved luminescence was measured at various emission energies. The rise and the decay of the luminescence was rapid close to the excitation energy and became longer as the detection energy was lowered. The carrier temperature was deduced from the high-energy tail of the luminescence spectrum as a function of time. The results for two densities are plotted in Fig. 9.63. *Maciel* et al. remarked that the cooling curve appear to be similar to those for quantum wells, although no direct comparison was made. The experimentally deduced cooling curves were compared with a hybrid Monte Carlo simulation with and without the inclusion of hot-phonon effects (Sect. 4.3.2). This comparison is shown in Fig. 9.63 as solid and dotted curves, respectively. A better agreement with the data is obtained when hot-phonon effects are included, showing that hot-phonon effects are also important in 1D systems.

9.4.4 Quantum Wires: High-Density Effects

Ultrafast optical spectroscopy is a powerful technique for investigating high-density effects in semiconductors. It provides a means for generating high excitation densities while minimizing lattice heating, and to follow the dynamics at very short times. We have seen many examples of this in Chaps. 3–5. As a result of many-body interactions between the photoexcited carriers, the electronic states and the interband matrix elements undergo profound changes (see, for example, [9.187]). Ultrafast spectroscopy provides a means for probing such effects. One high-density effect that has received considerable attention in bulk and reduced-dimensional semiconductors is the renormalization of the semiconductor bandgap as a result of a high density of photoexcited electron-hole pairs [9.466]. In indirect-bandgap semiconductors like Si and Ge, this leads to the phenomenon of electron-hole liquid or droplets [9.467–470], whereas in direct-bandgap semiconductors like GaAs, it leads to effects such as optical gain and lasing [9.187]. These phenomena have been extensively investigated in bulk and 2D semiconductors, and similar investigations of quantum wires have been reported in recent years. Quantum wires bring many new features to these investigations: an increased binding energy of excitons, a inverse square root density of states in the ideal case, and a *reduction* in the interband matrix elements near the bottom of the band resulting from Coulomb interactions [9.30, 31, 428]. Some of these effects in quantum wires are not yet fully understood, and some issues are still being debated. We discuss below some of the results on bandgap renormalization in quantum wires.

The topic of bandgap renormalization in quantum wires has generated considerable controversy. Theoretical analyses have predicted large bandgap renormalization in quantum wires [9.471–473]. Experimentally, time-resolved and cw luminescence spectroscopy has been used as the primary tool to investigate bandgap renormalization in quantum wires. A number of studies have been reported. *Cingolani* et al. [9.463, 464, 474] reported some of the first results on high-density electron-hole plasma in quantum wires. They investigated time-resolved luminescence from rectangularly etched, 600 Å wide GaAs quantum wires, the same as those employed for the investigation of dephasing of excitons by FWM discussed in Sect. 9.2.13, using 25 ps pulses from a low-repetition rate (2 Hz) Nd:YAG laser. *Wang* et al. [9.475] reported on luminescence from rectangularly etched, 300 Å wide $In_{0.53}Ga_{0.47}As/InP$ quantum wires using cw luminescence spectroscopy with low average power but high density. *Grundmann* et al. [9.16] have reported on cathodoluminescence study of single, stacked and array of V-groove GaAs quantum wires, similar to those discussed in Sect. 9.1.1 a. *Wegscheider* et al. [9.476] have studied lasing in T-wires (Sect. 9.1.1 a). *Ambigapathy* et al. [9.477] investigated GaAs V-groove quantum wires excited with 1–3 ps dye laser pulses at 5680 Å using time-correlated photon counting. There are many common features to the data reported by these and other researchers. The luminescence spectrum broadens on the high-energy side as the density of carriers is increased. In cw experiments, this occurs as one increases the pump power. In pulsed experiments, the broadening in the emission spectrum is evident at high densities and short times, but diminishes at longer times as the density is reduced. Another feature common to these data (except for those at the highest densities in [9.475]) is that the peak position of the emission remains relatively independent of the excitation density even as the emission spectrum broadens on the high-energy side. The experimental results from various groups are therefore in substantial agreement.

In spite of the apparent similarity in the data, the interpretations vary considerably. Large bandgap renormalization were deduced in some cases (see, for example, [9.464, 475]) whereas it was argued that bandgap renormalization is nonexistent in others (see, for example, [9.16, 477]). These controversies result from a number of factors, but boil down to the fact that the conclusions are based on an analysis of the emission lineshape. Determination of the experimental emission lineshape (spectrum) may be difficult in some cases if the emission of interest is superposed on strong background that has to be subtracted out. Once the lineshape is determined, bandgap renormalization is deduced by comparison with theoretical curves calculated with various degrees of sophistication. The theory contains renormalized bandgap as one of the parameters, and various other parameters such as the carrier density, carrier temperature, and low- and high-energy broadening. More important, some theoretical fits are made assuming simple free-electron-hole lineshape, whereas others consider sophisticated models including excitonic and many-body effects in varying degrees. With these uncertainties, it is not surprising that, although the results appear to be similar, the interpretations from different groups are rather different, and bandgap renormalization in quantum wires remains a controversial topic.

All Coulomb effects increase with increasing confinement and scale roughly with the exciton binding energy which is enhanced in 1D compared to 2D systems. This means that excitonic and many-body effects have to be incorporated in the lineshape analysis. At a low density such a model gives an excitonic emission line. With increasing density, the exciton binding energy decreases and the bandgap is renormalized to lower values. As in higher dimensions, one expects a good cancellation between the two effects, leaving the spectral position of the emission line relatively unaffected. Therefore, a careful comparison with a realistic theory is essential in deducing correct information from the experimental curves.

The most rigorous comparison of experiments and theory appears to have been made by *Wang* et al. [9.475] for the case of no magnetic field, and by *Bayer* et al. [9.478] for the case with applied magnetic fields. Starting with a single quantum-well structure with 50 Å $In_{0.53}Ga_{0.47}As$ well and InP barriers, they fabricated high-quality, $\approx$300 Å-wide quantum-wire structures by lithography and wet chemical etching. The samples were cooled to 1.8 K and excited by a cw Ar^+ laser at 5145 Å with an excitation density varying from 0.01 to 18 kW/cm^2. The average laser power was limited to <5 mW

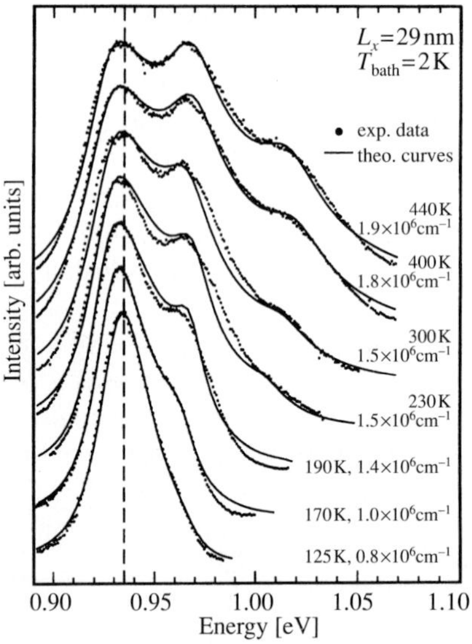

Fig. 9.64. Luminescence spectra for an array of 290 Å-wide etched $In_{0.53}Ga_{0.47}As/InP$ quantum wires at 2 K. The solid lines represent the line-shape fits to the spectra with a theoretical model that includes excitonic and many-body effects in the Hartree-Fock approximations. The carrier densities and temperatures for each curve obtained by the fit are indicated in the figure. The curves are displaced vertically for clarity [9.475]

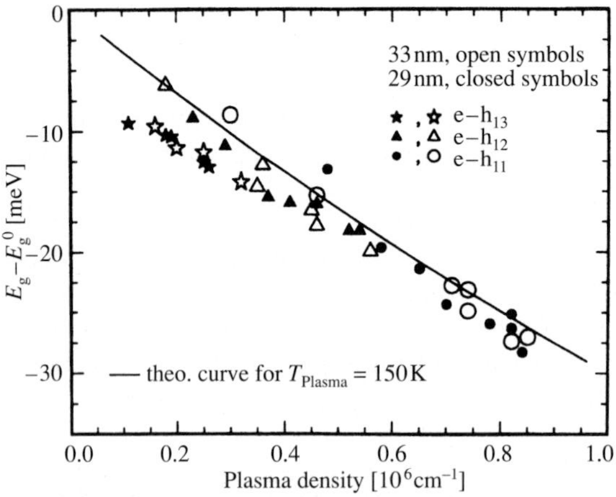

Fig. 9.65. Bandgap renormalization as function of the individual density in each of the respective subbands. Points are experimental, the line was calculated for the 330 Å wire at a constant plasma temperature of 150 K. Note that large values ($\approx$30 meV) of the bandgap renormalization are deduced even though the shift in the luminescence peaks in Fig. 9.64 is negligible [9.475]

to minimize heating. Figure 9.64 depicts the experimental spectra and the theoretical fits at various pump powers for a 290 Å wide quantum-wire structure, with the highest excitation intensity limited to 2 kW/cm^2 because the theoretical model needs to be refined at higher densities. A comparison of theoretical and experimental spectral shapes yields information about the plasma density, temperature and renormalized bandgaps of each of the three subband pairs at each excitation density. The bandgap renormalization of 25–30 meV was deduced from these results. An interesting aspect of the results was that the bandgap renormalization seemed to saturate at a density above 1.5×10^{6} cm^{-1}. This effect was attributed to a redistribution of the densities between subbands as the higher excitation intensity increased the plasma temperature. When the bandgap renormalization of a given subband is plotted as a function of the density in that subband, the bandgap renormalization of different subbands follow a continuous dependence on density without any abrupt changes between the subbands (Fig. 9.65). *Wang* et al. [9.475] concluded that the bandgap renormalization for a given subband depends only on the density in that subband. Since higher subbands are less densely populated, they exhibit a smaller bandgap renormalization. In contrast to these results, the bandgap renormalization at high magnetic fields (when the cyclotron energy exceeds the subband separation) was found to be quite small and depended mainly on the density in *other* subbands [9.478]. It was argued that this is because the magnetic field suppresses the motion of the free carriers along the wire so that the excitons in one band form an ideal gas, to a good approximation.

The results discussed above deduce a large bandgap renormalization in quantum wires at zero magnetic field. However, it is clear that there is yet no consensus in the literature on this issue. Perhaps use of other experimental techniques (e.g., pump-probe transmission spectroscopy to determine the gain spectrum) can shed further light on this controversial subject.

9.4.5 Quantum Dots: Relaxation Dynamics

Quantum dots are the natural culmination of the efforts to confine the carriers and excitons in semiconductors to lower dimensions. Confinement in one dimension leads to quantum wells (2D systems), in two dimensions to quantum wires (1D systems), and finally in all three dimensions to quantum dots (0D systems). Quantum dots with sufficiently small dimensions are expected to have a few discrete atomic-like energy levels. The interest in Quantum Dots (QD) stems from interest in fundamental physics of such 0D systems, as well as from their potential device applications. The 0D systems may be very attractive for semiconductor lasers and other devices [9.479] because of enhanced oscillator strengths and optical nonlinearities. Indeed, laser action employing quantum dots as the active medium has already been demonstrated [9.32–39]. The fundamental question of how carriers relax in a system of discrete energy levels has attracted much attention from theoretical [9.431, 480–494] and experimental [9.495–507] points of views. Relaxation by optical phonons, generally a dominant process in higher dimensional semiconductors, can occur only if the energy difference between the energy levels equals or is close to the optical phonon energy. Relaxation by acoustic phonons can occur when the conditions for relaxation by optical phonons are not satisfied. However, it was shown theoretically [9.431, 480,

489] that relaxation by acoustic phonons becomes less efficient as the energy difference between the initial and final states becomes large. As a result, there is considerable discussion in the literature of the idea of a "phonon bottleneck". It was also proposed that while phonon-assisted relaxation may slow down considerable, because of these considerations, other mechanisms such as Auger processes may still permit fast relaxation. Electron relaxation by Auger interaction of electrons in quantum dots with electrons in 2D quantum-well states [9.482], and with holes in the valence bands [9.488] has been investigated theoretically. In view of these developments, it is not surprising that most of the investigations of the dynamics of quantum dots have focused on the question of a phonon bottleneck, and carrier relaxation processes and rates [9.23, 486, 490, 491, 495, 497, 504–519]. Nonlinear optics of quantum dots is also beginning to be investigated [9.348], as discussed in Sect. 9.2.13. Note that quantum dots of II–VI semiconductors in glass [9.520] have been analysed extensively with many different techniques, including ultrafast spectroscopy [9.17]. We will restrict our discussion to quantum dots in III–V semiconductors.

In narrow quantum wells, the monolayer fluctuations in the quantum well width can lead to energy fluctuations comparable to or larger than the exciton binding energy. Under these conditions, luminescence from a very small region of the quantum well sample monitored with near-field techniques [9.26, 521] or with submicrometer apertures [9.13, 27, 28, 346, 522–525] allows investigation of a *single* quantum dot and leads to the observation of extremely sharp lines. Extremely sharp lines have also been observed by other techniques [9.526, 527]. Measurement of single quantum dot implies that there is no inhomogeneous broadening and that the observed linewidth is the homogeneous linewidth of the quantum dot. *Gammon* et al. [9.27, 28] reported luminescence linewidths as narrow as 23 μeV at ≈ 5 K. They concluded that the homogeneous linewidth is due to the intrinsic radiative recombination of the electron-hole pair in its lowest state, and deduced a value of 29 ps for the radiative lifetime (3.5×10^{10} s^{-1} for the radiative rate). Comparing this with the linewidth of 35 μeV for the first excited state of the quantum dot in the PL excitation spectrum, they concluded that the relaxation time between the first excited and the ground state of the quantum dot is 19 ps (relaxation rate of 5.3×10^{10} s^{-1}) in their case. This result indicates the *absence* of a significant phonon bottleneck, in agreement with the calculations for dots of this dimension [9.483]. (Sharp excited-state transitions have also been reported in microphotoluminescence studies by *Notomi* et al. [9.525]). *Marzin* et al. [9.522] have investigated single InAs quantum dots obtained by self-organized growth on GaAs [9.528]. Single quantum dots were isolated by etching square mesas with sides as small as 1000 Å. They observed sharp emission lines associated with single quantum dots. They deduce short capture and relaxation times into the dots.

More insight into the question of carrier relaxation in quantum dots can be obtained by time-resolved studies [9.497, 498, 504, 505, 510, 513]. *Bockelmann* et al. [9.504, 510] have reported time-resolved studies of a single quantum dot fabricated by laser-induced thermal interdiffusion in an undoped, 30 Å wide $GaAs/Al_{0.35}Ga_{0.65}As$ quantum well [9.346], using the technique of microphotoluminescence. An individual dot structure is defined by drawing a square frame of size w with a focused Ar^+ laser on the sample surface. An area of 6×6 μm^2 around the dot is interdiffused by scanning the laser beam continuously. Model calculations indicate that the lateral potentials are parabolic near the dot center with a barrier height of $\approx$35 meV for both electrons and holes. A series of quantum dots with various splittings in the cw PL were investigated with the technique of microphotoluminescence. We concentrate here on the results for the structure with w=450 nm, which showed the largest splitting (10 meV) between the two lowest quantum dot states. The structure was cooled to 7 K and excited by 1.5 ps pulses from a Ti:Sapphire laser focused to a diameter of 1.5 μm. The Time-Resolved PhotoLuminescence (TR-PL) was detected with a streak camera having 10 ps time resolution. The excitation photon energy was varied such that the excess energy of the photoexcited electron-hole pair was either less than or higher than the LO-phonon energy, and the excitation power was varied by two orders of magnitude. The TR-PL spectra recorded in various time windows for various excitation conditions are shown in Fig. 9.66. The PL signal appears with a rise time of 10 ps (the time resolution of the experiment) regardless of the excitation photon energy or power. The broadening of the spectrum at short delays increases as the excitation density is increased. From these observations *Bockelmann* et al. [9.504, 510] concluded that carrier relaxation in quantum dots proceeds via Coulomb interaction down to the lowest excitation power investigated (140 nW focused to 1.5 μm diame-

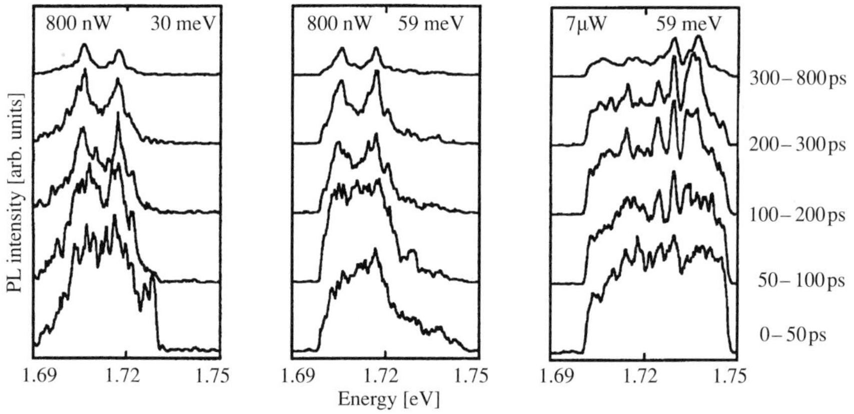

Fig. 9.66. Time-resolved PL spectra of a single quantum dot of the lateral dimension 4500 Å exhibiting a splitting of 10 meV in cw luminescence for excitation within the dot either 30 meV (less than the LO-phonon energy) or 59 meV (more than the LO-phonon energy) above the lowest PL peak. The exciting laser is focused to 1.5 μm, with an excitation power indicated on the top left of each figure. Higher-energy PL lines appear and persist to long delays at higher excitation densities [9.504, 510]

ter spot). Furthermore, they found that the TR-PL at long delays exhibits a series of peaks with the higher-energy peaks becoming more prominent at higher excitations. The time-integrated PL also shows a shift to higher energy as the excitation power is increased. These observations lead to the conclusion that Pauli blocking of the lower states becomes important with increasing density, and that a consistent picture of the dynamics in quantum dots is obtained by including the effect of Coulomb interaction for the carrier relaxation and Pauli blocking of the low-energy states.

Grosse et al. [9.505] have reported on an investigation of the carrier relaxation dynamics in strain-induced InGaAs quantum dots. These quantum dots were fabricated by the technique reported by *Lipsanen* et al. [9.25], and exhibited low inhomogeneous broadening effects. The structure investigated consists of a 70 Å $In_{0.1}Ga_{0.9}As$ quantum-well layer surrounded by GaAs barriers. The top barrier has a thickness of 50 Å. 750 Å wide and 220 Å high InP islands, grown in situ on this top layer, acted as stressors and produced a strain-induced parabolic lateral potential with a calculated confinement energy of 64 meV (Fig. 9.67). The degeneracy of each level [9.529] is indicated by the number of filled dots on each level in Fig. 9.67. The 10 K PL spectra for two different excitation densities (50 and 250 W/cm^2, dashed and solid curves, respectively) show well-defined peaks corresponding to interband transitions between levels with the same quantum number [9.25, 529] (Fig. 9.67). The higher-energy levels and the 2D quantum-well states begin to get occupied with increasing density. The linewidth of each transition is ≈ 5 nm or 10 meV. Absorption and PLE of the 1-1 transition show no detectable absorption for the quantum dot states, presumably because the holes from the unintentionally-doped quantum well occupy the quantum dot states.

For time-resolved studies, the sample was excited by 110 fs pulses from a Ti:Sapphire laser operating at 8000 Å and focused to ≈ 100 μm. With an average quantum dot density of 2×10^9 cm^{-2}, the excitation spot excited $\approx 20\,000$ quantum dots. The excitation created electron-hole pairs high in the quantum-well states. The excitation density was equivalent to the 250 W/cm^2 used for the high-intensity curve in Fig. 9.67. Time-resolved PL (TR-PL) at 10 K, recorded at different transition peaks using a streak camera with 13 ps time resolution, is depicted in Fig. 9.68. The rise times of all transitions are limited by the time resolution of the detection system. The ordinate in Fig. 9.68 is adjusted to be 2 for the 1-1 transition after the initial rise, on the assumption that the lowest electron state (1) is occupied by two electrons. The observation that the rise times of all transitions are limited by the time resolution of the detection system, and that the initial quantum-well PL is weaker than the quantum-dot PL show that the capture of carriers into the quantum-dot states, and their relaxation within the quantum-dot states, are extremely rapid (<13 ps). Secondly, luminescence transients exhibit a plateau-like behavior for 1.5 to 3.5 ns, with the higher-lying levels showing shorter plateaus. The plateaus indicate that the levels are fully occupied for a time that decreases for the higher-lying states. The decay of the PL from

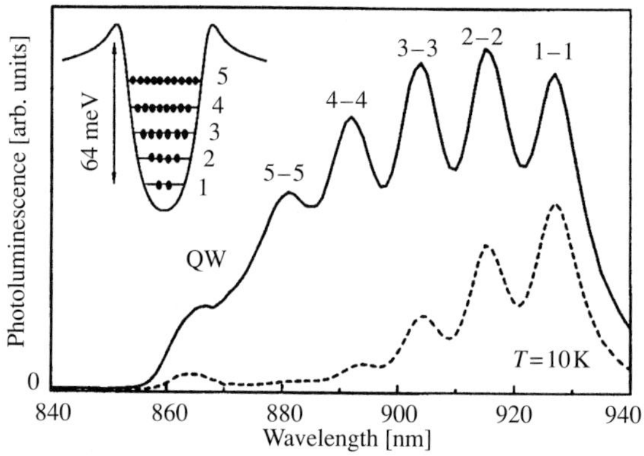

Fig. 9.67. PL spectra of strain-induced InGaAs quantum dots at 10 K for low (*dashed line*) and high (*solid line*) excitation intensities. The inset shows the strain-induced lateral potential for electrons in the conduction band. The number of filled circles on each quantum-dot energy level i gives the degeneracy of the level. The well-defined peaks correspond to the transitions between electron and hole levels of the same i, as indicated. PL from higher-energy levels and the quantum well is observable at high intensities [9.505]

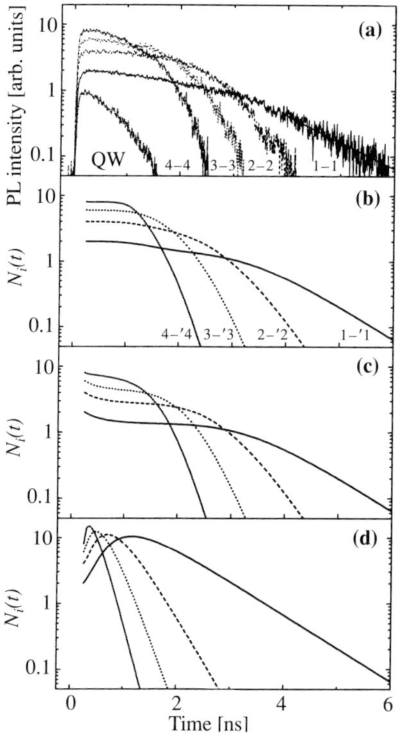

Fig. 9.68. (**a**) Time-resolved PL spectra of the four lowest quantum dot transitions and of the reference (uncovered) quantum well for a high excitation density condition corresponding to the *solid lines* in Fig. 9.67. (**b**) Calculated temporal evolution of the mean number of electrons occupying the four lowest states for a model that includes state filling, Coulomb scattering and acoustic phonon scattering, (**c**) same as (**b**) but neglecting Coulomb scattering, (**d**) same as (**c**) but neglecting state-filling effects [9.505]

higher states is faster than the decay of the PL from the 1-1 transition, indicating that transfer of carriers to the lower states is an important decay mechanism.

The fast rise time of PL was interpreted in terms of the 0D-2D Auger interaction model by *Bockelmann* and *Egeler* [9.482], as discussed above. The decay time constant at long times of the 1-1 transition (860±25 ps) is attributed to the radiative recombination time; non-radiative recombination is considered to be negligible for a number of reasons. The relaxation of electrons from higher states to the ground state is assumed to be a combination of an Auger interaction and acoustic phonon scattering. *Grosse* et al. have quantitatively analyzed the results with a rate-equation model which includes Pauli blocking, Auger interaction between 0D electrons and carriers (electrons and holes) in the 2D states and acoustic phonon scattering. The rates of Auger interaction and acoustic phonon scattering are treated as parameters. The results of the full calculation with an acoustic phonon scattering time of 570 ps and an Auger scattering time of 1 ps (Fig. 9.68 b) provide good agreement with the data. The researchers found that the fit is quite sensitive to the value of the acoustic phonon scattering time. Also, neglecting either the Auger process (Fig. 9.68 c), or both the Auger process and the Pauli blocking (Fig. 9.68 d) gives distinctly poorer agreement with the results. The results by *Bockelmann* et al. [9.510] and *Grosse* et al. [9.505] thus show that Auger processes do indeed play an important role in the carrier capture and relaxation of quantum dots at high densities, and remove the problem of phonon bottleneck expected in the absence of such processes.

Ohnesorge et al. [9.513] have reported on a detailed investigation of carrier capture, relaxation and recombination dynamics in quantum dots as a function of density, excitation energy and temperature also utilizing the technique of TR-PL. The structures they analyzed were grown by metal-organic chemical vapor deposition on a GaAs substrate [9.12]. $In_{0.5}Ga_{0.5}As$ was deposited on top of a 200 nm GaAs buffer layer. Large lattice mismatch transforms the surface into three-dimensional islands or quantum dots (Stranski-Krastanow growth mode) after the deposition of 1.5 monolayers. The quantum-dot diameter was ≈ 150 Å, the dot height was 50 Å before overgrowth (of 300 Å GaAs cap layer) and the dot density was $\approx 2\times10^9$ cm^{-2}. The cw PL studies revealed no peaks associated with the InGaAs wetting layer, and the integrated intensity of the quantum-dot luminescence to be comparable to that of a reference quantum well. The PLE spectrum of the quantum-dot luminescence exhibited two broad peaks presumably due to excited states in the dot, and a strong peak due to the GaAs barrier. Time-resolved PL measurements were performed using either a mode-locked, frequency-doubled, pulse-compressed Nd:YAG laser (3 ps pulses at 5320 Å) or a tunable Ti:Sapphire laser (1.5 ps in the 7200–10 800 Å range). Detection with a streak camera provided a time resolution of ≈ 10 ps.

The results were analyzed using a three-level model with τ_R as the time constant for capture and relaxation of carriers to the lowest quantum-dot

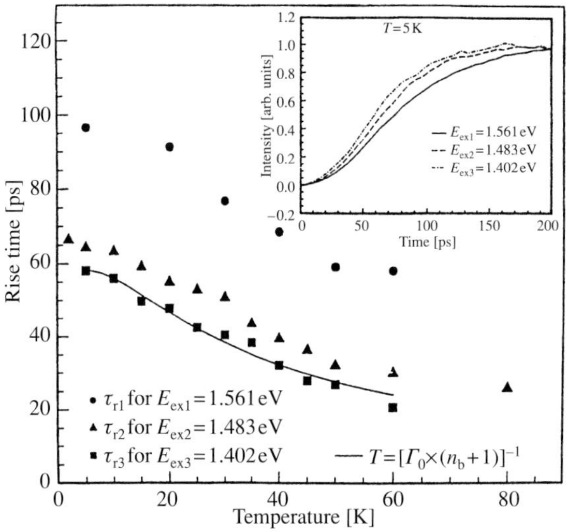

Fig. 9.69. Temperature dependence of the rise time of luminescence at 1.33 eV from self-assembled InGaAs/GaAs quantum dots as a function of temperature. Excitation produces less than one electron-hole pair per dot. *Points* are experimental, the *solid curve* represents the fit to a multi-acoustic-phonon relaxation model. The *inset* shows the rise curves for luminescence at 5 K for the three excitation energies [9.513]

state, and with τ_x as the excitonic lifetime. The values for τ_R obtained for two resonant (within the quantum-dot states) excitation energies and one non-resonant excitation in the GaAs barrier are plotted in Fig. 9.69 as a function of the sample temperature. The density of excitation was estimated to be ≈ 0.2 electron-hole pairs per dot per pulse. The results show that τ_R for non-resonant excitation is ≈ 30 ps longer than that for resonant excitation, a difference that is preserved as τ_R decreases with increasing temperature. Furthermore, *Ohensorge* et al. [9.513] found that the barrier luminescence also decays with a time constant of ≈ 30 ps. They concluded that the carriers excited in the GaAs barrier region diffuse and capture into high states of the quantum dots is ≈ 30 ps, and the carrier relaxation to the ground state of the quantum dots takes ≈ 60 ps at 5 K to ≈ 25 ps at 60 K. In order to understand the physical origin of capture and relaxation processes in quantum dots, they investigated the intensity dependence of the dynamics and noted that τ_R for non-resonant excitation decreases strongly with increasing intensity. On the basis of these studies, the authors concluded that at low densities (less than 1 electron-hole pair per dot per pulse) the relaxation of carriers in quantum dots is dominated by multiphonon relaxation process involving LA phonons of 2.7 meV energy. They further proposed that, at high excitation densities, the capture and relaxation of carriers in quantum dots is dominated by Auger processes [9.482, 488], and in agreement with the results [9.505, 510] discussed above. We note that *Raymond* et al. [9.498] have also investigated time-resolved PL from self-assembled quantum dots grown by MBE utilizing Stranski-Krastanow growth mode. Decay times of several hundred picoseconds were deduced for excited states, limited by relaxation to lower states and state-filling effects, in general agreement with the above discussion [9.505, 510]. An interesting aspect of the results is that the rise time of luminescence at various transitions was found to be 35±5 ps, relatively independent of the excitation level. It was argued that carrier diffusion limited this rise time, and that the carrier capture is much faster.

Multiple *optical* phonon replicas have been noted in cw and time-resolved studies of quantum dots [9.495, 497, 501, 502, 506, 523, 530, 531]. They have been associated with hot-exciton (or geminate electron-hole pair) luminescence, a phenomenon well known in bulk semiconductors [9.532]. In bulk semiconductors, hot-exciton luminescence is explained [9.532] by arguing that efficient channels of radiative or nonradiative recombination is available to all excited carriers. Therefore, exciton luminescence is observable only when photoexcited carriers can form excitons at a rate that can compete with the rates for these alternate channels. For carrier pairs at energies that are multiple optical phonon energy above the exciton energy, optical phonon emission can provide such an efficient channel of formation of excitons. This explains why hot-exciton luminescence is observed at excitation energies which are multiple optical phonon energies above the exciton energy. Since this argument can be extended to 2D and 1D systems, but not to 0D systems with discrete energy levels, an inhomogeneous size distribution of quantum dots and hence an inhomogeneous energy distribution, has been invoked to interprete the observation of multiple phonon features in the photoluminescence excitation (PLE) spectra. However, the observation of phonon replicas is still difficult to explain in quantum dots, and the nature of PLE and its explanation are still being debated in the literature (for a recent discussion see [9.502, 503, 506, 507]).

Notwithstanding the controversy, the dynamics of these replicas has been investigated [9.497, 506]. *Vollmer* et al. [9.497] investigated self-organized InP and observed up to three LO-phonon replicas, with phonon energy of 46 meV. They found that the rise time of the phonon replicas is always less than 10 ps, the time resolution of the experiment. The decay time of the phonon replicas is fast if further emission of LO phonons is possible, but slow if further relaxation can proceed only via acoustic phonon emission. Furthermore, the rise time of the ground-state emission is slow if it can be populated only via acoustic phonon emission. An acoustic phonon emission time of ≈ 60 ps was deduced from these studies. *Heitz* et al. have advanced a detailed study of energy relaxation processes in a single layer of InAs/GaAs quantum dots [9.506], as well as in stacked layers of InAs/GaAs quantum dots [9.507]. They found multiphonon emission in the former but not in the latter. For the single-layer case [9.506], they noted that the rise time of the emission is 28±5 ps for the ground state and <10 ps for the excited state, independent of the emission energy and whether the excitation is due to multiple phonon energies above the emission energy, or in the wetting layer or in the barrier layer. Also, the excited state decays with a time constant of ≈ 100 ps, independent of the emission energy. The recombination time constant of the ground state was found to depend on the emission energy, decaying from about 1000 ps at 1 eV to 300 ps at 1.25 eV. They noticed that a hole state presents a relaxation bottleneck that determines the ground-state occupation time after non-resonant excitation. For small quantum dots, *Heitz* et al. [9.507] noted that intradot relaxation is faster than the radiative and non-radiative recombination rates, thus explaining the absence of a phonon bottleneck. They also found no evidence of diffusion-limited dynamics in their samples.

The studies discussed here have provided considerable information on the nature of electronic states and the relaxation dynamics in quantum dots. The important roles played by Coulomb-induced relaxation (Auger-like processes) and state-filling have been established. However, the nature of electronic states and the dynamics of relaxation under the conditions when multiple phonon peaks are observed in photoluminescence-excitation spectra is still being debated and not fully understood. Further investigations will be required to clarify these controversial issues. It is clear that the studies on relaxation dynamics in quantum dots are just beginning, and will be an important part of the ultrafast studies of semiconductors in the near future.

9.4.6 Carrier-Transport Dynamics

We have discussed the carrier-capture dynamics in quantum wires and quantum dots in Sects. 9.4.3 and 9.4.5 respectively. In some cases, carrier capture into a low-dimensional semiconductor nanostructure involves spatial transport of carriers as discussed for the case of quantum wells in Sect. 8.4 and for the case of quantum wires in Sect. 9.4.3. In this subsection we discuss a recent measurement of femtosecond transport dynamics of carriers in bulk semiconductors under the influence of high electric fields.

High-field transport in semiconductors is of interest from fundamental and applied points of views. Transport of electrons and holes under high electric field determines the speed of electronic devices in many cases. For example, the transit time of carriers in the base-collector depletion region is an important parameter determining the speed of a Heterojunction Bipolar Transistor (HBT) [9.533]. With the shrinking of the device dimensions, the electric field in electronic devices can be extremely high, approaching 100 kV/cm. Predictions of device performance under these conditions is often based on an extrapolation of our understanding of the physics under steady-state conditions. Velocity-overshoot phenomena on the ≈ 100 fs time scale have been investigated using pump-probe differential absorption [9.534–536] and THz transients with a few THz bandwidth [9.537], and with 10 fs time resolution using a THz pump-probe technique [9.538], briefly mentioned in Sect. 8.1. We discuss here the first direct measurements of transient high-field transport using 10 fs pulses and electro-optic detection with ultrabroad bandwidth (>50 THz, Sect. 9.1.3 c).

The basic idea behind the technique is to photoexcite electron-hole pairs in an intrinsic region of the semiconductor under a high electric field with ≈ 10 fs pulses. The carriers accelerating under the influence of the electric fields emit electromagnetic radiation whose electric field is proportional to the acceleration of the carrier under certain conditions. If the emitted field can be detected with a THz detector with a bandwidth comparable to the optical pulse width, then the acceleration of the carriers can be determined on the ≈ 10 fs time scale. As discussed in Sect. 9.1.3 c, electro-optic detectors provide bandwidths exceeding 50 THz. *Leitenstorfer* et al. [9.539] have investigated high-field nonequilibrium transport in GaAs and InP with this technique. They fabricated large-area (5×8 mm^2) p-i-n^+ diodes with thin (20 nm) top p^+ layers. Electric fields in the i-region were calibrated by measuring Franz-Keldysh oscillations in the reflectivity spectra, and electric fields as high as 130 kV/cm were determined. The samples at room temperature were excited by 12 fs pulses from a Ti:Sapphire laser operating at 1.52 eV. The laser was incident at a large oblique angle and was weakly focused such that the focus was 12 mm from the sample surface. The excitation density was $\approx 1\times10^{15}$ cm^{-3}. An electro-optic detector was at the focus of the reflected laser beam (12 mm from the surface). The emitted THz radiation in the reflection direction was incident on the electro-optic detector after passing through a filter to remove the pump laser. Variably delayed pulses from the Ti:Sapphire laser were also incident on the electro-optic detector. The two linear polarizations of the transmitted Ti:Sapphire laser were separated and were incident on a pair of balanced photodetectors. The optics was arranged such that the difference signal from the balanced photodetectors is zero in the absence of the THz radiation on the electro-optic crystal. When the THz radiation is present, the electro-optic effect in the crystal rotates the polarization of the Ti:Sapphire beam, generating a difference signal in the balanced photodetectors (Sect. 9.1.3 c). Thus the difference signal reproduces the temporal variation in the amplitude of the THz field as the probe-laser delay is varied. 6–15 μm thick (110) GaP and ZnTe electro-optic crystals were mounted on (100) substrates that exhibit no electro-optic effect. The use of different electro-optic crystals allowed a cross-check on the results and analysis. The thin electro-optic crystals are necessary to achieve the desired detection bandwidth of >50 THz.

The results obtained by *Leitenstorfer* et al. are depicted in Fig. 9.70 for GaAs and InP at four different electric fields. Note that the noise in the data correspond to a few times 10^{-8} of the unmodulated signal, close to the shot noise limit of the photodetectors. The results can be qualitatively understood as follows: The electrons and holes are excited with some kinetic energy in the conduction and valence bands. Since the holes are heavier and accelerate to their saturation velocity in a very short time (≈ 30 fs at 100 kV/cm), most of the signal comes from the electrons. At the lowest electric fields intervalley transfer of electrons from the Γ to the L and the X valleys is not very effective, so the dynamics is governed by the acceleration of electrons, determined by the scattering processes in the Γ valley, and by the reduction

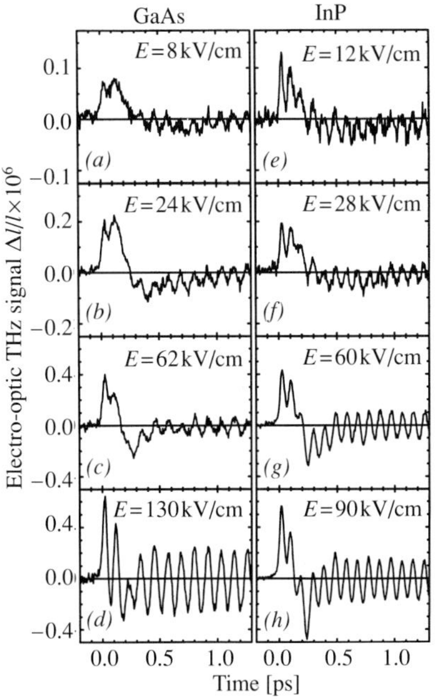

Fig. 9.70. THz electric field emitted by photoexcited electrons and holes accelerating under the influence of an electric field following the excitation of the i-region of a p-i-n structure by 12 fs pulses from a Ti:Sapphire laser at 1.52 eV. *Left panel*: GaAs, *right panel*: InP, both at 300 K. Data are presented for four different electric fields for each sample [9.539]

in the current as the carriers leave the i-region. At 24 kV/cm in GaAs, substantial transfer of electrons to the subsidiary valley leads to a peak and subsequent decrease in the average drift velocity of electrons (velocity overshoot phenomenon). Thus the signal decreases rapidly, goes through zero and develops a strong negative peak. The time at which the signal crosses zero indicates the time of zero acceleration, and hence the time to reach the peak velocity. At 28 kV/cm, InP does not exhibit a strong velocity overshoot behavior. At fields near 60 kV/cm, both GaAs and InP reveal strong velocity overshoot behavior, and an oscillatory component appears in the signal. This component becomes quite strong for GaAs at 130 kV/cm but not quite as strong, but still quite well-developed, in InP at 100 kV/cm. The oscillation frequency corresponds to the LO-phonon frequency in each material leading to the conclusion that these are coherent phonon oscillations (Sect. 2.11) induced by the polar lattice attempting to screen out the rapid motion of the carriers [9.257].

In order to obtain quantitative information about femtosecond carrier transport under these extreme non-equilibrium conditions, one must carefully consider all the corrections that need to be applied to the THz signals because of the sample structure, the frequency-dependent dielectric function of the sample, and the frequency-dependent linear dielectric function of the detector as well as the frequency dependence of the electro-optic coefficient of the detector. Once these corrections are determined, the double integral of the corrected signal corresponds to the distance traveled, and thus saturates at a value that can be associated with the thickness of the intrinsic i-region. *Leitenstorfer* et al. [9.539] argued that this provides an absolute scale to the signal from which the temporal variation of acceleration, velocity and distance can be determined. On the basis of such analyses, *Leitenstorfer* et al. [9.539] have deduced the electric field and material (GaAs vs. InP) dependence of a number of important parameters such as the peak velocity, the maximum acceleration, the distance traveled in the non-equilibrium regime and the time to reach the peak velocity as a function of the electric field. These first direct measurements of femtosecond high-field transport provide new insights into the dynamics under a high field. They have the potential to provide useful information about optimizing device performance in the ultrafast regime.

9.5 Epilogue

This chapter provided an overview of the recent developments in the field of ultrafast spectroscopy of semiconductors and semiconductor nanostructures, and described in detail some of the many exciting new developments in this field. The intent was not to provide a discussion of all new developments in topics that were treated in the first edition of the book (the first eight chapters), but to present a flavor of how the field is developing by focusing on the most exciting developments, albeit from one persons viewpoint. One can discern many trends as one compares the content of this chapter with that of the first eight chapters. I will not discuss these trends here, but end with the comment that changes in direction are signs of a vibrant field. I have no doubt that ultrafast spectroscopy of semiconductors will continue to be an exciting field of research in coming years.

References

Chapter 1

1.1 C. Kittel: *Introduction to Solid State Physics* (Wiley, New York 1956)
C. Kittel: *Quantum Theory of Solids* (Wiley, New York 1963)

1.2 H. Ibach, H. Lüth: *Solid-State Physics*, 2nd edn. (Springer, Berlin, Heidelberg 1995)

1.3 E.M. Conwell: *High Field Transport in Semiconductors* (Academic, New York 1967)

1.4 N.W. Ashcroft, N.D. Mermin: *Solid State Physics* (Saunders, Philadelphia 1976)

1.5 B.K. Ridley: *Quantum Processes in Semiconductors* (Clarendon, Oxford 1982)

1.6 K. Seeger: *Semiconductor Physics*, 6th edn., Springer Ser. Solid-State Sci., Vol. 40 (Springer, Berlin, Heidelberg 1997)

1.7 D.K. Ferry: *Semiconductors* (Macmillan, New York 1991)

1.8 P.Y. Yu, M. Cardona: *Fundamentals of Semiconductors: Physics and Material Properties*, 2nd edn. (Springer, Berlin, Heidelberg 1999)

1.9 J. Callaway: *Energy Band Theory* (Academic, Boston 1964)

1.10 M.L. Cohen, J.R. Chelikowsky: *Electronic Structure and Optical Properties of Semiconductors*, 2nd edn., Springer Ser. Solid-State Sci., Vol 75 (Springer, Berlin, Heidelberg 1989)

1.11 J.I. Pankove: *Optical Processes in Semiconductors* (Prentice Hall, New York 1971)

1.12 J.R. Chelikowsky, M.L. Cohen: Phys. Rev. B **14**, 556 (1976)

1.13 J.S. Blakemore: J. Appl. Phys. **53**, R123 (1982)

1.14 S. Adachi: J. Appl. Phys. **58**, R1 (1985)

1.15 A.Y. Cho, J.R. Arthur: Prog. Solid State Chem. **10**, 157 (1975)
A.Y. Cho: Thin Solid Films **100**, 291 (1983)

1.16 K. Ploog: In *Crystals: Growth, Properties and Applications*, ed. by H. C. Freyhardt (Springer, Berlin, Heidelberg 1980) p. 75

1.17 W.T. Tsang: Molecular beam epitaxy of III-V compound semiconductors. *Semiconductors and Semimetals* **22**, 95 (Academic, New York 1985)

1.18 M.A. Herman, H. Sitter: *Molecular Beam Epitaxy*, 2nd edn., Springer Ser. Mater. Sci., Vol. 7 (Springer, Berlin, Heidelberg 1996)

1.19 G. Bastard, J.A. Brum, R. Ferreira: Electronic states in semiconductor heterostructures. *Solid State Physics* **44**, 229-415 (Academic, Boston 1991)

1.20 C. Weisbuch: Applications of multiquantum wells, selective doping and superlattices. *Semiconductors and Semimetals* **24**, 1-133 (Academic, Boston 1987)

1.21 C. Weisbuch, B. Vinter: *Quantum Semiconductor Structures: Fundamentals and Applications* (Academic, Boston 1991)

1.22 M. Jaros: *Physics and Applications of Semiconductor Microstructures* (Clarendon, Oxford, UK 1978)

1.23 E.I. Ivchenko, P.E. Pikus: *Superlattices and Other Heterostructures: Symmetry and Optical Phenomena*, 2nd edn., Springer Ser. Solid-State Sci., Vol. 110 (Springer, Berlin, Heidelberg 1997)
1.24 Y.-C. Chang, J.N. Shulman: Phys. Rev. B **31**, 2069 (1985)
1.25 R. Dingle, H. Stormer, A.C. Gossard, W. Wiegmann: Appl. Phys. Lett. **33**, 665 (1978)
1.26 A.C. Gossard, A. Pinczuk: In *Synthetic Modulated Structures*, ed. by L. Chang, B. Giessen (Academic, Boston, MA 1985) pp. 215-255
1.27 R.E. Prange, S.M. Girvin (eds.): *The Quantum Hall Effect* (Springer, New York 1987)
T. Chakraborty, P. Pietiläinen: *The Quantum Hall Effects – Fractional and Integral*, 2nd edn., Springer Ser. Solid-State Sci., Vol. 85 (Springer, Berlin, Heidelberg 1995)
1.28 R.J. Elliott: Phys. Rev. **108**, 1384 (1957)
1.29 R.J. Elliott: In *Polarons and Excitons*, ed. by C. G. Kuper, G. D. Whitfield (Plenum, New York 1963) pp. 269-293
1.30 R.S. Knox: *Theory of Excitons* (Academic, New York 1963)
1.31 D.L. Dexter, R.S. Knox: *Excitons* (Interscience, New York 1965)
1.32 J.O. Dimmock: In *Optical Properties of III-V Compounds. Semiconductors and Semimetals* **3**, 259-319 (Academic, New York 1967
1.33 E.J. Johnson: In *Optical Properties of III-V Compounds. Semiconductors and Semimetals* **3**, 153-258 (Academic, New York 1967)
1.34 H. Haken, S. Nikitine (eds.): *Excitons at High Density*, Springer Tracts Mod. Phys., Vol. 73 (Springer, Berlin, Heidelberg 1975)
1.35 K. Cho: *Excitons*, Topics Curr. Phys., Vol. 14 (Springer, Berlin, Heidelberg 1979)
1.36 D.C. Reynolds, T.C. Collins: *Excitons: Their Properties and Uses* (Academic, New York 1981)
1.37 E.I. Rashba, M.D. Sturge (eds.): *Excitons* (North-Holland, Amsterdam 1982)
1.38 J.L.T. Waugh, G. Dolling: Phys. Rev. **132**, 2410 (1963)
1.39 B.K. Ridley: In *Hot Carriers in Semiconductor Nanostructures: Physics and Applications*, ed. by J. Shah (Academic, Boston 1992) pp. 17-51
1.40 B. Jusserand, M. Cardona: In *Light Scattering in Solids V: Superlattices and Other Microstructures*, ed. by M. Cardona, G. Güntherodt, Topics Appl. Phys., Vol. 66 (Springer, Berlin, Heidelberg 1989) pp. 49-152
1.41 B.R. Nag: *Theory of Electrical Transport in Semiconductors* (Pergamon, Oxford, U.K. 1972)
1.42 B.R. Nag: *Electron Transport in Compound Semiconductors*, Springer Ser. Solid-State Sci., Vol. 11 (Springer, Berlin, Heidelberg 1980)
1.43 L. Reggiani: *Hot-Electron Transport in Semiconductors*, Topics Appl. Phys., Vol. 58 (Springer, Berlin, Heidelberg 1985)
1.44 C. Jacoboni, P. Lugli: *The Monte Carlo Method for Semiconductor Device Simulation* (Springer, Vienna 1989)
1.45 M. Costato, L. Reggiani: Phys. Status Solidi (b) **58**, 471 (1973)
1.46 M. Costato, G. Gagliani, C. Jacoboni, L. Reggiani: J. Phys. Chem. Solids **35**, 1605 (1974)
1.47 J.D. Wiley, M. DiDomenico Jr.: Phys. Rev. B **2**, 427 (1970)

1.48 D.L. Rode: In *Transport Phenomena. Semiconductor and Semimetals* **10**, 1-90 (Academic, New York 1975)
1.49 J.D. Wiley: In *Transport Phenomena. Semiconductor and Semimetals* **10**, 91-174 (Academic, New York 1975)
1.50 S. Das Sarma: In *Hot Carriers in Semiconductor Nanostructures: Physics and Applications*, ed. by J. Shah (Academic, Boston 1992) pp. 53-85
1.51 J. Shah (ed.): *Hot Carriers in Semiconductor Nanostructures: Physics and Applications* (Academic, Boston 1992)
1.52 J.F. Ryan, A.C. Maciel (eds.): *Hot Carriers in Semiconductors* (HCIS-8), Semicond. Sci. Technol. Vol. 9 (Institute of Physics, Bristol 1994)
1.53 J. Shah: In *Hot Carriers in Semiconductor Nanostructures: Physics and Applications*, ed. by J. Shah (Academic, Boston 1992) pp. 3-14
1.54 D.W. Pohl, W. Denk, M. Lantz: Appl. Phys. Lett. **44**, 651 (1984)
D.W. Pohl: In *Scanning Tunneling Microscopy II*, 2nd edn., ed. by R. Wiesendanger, H.-J. Güntherodt, Springer Ser. Surf. Sci., Vol. 28 (Springer, Berlin, Heidelberg 1995) Chap. 7 and Sect. 9.4
1.55 E. Betzig, J.K. Trautman: Science **257**, 189 (1992)
1.56 B. Deveaud, J. Shah, T.C. Damen, B. Lambert, A. Chomette, A. Regreny: IEEE J. QE-**24**, 1641 (1988)
1.57 J.-L. Martin, A. Migus, G.A. Mourou, A.H. Zewail (eds.): *Ultrafast Phenomena VIII*, Springer Ser. Chem. Phys., Vol. 55 (Springer, Berlin, Heidelberg 1993)
1.58 P.F. Barbara, W.H. Knox, G.A. Mourou, A.H. Zewail (eds.): *Ultrafast Phenomena IX*, Springer Ser. Chem. Phys. Vol. 60 (Springer, Berlin, Heidelberg 1995)
1.59 W. Kaiser (ed.): *Ultrashort Laser Pulses*, 2nd edn., Topics Appl. Phys., Vol. 60 (Springer, Berlin, Heidelberg 1993)
1.60 A.E. Siegman: *Lasers* (University Science Books, Mill Valley, CA 1986)
1.61 C.V. Shank: In *Ultrashort Laser Pulses*, 2nd edn., ed. by W. Kaiser, Topics Appl. Phys., Vol. 60 (Springer, Berlin, Heidelberg 1993) Chap. 2
1.62 N. Sarukara, Y. Ishida, H. Nakano: Opt. Lett. **16**, 153 (1991)
1.63 D.E. Spence, W. Sibbett: J. Opt. Soc. Am. B **8**, 2053 (1991)
1.64 D.E. Spence, P.N. Kean, W. Sibbett: Opt. Lett. **16**, 42 (1991)
1.65 D.E. Spence, J.M. Evans, W.E. Sleat, W. Sibbett: Opt. Lett. **16**, 1762 (1991)
1.66 C. Spielman, F. Krausz, T. Brabec, E. Wintner, A.J. Schmidt: Opt. Lett. **16**, 1180 (1991)
1.67 L. Spinelli, B. Coulliard, N. Goldblatt, D. K. Negus: CLEO'91 (Baltimore, MD) Paper FDF 7
1.68 C.-P. Huang, H.C. Kapteyn, J.M. McIntosh, M.M. Murnane: Opt. Lett. **17**, 139 (1992)
1.69 F. Krausz, C. Spielman, T. Brabec, E. Wintner, A.J. Schmidt: Opt. Lett. **17**, 204 (1992)
1.70 N.H. Rizvi, P.M.W. French, J.R. Taylor: Opt. Lett. **17**, 279 (1992)
1.71 E.P. Ippen: Appl. Phys. B **58**, 159 (1994)
1.72 H.C. Kapteyn, M.M. Murnane: Opt. Photon. News **5**, 20 (1994)
1.73 Q. Fu, G. Mak, H.M. van Driel: Opt. Lett. **17**, 1006 (1992)
1.74 W.S. Pelouch, P.E. Powers, C.L. Tang: Opt. Lett. **17**, 1070 (1992)

1.75 C.L. Tang, W.R. Bosenberg, T. Ukachi, R.J. Lane, L.K. Cheng: IEEE Proc. **80**, 365 (1992)
1.76 H.M. van Driel, G. Mak: Canad. J. Physics **71**, 47 (1993)
1.77 J. D. Kafka, M. L. Watts, J. W. Pieterse: In *Ultrafast Phenomena IX*, ed. by P. F. Barbara, W. H. Knox, G. A. Mourou, A. H. Zewail, Springer Ser. Chem. Phys., Vol. 60 (Springer, Berlin, Heidelberg 1994) p. 185
1.78 D.T. Reid, J.M. Dudley, M. Ebrahimzadeh, W. Sibbett: In *Ultrafast Phenomena IX*, ed. by P.F. Barbara, W.H. Knox, G.A. Mourou, A.H. Zewail, Springer Ser. Chem. Phys., Vol. 60 (Springer, Berlin, Heidelberg 1994) pp. 205-207
1.79 L. Min, R.J.D. Miller: Appl. Phys. Lett. *56*, 524 (1990)
1.80 W.A. Kütt, G.C. Cho, M. Strahnen, H. Kurz: Appl. Surf. Sci. **50**, 325 (1990)
1.81 D. von der Linde, J. Kuhl, H. Klingenberg: Phys. Rev. Lett. **44**, 1505 (1980)
1.82 D.Y. Oberli, D.R. Wake, M.V. Klein, J. Klem, T. Henderson, H. Morkoc: Phys. Rev. Lett. **59**, 696 (1987)
1.83 H.J. Eichler, P. Günter, D.W. Pohl: *Laser-Induced Dynamics Gratings*, Springer Ser. Opt. Sci. Vol. 50 (Springer, Berlin, Heidelberg 1986)
1.84 M.D. Webb, S.T. Cundiff, D.G. Steel: Phys. Rev. Lett. **66**, 934 (1991)
1.85 A.M. Weiner, S. De Silvestri, E.P. Ippen: J. Opt. Soc. Am. B **2**, 645 (1985)
1.86 J.-Y. Bigot, M.T. Portella, R.W. Schoenlein, C.J. Bardeen, A. Migus, C.V. Shank: Phys. Rev. Lett. **66**, 1138 (1991)
1.87 E.T.J. Nibbering, D.A. Wiersma, K. Duppen: Phys. Rev. Lett. **66**, 2464 (1991)
1.88 E.T.J. Nibbering, D.A. Wiersma, K. Duppen: In *Coherent Optical Interactions in Solids*, ed. by R. T. Phillips (Plenum, New York 1994) pp. 181-198
1.89 H.J. Bakker, K. Leo, J. Shah, K. Kohler: Phys. Rev. B **49**, 8249 (1994)
1.90 N.A. Kurnit, I.D. Abella, S.R. Hartman: Phys. Rev. Lett. **13**, 567 (1964)
1.91 J. Shah: IEEE J. QE-**24**, 276 (1988)
1.92 G. Noll, U. Siegner, S.G. Shevel, E.O. Göbel: Phys. Rev. Lett. **64**, 792 (1990)
1.93 D.S. Kim, J. Shah, T.C. Damen, W. Schäfer, F. Jahnke, S. Schmitt-Rink, K. Köhler: Phys. Rev. Lett. **69**, 2725 (1992)
1.94 J.-Y. Bigot, M.-A. Mycek, S. Weiss, R.G. Ulbrich, D.S. Chemla: Phys. Rev. Lett. **70**, 3307 (1993)
1.95 D.S. Kim, J. Shah, D.A.B. Miller, T.C. Damen, W. Schäfer, L. Pfeiffer: Phys. Rev. B **48**, 17902 (1993)
1.96 Y.R. Shen: *The Principles of Nonlinear Optics* (Wiley, New York 1984)
D.L. Mills: *Nonlinear Optics*, 2nd edn. (Springer, Berlin, Heidelberg 1998)
1.97 M.D. Levenson, S.S. Kano: *Introduction to Nonlinear Laser Spectroscopy* (Academic, Boston 1988)
W. Demtröder: *Laser Spectroscopy*, 2nd edn. (Springer, Berlin, Heidelberg 1996)
1.98 W.E. Bron: In *Ultrashort Processes in Condensed Matter*, ed. by W. E. Bron (Plenum, New York 1993) pp. 101-142
1.99 W.E. Bron: In *Coherent Optical Interactions in Solids*, ed. by R. T. Phillips (Plenum, New York 1994) pp. 199-222
1.100 A. Honold, L. Schultheis, J. Kuhl, C.W. Tu: Appl. Phys. Lett. **52**, 2105 (1988)
1.101 D. Rosen, A.G. Doukas, Y. Budansky, A. Katz, R.R. Alfano: Appl. Phys. Lett. **39**, 935 (1981)

1.102 A. Olsson, D.J. Erskine, Z.Y. Xu, A. Schremer, C.L. Tang: Appl. Phys. Lett. **41**, 659 (1982)
1.103 M.B. Johnson, T.C. McGill, A.T. Hunter: J. Appl. Phys. **63**, 2077 (1988)
1.104 D. von der Linde, J. Kuhl, E. Rosengart: J. Lumin. **24/25**, 675 (1981)
1.105 H. Mahr, M.D. Hirsch: Opt. Commun. **13**, 96 (1975)
1.106 G. Fleming: *Chemical Applications of Ultrafast Spectroscopy* (Oxford Univ. Press, Oxford, UK 1986)
1.107 J. Shah: In *Spectroscopy of Nonequilibrium Electrons and Phonons*, ed. by C.V. Shank, B.P. Zakharchenya (Elsevier, Amsterdam 1992) pp. 57-112
1.108 J. Shah: In *Hot Carriers in Semiconductor Nanostructures: Physics and Applications*, ed. by J. Shah (Academic, Boston 1992) pp. 279-312
1.109 M.R. Freeman, D.D. Awschalom, J.M. Hong: Appl. Phys. Lett. **57**, 704 (1990)
1.110 A. Vinattieri, J. Shah, T.C. Damen, D.S. Kim, L.N. Pfeiffer, L.J. Sham: Solid State Commun. **88**, 189 (1993)
1.111 H. Wang, J. Shah, T.C. Damen, L.N. Pfeiffer: IQE'94, Techn. Digest Ser., Vol. 7 (Optical Society of America, Washington, DC 1994) postdeadline paper
1.112 P.R. Smith, D.H. Auston, M.C. Nuss: IEEE J. QE-**24**, 255 (1988)
1.113 C.H. Fattinger, D. Grischkowsky: Appl. Phys. Lett. **54**, 490 (1989)
1.114 T. Kuhn, F. Rossi: Phys. Rev. Lett. **69**, 977 (1992)
1.115 T. Kuhn, F. Rossi: Phys. Rev. B **46**, 7496 (1992)
1.116 F. Rossi, S. Haas, T. Kuhn: Semicond. Sci. Technol. **9**, 411 (1994)
1.117 X.-C. Zhang, B.B. Hu, J.T. Darrow, D.H. Auston: Appl. Phys. Lett. **56**, 1011 (1990)
1.118 B.B. Hu, J.T. Darrow, X.-C. Zhang, D.H. Auston, P.R. Smith: Appl. Phys. Lett. **56**, 886 (1990)
1.119 H.G. Roskos, M.C. Nuss, J. Shah, K. Leo, D.A.B. Miller, A.M. Fox, S. Schmitt-Rink, K. Köhler: Phys. Rev. Lett. **68**, 2216 (1992)
1.120 M.C. Nuss, P.C.M. Planken, I. Brener, H.G. Roskos, M.S.C. Luo, S.L. Chuang: Appl. Phys. B **58**, 249 (1994)
1.121 T.C.L.G. Sollner, J. Shah (eds.): *Picosecond Electronics and Optoelectronics*, OSA Proc. on Picosec. Electro. Optoelectron., Vol. 9 (Optical Society of America, Washington, DC 1991)
1.122 J. Shah, U. Mishra (eds.): *Ultrafast Electronics and Optoelectronics*, OSA Proc. on Ultrafast Electron. Optoelectron., Vol. 14 (Optical Society of America, Washington, DC 1993)
1.123 L.P. Kadanoff, G. Baym: *Quantum Statistical Mechanics* (Benjamin, New York 1962)
1.124 L.V. Keldysh: Sov. Phys. JETP **20**, 4 (1965)
1.125 A.L. Ivanov, L.V. Keldysh: Sov. Phys. JETP **57**, 234 (1983)
1.126 K. Henneberger, G. Manzke, V. May, R. Zimmermann: Physica A **138**, 557 (1986)
1.127 L. Hitzchke, G. Ropke, T. Seifert, R. Zimmermann: J. Phys. B **19**, 2443 (1986)
1.128 S. Schmitt-Rink, D.S. Chemla, H. Haug: Phys. Rev. B **37**, 941 (1988)
1.129 W. Schäfer: *Festkörperprobleme/Adv. Solid State Phys.* **28**, 63 (Vieweg, Braunschweig 1988)
1.130 R. Zimmermann: *Many-Particle Theory of Highly Excited Semiconductors* (Teubner, Leipzig 1988)

1.131 M. Hartmann, W. Schäfer: Phys. Status Solidi (b) **173**, 165 (1992)

1.132 H. Haug, S.W. Koch: *Quantum Theory of the Optical and Electronic Properties of Semiconductors* (World Scientific, Singapore 1993)

1.133 L. Allen, J.H. Eberly: *Optical Resonance and Two-Level Atoms* (Wiley, New York 1975)

1.134 M. Sargent III, M.O. Scully, W.E. Lamb Jr.: *Laser Physics* (Addison-Wesley, New York 1977)
P. Meystre, M. Sargent III: *Elements of Quantum Optics*, 3rd edn. (Springer, Berlin, Heidelberg 1999)

1.135 A.V. Kuznetsov: Phys. Rev. B **44**, 8721 (1991)

1.136 A.V. Kuznetsov: Phys. Rev. B **44**, 13381 (1991)

1.137 T. Kuhn, F. Rossi: Phys. Rev. B **46**, 7496 (1992)

1.138 A.V. Kuznetsov, C.J. Stanton: Phys. Rev. B **48**, 10828 (1993)

1.139 A. Leitenstorfer, A. Lohner, T. Elsaesser, S. Haas, F. Rossi, T. Kuhn, W. Klein, G. Boehm, G. Traenkle, G. Weimann: Phys. Rev. Lett. **73**, 1687 (1994)

1.140 F. Rossi, S. Haas, T. Kuhn: Phys. Rev. Lett. **72**, 152 (1994)

1.141 S.M. Kogan: Sov. Physics – Solid State **4**, 1813 (1963)

1.142 V.V. Paranjape, B.V. Paranjape: Phys. Rev. **166**, 757 (1968)

1.143 N. Perrin, H. Bud: Phys. Rev. B **6**, 1359 (1972)

1.144 P. Kocevar: J. Phys. C **5**, 3349 (1972)

1.145 J.H. Collet, A. Cornet, M. Pugnet, T. Amand: Solid State Commun. **42**, 883 (1982)

1.146 W. Pötz, P. Kocevar: Phys. Rev. B **28**, 7040 (1983)

1.147 J.H. Collet, T. Amand: J. Phys. Chem. Solids **47**, 153 (1986)

1.148 W. Pötz, P. Kocevar: In *Hot Carriers in Semiconductor Nanostructures: Physics and Applications*, ed. by J. Shah (Academic, Boston 1992) pp. 87-120

1.149 S.M. Goodnick, P. Lugli: In *Hot Carriers in Semiconductor Nanostructures: Physics and Applications*, ed. by J. Shah (Academic, Boston 1992) pp. 191-234

1.150 K. Binder, D.W. Heermann: *Monte Carlo Simulation in Statistical Physics*, 2nd edn., Springer Ser. Solid-State Sci., Vol. 80 (Springer, Berlin, Heidelberg 1992)

1.151 P. Lugli, D.K. Ferry: Phys. Rev. Lett. **56**, 1295 (1986)

1.152 D. Heermann: *Computer Simulation Methods in Theoretical Physics*, 2nd edn. (Springer, Berlin, Heidelberg 1990)

Chapter 2

2.1 L. Allen, J.H. Eberly: *Optical Resonance and Two-Level Atoms* (Interscience, New York 1975)

2.2 M. Sargent III, M.O. Scully, W.E. Lamb Jr.: *Laser Physics* (Addison-Wesley, New York 1977)
P. Meystre, M. Sargent III: *Elements of Quantum Optics*, 3rd edn. (Springer, Berlin, Heidelberg 1999)

2.3 Y.R. Shen: *The Principles of Nonlinear Optics* (Wiley, New York 1984)
D.L. Mills: *Nonlinear Optics*, 2nd edn. (Springer, Berlin, Heidelberg 1998)

2.4 M.D. Levenson, S.S. Kano: *Introduction to Nonlinear Laser Spectroscopy* (Academic, Boston 1988)
W. Demtröder: *Laser Spectroscopy*, 2nd edn. (Springer, Berlin, Heidelberg 1996)
2.5 F. Henneberger, S. Schmitt-Rink, E.O. Göbel (eds.): *Optics of Semiconductor Nanostructures* (Akademie Verlag, Berlin 1993)
R.T. Phillips (ed.): *Coherent Optical Processes in Semiconductors*, NATO ASI Series B: Physics, Vol. 30 (Plenum, New York 1994)
2.6 M. Ueta, H. Kanzaki, K. Kobayashi, Y. Toyozawa, E. Hanamura: *Excitonic Processes in Solids*, Springer Ser. Solid-State Sci., Vol. 60 (Springer, Berlin, Heidelberg 1986)
2.7 L.I. Schiff: *Quantum Mechanics* (McGraw Hill, New York 1955)
2.8 A. Messiah: *Quantum Mechanics* (North-Holland, Amsterdam 1961)
2.9 L.P. Kadanoff, G. Baym: *Quantum Statistical Mechanics* (Benjamin, New York 1962)
2.10 L.V. Keldysh: Sov. Phys. JETP **20**, 4 (1965)
2.11 A.L. Ivanov, L.V. Keldysh: Sov. Phys. JETP **57**, 234 (1983)
2.12 H. Haug, S. Schmitt-Rink: Progr. Quantum Electron. **9**, 3 (1984)
2.13 H. Haug: J. Lumin. **30**, 171 (1985)
2.14 K. Henneberger, G. Manzke, V. May, R. Zimmermann: Physica A **138**, 557 (1986)
2.15 K. Henneberger, V. May: Physica A **138**, 537 (1986)
2.16 W. Schäfer, J. Treusch: Z. Physik **63**, 407 (1986)
2.17 M. Lindberg, S.W. Koch: Phys. Rev. B **38**, 3342 (1988)
2.18 S. Schmitt-Rink, D.S. Chemla, H. Haug: Phys. Rev. B **37**, 941 (1988)
2.19 R. Zimmermann: *Many-Particle Theory of Highly Excited Semiconductors* (Teubner, Leipzig 1988)
2.20 H. Haug, S.W. Koch: *Quantum Theory of the Optical and Electronic Properties of Semiconductors* (World Scientific, Singapore 1993)
2.21 W. Schäfer: In *Optics of Semiconductor Nanostructures*, ed. by H. Henneberger, S. Schmitt-Rink, E.O. Göbel (Akademie, Berlin 1993) pp. 21-50
2.22 W. Schäfer: *Festkörperprobleme/Adv. Solid State Phys.* **28**, 63 (Vieweg, Braunschweig 1988)
2.23 R. Binder, S.W. Koch, M. Lindberg, W. Schäfer, F. Jahnke: Phys. Rev. B **43**, 6520 (1991)
2.24 M. Hartmann, W. Schäfer: Phys. Status Solidi (b) **173**, 165 (1992)
2.25 H. Haug (guest ed.): Proc. 3rd Conf. on Nonlinear Optics and Excitation Kinetics in Semiconductors. Phys. Status Solidi (b) **173** (1992)
2.26 R. Zimmermann (guest ed.): Proc. 4th Conf. on Nonlinear Optics and Excitation Kinetics in Semiconductors. Phys. Status Solidi (b) **188** (1995)
2.27 A.V. Kuznetsov: Phys. Rev. B **44**, 13381 (1991)
2.28 A. Tomita, A. Suzuki: IEEE J. QE-**27**, 1630 (1991)
2.29 S. Saikan, J.W. Lin, H. Nemoto: Phys. Rev. A **46**, 7123 (1992)
2.30 R. Zimmermann, J. Wauer: J. Lumin. **58**, 271 (1994)
2.31 J.-Y. Bigot, M.T. Portella, R.W. Schoenlein, C.J. Bardeen, A. Migus, C.V. Shank: Phys. Rev. Lett. **66**, 1138 (1991)
2.32 E.T.J. Nibbering, K. Duppen, D.A. Wiersma: J. Chem. Phys. **93**, 5477 (1990)
2.33 E.T.J. Nibbering, D.A. Wiersma, K. Duppen: Phys. Rev. Lett. **66**, 2464 (1991)

2.34 E.T.J. Nibbering, D.A. Wiersma, K. Duppen: In *Coherent Optical Interactions in Solids*, ed. by R.T. Phillips (Plenum, New York 1994) pp. 181-198
2.35 S. Mukamel: Adv. Chem. Phys. **70**, 165 (1988)
2.36 S. Schmitt-Rink, S. Mukamel, K. Leo, J. Shah, D.S. Chemla: Phys. Rev. A **44**, 2124 (1991)
2.37 N. Bloembergen, E.M. Percell, R.V. Pound: Phys. Rev. **73**, 679 (1948)
2.38 P.W. Anderson, P.R. Weiss: Rev. Mod. Phys. **25**, 269 (1953)
2.39 R. Kubo: J. Phys. Soc. Jpn. **9**, 935 (1954)
2.40 R. Kubo: In *Fluctuations, Relaxation and Resonance in Magentic Systems*, ed. by D. ter Haar (Plenum, New York 1962) pp. 23-68
2.41 R. Kubo: Adv. Chem. Phys. **15**, 101 (1969)
2.42 S. Schmitt-Rink, S. Mukamel, K. Leo, J. Shah, D.S. Chemla: Phys. Rev. A **44**, 2124 (1991)
2.43 W.B. Bosma, Y. Yan, S. Mukamel: Phys. Rev. A **42**, 6920 (1990)
2.44 F. Bloch: Phys. Rev. **70**, 460 (1946)
2.45 T. Yajima, Y. Taira: J. Phys. Soc. Jpn. **47**, 1620 (1979)
2.46 N.A. Kurnit, I.D. Abella, S.R. Hartman: Phys. Rev. Lett. **13**, 567 (1964)
2.47 A.M. Weiner, S. De Silvestri, E.P. Ippen: J. Opt. Soc. Am. B **2**, 645 (1985)
2.48 R. Loudon: *The Quantum Theory of Light* (Oxford Univ. Press, New York 1983)
2.49 P.C. Becker, H.L. Fragnito, C.H. Brito Cruz, R.L. Fork, J.E. Cunningham, J.E. Henry, C.V. Shank: Phys. Rev. Lett. **61**, 1647 (1988)
2.50 R.J. Elliott: Phys. Rev. **108**, 1384 (1957)
2.51 R.J. Elliott: In *Polarons and Excitons*, ed. by C.G. Kuper, G.D. Whitfield (Plenum, New York 1963) pp. 269-293
2.52 R.S. Knox: *Theory of Excitons* (Academic, New York 1963)
2.53 D.L. Dexter, R.S. Knox: *Excitons* (Interscience, New York 1965)
2.54 S. Schmitt-Rink, D.S. Chemla, D.A.B. Miller: Phys. Rev. B **32**, 6601 (1985)
2.55 S. Schmitt-Rink, D.A.B. Miller, D.S. Chemla: Phys. Rev. B **35**, 8113 (1987)
2.56 S. Schmitt-Rink, D.S. Chemla, D.A.B. Miller: Adv. Phys. **38**, 89 (1989)
2.57 W. Schäfer, K.-H. Schuldt, J. Treusch: Phys. Status Solidi (b) **147**, 699 (1988)
2.58 R. Binder, S.W. Koch, M. Lindberg, N. Peyghambarian, W. Schäfer: Phys. Rev. Lett. **65**, 899 (1990)
2.59 W. Schäfer, S. Schmitt-Rink, C. Stafford: SPIE Proc. **1280**, 24 (1990)
2.60 R. Binder, D. Scott, A.E. Paul, M. Lindberg, K. Henneberger, S.W. Koch: Phys. Rev. B **45**, 1107 (1992)
2.61 S.W. Koch, A. Knorr, R. Binder, M. Lindberg: Phys. Status Solidi (b) **173**, 177 (1992)
2.62 M. Lindberg, R. Binder, S.W. Koch: Phys. Rev. A **45**, 1865 (1992)
2.63 S.W. Koch, Y.Z. Hu, R. Binder: Physica B **189**, 176 (1993)
2.64 W. Schäfer, F. Jahnke, S. Schmitt-Rink: Phys. Rev. B **47**, 1217 (1993)
2.65 S. Schmitt-Rink, D.S. Chemla: Phys. Rev. Lett. **57**, 2752 (1986)
2.66 M. Wegener, D.S. Chemla, S. Schmitt-Rink, W. Schäfer: Phys. Rev. A **42**, 5675 (1990)
2.67 S. Saikan, H. Miyamoto, Y. Tosaki, A. Fujiwara: Phys. Rev. B **36**, 5074 (1987)
2.68 R.C.C. Leite, J. Shah, J.P. Gordon: Phys. Rev. Lett. **23**, 1332 (1969)

2.69 L. Schultheis, A. Honold, J. Kuhl, K. Köhler, C.W. Tu: Phys. Rev. B **34**, 9027 (1986)
2.70 A. Honold, L. Schultheis, J. Kuhl, C.W. Tu: Phys. Rev. B **40**, 6442 (1989)
2.71 H. Wang, K.B. Ferrio, D.G. Steel, Y.Z. Hu, R. Binder, S.W. Koch: Phys. Rev. Lett. **71**, 1261 (1993)
2.72 D.G. Steel, H. Wang, M. Jiang, K.B. Ferrio, S.T. Cundiff: In *Coherent Optical Interactions in Solids*, ed. by R.T. Phillips (Plenum, New York 1994) pp. 157-180
2.73 H. Wang, K.B. Ferrio, D.G. Steel, P.R. Berman, Y.Z. Hu, R. Binder, S.W. Koch: Phys. Rev. A **49**, R1551 (1994)
2.74 T.F. Heinz, S.L. Palfrey: J. Opt. Soc. Am. B **2**, 674 (1985)
2.75 Z. Vardeny, J. Tauc: Opt. Commun. **39**, 396 (1981)
2.76 H. Stolz: *Time-Resolved Light Scattering from Excitons*, Springer Tracts Mod. Phys., Vol. 130 (Springer, Berlin, Heidelberg 1994)
2.77 W. von der Osten, V. Langer, H. Stolz: In *Coherent Optical Interactions in Solids*, ed. by R.T. Phillips (Plenum, New York 1994) pp. 111-136
2.78 Y.R. Shen: Phys. Rev. B **14**, 1772 (1976)
2.79 Y. Toyozawa: J. Phys. Soc. Jpn. **41**, 400 (1976)
2.80 B.R. Mollow: *Progr. Optics* **19**, 1-43 (North-Holland, Amsterdam 1981)
2.81 J. Hegarty, M.D. Sturge, C. Weisbuch, A.C. Gossard, W. Wiegmann: Phys. Rev. Lett. **49**, 930 (1982)
2.82 L. Schultheis, M.D. Sturge, J. Hegarty: Appl. Phys. Lett. **47**, 995 (1985)
2.83 L. Schultheis, A. Honold, J. Kuhl, K. Köhler, C.W. Tu: Superlattices and Microstructures **2**, 441 (1986)
2.84 L. Schultheis, J. Kuhl, A. Honold, C.W. Tu: Phys. Rev. Lett. **57**, 1797 (1986)
2.85 L. Schultheis, J. Kuhl, A. Honold, C.W. Tu: Phys. Rev. Lett. **57**, 1635 (1986)
2.86 J. Hegarty: Phys. Rev. B **38**, 7843 (1988)
2.87 T.F. Albrecht: Solid State Electron. **37**, 1327 (1994)
2.88 L. Schultheis, J. Kuhl, A. Honold, C.W. Tu: In *Proc. 18th Int'l Conf. on the Physics of Semiconductors*, ed. by O. Engstrom (World Scientific, Singapore 1987) p.2
2.89 A. Honold, L. Schultheis, J. Kuhl, C. W. Tu: In *Ultrafast Phenomena VI*, ed. by T. Yajima, K. Yoshihara, C. B. Harris, S. Shionoya, Springer Ser. Chem. Phys., Vol. 48 (Springer, Berlin, Heidelberg 1989) pp. 307-309
2.90 D.G. Steel: Unpublished (1995)
2.91 D.S. Kim, J. Shah, J.E. Cunningham, T.C. Damen, W. Schäfer, M. Hartmann, S. Schmitt-Rink: Phys. Rev. Lett. **68**, 1006 (1992)
2.92 J. Lee, E.S. Koteles, M.O. Vassell: Phys. Rev. B **33**, 5512 (1986)
2.93 D.S. Chemla, D.A.B. Miller, P.W. Smith, A.C. Gossard, W. Wiegmann: IEEE J. QE-**20**, 265 (1984)
2.94 P.W. Anderson: Phys. Rev. **109**, 1492 (1958)
2.95 J. Hegarty, M.D. Sturge, A.C. Gossard, W. Wiegmann: Appl. Phys. Lett. **40**, 132 (1982)
2.96 J. Hegarty: Phys. Rev. B **25**, 4324 (1982)
2.97 J. Hegarty, M.D. Sturge: J. Lumin. **31 & 32**, 494 (1984)
2.98 J. Hegarty, L. Goldner, M.D. Sturge: Phys. Rev. B **30**, 7346 (1984)
2.99 J. Hegarty, M.D. Sturge: J. Opt. Soc. Am. B **2**, 1143 (1985)

2.100 J.T. Remillard, H. Wang, M.D. Webb, D.G. Steel, J. Oh, J. Pamulapati, P.K. Bhattacharya: Opt. Lett. **14**, 1131 (1989)
2.101 J.T. Remillard, H. Wang, M.D. Webb, D.G. Steel: IEEE J. QE-**25**, 408 (1989)
2.102 J.T. Remillard, H. Wang, M.D. Webb, D.G. Steel: J. Opt. Soc. Am. B **7**, 897 (1990)
2.103 H. Wang, J.T. Remillard, M.D. Webb, D.G. Steel, J. Pamulapati, J. Oh, P.K. Bhattacharya: Surf. Sci. **228**, 69 (1990)
2.104 M.D. Webb, S.T. Cundiff, D.G. Steel: Phys. Rev. B **43**, 12658 (1991)
2.105 M.D. Webb, S.T. Cundiff, D.G. Steel: Phys. Rev. Lett. **66**, 934 (1991)
2.106 D.G. Steel, H. Wang, S.T. Cundiff: In *Optics of Semiconductor Nanostructures*, ed. by H. Henneberger, S. Schmitt-Rink, E. O. Göbel (Akademie, Berlin 1993) pp. 75-125
2.107 S.T. Cundiff, D.G. Steel: IEEE J. QE-**28**, 2423 (1992)
2.108 S.T. Cundiff, H. Wang, D.G. Steel: Phys. Rev. B **46**, 7248 (1992)
2.109 H. Stolz, D. Schwarze, W. von der Osten, G. Weimann: Phys. Rev. B **47**, 9669 (1993)
2.110 G. Noll, U. Siegner, S.G. Shevel, E.O. Göbel: Phys. Rev. Lett. **64**, 792 (1990)
2.111 R.P. Stanley, J. Hegarty: In *Optics of Semiconductor Nanostructures*, ed. by H. Henneberger, S. Schmitt-Rink, E. O. Göbel (Akademie, Berlin 1993) pp. 263-290
2.112 U. Siegner, D. Weber, E.O. Göbel, D. Bennhardt, V. Heuckeroth, R. Saleh, S.D. Baranovskii, P. Thomas, H. Schwab, C. Klingshirn, J.M. Hvam, V.G. Lyssenko: Phys. Rev. B **46**, 4564 (1992)
2.113 C. Cohen-Tannoudji, J. Dupont-Roc, G. Grynberg: *Atom-Photon Interactions: Basic Processes and Applications* (Interscience, New York 1992)
2.114 D. Fröhlich, A. Nothe, K. Reimann: Phys. Rev. Lett. **55**, 1335 (1985)
2.115 D. Fröhlich, R. Wille, W. Schlapp, G. Weimann: Phys. Rev. Lett. **59**, 1748 (1987)
2.116 A. Mysyrowicz, D. Hulin, A. Antonetti, A. Migus, W.T. Masselink, H. Morkoç: Phys. Rev. Lett. **56**, 2748 (1986)
2.117 A. Von Lehmen, D.S. Chemla, J.E. Zucker, J.P. Heritage: Opt. Lett. **11**, 609 (1986)
2.118 M. Combescot, R. Combescot: Phys. Rev. Lett. **61**, 117 (1988)
2.119 R. Zimmermann: Phys. Status Solidi (b) **146**, 545 (1988)
2.120 R. Zimmermann: Phys. Status Solidi (b) **159**, 317 (1990)
2.121 R. Zimmermann, D. Fröhlich: In *Optics of Semiconductor Nanostructures*, ed. by H. Henneberger, S. Schmitt-Rink, E. O. Göbel (Akademie Verlag, Berlin 1993) pp. 51-74
2.122 N. Peyghambarian, S.W. Koch, M. Lindberg, B.D. Fluegel, M. Joffre: Phys. Rev. Lett. **62**, 1185 (1989)
2.123 W.H. Knox, D.S. Chemla, D.A.B. Miller, J.B. Stark, S. Schmitt-Rink: Phys. Rev. Lett. **62**, 1189 (1989)
2.124 D.S. Chemla, W.H. Knox, D.A.B. Miller, S. Schmitt-Rink, J.B. Stark, R. Zimmermann: J. Lumin. **44**, 233 (1989)
2.125 N. Peyghambarian, H.M. Gibbs, G. Khitrova, S.W. Koch: In *Nonlinear Optics. Fundamentals, Materials and Devices*, ed. by S. Miyata, Proc. 5th Toyota Conf. on Nonlinear Optical Materials (North-Holland, Amsterdam 1992) pp. 532-536

2.126 S.T. Cundiff, A. Knorr, J. Feldmann, S.W. Koch, E.O. Göbel, H. Nickel: Phys. Rev. Lett. **73**, 1178 (1994)

2.127 B.D. Fluegel, N. Peyghambarian, G. Olbright, M. Lindberg, S.W. Koch, M. Joffre, D. Hulin, A. Migus, A. Antonetti: Phys. Rev. Lett. **59**, 2588 (1987)

2.128 C.H. Brito Cruz, J.P. Gordon, P.C. Becker, R.L. Fork, C.V. Shank: IEEE J. QE-**24**, 261 (1988)

2.129 J.P. Sokoloff, M. Joffre, B.D. Fluegel, D. Hulin, M. Lindberg, S.W. Koch, A. Migus, A. Antonetti, N. Peyghambarian: Phys. Rev. B **38**, 7615 (1988)

2.130 M. Lindberg, S.W. Koch: J. Opt. Soc. Am. B **5**, 139 (1988)

2.131 M. Joffre, D. Hulin, A. Migus, A. Antonetti, C. Benoit a la Guillaume, N. Peyghambarian, M. Lindberg, S.W. Koch: Opt. Lett. **13**, 276 (1988)

2.132 D.A.B. Miller, D.S. Chemla, P.W. Smith, A.C. Gossard, W. Wiegmann: Opt. Lett. **8**, 477 (1983)

2.133 D.A.B. Miller, D.S. Chemla, D.J. Eilenberger, P.W. Smith, A.C. Gossard, W. Wiegmann: Appl. Phys. Lett. **42**, 925 (1983)

2.134 D.S. Chemla, D.A.B. Miller: J. Opt. Soc. Am. B **2**, 1155 (1985)

2.135 W. Schäfer, S. Schmitt-Rink: Unpublished (1992)

2.136 W. Zinth, W. Kaiser: In *Ultrashort Laser Pulses*, 2nd edn., ed. by W. Kaiser, Topics Appl. Phys., Vol. 60 (Springer, Berlin, Heidelberg 1993) pp. 235-278

2.137 R.L. Shoemaker, R.G. Brewer: Phys. Rev. Lett. **28**, 1430 (1972)

2.138 E. Gerdau, H. Rüffer, R. Hollatz, J.P. Hannon: Phys. Rev. Lett. **57**, 1141 (1986)

2.139 A. Laubereau: In *Ultrashort Laser Pulses*, 2nd edn., ed. by W. Kaiser, Topics Appl. Phys., Vol. 60 (Springer, Berlin, Heidelberg 1993) pp. 35-112

2.140 V. Langer, H. Stolz, W. von der Osten: Phys. Rev. Lett. **64**, 854 (1990)

2.141 E.O. Göbel, K. Leo, T.C. Damen, J. Shah, S. Schmitt-Rink, W. Schäfer, J.F. Muller, K. Köhler: Phys. Rev. Lett. **64**, 1801 (1990)

2.142 L. Goldstein, Y. Horikoshi, S. Tarucha, H. Okamoto: Jpn. J. Appl. Phys. **22**, 1489 (1983)

2.143 D. Bimberg, J. Christen, T. Fukunaga, H. Nakashima, D.E. Mars, J.N. Miller: J. Vac. Sci. Technol. B **5**, 1191 (1987)

2.144 B. Deveaud, T.C. Damen, J. Shah, C.W. Tu: Appl. Phys. Lett. **51**, 828 (1987)

2.145 A. Ourmazd, D.W. Taylor, J.E. Cunningham, C.W. Tu: Phys. Rev. Lett. **62**, 933 (1989)

2.146 M. Koch, J. Feldmann, E.O. Göbel, P. Thomas, J. Shah, K. Köhler: Phys. Rev. B **48**, 11480 (1993)

2.147 E.O. Göbel, M. Koch, J. Feldmann, G. von Plessen, T. Meier, A. Schulze, P. Thomas, S. Schmitt-Rink, K. Köhler, K. Ploog: Phys. Status Solidi (b) **173**, 21 (1992)

2.148 M. Koch, J. Feldmann, G. von Plessen, E.O. Göbel, P. Thomas, K. Köhler: Phys. Rev. Lett. **69**, 3631 (1992)

2.149 V.G. Lyssenko, J. Erland, I. Balslev, K.-H. Pantke, B.S. Razbirin, J.M. Hvam: Phys. Rev. B **48**, 5720 (1993)

2.150 K. Leo, T.C. Damen, J. Shah, E.O. Göbel, K. Köhler: Appl. Phys. Lett. **57**, 19 (1990)

2.151 B.F. Feuerbacher, J. Kuhl, R. Eccleston, K. Ploog: Solid State Commun. **74**, 1279 (1990)

2.152 K. Leo, E.O. Göbel, T.C. Damen, J. Shah, S. Schmitt-Rink, W. Schäfer, J.F. Muller, K. Köhler, P. Ganser: Phys. Rev. B **44**, 5726 (1991)

2.153 S. Schmitt-Rink, D. Bennhardt, V. Heuckeroth, P. Thomas, P. Haring, G. Maidorn, H.J. Bakker, K. Leo, D.S. Kim, J. Shah, K. Köhler: Phys. Rev. B **46**, 10460 (1992)

2.154 J. Kuhl, E.J. Mayer, G. Smith, R. Eccleston, D. Bennhardt, P. Thomas, K. Bott, O. Heller: In *Coherent Optical Interactions in Solids*, ed. by R.T. Phillips (Plenum, New York 1994) pp. 1-32

2.155 S. Bar-Ad, I. Bar-Joseph: Phys. Rev. Lett. **66**, 2491 (1991)

2.156 H. Wang, M. Jiang, R. Merlin, D.G. Steel: Phys. Rev. Lett. **69**, 804 (1992)

2.157 L.Q. Lambert, A. Compaan, I.D. Abella: Phys. Rev. A **4**, 2022 (1971)

2.158 A. Laubereau, W. Kaiser: Rev. Mod. Phys. **50**, 607 (1978)

2.159 A. Puri, J.L. Birman: Phys. Rev. A **27**, 1044 (1983)

2.160 H.J. Hartmann, A. Laubereau: Opt. Commun. **47**, 117 (1983)

2.161 J.J. Hopfield: Phys. Rev. **112**, 1555 (1958)

2.162 S.I. Pekar: Sov. Phys. JETP **6**, 785 (1958)

2.163 S.I. Pekar: Sov. Phys. JETP **7**, 813 (1958)

2.164 D. Fröhlich, A. Kulik, B. Uebbing: Phys. Rev. Lett. **67**, 2343 (1991)

2.165 D. Fröhlich, A. Kulik, B. Uebbing, V. Langer, H. Stolz, W. von der Osten: Phys. Status Solidi (b) **173**, 31 (1992)

2.166 K.-H. Pantke, P. Shillak, B.S. Razbirin, V.G. Lyssenko, J.M. Hvam: Phys. Rev. Lett. **70**, 327 (1993)

2.167 D.S. Kim, J. Shah, D.A.B. Miller, T.C. Damen, W. Schäfer, L. Pfeiffer: Phys. Rev. B **48**, 17902 (1993)

2.168 D.S. Kim, J. Shah, D.A.B. Miller, T.C. Damen, A. Vinattieri, W. Schäfer, L.N. Pfeiffer: Phys. Rev. B **50**, 18240 (1994)

2.169 J. Shah, D.S. Kim: J. Physique IV C **3**, 215 (1993)

2.170 H. Stolz, V. Langer, E. Schreiber, S.A. Permogorov, W. von der Osten: Phys. Rev. Lett. **67**, 679 (1991)

2.171 A.P. Heberle, W.W. Rühle, K. Ploog: Phys. Rev. Lett. **72**, 3887 (1994)

2.172 H. Wang, J. Shah, T. C. Damen, L. N. Pfeiffer: IQEC'94 (Anaheim, CA) OSA Techn. Digest Ser., Vol. 7 (Optical Society of America, Washington, DC 1994) postdeadline paper

2.173 H. Wang, J. Shah, T.C. Damen, L.N. Pfeiffer: Phys. Rev. Lett. **74**, 3061 (1995)

2.174 K. Leo, M. Wegener, J. Shah, D.S. Chemla, E.O. Göbel, T.C. Damen, S. Schmitt-Rink, W. Schäfer: Phys. Rev. Lett. **65**, 1340 (1990)

2.175 D.S. Kim, J. Shah, T.C. Damen, W. Schäfer, F. Jahnke, S. Schmitt-Rink, K. Köhler: Phys. Rev. Lett. **69**, 2725 (1992)

2.176 K. Leo, J. Shah, S. Schmitt-Rink, K. Köhler: In *Ultrafast Processes in Spectroscopy 1991*, ed. by A. Laubereau, A. Seilmeier (IoP, Bristol, UK 1992)

2.177 D. Bennhardt, P. Thomas, R. Eccleston, E.J. Mayer, J. Kuhl: Phys. Rev. B **47**, 13485 (1993)

2.178 R. Eccleston, J. Kuhl, D. Bennhardt, P. Thomas: Solid State Commun. **86**, 93 (1993)

2.179 J.-Y. Bigot, M.-A. Mycek, S. Weiss, R.G. Ulbrich, D.S. Chemla: Phys. Rev. Lett. **70**, 3307 (1993)

2.180 D.S. Chemla, J.-Y. Bigot, M.-A. Mycek, S. Weiss, W. Schäfer: Phys. Rev. B **50**, 8439 (1994)

2.181 J.-Y. Bigot, M.-A. Mycek, S. Weiss, R.G. Ulbrich, D.S. Chemla: In *Coherent Optical Interactions in Solids*, ed. by R.T. Phillips (Plenum, New York 1994) pp. 245-260

2.182 Y.Z. Hu, R. Binder, S.W. Koch: Phys. Rev. B **47**, 15679 (1993)

2.183 C. Stafford, S. Schmitt-Rink, W. Schäfer: Phys. Rev. B **41**, 10000 (1990)

2.184 D.S. Kim, J. Shah, W. Schäfer, S. Schmitt-Rink: Phys. Status Solidi (b) **173**, 11 (1992)

2.185 S. Weiss, M.-A. Mycek, J.-Y. Bigot, S. Schmitt-Rink, D.S. Chemla: Phys. Rev. Lett. **69**, 2685 (1992)

2.186 J. Paye: IEEE J. QE-**28**, 2262 (1992)

2.187 D.J. Kane, R. Trebino: IEEE J. QE-**29**, 571 (1993)

2.188 H. Wang, J. Shah, T.C. Damen, L.N. Pfeiffer: Solid State Commun. **91**, 869 (1994)

2.189 E.J. Mayer, G.O. Smith, V. Heuckeroth, J. Kuhl, K. Bott, A. Schulze, T. Meier, D. Bennhardt, S.W. Koch, P. Thomas, R. Hay, K. Ploog: Phys. Rev. B **50**, 14730 (1994)

2.190 E.A. Hylleraas, A. Ore: Phys. Rev. **71**, 493 (1947)

2.191 W.F. Brinkman, T.M. Rice, B. Bell: Phys. Rev. B **8**, 1570 (1973)

2.192 D.A. Kleinman: Phys. Rev. B **28**, 871 (1983)

2.193 O. Akimoto, E. Hanamura: J. Phys. Soc. Jpn. **33**, 1537 (1972)

2.194 E. Hanamura: J. Phys. Soc. Jpn. **39**, 1516 (1975)

2.195 D.S. Chemla and A. Maruani: Progr. Quantum Electron. **8**, 1 (1982)

2.196 I. Abram, A. Maruani, S. Schmitt-Rink: J. Phys. C **17**, 5163 (1984)

2.197 I. Abram: J. Opt. Soc. Am. B **2**, 1204 (1985)

2.198 M. Kuwata, N. Nagasawa: J. Phys. Soc. Jpn. **56**, 39 (1987)

2.199 R. Levy, B. Hönerlage, J.B. Grun: In *Optical Nonlinearities and Instabilities in Semiconductors*, ed. by H. Haug (Academic, Boston 1988)

2.200 C. Dörnfeld, J.M. Hvam: IEEE J. QE-**25**, 904 (1989)

2.201 M. Kuwata, T. Itoh, E. Hanamura, N. Nagasawa, A. Mysyrowicz: SPIE Proc. **1127**, 55 (1989)

2.202 H. Akiyama, T. Kuga, M. Matsuoka, M. Kuwata-Gonokami: Phys. Rev. B **42**, 5621 (1990)

2.203 K.-H. Pantke, J.M. Hvam: Int'l J. Mod. Phy. B **8**, 73 (1994)

2.204 R.C. Miller, D.A. Kleinman, A.C. Gossard, O. Munteanu: Phys. Rev. B **25**, 6545 (1982)

2.205 S. Charbonneau, T. Steiner, M.L.W. Thewalt, E.S. Koteles, J.Y. Chi, B. Elman: Phys. Rev. B **38**, 3583 (1988)

2.206 D.C. Reynolds, K.K. Bajaj, C.E. Stutz, R.L. Jones, W.M. Theis, P.W. Yu, K.R. Evans: Phys. Rev. B **40**, 3340 (1989)

2.207 T.W. Steiner, A.G. Steele, S. Charbonneau, M.L.W. Thewalt, E.S. Koteles, B. Elman: Solid State Commun. **69**, 1139 (1989)

2.208 R. Cingolani, K. Ploog, G. Peter, R. Hahn, E.O. Göbel, C. Moro, A. Cingolani: Phys. Rev. B **41**, 3272 (1990)

2.209 H. Wang, J. Shah, T. C. Damen, L. N. Pfeiffer: IQEC'94 (Anaheim, CA) OSA Techn. Digest Ser., Vol. 7 (Optical Society of America, Washington, DC 1994) paper QPD22
2.210 H. Wang, J. Shah, T.C. Damen, A.L. Ivanov, H. Haug, L.N. Pfeiffer: Unpublished (1995)
2.211 B.F. Feuerbacher, J. Kuhl, K. Ploog: Phys. Rev. B **43**, 2439 (1991)
2.212 K.-H. Pantke, D. Oberhauser, V.G. Lyssenko, J.M. Hvam, G. Weimann: Phys. Rev. B **47**, 2413 (1993)
2.213 S. Bar-Ad, I. Bar-Joseph: Phys. Rev. Lett. **68**, 349 (1992)
2.214 D.J. Lovering, R.T. Phillips, G.J. Denton, G.W. Smith: Phys. Rev. Lett. **68**, 1880 (1992)
2.215 D. Oberhauser, K.-H. Pantke, W. Langbein, V.G. Lyssenko, H. Kalt, J.M. Hvam, G. Weimann, C. Klingshirn: Phys. Status Solidi (b) **173**, 53 (1992)
2.216 G.O. Smith, E.J. Mayer, J. Kuhl, K. Ploog: Solid State Commun. **92**, 325 (1994)
2.217 H. Wang, J. Shah, T. C. Damen, L. N. Pfeiffer: In *Ultrafast Phenomena IX*, ed. by P. F. Barbara, W.H. Knox, G. A. Mourou, A. H. Zewail, Springer Ser. Chem. Phys., Vol. 60 (Springer, Berlin, Heidelberg 1994) pp. 395-396
2.218 T. Saiki, M. Kuwata-Gonokami, T. Matsusue, H. Sakaki: Phys. Rev. B **49**, 7817 (1994)
2.219 Y.Z. Hu, S.W. Koch, M. Lindberg, N. Peyghambarian, E.L. Pollock, F.F. Abraham: Phys. Rev. Lett. **64**, 1805 (1990)
2.220 E. Hanamura: Phys. Rev. B **44**, 8514 (1991)
2.221 G. Finkelstein, S. Bar-Ad, O. Carmel, I. Bar-Joseph, Y.B. Levinson: Phys. Rev. B **47**, 12964 (1993)
2.222 G.O. Smith, E.J. Mayer, V. Heukeroth, J. Kuhl, K. Bott, T. Meier, A. Schulze, D. Bennhardt, S.W. Koch, P. Thomas, R. Hey, K. Ploog: Solid State Commun. **94**, 373 (1995)
2.223 E.J. Mayer, G.O. Smith, V. Heuckeroth, J. Kuhl, K. Bott, A. Schulze, T. Meier, S.W. Koch, P. Thomas, R. Hey, K. Ploog: Phys. Rev. B **51**, 10909 (1995)
2.224 M.Z. Maialle, L.J. Sham: Phys. Rev. Lett. **73**, 3310 (1994)
2.225 A.L. Ivanov, H. Haug: Phys. Rev. Lett. **74**, 438 (1995)
2.226 H.H. Yaffe, Y. Prior, J.P. Harbison, L.T. Florez: J. Opt. Soc. Am. B **10**, 578 (1993)
2.227 K.B. Ferrio, D.G. Steel: Unpublished (1995)
2.228 K. Leo, J. Shah, E.O. Göbel, T.C. Damen, S. Schmitt-Rink, W. Schäfer, K. Köhler: Phys. Rev. Lett. **66**, 201 (1991)
2.229 G. Bastard, C. Delalande, R. Ferreira, H.U. Liu: J. Lumin. **44**, 247 (1989)
2.230 R. Ferreira, C. Delalande, H.W. Liu, G. Bastard, B. Etienne, J.F. Palmier: Phys. Rev. B **42**, 9170 (1990)
2.231 A.M. Fox, D.A.B. Miller, G. Livescu, J.E. Cunningham, J.E. Henry, W.Y. Jan: Phys. Rev. B **42**, 1841 (1990)
2.232 A.M. Fox, D.A.B. Miller, G. Livescu, J.E. Cunningham, W.Y. Jan: Phys. Rev. B **44**, 6231 (1991)
2.233 R. Binder, T. Kuhn, G. Mahler: Phys. Rev. B **50**, 18319 (1994)
2.234 A.J. Leggett, S. Chakravarty, A.T. Dorsey, P.A. Fisher, A. Garg, W. Zwerge: Rev. Mod. Phys. **59**, 1 (1987)

2.235 K. Leo, J. Shah, J.P. Gordon, T.C. Damen, D.A.B. Miller, C.W. Tu, J.E. Cunningham: Phys. Rev. B **42**, 7065 (1990)
2.236 J. Shah, K. Leo, D.Y. Oberli, T.C. Damen: In *Ultrashort Processes in Condensed Matter*, ed. by W. E. Bron (Plenum, New York 1993) pp. 53-100
2.237 K. Leo, J. Shah, T.C. Damen, A. Schulze, T. Meier, S. Schmitt-Rink, P. Thomas, E.O. Göbel, S.L. Chuang, M.S.C. Luo, W. Schäfer, K. Köhler, P. Ganser: IEEE J. QE-**28**, 2498 (1992)
2.238 J. Feldmann, T. Meier, G. von Plessen, M. Koch, E.O. Göbel, P. Thomas, G. Bacher, C. Hartmann, H. Schweizer, W. Schäfer, H. Nickel: Phys. Rev. Lett. **70**, 3027 (1993)
2.239 F. Bloch: Z. Physik **52**, 555 (1928)
2.240 G.H. Wannier: Phys. Rev. **52**, 191 (1937)
2.241 G.H. Wannier: Rev. Mod. Phys. **34**, 645 (1962)
2.242 J.B. Krieger, G.J. Iafrate: Physica B & C **134**, 228 (1985)
2.243 J.B. Krieger, G.J. Iafrate: Phys. Rev. B **33**, 5494 (1986)
2.244 D. Emin, C.F. Hart: Phys. Rev. B **36**, 7353 (1987)
2.245 J.B. Krieger, G.J. Iafrate: Phys. Rev. B **38**, 6324 (1988)
2.246 D. Emin, C.F. Hart: Phys. Rev. B **41**, 3859 (1990)
2.247 G. Bastard, R. Ferreira: In *Spectroscopy of Semiconductor Microstructures*, ed. by G. Fasol, A. Fasolino, P. Lugli (Plenum, New York 1989) pp. 333-346
2.248 P. Feuer: Phys. Rev. **88**, 92 (1952)
2.249 M. Dignam, J.E. Sipe, J. Shah: Phys. Rev. B **49**, 10502 (1994)
2.250 E.E. Mendez, F. Agullo-Rueda, J.M. Hong: Phys. Rev. Lett. **60**, 2426 (1988)
2.251 P. Voisin, J. Bleuse, C. Bouch, S. Gaillard, C. Alibert, A. Regreny: Phys. Rev. Lett. **61**, 1639 (1988)
2.252 M. Dignam, J.E. Sipe: Phys. Rev. Lett. **64**, 1797 (1990)
2.253 A.M. Bouchard, M. Luban: Phys. Rev. B **47**, 6815 (1993)
2.254 S.M. Zakharov, E.A. Manykin: Izvestiya Akademii Nauk SSSR **37**, 136 (1973)
2.255 G. von Plessen, P. Thomas: Phys. Rev. B **45**, 9185 (1992)
2.256 J. Feldmann, K. Leo, J. Shah, D.A.B. Miller, J.E. Cunningham, T. Meier, G. von Plessen, A. Schulze, P. Thomas, S. Schmitt-Rink: Phys. Rev. B **46**, 7252 (1992)
2.257 K. Leo, P.H. Bolivar, F. Brüggemann, R. Schwedler: Solid State Commun. **84**, 943 (1992)
2.258 P. H. Bolivar, P. Leisching, K. Leo, J. Shah, K. Köhler: In *Ultrafast Electronics and Optoelectronics*, ed. by J. Shah, U. Mishra, OSA Proc. Ultrafast Electron. Optoelectron., Vol. 14 (Optical Society of America, Washington, DC 1993) pp. 142-146
2.259 P. Leisching, P. Haring Bolivar, W. Beck, Y. Dhaibi, F. Bruggemann, R. Schwedler, H. Kurz, K. Leo, K. Kohler: Phys. Rev. B **50**, 14389 (1994)
2.260 G. von Plessen, T. Meier, J. Feldmann, E.O. Göbel, P. Thomas, K.W. Goossen, J.M. Kuo, R.F. Kopf: Phys. Rev. B **49**, 14058 (1994)
2.261 T. Dekorsy, P. Leisching, K. Kohler, H. Kurz: Phys. Rev. B **50**, 8106 (1994)
2.262 T. Dekorsy, R. Ott, H. Kurz, K. Kohler: Phys. Rev. B **51**, 17275 (1995)
2.263 M. Dignam, J.E. Sipe: Phys. Rev. B **43**, 4097 (1991)
2.264 T. Meier, G. von Plessen, P. Thomas, S.W. Koch: Phys. Rev. Lett. **73**, 902 (1994)

2.265 P. Leisching, T. Dekorsy, H.J. Bakker, H. Kurz, K. Kohler: Phys. Rev. B **51**, 18015 (1995)
2.266 F. Rossi, T. Meier, P. Thomas, S.W. Koch, P.E. Selbmann, E. Molinari: Phys. Rev. B **51**, 16943 (1995)
2.267 C. Waschke, H.G. Roskos, R. Schwedler, K. Leo, H. Kurz, K. Köhler: Phys. Rev. Lett. **70**, 3319 (1993)
2.268 J.L. Oudar, D. Hulin, A. Migus, A. Antonetti, F. Alexandre: Phys. Rev. Lett. **55**, 2074 (1985)
2.269 M.T. Portella, J.-Y. Bigot, R.W. Schoenlein, J.E. Cunningham, C.V. Shank: Appl. Phys. Lett. **60**, 2123 (1992)
2.270 J.-Y. Bigot, M.T. Portella, R.W. Schoenlein, J.E. Cunningham, C.V. Shank: Phys. Rev. Lett. **67**, 636 (1991)
2.271 A.C. Gossard, A. Pinczuk: In *Synthetic Modulated Structures*, ed. by L. Chang, B. Giessen (Academic, Boston 1985) pp. 215-255
2.272 D. Pines, P. Nozieres: *The Theory of Quantum Liquids* (Benjamin, New York 1966)
2.273 G. Livescu, D.A.B. Miller, D.S. Chemla, M. Ramaswamy, T.Y. Chang, N. Sauer, A.C. Gossard, J.H. English: IEEE J. QE-**24**, 1677 (1988)
2.274 M.S. Skolnick, J.M. Rorison, K.J. Nash, D.J. Mowbray, P.R. Tapster, S.J. Bass, A.D. Pitt: Phys. Rev. Lett. **58**, 2130 (1987)
2.275 D.S. Kim, J. Shah, J.E. Cunningham, T.C. Damen, S. Schmitt-Rink, W. Schäfer: Phys. Rev. Lett. **68**, 2838 (1992)
2.276 P. Hawrylak: Phys. Rev. Lett. **59**, 485 (1987)
2.277 P. Hawrylak, J.F. Young, P. Brockmann: Semicond. Sci. Technol. **9**, 432 (1994)
2.278 J. Shah, D.S. Kim: In *Coherent Optical Interactions in Solids*, ed. by R.T. Phillips (Plenum, New York 1994) pp. 137-155
2.279 W.E. Bron: In *Ultrashort Processes in Condensed Matter*, ed. by W.E. Bron (Plenum, New York 1993) pp. 101-142
2.280 W.E. Bron: In *Coherent Optical Interactions in Solids*, ed. by R.T. Phillips (Plenum, New York 1994) pp. 199-222
2.281 S. Ruhman, A.G. Joly, K.A. Nelson: IEEE J. QE-**24**, 460 (1988)
2.282 A.M. Weiner, D.E. Leaird, G.P. Wiederrecht, K.A. Nelson: Science **247**, 1317 (1990)
2.283 A.M. Weiner, D.E. Leaird, G.P. Weiderrecht, K.A. Nelson: J. Opt. Soc. Am. B **8**, 1264 (1991)
2.284 G.C. Cho, W.A. Kütt, H. Kurz: Phys. Rev. Lett. **65**, 764 (1990)
2.285 W.A. Kütt, W. Albrecht, H. Kurz: IEEE J. QE-**28**, 2434 (1992)
2.286 T. Pfeifer, W.A. Kütt, H. Kurz, R. Scholz: Phys. Rev. Lett. **69**, 3248 (1992)
2.287 T. Pfeifer, T. Dekorsy, W.A. Kütt, H. Kurz: Appl. Phys. A **55**, 482 (1992)
2.288 T. Dekorsy, W.A. Kütt, T. Pfeifer, H. Kurz: Europhys. Lett. **23**, 223 (1993)
2.289 R. Scholz, T. Pfeifer, H. Kurz: Phys. Rev. B **47**, 16229 (1993)
2.290 T.K. Cheng, S.D. Brorson, A.S. Kazeroonian, J.S. Moodera, G. Dresselhaus, M.S. Dresselhaus, E.P. Ippen: Appl. Phys. Lett. **57**, 1004 (1990)
2.291 T.K. Cheng, S. Vidal, M.J. Zeiger, G. Dresselhaus, M.S. Dresselhaus, E.P. Ippen: Appl. Phys. Lett. **59**, 1923 (1990)
2.292 M.J. Zeiger, S. Vidal, T.K. Cheng, E.P. Ippen, G. Dresselhaus, M.S. Dresselhaus: Phys. Rev. B **45**, 768 (1992)

2.293 J.M. Chwalek, C. Uher, J.F. Whitaker, G.A. Mourou: Appl. Phys. Lett. **58**, 980 (1991)

2.294 L. Min, R.J.D. Miller: Appl. Phys. Lett. **56**, 524 (1990)

2.295 W.A. Kütt, G.C. Cho, M. Strahnen, H. Kurz: Appl. Surf. Sci. **50**, 325 (1990)

2.296 A.V. Kuznetsov, C.J. Stanton: Phys. Rev. Lett. **73**, 3243 (1994)

2.297 F. Vallee, F. Bogani: Phys. Rev. B **43**, 12049 (1991)

2.298 D.H. Auston, M.C. Nuss: IEEE J. QE-**24**, 184 (1988)

2.299 X.-C. Zhang, X.F. Ma, Y. Jin, T.-M. Lu, E.P. Boden, P.D. Phelps, K.R. Stewart, C.P. Yakymyshen: Appl. Phys. Lett. **61**, 3080 (1993)

2.300 D.H. Auston: Appl. Phys. Lett. **26**, 101 (1974)

2.301 P.R. Smith, D.H. Auston, M.C. Nuss: IEEE J. QE-**24**, 255 (1988)

2.302 C.H. Fattinger, D. Grischkowsky: Appl. Phys. Lett. **54**, 490 (1989)

2.303 X.-C. Zhang, B.B. Hu, J.T. Darrow, D.H. Auston: Appl. Phys. Lett. **56**, 1011 (1990)

2.304 B.B. Hu, J.T. Darrow, X.-C. Zhang, D.H. Auston, P.R. Smith: Appl. Phys. Lett. **56**, 886 (1990)

2.305 X.-C. Zhang, B.B. Hu, S.H. Xin, D.H. Auston: Appl. Phys. Lett. **57**, 753 (1990)

2.306 T.C.L.G. Sollner, J. Shah (eds.): *Picosecond Electroinics and Optoelectronics*, OSA Proc. Picosec. Electron. Optoelectron., Vol. 9 (Optical Society of America, Washington, DC 1991)

2.307 J. Shah, U. Mishra (eds.): *Ultrafast Electronics and Optoelectronics*, OSA Proc. Ultrafast Electron. Optoelectron., Vol. 14 (Optical Society of America, Washington, DC 1993)

2.308 J.-L. Martin, A. Migus, G.A. Mourou, A.H. Zewail (eds.): *Ultrafast Phenomena VIII*, Springer Ser. Chem. Phys. Vol. 55 (Springer, Berlin, Heidelberg 1993)
P.F. Barbara, W.H. Knox, G.A. Mourou, A.H. Zewail (eds.): *Ultrafast Phenomena IX*, Springer Ser. Chem. Phys. Vol. 60 (Springer, Berlin, Heidelberg 1994)

2.309 H.G. Roskos, M.C. Nuss, J. Shah, K. Leo, D.A.B. Miller, A.M. Fox, S. Schmitt-Rink, K. Köhler: Phys. Rev. Lett. **68**, 2216 (1992)

2.310 P.C.M. Planken, M.C. Nuss, I. Brener, K.W. Goossen, M.S.C. Luo, S.L. Chuang, L. Pfeiffer: Phys. Rev. Lett. **69**, 3800 (1992)

2.311 M.C. Nuss, P.C.M. Planken, I. Brener, H.G. Roskos, M.S.C. Luo, S.L. Chuang: Appl. Phys. B **58**, 249 (1994)

2.312 M. van Exter, D. Grischkowsky: IEEE Trans. MTT-**38**, 1684 (1990)

2.313 S.L. Chuang, S. Schmitt-Rink, B. Greene, P.N. Saeta, A.F.J. Levi: Phys. Rev. Lett. **68**, 102 (1992)

2.314 M.S.C. Luo, S.L. Chuang, P.C.M. Planken, I. Brener, H.G. Roskos, M.C. Nuss: IEEE J. QE-**30**, 1478 (1994)

2.315 M.S.C. Luo, S.L. Chuang, P.C.M. Planken, I. Brener, M.C. Nuss: Phys. Rev. B **48**, 11043 (1993)

2.316 M.Z. Maialle, E.A. de Andrada e Silva, L.J. Sham: Phys. Rev. B **47**, 15776 (1993)

2.317 C. Waschke, H.G. Roskos, K. Leo, H. Kurz, K. Köhler: Semicond. Sci. Technol. **9**, 416 (1994)

2.318 C. Waschke, P. Leisching, P.H. Bolivar, R. Schwedler, F. Bruggemann, H.G. Roskos, K. Leo, H. Kurz, K. Köhler: Solid State Electron. **37**, 1321 (1994)

2.319 P. Leisching, C. Waschke, P.H. Bolivar, W. Beck, H.G. Roskos, K. Leo, H. Kurz, K. Köhler, P. Ganser: In *Coherent Optical Interactions in Solids*, ed. by R.T. Phillips (Plenum, New York 1994) pp. 325-331
2.320 J. Shah: In *Nonlinear Optics for High-Speed Electronics and Optical Frequency Conversion*, ed. by N. Peyghambarian, H. Everitt, R.C. Eckardt, D.D. Lowenthal. SPIE Proc. **2145**, 144-154 (SPIE, Bellingham, WA 1994)
2.321 H.G. Roskos: *Festkörperprobleme/Adv. Solid State Phys.* **34**, 297 (Vieweg, Braunschweig 1994)
2.322 H.G. Roskos, C. Waschke, R. Schwedler, P. Leisching, Y. Dhaibi, H. Kurz, K. Köhler: Superlattices and Microstructures **15**, 281 (1994)
2.323 H.G. Roskos, C. Waschke, K. Victor, K. Köhler, H. Kurz: Jpn. J. Appl. Phys. Pt.1 **34**, 1370 (1995)
2.324 A. Sibille, J.F. Palmier, H. Wang, J.C. Esnault, F. Mollot: Solid State Electron. **32**, 1461 (1989)
2.325 A. Sibille, J.F. Palmier, F. Mollot, H. Wang, J.C. Esnault: Phys. Rev. B **39**, 6272 (1989)
2.326 A. Sibille, J.F. Palmier, H. Wang, J.C. Esnault, F. Mollot: Appl. Phys. Lett. **56**, 256 (1990)
2.327 A. Sibille, J.F. Palmier, H. Wang, F. Mollot: Phys. Rev. Lett. **64**, 52 (1990)
2.328 W.S. Warren, A.H. Zewail: J. Chem. Phys. **78**, 3583 (1983)
2.329 A.H. Zewail, R. Bernstein: Chem. Eng. News **66**, 24 (1988)
2.330 H. Rabitz, S.H. Shi: In *Advances in Molecular Vibrations and Collision Dynamics*, ed. by J. Bowman (Jai, Greenwich, CT 1991) pp. 187
2.331 S. Rice: Science **258**, 412 (1992)
2.332 N.F. Scherer, A. Matro, L.D. Ziegler, M. Du, R.J. Carlson, J.A. Cina, G.R. Fleming: J. Chem. Phys. **96**, 4180 (1992)
2.333 W.S. Warren, H. Rabitz, M. Dahleh: Science **259**, 1581 (1993)
2.334 J.P. Heritage, A.M. Weiner, R.N. Thurston: Opt. Lett. **10**, 609 (1985)
2.335 J.P. Heritage, R.N. Thurston, W.J. Tomlinson, A.M. Weiner, R.H. Stolen: Appl. Phys. Lett. **47**, 87 (1985)
2.336 R.N. Thurston, J.P. Heritage, A.M. Weiner, W.J. Tomlinson: IEEE J. QE-**22**, 682 (1986)
2.337 A.M. Weiner, J.P. Heritage: Rev. Physique Appl. **22**, 1619 (1987)
2.338 A.M. Weiner, J.P. Heritage, E.M. Kirschner: J. Opt. Soc. Am. B **5**, 1563 (1988)
2.339 A.M. Weiner, J.P. Heritage, J.A. Salehi: Opt. Lett. **13**, 300 (1988)
2.340 A.M. Weiner, D.E. Leaird, J.S. Patel, J.R. Wullert: Opt. Lett. **15**, 326 (1990)
2.341 A.M. Weiner, D.E. Leaird: Opt. Lett. **15**, 51 (1990)
2.342 A.M. Weiner, D.E. Leaird, D.H. Reitze, E.G. Paek: IEEE J. QE-**28**, 2251 (1992)
2.343 A.M. Weiner, D.E. Leaird, J.S. Patel, J.R. Wullert: IEEE J. QE-**28**, 908 (1992)
2.344 A.M. Weiner, D.E. Leaird, D.H. Reitze, E.G. Paek: Opt. Lett. **17**, 224 (1992)
2.345 A.M. Weiner, S. Oudin, D.E. Leaird, D.H. Reitze: J. Opt. Soc. Am. A **10**, 1112 (1993)
2.346 I. Brener, P.C.M. Planken, M.C. Nuss, L. Pfeiffer, D.E. Leaird, A.M. Weiner: Appl. Phys. Lett. **63**, 2213 (1993)
2.347 P.C.M. Planken, I. Brener, M.C. Nuss, M.S.C. Luo, S.L. Chuang: Phys. Rev. B **48**, 4903 (1993)

Chapter 3

3.1 J.H. Collet, T. Amand: J. Phys. Chem. Solids **47**, 153 (1986)

3.2 A.V. Kuznetsov: Phys. Rev. B **44**, 8721 (1991)

3.3 A.V. Kuznetsov: Phys. Rev. B **44**, 13381 (1991)

3.4 T. Kuhn, F. Rossi: Phys. Rev. B **46**, 7496 (1992)

3.5 F. Rossi, S. Haas, T. Kuhn: Phys. Rev. Lett. **72**, 152 (1994)

3.6 A.V. Kuznetsov, C.J. Stanton: Phys. Rev. B **48**, 10828 (1993)

3.7 T. Kuhn, F. Rossi: Phys. Rev. Lett. **69**, 977 (1992)

3.8 F. Rossi, S. Haas, T. Kuhn: Semicond. Sci. Technol. **9**, 411 (1994)

3.9 A. Lohner, K. Rick, P. Leisching, A. Leitenstorfer, T. Elsaesser, T. Kuhn, F. Rossi, W. Stolz: Phys. Rev. Lett. **71**, 77 (1993)

3.10 A. Leitenstorfer, A. Lohner, T. Elsaesser, S. Haas, F. Rossi, T. Kuhn, W. Klein, G. Boehm, G. Traenkle, G. Weimann: Phys. Rev. Lett. **73**, 1687 (1994)

3.11 J.L. Oudar, D. Hulin, A. Migus, A. Antonetti, F. Alexandre: Phys. Rev. Lett. **55**, 2074 (1985)

3.12 W.Z. Lin, J.G. Fujimoto, E.P. Ippen, R.A. Logan: Appl. Phys. Lett. **51**, 161 (1987)

3.13 W.Z. Lin, J.G. Fujimoto, E.P. Ippen, R.A. Logan: Appl. Phys. Lett. **50**, 124 (1987)

3.14 W.Z. Lin, R.W. Schoenlein, J.G. Fujimoto, E.P. Ippen: IEEE J. QE-**24**, 267 (1988)

3.15 D.J. Erskine: Phys. Rev. Lett. **51**, 840 (1983)

3.16 D.J. Erskine, A.J. Taylor, C.L. Tang: Appl. Phys. Lett. **45**, 54 (1984)

3.17 A.J. Taylor, D.J. Erskine, C.L. Tang: J. Opt. Soc. Am. B **2**, 663 (1985)

3.18 M.J. Rosker, F.W. Wise, C.L. Tang: Appl. Phys. Lett. **49**, 1726 (1986)

3.19 C.W.W. Bradley, R.A. Taylor, J.F. Ryan: Solid State Electron. **32**, 1173 (1989)

3.20 R.A. Taylor, C.W.W. Bradley, N. Mayhew, T.N. Thomas, J.F. Ryan: J. Lumin. **53**, 321 (1992)

3.21 J. Nunnenkamp, J.H. Collet, J. Klebniczki, J. Kuhl, K. Ploog: Phys. Rev. B **43**, 14047 (1991)

3.22 J.-P. Foing, D. Hulin, M. Joffre, M.K. Jackson, J.L. Oudar, C. Tanguy, M. Combescot: Phys. Rev. Lett. **68**, 110 (1992)

3.23 J.-P. Foing, M. Joffre, M.K. Jackson, J.L. Oudar, D. Hulin: Phys. Status Solidi (b) **173**, 281 (1992)

3.24 W.H. Knox, C. Hirlimann, D.A.B. Miller, J. Shah, D.S. Chemla, C.V. Shank: Phys. Rev. Lett. **56**, 1191 (1986)

3.25 A.E. Ruckenstein, S. Schmitt-Rink: Phys. Rev. B **35**, 7551 (1987)

3.26 G. Livescu, D.A.B. Miller, D.S. Chemla, M. Ramaswamy, T.Y. Chang, N. Sauer, A.C. Gossard, J.H. English: IEEE J. QE-**24**, 1677 (1988)

3.27 M.S. Skolnick, J.M. Rorison, K.J. Nash, D.J. Mowbray, P.R. Tapster, S.J. Bass, A.D. Pitt: Phys. Rev. Lett. **58**, 2130 (1987)

3.28 G.D. Mahan: Phys. Rev. 153, 882 (1967)

3.29 V.M. Asnin, V.I. Stepanov, R. Zimmermann, M. Rosler: Solid State Commun. **47**, 655 (1983)

3.30 R. Zimmermann: Phys. Status Solidi (b) **146**, 371 (1988)
3.31 C. Tanguy, M. Combescot: Phys. Rev. Lett. **68**, 1935 (1992)
3.32 J. Shah, G.J. Iafrate (guest eds.): Hot Carriers in Semiconductors: Proc. HCIS-5. Solid State Electron. **31**, Nos. 3 and 4 (1988)
3.33 D.K. Ferry, L.A. Akers (guest eds.): Hot Carriers in Semiconductors: Proc. HCIS-6. Solid State Electron. **32**, No. 12 (1989)
3.34 C. Hamaguchi, M. Inoue (guest eds.): Hot Carriers in Semiconductors: Proc. HCIS-7. Semicond. Sci. Technol. **7**, B (1992)
3.35 J.F. Ryan, A.C. Maciel (guest eds.): Hot Carriers in Semiconductors: Proc. HCIS-8. Semicond. Sci. Technol. **9**, 5S (1994)
3.36 T. Elsaesser, J. Shah, L. Rota, P. Lugli: Phys. Rev. Lett. **66**, 1757 (1991)
3.37 P. Lugli, D.K. Ferry: Phys. Rev. Lett. **56**, 1295 (1986)
3.38 D.K. Ferry: *Semiconductors* (Macmillan, New York 1991)
3.39 D.K. Ferry, A.M. Kriman, M.J. Kann, R.P. Joshi: Comp. Phys. Commun. **67**, 119 (1991)
3.40 J.F. Young, P.J. Kelly: Phys. Rev. B **47**, 6316 (1993)
3.41 T. Gong, P.M. Fauchet, J.F. Young, P.J. Kelly: Appl. Phys. Lett. **62**, 522 (1993)
3.42 R.G. Ulbrich: Phys. Rev. B **8**, 5719 (1973)
3.43 B.P. Zakharchenya, V.I. Zemskii, D.N. Mirlin: Sov. Phys. JETP **43**, 569 (1976)
3.44 D.N. Mirlin, I.Y. Karlik, L.P. Nikitin, I.J. Reshina, V.F. Sapega: Solid State Commun. **37**, 757 (1981)
3.45 G. Fasol, H.P. Hughes: Phys. Rev. B **33**, 2953 (1986)
3.46 S.A. Lyon: J. Lumin. **35**, 121 (1986)
3.47 R.G. Ulbrich, J.A. Kash, J.C. Tsang: Phys. Rev. Lett. **62**, 949 (1989)
3.48 G. Fasol, W. Hackenburg, H.P. Hughes, K. Ploog, E. Bauer, H. Kano: Phys. Rev. B **41**, 1461 (1990)
3.49 D.N. Mirlin, V.I. Perel′: In *Spectroscopy of Nonequilibrium Electrons and Phonons*, ed. by C.V. Shank, B.P. Zakharchenya (North-Holland, Amsterdam 1992) pp. 269-325
3.50 M.A. Alekseev, D.N. Mirlin, J.A. Kash, J.C. Tsang, R.G. Ulbrich: Phys. Rev. Lett. **65**, 274 (1990)
3.51 D.W. Snoke, W.W. Rühle, Y.-C. Lu, E. Bauser: Phys. Rev. Lett. **68**, 990 (1992)
3.52 D.W. Snoke, W.W. Rühle, Y.-C. Lu, E. Bauser: Phys. Rev. B **45**, 10979 (1992)
3.53 D.W. Snoke: Phys. Rev. B **47**, 13346 (1993)
3.54 K. Kash, P.A. Wolff, W.A. Bonner: Appl. Phys. Lett. **42**, 173 (1983)
3.55 C.L. Collins, P.Y. Yu: Phys. Rev. B **30**, 4501 (1984)
3.56 J.A. Kash, J.C. Tsang: In *Light Scattering in Solids VI*, ed. by M. Cardona, G. Güntherodt, Topics Appl. Phys., Vol. 68 (Springer, Berlin, Heidelberg 1990) Chap. 8
3.57 J.A. Kash, J.C. Tsang: In *Spectroscopy of Nonequilibrium Electrons and Phonons*, ed. by C.V. Shank, B.P. Zakharchenya (North-Holland, Amsterdam 1992) pp. 113-168
3.58 P.C. Becker, H.L. Fragnito, C.H.B. Cruz, J. Shah, R.L. Fork, J.E. Cunningham, J.E. Henry, C.V. Shank: Appl. Phys. Lett. **53**, 2089 (1988)
3.59 J.-Y. Bigot, M.T. Portella, R.W. Schoenlein, J.E. Cunningham, C.V. Shank: Phys. Rev. Lett. **65**, 3429 (1990)

3.60 S. Zollner, S. Gopalan, M. Cardona: J. Appl. Phys. **68**, 1682 (1990)
3.61 S. Zollner, S. Gopalan, M. Cardona: Phys. Rev. B **44**, 13446 (1991)
3.62 S. Krishnamurthy, M. Cardona: J. Appl. Phys. **74**, 2117 (1993)
3.63 R.R. Alfano (ed.): *Ultrafast Laser Probe Phenomena in Bulk and Microstructure Semiconductors II*. SPIE Proc. **942** (SPIE, Bellingham, WA 1988)
3.64 J. Shah, B. Deveaud, T.C. Damen, W.T. Tsang, A.C. Gossard, P. Lugli: Phys. Rev. Lett. **59**, 2222 (1987)
3.65 S. Zollner, S. Gopalan, M. Cardona: In *Ultrafast laser Probe Phenomena in Bulk and Microstructure Semiconductors III*, ed. by R.R. Alfano. SPIE Proc. **1282**, 78-85 (SPIE, Bellingham, WA 1990)
3.66 D.Y. Oberli, J. Shah, T.C. Damen: Phys. Rev. B **40**, 1323 (1989)
3.67 C.J. Stanton, D.W. Bailey, K. Hess: IEEE J. QE-**24**, 1614 (1988)
3.68 F.W. Wise, I.A. Walmsley, C.L. Tang: Appl. Phys. Lett. **51**, 605 (1987)
3.69 W.H. Knox, D.S. Chemla, G. Livescu, J.E. Cunningham, J.E. Henry: Phys. Rev. Lett. **61**, 1290 (1988)
3.70 W.H. Knox: In *Hot Carriers in Semiconductor Nanostructures: Physics and Applications*, ed. by J. Shah (Academic, Boston 1992) pp. 313-344
3.71 J.A. Kash: Phys. Rev. B **48**, 18336 (1993)
3.72 J.A. Kash: Phys. Rev. B **40**, 3455 (1989)
3.73 R.A. Höpfel, J. Shah, P.A. Wolff, A.C. Gossard: Appl. Phys. Lett. **49**, 572 (1986)
3.74 R.A. Höpfel, J. Shah, P.A. Wolff, A.C. Gossard: Phys. Rev. Lett. **56**, 2736 (1986)
3.75 R.A. Höpfel, J. Shah, P.A. Wolff, A.C. Gossard: Phys. Rev. B **37**, 6941 (1988)
3.76 S.M. Goodnick, P. Lugli: Phys. Rev. B **38**, 10135 (1988)
3.77 S.M. Goodnick, P. Lugli: Phys. Rev. B **37**, 2578 (1988)
3.78 D.S. Kim, J. Shah, J.E. Cunningham, T.C. Damen, S. Schmitt-Rink, W. Schäfer: Phys. Rev. Lett. **68**, 2838 (1992)
3.79 A. Tomita, J. Shah, J.E. Cunningham, S.M. Goodnick, P. Lugli, S.L. Chuang: Phys. Rev. B **48**, 5708 (1993)
3.80 X.Q. Zhou, K. Leo, H. Kurz: Phys. Rev. B **45**, 3886 (1992)
3.81 A. Chebira, J. Chesnoy, G.M. Gale: Phys. Rev. B **46**, 4559 (1992)
3.82 M. Woerner, W. Frey, M.T. Portella, C. Ludwig, T. Elsaesser, W. Kaiser: Phys. Rev. B **49**, 17007 (1995)
3.83 A. Seilmeier, H.J. Hubner, G. Abstreiter, G. Weimann, W. Schlapp: Phys. Rev. Lett. **59**, 1345 (1987)
3.84 B.K. Ridley: J. Phys. C **15**, 5899 (1982)
3.85 B.K. Ridley: In *Hot Carriers in Semiconductor Nanostructures: Physics and Applications*, ed. by J. Shah (Academic, Boston 1992) pp. 17-51
3.86 P. Lugli, S.M. Goodnick: Phys. Rev. Lett. **59**, 716 (1987)
3.87 S.M. Goodnick, J.E. Lary, P. Lugli: Superlattices and Microstructures **10**, 461 (1991)
3.88 S.M. Goodnick, P. Lugli: In *Hot Carriers in Semiconductor Nanostructures: Physics and Applications*, ed. by J. Shah (Academic, Boston 1992) pp. 191-234
3.89 J.A. Levenson, G. Dolique, J.L. Oudar, I.A. Abram: Solid State Electron. **32**, 1869 (1989)

3.90 H. Lobentanzer, W. Stolz, K. Ploog, R.J. Bauerle, T. Elsaesser: Solid State Electron. **32**, 1875 (1989)

3.91 T. Elsaesser: Phys. Rev. B **38**, 4307 (1988)

3.92 A. Mooradian, A. L. McWhorter: In *Proc. 10th Int'l Conf. on the Physics of Semiconductors*, ed. by S.P. Keller, J.C. Hensel, F. Stern (Atomic Energy Commission, Oak Ridge, TN 1970) pp. 380-386

3.93 G. Abstreiter, M. Cardona, A. Pinczuk: In *Light Scattering in Solids IV*, ed. by M. Cardona, G. Güntherodt, Topics Appl. Phys., Vol. 54 (Springer, Berlin, Heidelberg 1984) Chap. 2

3.94 D.Y. Oberli, D.R. Wake, M.V. Klein, J. Klem, T. Henderson, H. Morkoç: Phys. Rev. Lett. **59**, 696 (1987)

3.95 M.C. Tatham, J.F. Ryan, C.T. Foxon: Phys. Rev. Lett. **63**, 1637 (1989)

3.96 B.K. Ridley: Phys. Rev. B **39**, 5282 (1989)

3.97 B.K. Ridley: Sci. Progr. **74**, 465 (1990)

3.98 M.P. Chamberlain, M. Babiker, B.K. Ridley: Superlattices and Microstructures **9**, 227 (1991)

Chapter 4

4.1 E.M. Conwell: *High Field Transport in Semiconductors* (Academic, New York 1967)

4.2 B.R. Nag: *Theory of Electrical Transport in Semiconductors* (Pergamon, Oxford, UK 1972)

4.3 B.R. Nag: *Electron Transport in Compound Semiconductors*, Springer Ser. Solid-State Sci., Vol. 11 (Springer, Berlin, Heidelberg 1980)

4.4 B.K. Ridley: *Quantum Processes in Semiconductors* (Clarendon, Oxford, UK 1982)

4.5 K. Seeger: *Semiconductor Physics*, 6th edn., Springer Ser. Solid-State Sci., Vol. 40 (Springer, Berlin, Heidelberg 1991)

4.6 D.K. Ferry: *Semiconductors* (Macmillan, New York 1997)

4.7 J. Shah, R.C.C. Leite: Phys. Rev. Lett. **22**, 1304 (1969)

4.8 G. Bauer: *Springer Tracks Mod. Phys.* **74**, 1-6 (Springer, Berlin, Heidelberg 1974)

4.9 M. Voos, R.F. Leheny, J. Shah: In *Handbook on Semiconductors Vol II: Optical Properties of Semiconductors*, ed. by M. Balkanski (North-Holland, Amsterdam 1980) pp. 329-416

4.10 J. Shah, R.F. Leheny: In *Semiconductors Probed by Ultrafast Laser Spectroscopy*, ed. by R.R. Alfano (Academic, Orlando, FL 1984) pp. 45-75

4.11 J. Shah: In *Hot Carriers in Semiconductor Nanostructures: Physics and Applications*, ed. by J. Shah (Academic, Boston 1992) pp. 279-312

4.12 J. Shah, A. Pinczuk, A.C. Gossard, W. Wiegmann: Phys. Rev. Lett. **54**, 2045 (1985)

4.13 J. Shah: IEEE J. QE-**22**, 1728 (1986)

4.14 J. Shah: In *Physics of the Two-Dimensional Electron Gas*, ed. by J.T. Devreese, F. M. Peeters (Plenum, New York 1987) pp. 183-225

4.15 S.M. Kogan: Sov. Phys. – Solid State **4**, 1813 (1963)

4.16 M. Costato, L. Reggiani: Phys. Status Solidi (b) **58**, 471 (1973)

4.17 M. Costato, G. Gagliani, C. Jacoboni, L. Reggiani: J. Phys. Chem. Solids **35**, 1605 (1974)

4.18 J. Shah: Solid State Electron. **21**, 43 (1978)

4.19 E.O. Göbel, O. Hildebrand: Phys. Status Solidi (b) **88**, 645 (1978)

4.20 S. Das Sarma: In *Hot Carriers in Semiconductor Nanostructures: Physics and Applications*, ed. by J. Shah (Academic, Boston 1992) pp. 53-85

4.21 S. Das Sarma, J.K. Jain, R. Jalabert: Phys. Rev. B **37**, 4560 (1988)

4.22 J. Shah: Appl. Phys. Lett. **20**, 479 (1972)

4.23 J. Shah: Phys. Rev. B **9**, 562 (1974)

4.24 J. Shah, R.F. Leheny, R.E. Nahory, M.A. Pollack: Appl. Phys. Lett. **37**, 475 (1980)

4.25 J. Shah, R.F. Leheny, R.E. Nahory, H. Temkin: Appl. Phys. Lett. **39**, 618 (1981)

4.26 J. Shah, B. Etienne, R.F. Leheny, R.E. Nahory, A.E. DiGiovanni: J. Appl. Phys. **53**, 9224 (1982)

4.27 R.S. Turtelli, A.R.B. Castro, R.C.C. Leite: Solid State Commun. **16**, 969 (1975)

4.28 A. Mooradian, A. L. McWhorter: In *Proc. 10th Int'l Conf. on the Physics of Semiconductors*, ed. by S.P. Keller, J.C. Hensel, F. Stern (Atomic Energy Commission, Oak Ridge, TN 1970) pp. 380-386

4.29 P.D. Southgate, D.S. Hall, A.B. Dreeben: J. Appl. Phys. **42**, 2868 (1971)

4.30 N. Takenaka, M. Inoue, J. Shirafuji, Y. Inuishi: J. Phys. Soc. Jpn. **45**, 1630 (1978)

4.31 A. Pinczuk, J. Shah, A.C. Gossard: Solid State Electron. **31**, 477 (1988)

4.32 J. Shah, C. Lin, R.F. Leheny, A.E. DiGiovanni: Solid State Commun. **18**, 487 (1976)

4.33 B. Etienne, J. Shah, R.F. Leheny, R.E. Nahory: Appl. Phys. Lett. **41**, 1018 (1982)

4.34 J. Shah: Solid State Electron. **32**, 1051 (1989)

4.35 R.F. Leheny, J. Shah, R.L. Fork, C.V. Shank, A. Migus: Solid State Commun. **31**, 809 (1979)

4.36 C.V. Shank, R.L. Fork, R.F. Leheny, J. Shah: Phys. Rev. Lett. **42**, 112 (1979)

4.37 D. von der Linde, R. Lambrich: Phys. Rev. Lett. **42**, 1090 (1989)

4.38 S. Tanaka, H. Kobayashi, H. Sato, H. Shionoya: J. Phys. Soc. Jpn. **49**, 1051 (1980)

4.39 J.H. Collet, W.W. Rühle, M. Pugnet, K. Leo, A. Million: Phys. Rev. B **40**, 12296 (1989)

4.40 K. Kash, J. Shah: Appl. Phys. Lett. **45**, 401 (1984)

4.41 E.J. Yoffa: Phys. Rev. B **23**, 1909 (1981)

4.42 H.M. van Driel: Phys. Rev. B **19**, 5928 (1979)

4.43 M. Pugnet, J.H. Collet, A. Cornet: Solid State Commun. **38**, 531 (1981)

4.44 J.H. Collet, A. Cornet, M. Pugnet, T. Amand: Solid State Commun. **42**, 883 (1982)

4.45 W. Pötz, P. Kocevar: Phys. Rev. B **28**, 7040 (1983)

4.46 W. Pötz, P. Kocevar: In *Hot Carriers in Semiconductor Nanostructures: Physics and Applications*, ed. by J. Shah (Academic, Boston 1992) pp. 87-120

4.47 D. von der Linde, J. Kuhl, H. Klingenberg: Phys. Rev. Lett. **44**, 1505 (1980)

4.48 J.A. Kash, C. Tsang, J.M. Hvam: Phys. Rev. Lett. **54**, 2151 (1985)

4.49 J.A. Kash, J.C. Tsang: In *Light Scattering in Solids VI*, ed. by M. Cardona, G. Güntherodt, Topics Appl. Phys., Vol. 68 (Springer, Berlin, Heidelberg 1990) Chap. 8

4.50 J.A. Kash, J.C. Tsang: In *Spectroscopy of Nonequilibrium Electrons and Phonons*, ed. by C V. Shank, B.P. Zakharchenya (Elsevier, Amsterdam 1992) pp. 113-168
4.51 K.T. Tsen: Int'l J. Mod. Phy. B **7**, 4165 (1993)
4.52 J. Shah, R.C.C. Leite, J.F. Scott: Solid State Commun. **8**, 1089 (1970)
4.53 J.H. Collet, J.L. Oudar, T. Amand: Phys. Rev. B **34**, 26 (1986)
4.54 J.H. Collet, T. Amand: J. Phys. Chem. Solids **47**, 153 (1986)
4.55 J.H. Collet, T. Amand, M. Pugnet: Phys. Lett. A **96**, 368 (1994)
4.56 P. Lugli. S.M. Goodnick: Phys. Rev. Lett. **59**, 716 (1987)
4.57 C. Jacoboni, P. Lugli: *The Monte Carlo Method for Semiconductor Device Simulation* (Springer, Vienna 1989)
4.58 P. Lugli, P. Bordone, L. Reggiani, M. Rieger, P. Kocevar, S.M. Goodnick: Phys. Rev. B **39**, 7852 (1989)
4.59 S.M. Goodnick, P. Lugli: In *Hot Carriers in Semiconductor Nanostructures: Physics and Applications*, ed. by J. Shah (Academic, Boston 1992) pp. 191-234
4.60 P. Lugli: In *Spectroscopy of Nonequilibrium Electrons and Phonons*, ed. by C.V. Shank, B.P. Zakharchenya (Elsevier, Amsterdam 1992) pp. 1-56
4.61 M.C. Marchetti, W. Pötz: Phys. Rev. B **40**, 12391 (1989)
4.62 M.C. Marchetti, W. Pötz: J. Vac. Sci. & Technol. B **6**, 1341 (1988)
4.63 W. Pötz, M.C. Marchetti: SPIE Proc. **942**, 100 (1988)
4.64 D. Pines, P. Nozieres: *The Theory of Quantum Liquids* (Benjamin, New York 1966)
4.65 J. Lindhard: Kgl. Danske Videnskab Mat.-Fys. Medd. **28**, 1 (1954)
4.66 H. Haug, S.W. Koch: *Quantum Theory of the Optical and Electronic Properties of Semiconductors* (World Scientific, Singapore 1993)
4.67 J.R. Hayes, A.F.J. Levi: IEEE J. QE-**22**, 1744 (1986)
4.68 J.F. Young, N.L. Henry, P.J. Kelly: Solid State Electron. **32**, 1567 (1989)
4.69 J.F. Young, P.J. Kelly: Phys. Rev. B **47**, 6316 (1993)
4.70 P. Lugli, D.K. Ferry: Phys. Rev. Lett. **56**, 1295 (1986)
4.71 S.M. Goodnick, P. Lugli: Phys. Rev. B **38**, 10135 (1988)
4.72 S.M. Goodnick, P. Lugli: Phys. Rev. B **37**, 2578 (1988)
4.73 D.K. Ferry, A.M. Kriman, M.J. Kann, R.P. Joshi: Comp. Phys. Commun. **67**, 119 (1991)
4.74 J.H. Collet: Phys. Rev. B **47**, 10279 (1993)
4.75 S. Das Sarma, B.A. Mason: Phys. Rev. B **31**, 5536 (1985)
4.76 J.R. Senna, S. Das Sarma: Solid State Commun. **64**, 1397 (1987)
4.77 S. Das Sarma, J.K. Jain, R. Jalabert: Phys. Rev. B **37**, 6290 (1988)
4.78 J.K. Jain, R. Jalabert, S. Das Sarma: Phys. Rev. Lett. **60**, 353 (1988)
4.79 J.F. Young, T. Gong, P.J. Kelly, P.M. Fauchet: Phys. Rev. B **50**, 2208 (1994)
4.80 S. Das Sarma, J.K. Jain, R. Jalabert: Phys. Rev. B **37**, 1228 (1988)
4.81 M. Asche and O.G. Sarbei: Phys. Status Solidi (b) **126**, 883 (1984)
4.82 W. Pötz: Phys. Rev. B **36**, 5016 (1987)
4.83 H. Lobentanzer, H.-J. Polland, W.W. Rühle, W. Stolz, K. Ploog: Appl. Phys. Lett. **51**, 673 (1987)
4.84 H. Lobentanzer, W.W. Rühle, H.-J. Polland, W. Stolz, K. Ploog: Phys. Rev. B **36**, 2954 (1987)
4.85 H. Lobentanzer, H.-J. Polland, W.W. Rühle, W. Stolz, K. Ploog: Phys. Rev. B **36**, 1136 (1987)

4.86 H. Lobentanzer, W.W. Rühle, W. Stolz, K. Ploog: Solid State Commun. **62**, 53 (1987)

4.87 A.S. Vengurlekar, S.S. Prabhu, S.K. Roy, J. Shah: Phys. Rev. B **50**, 8348 (1994)

4.88 R. Dingle, H. Stormer, A.C. Gossard, W. Wiegmann: Appl. Phys. Lett. **33**, 665 (1978)

4.89 A.C. Gossard, A. Pinczuk: In *Synthetic Modulated Structures*, ed. by L. Chang, B. Giessen (Academic, Boston 1985) pp. 215-255

4.90 C.H. Yang, J.M. Carson-Swindle, S.A. Lyon, J.M. Worlock: Phys. Rev. Lett. **55**, 2359 (1985)

4.91 K. Kash, J. Shah, D. Block, A.C. Gossard, W. Weigmann: Physica B & C **134**, 189 (1985)

4.92 J. Shah, A. Pinczuk, A.C. Gossard, W. Wiegmann, K. Kash: Surf. Sci. **174**, 363 (1986)

4.93 J. A. P. Da Costa, R. A. Taylor, A. J. Turberfield, J. F. Ryan, W. I. Wang: In *Proc. 18th Int'l Conf. on the Physics of Semiconductors*, ed. by O. Engstrom (World Scientific, Singapore 1987) pp. 1327-1330

4.94 K. Leo, W.W. Rühle: SPIE Proc. **942**, 231 (1988)

4.95 J.F. Ryan, M.C. Tatham, D.J. Westland, C.T. Foxon, M.D. Scott, W.I. Wang: SPIE Proc. **942**, 256 (1988)

4.96 J. Shah, G.J. Iafrate (guest eds.): Hot Carriers in Semiconductors: Proc. HCIS-5. Solid State Electron. **31**, Nos. 3 and 4 (1988)

4.97 M.C. Tatham, R.A. Taylor, J.F. Ryan, W.I. Wang, C.T. Foxon: Solid State Electron. **31**, 459 (1988)

4.98 D.K. Ferry, L.A. Akers (guest eds.): Hot Carriers in Semiconductors: Proc. HCIS-6. Solid State Electron. **32**, No. 12 (1989)

4.99 C. Hamaguchi, M. Inoue (guest eds.): Hot Carriers in Semiconductors: Proc. HCIS-7. Semicond. Sci. Technol. **7**, B (1992)

4.100 A. Tomita, J. Shah, J.E. Cunningham, S.M. Goodnick, P. Lugli, S.L. Chuang: Phys. Rev. B **48**, 5708 (1993)

4.101 J.F. Ryan, A.C. Maciel (guest eds.): Hot Carriers in Semiconductors: Proc. HCIS-8. Semicond. Sci. Technol. **9**, 5S (1994)

4.102 W.W. Rühle, H.-J. Polland: Phys. Rev. B **36**, 1683 (1987)

4.103 W.W. Rühle, K. Leo, E. Bauser: Phys. Rev. B **40**, 1756 (1989)

4.104 S. Tanaka, H. Kobayashi, H. Saito, S. Shionoya: Solid State Commun. **33**, 167 (1980)

4.105 C.V. Shank, R.L. Fork, R. Yen, J. Shah, B.I. Greene, A.C. Gossard, C. Weisbuch: Solid State Commun. **47**, 981 (1983)

4.106 J.F. Ryan, R.A. Taylor, A.J. Turberfield, A. Maciel, J.M. Worlock, A.C. Gossard, W. Wiegmann: Phys. Rev. Lett. **53**, 1841 (1984)

4.107 Z.Y. Xu, C.L. Tang: Appl. Phys. Lett. **44**, 692 (1984)

4.108 D.K. Ferry: Surf. Sci. **75**, 86 (1978)

4.109 K. Hess: Appl. Phys. Lett. **35**, 484 (1979)

4.110 P.J. Price: Ann. Phys. **133**, 217 (1981)

4.111 P.J. Price: J. Vac. Sci. Technol. **19**, 599 (1981)

4.112 F.A. Riddoch, B.K. Ridley: J. Phys. C **16**, 6971 (1983)

4.113 J.P. Leburton: J. Appl. Phys. **56**, 2850 (1984)

4.114 F.A. Riddoch and B.K. Ridley: Surf. Sci. **142**, 260 (1984)
4.115 B.K. Ridley: In *Hot Carriers in Semiconductor Nanostructures: Physics and Applications*, ed. by J. Shah (Academic, Boston 1992) pp. 17-51
4.116 Z.Y. Xu, V.G. Kreismanis, C.L. Tang: Appl. Phys. Lett. **43**, 415 (1983)
4.117 R.W.J. Hollering, T.T.J.M. Berendtschot, H.J.A. Bluyssen, P. Wyder, M.R. Leys, J.H. Wolter: Solid State Commun. **57**, 527 (1986)
4.118 M.J. Rosker, F.W. Wise, C.L. Tang: Appl. Phys. Lett. **49**, 1726 (1986)
4.119 J.F. Ryan, R.A. Taylor, A.J. Turberfield, J.M. Worlock: Surf. Sci. **170**, 511 (1986)
4.120 K. Leo, W.W. Rühle, K. Ploog: Phys. Rev. B **38**, 1947 (1988)
4.121 K. Leo, W.W. Rühle, H.J. Queisser, K. Ploog: Phys. Rev. B **37**, 7121 (1988)
4.122 K. Leo, W.W. Rühle, H.J. Queisser, K. Ploog: Appl. Phys. A **45**, 35 (1988)
4.123 W.W. Rühle, H.-J. Polland, E. Bauser, K. Ploog, C.W. Tu: Solid State Electron. **31**, 407 (1988)
4.124 D.J. Westland, J.F. Ryan, M.D. Scott, J.I. Davies, J.R. Riffat: Solid State Electron. **31**, 431 (1988)
4.125 K. Leo, W.W. Rühle, K. Ploog: Solid State Electron. **32**, 1863 (1989)
4.126 K. Leo, W.W. Rühle, K. Ploog: Solid State Commun. **71**, 101 (1989)
4.127 J.F. Ryan, M.C. Tatham: Solid State Electron. **32**, 1429 (1989)
4.128 K. Leo, W.W. Rühle, E. Bauser, K. Ploog: SPIE Proc. **1282**, 134 (1990)
4.129 W.S. Pelouch, R.J. Ellingson, P.E. Powers, C.L. Tang, D.H. Levi, A.J. Nozik: SPIE Proc. **1677**, 260 (1992)
4.130 W.S. Pelouch, R.J. Ellingson, P.E. Powers, C.L. Tang, D.M. Szmyd, A.J. Nozik: Phys. Rev. B **45**, 1450 (1992)
4.131 X.Q. Zhou, H.M. van Driel, W.W. Rühle, K. Ploog: Phys. Rev. B **46**, 16148 (1992)
4.132 Y. Rosenwaks, M.C. Hanna, D.H. Levi, D.M. Szmyd, R.K. Ahrenkiel, A.J. Nozik: Phys. Rev. B **48**, 14675 (1993)
4.133 J. Shah, A. Pinczuk, A.C. Gossard, W. Wiegmann: Physica B & C **134**, 174 (1985)

Chapter 5

5.1 R.E. Nahory: Phys. Rev. **178**, 1293 (1969)
5.2 R.W. Shaw: Phys. Rev. B **3**, 3283 (1971)
5.3 C. Weisbuch: Solid State Electron. **21**, 179 (1978)
5.4 R.E. Nahory, H.Y. Fan: Phys. Rev. Lett. **17**, 251 (1966)
5.5 J. Conradi, R.R. Haering: Phys. Rev. Lett. **20**, 1344 (1968)
5.6 R.C.C. Leite, J.F. Scott, T.C. Damen: Phys. Rev. Lett. **22**, 780 (1969)
5.7 E.F. Gross, S.A. Permogorov, V. Travnikov, A. Selkin: J. Phys. Chem. Solids **31**, 2595 (1970)
5.8 S.A. Permogorov: Phys. Status Solidi (b) **68**, 9 (1975)
5.9 D. von der Linde, J. Kuhl, H. Klingenberg: Phys. Rev. Lett. **44**, 1505 (1980)
5.10 J. Kuhl, D. von der Linde: In *Picosecond Phenomena III*, ed. by K.B. Eisenthal, R.M. Hochstrasser, W. Kaiser, A. Laubereau, Springer Ser. Chem. Phys., Vol. 23 (Springer, Berlin, Heidelberg 1982) pp. 336-340

5.11 K.T. Tsen, D.A. Abramsohn, R. Bray: Phys. Rev. B **26**, 4770 (1982)
5.12 C.L. Collins, P.Y. Yu: Phys. Rev. B **30**, 4501 (1984)
5.13 C.L. Collins, P.Y. Yu: Solid State Commun. **51**, 123 (1984)
5.14 J. Kuhl, B. K. Rhee, W. E. Bron: In *Phonon Scattering in Condensed Matter*, ed. by W. Eisenmenger, K. Lassmann, S. Dottinger, Springer Ser. Solid-State Sci., Vol. 51 (Springer, Berlin, Heidelberg 1984) pp. 106-108
5.15 J. Kuhl, W.E. Bron: Solid State Commun. **49**, 935 (1984)
5.16 W.E. Bron: In *Nonequilibrium Phonon Dynamics*, ed. by W.E. Bron (Plenum, New York 1985) pp. 1-57
5.17 J.A. Kash, C. Tsang, J.M. Hvam: Phys. Rev. Lett. **54**, 2151 (1985)
5.18 W.E. Bron, J. Kuhl, B.K. Rhee: Phys. Rev. B **34**, 6961 (1986)
5.19 K.T. Tsen, H. Morkoç: Phys. Rev. B **34**, 4412 (1986)
5.20 J.A. Kash, S.S. Jha, J.C. Tsang: Phys. Rev. Lett. **58**, 1869 (1987)
5.21 A.C. Maciel, L.C. Campelo Cruz, J.F. Ryan: J. Phys. C **20**, 3041 (1987)
5.22 K.T. Tsen, G. Halama, F. Luty: Phys. Rev. B **36**, 9247 (1987)
5.23 J.A. Kash: SPIE Proc. **942**, 138 (1988)
5.24 J.A. Kash, J.M. Hvam, J.C. Tsang, T.F. Kuech: Phys. Rev. B **38**, 5776 (1988)
5.25 J.F. Ryan, M.C. Tatham, D.J. Westland, C.T. Foxon, M.D. Scott, W.I. Wang: SPIE Proc. **942**, 256 (1988)
5.26 K.T. Tsen, H. Morkoç: Phys. Rev. B **38**, 5615 (1988)
K.T. Tsen, H. Morkoç: Phys. Rev. B **37**, 7137 (1988)
5.27 P. Yu, M. Cardona: *Fundamentals of Semiconductors* (Springer, Berlin, Heidelberg 1996)
5.28 D. von der Linde: In *Ultrashort Laser Pulses*, 2nd edn., ed. by W. Kaiser, Topics Appl. Phys., Vol. 60 (Springer, Berlin, Heidelberg 1988) Chap. 4
5.29 W.E. Bron, S. Mehta, J. Kuhl, M. Klingenstein: Phys. Rev. B **39**, 12642 (1989)
5.30 A.M. de Paula, R.A. Taylor, C.W.W. Bradley, A.J. Turberfield, J.F. Ryan: Superlattices and Microstructures **6**, 199 (1989)
5.31 M.C. Tatham, J.F. Ryan, C.T. Foxon: Phys. Rev. Lett. **63**, 1637 (1989)
5.32 K.T. Tsen, R.P. Joshi, D.K. Ferry, H. Morkoç: Phys. Rev. B **39**, 1446 (1989)
5.33 J.A. Kash, J.C. Tsang: In *Light Scattering in Solids VI*, ed. by M. Cardona, G. Güntherodt, Topics Appl. Phys., Vol. 68 (Springer, Berlin, Heidelberg 1990) Chap. 8
5.34 D.S. Kim, P.Y. Yu: Phys. Rev. B **43**, 4158 (1991)
5.35 K.T. Tsen, K.R. Wald, T. Ruf, P.Y. Yu, H. Morkoç: Phys. Rev. Lett. **67**, 2557 (1991)
5.36 K.T. Tsen, D.J. Smith, S.-C.Y. Tsen, N.S. Kumar, H. Morkoç: J. Appl. Phys. **70**, 418 (1991)
5.37 G. Weber, A.M. de Paula, J.F. Ryan: Semicond. Sci. Technol. **6**, 397 (1991)
5.38 C. Chia, O.F. Sankey, K.T. Tsen: Phys. Rev. B **45**, 6509 (1992)
5.39 J.A. Kash, J.C. Tsang: In *Spectroscopy of Nonequilibrium Electrons and Phonons*, ed. by C.V. Shank, B.P. Zakharchenya (Elsevier, Amsterdam 1992) pp. 113-168
5.40 P. Lugli, P. Bordone, E. Molinari, H. Rücker, A.M. de Paula, A.C. Maciel, J.F. Ryan, M. Shayegan: Semicond. Sci. Technol. **7**, B116 (1992)
5.41 J.F. Ryan, M.C. Tatham: In *Hot Carriers in Semiconductor Nanostructures: Physics and Applications*, ed. by J. Shah (Academic, Boston 1992) pp. 345-378

5.42 G.O. Smith, T. Juhasz, W.E. Bron, Y.B. Levinson: Phys. Rev. Lett. **68**, 2366 (1992)
5.43 G. Weber, J.F. Ryan: Phys. Rev. B **45**, 11202 (1992)
5.44 W.E. Bron: In *Ultrashort Processes in Condensed Matter*, ed. by W.E. Bron (Plenum, New York 1993) pp. 101-142
5.45 T. Ruf, K. Wald, P.Y. Yu, K.T. Tsen, H. Morkoc, K.T. Chan: Superlattices and Microstructures **13**, 203 (1993)
5.46 K.T. Tsen: Int'l J. Mod. Phy. B **7**, 4165 (1993)
5.47 K.T. Tsen, C. Chia, J. West, H. Morkoc: Mod. Phys. Lett. B **7**, 887 (1993)
5.48 K.T. Tsen, R. Joshi, H. Morkoc: Appl. Phys. Lett. **62**, 2075 (1993)
5.49 W.E. Bron: In *Coherent Optical Interactions in Solids*, ed. by R.T. Phillips (Plenum, New York 1994) pp. 199-222
5.50 S.E. Esipov, Y.B. Levinson: Adv. Phys. **36**, 331 (1987)
5.51 J. Shah, R.F. Leheny, A.H. Dayem: Phys. Rev. Lett. **33**, 818 (1974)
5.52 R.G. Ulbrich: In *Nonequilibrium Phonon Dynamics*, ed. by W.E. Bron (Plenum, New York 1985) pp. 101-127
5.53 G.A. Northrop, J.P. Wolfe: In *Nonequilibrium Phonon Dynamics*, ed. by W.E. Bron (Plenum, New York 1985) pp. 165-242
5.54 J. Shah, R.C.C. Leite, J.F. Scott: Solid State Commun. **8**, 1089 (1970)
5.55 Y.B. Levinson, B.N. Levinsky: Solid State Commun. **16**, 713 (1975)
5.56 W. Pötz, P. Kocevar: Phys. Rev. B **28**, 7040 (1983)
5.57 J.A. Kash, J.C. Tsang: Solid State Electron. **31**, 419 (1988)
5.58 G.C. Cho, W.A. Kütt, H. Kurz: Phys. Rev. Lett. **65**, 764 (1990)
5.59 W.A. Kütt, W. Albrecht, H. Kurz: IEEE J. QE-**28**, 2434 (1992)
5.60 F. Vallee, F. Bogani: Phys. Rev. B **43**, 12049 (1991)
5.61 T. Elsaesser, J. Shah, L. Rota, P. Lugli: Phys. Rev. Lett. **66**, 1757 (1991)
5.62 L. Rota, P. Lugli, T. Elsaesser, J. Shah: Phys. Rev. B **47**, 4226 (1993)
5.63 P. Lugli, P. Bordone, L. Reggiani, M. Rieger, P. Kocevar, S.M. Goodnick: Phys. Rev. B **39**, 7852 (1989)
5.64 W.G. Spitzer: *Optical Properties of III-V Compounds. Semiconductors and Semimetals* **3**, 17-69 (Academic, New York 1967)
5.65 Y.R. Shen: *The Principles of Nonlinear Optics* (Wiley, New York 1984)
D.L. Mills: *Nonlinear Optics*, 2nd edn. (Springer, Berlin, Heidelberg 1998)
5.66 J. Kuhl, W.E. Bron: Physica B & C **117-118**, 532 (1983)
5.67 T. Juhasz, W.E. Bron: Phys. Rev. Lett. **63**, 2385 (1989)
5.68 R. Fuchs, K.L. Kliewer: Phys. Rev. **140**, 2076 (1965)
5.69 B.K. Ridley: J. Phys. C **15**, 5899 (1982)
5.70 M. Babiker: J. Phys. C **19**, 683 (1985)
5.71 M. Babiker: J. Phys. C **19**, 339 (1986)
5.72 L. Wendler, R. Haupt: Phys. Status Solidi (b) **143**, 487 (1987)
5.73 L. Wendler, R. Haupt: Phys. Status Solidi (b) **141**, 493 (1987)
5.74 K. Huang, B. Zhu: Phys. Rev. B **38**, 13377 (1988)
5.75 K. Huang, Bang-Fen Zhu: Phys. Rev. B **38**, 2183 (1988)
5.76 L. Wendler, R. Haupt, F. Bechstedt, H. Rücker, R. Enderlein: Superlattices and Microstructures **4**, 577 (1988)
5.77 S. Baroni, P. Giannozzi, E. Molinari: Phys. Rev. B **41**, 3870 (1990)

5.78 S. Rudin, T.L. Reincke: Phys. Rev. B **41**, 7713 (1990)
5.79 P. Lugli, E. Molinari, H. Rücker: Superlattices and Microstructures **10**, 471 (1991)
5.80 B.K. Ridley, M. Babiker: Phys. Rev. **41**, 9096 (1991)
5.81 H. Rücker, E. Molinari, P. Lugli: Phys. Rev. B **44**, 3463 (1991)
5.82 E. Molinari, C. Bungaro, M. Gulia, P. Lugli, H. Rücker: Semicond. Sci. Technol. **7**, B67 (1992)
5.83 B.K. Ridley: SPIE Proc. **1675**, 492 (1992)
5.84 B.K. Ridley: In *Hot Carriers in Semiconductor Nanostructures: Physics and Applications*, ed. by J. Shah (Academic, Boston 1992) pp. 17-51
5.85 H. Rücker, E. Molinari, P. Lugli: Phys. Rev. B **45**, 6747 (1992)
5.86 N.C. Constantinou, O. Al-Dossary, B.K. Ridley: Solid State Commun. **86**, 191 (1993)
5.87 B.K. Ridley: Phys. Rev. B **47**, 4592 (1993)
5.88 C. Colvard, T.A. Gant, M.V. Klein, R. Merlin, R. Fischer, H. Morkoç, A.C. Gossard: Phys. Rev. B **31**, 2080 (1985)
5.89 M.V. Klein: IEEE J. QE-**22**, 1760 (1986)
5.90 B. Jusserand, M. Cardona: In *Light Scattering in Solids V*, ed. M. Cardona, G. Güntherodt, Topics Appl. Phys., Vol. 66 (Springer, Berlin, Heidelberg 1989) Chap. 3
5.91 A.K. Sood, J. Menendez, M. Cardona, K. Ploog: Phys. Rev. Lett. **54**, 2111 (1985)
5.92 A.K. Sood, J. Menendez, M. Cardona, K. Ploog: Phys. Rev. Lett. **54**, 2115 (1985)
5.93 T.A. Gant, M. Delaney, M.V. Klein, R. Houdre, H. Morkoç: Phys. Rev. B **39**, 1696 (1989)
5.94 M.C. Tatham, J.F. Ryan: Semicond. Sci. Technol. **7**, B102 (1992)
5.95 B.K. Ridley: Phys. Rev. B **39**, 5282 (1989)
5.96 R. Loudon: Proc. Roy. Soc. A **275**, 218 (1963)
5.97 D.A. Abramsohn, K.T. Tsen, R. Bray: Phys. Rev. B **26**, 6571 (1982)
5.98 K.T. Tsen, D.A. Abramsohn, R. Bray: Phys. Rev. B **26**, 47709 (1982)
5.99 J. Shah, R.F. Leheny, W.F. Brinkman: Phys. Rev. B **10**, 659 (1974)
5.100 A.M. Weiner, D.E. Leaird, G.P. Wiederrecht, K.A. Nelson: Science **247**, 1317 (1990)
5.101 A.M. Weiner, D.E. Leaird, G.P. Weiderrecht, K.A. Nelson: J. Opt. Soc. Am. B **8**, 1264 (1991)
5.102 A.J. Shields, M.P. Chamberlain, M. Cardona, K. Ebert: Phys. Rev. B **51**, 17728 (1995)

Chapter 6

6.1 J. Frenkel: Phys. Rev. **37**, 17 (1931)
J. Frenkel: Phys. Rev. **37**, 1276 (1931)
6.2 R.E. Peierls: Ann. Physik **13**, 905 (1932)
6.3 G.H. Wannier: Phys. Rev. **52**, 191 (1937)
6.4 M. Ueta, H. Kanzaki, K. Kobayashi, Y. Toyozawa, E. Hanamura: *Excitonic Processes in Solids*, Springer Ser. Solid-State Sci., Vol. 60 (Springer, Berlin, Heidelberg 1986)

6.5 R.S. Knox: *Theory of Excitons* (Academic, New York 1963)

6.6 D.L. Dexter, R.S. Knox: *Excitons* (Interscience, New York 1965)

6.7 J.O. Dimmock: *Optical Properties of III-V Compounds. Semiconductors and Semimetals* **3**, 259-319 (Academic, New York 1967)

6.8 H. Haken, S. Nikitine (eds.): *Excitons at High Density*, Springer Tracts Mod. Phys., Vol. 73 (Springer, Berlin, Heidelberg 1975)

6.9 K. Cho (ed.): *Excitons*, Topics Curr. Phys., Vol. 14 (Springer, Berlin, Heidelberg 1979)

6.10 D.C. Reynolds, T.C. Collins: *Excitons: Their Properties and Uses* (Academic, New York 1981)

6.11 E.I. Rashba, M.D. Sturge (eds.): *Excitons* (North-Holland, Amsterdam 1982)

6.12 R.J. Elliott: Phys. Rev. **108**, 1384 (1957)

6.13 R.J. Elliott: In *Polarons and Excitons*, ed. by C.G. Kuper, G.D. Whitfield (Plenum, New York 1963) pp. 269-293

6.14 S. Schmitt-Rink, D.S. Chemla, D.A.B. Miller: Adv. Phys. **38**, 89 (1989)

6.15 J.J. Hopfield: Phys. Rev. **112**, 1555 (1958)

6.16 Y. Toyozawa: Prog. Theor. Phys. Suppl. **12**, 111 (1959)

6.17 K. Cho: In *Excitons*, ed. by K. Cho, Topics Curr. Phys., Vol. 14 (Springer, Berlin, Heidelberg 1979) Chap. 2

6.18 B. Fischer, J. Lagois: In *Excitons*, ed. by K. Cho, Topics Curr. Phys., Vol. 14 (Springer, Berlin, Heidelberg 1979) Chap. 4

6.19 P.Y. Yu: In *Excitons*, ed. by K. Cho, Topics Curr. Phys., Vol. 14 (Springer, Berlin, Heidelberg 1979) Chap. 5

6.20 J.L. Birman: In *Excitons*, ed. by E.I. Rashba, M.D. Sturge (North-Holland, Amsterdam 1982) pp. 83-140

6.21 E.S. Koteles: In *Excitons*, ed. by E.I. Rashba, M.D. Sturge (North-Holland, Amsterdam 1982) pp. 83-140

6.22 C. Weisbuch, R.G. Ulbrich: In *Light Scattering in Solids III*, ed. by M. Cardona, G. Güntherodt, Topics Appl. Phys., Vol. 51 (Springer, Berlin, Heidelberg 1982) Chap. 7

6.23 F. Bassani: In *Optical Properties of Semiconductors*, ed. by G. Martinez (Kluwer, Dordrecht, Netherlands 1993) pp. 324-364

6.24 H. Haug, S.W. Koch: Quantum Theory of the Optical and Electronic Properties of Semiconductors (World Scientific, Singapore 1993)
C.F. Klingshirn: *Semiconductor Optics* (Springer, Berlin, Heidelberg 1995)

6.25 V.M. Agranovich, V.L. Ginzburg: *Spatial Dispersion in Crystal Optics and Theory of Excitons* (Interscience, London 1966)
V.M. Agranovich, V.L. Ginzburg: *Crystal Optics with Spatial Dispersion, and Excitons*, 2nd edn., Springer Ser. Solid-State Sci., Vol. 42 (Springer, Berlin, Heidelberg 1984)

6.26 S.I. Pekar: Sov. Phys. JETP **6**, 785 (1958)

6.27 S.I. Pekar: Sov. Phys. JETP **7**, 813 (1958)

6.28 V.M. Agranovich, V.O. Dubovskii: JETP Lett. **3**, 223 (1966)

6.29 A. Quattropani, L.C. Andreani, F. Bassani: Nuovo Cimento D **7**, 55 (1986)

6.30 L.C. Andreani, F. Bassani: Phys. Rev. B **41**, 7536 (1990)

6.31 F. Tassone, F. Bassani, L.C. Andreani: Nuovo Cimento D **12**, 1673 (1990)

6.32 V. M. Agranovich: In *Proc. Int'l Conf. on Opticsl of Excitons in Confined Systems*, ed. by A. D'Andrea, R. Del Sole, R. Girlanda, A. Quattropani, IoP Conf. Ser., Vol. 123 (IoP, Bristol, UK 1992) p. 1

6.33 E. Hanamura: SPIE Proc. **1268**, 96 (1990)

6.34 E. Hanamura: Phys. Rev. B **38**, 1228 (1988)

6.35 F. Tassone, F. Bassani, L.C. Andreani: Phys. Rev. B **45**, 6023 (1992)

6.36 R.T. Harley: In *Coherent Optical Interactions in Solids*, ed. by R.T. Phillips (Plenum, New York 1994) pp. 91-109

6.37 J. Lee, E.S. Koteles, M.O. Vassell: Phys. Rev. B **33**, 5512 (1986)

6.38 H. Stolz, D. Schwarze, W. von der Osten, G. Weimann: Superlattices and Microstructures **6**, 271 (1989)

6.39 S. Rudin, T.L. Reinecke, B. Segall: Phys. Rev. B **42**, 11218 (1990)

6.40 S. Rudin, T.L. Reinecke: Phys. Rev. B **41**, 3017 (1990)

6.41 F. Meier, B.P. Zakharchenya (eds.): *Optical Orientation* (North-Holland, Amsterdam 1984)

6.42 E.J. Johnson, R.J. Seymour, R.R. Alfano: In *Semiconductors Probed by Ultrafast Laser Spectroscopy*, ed. by R.R. Alfano (Academic, Orlando, FL 1984) pp. 199-241

6.43 M.Z. Maialle, E.A. de Andrada e Silva, L.J. Sham: Phys. Rev. B **47**, 15776 (1993)

6.44 M. Z. Maialle, L. J. Sham: Surf. Sci. **305**, 256 (1994)

6.45 W. Ekardt, K. Losch, D. Bimberg: Phys. Rev. B **20**, 3303 (1979)

6.46 Y. Shinozuka, M. Matsuura: Phys. Rev. B **28**, 4878 (1983)

6.47 G.E. Bauer, T. Ando: Phys. Rev. B **31**, 8321 (1985)

6.48 M. Matsuura, T. Kamizato: Surf. Sci. **174**, 183 (1986)

6.49 U. Rossler, S. Jorda, D. Broido: Solid State Commun. **73**, 209 (1990)

6.50 F. Askary, P.Y. Yu: Phys. Rev. B **31**, 6643 (1985)

6.51 G.W. 't Hooft, W.A.J.A. van der Poel, L.W. Molenkamp, C.T. Foxon: Phys. Rev. B **35**, 8281 (1987)

6.52 W.J. Rappel, L.F. Feiner, M.F.H. Schuurmans: Phys. Rev. B **38**, 7874 (1988)

6.53 L.C. Andreani, F. Tassone, F. Bassani: Solid State Commun. **77**, 641 (1991)

6.54 D.S. Citrin: Phys. Rev. B **47**, 3832 (1993)

6.55 D.S. Citrin: Solid State Commun. **84**, 281 (1992)

6.56 L.C. Andreani, A. Pasquarello: Phys. Rev. B **42**, 8928 (1990)

6.57 L.C. Andreani, A. Pasquarello: Superlattices and Microstructures **5**, 59 (1989)

6.58 L.C. Andreani, A. Pasquarello: Europhys. Lett. **6**, 259 (1988)

6.59 T. Tokihiro, Y. Manabe, E. Hanamura: Phys. Rev. B **47**, 2019 (1993)

6.60 D.S. Citrin: Solid State Commun. **89**, 139 (1994)

6.61 F.C. Spano, J.R. Kuklinski, S. Mukamel: Phys. Rev. Lett. **65**, 211 (1990)

6.62 H. Stolz, D. Schwarze, W. von der Osten, G. Weimann: Phys. Rev. B **47**, 9669 (1993)

6.63 H. Stolz, D. Schwarze, W. von der Osten, G. Weimann, W. Schottky: Superlattices and Microstructures **9**, 511 (1991)

6.64 R. Zimmermann: Unpublished (1994)

6.65 J. Kusano, Y. Segawa, T. Aoyagi, S. Namba, H. Okamoto: Phys. Rev. B **40**, 1685 (1989)

6.66 T.C. Damen, J. Shah, D.Y. Oberli, D.S. Chemla, J.E. Cunningham, J.M. Kuo: Phys. Rev. B **42**, 7434 (1990)
6.67 P.W.M. Blom, P.J. van Hall, C. Smit, J.P. Cuypers, J.H. Wolter: Phys. Rev. Lett. **71**, 3878 (1993)
6.68 J. Shah: IEEE J. QE-**24**, 276 (1988)
6.69 P.W.M. Blom, J. Claes, J.E.M. Haverkort, J.H. Wolter: Opt. Quantum Electron. **26**, S667 (1994)
6.70 P. Roussignol, C. Delalande, A. Vinattieri, L. Carraresi, M. Colocci: Phys. Rev. B **45**, 6965 (1992)
6.71 B. Deveaud, F. Clerot, N. Roy, K. Satzke, B. Sermage, D.S. Katzer: Phys. Rev. Lett. **67**, 2355 (1991)
6.72 S. Nikitine: In *Optical Properties of Solids*, ed. by S. Nudelman, S.S. Mitra (Plenum, New York 1969) pp. 197-237
6.73 A.T. Agekyan: Phys. Status Solidi (a) **43**, 11 (1977)
6.74 D.W. Snoke, D. Braun, M. Cardona: Phys. Rev. B **44**, 2991 (1991)
6.75 J. Shah, R.F. Leheny, W.F. Brinkman: Phys. Rev. B **10**, 659 (1974)
6.76 W.E. Bron: In *Nonequilibrium Phonon Dynamics*, ed. by W.E. Bron (Plenum, New York 1985) pp. 1-57
6.77 R.J. Seymour, R.R. Alfano: Appl. Phys. Lett. **37**, 231 (1980)
6.78 R.J. Seymour, M.R. Junnarkar, R.R. Alfano: Phys. Rev. B **24**, 3623 (1981)
6.79 G. Bastard, L.L. Chang: Phys. Rev. B **41**, 7899 (1990)
6.80 M.R. Freeman, D.D. Awschalom, J.M. Hong, L.L. Chang: Phys. Rev. Lett. **64**, 2430 (1990)
6.81 R. Ferreira, G. Bastard: Phys. Rev. B **43**, 9687 (1991)
6.82 G. Bastard: Phys. Rev. B **46**, 4253 (1992)
6.83 G. Bastard, R. Ferreira: Surf. Sci. **267**, 335 (1992)
6.84 P. Roussignol, P. Rolland, R. Ferreira, C. Delalande, G. Bastard, A. Vinattieri, J. Martinez-Pastor, L. Carraresi, M. Colocci, J.F. Palmier, B. Etienne: Phys. Rev. B **46**, 7292 (1992)
6.85 P. Roussignol, P. Rolland, R. Ferreira, C. Delalande, G. Bastard, A. Vinattieri, L. Carraresi, M. Colocci, B. Etienne: Surf. Sci. **267**, 360 (1992)
6.86 D.D. Awschalom and N. Samarth: In *Optics of Semiconductor Nanostructures*, ed. by H. Henneberger, S. Schmitt-Rink, E.O. Göbel (Akademie, Berlin 1993) pp. 291-311
6.87 T.C. Damen, L. Viña, J.E. Cunningham, J. Shah, L.J. Sham: Phys. Rev. Lett. **67**, 3432 (1991)
6.88 T.C. Damen, K. Leo, J. Shah, J.E. Cunningham: Appl. Phys. Lett. **58**, 1902 (1991)
6.89 L. Viña, T.C. Damen, J.E. Cunningham, J. Shah, L.J. Sham: Superlattices and Microstructures **12**, 379 (1992)
6.90 P.J. Dean, D.C. Herbert: In *Excitons*, ed. by K. Cho, Topics Curr. Phys., Vol. 14 (Springer, Berlin, Heidelberg 1979) Chap. 3
6.91 M. Voos, R.F. Leheny, J. Shah: In *Handbook on Semiconductors: II. Optical Properties of Semiconductors*, ed. by M. Balkanski (North-Holland, Amsterdam 1980) pp. 329-416

6.92 P.T. Landsberg: *Recombination in Semiconductors* (Cambridge Univ. Press, Cambridge 1991)
6.93 J. Feldmann, G. Peter, E.O. Göbel, P. Dawson, K. Moore, C.T. Foxon, R.J. Elliott: Phys. Rev. Lett. **59**, 2337 (1987)
6.94 M. Colocci, M. Gurioli, J. Martinez-Pastor: J. Physique IV C **3**, 3 (1993)
6.95 H. Akiyama, S. Koshiba, T. Someya, K. Wada, H. Noge, Y. Nakamura, T. Inoshita, A. Shimizu, H. Sakaki: Phys. Rev. Lett. **72**, 924 (1994)
6.96 D. Gershoni, M. Katz, W. Wegscheider, L.N. Pfeiffer, R.A. Logan, K. West: Unpublished (1993)
6.97 R. Eccleston, B.F. Feuerbacher, J. Kuhl, W.W. Rühle, K. Ploog: Phys. Rev. B **45**, 11403 (1992)
6.98 H. Haug, S. Schmitt-Rink: Progr. Quantum Electron. **9**, 3 (1984)
6.99 D.S. Chemla, D.A.B. Miller, P.W. Smith: Opt. Eng. **24**, 556 (1985)
6.100 S. Schmitt-Rink, D.S. Chemla, D.A.B. Miller: Phys. Rev. B **32**, 6601 (1985)
6.101 D.S. Chemla: *Nonlinear Optics: Materials and Devices*, ed. by C. Flytzanis, J.L. Oudar, Springer Proc. Phys. **7**, 65-78 (Springer, Berlin, Heidelberg 1986)
6.102 D.S. Chemla: In *Laser Optics of Condensed Matter*, ed. by J.L. Birman (Plenum, New York 1988)
6.103 S. Schmitt-Rink, D.S. Chemla, H. Haug: Phys. Rev. B **37**, 941 (1988)
6.104 W.H. Knox, R.L. Fork, M.C. Downer, D.A.B. Miller, D.S. Chemla, C.V. Shank, A.C. Gossard, W. Wiegmann: Phys. Rev. Lett. **54**, 1306 (1985)
6.105 D.S. Kim, J. Shah, J.E. Cunningham, T.C. Damen, W. Schäfer, M. Hartmann, S. Schmitt-Rink: Phys. Rev. Lett. **68**, 1006 (1992)
6.106 A.V. Nurmikko, R.L. Gunshor: In *Optics of Semiconductor Nanostructures*, ed. by H. Henneberger, S. Schmitt-Rink, E. O. Göbel (Akademie, Berlin 1993) pp. 231-261
6.107 P.C. Becker, D. Lee, A.M. Johnson, A.G. Prosser, R.D. Feldman, R.F. Austin, R.E. Behringer: Phys. Rev. Lett. **68**, 1876 (1992)
6.108 P.C. Becker, D. Lee, M.R.X. Barros, A.M. Johnson, A.G. Prosser, R.D. Feldman, R.F. Austin, R.E. Behringer: IEEE J. QE-**28**, 2535 (1992)
6.109 J.B. Stark, W.H. Knox, D.S. Chemla, W. Schäfer, S. Schmitt-Rink, C. Stafford: Phys. Rev. Lett. **65**, 3033 (1990)
6.110 J.B. Stark, W.H. Knox, D.S. Chemla: Phys. Rev. B **46**, 7919 (1992)
6.111 J.B. Stark, W.H. Knox, D.S. Chemla: Phys. Rev. Lett. **68**, 3080 (1992)
6.112 J.B. Stark, W.H. Knox, D.S. Chemla: In *Optics of Semiconductor Nanostructures*, ed. by H. Henneberger, S. Schmitt-Rink, E.O. Göbel (Akademie, Berlin 1993) pp. 399-445
6.113 M.Z. Maialle, L.J. Sham: Phys. Rev. Lett. **73**, 3310 (1994)
6.114 S. Bar-Ad, I. Bar-Joseph: Phys. Rev. Lett. **68**, 349 (1992)
6.115 I. Brener, W.H. Knox, K.W. Goossen, J.E. Cunningham: Phys. Rev. Lett. **70**, 319 (1993)
6.116 J. Hegarty, M.D. Sturge, C. Weisbuch, A.C. Gossard, W. Wiegmann: Phys. Rev. Lett. **49**, 930 (1982)
6.117 H. Stolz, V. Langer, W. von der Osten: J. Lumin. **48** & **49**, 72 (1991)
6.118 H. Stolz: Phys. Status Solidi (b) **173**, 99 (1992)

6.119 H. Stolz: *Time-Resolved Light Scattering from Excitons*, Springer Tracts Mod. Phys., Vol. 130 (Springer, Berlin, Heidelberg 1994)
6.120 M.R. Freeman, D.D. Awschalom, J.M. Hong: Appl. Phys. Lett. **57**, 704 (1990)
6.121 A. Vinattieri, J. Shah, T.C. Damen, D.S. Kim, L.N. Pfeiffer, L.J. Sham: Solid State Commun. **88**, 189 (1993)
6.122 B. Sermage, B. Deveaud, K. Satzke, F. Clerot, C. Dumas, N. Roy, D.S. Katzer, F. Mollot, R. Planel, M. Berz, J.L. Oudar: Superlattices and Microstructures **13**, 271 (1993)
6.123 B. Sermage, S. Long, B. Deveaud, D.S. Katzer: J. Physique IV C5, **3**, 19 (1993)
6.124 A. Vinattieri, J. Shah, T.C. Damen, D.S. Kim, L.N. Pfeiffer, L.J. Sham: J. Physique IV C5, **3**, 27 (1993)
6.125 A. Vinattieri, J. Shah, T.C. Damen, K.W. Goossen, L.N. Pfeiffer, M.Z. Maialle, L.J. Sham: Appl. Phys. Lett. **63**, 3164 (1993)
6.126 A. Vinattieri, J. Shah, T. C. Damen, L. N. Pfeiffer, L. J. Sham, M. Z. Maialle: *Ultrafast Phenomena in Semiconductors*, ed. by D. K. Ferry and H. M. van Driel, SPIE Proc. **2142**, 2-13 (SPIE, Bellingham, WA 1994)
6.127 A. Vinattieri, J. Shah, T.C. Damen, D.S. Kim, L.N. Pfeiffer, M.Z. Maialle, L.J. Sham: Phys. Rev. B **50**, 10868 (1994)
6.128 A. Lohner, P. Michler, W.W. Rühle: IQEC'95, OSA Techn. Digest Ser., Vol. 14 (Optical Society of America, Washington, DC 1995) pp. 108-110
6.129 E.O. Göbel, H. Jung, J. Kuhl, K. Ploog: Phys. Rev. Lett. **51**, 1588 (1983)
6.130 H. Wang, J. Shah, T. C. Damen, L. N. Pfeiffer: IQEC'94, OSA Techn. Digest Ser., Vol. 7 (Optical Society of America, Washington, DC 1994) paper QPD22
6.131 H. Wang, J. Shah, T.C. Damen, L.N. Pfeiffer: Phys. Rev. Lett. **74**, 3061 (1995)
6.132 H. Wang, J. Shah, T.C. Damen, L. N. Pfeiffer: *Ultrafast Phenomena IX*, ed. by P.F. Barbara, W.H. Knox, G.A. Mourou, A.H. Zewail, Springer Ser. Chem. Phys., Vol. 60 (Springer, Berlin, Heidelberg 1994) pp. 362-363
6.133 H. Wang, J. Shah, T.C. Damen, L.N. Pfeiffer: Solid State Commun. **91**, 869 (1994)
6.134 The decay is non-exponential and becomes slower at longer times. The initial decay rate is consistent with earlier measurements [6.127]
6.135 S. Haroche, J.A. Paisner, A.L. Schawlow: Phys. Rev. Lett. **30**, 948 (1973)
6.136 H. Stolz, V. Langer, E. Schreiber, S.A. Permogorov, W. von der Osten: Phys. Rev. Lett. **67**, 679 (1991)
6.137 V. Langer, H. Stolz, W. von der Osten: Phys. Rev. Lett. **64**, 854 (1990)
6.138 A.P. Heberle, W.W. Rühle, K. Ploog: Phys. Rev. Lett. **72**, 3887 (1994)
6.139 P.R. Berman: J. Opt. Soc. Am. B **4**, 564 (1986)
6.140 J. Hegarty, L. Goldner, M.D. Sturge: Phys. Rev. B **30**, 7346 (1984)
6.141 S.A. Permogorov, A. Reznitsky, A. Naumov, H. Stolz, W. von der Osten: J. Lumin. **40 & 41**, 483 (1988)
6.142 H. Stolz: Adv. Solid State Phys. **31**, 219 (1991)
6.143 H. Wang, J. Shah, T.C. Damen, L.N. Pfeiffer, J.E. Cunningham: Phys. Status Solidi (b) **188**, 381 (1995)

Chapter 7

7.1 E.R. Brown: In *Hot Carriers in Semiconductor Nanostructures: Physics and Applications*, ed. J. Shah (Academic, Boston 1992) pp. 469-498

7.2 C. Minot, H. LePerson, F. Alexandre, J.F. Palmier: Appl. Phys. Lett. **51**, 1626 (1987)

7.3 S. Tarucha, K. Ploog, K. von Klitzing: Phys. Rev. B **36**, 4558 (1987)

7.4 H. Schneider, K. von Klitzing, K. Ploog: Superlattices and Microstructures **5**, 383 (1989)

7.5 H. Schneider, H.T. Grahn, K. von Klitzing, K. Ploog: Phys. Rev. B **40**, 10040 (1989)

7.6 H. Schneider, W.W. Rühle, K. von Klitzing, K. Ploog: Appl. Phys. Lett. **54**, 2656 (1989)

7.7 S. Tarucha, K. Ploog: Phys. Rev. B **39**, 5353 (1989)

7.8 T.C.L.G. Sollner, W.D. Goodhue, P.E. Tannenwald, C.D. Parker, D.D. Peck: Appl. Phys. Lett. **43**, 588 (1983)

7.9 J.F. Whitaker, G.A. Mourou, T.C.L.G. Sollner, W.D. Goodhue: Appl. Phys. Lett. **53**, 385 (1988)

7.10 M. Tsuchiya, T. Matsusue, H. Sakaki: Phys. Rev. Lett. **59**, 2356 (1987)

7.11 T.B. Norris, X.J. Song, W.J. Schaff, L.F. Eastman, G. Wicks, G.A. Mourou: Appl. Phys. Lett. **54**, 60 (1989)

7.12 M.K. Jackson, M.B. Johnson, D.H. Chow, T.C. McGill, C.W. Nieh: Appl. Phys. Lett. **54**, 552 (1989)

7.13 J. Shah, K. Leo, D.Y. Oberli, T.C. Damen: In *Ultrashort Processes in Condensed Matter*, ed. W.E. Bron (Plenum, New York 1993) pp. 53-100

7.14 A.P. Heberle, W.W. Rühle, K. Köhler: Phys. Status Solidi (b) **173**, 381 (1992)

7.15 M. Büttiker, R. Landauer: Phys. Rev. Lett. **49**, 1739 (1982)

7.16 M. Büttiker, R. Landauer: Physica Scripta **32**, 429 (1985)

7.17 S. Collins, D. Lowe, J.R. Barker: J. Phys. C **20**, 6213 (1987)

7.18 A.P. Jauho: In *Hot Carriers in Semiconductor Nanostructures: Physics and Applications*, ed. J. Shah (Academic, Boston (1992) pp. 121-151

7.19 C.R. Leavens and G.C. Aers: In *Scanning Tunneling Microscopy III*, ed. by R. Wiesendanger, H.-J. Güntherodt, Springer Ser. Surf. Sci., Vol. 29 (Springer, Berlin, Heidelberg 1992) Chap. 7

7.20 H.C. Liu: Appl. Phys. Lett. **50**, 1019 (1987)

7.21 M.G.W. Alexander, M. Nido, K. Reimann, W.W. Rühle, K. Köhler: Appl. Phys. Lett. **55**, 2517 (1989)

7.22 G. Bastard, J.A. Brum, R. Ferreira: Advances in research and applications. *Solid State Physics* **44**, 229-415 (Academic, Boston 1991)

7.23 E.O. Kane: In *Solid State Physics: Advances in Research and Applications*, ed. by C. B. Duke, Solid State Physics, Suppl. 10 (Academic Press, New York 1969)

7.24 S. Luryi: Appl. Phys. Lett. **47**, 490 (1985)

7.25 S. Luryi: Superlattices and Microstructures **5**, 375 (1989)

7.26 S. Luryi: IEEE J. QE-**27**, 53 (1991)

7.27 J. Shah: In *Spectroscopy of Semiconductor Microstructures*, ed. by G. Fasol, A. Fasolino, P. Lugli (Plenum, New York 1989) pp. 535-560

7.28 K. Leo, J. Shah, E.O. Göbel, T.C. Damen, S. Schmitt-Rink, W. Schäfer, K. Köhler: Phys. Rev. Lett. **66**, 201 (1991)
7.29 A.J. Leggett, S. Chakravarty, A.T. Dorsey, P.A. Fisher, A. Garg, W. Zwerge: Rev. Mod. Phys. **59**, 1 (1987)
7.30 K. Leo, J. Shah, J.P. Gordon, T.C. Damen, D.A.B. Miller, C.W. Tu, J.E. Cunningham: Phys. Rev. B **42**, 7065 (1990)
7.31 K. Leo, J. Shah, J.P. Gordon, T.C. Damen, D.A.B. Miller, C.W. Tu, J.E. Cunningham, J.E. Henry: SPIE Proc. **1283**, 35 (1990)
7.32 S.A. Gurvitz, I. Bar-Joseph, B. Deveaud: Phys. Rev. B **43**, 14703 (1991)
7.33 G. Cohen, S.A. Gurvitz, I. Bar-Joseph, B. Deveaud, P. Bergman, A. Regreny: Phys. Rev. B **47**, 16012 (1993)
7.34 A.D. Stone, P.A. Lee: Phys. Rev. Lett. **54**, 1196 (1985)
7.35 L.L. Chang, L. Esaki, R. Tsu: Appl. Phys. Lett. **24**, 593 (1974)
7.36 F. Capasso, K. Mohammed, A.Y. Cho: IEEE J. QE-**22**, 1853 (1986)
7.37 W.D. Goodhue, T.C.L.G. Sollner, H.Q. Le, E.R. Brown, B.A. Vojak: Appl. Phys. Lett. **49**, 1088 (1986)
7.38 E.R. Brown, T.C.L.G. Sollner, W.D. Goodhue, C.D. Parker, C.L. Chen: Appl. Phys. Lett. **55**, 1777 (1989)
7.39 E.R. Brown, C.D. Parker, T.C.L.G. Sollner: Appl. Phys. Lett. **54**, 934 (1989)
7.40 V.J. Goldman, D.C. Tsui, J.E. Cunningham: Phys. Rev. Lett. **58**, 1256 (1987)
7.41 V.J. Goldman, D.C. Tsui, J.E. Cunningham: Phys. Rev. B **35**, 9387 (1987)
7.42 J.F. Young, B.M. Wood, G.C. Aers, R.L.S. Devine, H.C. Liu, D. Landheer, Buchanan: Phys. Rev. Lett. **60**, 2085 (1988)
7.43 I. Bar-Joseph, T.K. Woodward, D.S. Chemla, D. Sivco, A.Y. Cho: Phys. Rev. B **41**, 3264 (1990)
7.44 I. Bar-Joseph, T.K. Woodward, D.S. Chemla, Y. Gedalyahu, A. Yacoby, D. Sivco, A.Y. Cho: Superlattices and Microstructures **8**, 409 (1990)
7.45 H. Yoshimura, J.L. Schulman, H. Sakaki: Phys. Rev. Lett. **64**, 2422 (1990)
7.46 I. Bar-Joseph, Y. Gedalyahu, A. Yacoby, T.K. Woodward, D.S. Chemla, D.L. Sivco, A.Y. Cho: Phys. Rev. B **44**, 8361 (1991)
7.47 T.K. Woodward, D.S. Chemla, I. Bar-Joseph, D.L. Sivco, A.Y. Cho: In *Picosecond Electronics and Optoelectronics*, ed. by T.C.L.G. Sollner, J. Shah, OSA Proc. Picosec. Electron. Optoelectron., Vol. 9 (Optical Society of America, Washington, DC 1991) p. 198
7.48 T.K. Woodward, D.S. Chemla, I. Bar-Joseph, H.U. Baranger, D.L. Sivco, A.Y. Cho: Phys. Rev. B **44**, 1353 (1991)
7.49 T. Tada, A. Yamaguchi, T. Ninomiya, H. Uchiki: J. Appl. Phys. **63**, 5491 (1988)
7.50 M. Tsuchiya, T. Matsusue, H. Sakaki: In *Ultrafast Phenomena VI*, ed. by T. Yajima, K. Yoshihara, C. B. Harris, S. Shionoya, Springer Ser. Chem. Phys., Vol. 48 (Springer, Berlin, Heidelberg 1988) pp. 304-307
7.51 H. Sakaki, T. Matsusue, M. Tsuchiya: IEEE J. QE-**25**, 2498 (1989)
7.52 T.B. Norris, V. Vodjdani, B. Vinter, C. Weisbuch, G.A. Morou: Phys. Rev. B **40**, 1392 (1989)
7.53 D.Y. Oberli, J. Shah, T.C. Damen, C.W. Tu, T.Y. Chang, D.A.B. Miller, J.E. Henry, R.F. Kopf, N. Sauer, A.E. DiGiovanni: Phys. Rev. B **40**, 3028 (1989)

7.54 D.Y. Oberli, J. Shah, B. Deveaud, T.C. Damen: In *Picosecond Electronics and Optoelectronics*, ed. by T.C.L.G. Sollner, D.M. Bloom, OSA Proc. Picosec. Electron. Optoelectron., Vol. 4 (Optical Society of America, Washington, DC 1989) p. 94

7.55 D.Y. Oberli, J. Shah, T.C. Damen, J.M. Kuo, J.E. Henry, J. Lary, S.M. Goodnick: Appl. Phys. Lett. **56**, 1239 (1990)

7.56 M.G.W. Alexander, W.W. Rühle, R. Sauer, W.T. Tsang: Appl. Phys. Lett. **55**, 885 (1989)

7.57 N. Sawaki, R.A. Höpfel, E. Gornik, H. Kano: Appl. Phys. Lett. **55**, 1996 (1989)

7.58 A.P. Heberle, W.W. Rühle, K. Köhler: In *Ultrafast Processes in Spectroscopy*, ed. by A. Lauberau, A. Seilmeir (IoP, Bristol 1992) p. 367

7.59 P. Roussignol, M. Gurioli, L. Carraresi, M. Colocci, A. Vinattieri, C. Deparis, J. Massies, G. Neu: Superlattices and Microstructures **9**, 151 (1991)

7.60 B. Deveaud, A. Chomette, F. Clerot, P. Auvray, A. Regreny, R. Ferreira, G. Bastard: Phys. Rev. B **42**, 7021 (1990)

7.61 S. Muto, T. Inata, A. Tackeuchi, Y. Sugiyama, T. Fujii: Appl. Phys. Lett. **58**, 2393 (1991)

7.62 R. Sauer, K. Thonke, W.T. Tsang: Phys. Rev. Lett. **61**, 609 (1988)

7.63 H.W. Liu, R. Ferreira, G. Bastard, C. Delalande, J.F. Palmier, B. Etienne: Surf. Sci. **228**, 383 (1990)

7.64 S. Charbonneau, M.L.W. Thewalt, E.S. Koteles, B. Elman: Phys. Rev. B **38**, 6287 (1988)

7.65 R. Ferreira, P. Rolland, P. Roussignol, C. Delalande, A. Vinattieri, L. Carraresi, M. Colocci, N. Roy, B. Sermage, J.F. Palmier, B. Etienne: Phys. Rev. B **45**, 11782 (1992)

7.66 Y.J. Chen, E.S. Koteles, B.S. Elman, C. Armiento: Phys. Rev. B **36**, 4562 (1987)

7.67 E.S. Koteles, B. Elman, J. Lee, S. Charbonneau, M.W.L. Thewalt: SPIE Proc. **1283**, 143 (1990)

7.68 T. Weil, B. Vinter: Appl. Phys. Lett. **50**, 1281 (1987)

7.69 S. Luryi: Solid State Commun. **65**, 787 (1988)

7.70 R. Ferreira and G. Bastard: Surf. Sci. **229**, 165 (1990)

7.71 G. Bastard, C. Delalande, R. Ferreira, H.U. Liu: J. Lumin. **44**, 247 (1989)

7.72 J. Lary, S.M. Goodnick, P. Lugli, D.Y. Oberli, J. Shah: Solid State Electron. **32**, 1283 (1989)

7.73 N.S. Wingreen, K.W. Jacobsen, J.W. Wilkins: Phys. Rev. Lett. **61**, 1396 (1988)

7.74 M. Büttiker: IBM J. Res. Develop. **32**, 63 (1988)

7.75 B.Y. Gelfand, S. Schmitt-Rink, A.F.J. Levi: Phys. Rev. Lett. **62**, 1683 (1989)

7.76 W.W. Rühle, M.G.W. Alexander, M. Nido: In *Proc. 20th Int'l Conf. on the Physics of Semiconductors*, ed. by E. M. Anastassakis, J. D. Joannopoulos (World Scientific, Singapore 1990)

7.77 D.Y. Oberli, J. Shah, T.C. Damen, J.M. Kuo, R.F. Kopf, J.E. Henry: Surf. Sci. **229**, 189 (1990)

7.78 S.M. Goodnick, J.E. Lary, P. Lugli: Superlattices and Microstructures **10**, 461 (1991)

7.79 H. Rücker, P. Lugli, S.M. Goodnick, J.E. Lary: Semicond. Sci. Technol. **7**, B98 (1992)

7.80 P. Roussignol, A. Vinattieri, L. Carraresi, M. Colocci, A. Fasolino: Phys. Rev. B **44**, 8873 (1991)
7.81 T.C. Damen, J. Shah, D.Y. Oberli, D.S. Chemla, J.E. Cunningham, J.M. Kuo: Phys. Rev. B **42**, 7434 (1990)
7.82 R. Wessel, M. Altarelli: Phys. Rev. B **39**, 12802 (1989)
7.83 E.T. Yu, M.K. Jackson, T.C. McGill: Appl. Phys. Lett. **55**, 744 (1989)
7.84 M.G.W. Alexander, M. Nido, W.W. Rühle, K. Köhler: Phys. Rev. B **41**, 12295 (1990)
7.85 M. Nido, M.G.W. Alexander, W.W. Rühle, K. Köhler: Phys. Rev. B **43**, 1839 (1991)
7.86 T.B. Norris, N. Vodjdani, B. Vinter, E. Costard, E. Bockenhoff: Phys. Rev. B **43**, 1867 (1991)
7.87 A.P. Heberle, W.W. Rühle, K. Köhler: SPIE Proc. **1677**, 234 (1992)
7.88 M. Nido, M.G.W. Alexander, W.W. Rühle, K. Köhler: SPIE Proc. **1268**, 177 (1990)
7.89 M. Nido, M.G.W. Alexander, W.W. Rühle, T. Schweizer, K. Köhler: Appl. Phys. Lett. **56**, 355 (1990)
7.90 M.G.W. Alexander, M. Nido, W.W. Rühle, K. Köhler: Superlattices and Microstructures **9**, 83 (1991)
7.91 W.W. Rühle, A.P. Heberle, M.G.W. Alexander, M. Nido, K. Köhler: Physica Scripta T **39**, 278 (1991)
7.92 K. Leo, J. Shah, T.C. Damen, A. Schulze, T. Meier, S. Schmitt-Rink, P. Thomas, E.O. Göbel, S.L. Chuang, M.S.C. Luo, W. Schäfer, K. Köhler, P. Ganser: IEEE J. QE-**28**, 2498 (1992)

Chapter 8

8.1 J.R. Haynes, W. Shockley: Phys. Rev. **81**, 835 (1951)
8.2 L. Reggiani: In *Physics of Nonlinear Transport*, ed. by D.K. Ferry, J.R. Barker, C. Jacoboni (Plenum, New York 1980) pp. 243-254
8.3 C. Canali, F. Nava, L. Reggiani: In *Hot Electron Transport in Semiconductors*, ed. by L. Reggiani, Topics Appl. Phys., Vol. 58 (Springer, Berlin, Heidelberg 1985) Chap. 3
8.4 D.F. Nelson, J.A. Cooper Jr., A.R. Tretola: Appl. Phys. Lett. **41**, 857 (1982)
8.5 R.A. Höpfel, J. Shah, D. Block, A.C. Gossard: Appl. Phys. Lett. **48**, 148 (1986)
8.6 D.W. Pohl, W. Denk, M. Lantz: Appl. Phys. Lett. **44**, 651 (1984)
D.W. Pohl: In *Scanning Tunneling Microscopy II*, 2nd edn., ed. R. Wiesendanger, H.-J. Güntherodt, Springer Ser. Surf. Sci., Vol. 28 (Springer, Berlin, Heidelberg 1995) Chap. 7 and Sect. 9.4
8.7 E. Betzig, J.K. Trautman: Science **257**, 189 (1992)
8.8 G. Binnig, H. Rohrer: Helvetica Physica Acta **55**, 726 (1982)
H.-J. Güntherodt, R. Wiesendanger (eds.): *Scanning Tunneling Microscopy I*, 2nd edn., Springer Ser. Surf. Sci., Vol. 20 (Springer, Berlin, Heidelberg 1994)
8.9 G. Binnig, C.F. Quate, C. Gerber: Phys. Rev. Lett. **56**, 726 (1986)

8.10 J.B. Stark, U. Mohideen, E. Betzig, R.E. Slusher: In *Ultrafast Phenomena IX*, ed. by P.F. Barbara, W.H. Knox, G.A. Mourou, A.H. Zewail, Springer Ser. Chem. Phys., Vol. 60 (Springer, Berlin, Heidelberg 1994) pp. 349-350

8.11 S. Weiss, D.F. Ogletree, D. Botkin, M. Salmeron, D.S. Chemla: Appl. Phys. Lett. **63**, 2567 (1993)

8.12 S. Weiss, D. Botkin, D.F. Ogletree, M. Salmeron, D.S. Chemla: In *Ultrafast Phenomena IX*, ed. by P.F. Barbara, W.H. Knox, G.A. Mourou, A.H. Zewail, Springer Ser. Chem. Phys., Vol. 60 (Springer, Berlin, Heidelberg 1994) pp. 347-348

8.13 G. Nunes and M.R. Freeman: Science **262**, 1029 (1993)

8.14 D.M. Bloom, A.S. Hou, F. Ho, B.A. Nechay: In *Ultrafast Phenomena IX*, ed. by P.F. Barbara, W.H. Knox, G.A. Mourou, A.H. Zewail, Springer Ser. Chem. Phys., Vol. 60 (Springer, Berlin, Heidelberg 1994) pp. 17-21
A.S. Hou, F. Ho, D.M. Bloom: Electron. Lett. **28**, 2302 (1992)

8.15 C.J. Chen: In *Scannung Tunneling Microscopy III*, ed. by R. Wiesendanger, H.-J. Güntherodt, Springer Ser. Surf. Sci., Vol. 29 (Springer, Berlin, Heidelberg 1993) Chap. 7

8.16 J. Nees, S. Wakana, C.Y. Chen: In *Ultrafast Phenomena IX*, ed. by P.F. Barbara, W.H. Knox, G.A. Mourou, A.H. Zewail, Springer Ser. Chem. Phys., Vol. 60 (Springer, Berlin, Heidelberg 1994) pp. 139-140

8.17 C. Bai: *Scanning Tunneling Microscopy and its Applications*, Springer Ser. Surf. Sci., Vol. 32 (Springer, Berlin, Heidelberg 1995)

8.18 W. Franz: Z. Naturforsch. **13**, 484 (1958)
L.V. Keldysh: Sov. Phys. JETP **7**, 788 (1958)

8.19 D.A.B. Miller, D.S. Chemla, S. Schmitt-Rink: Phys. Rev. B **33**, 6976 (1986)

8.20 B.B. Hu, E.A. de Souza, W.H. Knox, J.E. Cunningham, M.C. Nuss, A.V. Kuznetsov, S.L. Chuang: Phys. Rev. Lett. **74**, 1689 (1995)

8.21 H. Hillmer, G. Mayer, A. Forchel, K.S. Lochner, E. Bauser: Appl. Phys. Lett. **49**, 948 (1986)

8.22 A. Chomette, B. Deveaud, J.Y. Emery, A. Regreny: Superlattices and Microstructures **1**, 201 (1985)

8.23 B. Deveaud, J. Shah, T.C. Damen, B. Lambert, A. Chomette, A. Regreny: IEEE J. QE-**24**, 1641 (1988)

8.24 J. Shah: In *Spectroscopy of Nonequilibrium Electrons and Phonons*, ed. by C.V. Shank, B.P. Zakharchenya (Elsevier, Amsterdam 1992) pp. 57-112

8.25 J. Shah: In *Hot Carriers in Semiconductor Nanostructures: Physics and Applications*, ed. by J. Shah (Academic, Boston 1992) pp. 279-312

8.26 J. Shah: In *Spectroscopy of Semiconductor Microstructures*, ed. by G. Fasol, A. Fasolino, P. Lugli (Plenum, New York 1989) pp. 535-560

8.27 D.Y. Oberli, J. Shah, B. Deveaud, T.C. Damen: In *Picosecond Electronics and Optoelectronics*, ed. by T.C.L.G. Sollner, D.M. Bloom, OSA Proc. Picosec. Electron. Optoelectron., Vol. 4 (Optical Society of America, Washington, DC 1989) p. 94

8.28 R.A. Höpfel, J. Shah, P.A. Wolff, A.C. Gossard: Phys. Rev. Lett. **56**, 2736 (1986)

8.29 R.A. Höpfel, J. Shah, P.A. Wolff, A.C. Gossard: Phys. Rev. B **37**, 6941 (1988)

8.30 H. Hillmer, S. Hansmann, A. Forchel, M. Morohashi, E. Lopez, H.P. Meier, K. Ploog: Appl. Phys. Lett. **53**, 1937 (1988)
8.31 H. Hillmer, A. Forchel, C.W. Tu: J. Phys. C **5**, 5563 (1993)
8.32 H. Hillmer, C.W. Tu: Appl. Phys. A **56**, 445 (1994)
8.33 H. Hillmer, H.L. Zhu, H. Burkhard: J. Appl. Phys. **73**, 1035 (1993)
8.34 A. Forchel, B. Laurich, H. Hillmer, G. Tränkle, M. Pilkuhn: J. Lumin. **30**, 67 (1985)
8.35 B.F. Levine, W.T. Tsang, C.G. Bethea, F. Capasso: Appl. Phys. Lett. **41**, 470 (1982)
8.36 D.J. Westland, D. Mihailovic, J.F. Ryan, M.D. Scott: Appl. Phys. Lett. **51**, 590 (1987)
8.37 H. Hillmer, A. Forchel, T. Kuhn, G. Mahler, H.P. Meier: Phys. Rev. B **43**, 13992 (1991)
8.38 B. Deveaud, A. Chomette, B. Lambert, A. Regreny, R. Romestain, P. Edel: Solid State Commun. **57**, 885 (1986)
8.39 C.V. Shank, R.L. Fork, B.I. Greene, F.K. Reinhart, R.A. Logan: Appl. Phys. Lett. **38**, 104 (1981)
8.40 T.J. Maloney, J. Frey: J. Appl. Phys. **48**, 781 (1977)
8.41 W. Sha, T.B. Norris, W.J. Schaff, K.E. Meyer: Phys. Rev. Lett. **67**, 2553 (1991)
8.42 J. Son, W. Sha, J. Kim, T.B. Norris, J.F. Whitaker, G.A. Mourou: Appl. Phys. Lett. **63**, 923 (1993)
8.43 T.B. Norris, W. Sha, W.J. Schaff: SPIE Proc. **1677**, 241 (1992)
8.44 F. Capasso, K. Mohammed, A.Y. Cho: IEEE J. QE-**22**, 1853 (1986)
8.45 P.W. Anderson: Phys. Rev. **109**, 1492 (1958)
8.46 A. Chomette, B. Deveaud, A. Regreny, G. Bastard: Phys. Rev. Lett. **57**, 1464 (1986)
8.47 D. Calecki, J.F. Palmier, A. Chomette: J. Phys. C **17**, 5017 (1984)
8.48 J.F. Palmier, H. Le Person, C. Minot, A. Chomette, A. Regreny, D. Calecki: Superlattices and Microstructures **1**, 67 (1985)
8.49 J.F. Palmier and A. Chomette: J. Physique **43**, 381 (1982)
8.50 G.J. Warren and P.N. Butcher: Semicond. Sci. Technol. **1**, 133 (1986)
8.51 B. Deveaud, J. Shah, T.C. Damen, B. Lambert, A. Regreny: Phys. Rev. Lett. **58**, 2582 (1987)
8.52 E. Yang, S. Das Sarma: Phys. Rev. B **37**, 10090 (1988)
8.53 B. Lambert, F. Clerot, B. Deveaud, A. Chomette, G. Talalaeff, A. Regreny: J. Lumin. **44**, 277 (1989)
8.54 B. Deveaud, F. Clerot, A. Chomette, B. Lambert, P. Auvray, M. Gauneau: Appl. Phys. Lett. **59**, 2168 (1991)
8.55 D.A.B. Miller, D.S. Chemla, T.C. Damen, A.C. Gossard, W. Wiegmann, T.H. Wood, C.A. Burrus: Appl. Phys. Lett. **45**, 13 (1984)
D.A.B. Miller: Opt. Quantum Electron. **22**, S61 (1990)
8.56 C.F. Klingshirn: *Semiconductor Optics* (Springer, Berlin, Heidelberg 1995)
R.G. Hunsperger: *Integrated Optics*, 4th edn. (Springer, Berlin, Heidelberg 1995)
8.57 F. Capasso, K. Mohammed, A.Y. Cho: Appl. Phys. Lett. **48**, 478 (1986)
8.58 H. Schneider, K. von Klitzing, K. Ploog: Superlattices and Microstructures **5**, 383 (1989)

8.59 J. Callaway: Phys. Rev. **130**, 549 (1963)

8.60 D.A.B. Miller, D.S. Chemla, T.C. Damen, A.C. Gossard, W. Wiegmann, T.H. Wood, C.A. Burrus: Phys. Rev. B **32**, 1043 (1985)

8.61 G. Livescu, A.M. Fox, D.A.B. Miller, T. Sizer, W.H. Knox, A.C. Gossard, J.H. English: Phys. Rev. Lett. **63**, 438 (1989)

8.62 A.M. Fox, D.A.B. Miller, G. Livescu, J.E. Cunningham, W.Y. Jan: IEEE J. QE-**27**, 2281 (1991)

8.63 A.M. Fox, D.A.B. Miller, G. Livescu, J.E. Cunningham, W.Y. Jan: In *Picosecond Electronics and Optoelectronics*, ed. by T.C.L.G. Sollner, J. Shah, OSA Proc. Picosec. Electron. Optoelectron., Vol.9 (Optical Society of America, Washington, DC 1991) p.210

8.64 A. Larsson, P.A. Andrekson, S.T. Eng, A. Yariv: IEEE J. QE-**24**, 787 (1988)

8.65 K. Hess, B.A. Bojak, N. Holonyak, Jr., R. Chin, P.D. Dapkus: Solid State Electron. **23**, 585 (1980)

8.66 S.V. Kozyrev, Y. Shik: Sov. Phys. – Semicond. **19**, 1024 (1985)

8.67 M. Babiker, B.K. Ridley: Superlattices and Microstructures **2**, 287 (1986)

8.68 J.A. Brum, G. Bastard: Phys. Rev. B **33**, 1420 (1986)

8.69 B. Deveaud, A. Chomette, D. Morris, A. Regreny: Solid State Commun. **85**, 367 (1993)

8.70 P.W.M. Blom, J. Claes, J.E.M. Haverkort, J.H. Wolter: Opt. Quantum Electron. **26**, S667 (1994)

8.71 P.J. van Hall, P.W.M. Blom: Superlattices and Microstructures **13**, 329 (1993)

8.72 M. Babiker, A. Ghosal, B.K. Ridley: Superlattices and Microstructures **5**, 133 (1989)

8.73 P.W.M. Blom, R.F. Mols, J.E.M. Haverkort, M.R. Leys, J.H. Wolter: ECOC'90, 16th Europ. Conf. on Opt. Commun. (1990) p.3

8.74 P.W.M. Blom, J.E.M. Haverkort, J.H. Wolter: Appl. Phys. Lett. **58**, 2767 (1991)

8.75 S. Morin, B. Deveaud, F. Clerot, K. Fujiwara, K. Mitsunaga: IEEE J. QE-**27**, 1669 (1991)

8.76 P.W.M. Blom, C. Smit, J.E.M. Haverkort, J.H. Wolter: Phys. Rev. B **47**, 2072 (1993)

8.77 J.E.M. Haverkort, P.W.M. Blom, J.H. Wolter: SPIE Proc. **1985**, 705 (1993)

8.78 D. Morris, B. Deveaud, A. Regreny, P. Auvray: Phys. Rev. B **47**, 6819 (1993)

8.79 B. Deveaud, D. Morris, A. Regreny, M.R.X. Barros, P.C. Becker, J.M. Gerard: Opt. Quantum Electron. **26**, S679 (1994)

8.80 M.R.X. Barros, P.C. Becker, D. Morris, B. Deveaud, A. Regreny, F.A. Beisser: Phys. Rev. B **47**, 10951 (1993)

8.81 J. Feldmann, G. Peter, E.O. Göbel, K. Leo, H.-J. Polland, K. Ploog, K. Fujiwara, T. Nakayama: Appl. Phys. Lett. **51**, 226 (1987)

8.82 D.J. Westland, D. Mihailovic, J.F. Ryan, M.D. Scott: Surf. Sci. **196**, 399 (1988)

8.83 D. Bimberg, J. Christen, A. Steckenborn, G. Weimann, W. Schlapp: J. Lumin. **30**, 562 (1985)

8.84 D.J. Westland, D. Mihailovic, J.F. Ryan, M.D. Scott: Appl. Phys. Lett. **51**, 590 (1987)

8.85 B. Deveaud, J. Shah, T.C. Damen, W.T. Tsang: Appl. Phys. Lett. **52**, 1886 (1988)

8.86 R. Kersting, X.Q. Zhou, K. Wolter, D. Grutzmacher, H. Kurz: Superlattices and Microstructures **7**, 345 (1990)

8.87 Y. Shiraki, S. Fukatsu, Y. Katayama, R. Ito, A. Fujiwara, N. Ogasawara: In *Proc. 20th Int'l Conf. on the Physics of Semiconductors*, ed. by E.M. Anastassakis, J.D. Joannopoulos (World Scientific, Singapore 1990) pp. 1573-1576

8.88 I. Suemune: IEEE J. QE-**27**, 1149 (1991)

8.89 R. Nagarajan, T. Fukushima, S.W. Corzine, J.E. Bowers: Appl. Phys. Lett. **59**, 1835 (1991)

8.90 R. Nagarajan: In *Ultrafast Electronics and Optoelectronics*, ed. by J. Shah, U. Mishra, OSA Proc. Ultrafast Electron. Optoelectron., Vol. 14 (Optical Society of America, Washington, DC 1993) p. 13-16

8.91 H. Sakaki, H. Noge (eds.): *Nanostructures and Quantum Effects*, Springer Ser. Mater. Sci., Vol. 30 (Springer, Berlin, Heidelberg 1994)

8.92 S. Weiss, J.M. Wiesenfeld, D.S. Chemla, G. Raybon, G. Sucha, M. Wegener, G. Eisenstein, C.A. Burrus, A.G. Dentai, U. Koren, B.I. Miller, H. Temkin, R.A. Logan, T. Tanban-Ek: Appl. Phys. Lett. **60**, 9 (1992)

8.93 J.E.M. Haverkort, P.W.M. Blom, P.J. van Hall, J. Claes, J.H. Wolter: Phys. Status Solidi (b) **188**, 139 (1995)

Chapter 9

9.1 H. Haug, A.P. Jauho: *Quantum Kinetics in Transport and Optics of Semiconductors*, 2nd edn., Springer Ser. Solid-State Sci., Vol. 123 (Springer, Berlin, Heidelberg, 1998)

9.2 S. Mukamel: *Principles of Nonlinear Optical Spectroscopy* (Oxford Univ. Press, New York 1995)

9.3 D.S. Chemla: In *Nonlinear Optics in Semiconductors*, ed. by E. Garmire, A.R. Kost (Academic, San Diego 1998)

9.4 P.M. Petroff, A.C. Gossard; W. Wiegmann: Appl. Phys. Lett. **45**, 620 (1984)

9.5 B. Etienne. F. Laurelle: J. Cryst. Growth **127**, 1056 (1993)

9.6 R. Nötzel, N.N. Ledentsov, L. Däweritz, M. Hohnstein, K. Ploog: Phys. Rev. Lett. **67**, 3812 (1991)

9.7 E. Kapon, D.M. Hwang, R. Bhat: Phys. Rev. Lett. **63**, 430 (1989)

9.8 L. Pfeiffer, H.L. Stormer, K.W. Baldwin, K.W. West, A.R. Goni, A. Pinczuk, R.C. Ashoori, M. Dignam, W. Wegscheider: J. Cryst. Growth **127**, 849 (1993)

9.9 S. Guha, A. Madhukar, K.C. Rajkumar: Appl. Phys. Lett. **57**, 2110 (1990)

9.10 D. Leonard, M. Krishnamurthy, C.M. Reaves, S. DenBaars, P.M. Petroff: Appl. Phys. Lett. **63**, 3203 (1993)

9.11 R. Nötzel, J. Temmyo, T. Tamamura: Nature **369**, 131 (1994)

9.12 J. Oshinowo, M. Nishioka, S. Ishida, Y. Arakawa: Appl. Phys. Lett. **65**, 1421 (1994)

9.13 M. Grundmann, J. Christen, N.N. Ledentsov, J. Bohrer, D. Bimberg, S.S. Ruvimov, P. Werner, U. Richter, U. Gösele, J. Heydenreich, V.M. Ustinov, A.Y. Egorov, E.E. Zhukov, P.S. Kop′eV, Z. Alferov: Phys. Rev. Lett. **74**, 4043 (1995)

9.14 Y. Arakawa, M. Nishioka, H. Nakayama, M. Kitamura: IEICE Trans. Electron. E **79**-C, 1487 (1996)

9.15 V. Berger, G. Vermeire, P. Demeester, C. Weisbuch: Superlattices, Microstructures **20**, 125 (1996)

9.16 M. Grundmann, J. Christen, D. Bimberg, E. Kapon: J. Nonlin. Opt. Phys. Mater. **4**, 99 (1995)

9.17 U. Woggon: *Optical Properties of Semiconductor Quantum Dots*. Springer Tracts Mod. Phys. Vol. 136 (Springer, Berlin, Heidelberg 1997)

9.18 D.J. Eaglesham, M. Cerrulo: Phys. Rev. Lett. **64**, 1943 (1990)

9.19 G. Schedelbeck, W. Wegscheider, M. Bichler, G. Abstreiter: Science **278**, 1792 (1997)

9.20 W. Wegscheider, G. Schedelbeck, G. Abstreiter, M. Rother, M. Bichler: Phys. Rev. Lett. **79**, 1917 (1997)

9.21 W. Wegscheider, G. Schedelbeck, M. Bichler, G. Abstreiter: Phys. Status Solidi (a) **164**, 601 (1997)

9.22 M. Grundmann, D. Bimberg: Phys. Rev. B **55**, 4054 (1997)

9.23 A. Forchel, P. Ils, K. H. Wang, O. Schilling, R. Steffen, J. Oshinowo: Microelectronic Eng. **32**, 317 (1996)

9.24 K. Kash, R. Bhat, D.D. Mahoney, P.S.D Lin, A. Scherer, J.M. Worlock, B.P. Van der Gaag, M. Koza, P. Grabbe: Appl. Phys. Lett. **55**, 681 (1989)

9.25 H. Lipsanen, M. Sopanen, J. Ahopelto: Phys. Rev. B **51**, 13868 (1995)

9.26 H.F. Hess, E. Betzig, T.D. Harris, L.N. Pfeiffer, K.W. West: Science **264**, 1740 (1994)

9.27 D. Gammon, E.S. Snow, B.V. Shanabrook, D.S. Katzer, D. Park: Science **273**, 87 (1996)

9.28 D. Gammon, E.S. Snow, B.V. Shanabrook, D.S. Katzer, D. Park: Phys. Rev. Lett. **76**, 3005 (1996)

9.29 Y. Arakawa, H. Sakaki: Appl. Phys. Lett. **40**, 939 (1982)

9.30 T. Ogawa, T. Takagahara: Phys. Rev. B **43**, 14325 (1991)

9.31 F. Rossi, E. Molinari: Phys. Rev. Lett. **76**, 3642 (1996)

9.32 Z. Alferov, N.Y. Gordeev, S.V. Zaitsev, K.L.V. Kochnev, V.V. Komin, I.L. Krestnikov, N.N. Ledentsov, A.V. Lunev, M.V. Maksimov, S.S. Ruvimov, A.V. Sakharov, A.V. Tsapui′nikov, Y. Shenyakov, D. Bimberg: Semiconductors **30**, 197 (1996)

9.33 D. Bimberg, N.N. Ledentsov, M. Grundmann, N. Kirstaedter, O.G. Schmidt, M.H. Mao, V.M. Ustinov, A.Y. Egorov, A.E. Zhukov, P.S. Kopev, Z. Alferov, S.S. Ruvimov, U. Gösele, J. Heydenreich: Phys. Status Solidi (b) **194**, 159 (1996)

9.34 D. Bimberg, N. Kirstaedter, N.N. Ledentsov, Zh.I. Alferov, P.S. Kop′ev, V.M. Ustinov: IEEE J. Sel. Top. Quant. Electron. **3**, 196 (1997)

9.35 Yu.M. Shernyakov, A.Yu. Egorov, A.E. Zhukov, S.V. Zaitsev, A.R. Kovsh, I.L. Krestnikov, A.V. Lunev, N.N. Ledentsov, M.V. Maksimov, A.V. Sakharov, V.M. Ustinov, C. Chen, P.S. Kop′ev, Z.I. Alferov, D. Bimberg: Tech. Phys. Lett. **23**, 149 (1997)

9.36 S.V. Zaitsev, N.Y. Gordeev, V.I. Kopchatov, A.M. Georgievskii, V.M. Ustinov, A.E. Zhukov, A.Y. Egorov, A.R. Kovsh, N.N. Ledentsov, P.S. Kop′ev, Z.I. Alferov, D. Bimberg: Semiconductors **31**, 947 (1997)

9.37 M.V. Maximov, A.Y. Shernyakov, A.F. Tsatsulnikov, A.V. Lunev, A.V. Sakharov, V.M. Ustinov, A.Y. Egorov, A.E. Zhukov, A.R. Kovsh, P.S. Kopev, L.V. Asryan, Z.I. Alferov, N.N. Ledentsov, D. Bimberg, A.O. Kosogov, P. Werner: J. Appl. Phys. **83**, 5561 (1998)

9.38 V.M. Ustinov, A.E. Zhukov, A.Y. Egorov, A.R. Kovsh, S.V. Zaitsev, N.Y. Gordeev, V.I. Kopchatov, N.N. Ledentsov, A.F. Tsatsulnikov, B.V. Volovik, P.S. Kopev, Z.I. Alferov, S.S. Ruvimov, Z. Liliental-Weber, D. Bimberg: Electron. Lett. **34**, 670 (1998)

9.39 A.E. Zhukov, V.M. Ustinov, A.Y. Egorov, A.R. Kovsh, A.F. Tratsul'nikov, M.V. Maximov, N.N. Ledentsov, S.V. Zaitsev, N.Y. Gordeev, V.I. Kopchatov, Y.M. Shernyakov, P.S. Kop'ev, D.Bimberg, Z.I. Alferov: J. Electron. Mater. **27**, 106 (1998)

9.40 Y. Yamamoto: Jpn. J. Appl. Phys. Lett. **30**, L2039 (1991)

9.41 J.L. Oudar, R. Kuszelewicz, B. Sfez, D. Pella, R. Azoulay: Superlattices, Microstructures **12**, 89 (1992)

9.42 M.S. Ü nlü, K. Kishino, H.J. Liaw, H. Morkoc: J. Appl. Phys. **71**, 4049 (1992)

9.43 Y. Yamamoto, S. Machida, G. Björk: Opt. Quant. Electron. **24**, S215 (1992)

9.44 H. Yokoyama: Science **256**, 66 (1992)

9.45 K. Nishioka, K. Tanaka, I. Nakamura, Y. Lee, M. Yamanishi: Appl. Phys. Lett. **63**, 2944 (1993)

9.46 R.P. Stanley, R. Houdrι, U. Oersterle, M. Ilegems, C. Weisbuch: Phys. Rev. A **48**, 2246 (1993)

9.47 Y. Yamamoto, G. Björk, H. Heitmann, R. Horowicz: In *Optics of Semiconductor Nanostructures*, ed. by H. Henneberger, S. Schmitt-Rink, E.O. Göbel (Akademie, Berlin 993)

9.48 H. Yokoyama, S.D. Brorson, E.P. Ippen, K. Nishi, T. Anan, M. Suzuki, Y. Nambu: In *Semiconductor Interfaces, Microstructures, Devices: Properties, Applications*, ed. by Z.C. Feng (IoP, Bristol, UK 1993)

9.49 L.C. Andreani, V. Savona, P. Schwendimann, A. Quattropani: Superlattices, Microstructures **15**, 453 (1994)

9.50 D.S. Citrin: IEEE J. QE-**30**, 997 (1994)

9.51 R. Houdrι, C. Weisbuch, R.P. Stanley, U. Oesterle, P. Pellandini, M. Ilegems: Phys. Rev. Lett. **73**, 2043 (1994)

9.52 V. Savona, Z. Hradil, A. Quattropani, P. Schwendimann: Phys. Rev. B **49**, 8774 (1994)

9.53 R.E. Slusher, C. Weisbuch: Sol. State Commun. **92**, 149 (1994)

9.54 R.P. Stanley, R. Houdrι, U. Oesterle, M. Ilegems, C. Weisbuch: Appl. Phys. Lett. **65**, 2093 (1994)

9.55 J.H. Collet, R. Buhleier, J.O. White: J. Opt. Soc. Am. B **12**, 2439 (1995)

9.56 R. Houdré, J.L. Gibernon, P. Pellandini, R.P. Stanley, U. Oesterle, C. Weisbuch, J, O'Gorman, B. Roycroft, M. Ilegems: Phys. Rev. B **52**, 7810 (1995)

9.57 T.B. Norris: In *Confined Electrons, Photons: New Physics, Applications*, ed. by E. Burstein, C. Weisbuch (Plenum, New York 1995)

9.58 R.J. Ram, D.I. Babid, Y.A. York, J.E. Bowers: IEEE J. QE-**31**, 399 (1995)

9.59 V. Savona, F. Tassone: Sol. State Commun. **95**, 673 (1995)

9.60 V. Savona, L.C. Andreani, P. Schwendimann, A. Quattropania: Sol. State Commun. **93**, 733 (1995)

9.61 H. Wang, J. Shah, T.C. Damen, W.Y. Jan, J.E. Cunningham, M.H. Hong: Phys. Rev. B **51**, 14713 (1995)

9.62 H. Yokoyama, K. Ujihara (eds.): *Spontaneous Emission, Laser Oscillation in Microcavities* (CRC, Boca Raton, FL 1995)

9.63 M. Baba, K. Iga: In *Spontaneous Emission, Laser Oscillations in Microcavities*, ed. by H. Yokoyama, K. Ujihara (CRC, Boca Raton, FL 1995)

9.64 C. Weisbuch, R. Houdrι, R. P. Stanley: In *Spontaneous Emission, Laser Oscillations in Microcavities*, ed. by H. Yokoyama, K. Ujihara (CRC, Boca Raton, FL 1995)

9.65 H. Yokoyama: In *Spontaneous Emission, Laser Oscillations in Microcavities*, ed. by H. Yokoyama, K. Ujihara, (CRC, Boca Raton, FL 1995) Chap.8

9.66 H. Yokoyama: In *Spontaneous Emission, Laser Oscillations in Microcavities*, ed. by H. Yokoyama, K. Ujihara (CRC, Boca Raton, FL 1995) Chap.9

9.67 R. Buhleier, V. Bardinal, J. H. Collet, C. Fontaine, M. Hübner, J. Kuhl: Appl. Phys. Lett. **69**, 2240 (1996)

9.68 D.S. Citrin: Phys. Rev. Lett. **77**, 4596 (1996)

9.69 D.S. Citrin, T.B. Norris: IEEE J. Sel. Top. Quant. Electron. **2**, 401 (1996)

9.70 D.S. Citrin, M. Yamanishi, Y. Kadoya: IEEE J. Sel. Top. Quant. Electron. **2**, 720 (1996)

9.71 F. Jahnke, M. Kira, S.W. Koch: Phys. Rev. Lett. **77**, 5257 (1996)

9.72 E.K. Lindmark, T.R. Nelson Jr., G. Khitrova, H.M. Gibbs, A.V. Kavokin, M.A. Kaliteevskii: Opt. Lett. **21**, 994 (1996)

9.73 P. Michler, M. Hilpert, H.C. Schneider, F. Jahnke, S.W. Koch, W.W. Ruhle: *Physics of Semiconductors*, 23rd Int'l Conf., ed. by M. Scheffler, R. Zimmermann (World Scientific, Singapore 1996) p.3095

9.74 T.B. Norris, J.K. Rhee, R. Lai, D.S. Citrin, M. Nishioka, Y. Arakawa: Progr. Crys. Growth, Charac. Mater. **33**, 155 (1996)

9.75 J. Rarity, C. Weisbuch (eds.): *Microcavities, Photonic Bandgaps: Physics, Applications* (Kluwer, Dordrecht, Netherlands 1996)

9.76 V. Savona, F. Tassone, C. Piermarocchi, A. Quattropani, P. Schwendimann: Phys. Rev. B **53**, 13051 (1996)

9.77 R.P. Stanley, R. Houdrι, C. Weisbuch, U. Oesterle, M. Ilegems: Phys. Rev. B **53**, 10995 (1996)

9.78 F. Tassone, C. Piermarocchi, V. Savona, A. Quattropani, P. Schwendimann: Phys. Rev. B **53**, R7642 (1996)

9.79 P.R. Villeneuve, S. Fan, J.D. Joannopoulos: Phys. Rev. B **54**, 7837 (1996)

9.80 C. Weisbuch: Phys. Scripta T **68**, 102 (1996)

9.81 C. Weisbuch, R. Houdrι, R.P. Stanley: Phys. Scripta T **66**, 121 (1996)

9.82 D.M. Whittaker, P. Kinsler, T.A. Fisher, M.S. Skolnick, A. Armitage, A.M. Afshar, M.D. Sturge, J.S. Roberts: Phys. Rev. Lett. **77**, 4792 (1996)

9.83 J. Bloch, R. Planel, V. Thierry-Mieg, J.-M. Gèrard, D. Barrier, J.-Y. Marzin, E. Costard: Superlattices, Microstructures **22**, 371 (1997)

9.84 G. Bongiovanni, A. Mura, F. Quochi, S. Gurtler, J.L. Staehli, F. Tassone, R.P. Stanley, U. Oesterle, R. Houdré: Phys. Rev. B **55**, 7084 (1997)

9.85 H. Cao, S. Jiang, S. Machida, Y. Takiguchi, Y. Yamamoto: Appl. Phys. Lett. **71**, 1461 (1997)

9.86 A. Fainstein, B. Jusserand, V. Thierry-Mieg: Phys. Rev. Lett. **78**, 1576 (1997)

9.87 Y. Fu, M. Willander, E.L. Ivchenko, A.A. Kiselev: Phys. Rev. B **55**, 9872 (1997)

9.88 L.A. Graham, Q. Deng, D.G. Deppe, D.L. Huffaker: Appl. Phys. Lett. **70**, 814 (1997)

9.89 S. Hallstein, J.D. Berger, M. Hilpert, H.C. Schneider, W.W. Rühle, F. Jahnke, S.W. Koch, H.M. Gibbs, G. Khitrova, M. Östreich: Phys. Rev. B **56**, R7076 (1997)

9.90 Y. Hanamaki, H. Kinoshita, H. Akiyama, N. Ogasawara, Y. Shiraki: Phys. Rev. B **56**, R4379 (1997)

9.91 F. Jahnke, M. Kira, S.W. Koch: Z. Phys. B **104**, 559 (1997)

9.92 M. A. Kaliteevskii: Tech. Phys. Lett. **23**, 120 (1997)

9.93 M. Kira, F. Jahnke, S.W. Koch: Solid State Commun. **102**, 703 (1997)

9.94 M. Kuwata-Gonokami, S. Inouye, H. Suzuura, M. Shirane, R. Shimano, T. Someya, H. Sakaki: Phys. Rev. Lett. **79**, 1341 (1997)

9.95 V. Savona, C. Piermarocchi, A. Quattropani, F. Tassone, P. Schwendimann: Phys. Rev. Lett. **78**, 4470 (1997)

9.96 O. Lyngnes, J.D. Berger, J.P. Prineas, S. Park, G. Khitrova, H.M. Gibbs, F. Jahnke, M. Kira, S.W. Koch: Solid State Commun. **104**, 297 (1997)

9.97 B. Ohnesorge, M. Bayer, A. Forchel, J.P. Reithmaier, N.A. Gippius, S.G. Tikhodeev: Phys. Rev. B **56**, R4367 (1997)

9.98 P. Pellandini, R.P. Stanley, R. Houdrι, U. Oesterle, M. Ilegems, C. Weisbuch: Appl. Phys. Lett. **71**, 864 (1997)

9.99 B. Sermage, S. Long, H. Eskinazi: Superlattices, Microstructures **22**, 375 (1997)

9.100 R. Shimano , J. Lumin. **72-74**, 297 (1997)

9.101 R.P. Stanley, S. Pau, U. Oesterle, R. Houdrι, M. Ilegems: Phys. Rev. B **55**, R4867 (1997)

9.102 F. Tassone, C. Piermarocchi, V. Savona, A. Quattropani, P. Schwendimann: Phys. Rev. B **56**, 7554 (1997)

9.103 J. Tignon, R. Ferreira, J. Wainstain, C. Delalande, P. Voisin, M. Voos, R. Houdré, U. Oesterle, R.P. Stanley: Phys. Rev. B **56**, 4068 (1997)

9.104 J. Wainstain, G. Cassabois, P. Roussignol: Superlattices, Microstructures **22**, 389 (1997)

9.105 E. Yamanishi, Y. Kadoya: Superlattices, Microstructures **22**, 97 (1997)

9.106 G. Khitrova, H.M. Gibbs, F. Jahnke, M. Kira, S.W. Koch: Rev. Mod. Phys. (1998), in press

9.107 M. S. Skolnick, T.A. Fisher, D.M. Whittaker: Sem. Sci. Technol. **13**, 645 (1998)

9.108 E. Burstein, C. Weisbuch (eds.): *Confined Electrons, Photons: New Physics, Applications*, NATO ASI Ser. B, Vol.340 (Plenum, New York 1995)

9.109 P.R. Berman (ed.): *Cavity Quantum Electrodynamics* (Academic, Boston 1994)

9.110 Y. Yamamoto, R.E. Slusher: Phys. Today **46**, 66 (June 1993)

9.111 E.M. Purcell: Phys. Rev. **69**, 681 (1946)

9.112 E.T. Jaynes, F.W. Cummings: Proc. IEEE **51**, 89 (1963)

9.113 M. Brune, F. Schmidt-Kaler, A. Maali, E. Hagley, J.M. Raimond, S. Haroche: Phys. Rev. Lett. **76**, 1800 (1996)

9.114 C. Weisbuch, M. Nishioka, A. Ishikawa, Y. Arakawa: Phys. Rev. Lett. **69**, 3314 (1992)

9.115 J.J. Hopfield: Phys. Rev. **112**, 1555 (1958)

9.116 Y. Toyozawa: Prog. Theor. Phys. Suppl. **12**, 111 (1959)

9.117 L.C. Andreani, F. Bassani: Phys. Rev. B **41**, 7536 (1990)

9.118 L.C. Andreani: Phys. Status Solidi (b) **188**, 29 (1995)

9.119 J.L. Jewell, S.L. McCall, Y.H. Lee, A. Scherer, A.C. Gossard, J.H. English: Appl. Phys. Lett. **54**, 1400 (1989)

9.120 J.L. Jewell, A. Scherer, S.L. McCall, Y.H. Lee, S. Walker, J.P. Harbison, L.T. Florez: Electron. Lett. **25**, 1123 (1989)

9.121 J.L. Jewell, Y.H. Lee, A. Scherer: Opt. Eng. **29**, 210 (1990)

9.122 J.L. Jewell, J.P. Harbison, A. Scherer: Sci. Am. (Int'l Edn.) **265**, 56 (1991)

9.123 J.L. Jewell, J.P. Harbison, A. Scherer, Y.H. Lee, L.T. Florez: IEEE J. QE-**27**, 1332 (1991)

9.124 W.S. Hobson, U. Mohideen, S.J. Pearton, R.E. Slusher, F. Ren: Electron. Lett. **29**, 2199 (1993)

9.125 R.E. Slusher, A.F. Levi, U. Mohideen, S.L. McCall, S.J. Pearton, R.A. Logan: Appl. Phys. Lett. **63**, 1310 (1993)

9.126 S.L. McCall, A.F. Levi, R.E. Slusher, S.J. Pearton, R.A. Logan: Appl. Phys. Lett. **60**, 289 (1992)

9.127 U. Mohideen, R.E. Slusher, F. Johnke, S.W. Koch: Phys. Rev. Lett. **73**, 1785 (1994)

9.128 U. Mohideen, W.S. Hobson, S.J. Pearton, F. Ren, R.E. Slusher: Appl. Phys. Lett. **64**, 1911 (1994)

9.129 R.E. Slusher, S.L. McCall, U. Mohideen, A.F. Levi: SPIE Proc. **2145**, 36 (1994)

9.130 J.B. Stark, U. Mohideen, E. Betzig, R.E. Slusher: *Ultrafast Phenomena IX*, ed. by P.F. Barbara, W.H. Knox, G.A. Mourou, A.H. Zewail, Springer Ser. Chem. Phys., Vol.60 (Springer, Berlin, Heidelberg 1994) p.310

9.131 E. Yablonovitch, T.J. Gmitter, R. Bhat: Phys. Rev. Lett. **61**, 2546 (1988)

9.132 J.D. Joannopoulos, R.D. Meade, J.N. Winn: *Photonic Crystal: Molding the Flow of Light* (Princeton Univ. Press, Princeton, NJ 1995)

9.133 C.C. Cheng, V. Arbet-Engels, A. Scherer, E. Yablonovitch: Phys. Scripta T **68**, 17 (1996)

9.134 C.C. Cheng, A. Scherer, T. Rong-Chung, Y. Fainman, G. Witzgall, E. Yablonovitch: J. Vac. Sci. Tech. B **15**, 2764 (1997)

9.135 R.L. Fork, C.H. Brito Cruz, P.C. Becker, C.V. Shank: Opt. Lett. **12**, 483 (1987)

9.136 I.D. Jung, F.X. Kartner, N. Matuschek, D.H. Sutter, F. Morier-Genoud, G. Zhang, U. Keller, V. Scheuer, M. Tilsch, T. Tschudi: Opt. Lett. **22**, 1009 (1997)

9.137 A. Baltuska, Z. Wei, M.S. Pshenichnikov, D.A. Wiersma, R. Szipocs: Appl. Phys. B **65**, 175 (1997)

9.138 A. Baltuska, Z. Wei, M.S. Pshenichnikov, D.A. Wiersma: Opt. Lett. **22**, 102 (1997)

9.139 M. Nisoli, S. Stagira, S. De Silvestri, O. Svelto, S. Sartania, Z. Cheng, M. Lenzner, C. Spielmann, F. Krausz: Appl. Phys. B **65** 189 (1997)

9.140 M. Nisoli, S. De Silvestri, O. Svelto, R. Szipocs, K. Ferencz, C. Spielmann, S. Sartania, F. Krausz: Opt. Lett. **22**, 522 (1997)

9.141 A. Antonetti, F. Blasco, J.P. Chambaret, G. Cheriaux, G. darpentigny, C. Le Blan, P. Rousseau, S. Ranc, G. Rey, F. Salin: Appl. Phys. B **65**, 197 (1997)

9.142 G. Mourou: Appl. Phys. B **65**, 205 (1997)

9.143 W.P. Leemans, R.W. Schoenlein, P. Volfbeyn, A.H. Chin, T.E. Glover, P. Balling, M. Zolotorev, K.J. Kim, S. Chattopadhyay, C.V. Shank: Phys. Rev. Lett. **77**, 4182 (1996)

9.144 R.W. Schoenlein, W.P. Leemans, A.H. Chin, P. Volfbeyn, T.E. Glover, P. Balling, M. Zolotorev, K.J. Kim, S. Chattopadhyay, C.V. Shank: Science **274**, 236 (1996)

9.145 W.P. Leemans, R.W. Schoenlein, P. Volfbeyn, A.H. Chin, T.E. Glover, P. Bailing, M. Zolotorev, K.J. Kim, S. Chattopadhyay, C.V. Shank: IEEE J. QE-**33**, 1925 (1997)

9.146 L.E. Nelson, D.J. Jones, K. Tamura, H.A. Haus, E.P. Ippen: Appl. Phys. B **65**, 277 (1997)

9.147 M.E. Fermann, A. Galvanauskas, G. Sucha, D. Harter: Appl. Phys. B **65**, 259 (1997)

9.148 D.M. Mittleman, S. Hunsche, L. Boivin, M.C. Nuss: Opt. Lett. **22**, 904 (1997)

9.149 D.M. Mittleman, J. Cunningham, M.C. Nuss, M. Geva: Appl. Phys. Lett. **71**, 16 (1997)

9.150 S. Hunsche, D.M. Mittleman, M. Koch, M.C. Nuss: IEICE Trans. Electron. E **81**-C, 269 (1998)

9.151 E.A. de Souza, M.C. Nuss, W.H. Knox, D.A. Miller: Opt. Lett. **20**, 1166 (1995)

9.152 M.C. Nuss, W.H. Knox, U. Koren: Electron. Lett. **32**, 1311 (1996)

9.153 J.B. Stark, M.C. Nuss, W.H. Knox, S.T. Cundiff, L. Boivin, S.G. Grubb, D. Tipton, D. DiGiovanni, U. Koren, K. Dreyer: IEEE Photon. Tech. Lett. **9**, 1170 (1997)

9.154 J. Paye, M. Ramaswamy, J.G. Fujimoto, E.P. Ippen: Opt. Lett. **18**, 1946 (1993)

9.155 J. Paye: IEEE J. QE-**28**, 2262 (1992)

9.156 D.J. Kane, R. Trebino: Opt. Lett. **18**, 823 (1993)

9.157 D.J. Kane, R. Trebino: IEEE J. QE-**29**, 571 (1993)

9.158 R. Trebino, D.J. Kane: J. Opt. Soc. Am. A **10**, 1101 (1993)

9.159 K.W. DeLong, R. Trebino, J. Hunter, W.E. White: J. Opt. Soc. Am. B **11**, 2206 (1994)

9.160 K.W. DeLong, D.N. Fittinghoff, R. Trebino: IEEE J. QE-**32**, 1253 (1996)

9.161 K. Naganuma, K. Mogi, H. Yamada: IEEE J. QE-**25**, 1225 (1989)

9.162 A.C. Albrecht, K. Seibert, H. Kurz: Opt. Commun. **84**, 223 (1991)

9.163 J.L. A. Chilla, O.E. Martinez: IEEE J. QE-**27**, 1228 (1991)

9.164 J.-P. Foing, J.P. Likforman, M. Joffre, A. Migus: IEEE J. QE-**28**, 2285 (1992)

9.165 L. Lepetit, G. Cheriaux, M. Joffre: J. Opt. Soc. Am. B **12**, 2467 (1995)

9.166 D.N. Fittinghoff, R. Trebino, J. Bowie, J.N. Sweetser, M.A. Krumbugel, K.W. DeLong, I.A. Walmsley: Generation, Amplification,, Measurement of Ultrashort Laser Pulses III, Proc. SPIE 2701 (1996); p. 118

9.167 D.N. Fittinghoff, J.L. Bowie, J.N. Sweetser: Opt. Lett. **21**, 884 (1996)

9.168 L. Lepetit, M. Joffre: Opt. Lett. **21**, 564 (1996)

9.169 L. Lepetit, G. Cheriaux, M. Joffre: J. Nonlin. Opt. Phys. Mater. **5**, 465 (1996)
9.170 W. J. Walecki, D.N. Fittinghoff, A.L. Smirl, R. Trebino: Opt. Lett. **22**, 81 (1997)
9.171 J.J. Baumberg, A.P. Heberle, K. Köhler, A.V. Kavokin: HCIS-10, Phys. Status Solidi **204**, 9 (1997)
9.172 Q. Wu, X.-C. Zhang: Appl. Phys. Lett. **70**, 1784 (1997)
A. Leitenstorfer, S. Hunsche, J. Shah, M.C. Nuss, W.H. Knox: Appl. Phys. Lett. **74**, 1516 (1999)
9.173 W.L. Faust, C.H. Henry, R.H. Eick: Phys. Rev. **173**, 781 (1968)
9.174 A. Schmenkel, L. Banyai, H. Haug: J. Lumin. **76-77**, 134 (1998)
9.175 C. Fürst, A. Leitenstorfer, A. Laubereau, R. Zimmermann: Phys. Rev. Lett. **78**, 3733 (1997)
9.176 K. El Sayed, C.J. Stanton: Phys. Rev. B **55**, 9671 (1997)
9.177 L. Banyai, D.B.T. Thoai, E. Reitsamer: Phys. Rev. Lett. **75**, 2188 (1995)
9.178 D.B. Tran Thoai, H. Haug: Phys. Rev. B **47**, 3574 (1993)
9.179 D.B. Tran Thoai, L. Banyai, E. Reitsamer, H. Haug: Phys. Status Solidi (b) **188**, 387 (1995)
9.180 J.-P. Foing, D. Hulin, M. Joffre, J.L. Oudar, C. Tanguy, M. Combescot: Phys. Rev. Lett. **68**, 110 (1992)
9.181 M.U. Wehner, M.H. Ulm, D.S. Chemla, M. Wegener: Phys. Rev. Lett. **80**, 1992 (1998)
9.182 M.U. Wehner, M.H. Ulm, M. Wegener: Opt. Lett. **22**, 1455 (1997)
9.183 M.U. Wehner, J. Hetzler, M. Wegener: Phys. Rev. B **55**, 4031 (1997)
9.184 F.X. Camescasse, A. Alexandrou, D. Hulin, L. Banyai, D.B.T. Thoai, H. Haug: Phys. Rev. Lett. **77**, 5429 (1996)
9.185 S. Bar-Ad, P. Kner, M.V. Marquezini, D.S. Chemla, K. El Sayed: Phys. Rev. Lett. **77**, 3177 (1996)
9.186 T. Elsaesser, J. Shah, L. Rota, P. Lugli: Phys. Rev. Lett. **66**, 1757 (1991)
9.187 H. Haug, S.W. Koch: *Quantum Theory of the Optical, Electronic Properties of Semiconductors* (World Scientific, Singapore 1993)
9.188 F.C. Spano, S. Mukamel: Phys. Rev. A **40**, 5783 (1989)
9.189 S. Mukamel: In *Molecular Nonlinear Optics*, ed. by J. Zyss (Academic, San Diego, CA 1994)
9.190 M.Z. Maialle, L.J. Sham: Phys. Rev. Lett. **73**, 3310 (1994)
9.191 T. Östreich, K. Schönhammer, L. J. Sham: Phys. Rev. Lett. **74**, 4698 (1995)
9.192 V.M. Axt, A. Stahl: Z. Phys. B **93**, 195 (1994)
9.193 P. Kner, S. Bar-Ad, M.V. Marquezini, D.S. Chemla, W. Schäfer: Phys. Status Solidi (a) **164**, 957 (1997)
9.194 P. Kner, S. Bar-Ad, M.V. Marquezini, D.S. Chemla, W. Schäfer: Phys. Rev. Lett. **78**, 1319 (1997)
9.195 V.M. Axt, G. Bartels, A. Stahl: Phys. Rev. Lett. **76**, 2543 (1996)
9.196 W. Schäfer, D.S. Kim, J. Shah, T.C. Damen, J.E. Cunningham, K.W. Goossen, L.N. Pfeiffer, K. Köhler: Phys. Rev. B **53**, 16429 (1996)
9.197 M. Wegener, D.S. Chemla, S. Schmitt-Rink, W. Schäfer: Phys. Rev. A **42**, 5675 (1990)
9.198 M. Dignam, J.E. Sipe: Phys. Rev. Lett. **64**, 1797 (1990)
9.199 M. Dignam, J.E. Sipe: Phys. Rev. B **43**, 4097 (1991)

9.200 T. Meier, G. von Plessen, P. Thomas, S.W. Koch: Phys. Rev. Lett. **73**, 902 (1994)
9.201 M. Dignam, J.E. Sipe, J. Shah: Phys. Rev. B **49**, 10502 (1994)
9.202 P. Haring Bolivar, F. Wolter, A. Muller, H.G. Roskos, H. Kurz, K. Köhler: Phys. Rev. Lett. **78**, 2232 (1997)
9.203 P.C.M. Planken, I. Brener, M.C. Nuss, M.S.C. Luo, S.L. Chuang: Phys. Rev. B **48**, 4903 (1993)
9.204 D.S. Citrin, T.B. Norris: IEEE J. QE-**33**, 404 (1997)
9.205 W. Pötz: Phys. Rev. Lett. **79**, 3262 (1997)
9.206 W. Pötz: Appl. Phys. Lett. **71**, 395 (1997)
9.207 A.P. Heberle, J.J. Baumberg, E. Binder, T. Kuhn, K. Köhler, K.H. Ploog: IEEE J. Sel. Top. Quant. Electron. **2**, 769 (1996)
9.208 A.P. Heberle, J.J. Baumberg, K. Köhler: Phys. Rev. Lett. **75**, 2598 (1995)
9.209 J.J. Baumberg, A.P. Heberle, K. Köhler, K. Ploog: J. Opt. Soc. Am. B **13**, 1246 (1996)
9.210 A. Vinattieri, J. Shah, T. C. Damen, D.S. Kim, L.N. Pfeiffer, M.Z. Maialle, L.J. Sham: Phys. Rev. B **50**, 10868 (1994)
9.211 T. Amand, D. Robart, X. Marie, M. Brousseau, P. Le Jeune, J. Barrau: Phys. Rev. B **55**, 9880 (1997)
9.212 F. Meier, B.P. Zakharchenya (eds.): *Optical Orientation* (North-Holland, Amsterdam 1984)
9.213 X. Marie, P. Le Jeune, T. Amand, M. Brousseau, J. Berrau, M. Paillard, R. Planei: Phys. Rev. Lett. **79**, 3222 (1997)
9.214 H. Wang, J. Shah, T.C. Damen, L.N. Pfeiffer: Phys. Rev. Lett. **74**, 3061 (1995)
9.215 E. Dupont, P. B. Corkum, H.C. Liu, M. Buchanan, Z.R. Wasilewski: Phys. Rev. Lett. **74**, 3592 (1995)
9.216 R. Atanasov, A. Hache, J.L.P. Hughes, H.M. van Driel, J.E. Sipe: Phys. Rev. Lett. **76**, 1703 (1996)
9.217 A. Hache, Y. Kostoulas, R. Atanasov, J.L.P. Hughes, J.E. Sipe, H.M. van Driel: Phys. Rev. Lett. **78**, 306 (1997)
9.218 U. Fano: Phys. Rev. **124**, 1866 (1961)
9.219 X. Chen, W.J. Walecki, O. Buccafusca, D.N. Fittinghoff, A.L. Smirl: Phys. Rev. B **56**, 9738 (1997)
9.220 J.P. Likforman, M. Joffre, V. Thierry-Mieg: Opt. Lett. **22**, 1104 (1997)
9.221 A.L. Smirl, W.J. Walecki, X. Chen, O. Buccafusca: Phys. Status Solidi (b) **204**, 16 (1997)
9.222 O. Buccafusca, X. Chen, W.J. Walecki, A.L. Smirl: J. Opt. Soc. Am. B **15**, 1218 (1998)
9.223 K. W. DeLong, R. Trebino, D.J. Kane: J. Opt. Soc. Am. B **11**, 1595 (1995)
9.224 S. Schmitt-Rink, D. Bennhardt, V. Heuckeroth, P. Thomas, P.H. Bolivar, G. Maidorn, H.J. Bakker, K. Leo, D.S. Kim, J. Shah, K. Köhler: Phys. Rev. B **46**, 10460 (1992)
9.225 J. Kuhl, E. J. Mayer, G. Smith, R. Eccleston, D. Bennhardt, P. Thomas, K. Bott, O. Heller: In *Coherent Optical Interactions in Solids*, ed. by R.T. Phillips (Plenum, New York 1994)
9.226 D.G. Steel, H. Wang, M. Jiang, K.B. Ferrio, S.T. Cundiff: In *Coherent Optical Interactions in Solids*, ed. by R. T. Phillips (Plenum, New York 1994)

9.227 S. Patkar, A.E. Paul, W. Sha, J.A. Bolger, A.L. Smirl: Phys. Rev. B **51**, 10789 (1995)

9.228 J.A. Bolger, A.E. Paul, A.L. Smirl: Phys. Rev. B **54**, 11666 (1996)

9.229 A.E. Paul, J.A. Bolger, A.L. Smirl, J.G. Pellegrino: J. Opt. Soc. Am. B **13**, 1016 (1996)

9.230 D.S. Kim, J. Shah, T.C. Damen, W. Schäfer, F. Jahnke, S. Schmitt-Rink, K. Köhler: Phys. Rev. Lett. **69**, 2725 (1992)

9.231 J.B. Stark, W.H. Knox, D.S. Chemla: In *Optics of Semiconductor Nanostructures*, ed. by H. Henneberger, S. Schmitt-Rink, E.O. Göbel (Akademie, Berlin 1993)

9.232 J. Feldmann, T. Meier, G. von Plessen, M. Koch, E.O. Göbel, P. Thomas, G. Bacher, C. Hartmann, H. Schweizer, W. Schäfer, N. Nickel: Phys. Rev. Lett. **70**, 3027 (1993)

9.233 J.J. Hopfield, P.J. Dean, D.G. Thomas: Phys. Rev. **158**, 748 (1967)

9.234 H. Kleinpoppen, M.R.C. McDowell (eds.): *Electron, Photon Interaction with Atoms* (Plenum, New York 1976)

9.235 D.Y. Oberli, G. Böhm, G. Weimann, J.A. Brum: Phys. Rev. B **49**, 5757 (1994)

9.236 S. Glutsch, U. Siegner, M.-A. Mycek, D.S. Chemla: Phys. Rev. B **50**, 17009 (1994)

9.237 U. Siegner, M.-A. Mycek, S. Glutsch, D.S. Chemla: Phys. Rev. Lett. **74**, 470 (1995)

9.238 S. Bar-Ad, P. Kner, M.V. Marquezini, S. Mukamel, D.S. Chemla: Phys. Rev. Lett. **78**, 1363 (1997)

9.239 S.T. Cundiff, M. Koch, W.H. Knox, J. Shah, W. Stolz: Phys. Rev. Lett. **77**, 1107 (1996)

9.240 D.S. Kim, J. Shah, J.E. Cunningham, T.C. Damen, W. Schäfer, M. Hartmann, S. Schmitt-Rink: Phys. Rev. Lett. **68**, 1006 (1992)

9.241 M.U. Wehner, D. Steinbach, M. Wegener, T. Marschner, W. Stolz: J. Opt. Soc. Am. B **13**, 977 (1996)

9.242 M.U. Wehner, D. Steinbach, M. Wegener: Phys. Rev. B **54**, R5211 (1996)

9.243 D. Birkedal, V.G. Lyssenko, J.M. Hvam, K. El Sayed: Phys. Rev. B **54**, R14250 (1996)

9.244 H. Wang, K.B. Ferrio, D.G. Steel, Y.Z. Hu, R. Binder, S.W. Koch: Phys. Rev. Lett. **71**, 1261 (1993)

9.245 D.S. Kim, J.Y. Sohn, J.S. Yahng, Y.H. Ahm, K.J. Yee, D.S. Yee, Y.D. Jho, S.C. Hohng, D.H. Kim, W.S. Kim, J.C. Woo, T. Meier, S.W. Koch, D.H. Whoo, E.K. Kim, S.H. Kim, C.S. Kim: Phys. Rev. Lett. **80**, 4803 (1998)

9.246 V.G. Lyssenko, G. Valusis, F. Loser, T. Hasche, K. Leo, M.M. Dignam, K. Köhler: Phys. Rev. Lett. **79**, 301 (1997)

9.247 H. Kurz, H.G. Roskos, T. Dekorsy, K. Köhler: Phil. Trans. Royal Soc. (London) A **354**, 2295 (1996)

9.248 P. Leisching, T. Dekorsy, H.J. Bakker, H. Kurz, K. Köhler: Phys. Rev. B **51**, 18015 (1995)

9.249 P. Leisching, T. Dekorsy, H. J. Bakker, H.G. Roskos, H. Kurz, K. Köhler: J. Opt. Soc. Am. B **13**, 1009 (1996)

9.250 R. Martini, G. Klose, H.G. Roskos, H. Kurz, H.T. Grahn, R. Hey: Phys. Rev. B **54**, R14325 (1996)

9.251 G.C. Cho, T. Dekorsy, H.J. Bakker, H. Kurz, A. Kohl, B. Opitz: Phys. Rev. B **54**, 4420 (1996)
9.252 E. Peik, M. Ben Dahan, I. Bouchoule, Y. Castin, C. Salomon: Phys. Rev. A **55**, 2989 (1997)
9.253 C.F. Bharucha, K.W. Madison, P.R. Morrow, R.K. Willardson, B. Sundaram, M.G. Raizen: Phys. Rev. A **55**, R587 (1997)
9.254 S.K. Wilkinson, C.F. Bharucha, K.W. Madison, Q. Niu, M.G. Raizen: Phys. Rev. Lett. **76**, 4512 (1996)
9.255 M. Ben Dahan, E. Peik, J. Reichel, Y. Castin, C. Salomon: Phys. Rev. Lett. **76**, 4508 (1996)
9.256 M.G. Raizen, C. Salomon, Q. Niu: Phys. Today **50**, 30 (July 1997)
9.257 T. Dekorsy, H. Auer, C. Waschke, H.J. Bakker, H.G. Roskos, H. Kurz, V. Wagner, P. Grosse: Phys. Rev. Lett. **74**, 738 (1995)
9.258 T. Dekorsy, H. Auer, H.J. Bakker, H.G. Roskos, H. Kurz: Phys. Rev. B **53**, 4005 (1996)
9.259 G.A. Garrett, T.F. Albrecht, J.F. Whitaker, R. Merlin: Phys. Rev. Lett. **77**, 3661 (1996)
9.260 R. Merlin: Solid State Commun. **102**, 207 (1997)
9.261 J.J. Baumberg, D.A. Williams, K. Köhler: Phys. Rev. Lett. **78**, 3358 (1997)
9.262 T.D. Krauss, F.W. Wise: Phys. Rev. Lett. **79**, 5102 (1998)
9.263 W. Sha, A. L. Smirl, W. F. Tseng: Phys. Rev. Lett. 74, 4273 (1995)
9.264 R. Kersting, K. Unterrainer, G. Strasser, H.F. Kauffmann, E. Gornik: Phys. Rev. Lett. **79**, 3038 (1997)
9.265 A. Pinczuk, J. Shah, P.A. Wolff: Phys. Rev. Lett. **47**, 1487 (1981)
9.266 G.C. Cho, H.J. Bakker, T. Dekorsy, H. Kurz: Phys. Rev. B **53**, 6904 (1996)
9.267 T. Dekorsy, A.M.T. Kim, G.C. Cho, S. Hunsche, H.J. Bakker, H. Kurz, S.L. Chuang, K. Köhler: Phys. Rev. Lett. **77**, 3045 (1996)
9.268 K. Leo, J. Shah, T.C. Damen, A. Schulze, T. Meier, S. Schmitt-Rink, P. Thomas, E.O. Göbel, S.L. Chuang, M.S.C. Luo, W. Schäfer, K. Köhler, P. Ganser: IEEE J. QE-**28**, 2498 (1992)
9.269 K. Leo, E.O. Göbel, T.C. Damen, J. Shah, S. Schmitt-Rink, W. Schäfer, J.F. Muller, K. Köhler, P. Ganzer: Phys. Rev. B **44**, 5726 (1991)
9.270 E.O. Göbel, K. Leo, T.C. Damen, J. Shah, S. Schmitt-Rink, W. Schäfer, J.F. Muller, K. Köhler: Phys. Rev. Lett. **64**, 1801 (1990)
9.271 M. Joschko, M. Woerner, T. Elsaesser, E. Binder, T. Kuhn, Y. Hey, H. Kostial, K. Ploog: Phys. Rev. Lett. **78**, 737 (1997)
9.272 D.S. Kim, J. Shah, T.C. Damen, L.N. Pfeiffer, W. Schäfer: Phys. Rev. B **50**, 5775 (1994)
9.273 V.I. Klimov, S. Hunsche, H. Kurz: Phys. Rev. B **50**, 8110 (1994)
9.274 E.J. Mayer, G.O. Smith, V. Heuckeroth, J. Kuhl, K. Bott, A. Schulze, T. Meier, D. Bennhardt, S.W. Koch, P. Thomas, R. Hey, K. Ploog: Phys. Rev. B **50**, 14730 (1994)
9.275 S.S. Prabhu, A.S. Vengurlekar, S.K. Roy, J. Shah: Phys. Rev. B **50**, 18098 (1994)
9.276 G.O. Smith, E.J. Mayer, J. Kuhl, K. Ploog: Solid State Commun. **92**, 325 (1994)
9.277 H. Wang, J. Shah, T.C. Damen, L.N. Pfeiffer: Solid State Commun. **91**, 869 (1994)

9.278 E.J. Mayer, G.O. Smith, V. Heuckeroth, J. Kuhl, K. Bott, A. Schulze, T. Meier, S.W. Koch, P. Thomas, R. Hey, K. Ploog: Phys. Rev. B **51**, 10909 (1995)

9.279 J. Min, A.C. Schaefer, D.G. Steel: Phys. Rev. B **51**, 16714 (1995)

9.280 H. Wang, J. Shah, T.C. Damen, L.N. Pfeiffer: In *Ultrafast Phenomena IX*, ed. by P.F. Barbara, W.H. Knox, G.A. Mourou,, A.H. Zewail, Springer Ser. Chem. Phys., Vol. 60 (Springer, Berlin, Heidelberg 1994) p.362

9.281 H. Wang, J. Shah, T.C. Damen,, L.N. Pfeiffer: In *Ultrafast Phenomena IX*, ed. by P.F. Barbra, W.H. Knox, G.A. Mourou,, A.H. Zewail. Springer Ser. Chem. Phys., Vol. 60 (Springer, Berlin, Heidelberg 1994) p.395

9.282 D. Birkedal, J. Singh, V.G. Lyssenko, J. Erland, J. M. Hvam: Phys. Rev. Lett. **76**, 672 (1996)

9.283 J. Erland, D. Birkedal, V.G. Lyssenko, J.M. Hvam: J. Opt. Soc. Am. B **13**, 981 (1996)

9.284 K.B. Ferrio, D.G. Steel: Phys. Rev. B **54**, R5231 (1996)

9.285 H. Wang, J. Shah, T.C. Damen, A.L. Ivanov, H. Haug, L.N. Pfeiffer: Solid State Commun. **98**, 807 (1996)

9.286 A. Euteneuer, J. Mobius, R. Rettig, E.J. Mayer, M. Hofmann, W. Stolz, E.O. Göbel, W.W. Rühle: Phys. Rev. B **56**, R10028 (1997)

9.287 O. Heller, P. Lelong, G. Bastard: Phys. Rev. B **56**, 4702 (1997)

9.288 J. M. Hvam, D. Birkedal, V. Mizeikis, K. E. Sayed: J. Lumin. **72-74**, 25 (1997)

9.289 M. Ikezawa, Y. Masumoto, T. Takagahara, S.V. Nair: Phys. Rev. Lett. **79**, 3522 (1997)

9.290 V. Mizeikis, D. Birkedal, W. Langebein, V.G. Lyssenko, J.M. Hvam: Phys. Rev. B **55**, 5284 (1997)

9.291 V.M. Axt, A. Stahl: Z. Phys. B **93**, 205 (1994)

9.292 D.S. Citrin: Phys. Rev. B **50**, 17655 (1994)

9.293 A.L. Ivanov, H. Haug: Phys. Rev. Lett. **74**, 438 (1995)

9.294 A.L. Ivanov, H. Wang, J. Shah, T.C. Damen, L.V. Keldysh, H. Haug, L.N. Pfeiffer: Phys. Rev. B **56**, 3941 (1997)

9.295 D.A. Kleinman: Phys. Rev. B **28**, 871 (1983)

9.296 J. Singh, D. Birkedal, V.G. Lyssenko, J.M. Hvam: Phys. Rev. B **53**, 15909 (1996)

9.297 Y. Zhu, D.J. Gauthier, S.E. Morin, Q. Wu, H.J. Carmichael, T.S. Mossberg: Phys. Rev. Lett. **64**, 2499 (1990)

9.298 H. Wang, J. Shah, T.C. Damen, L.N. Pfeiffer, J.E. Cunningham: Phys. Status Solidi (b) **188**, 381 (1995)

9.299 P.V. Kelkar, V.G. Kozlov, A.V. Nurmikko, C.C. Chu, J. Han, R.L. Gunshor: Phys. Rev. B **56**, 7564 (1997)

9.300 M. Koch, J. Shah, T. Meier: Phys. Rev. B **57**, R2049 (1998)

9.301 M. Koch, J. Feldmann, G. von Plessen, E.O. Göbel, P. Thomas, K. Köhler: Phys. Rev. Lett. **69**, 3631 (1992)

9.302 J.-K. Rhee, D.S. Citrin, T.B. Norris, Y. Arakawa, M. Nishioka: Solid State Commun. **97**, 941 (1996)

9.303 R. Houdrι, R.P. Stanley, M. Ilegems: Phys. Rev. A **53**, 2711 (1996)

9.304 S. Pau, G. Bjor, H. Cao, E. Hanamura, Y. Yamamoto: Solid State Commun. **98**, 781 (1996)

9.305 V. Savona, C. Weisbuch: Phys. Rev. B **54**, 10835 (1996)

9.306 A.V. Kavokin: Phys. Rev. B **57**, 3757 (1998)
9.307 C. Ell, J.P. Prineas, T.R. Nelson Jr., S. Park, H.M. Gibbs, G. Khitrova, C. Koch, R. Houdré: Phys. Rev. Lett. **80**, 4795 (1998)
9.308 D.M. Whittaker: Phys. Rev. Lett. **80**, 4791 (1998)
9.309 H. Cao, S. Pau, J.M. Jacobson, G. Bjork, Y. Yamamoto, A. Imamoglu: Phys. Rev. A **55**, 4632 (1997)
9.310 M. Kira, F. Jahnke, S.W. Koch, J.D. Berger, D.V. Wick, T.R. Nelson, G. Khitrova, H.M. Gibbs: Phys. Rev. Lett. **79**, 5179 (1997)
9.311 F. Quochi, G. Bongiovanni, A. Mura, J.L. Staehli, B. Deveand, R.P. Stanley, U. Oesterle, R. Houdré: Phys. Rev. Lett. **80**, 4733 (1998)
9.312 S. Jiang, M. Degenais, R.A. Morgan: Appl. Phys. Lett. **65**, 1334 (1994)
9.313 R. Buhleier, J.H. Collet, V. Bardinal, C. Fontaine, R. Legros, M. Hübner, J. Kuhl: J. Nonlin. Opt. Phys. Mater. **5**, 637 (1996)
9.314 M. Tsuchiya, J. Shah, T.C. Damen, J.E. Cunningham: Appl. Phys. Lett. **71**, 2650 (1997)
9.315 M.S. Ünlü, S. Strite: J. Appl. Phys. **78**, 607 (1995)
9.316 I. Hayashi: In *Spontaneous Emission, Laser Oscillations in Microcavities*, ed. by H. Yokoyama, K. Ujihara (CRC, Boca Raton 1995)
9.317 D. Fröhlich, A. Kulik, B. Uebbing, A. Mysyrowicz, V. Langer, H. Stolz, W. von der Osten: Phys. Rev. Lett. **67**, 2343 (1991)
9.318 D.S. Kim, J. Shah, D.A.B. Miller, T.C. Damen, W. Schäfer, L. Pfeiffer: Phys. Rev. B **48**, 17902 (1993)
9.319 N. Wang, V. Chernyak, S. Mukamel: Phys. Rev. B **50**, 5609 (1994)
9.320 D. S. Citrin: Solid State Commun. **89**, 139 (1994)
9.321 T. Stroucken, A. Knorr, C. Anthony, A. Schulze, P. Thomas, S.W. Koch, M. Koch, S.T. Cundiff, J. Feldmann, E.O Göbel: Phys. Rev. Lett. **74**, 2391-2394 (1995)
9.322 V. Chernyak, N. Wang, S. Mukamel: Phys. Rept. **263**, 214 (1995)
9.323 T. Stroucken, A. Knorr, P. Thomas, S.W. Koch: Phys. Rev. B **53**, 2026 (1996)
9.324 T. Stroucken, S. Haas, B. Grote, S.W. Koch, M. Hübner, D. Ammerlahn, J. Kuhl: In *Festkörperprobleme / Adv. Solid State Phys.*, ed. by B. Kramer. **38**, 265-279 (Vieweg, Braunschweig 1999)
9.325 J. Kuhl, M. Hübner, D. Ammerlahn, T. Stroucken, B. Grote, S. Haas, S.W. Koch, G. Khitrova, H.M. Gibbs, R. Hey, K. Ploog: In *Festkörperprobleme / Adv. Solid State Phys.*, ed. by B. Kramer. **38**, 281-295 (Vieweg, Braunschweig 1999)
9.326 V. M. Agranovich, V.O. Dubovskii: JETP Lett. **3**, 223 (1966)
9.327 E. Hanamura: Phys. Rev. B **38**, 1228 (1988)
9.328 L. C. Andreani, F. Tassone, F. Bassani: Solid State Commun. **77**, 641 (1991)
9.329 D. S. Citrin: Solid State Commun. **84**, 281 (1992)
9.330 D. Weber, J. Feldmann, E.O. Göbel, T. Stroucken, A. Knorr, S.W. Koch, D.S. Citrin, K. Köhler: J. Opt. Soc. Am. B **13**, 1241 (1996)
9.331 D.S. Kim, J. Shah, D.A.B. Miller, T.C. Damen, A. Vinattieri, W. Schäfer, L.N. Pfeiffer: Phys. Rev. B **50**, 18240 (1994)
9.332 M. Hübner, J. Kuhl, T. Stroucken, A. Knorr, S.W. Koch, R. Hey, K. Ploog: Phys. Rev. Lett. **76**, 4199 (1996)

9.333 S. Haas, T. Stroucken, M. Hübner, J. Kuhl, B. Grote, A. Knorr, F. Jahnke, S.W. Koch, R. hey, K. Ploog: Phys. Rev. B **57**, 14860 (1998)
9.334 M. Hübner, J. Kuhl, B. Grote, T. Stroucken, S.W. Koch, R. Hey, K. Ploog: Unpublished (1998)
9.335 M. Hübner, J. Kuhl, B. Grote, T. Strouken, S. Haas, A. Knorr, S.W. Koch, G. Khitrova, H.M. Gibbs: Phys. Stat. Sol. (b) **206**, 333 (1998)
9.336 M. Hübner, J. Kuhl, S. Haas, T. Strouken, S.W. Koch, R. Hey, K. Ploog: Solid State Commun. **105**, 105 (1998)
9.337 J.J. Baumberg, A.P. Heberle, A.V. Kavokin, M.R. Vladimirova, K. Köhler: Phys. Rev. Lett. **80**, 3567 (1998)
9.338 A.V. Kavokin, J.J. Baumberg: Phys. Rev. B **57**, R12697 (1998)
9.339 L. Schultheis, J. Kuhl, A. Honold, C.W. Tu: Phys. Rev. Lett. **57**, 1635 (1986)
9.340 L. Schultheis, A. Honold, J. Kuhl, K. Köhler, C.W. Tu: Phys. Rev. B **34**, 9027 (1986)
9.341 A. Honold, L. Schultheis, J. Kuhl, C.W. Tu: Phys. Rev. B **40**, 6442 (1989)
9.342 E.J. Mayer, J.O. White, G.O. Smith, H. Lage, D. Heitmann, K. Ploog, J. Kuhl: Phys. Rev. B **49**, 2993 (1994)
9.343 W. Braun, M. Bayer, A. Forchel, H. Zull, J.P. Reithmaier, A.I.Filin, T.L. reinecke: Phys. Rev. B **56**, 12096 (1997)
9.344 W. Braun, M. Bayer, A. Forchel, H. Zull, J.P. Reithmaier, A.I. Filin, S.N. Walck, T.L. Reinecke: Phys. Rev. B **55**, 9290 (1997)
9.345 M. Bayer, S. Walck, T. L. Reinecke, A. Forchel: Europhys. Lett. **39**, 453 (1997)
9.346 K. Brunner, U. Bockelmann, G. Abstreiter, M. Walther, G. Bohm, G. Trankle, G. Weimann: Phys. Rev. Lett. **69**, 3216 (1992)
9.347 D. Gammon, E.S. Snow, B.V. Shanabrook, D.S. Katzer, D. Park: Phys. Rev. Lett. **76**, 3005 (1996)
9.348 N.H. Bonadeo, G. Chen, D. Gammon, D.G. Steel: Phys. Rev. Lett. **81**, 2959 (1998)
9.349 H. Stolz: *Time-Resolved Light Scattering from Excitons*. Springer Tracts Mod. Phys., Vol. 130 (Springer, Berlin, Heidelberg 1994)
9.350 R. Zimmermann: Il Nuovo Cimento D **17**, 1801 (1995)
9.351 D.S. Citrin: Phys. Rev. B **54**, 16425 (1996)
9.352 D.S. Citrin: Phys. Rev. B **54**, 14572 (1996)
9.353 S. Ceccherini, F. Bogani, M. Gurioli, M. Colocci: Opt. Commun. **132**, 77 (1996)
9.354 S. Haacke, R. A. Taylor, R. Zimmermann, I. Bar-Joseph, B. Deveaud: Phys. Rev. Lett. **78**, 2228 (1997)
9.355 M. Gurioli, F. Bogani, S. Ceccherini, M. Colocci: Phys. Rev. Lett. **78**, 3205 (1997)
9.356 R. Duer, I. Shtrichman, D. Gershoni, E. Ehrenfreund: Phys. Rev. Lett. **78**, 3919 (1997)
9.357 M. Woerner, J. Shah: Phys. Rev. Lett. **81**, 4208 (1998)
9.358 D. Birkedal, J. Shah: Phys. Rev. Lett. **81**, 2372 (1998)
9.359 R. Zimmermann: Unpublished (1998)
9.360 J.W. Strutt (Lord Rayleigh): Phil. Mag. **47**, 375 (1899)
9.361 B. Crosignani, P. Di Porto, M. Bartolotti: *Statistical Properties of Scattered Light* (Academic, New York 1975)

9.362 J.C. Dainty (ed.): *Laser Speckle, Related Phenomena*, 2nd edn., Topics Appl. Phys., Vol.9 (Springer, Berlin, Heidelberg 1984)
9.363 J.W. Goodman: *Statistical Optics*, (Wiley-Interscience, New York 1985)
9.364 A. Ourmazd, D.W. Taylor, J.E. Cunningham, C.W. Tu: Phys. Rev. Lett. **62**, 933 (1989)
9.365 J. Hegarty, M.D. Sturge, C. Weisbuch, A.C. Gossard, W. Wiegmann: Phys. Rev. Lett. **49**, 930 (1982)
9.366 T. B. Norris, J.-K. Rhee, C.-Y. Sung, Y. Arakawa, M. Nishioka, C. Weisbuch: Phys. Rev. B **50**, 14663 (1994)
9.367 P. Michler, W. W. Rühle, G. Reiner, K.J. Ebeling, A. Moritz: Appl. Phys. Lett. **67**, 1363 (1995)
9.368 P. Michler, A. Lohner, W.W. Rühle, G. Reiner: Appl. Phys. Lett. **66**, 1599 (1995)
9.369 P. Michler, M. Hilpert, W.W. Rühle, H.D. Wolf, D. Bernklau, H. Riechert: Appl. Phys. Lett. **68**, 156 (1996)
9.370 G. Bjork, S. Pau, J.M. Jacobson, Hu. Cao, Y. Yamamoto: J. Opt. Soc. Am. B **13**, 1069 (1996)
9.371 B. Sermage, S. Long, I. Abram, J.-Y. Marzin, J. Bloch, R. Planel, V. Thierry-Mieg: Phys. Rev. B **53**, 16516 (1996)
9.372 J.D. Berger, S. Hallstein, O. Lyngnes, W.W. Ruhle, G. Khitrova, H.M. Gibbs: Phys. Rev. B **55**, R4910 (1997)
9.373 R. Huang, C. Hui, Y. Yamamoto: Phys. Rev. B **56**, 9217 (1997)
9.374 P. Michler, M. Hilpert, G. Reiner: Appl. Phys. Lett. **70**, 2073 (1997)
9.375 J. Bloch, J.-Y. Marzin: Phys. Rev. B **56**, 2103 (1997)
9.376 M. Tsuchiya, M. Koch, J. Shah, T.C. Damen, W.Y. Jan, J.E. Cunningham: Appl. Phys. Lett. **71**, 1240 (1997)
9.377 M. Elsasser, S.G. Hense, M. Wegener: Appl. Phys. Lett. **70**, 853 (1997)
9.378 S.G. Hense, M. Wegener: Phys. Rev. B **55**, 9255 (1997)
9.379 Y. Yamamoto, F. Matinaga, S. Machida, A. Karlsson, J. Jacobson, G. Björk: J. Physique IV, C **3**, 39 (1993)
9.380 H. Cao, S. Pau, Y. Yamamoto, G. Bjork: Phys. Rev. B **54**, 8083 (1996)
9.381 S. Pau, J. Jacobson, G. Bjork, Y. Yamamoto: J. Opt. Soc. Am. B **13**, 1078 (1996)
9.382 S. Pau, G. Bjork, J. Jacobson, Y. Yamamoto: IEEE J. QE-**32**, 567 (1996)
9.383 J. Tignon, P. Voisin, C. Delalande, M. Voos, R. Houdré, U. Oesterle, R.P. Stanley: Phys. Rev. Lett. **74**, 3967 (1995)
9.384 T.A. Fisher, A.M. Afshar, M.S. Skolnick, D.M. Whittaker, J.S. Roberts: Phys. Rev. B **53**, R10469 (1996)
9.385 J.D. Berger, O. Lyngnes, H.M. Gibbs, G. Khitrova, T.R. Nelson, E.K. Lindmark, A.V. Kavokin, M.A. Kaliteevskii, V.V. Zapasskii: Phys. Rev. B **54**, 1975 (1996)
9.386 A.V. Kavokin, M.R. Vladimirova, M.A. Kaliteevskii, O. Lyngnes, J.D. Berger, H.M. Gibbs, G. Khitrova: Phys. Rev. B **56**, 1087 (1997)
9.387 G. Hasnain, J.M. Wiesenfeld, T.C. Damen, T.C. Damen, J. Shah, J.D. Wynn, Y.H. Wang, A.Y. Cho: IEEE Photon. Tech. Lett. **4**, 6 (1992)
9.388 J. Wainstain, C. Delalande, M. Voos, R. Houdrι, R.P. Stanley, U. Oesterle: Solid State Commun. **99**, 317 (1996)

9.389 J.P. Reithmaier, M. Rohner, H. Zull, F. Schafer, A. Forchel, P.A. Knipp, T.L. Reinecke: Phys. Rev. Lett. **78**, 378 (1997)
9.390 S. Lutgen, R.A. Kaindl, M. Woerner, T. Elsaesser, A. Hase, H. Kunzel, M. Gulia, D. Meglio, P. Lugli: Phys. Rev. Lett. **77**, 3657 (1996)
9.391 F. Capasso, J. Faist, C. Sirtori, D.L. Sivco, J.N. Baillargeon, A.L. Hutchinson, A.Y. Cho: Phil. Trans. Royal Soc. (London) A **354**, 2463 (1996)
9.392 C. Hartmann, G. Martinez, A. Fischer, W. Braun, K. Ploog: Phys. Rev. Lett. **80**, 810 (1998)
9.393 M. Hartig, S. Haacke, B. Deveaud, L. Rota: Phys. Rev. B **54**, R14269 (1996)
9.394 M. Hartig, S. Haacke, R.A. Taylor, L. Rota, B. Deveaud: Superlattices, Microstructures **21**, 77 (1997)
9.395 M. Hartig, S. Haacke, P.E. Selbmann, B. Deveaud, R.A. Taylor, L. Rota: Phys. Rev. Lett. **80**, 1940 (1998)
9.396 R.A. Kaindl, S. Lutgen, M. Woerner, T. Elsaesser, B. Nottlemann, V.M. Axt, T. Kuhn, A. Hase, H. Kunzel: Phys. Rev. Lett. **80**, 3575 (1998)
9.397 X. Q. Zhou, K. Leo, H. Kurz: Phys. Rev. B **45**, 3886 (1992)
9.398 A. Chebira, J. Chesnoy, G. M. Gale: Phys. Rev. B **46**, 4559 (1992)
9.399 A. Tomita, J. Shah, J. E. Cunningham, S.M. Goodnick, P. Lugli, S.L. Chuang: Phys. Rev. B **48**, 5708 (1993)
9.400 M. Woerner, W. Frey, M.T. Portella, C. Ludwig, T. Elsaesser, W. Kaiser: Phys. Rev. B **49**, 17007 (1995)
9.401 K. Tanaka, H. Ohtake, H. Nansei, T. Suemoto: Phys. Rev. B **52**, 10709 (1995)
9.402 W.W. Rühle, H.-J. Polland: Phys. Rev. B **36**, 1683 (1987)
9.403 R. Tommasi, P. Langot, F. Vallé: Appl. Phys. Lett. **66**, 1361 (1995)
9.404 P. Langot, R. Tommasi, F. Vallé: Phys. Rev. B **54**, 1775 (1996)
9.405 N. Del Fatti, P. Langot, R. Tommasi, F. Vallé: Appl. Phys. Lett. **71**, 75 (1997)
9.406 P. Langot, N. Del Fatti, R. Tommasi, F. Vallé: Opt. Commun. **137**, 285 (1997)
9.407 H. Wang, J. Shah, T. C. Damen, S.W. Pierson, T.L. Reinecke, L.N. Pfeiffer, K. West: Phys. Rev. B **52**, R17013 (1995)
9.408 A.M.T. Kim, S. Hunsche, T. Dekorsy, H. Kurz, K. Köhler: Appl. Phys. Lett. **68**, 2956 (1996)
9.409 R. Kumar, A.S. Vengurlekar, S.S. Prabhu, J. Shah, L.N. Pfeiffer: Phys. Rev. B **54**, 4891 (1996)
9.410 S.S. Prabhu, A.S. Vengurlekar, J. Shah: Phys. Rev. B **53**, R10465 (1996)
9.411 P. E. Selbmann, M. Gulia, F. Rossi, E. Molinari, P. Lugli: In *Hot Carriers in Semiconductors*, ed. by K. Hess, J.P. Leburton,, U. Ravaioli (Plenum, New York 1996) p. 19
9.412 M. Gulia, F. Rossi, E. Molinari, P.E. Selbmann, P. Lugli: Phys. Rev. B **55**, R16049 (1997)
9.413 M. Umlauff, J. Hoffmann, H. Kalt, W. Langbein, J.M. Hvam, M. Scholl, J. Sollner, M. Heuken, B. Jobst, D. Hommel: Phys. Rev. B **57**, 1390 (1998)
9.414 M. Aven, J.S. Prener (eds.): *Physics, Chemistry of II-VI Compounds* (North-Holland, Amsterdam 1967)
9.415 J. Shah, R.F. Leheny, A.H. Dayem: Phys. Rev. Lett. **33**, 818 (1974)
9.416 W.H. Knox, R.L. Fork, M.C. Downer, D.A.B. Miller, D.S. Chemia, C.V. Shank, A.C. Gossard, W. Wiegmann: Phys. Rev. Lett. **54**, 1306 (1985)

9.417 R. Kumar, A.S. Vengurlekar, A. Venu Gopal, F. Laurelle, B. Etienne, J. Shah: Phys. Rev. Lett. **81**, 2578 (1998)
9.418 B. Etienne, F. Laurelle, J. Bloch, L. Sfaxi, F. Lelarge: J. Cryst. Growth **150**, 336 (1995)
9.419 B. Etienne, F. Laurelle, Z. Wang, L. Sfaxi, F. Lelarge, F. Petit, T.C.A. Melin: Sem. Sci. Technol. **11**, 1534 (1996)
9.420 T.C. Damen, J. Shah, D.Y. Oberli, D.S. Chemla, J.E. Cunningham, J.M. Kuo: Phys. Rev. B **42**, 7434 (1990)
9.421 D.S. Citrin: Phys. Rev. Lett. **69**, 3393 (1992)
9.422 H. Akiyama, S. Koshiba, T. Someya, K. Wada, H. Noge, Y. Nakamura, T. Inishita, A. Shimizu, H. Sakaki: Phys. Rev. Lett. **72**, 924 (1994)
9.423 D. Gershoni, M. Katz, W. Wegscheider, L.N. Pfeiffer, R.A. Logan, K. West: Phys. Rev. B **50**, 8930 (1994)
9.424 M. Lomascolo, A.P. Ciccarese, R. Cingolani, R. Rinaldi, F.K. Reinhart: J. Appl. Phys. **83**, 302 (1998)
9.425 J. Feldmann, G. Peter, E.O. Göbel, P. Panson, K. Moors, C.T. Foxon, R.J. Elliott: Phys. Rev. Lett. **59**, 2337 (1987)
9.426 D.S. Citrin: J. Nonlin. Opt. Phys. Mater. **4**, 83 (1995)
9.427 C. Kiener, L. Rota, J. M. Freyland, K. Turner, A.C. Maciel, R.F. Ryan, U. Marti, D. Martin, F. Morier-Gemoud, F.K. Reinhart: Appl. Phys. Lett. **67**, 2851 (1995)
9.428 F. Rossi, E. Molinari: Phys. Rev. B **53**, 16462 (1996)
9.429 M. Yamauchi, Y. Nakamura, Y. Kadoya, H. Sugawara, H. Sakaki: Jpn. J. Appl. Phys. Pt.I, **35**, 1886 (1996)
9.430 G. Goldoni, F. Rossi, E. Molinari, A. Fasolino: Phys. Rev. B **55**, 7110 (1997)
9.431 U. Bockelmann, G. Bastard: Phys. Rev. B **42**, 8947 (1990)
9.432 N.C. Constantinou, B.K. Ridley: Phys. Rev. B **41**, 10622 (1990)
9.433 V.B. Campos, S. Das Sarma: Phys. Rev. B **45**, 3898 (1992)
9.434 P. Lugli, L. Rota, F. Rossi: Phys. Stat. Sol. (b) **173**, 229 (1992)
9.435 L. Rota, F. Rossi, S.M. Goodnick, P. Lugli, E. Molinari, W. Porod: Phys. Rev. B **47**, 1632 (1993)
9.436 L. Rota, F. Rossi, P. Lugli, E. Molinari: Phys. Rev. B **52**, 5183 (1995)
9.437 S. Das Sarma, V.B. Campos: Phys. Rev. B **49**, 1867 (1994)
9.438 L. Rota, J.F. Ryan, F. Rossi, P. Lugli, E. Molinari: Europhys. Lett. **28**, 277 (1994)
9.439 S. Das Sarma, E.H. Hwang: Phys. Rev. B **54**, 1936 (1996)
9.440 L. Zheng, S. Das Sarma: Phys. Rev. B **54**, 2751 (1996)
9.441 C. Ammann, M.A. Dupertuis, U. Bockelmann, B. Deveaud: Phys. Rev. B **55**, 2420 (1997)
9.442 J. Dreybrodt, F. Daiminger, J.P. Reithmaier, A. Forchel: Phys. Rev. B **51**, 4657 (1995)
9.443 A.C. Maciel, C. Kiener, L. Rota, J.F. Ryan, U. Marti, D. Martin, F. Morier-Gemoud, F.K. Reinhart: Appl. Phys. Lett. **66**, 3039 (1995)
9.444 J.M. Freyland, K. Turner, C. Kiener, L. Rota, A.C. Maciel, J.F. Ryan: In *Hot Carriers in Semiconductors*, ed. by K. Hess, J.P. Leburton, U. Ravaioli (Plenum, New York 1996) p.323

9.445 C. Kiener, L. Rota, A.C. Maciel, J.M. Freyland, J.F. Ryan: Appl. Phys. Lett. **68**, 2061 (1996)

9.446 J.F. Ryan, A.C. Maciel, C. Kiener, L. Rota, K. Turner, J.M. Freyland, U. Marti, D. Martin, F. Morier-Gemound, F.K. Reinhart: Phys. Rev. B **53**, R4225 (1996)

9.447 C.Y. Sung, T.B. Norris, X.K. Zhang, Y.L. Lam, I. Vurgaftman, J. Singh, P.K. Bhattacharya: Solid State Electronics **40**, 751 (1996)

9.448 A. Richter, G. Behme, M. Suptitz, C. Lienau, T. Elsaesser, M. Ramsteiner, R. Nötzel, K. Ploog: Phys. Rev. Lett. **79**, 2145 (1997)

9.449 K. Turner, J.M. Freyland, A.C. Maciel, C. Kiener, L. Rota, J.F. Ryan, U. Marti, D. Martin, F. Morier-Gemoud, F.K. Reinhart: In *Hot Carriers in Semiconductors*, ed. by K. Hess, J.P. Leburton, U. Ravaioli (Plenum, New York 1996) p.327

9.450 E. Kapon, D.M. Hwang, R. Bhat: Phys. Rev. Lett. **63**, 430 (1989)

9.451 J. Christen, M. Grundmann, D. Bimberg: J. Vac. Sci. Tech. B **9**, 2358 (1991)

9.452 J. Christen, E. Kapon, E. Colas, D.M. Hwang, L.M. Schiavone, M. Grundmann, D. Bimberg: Surf. Sci. **267**, 257 (1992)

9.453 R.D. Grober, T.D. Harris, J.K. Trautman, E. Betzig, W. Wegscheider, L. Pfeiffer, K. West: Appl. Phys. Lett. **64**, 1421 (1994)

9.454 H. Ghaemi, C. Cates, B.B. Goldberg: Ultramicrosc. **57**, 165 (1995)

9.455 H.F. Ghaemi, C. Cates, B.B. Goldberg: Superlattices, Microstructures **17**, 15 (1995)

9.456 T.D. Harris, D. Gershoni, L. Pfeiffer, M. Nirmal, J.K. Trautman, J.J. Macklin: Sem. Sci. Technol. **11**, 1569 (1996)

9.457 T.D. Harris, D. Gershoni, R.D. Grober, L. Pfeiffer, K. West, N. Chand: Appl. Phys. Lett. **68**, 988 (1996)

9.458 T. Someya, H. Akiyama, H. Sakaki: J. Appl. Phys. **79**, 2522 (1996)

9.459 D. Gershoni, T.D. Harris, L.N. Pfeiffer: Nanotechnology **8**, A44 (1997)

9.460 M. Katz, D. Gershoni, T. D. Harris, L. N. Pfeiffer: J. Lumin. **72-74**, 12 (1997)

9.461 E. Betzig, J.K. Trautman, T.D. Harris, J.S. Weiner, R.L. Kostelak: Science **251**, 1468 (1991)

9.462 G. Mayer, B.E. Maile, R. Germann, A. Forchel, P. Grambow, H.P. Meier: Appl. Phys. Lett. **56**, 2016 (1990)

9.463 R. Cingolani, H. Lage, L. Tapfer, H. Kalt, D. Heitmann, K. Ploog: Phys. Rev. Lett. **67**, 891 (1991)

9.464 R. Cingolani, R. Rinaldi, M. Ferrara, G.C. La Rocca, H. Lage, D. Heitmann, K. Ploog, H. Kalt: Phys. Rev. B **48**, 14331 (1993)

9.465 A.C. Maciel, J.F. Ryan, R. Rinaldi, R. Cingolani, M. Ferrara, U. Marti, D. Martin, F. Morier-Genoud, F.K. Reinhart: Sem. Sci. Technol. **9**, 893 (1994)

9.466 P. Vashishta, R.K. Kalia: Phys. Rev. B **25**, 6492 (1982)

9.467 T.M. Rice: *Electron-Hole Droplets: Theory*, Solid State Physics (Academic, New York 1978)

9.468 J.C. Hensel, T.G. Phillips, G.A. Thomas: *Electron-Hole Droplets: Experimental Aspects*. Solid State Phys., Vol.32 (Academic, New York 1977)

9.469 J. Shah: In *Encyclopedia of Physics*, ed. by R.G. Lerner, G.L. Trigg (Addison-Wesley, Reading, MA 1981) p.254

9.470 J. Shah: In *Concise Encyclopedia of Solid State Physics*, ed. by R. G. Lerner, G.L. Trigg (Addison-Wesley, Reading, MA 1983)

9.471 S. Benner, H. Haug: Europhys. Lett. **16**, 579 (1991)
9.472 S. Benner, H. Haug: Phys. Rev. B **47**, 15750 (1993)
9.473 B.Y.K. Hu, S. Das Sarma: Phys. Rev. B **48**, 5469 (1993)
9.474 R. Cingolani, R. Rinaldi: La Rivisita del Nuovo Cimento **16**, 1 (1993)
9.475 K. H. Wang, M. Bayer, A. Forchel, S. Branner, H. Hang, P. Pagnod-Rossiaux, L. Goldstein: Phys. Rev. B **53**, R10505 (1996)
9.476 W. Wegscheider, L.N. Pfeiffer, M. Dignam, A. Pinczuk, K.W. West, S.L. McCall, R. Hull: Phys. Rev. Lett. **71**, 4071 (1993)
9.477 R. Ambigapathy, I. Bar-Joseph, D.Y. Oberli, S. Haacke, M.J. Brasil, F.K. Reinhart, E. Kapon, B. Deveaud: Phys. Rev. Lett. **78**, 3579 (1997)
9.478 M. Bayer, C. Schlier, C. Greus, A. Forchel, S. Benner, H. Haug: Jpn. J. Appl. Phys. Pt.I **34**, 4408 (1995)
9.479 C. Weisbuch, B. Vinter: *Quantum Semiconductor Structures: Fundamentals, Applications* (Academic, Boston 1991)
9.480 U. Bockelmann, G. Bastard: Europhys. Lett. **15**, 215 (1991)
9.481 H. Benisty, C.M. Sotomayor-Torres, C. Weisbuch: Phys. Rev. B **44**, 10945 (1991)
9.482 U. Bockelmann, T. Egeler: Phys. Rev. B **46**, 15574 (1992)
9.483 U. Bockelmann: Phys. Rev. B **48**, 17637 (1993)
9.484 U. Bockelmann: Sem. Sci. Technol. **9**, 865 (1994)
9.485 U. Bockelmann: Phys. Rev. B **50**, 17271 (1994)
9.486 U. Bockelmann, K. Brunner, G. Abstreiter: Solid-State Electron. **37**, 1109 (1994)
9.487 H. Benisty: Phys. Rev. B **51**, 13281 (1995)
9.488 A. L. Efros, V.A. Kharchenko , M. Rosen: Solid State Commun. **93**, 281 (1995)
9.489 T. Inoshita, H. Sakaki: Physica B **227**, 373 (1996)
9.490 M. Grundmann, D. Bimberg: Phys. Rev. B **55**, 9740 (1997)
9.491 M. Grundmann, R. Heitz, D. Bimberg, J.H. Sandmann, J. Feldmann: Phys. Stat. Sol. (b) **203**, 121 (1997)
9.492 T. Inoshita, H. Sakaki: Phys. Rev. B **56**, R4355 (1997)
9.493 J.L. Pan: Phys. Rev. B **49**, 2536 (1994)
9.494 P.C. Sercel: Phys. Rev. B **53**, 14532 (1996)
9.495 P.D. Wang, C.M. Sotomayor-Torres: J. Appl. Phys. **74**, 5047 (1993)
9.496 S. Fafard, R. Leon, D. Leonard, J.L. Merz, P.M. Petroff: Phys. Rev. B **52**, 5752 (1995)
9.497 M. Vollmer, E.J. Mayer, W.W. Rühle, A. Kurtenbach, K. Eberl: Phys. Rev. B **54**, R17292 (1996)
9.498 S. Raymond, S. Fafard, P.J. Poole, A. Wojs, P. Hawrylak, S. Charbonneau, D. L.leonard, R. Leon, P.M. Petroff, J.L. Merz: Phys. Rev. B **54**, 11548 (1996)
9.499 M. Grundmann, N.N. Ledentsov, O. Stier, D. Bimberg, V.M. Ustinov, M.V. Kop'ev, Z.I. Alferov: Appl. Phys. Lett. **68**, 979 (1996)
9.500 M. Grundmann, R. Heitz, N. Ledentsov, O. Stier, D. Bimberg, V.M. Ustinov, M.V. Kop'ev, Z.I. Alferov, S.S. Ruvimov, P.Werner, U. Gösele, J. Heydenreich: Superlattices, Microstructures **19**, 81 (1996)
9.501 R. Heitz, M. Grundmann, N.N. Ledentsov, B.L. Eckey, M. Veit, D. Bimberg, V.M. Ustinov, A.Y. Egorov, A.E. Zhukov, M.V. Kop'ev, Z.I. Alferov: Appl. Phys. Lett. **68**, 361 (1996)

9.502 M.J. Steer, D.J. Mowbray, W.R. Tribe, M.S. Skolnick, M.D. Sturge, M. Hopkinson, A.G. Cullis, C.R. Whitehouse, R. Murray: Phys. Rev. B **54**, 17738 (1996)

9.503 K.H. Schmidt, G. Medeiros-Ribeiro, M. Östreich, P.M. Petroff, G.H. Döhler: Phys. Rev. B **54**, 11346 (1996)

9.504 U. Bockelmann, W. Heller, A. Filoramo, P. Roussignol: Phys. Rev. B **55**, 4456 (1997)

9.505 S. Grosse, J.H.H. Sandmann, G. von Plessen, J. Feldmann, H. Lipsanen, M. Sopanen, J. Tulkki, J. Ahopello: Phys. Rev. B **55**, 4473 (1997)

9.506 R. Heitz, M. Veit, N.N. Ledentsov, A. Hoffmann, D. Bimberg, V.M. Ustinov, P.S. Kop′ev, Z. Alferov: Phys. Rev. B **56**, 10435 (1997)

9.507 R. Heitz, A. Kalburge, Q. Xie, M. Grundmann, P. Chen, A. Hoffmann, A. Madhukar, D. Bimberg: Phys. Rev. B **57**, 9050 (1998)

9.508 F. Adler, M. Burkard, H. Schweizer, S. Benner, H. Haug, W. Klein, G. Traenkle, G. Weimann?: Phys. Status Solidi (b) **188**, 241 (1995)

9.509 U. Bockelmann, P. Roussignol, A. Filoramo, W. Heller, G. Abstreiter: Solid State Electron. **40**, 541 (1996)

9.510 U. Bockelmann, P. Roussignol, A. Filoramo, W. Heller, G. Abstreiter, K.Brunner, G. Böhm, G. Weimann: Phys. Rev. Lett. **76**, 3622 (1996)

9.511 F. Bogani, L. Carraresi, R. Mattolini, M. Colocci, A. Bosacchi, S. Franchi: Solid State Electron. **40**, 363 (1996)

9.512 R. Heitz, M. Grundmann, N.N. Ledentsov, L. Eckey, M. Veit, B. Bimberg, V.M. Ustinov, A. Yu, Egorov, A.E. Zhukov, P.S. Kop′ev, Z.I. Alferov: Surf. Sci. **361-362**, 770 (1996)

9.513 B. Ohnesorge, M. Albrecht, J. Oshinowo, A. Forchel, Y. Arakawa: Phys. Rev. B **54**, 11532 (1996)

9.514 C.-K. Sun, A.G. Wang, J.E. Bowers, B. Brar, H.-R. Blank, H. Kroemer, M.H. Pilkhuhn: Appl. Phys. Lett. **68**, 1543 (1996)

9.515 U. Bockelmann, W. Heller, G. Abstreiter: Phys. Rev. B **55**, 4469 (1997)

9.516 C. Guasch, C.M.S. Torres, N.N. Ledentsov, D. Bimberg, V.N. Ustinov, P.S. Kop′ev: Superlattices, Microstructures **21**, 509 (1997)

9.517 S. Marcinkevicius, R. Leon: Phys. Status Solidi (b) **204**, 290 (1997)

9.518 Y. Sugiyama, Y. Nakata, T. Futatsugi, M. Sugawara, Y. Awano, N. Yokoyama: Jan. J. Appl. Phys., Lett. **36**, L158 (1997)

9.519 F. Hatami, M. Grundmann, N.N. Ledentsov, F. Heinrichsdorff, R. Heitz, J. Bohrer, D. Bimberg, S.S. Ruvimov, P. Werner, V.M. Ustinov, P.S. Kop′ev, Z.I. Alferov: Phys. Rev. B **57**, 4635 (1998)

9.520 M.G. Bawendi: In *Confined Electrons, Photons: New Physics, Applications*, ed. by E. Burstein, C. Weisbuch (Plenum, New York 1995)

9.521 Y. Toda, M. Kourogi, M. Ohtsu, Y. Nagamune, Y. Arakawa: Appl. Phys. Lett. **69**, 827 (1996)

9.522 J.-Y. Marzin, J.-M. Gérard, A. Izraκl, D. Barrier, G. Bastard: Phys. Rev. Lett. **73**, 716 (1994)

9.523 S. Fafard, R. Leon, D. Leonard, J.L. Merz, P.M. Petroff: Phys. Rev. B **50**, 8086 (1994)

9.524 R. Leon, P.M. Petroff, D. Leonard, S. Fafard: Science **267**, 1966 (1995)

9.525 M. Notomi, T. Furuta, H. Kamada, J. Temmyo, T. Tamamura: Phys. Rev. B **53**, 15743 (1996)

9.526 K. Brunner, G. Abstreiter, G. Bohm, G. Trankle, G. Weimann: Appl. Phys. Lett. **64**, 3320 (1994)

9.527 K. Brunner, G. Abstreiter, G. Bohm, G. Trankle, G. Weimann: Phys. Rev. Lett. **73**, 1138 (1994)

9.528 J.-M. Gérard: In *Confined Electrons, Photons: New Physics, Applications*, ed. by E. Burstein, C. Weisbuch (Plenum, New York 1995)

9.529 J. Tulkki, A. Heinamaki: Phys. Rev. B **52**, 8329 (1995)

9.530 S. Raymond, S. Fafard, S. Charbonneau, R. Leon, D. Leonard, P.M. Petroff, J.L. Merz: Phys. Rev. B **52**, 17238 (1995)

9.531 P.D. Wang, N.N. Ledentsov, C.M. Sotomayor-Torres, A.E. Zhukov, P.S. Kopev, V. M. Ustinov: J. Appl. Phys. **79**, 7164 (1996)

9.532 S. A. Permogorov: Phys. Stat. Sol. (b) **68**, 9 (1975)

9.533 F. Ali, A. Gupta (eds.): *HEMTs & HBTs: Devices, Fabrication, Circuits* (Artech House, Boston 1991)

9.534 C.V. Shank, R.L. Fork, B.I. Greene, F.K. Reinhart, R.A. Logan: Appl. Phys. Lett. **38**, 104 (1981)

9.535 W. Sha, J.-K. Rhee, T.B. Norris, W.J. Schaff: IEEE J. QE-**28**, 2445 (1992)

9.536 H. Heesel, S. Hunsche, H. Mikkelsen, T. Dekorsy, K. Leo, H. Kurz: Phys. Rev. B **47**, 16000 (1993)

9.537 Joo-Hiuk Son, T.B. Norris, J.F. Whitaker: J. Opt. Soc. Am. B **11**, 2519 (1994)

9.538 B.B. Hu, E.A. de Souza, W.H. Knox, J.E. Cunningham, M.C. Nuss, A.V. Kuznetsov, S.L. Chuang: Phys. Rev. Lett. **74**, 1689 (1995)

9.539 A. Leitenstorfer, S. Hunsche, J. Shah, M.C. Nuss, W.H. Knox: Unpublished (1998)

Subject Index

D

Q

Springer Series in Solid-State Sciences

Editors: M. Cardona P. Fulde K. von Klitzing H.-J. Queisser

1 **Principles of Magnetic Resonance** 3rd Edition By C. P. Slichter

2 **Introduction to Solid-State Theory** By O. Madelung

3 **Dynamical Scattering of X-Rays in Crystals** By Z. G. Pinsker

4 **Inelastic Electron Tunneling Spectroscopy** Editor: T. Wolfram

5 **Fundamentals of Crystal Growth I** Macroscopic Equilibrium and Transport Concepts By F. E. Rosenberger

6 **Magnetic Flux Structures in Superconductors** By R. P. Huebener

7 **Green's Functions in Quantum Physics** 2nd Edition By E. N. Economou

8 **Solitons and Condensed Matter Physics** Editors: A. R. Bishop and T. Schneider

9 **Photoferroelectrics** By V. M. Fridkin

10 **Phonon Dispersion Relations in Insulators** By H. Bilz and W. Kress

11 **Electron Transport in Compound Semiconductors** By B. R. Nag

12 **The Physics of Elementary Excitations** By S. Nakajima, Y. Toyozawa, and R. Abe

13 **The Physics of Selenium and Tellurium** Editors: E. Gerlach and P. Grosse

14 **Magnetic Bubble Technology** 2nd Edition By A. H. Eschenfelder

15 **Modern Crystallography I** Fundamentals of Crystals Symmetry, and Methods of Structural Crystallography 2nd Edition By B. K. Vainshtein

16 **Organic Molecular Crystals** Their Electronic States By E. A. Silinsh

17 **The Theory of Magnetism I** Statics and Dynamics By D. C. Mattis

18 **Relaxation of Elementary Excitations** Editors: R. Kubo and E. Hanamura

19 **Solitons** Mathematical Methods for Physicists By. G. Eilenberger

20 **Theory of Nonlinear Lattices** 2nd Edition By M. Toda

21 **Modern Crystallography II** Structure of Crystals 2nd Edition By B. K. Vainshtein, V. L. Indenbom, and V. M. Fridkin

22 **Point Defects in Semiconductors I** Theoretical Aspects By M. Lannoo and J. Bourgoin

23 **Physics in One Dimension** Editors: J. Bernasconi and T. Schneider

24 **Physics in High Magnetics Fields** Editors: S. Chikazumi and N. Miura

25 **Fundamental Physics of Amorphous Semiconductors** Editor: F. Yonezawa

26 **Elastic Media with Microstructure I** One-Dimensional Models By I. A. Kunin

27 **Superconductivity of Transition Metals** Their Alloys and Compounds By S. V. Vonsovsky, Yu. A. Izyumov, and E. Z. Kurmaev

28 **The Structure and Properties of Matter** Editor: T. Matsubara

29 **Electron Correlation and Magnetism in Narrow-Band Systems** Editor: T. Moriya

30 **Statistical Physics I** Equilibrium Statistical Mechanics 2nd Edition By M. Toda, R. Kubo, N. Saito

31 **Statistical Physics II** Nonequilibrium Statistical Mechanics 2nd Edition By R. Kubo, M. Toda, N. Hashitsume

32 **Quantum Theory of Magnetism** 2nd Edition By R. M. White

33 **Mixed Crystals** By A. I. Kitaigorodsky

34 **Phonons: Theory and Experiments I** Lattice Dynamics and Models of Interatomic Forces By P. Brüesch

35 **Point Defects in Semiconductors II** Experimental Aspects By J. Bourgoin and M. Lannoo

36 **Modern Crystallography III** Crystal Growth By A. A. Chernov

37 **Modern Chrystallography IV** Physical Properties of Crystals Editor: L. A. Shuvalov

38 **Physics of Intercalation Compounds** Editors: L. Pietronero and E. Tosatti

39 **Anderson Localization** Editors: Y. Nagaoka and H. Fukuyama

40 **Semiconductor Physics** An Introduction 6th Edition By K. Seeger

41 **The LMTO Method** Muffin-Tin Orbitals and Electronic Structure By H. L. Skriver

42 **Crystal Optics with Spatial Dispersion, and Excitons** 2nd Edition By V. M. Agranovich and V. L. Ginzburg

43 **Structure Analysis of Point Defects in Solids** An Introduction to Multiple Magnetic Resonance Spectroscopy By J.-M. Spaeth, J. R. Niklas, and R. H. Bartram

44 **Elastic Media with Microstructure II** Three-Dimensional Models By I. A. Kunin

45 **Electronic Properties of Doped Semiconductors** By B. I. Shklovskii and A. L. Efros

46 **Topological Disorder in Condensed Matter** Editors: F. Yonezawa and T. Ninomiya

Springer Series in Solid-State Sciences

Editors: M. Cardona P. Fulde K. von Klitzing H.-J. Queisser

47 **Statics and Dynamics of Nonlinear Systems**
Editors: G. Benedek, H. Bilz, and R. Zeyher

48 **Magnetic Phase Transitions**
Editors: M. Ausloos and R. J. Elliott

49 **Organic Molecular Aggregates**
Electronic Excitation and Interaction Processes
Editors: P. Reineker, H. Haken, and H. C. Wolf

50 **Multiple Diffraction of X-Rays in Crystals**
By Shih-Lin Chang

51 **Phonon Scattering in Condensed Matter**
Editors: W. Eisenmenger, K. Laßmann, and S. Döttinger

52 **Superconductivity in Magnetic and Exotic Materials** Editors: T. Matsubara and A. Kotani

53 **Two-Dimensional Systems, Heterostructures, and Superlattices**
Editors: G. Bauer, F. Kuchar, and H. Heinrich

54 **Magnetic Excitations and Fluctuations**
Editors: S. W. Lovesey, U. Balucani, F. Borsa, and V. Tognetti

55 **The Theory of Magnetism II** Thermodynamics and Statistical Mechanics By D. C. Mattis

56 **Spin Fluctuations in Itinerant Electron Magnetism** By T. Moriya

57 **Polycrystalline Semiconductors**
Physical Properties and Applications
Editor: G. Harbeke

58 **The Recursion Method and Its Applications**
Editors: D. G. Pettifor and D. L. Weaire

59 **Dynamical Processes and Ordering on Solid Surfaces** Editors: A. Yoshimori and M. Tsukada

60 **Excitonic Processes in Solids**
By M. Ueta, H. Kanzaki, K. Kobayashi, Y. Toyozawa, and E. Hanamura

61 **Localization, Interaction, and Transport Phenomena** Editors: B. Kramer, G. Bergmann, and Y. Bruynseraede

62 **Theory of Heavy Fermions and Valence Fluctuations** Editors: T. Kasuya and T. Saso

63 **Electronic Properties of Polymers and Related Compounds**
Editors: H. Kuzmany, M. Mehring, and S. Roth

64 **Symmetries in Physics** Group Theory Applied to Physical Problems 2nd Edition
By W. Ludwig and C. Falter

65 **Phonons: Theory and Experiments II**
Experiments and Interpretation of Experimental Results By P. Brüesch

66 **Phonons: Theory and Experiments III**
Phenomena Related to Phonons
By P. Brüesch

67 **Two-Dimensional Systems: Physics and New Devices**
Editors: G. Bauer, F. Kuchar, and H. Heinrich

68 **Phonon Scattering in Condensed Matter V**
Editors: A. C. Anderson and J. P. Wolfe

69 **Nonlinearity in Condensed Matter**
Editors: A. R. Bishop, D. K. Campbell, P. Kumar, and S. E. Trullinger

70 **From Hamiltonians to Phase Diagrams**
The Electronic and Statistical-Mechanical Theory of sp-Bonded Metals and Alloys By J. Hafner

71 **High Magnetic Fields in Semiconductor Physics**
Editor: G. Landwehr

72 **One-Dimensional Conductors**
By S. Kagoshima, H. Nagasawa, and T. Sambongi

73 **Quantum Solid-State Physics**
Editors: S. V. Vonsovsky and M. I. Katsnelson

74 **Quantum Monte Carlo Methods in Equilibrium and Nonequilibrium Systems** Editor: M. Suzuki

75 **Electronic Structure and Optical Properties of Semiconductors** 2nd Edition
By M. L. Cohen and J. R. Chelikowsky

76 **Electronic Properties of Conjugated Polymers**
Editors: H. Kuzmany, M. Mehring, and S. Roth

77 **Fermi Surface Effects**
Editors: J. Kondo and A. Yoshimori

78 **Group Theory and Its Applications in Physics**
2nd Edition
By T. Inui, Y. Tanabe, and Y. Onodera

79 **Elementary Excitations in Quantum Fluids**
Editors: K. Ohbayashi and M. Watabe

80 **Monte Carlo Simulation in Statistical Physics**
An Introduction 3rd Edition
By K. Binder and D. W. Heermann

81 **Core-Level Spectroscopy in Condensed Systems**
Editors: J. Kanamori and A. Kotani

82 **Photoelectron Spectroscopy**
Principle and Applications 2nd Edition
By S. Hüfner

83 **Physics and Technology of Submicron Structures**
Editors: H. Heinrich, G. Bauer, and F. Kuchar

84 **Beyond the Crystalline State** An Emerging Perspective By G. Venkataraman, D. Sahoo, and V. Balakrishnan

85 **The Quantum Hall Effects**
Fractional and Integral 2nd Edition
By T. Chakraborty and P. Pietiläinen

86 **The Quantum Statistics of Dynamic Processes**
By E. Fick and G. Sauermann

87 **High Magnetic Fields in Semiconductor Physics II**
Transport and Optics Editor: G. Landwehr

88 **Organic Superconductors** 2nd Edition
By T. Ishiguro, K. Yamaji, and G. Saito

89 **Strong Correlation and Superconductivity**
Editors: H. Fukuyama, S. Maekawa, and A. P. Malozemoff

Springer Series in Solid-State Sciences

Editors: M. Cardona P. Fulde K. von Klitzing H.-J. Queisser

Managing Editor: H. K.V. Lotsch